16th European White Dwarfs Workshop 2008

Journal of Physics: Conference Series Volume 172

Barcelona, Spain
30 June – 4 July 2008

Editors:

Enrique García-Berro
Jordi Isern
Margarida Hernanz
Santiago Torres

ISBN: 978-1-61738-311-3

Printed from e-media with permission by:

Curran Associates, Inc.
57 Morehouse Lane
Red Hook, NY 12571

Some format issues inherent in the e-media version may also appear in this print version.

Printed by Curran Associates, Inc. (2010)

For permission requests, please contact the Institute of Physics
at the address below.

Institute of Physics
Dirac House, Temple Back
Bristol BS1 6BE UK

Phone: 44 1 17 929 7481
Fax: 44 1 17 920 0979

techtracking@iop.org

Additional copies of this publication are available from:

Curran Associates, Inc.
57 Morehouse Lane
Red Hook, NY 12571 USA
Phone: 845-758-0400
Fax: 845-758-2634
Email: curran@proceedings.com
Web: www.proceedings.com

TABLE OF CONTENTS

LUMINOSITY FUNCTION, MASS FUNCTION AND POPULATIONS

WHITE DWARF STRUCTURE AND EVOLUTION

WHITE DWARF AGES

WHITE DWARF CATALOGS AND SURVEYS

CENTRAL STARS OF PLANETARY NEBULAE

SUPERNOVAE PROGENITORS

WHITE DWARFS IN NOVAE AND CVS

PHYSICAL PROCESSES IN WHITE DWARFS AND MAGNETIC WHITE DWARFS

DISK, DUST AND PLANETS AROUND WHITE DWARFS

PULSATING WHITE DWARFS

PREFACE

The 16th European Workshop on White Dwarfs was held in Barcelona, Spain, from 30 June to 4 July 2008 at the premises of the UPC. Almost 120 participants from Europe (France, Germany, United Kingdom, Italy, and several others), America (USA, Canada, Argentina, Brazil, and Chile), and other continents (Australia, South Africa, . . .) attended the workshop. Among these participants were the most relevant specialists in the field. The topics covered by the conference were:

- White dwarf structure and evolution
- Progenitors and Planetary Nebulae
- White dwarfs in binaries: cataclysmic variables, double degenerates and other binaries
- White dwarfs, dust disks and planetary systems
- Atmospheres, chemical composition, magnetic fields
- Variable white dwarfs
- White dwarfs in stellar clusters and the halo
- White Dwarfs as SNIa progenitors

The programme included 54 talks, and 45 posters. The oral presentations were distributed into the following sessions:

- Luminosity function, mass function and populations
- White dwarf structure and evolution
- White dwarf ages
- White dwarf catalogs and surveys
- Central stars of planetary nebulae
- Supernovae progenitors
- White dwarfs in novae and CVs
- Physical processes in white dwarfs and magnetic white dwarfs
- Disks, dust and planets around white dwarfs
- Pulsating white dwarfs

Additionally we had a special open session about Spitzer and white dwarfs.

The Proceedings of the 16th European Workshop on White Dwarfs are representative of the current state-of-the-art of the research field and include new and exciting results. We acknowledge the very positive attitude of the attendants to the workshop, which stimulated very fruitful discussions that took place in all the sessions and after the official schedule. Also, the meeting allowed new collaborations tp start that will undoubtedly result in significant advances in the research field. We also acknowledge the willingness of the participants to deliver their contributions before the final deadline. We sincerely thank them.

The white dwarf community has been steadily growing since the first white dwarf workshop, held in Kiel (Germany) in 1974. Some of the participants in the first colloquium have already effectively retired; others – although officially retired – continue to attend successive workshops, Professor Weidemann, one of the first organizers, being a leading example. We hope we will be able to continue counting on them for many years. A very graphical view of the evolution of the field can be found in the homepage of Professor Detlev Koester, who has collected pictures of almost all the previous workshops:.

Additionally, several astronomers coming from related fields have joined our (not so) small community. Most importantly, several generations of young scientists gave their first talks in these workshops. In summary our community is an active one, and we have close, durable and solid ties of friendship. We are optimistic and we foresee that the spirit of the previous workshops will continue in future editions.

We would like to express our deepest gratitude to our sponsors: The Universitat Politècnica de Catalunya (UPC), the Institut de Ciències de l'Espai (CSIC), the Institute for Space Studies of Catalonia (IEEC), the Spanish Ministry of Education and Science, the Generalitat de Catalunya, the Ajuntament de Barcelona, the School of Civil Engineering of Barcelona and UPCnet.

Finally, the IEEC staff and our graduate students have enthusiastically supported the organization of the workshop in every single detail; without them we would have not succeeded. We thank them especially. Also, we acknowledge the task of the Scientific Organizing Committee, which gave their full support in all the scientific tasks.

Enrique García-Berro, UPC
Margarida Hernanz, ICE (CSIC)
Jordi Isern, ICE (CSIC)
Santiago Torres, UPC
Editors

Journal of Physics: Conference Series **172** (2009) 012001 doi:10.1088/1742-6596/172/1/012001

The Luminosity Function for White Dwarfs in the SuperCOSMOS Sky Survey.

N R Rowell[1], N C Hambly[1] and P Bergeron[2]

[1] Institue for Astronomy, University of Edinburgh, Blackford Hill, Edinburgh EH9 3HJ, UK
[2] Département de Physique, Université de Montréal, C.P. 6128, Succ. Centre-Ville, Montréal, Québec H3C 3J7, Canada

E-mail: nr@roe.ac.uk, bergeron@astro.umontreal.ca

Abstract. We present the techniques and early results of our program to measure the luminosity function for white dwarfs in the SuperCOSMOS Sky Survey. Our survey covers over three quarters of the sky to a mean depth of $I \sim 19.2$, and finds $\sim 9,500$ Galactic disk white dwarf candidates on applying a conservative lower tangential velocity cut of $30\,\mathrm{kms}^{-1}$. Novel techniques introduced in this survey include allowing the lower proper motion limit to vary according to apparent magnitude, fully exploiting the accuracy of proper motion measurements to increase the sample size. Our luminosity function shows good agreement with that measured in similar works. We find a pronounced drop in the local number density of WDs at a $M_{\mathrm{bol}} \sim 15.75$, and an inflexion in the luminosity function at $M_{\mathrm{bol}} \sim 12$.

1. Introduction

The potential use of White Dwarfs (WDs) as chronometers has long been recognised. The relative accuracy of stellar age estimates based on up-to-date models, combined with the long evolutionary timescales of WDs and their abundance in most Galactic systems, means this approach has been used to age a wide variety of stellar configurations. These include WD - subdwarf binaries (Monteiro et al. 2006), single burst populations such as globular and open clusters (Hansen et al. 2007; von Hippel & Gilmore 2000; Jeffery et al. 2007), and continuous populations such as the Galactic disk (Winget et al. 1987, Oswalt et al. 1995, Leggett et al. 1998). The standard tool used for this technique is the luminosity function (LF) for WDs, which contains information on both the absolute age and the star formation history of a stellar system. The star formation history is essentially imprinted on the entire LF, and authors working on parametric forms for the LF (e.g. Noh & Scalo 1990) predict bumps and inflexions corresponding to changes in the rate of star production in the past. The absolute age is constrained by the marked downturn in the LF at faint magnitudes, where the cooling ages of WDs converge on that of the population as a whole.

Modern surveys that apply this to the Galactic disk have suffered from small observational samples of faint WDs at the LF turning point. This is confounded by the fact that the atmospheric physics of such "ultracool" WDs is currently poorly understood, with only a handful of stars extensively studied (e.g. Bergeron & Leggett 2002). The largest LF measurement currently published is that of Harris et al. (2006), using a sample of 6000 WDs selected from a combination of the SDSS and USNO-B. The presence of only 4 stars beyond the turning point precludes any accurate constraint on the LF at these magnitudes.

We are working on a program to measure the WDLF by exploiting data from the all-sky SuperCOSMOS Sky Survey (SSS). Sensitive proper motion detection combined with accurate photometry over wide angles enables large samples of WDs to be catalogued using reduced proper motion methods. In section 2 we outline the SSS, and go on to descibe our survey procedures in sections 3, 4 and 5. We present our LF in section 6, and discuss the limitations and possible future refinements in section 7.

2. The data

The SuperCOSMOS Sky Survey was compiled by digitizing several generations of photographic Schmidt plate surveys. The photographic source material includes the POSS-I and POSS-II surveys in the northern hemisphere, and a combination of ESO and SERC surveys in the south. Over 1700 fields are required to cover the entire sky, with each field being observed in four photographic bands $B_J R_{59F} I_N$ and $R_{103aE/63F}$ in the north/south respectively. Only $B_J R_{59F}$ and I_N are normally used for photometry, $R_{103aE/63F}$ being generally of poorer quality and very similar in response to R_{59F}. $R_{103aE/63F}$ is used to provide an early epoch for improved astrometric solutions in each field. The subscripts refer to particular filter/emulsion combinations, and will be dropped throughout the rest of this paper. We will sometimes refer to $R_{103aE/63F}$ and R_{59F} as first- and second-epoch R respectively. Due to reasons outlined in section 4.1.1, we restrict our analysis to objects detected at all four epochs. The survey data is housed in relational databases available online at `http://surveys.roe.ac.uk/ssa/`. An introduction to the SSS and further technical details can be found in Hambly et al. (2001a). We avoid the Galactic plane by 10° and the centre by 20°. These regions are heavily affected by blending and poor photometric calibration due to differential extinction within the field. We also reject six fields centred on the Magellanic clouds for the same reasons. This gives a total sky coverage of 10.2 sr, or about 80% of the sky. Taking into account the regions lost around bright stars due to diffraction spikes and scattered light haloes, the total survey area is 9.4 sr, or just over 75% of the sky.

3. Sample selection

3.1. Image morphology criteria

The SuperCOSMOS Image Analysis Mode software measures 32 parameters for each object detection, some of which we restrict in order to ensure our final WD catalogue only contains objects with reliable, high quality images. We only accept objects with the *deblending flag* set to zero, meaning no image fragmentation has been detected. The *quality number* is an integer indicating situations encountered during image analysis, such as proximity to a bright star or plate boundary, and is set such that increasingly severe circumstances are assigned higher values. We restrict this parameter to values less than 128. The *profile classification statistic* η gives a magnitude-independent measure of the 'stellarity' of an image, by quantifying the deviation of the radial profile slope from that of the mean stellar template. It is scaled to a zero mean, unit variance Gaussian distribution. Objects with $|\eta| > 4\sigma$ are rejected.

Previous studies utilising digitized Schmidt plate data have placed cuts on the ellipticities of images, in order to limit contamination by faint galaxies and noise. However, we have found that the ellipticities measured by SuperCOSMOS are extremely noisy at intermediate to faint magnitudes, and that any intuitively sensible cut will result in a seriously incomplete sample of stars. This may be due in part to the eight-fold nearest neighbour pixel connectivity employed by the SuperCOSMOS image analysis software. We have decided to ignore this parameter.

3.2. Contamination in north due to repeated photographic copying

The northern half of the SSS data was digitized from photographic copies of POSS-I and POSS-II survey originals. Unfortunately, this seems to have degraded the quality of the data, resulting in erroneous astrometric fits for a fraction of objects. This may be due to the photographic copying process rendering blended objects harder to detect. Certainly, the distribution of the astrometric residuals for affected objects deviates significantly from the expected χ^2 form, indicating that something is going wrong with the centroiding. This problem first became obvious when we started to analyse the reduced proper motion (RPM) diagram for the northern hemisphere (see section 5), and noticed an extended population of objects at high values of RPM, in approximately the same region as cool WDs. This is what is expected if the proper motions of objects no longer correlate with distance — essentially, stars with bad proper motions are scattered from their usual locii and any discrimination between distinct Galactic and stellar components is lost. The main consequence for this work is that the highly numerous disk main sequence stars tend to scatter into the WD region, completely swamping the cool WD locus. Unfortunately, this contamination cannot adequately be removed by simply applying judicious selection cuts on the image parameters, and we have been forced to devise a new statistic to identify these objects. This statistic is calculated using the known distributions of χ^2 (astrometric fit residuals) and η for both reliable and unreliable objects. The main result is that bad objects are successfully removed by placing a selection cut on this new statistic, and that a certain fraction of good objects are also removed. This fraction can be precisely measured and corrected for in later analysis, given that this statistic is independent of magnitude, colour and proper motion.

4. Survey limits

4.1. Proper Motion

The upper limit on proper motions is set to $0\farcs18$ yr^{-1} for the entire survey. This arises from the mean epoch spread and reliable pairing radius between consecutive observations in each field. This limit is not so important for producing large stellar catalogues, due to the lower probability of stars having such motions. Note, however, that nearby ultracool WDs and halo stars are expected to show larger motions. To produce a large sample of stars, we reduce the lower proper motion limit as far as possible while maintaining a 5σ proper motion detection threshold on all objects. This level has been chosen to limit observational scatter in the RPM diagram. Our lower limit is thus dictated by the proper motion errors, which in general differ from field to field due to varying time baselines and plate quality, and within individual fields are a function of apparent magnitude, due mostly to centroiding precision as detailed in section 2.2.5 of Hambly et al. (2001c). It is natural therefore to allow our lower proper motion limit to vary from field to field, and to make it a function of apparent magnitude within each field. In general, there is no analytic form for the distribution of proper motion errors with apparent magnitude, and a numerical solution has to be found.

4.1.1. Proper motion limit as a function of apparent magnitude For every survey field, we plot the proper motion uncertainty σ_μ against B for all objects. Many fields share the same general form for this distribution, in particular a minimum of $\sigma_\mu \sim 5$ mas yr^{-1} at around $B \sim 17$, rising to considerably larger errors at brighter and fainter magnitudes. Objects detected on only three plates show a similar distribution offset to larger σ_μ by up to $0\farcs04$ yr^{-1}. Rather than attempting to define a separate lower proper motion limit for these objects, we simply discard them. We wish to find the upper boundary of this distribution, denoted $\sigma_\mu^{\mathrm{upper}}(B)$. We may then define our 5σ-threshold lower proper motion limit at every magnitude by simply multiplying this function by 5. $\sigma_\mu^{\mathrm{upper}}(B)$ is estimated by first binning objects according to magnitude, interactively so that each bin contains 100 points, which allows reasonable estimation of $\sigma_\mu^{\mathrm{upper}}(B)$ at the extremes where objects are of low density. This results in N bins, where $N = N_{\mathrm{objects}}/100$ and is typically of the order of 100. Each of the N points $\sigma_\mu^{\mathrm{upper}}(B_N)$ is then defined by the largest σ_μ at the mean B value in each bin, after rejecting the top 3% of points as outliers. The set of points $\sigma_\mu^{\rm upper}(B_N)$ shows considerable noise on small scales, and is processed through one stage of smoothing to remove this. A Savitzky-Golay filtering technique is used (Savitzky & Golay 1964), generalized to allow for non-regular data. This results in a final set of N points $\sigma_\mu^{\rm upper}(B_N)$, which are interpolated linearly to define $\sigma_\mu^{\rm upper}(B)$ at all magnitudes. Figure 1 shows an example of the proper motion error distribution and $\sigma_\mu^{\rm upper}(B)$ for one field.

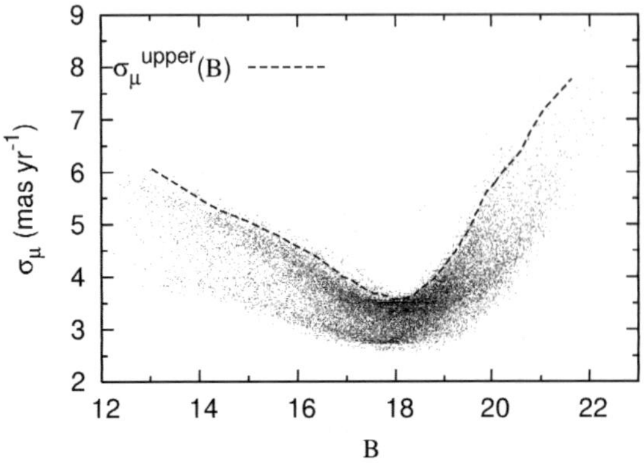

Figure 1: Proper motion error distribution for northern field 12, with the upper boundary set by $\sigma_\mu^{\rm upper}(B)$.

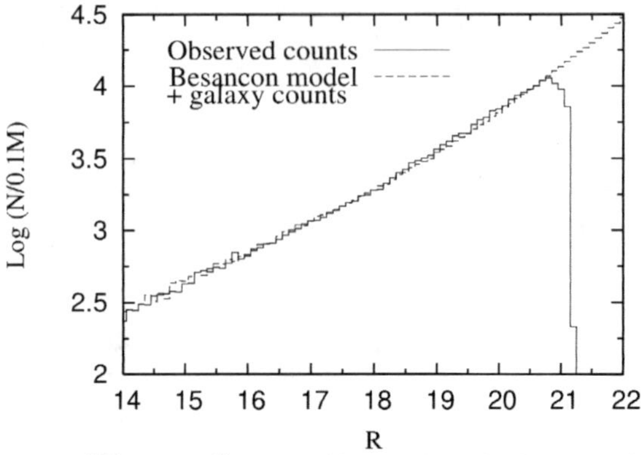

Figure 2: Observed and theoretical number count histograms for SERC-R survey field 149.

4.2. Magnitude

The magnitude completeness of catalogues derived from Schmidt plates has been analysed extensively in the literature. We follow a similar approach to Tinney, Reid and Mould (1993), except that with nearly 8000 Schmidt plates in 1700 fields it is unfeasible to model the differential stellar and galactic number counts in each. We proceed by assuming that within each of the eight photographic surveys compiled into the SSS the completeness of each plate is an identical function of the plate detection limit. This allows for the considerable variation in depth of plates within each survey. We estimate the *completeness function* for each survey, using a subset of five fields chosen at high Galactic latitude, in order to reduce the effects of image crowding and interstellar extinction. The completeness function is defined to be the fraction of objects recovered as a function of apparent magnitude, and relies on an estimate of the intrinsic stellar and galactic number counts in each field. Figure 2 shows the typical number count histograms for observed and modelled data in one field.

4.2.1. Theoretical source counts Intrinsic stellar number counts are modelled using the Besancon Galaxy model (Robin et al. 2003). These will generally be different in each field due to different lines of sight through the Galaxy. Conversely, galaxy number counts are assumed to be isotropic, with the large ($\sim 6^\circ \times 6^\circ$) fields smoothing out any small scale structure. Suitable empirical counts have been published by many authors (eg Jones et al. 2001), and in this study we use those compiled by the Durham Cosmology Group[1]. In log-number space the galaxy counts are well fitted by straight lines in each band over the range of interest.

4.2.2. Completeness of digitized Schmidt plates as a function of plate depth The completeness function in each field is given by $\mathcal{C}(m) = N_{\rm obs}(m)/N_{\rm intrinsic}(m)$. For each survey we obtain five measures of $\mathcal{C}(m)$, which are individually shifted according to the plate depth then averaged to obtain $\overline{\mathcal{C}}(m')$, where m' is magnitudes above the plate limit. Uncertainties are assigned from the sample variance of the individual measures. The resulting eight functions are used to

[1] Available online at `http://star-www.dur.ac.uk/~nm/pubhtml/counts/counts.html`

define dynamic magnitude limits for the remaining fields in each survey. One such function is presented in figure 3. The offset from the plate limit chosen as the completeness limit ($m'_{\rm lim}$) for each survey is given in table 1, along with the corresponding mean magnitude limits for each band.

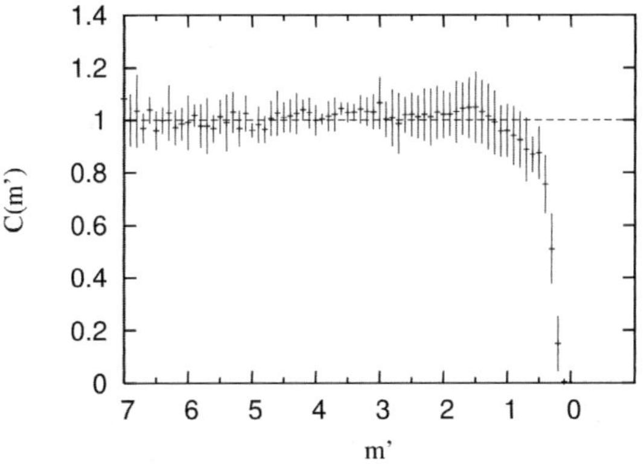

Figure 3: Mean completeness function for survey ESO-R. m' refers to magnitudes above the plate limit; in this example $m'_{\rm lim} = 0.6$.

Survey	m'_{lim}	Mean depth
Northern hemisphere:		
POSSI-E	0.8	20.2
POSSII-B	0.6	22.0
POSSII-R	0.7	20.5
POSSII-I	0.6	19.4
Southern hemisphere:		
ESO-R	0.6	20.3
SERC-J	0.5	22.2
SERC-R	0.5	20.5
SERC-I	0.6	19.0

Table 1: Completeness limit offsets and mean depth of surveys.

5. Methods

5.1. Reduced proper motion

The *reduced proper motion* statistic $H_m = m + 5\log(\mu) + 5$ uses the correlation between proper motion and distance to estimate the absolute magnitude of nearby stars, and is commonly employed for identifying extremely sub-luminous WDs in proper motion surveys. The fact that H_m can be expressed in quantities intrinsic to the star $H_m = M + 5\log(V_{\rm tan}) - 3.379$ suggests that we can use stellar and kinematic models to predict the range of H_m for WDs of various colour and velocity. This allows us to perform rigorous selections on H_m to produce well defined catalogues of WD candidates. This is done in the H_m-colour plane, where an appropriate colour-magnitude relation can be used to find the regions inhabited by WDs of various tangential velocity. Figure 4 shows the reduced proper motion diagram (RPMD) for our objects. This is topologically similar to the HR diagram. Cooling tracks for WDs with hydrogen and helium atmospheres and a range of $V_{\rm tan}$ are plotted. WDs of larger tangential velocity are located below these lines, and we select objects from below the DA track as our WD candidates.

5.1.1. Tangential velocity cut Cool WDs of low velocity share identical values of H_m with subdwarfs in the high velocity tail of the halo, so these two populations overlap to some extent in the RPMD. Contamination by subdwarfs can be reduced by applying a minimum tangential velocity threshold to the kinematic selection of WDs. This essentially selects WD candidates from regions more widely separated from the subdwarf locus. The fraction of WDs that fall below the chosen $V_{\rm tan}$ threshold can be corrected for by including a “discovery fraction”. This is calculated by projecting the WD velocity ellipsoid onto the sky, and integrating the resulting V_{tan} distribution over the range of allowed velocities, according to the prescription of Murray (1983). As each line of sight sees a slightly different projection of the velocity ellipsoid, this has to be done separately for each survey field. We assume the velocity ellipsoid for disk WDs matches that of local dM stars measured by Reid et al 1995, and set $\sigma_{u,v,w} =$ (43,31,25) km s^{-1} with an asymmetric drift of –22 km s^{-1} relative to the local standard of rest. The discovery fraction has a mean value of 0.85 (0.69) (0.52) for $V_{\rm tan} >$ 20(30)(40) km s^{-1}.

5.2. Photometric parallaxes
Photometric distances and temperatures are obtained by fitting the two SSS colours using the model atmospheres and cooling sequences described in Bergeron et al. (2001) and references therein (see also `http://www.astro.umontreal.ca/~bergeron/CoolingModels`), provided in the SuperCOSMOS passbands and interpolated at 10K intervals.

The appropriate two-colour diagram and model tracks are presented in figure 6. We use a simple variance-weighted least-squares fitting procedure, and reject objects with $\chi^2 > 5$. Such objects are mostly unresolved binaries, as detailed in section 5.2.2. With only one degree of freedom we cannot measure the gravity or atmosphere type of each star, so we assume $\log(g) = 8$ for all stars and fit both hydrogen (DA) and helium (DB) atmospheres. Over- and under-massive WDs exist in roughly equal proportion, and fitting them with normal-mass atmospheres has opposite effects and so should (roughly) cancel out. We make no correction for reddening, as we expect the effect to be very small due to the proximity of our stars.

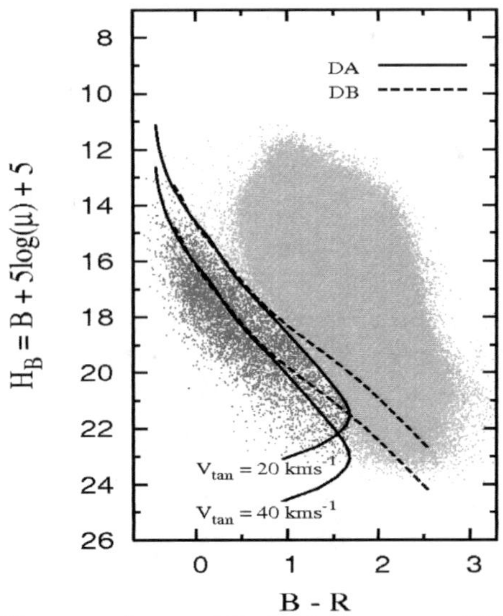

Figure 4: RPMD for stars in SSS. Light grey points are subdwarfs and disk main sequence stars, dark grey points are stars selected as WD candidates.

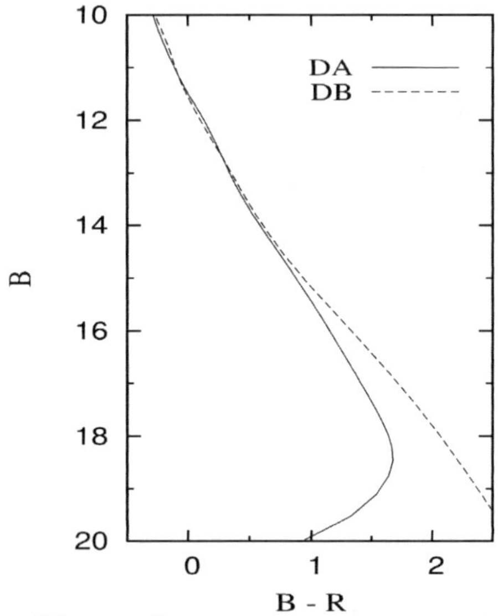

Figure 5: Theoretical colour-magnitude relations for hydrogen-rich (DA) and helium-rich (DB) WDs, assuming a normal surface gravity of $\log(g) = 8$.

5.2.1. WD atmosphere types We allow WD candidates to contribute to the LF as both hydrogen and helium atmosphere WDs, with a weight for each type set according to the observed stellar abundances. For WDs bluer than $B - R \sim 1$, corresponding to $T_{\mathrm{eff}} \gtrsim 5250$K, the particular choice of atmosphere has little effect on the fitted luminosity, so we simply assume all stars are hydrogen types. At lower effective temperatures helium atmosphere stars are more luminous, by around 1 mag at $B - R \sim 1.5$ (see figure 5), and we therefore expect helium WDs to be over-represented at the reddest colours. We shift the weight in favour of helium WDs in this range, and set it according to the fitted hydrogen atmosphere absolute magnitude. The values used are given in table 2.

5.2.2. WD+dM binaries WDs with unresolved cool companions will tend to scatter out of the WD locus in the colour-colour plane, due to the companion star contributing flux to the I band. Harris et al. (2006) include these objects in their survey by accepting only the bluest bands, however with only two colours we cannot do this and so must reject these objects. This to some extent represents an incompleteness in our survey, as around 10% of WD candidates are rejected in this way.

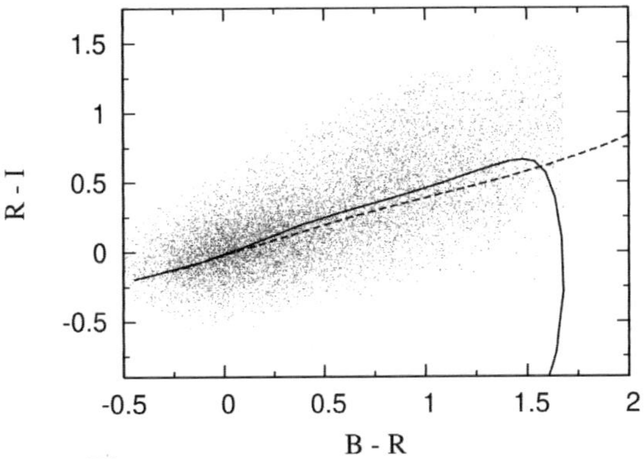

Figure 6: Two-colour plot for WD candidates and synthetic WD models. The dashed line is for helium-rich DB models, the solid line is for hydrogen-rich DAs.

$M_{bol,\,H}$	w_H	w_{He}
< 15.2	1.0	0
$15.2 \rightarrow 15.5$	0.5	0.5
$15.5 \rightarrow 16.0$	0.2	0.8
> 16.0	0	1.0

Table 2: Weights assigned to hydrogen and helium atmosphere WDs.

6. Luminosity function methods

6.1. The generalized V_{max} technique

The spatial density of objects retrieved in a survey with both magnitude and proper motion limits can be most readily obtained using Schmidt's $1/V_{max}$ method. This involves summing the inverse of the *maximum* volume in which each object could have been observed, given it's absolute magnitude and tangential velocity, and the survey limits. This method has since been generalised to non-uniformly distributed populations, in particular to allow for the exponential density profile of the Galactic disk. V_{max} is obtained by integrating the density law along the line of sight to the star, over the distance range in which the star passes the survey selection criteria. Errors are assigned to each luminosity bin by assuming Poisson statistics.

6.1.1. V_{max} with dynamic proper motion limits A problem arises when the proper motion limit is allowed to vary with apparent magnitude in a non-analytic way. The distances d_{min} and d_{max} at which the star's proper motion crosses the survey limits can no longer be calculated analytically, and so this integral is evaluated in the following manner. The limiting distances due to the magnitude limits are calculated as normal, then the integral is evaluated numerically between these limits in steps of finite size. At each distance step, the star's proper motion is calculated and compared to the limits appropriate for the new apparent magnitude, and if the star is within the limits then the volume associated with this step is added to the integral. The final stage is to select an appropriate step size for this calculation. We have chosen to use steps of constant *generalized volume*, taking into account the disk density profile. This ensures that each step makes an identical contribution to the sum. Note that calculating the distance interval necessary to keep the generalized volume of each step constant is also non-trivial, and leads ultimately to a transcendental equation that must be solved numerically.

6.2. The white dwarf luminosity function

The LF derived for the 9480 WDs with $V_{tan} > 30$ km s^{-1} is presented in figure 7. A scaleheight of 250 pc has been assumed for the density profile, to allow comparison with other works. The LF is remarkably featureless on the rising side, apart from a slight inflexion at $M_{bol} \sim 12$. The faint end shows a drop in density at $M_{bol} = 15.75$, in agreement with that reported by other authors. Integrating the LF gives a total WD number density in the solar neighbourhood of

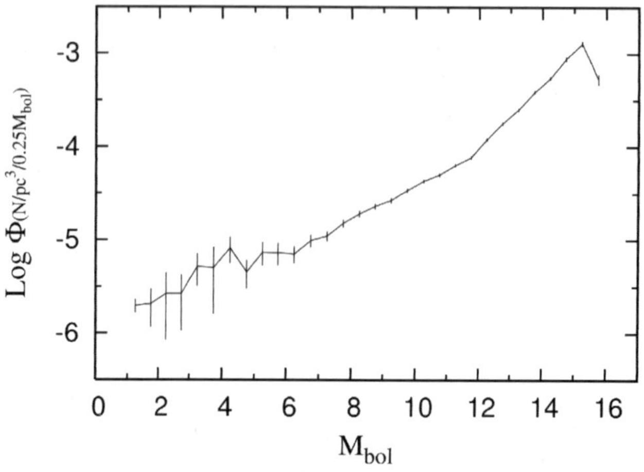

Figure 7. LF for WDs with $V_{\rm tan} > 30\,{\rm kms}^{-1}$.

2.3×10^{-3} pc^{-3}. $\langle V/V_{\rm max}\rangle = 0.505 \pm 0.003$ for this survey, somewhat larger than that for a complete survey ($\langle V/V_{\rm max}\rangle = 0.5$) but it should be noted that the assumed scaleheight affects this calculation, and increasing it to 350 pc results in $\langle V/V_{\rm max}\rangle = 0.498 \pm 0.003$.

7. Future work

Older stars are expected to have larger scaleheights due to kinematic heating, and $V_{\rm max}$ will be underestimated for these objects by assuming a value of 250 pc. We intend to improve our LF by measuring the variation in scaleheight with $M_{\rm bol}$ for our objects. Optical spectra have been obtained for a subsample of WD candidates, which we will use to assess the reliability of our method and contamination from non- degenerate stars. We also have a high proper motion catalogue of SSS objects with motions in the range $0.''18 < \mu < 10.''0$ yr^{-1}, which contains most of the interesting ultracool WDs featured in the literature. Including this will push our disk LF to fainter magnitudes, and will also possibly provide enough halo WD candidates to measure a LF for this component.

References

Bergeron P and Leggett S K 2002 *ApJ* **580** 1070
Bergeron P Leggett S K & Ruiz MT 2001 *ApJ Suppl.* **133** 413
Hambly N C *et al* 2001a *MNRAS* **326** 1279
Hambly N C Davenhall A C Irwin M J & MacGillivray H T 2001c *MNRAS* **326** 1315
Hansen B *et al* 2007 *ApJ* **671** 380
Harris H C *et al* 2006 *AJ* **131** 571
von Hippel T & Gilmore G 2000 *AJ* **120** 1384
Jeffery E von Hippel T Jefferys W Winget D Stein N & DeGennaro S 2007 *ApJ* **658** 391
Jones L R Fong R Shanks T Ellis R S & Peterson B A 1991 *MNRAS* **249** 481
Leggett S K Ruiz M T & Bergeron P 1998 *ApJ* **497** 294
Monteiro H Jao W C Henry T Subasavage J & Beaulieu T 2006 *ApJ* **638** 446
Murray C A 1983 *Vectorial Astrometry* 284
Noh H & Scalo J 1990 *ApJ* **352** 605
Oswalt T Smith J Wood M & Hintzen P 1995 *Nat.* **382** 692
Reid I N Hawley S L & Gizis J E 1995 *AJ* **110** 1838
Robin A C Reylé C Derriére S & Picaud S 2003 *A&A* **409** 523
Savitzky A & Golay M 1964 *Analytical Chemistry* **36** 1627
Tinney C G Reid I N & Mould J R 1993 *ApJ* **414** 254
Winget D Hansen C Liebert J Van Horn H Fontaine G Nather R Kepler S & Lamb D 1987 *ApJ* **315** L77

SDSS DR4: progress on the hot white dwarf luminosity function

J Krzesinski[1,2], A Nitta[3], S J Kleinman[3], S Hügelmeyer[4], S Dreizler[4], J Liebert[5] and H Harris[6]

[1] Mt. Suhora Observatory, Cracow Pedagogical University, ul. Podchorazych 2, 30-084 Cracow, Poland

[2] Apache Point Observatory, PO Box 59, Sunspot, NM 88349, USA

[3] Gemini Telescope, 670 N. A'Ohoku Place, Hilo, HI 96720, USA

[4] Institut für Astrophysik, Georg-August-Universität Göttingen, Friedrich-Hund-Platz 1, D-37077 Göttingen, Germany

[5] Steward Observatory, University of Arizona, Tucson, AZ 85726, USA

[6] U.S. Naval Observatory, PO Box 1149, Flagstaff, AZ 86002-1149, USA

E-mail: jk@astro.as.ap.krakow.pl

Abstract. A large collection of white dwarf (WD) spectra from Sloan Digital Sky Survey (SDSS) data release 4 (DR4) WD catalog data allowed us to obtain a luminosity function (LF) for the hottest WDs. The LF was calculated basing on spectra of the WDs from a special class of SDSS objects called hot standards (HS), for which the WD sample completeness can be estimated. WD distances were determined from the observed and absolute SDSS g filter magnitudes derived from WD stellar atmosphere model fits to the SDSS spectra. The result LF covers -0.5 up to 7 absolute bolometric magnitude range. It shows a slight disagreement at the cold end comparing to the literature LFs but it can be due to some problems with HS WD sample completeness estimations. It is clearly visible that the LF has unexpected plateau between 1.5 and 4 absolute bolometric magnitudes. There is nothing similar visible in a LF built only for non-DA WDs. The plateau might be real but we are concern that its existence can be connected to the problems with spectra calibration of the blue objects in SDSS DR4 data and lack of good stellar atmosphere models for the hottest DA white dwarfs.

A sample of WDs used to calculate hot WD LF was selected from DR4 WD catalog (Eisenstein et al. 2006). The catalog contains spectroscopically identified WDs from SDSS (York et al. 2000) DR4 data (Adelman-McCarthy et al. 2006), which covers 4783 square degrees of the sky with spectroscopic observations. However not all WDs from the catalog were used to build LF but only those marked as a special SDSS class of objects called HS (Krzesinski et al. 2004), which SDSS uses to calibrate the observations. Shortly, the HSs are a subclass of SDSS objects which fulfil the following criteria: their observed g filter magnitude is greater than 14 mag and dereddened g magnitude is less than 19 mag (see Schlegel et al. 1998). Their dereddened color differences $(u-g)_0$ and $(g-r)_0$ falls between –1.5 and 0 magnitudes. Objects fulfilling these requirements are allocated fibers for spectroscopy. In addition to above HS criteria, a candidate for HS has to pass some isolation and quality requirements to receive a fiber for spectroscopic observations. These requirements are very restrictive and largely decrease HS completeness. Never-less, HSs are the most complete of all stars observed in SDSS survey because of their priorities to be observed over all other star categories and its completeness can be estimated.

The SDSS HS color–color constrains restricting the sample impose a WD selection effect by cutting out more DA than DB class of WDs from the sample. In order to avoid that effect we choose to restrict our HS WD sample in the temperature. Therefore, for the analysis only WDs with $T_{\rm eff}$ greater than 23500 K and $\log g > 7.0$ were used.

To calculate our WD LF a general inverse maximum volume ($1/V_{\rm max}$) formula introduced by Schmidt 1968 was used. In the formula $V_{\rm max}$ describes a maximum volume in which a star can contribute to the LF:

$$V_{\rm max} = \frac{4\pi}{3}\beta c_b c_g c_{\rm HS}(r^3_{\rm max} - r^3_{\rm min})e^{-\frac{|z|}{250}}$$

where, β is the ratio of the SDSS DR4 spectral observation coverage of the sky (equals 4783, Adelman-McCarthy et al. 2006) to the total sky surface in square degrees. Since SDSS HS fiber allocation procedure strongly depends on the object Galactic latitude b and object brightness in SDSS g filter (Eisenstein et al. 2006, section 5.6) our WD sample completeness coefficient will not be a constant value. Therefore two coefficients c_b and c_g defined as linear functions of the object latitude b (c_b) and its brightness (c_g) were introduced to take care of the incompleteness due to the fiber allocation. An additional restriction comes from HS image quality. Eisenstein et al. 2006 estimates that only 77% of photometricaly targeted HS objects pass the stringent HS photometric quality flags and the third completeness coefficient $c_{\rm HS} = 0.77$ in the equation above reflects that. Finally, the $e^{-\frac{|z|}{250}}$ factor in the formula takes into account exponentially decreasing WD density with their distance z (in parsecs) from the Galactic plane. The Galactic scale height was assumed to equal 250 pc (the same as in Harris et al. 2006).

The distances to WDs were determined from observed and absolute SDSS g filter magnitudes derived from WD stellar atmosphere model fits to the SDSS spectra. However, one has to keep in mind that the DR4 WD catalog model fits show systematic errors which affect DAs with effective temperatures greater than 30 000 K (Eisenstein et al. 2006). The trend is that these model fits to SDSS spectra tend to be higher in temperature than earlier literature fits and that converts into brighter object in bolometric magnitudes ($M_{\rm bol}$). Therefore we derived a parabolic temperature correction function for the DR4 WD catalog $T_{\rm eff}$ and apply it to the whole HS DA sample. Also, since DR4 WD catalog $T_{\rm eff}$ and $\log g$ fits were calculated with LTE models, we used Napiwotzki et al. 1999 NLTE corrections to the $\log g$ and $T_{\rm eff}$ of all DAs with $T_{\rm eff} >$ 40 000 K. Taking both, the temperature and NLTE corrections into account, we calculated the LF shown in figure 1.

In the figure, bolometric magnitude bins are centered on half magnitudes and bin widths are equal 1 magnitude. For comparison we plotted WD LF by Harris et al. (2006) based on SDSS Data Release 3 (Abazajian et al. (2005) in the same figure. As one can see, the cold end of our LF is slightly above the hot end of Harris et al. LF and the displacement is greater than the error bars. The largest influence on the LF position in such diagrams is due to the WD sample completeness estimation. If that were the case, then our completeness, which basically follows the recipe given by Eisenstein et al. (2006), would be underestimated. To investigate the problem we had undertaken our own HS WD sample completeness analysis, independent on Eisenstein et al. (2006). At this moment however, we are not able to draw any certain conclusions about the completeness estimations of the HS sample.

Farther, hotter part of our LF in figure 1 goes gradually down, then became horizontal between 1.5 and 4 magnitudes which is rather an unexpected behavior. Since the LF is mostly populated by DA WDs, the flat part of the LF (or a plateau) is likely intrinsic to DAs only. To make sure that this is the case we made separate plots (not shown here) of DA and non–DA LFs, which confirmed that the plateau exists only in DA LF and not in non–DA one. By separating DA and non–DA LFs we could also demonstrate that there is a sudden drop in the space density of non-DA WDs near $M_{\rm bol} = 2$ magnitudes (which corresponds to 80 000 – 90 000 K) where the hottest part of DA LF begins. The meaning of that is such that we actually see the evolutionary

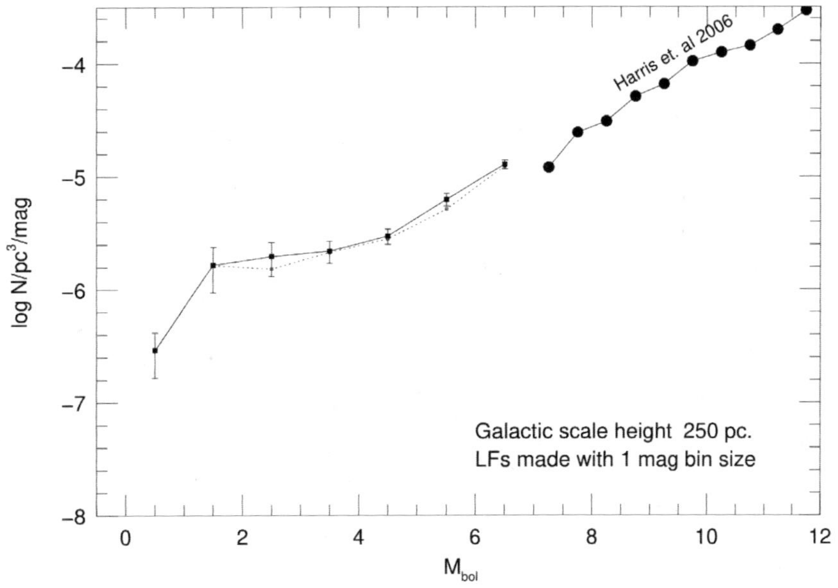

Figure 1. Hot WD LF plot for the HS sample and a bin width of 1 mag centered on a half magnitudes. The galactic scale height: 250 parsecs. Note that Harris et al. (2006) LF bin width is 0.5 magnitude.

transformation of non-DA WDs into DA ones. During the evolution temperatures of the WDs are gradually decreasing. As a consequence, approximately at 90 000 K, the effectiveness of the convective mixing in their atmospheres decreases allowing for heavier elements to settle down. That way a pure hydrogen WD atmosphere is created and a non–DA WD becomes a DA one. We start to see first DA WDs in the DA LF while at the same time the space density of non-DAs drops rapidly in the non–DA LF.

At the cold end of the non–DA LF we do not see any unexpected variations but it is worth to notice that there is no gap or any significant drop of non–DB LF within DB gap temperature region. This is due to recently found hot DB WDs (Eisenstein et al. 2006) which filled up the gap.

In a summary, the small discrepancies between our and literature LFs can be easily corrected by changes in the HS sample completeness estimation. From the other hand our LF was calculated based on distances determined from the atmosphere model fits to the SDSS spectra. That seems to be more accurate approach then any statistical proper motion method used to built the LFs in the literature. Therefore further analysis of that problem is required. A new light might bring our own (independent from Eisenstein et al. 2006) HS sample completeness estimation from SDSS targeting code but it is still under investigation.

For the moment we failed to explain the reason for the DA LF plateau presence with any physical cause but we incline to say, that the temperature and NLTE corrections we applied should have removed most of the problems with DA WD temperature and $\log g$ determinations. In that case the plateau we see in the LF is real. However, one has also keep in mind that there are other difficulties connected to the temperature determination of the hottest DAs from the atmosphere model fits. For example, metals are not included in the NLTE DA atmosphere models and their influence on the temperature determination from the models might be of order of NLTE corrections (Barstow et al. 1998). Therefore, farther investigation of the hottest part of the DA LF is required.

Acknowledgments

Funding for the SDSS and SDSS-II has been provided by the Alfred P. Sloan Foundation, the Participating Institutions, the National Science Foundation, the U.S. Department of Energy, the National Aeronautics and Space Administration, the Japanese Monbukagakusho, the Max Planck Society, and the Higher Education Funding Council for England. The SDSS Web Site is http://www.sdss.org/. The SDSS is managed by the Astrophysical Research Consortium for the Participating Institutions. The Participating Institutions are the American Museum of Natural History, Astrophysical Institute Potsdam, University of Basel, University of Cambridge, Case Western Reserve University, University of Chicago, Drexel University, Fermilab, the Institute for Advanced Study, the Japan Participation Group, Johns Hopkins University, the Joint Institute for Nuclear Astrophysics, the Kavli Institute for Particle Astrophysics and Cosmology, the Korean Scientist Group, the Chinese Academy of Sciences (LAMOST), Los Alamos National Laboratory, the Max-Planck-Institute for Astronomy (MPIA), the Max-Planck-Institute for Astrophysics (MPA), New Mexico State University, Ohio State University, University of Pittsburgh, University of Portsmouth, Princeton University, the United States Naval Observatory, and the University of Washington.

References

Abazajian K et al. 2005 *AJ* **129** 1755
Adelman-McCarthy J K et al. 2006 *ApJS* **162** 38
Barstow M A Hubeny I & Holberg J B 1998 *MNRAS* **299** 520
Eisenstein D J et al. 2006 *AJ* **132** 676
Eisenstein D J et al. 2006 *ApJ* **167** 40
Harris H C et al. 2006 *AJ* **131** 571
Krzesiński J Nitta A Kleinman S J Harris H C Liebert J Schmidt G Lamb D Q & Brinkmann J 2004 *A&A* **417** 1093
Napiwotzki R Green P J & Saffer R A 1999 *ApJ* **517** 399
Schlegel D J Finkbeiner D P & Davis M 1998 *ApJ* **500** 525
Schmidt M 1968 *ApJ* **151** 393
York D G et al. 2000 *AJ* **120** 1579

Journal of Physics: Conference Series **172** (2009) 012003 doi:10.1088/1742-6596/172/1/012003

White dwarfs, red dwarfs and halo dark matter

E García–Berro[1,2], S Torres[1,2], J Camacho[1] and J Isern[3,2]

[1]Departament de Física Aplicada, Escola Politècnica Superior de Castelldefels, Universitat Politècnica de Catalunya, Av. del Canal Olímpic, s/n, 08860 Castelldefels, Spain

[2]Institute for Space Studies of Catalonia, c/Gran Capità 2–4, Edif. Nexus 104, 08034 Barcelona, Spain

[3]Institut de Ciències de l'Espai, CSIC, Campus UAB, Facultat de Ciències, Torre C-5, 08193 Bellaterra, Spain

E-mail: garcia@fa.upc.edu

Abstract. The nature of the microlensing events observed by the MACHO team towards the LMC still remains controversial. Low-mass substellar objects and stars with masses larger than $\sim 1\,M_{\odot}$ have been ruled out, while stars of $\sim 0.5\,M_{\odot}$ are the most probable candidates. This means that the microlenses should be either red or white dwarfs. Consequently, we assess jointly the relative contributions of both types of stars to the mass budget of the Galactic halo. We use a Monte Carlo code that incorporates up-to-date evolutionary sequences of both red dwarfs and white dwarfs as well as detailed descriptions of both our Galaxy and the LMC and we compare the synthetic populations obtained with our simulator with the results obtained by the MACHO and EROS experiments. We find that the contribution of the red dwarf population is not enough to explain the number of events measured by the MACHO team. Even though, the optical depth obtained in our simulations almost doubles that obtained when taking into account the white dwarf population alone. Finally, we also find that the contribution to the halo dark matter of the entire population under study is smaller than 10%, at the 95% confidence level.

1. Introduction

Since the pioneering proposal of Paczyński (1986) that gravitational microlensing could be an useful tool to study Galactic dark matter, considerable observational and theoretical efforts have been invested in this issue. The most likely candidates for building-up the baryonic dark matter are massive compact halo-objects (MACHOS). It has been suggested that MACHOs could be planets, brown and red dwarfs, primordial black holes, molecular clumps, and old white dwarfs. On the other hand the MACHO (Alcock et al. 1997, 2000), EROS (Lasserre et al. 2001; Goldman et al. 2002; Tisserand et al. 2007), OGLE (Udalski et al. 1994), MOA (Muraki et al. 1999), and SuperMACHO (Becker et al. 2005) teams have monitored millions of stars during several years, in both the Large Magellanic Cloud (LMC) and the Small Magellanic Cloud (SMC), to search for microlensing events. However, the MACHO and EROS teams have been the only ones that have published their results. Although there exist notable differences between the results reported by both teams there is as well agreement in some of the most important points. In particular, no microlensing candidates have been found by the MACHO team or the EROS group with event durations between a few hours and 20 days. This implies that the Galactic halo cannot contain more than a 10% of dark objects in the mass range $10^{-7} < M/M_{\odot} < 10^{-3}$, thus ruling out planets and brown dwarfs. Moreover, the MACHO collaboration detected ~ 17 microlensing events during their 5.7 yr analysis of 11.9 million stars in the LMC (Alcock et al. 2000). However, the number of microlensing events has been the subject of some debate. After the publication of the first results (Alcock et al. 2000) in which 17 microlensing events were reported, it was discovered by the EROS collaboration that one of the MACHO events was caused by stellar variability rather than microlensing. This suggested that some of the other events might be variable stars, as well. Later, it was shown that most of the MACHO microlensing candidates were likely to be actual microlensing events (Bennett et al. 2005) and that 13 events were problably genuine events. Nevertheless, later on it has been shown that of these 13 events one is known to be a variable star , and it is most likely that another event is a variable star, as well. Thus, we are left with 11 microlensing events. Using their original data Alcock et al. (2000) derived an optical depth towards the LMC of $\tau = 1.2^{+0.4}_{-0.3} \times 10^{-7}$ or, equivalently, a halo fraction $0.08 < f < 0.50$, at the 95% confidence level, with a MACHO mass in the range $0.15 \leq M/M_{\odot} \leq 0.50$, depending on the halo model. More recent calculations suggest (Bennett et al. 2005) that the optical depth is $\tau = (1.0 \pm 0.3) \times 10^{-7}$. On the other hand, the non-detections reported by the EROS collaboration have provided an upper limit (Tisserand et al. 2007). Their results imply that, at the 95% confidence level, that the contribution of stars of $\sim 0.5\, M_{\odot}$ is less than 7%. All these findings have estimulated a vivid discussion about the location and nature of the lenses. A full variety of possible explanations have been proposed (and discarded) to reproduce the microlensing events. Some of these explanations are self-lensing by the LMC itself, tidal debris or a dwarf galaxy toward the LMC (Zhao 1998), a Galactic-extended shroud population of white dwarfs (Gates & Gyuk 2001), blending effects (Belokurov, Evans & Le Du 2003, 2004), biased mass-functions (Adams & Laughlin 1996) and spatially-varying mass-functions (Kerins & Evans 1998; Rahvar 2005). All in all, in spite of the numerous theoretical studies (Isern et al. 1998; Reylé et al. 2001; Flynn et al. 2003; García-Berro et al. 2004; Camacho et al. 2007) the mistery still remains unsolved. In this work, we analyze, in a comprehensive manner, a significant range of masses $0.08 < M/M_{\odot} < 10$ likely to reproduce the observational microlensing results and, thus, to contribute to the halo dark matter. This mass range represents almost 90% of the stellar content. The paper is organized as follows. In Sect. 2 we describe in detail how the sample of microlensing events is built. In Sect. 3 we present our main results and we compare them with the observational results of the MACHO and EROS teams, whereas in Sect. 4 we discuss the significance of our results. Finally in Sect. 5 we summarize our major findings and we present our conclusions.

2. Building the sample

A detailed description of our Monte Carlo simulator has been presented in Torres et al. (2002), García–Berro et al. (2004), and Camacho et al. (2007). Consequently, here we only summarize its most salient features. The Galactic halo was modeled assuming a spherically-symmetric halo. In particular, the model used here is an isothermal sphere of radius 5 kpc. The position of each synthetic star was randomly chosen according to this density profile. The mass distribution of synthetic stars was computed using two different initial mass functions, the standard initial mass function of Scalo (1998) and the biased log-normal initial mass function proposed by Adams & Laughlin (1996), which is representative of other non-conventional initial mass functions. We note, however, that these biased initial mass functions appear to be incompatible with the observed properties of the halo white dwarf population (Isern et al. 1998; García–Berro et al. 2004), and with other observations. The main-sequence mass is obtained by drawing a pseudo-random number, according to the adopted initial mass function. The time at which a star was born is randomly choosen assuming that the halo was formed 14 Gyr ago in an intense burst of star formation of duration ~ 1 Gyr. The main-sequence lifetime as a function of the main sequence mass is that of Iben & Laughlin (1989). Once the mass of a synthetic star is known, its main-sequence lifetime is inferred and we can determine which stars evolve into white dwarfs and which ones remain on the main sequence as red dwarfs. We considered red dwarfs to have masses in the range $0.08 < M/M_{\odot} < 1$. For these stars, we have adopted the evolutionary models of Baraffe et al. (1998). For stars that had time to enter the white dwarf cooling track effective temperatures, luminosities, and colors are derived using a set of theoretical cooling sequences and an initial-to-final-mass relationship (Iben & Laughlin, 1989). White dwarfs with masses smaller than $M_{\rm WD} = 1.1\,M_{\odot}$ are expected to have a CO core and for these we adopt the cooling tracks of Salaris et al. (2000). White dwarfs with masses larger than $M_{\rm WD} = 1.1\,M_{\odot}$ have ONe cores and for these, we adopt the cooling sequences of Althaus et al. (2007).

The kinematical properties of the halo population were modeled using Gaussian laws (Binney & Tremaine 1987) with radial and tangential velocity dispersions that reproduce the flat rotation curve of our Galaxy. We have adopted standard values for the circular velocity $V_{\rm c} = 220$ km/s and the peculiar velocity of the Sun $(U_{\odot}, V_{\odot}, W_{\odot}) = (10.0, 15.0, 8.0)$ km/s (Dehnen & Binney 1998). We have discarded stars that had velocities smaller than 250 km/s, since these would not be considered halo members. We also have rejected stars of velocities larger than 750 km/s, because these velocities exceed 1.5 times the escape velocity. To compare simulated results with observations, a normalization criterion should be used. We have normalized our simulations to the local density of halo white dwarfs obtained from the halo white dwarf luminosity function of Torres et al. (1998), but taking into account the new halo white dwarf candidates reported by Vidrih et al. (2007).

To reproduce the microlensing experiments towards the LMC, we have followed the detailed LMC descriptions of Gyuk et al. (2000) and Kallivayalil et al. (2006). This model provides a synthetic population of stars representative of the monitored point sources. After each simulation was completed, we evaluate which star of the Galactic halo could be responsible for a microlensing event. We have considered only stars that fulfilled certain criteria. First, the lensing star should be fainter than a given magnitude limit. Secondly, we also checked whether the lens was inside the Einstein tube of the monitored star. Finally, we filter those stars that are candidates to produce a microlensing event with the detection efficiency function, which depends on the particular characteristics of the experiment. In the case of the MACHO project, we have adopted 1.1×10^7 stars during 5.7 yr over 13.4 deg^2, whereas the detection efficiency has been modeled using the fit of Alcock et al. (2000). For the EROS experiment, we have used 0.7×10^7 stars over a wider field of 84 deg^2 and a period of 6.7 yr. To evaluate the detection efficiency, we have adopted a numerical fit to the results of Tisserand et al. (2007).

3. Results

3.1. The MACHO experiment

In the left panels of Fig. 1, we show the contribution of the different simulated populations to the optical depth, for the two initial mass functions studied here, as a function of the adopted magnitude cut. The value of the optical depth obtained in our simulations has been normalized to the value of the optical depth derived by Alcock et al. (2000), $\tau_0 = 1.0 \times 10^{-7}$. The white dwarf populations are represented by solid and open squares for the CO and ONe white dwarfs, respectively. Open triangles indicate the contribution of red dwarfs, while the contribution of the three populations is represented by open circles. The combined contribution of red dwarfs and white dwarfs is at most one third of the observed optical depth, when a totally-unrealistic magnitude cut is adopted, $m_V \sim 15^{\rm mag}$. In addition, it shoud be noted that there is a clearly decreasing trend as the adopted magnitude cut is increased. This is because fewer objects contribute to the optical depth as the magnitude cut increases. Furthermore, the slope of the distributions is different for the three types of objects considered. For instance, the contribution of red dwarfs decreases faster than the contribution of CO white dwarfs as the magnitude cut increases, which, in turn, decreases faster than the contribution of ONe white dwarfs. This reflects the fact that, in general, red dwarfs are brighter than regular, CO white dwarfs. Finally,

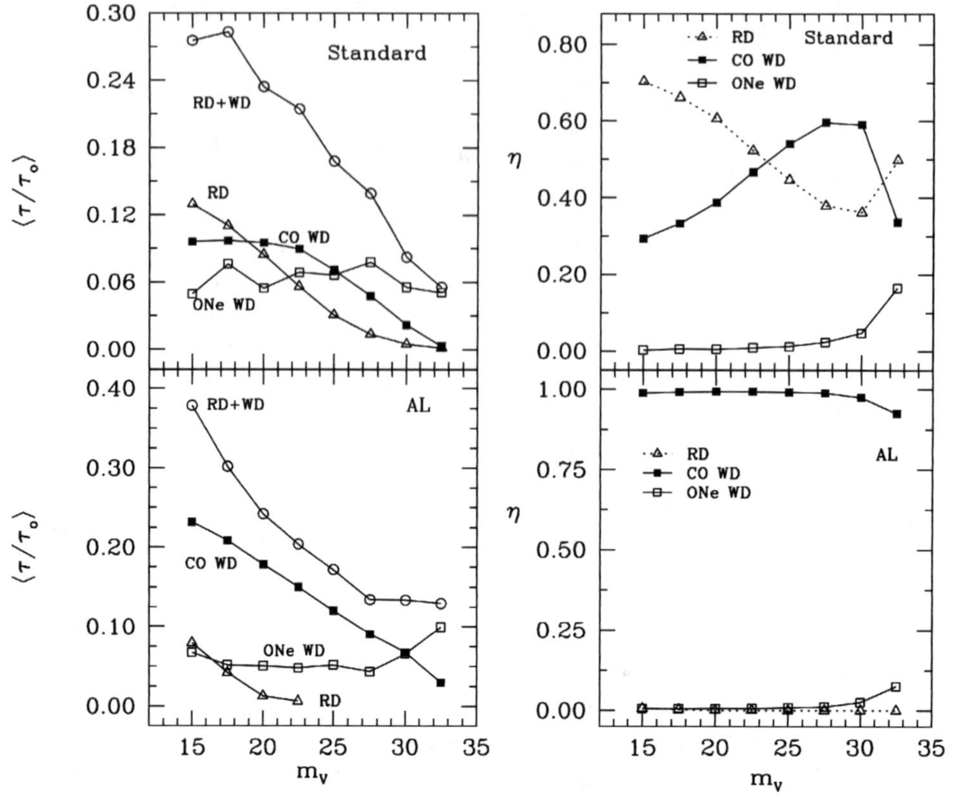

Figure 1. Left panels: microlensing optical depth towards the LMC as a function of the limiting magnitude. Solid and open squares represent the CO and ONe white-dwarf populations, respectively. Red dwarfs are represented using open triangles, while the entire population is shown using open circles. Right panels: fraction of microlenses with respect to the entire population, as a function of the magnitude cut.

ONe white dwarfs cool rapidly (Althaus et al. 2007) and thus, for realistic halo ages, are faint objects. Consequently, their contribution remains almost constant.

In Table 1 we summarize the average values of parameters of the entire population for two initial mass functions and four magnitude cuts, that is, present the number of microlensing events, the average mass of the microlenses, their average proper motion, distance and tangential velocity, the corresponding Einstein crossing times and, finally, the contribution to the microlensing optical depth. As can be seen, the average distance of the sample increases as the magnitude cut increases. This is a natural consequence of selecting more distant objects which, in turn, implies longer Einstein crossing times. This behavior is independent of the assumed initial mass function. Note as well that the expected number of events or the average mass of the sample do not depend on the adopted magnitude cut. In the case of a standard initial mass function, no more than one microlensing event should be expected at the 1σ confidence level, while for a log-normal mass function up to 5 events might be expected. In any case, the expected number of microlensing events is far from the 11 events claimed by the MACHO experiment.

We have also evaluated the fraction of microlenses for the different populations as a function of the adopted magnitude cut. The results are shown in the right panels of Fig. 1 for both the standard initial mass function (top panel) and the log-normal initial mass function (bottom panel). For the standard initial mass function, the relative contribution of red dwarfs decreases with increasing value of magnitude cuts, while that of the CO white dwarfs increases. Both

Journal of Physics: Conference Series **172** (2009) 012003 doi:10.1088/1742-6596/172/1/012003

Table 1. Summary of the results obtained for the entire simulated population of microlenses in the direction of the LMC for a halo age of of 14 Gyr, different model initial mass functions, and several magnitude cuts.

	Standard				AL			
Magnitude	17.5	22.5	27.5	32.5	17.5	22.5	27.5	32.5
$\langle N_{\rm WD}\rangle$	0 ± 1	0 ± 1	0 ± 1	0 ± 1	3 ± 2	2 ± 1	1 ± 1	0 ± 1
$\langle m\rangle$ $(M/M_\odot)$	0.421	0.411	0.427	0.443	0.638	0.636	0.640	0.684
$\langle\mu\rangle$ $('' \, {\rm yr}^{-1})$	0.020	0.017	0.008	0.004	0.036	0.025	0.011	0.003
$\langle d\rangle$ (kpc)	2.48	3.79	6.62	13.08	1.39	2.15	5.13	19.6
$\langle V_{\rm tan}\rangle$ $({\rm km\,s^{-1}})$	240	247	262	241	240	252	263	261
$\langle\hat{t}_{\rm E}\rangle$ (d)	41.2	49.3	63.3	82.8	34.7	46.6	76.4	126.8
$\langle\tau/\tau_0\rangle$	0.283	0.214	0.139	0.055	0.302	0.204	0.140	0.129

contributions are equal for a magnitude cut of $\approx 24^{\rm mag}$. Finally, the contribution of ONe white dwarfs remains roughly constant and only becomes significant for large magnitude cuts. These trends can be attributed to the fact that red dwarfs are more numerous at bright magnitudes than white dwarfs, for which typical luminosities are of the order of $\log(L/L_\odot) \simeq -3.5$. The situation is completely different when the log-normal initial mass function of Adams & Laughlin (1996) is used. As can be seen the number of microlenses is practically dominated by the CO white dwarf contribution, while the contribution of red dwarfs and ONe white dwarfs is negligible.

We have also performed a Z^2 statistical test of the compatibility of the different populations with the observed data. The Z^2 statistical test (Lucy 2000) is an improvement to the standard χ^2 statistical test, especially designed for meagre data sets. In Table 2, we show the Z^2 probability that the different simulated populations are compatible with the distribution of Einstein times obtained by the MACHO experiment. It is convenient to clarify that this probability is an estimate of the degree to which the observed event-rate distribution can be derived from a single population of stars. As can be seen in Table 2, the CO white dwarf population provides the most appropriate description of the observational data, given that its compatibility is as high as 0.90 for the faintest magnitude cut. Moreover, the compatibility of this population with the observational data increases at fainter magnitude bins. In sharp contrast, the population of red dwarfs presents a decreasing trend as the magnitude cut increases and, additionally, the compatibility with the observational data is, at most, 0.70. With regard to the ONe white dwarf population, the compatibility presents an almost constant value of around 0.70, independently of the magnitude cut. These results indicate that the CO white-dwarf population can reproduce the observed distribution of microlensing event-rates. Even more, they dominate the behavior of the entire population, as can be seen from the final row of Table 2, in which we analyze the compatibility of the entire population of simulated stars.

3.2. The EROS experiment

While the MACHO team has claimed the identification of up to 11 observed events, the EROS collaboration has not found any microlensing event towards the LMC and only one candidate event towards the SMC. Adopting a standard halo model and assuming $\tau_{\rm SMC} = 1.4\tau_{\rm LMC}$, the EROS results imply an optical depth $\tau_0 = 0.36 \times 10^{-7}$ (Tisserand et al. 2007), which is four times smaller than that obtained by the MACHO team. We have performed a set of simulations emulating the conditions of the EROS experiment with inputs similar to those of the MACHO experiment. Our simulations show that, for the standard initial mass function, the expected

Table 2. Compatibility, as obtained using the Z^2 statistical test, of the observed MACHO distribution of Einstein crossing times and those of the different simulated populations.

Magnitude	$17.5^{\rm mag}$	$22.5^{\rm mag}$	$25.0^{\rm mag}$	$27.5^{\rm mag}$	$30.0^{\rm mag}$
Red dwarfs	0.58	0.67	0.65	0.55	0.40
CO white dwarfs	0.61	0.70	0.80	0.87	0.90
ONe white dwarfs	0.76	0.55	0.68	0.68	0.72
Whole population	0.59	0.69	0.75	0.82	0.89

optical depth could be 70% of the value found by the EROS team. The value obtained when only the white-dwarf population was considered was previously found to be 50% (Camacho et al. 2007). The simulations presented here show a better agreement with the results of the EROS experiment. Obviously, the reason for this is the inclusion of the red dwarf population. However, when a non-standard initial mass function is considered, the results are only marginally different of those obtained for a white dwarf population, given the fact that in this case the role of red dwarfs is limited. In summary, our results are in fair agreement with those obtained by the EROS experiment, and appear to indicate that the microlensing optical depth, obtained by the MACHO collaboration, is probably an overestimate.

4. Discussion

Based on their ~ 15 microlensing events, the MACHO collaboration has derived an estimate of the halo fraction of dark matter f, as well as the MACHO mass m, using maximum-likelihood techniques. A similar analysis was completed by the EROS team, but with the significant difference that, in this case, no event was reported for the LMC, which implies that only an upper limit to the halo mass fraction can be obtained. To compare the results of the MACHO and EROS collaborations with our Monte Carlo simulations we have adopted as our reference model the isothermal sphere of core radius 5 kpc, with a value of $\rho_0 = 0.0079\,M_\odot$ pc^{-3} for the local dark matter density and disregarding the contribution of the LMC halo. Using this model, we have obtained that the optical depth towards the LMC is $\tau_{\rm LMC} = 5.1 \times 10^{-7} f$. The different estimates of the halo mass fraction f, as a function of mass, are plotted in Fig. 2. We show using a solid line the curve of the MACHO 95% confidence level, as taken from Alcock et al. (2000), and the EROS 95% confidence-level upper-limit, based on no observed events in the EROS-1 and EROS-2 data (Tisserand et al. 2007). We also represent the individual contributions of each population studied and the entire population, in addition to the corresponding 95% confidence level error bars. It is noticeable that the value obtained for the entire halo simulated population agrees within the 95% confidence level curves of both observational estimates. Our results therefore predict that the range of stellar masses within 0.08 and $10\,M_\odot$, provides $f = 0.05$ and an average mass of $0.411\,M_\odot$ to the halo dark matter, in agreement with the observational data.

5. Conclusions

We have extended our previous studies of the contribution to the halo dark matter of the white dwarf population and included the Galactic population of red dwarfs, we have estimated as well the contribution of these objects to the microlensing optical depth towards the LMC and compared our estimate with the measurements of the MACHO and EROS collaborations. Our estimate is based on a series of Monte Carlo simulations that incorporate the most up-to-date evolutionary tracks for red dwarfs, CO white dwarfs, and ONe white dwarfs, and reliable models of our Galaxy and the LMC. In a first set of simulations, we have found that the contribution of the red dwarf population practically doubles the contribution found so far for the white

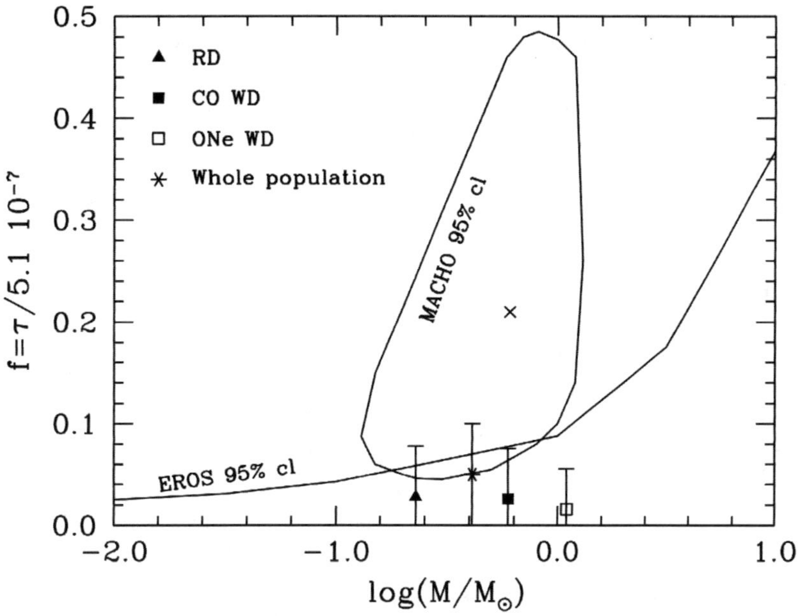

Figure 2. Halo dark matter fraction as a function of the mass of the lens. We plot, using a solid line, the 95% confidence-level curve for the MACHO experiment and the 95% confidence-level upper limit for the EROS experiment.

dwarf population. Our results indicate that the entire population of these stars can account for at most ~ 0.3 of the optical depth found by the MACHO team. This value implies that the contribution of the full range of masses between 0.08 and $10\,M_{\odot}$ represents 5% of the halo dark matter with an average mass of $0.4\,M_{\odot}$. Although this result is in partial agreement with the 95% confidence level MACHO estimate for a standard isothermal sphere and no halo LMC contribution, the expected number of events obtained by our simulations (3 events at the 95% confidence level) is substantially below the 11 observed MACHO events. These arguments reinforce the idea, previously pointed out by other studies, that the optical depth found by the MACHO team is likely an overstimate, probably due to contamination of self-lensing objects, variable stars and others. Moreover, we have assessed the compatibility between the observed event rate distribution and the ones obtained for the different populations under study. Our results show that the CO white dwarf population can reproduce fairly well the observed event-rate distribution although, as mentioned earlier, the expected number of events is considerable smaller. On the other hand, the negative results obtained by the EROS team towards the LMC are in agreement with our standard halo simulation. Finally, and for the sake of completeness, we have studied the effects of a log-normal biased initial mass function. In this case, the contribution of the red dwarf population is only marginal given that the production of low-mass stars is strongly inhibited. Accordingly, the total contribution to the microlensing optical depth is not different from that found in previous studies of the white dwarf contribution.

Acknowledgments
Part of this work was supported by MEC grants AYA05–08013–C03–01 and 02, ESP2007–61593, by the European Union FEDER funds and by the AGAUR.

References

Adams F C & Laughlin G 1996 *ApJ* **468** 686
Alcock C Allsman R A Alves D R et al. 1997 *ApJ* **486** 69
Alcock C Allsman R A Alves D R et al. 2000 *ApJ* **542** 281
Althaus L G García–Berro E Isern J Córsico A H & Rohrmann R D 2007 *A&A* **465** 249
Baraffe I Chabrier G Allard F & Hauschildt P H 1998 *A&A* **337** 403
Becker A C Rest A Stubbs C et al., 2005, *IAU Symposium* **225** 357
Belokurov V Evans N W Le Du Y 2003 *M.N.R.A.S.* **341** 1373
Belokurov V Evans N W Le Du Y 2004 *M.N.R.A.S.* **352** 233
Bennet D P Becker A C & Tomaney 2005 *ApJ* **631** 301
Binney J & Tremaine H 1987 *Galactic Dynamics* (Princeton: Princeton Univ. Press)
Camacho J Torres S Isern J Althaus L G & García–Berro E 2007 *A&A* **471** 151
Dehnen W & Binney J 1998 *M.N.R.A.S.* **298** 387
Flynn C Holopainen J & Holmberg J 2003 *M.N.R.A.S.* **339** 817
García–Berro E Torres S Isern J & Burkert A 2004 *A&A* **418** 53
Gates E I & Gyuk G 2001 *ApJ* **547** 786
Goldman B Afonso A Alard C et al. 2002 *A&A* **389** 69
Gyuk G Dalal N & Griest K 2000 *ApJ* **535** 90
Iben I & Laughlin G 1989 *ApJ* **341** 312
Isern J García–Berro E Hernanz M Mochkovitch R & Torres S 1998 *ApJ* **503** 239
Kallivayalil N van der Marel R P Alcock C Axelrod T Cook K H Drake A J & Geha M 2006 *ApJ* **638** 772
Kerins E & Evans N W 1998 *ApJ* **503** 75
Lasserre T Afonso C Albert J N et al. 2001 *A&A* **355** L39
Lucy L 2000 *M.N.R.A.S.* **318** 92
Muraki Y Sumi T Abe F et al. 1997 *ApJ* **490** 493
Paczyński B 1986 *ApJ* **304** 1
Rahvar S 2005 *M.N.R.A.S.* **356** 1127
Reylé C Robin A C & Créze M 2001 *A&A* **378** L53
Salaris M García–Berro E Hernanz M Isern J & Saumon D 2000 *ApJ* **544** 1036
Scalo J 1998 in *"The Stellar Initial Mass Function"* Eds.: G Gilmore & D Howell (San Francisco: PASP Conference Series), **142**, 201
Tisserand P Le Guillou L Afonso et al. 2007 *A&A* **469** 387
Torres S García–Berro E & Isern J 1998 *ApJ* **508** L71
Torres S García–Berro E Burkert A & Isern J 2002 *M.N.R.A.S.* **336** 971
Udalski A Szymanski M Kaluzny J et al. 1994 *Acta Astron.* **44** 1
Vidrih S Bramich D M Hewett et al. 2007 *M.N.R.A.S.* **382** 515
Zhao H S 1998 *M.N.R.A.S.* **294** 139

Journal of Physics: Conference Series **172** (2009) 012004 doi:10.1088/1742-6596/172/1/012004

The galactic population of white dwarfs

Ralf Napiwotzki

Centre for Astrophysics Research, University of Hertfordshire, College Lane, Hatfield AL10 9AB, UK

E-mail: r.napiwotzki@herts.ac.uk

Abstract. The contribution of white dwarfs of the different Galactic populations to the stellar content of our Galaxy is only poorly known. Some authors claim a vast population of halo white dwarfs, which would be in accordance with some investigations of the early phases of Galaxy formation claiming a top-heavy initial–mass–function. Here, I present a model of the population of white dwarfs in the Milky Way based on observations of the local white dwarf sample and a standard model of Galactic structure. This model will be used to estimate the space densities of thin disc, thick disc and halo white dwarfs and their contribution to the baryonic mass budget of the Milky Way. One result of this investigation is that white dwarfs of the halo population contribute a large fraction of the Galactic white dwarf number count, but they are not responsible for the lion's share of stellar mass in the Milky Way. Another important result is the substantial contribution of the – often neglected – population of thick disc white dwarfs. Misclassification of thick disc white dwarfs is responsible for overestimates of the halo population in previous investigations.

1. Introduction

The stellar populations of our Galaxy are divided into the thin disc, thick disc, halo and Galactic bulge. Only the thin disc still produces young stars. Star formation has ceased in all other populations at least 10 Gyr ago and all stars born more massive than the Sun have already evolved away from the main sequence, producing faint remnants: mostly white dwarfs plus some neutron stars and black holes. If the initial–mass–function in the early phases of galaxy formation was as top heavy as claimed by some high redshift investigations (e.g. Baugh et al. 2005) white dwarfs could be responsible for an important fraction of the Galactic mass budget.

Observations by the MACHO project of microlensing events in the Magellanic Clouds attributed to white dwarfs (Alcock et al. 2000) and a deep proper motion survey for high-velocity white dwarfs by Oppenheimer et al. (2001) appeared to confirm a high density of white dwarfs in the Galactic halo. Both results remain controversial. A recent analysis and discussion of the MACHO results is provided by Torres et al. (2008). The Oppenheimer et al. (2001) results have been criticised by Reid et al. (2001) and Pauli et al. (2003), but open questions remain. Space densities of thick disc and halo white dwarfs and their contribution to the Milky Way mass budget are still open questions.

I will present a Monte Carlo simulation of the white dwarf populations in our Galaxy. The simulation models the complete Galactic content of white dwarfs and post-AGB stars. Simulated stars can be selected according to criteria resembling the selection of observed samples. After a short description of the model I will describe how it is calibrated using work based on observations of white dwarfs with the Supernova type Ia Progenitor surveY (SPY; Napiwotzki et al. 2003).

This population model will be used to estimate the local space density of the three white dwarf populations and their contribution to the Galactic mass. Sect. 4 compares results with the Oppenheimer et al. (2001) observations. This contribution finishes with a discussion in Sect. 5.

2. Population model and calibration

The simulations are based on the model of the Galactic structure presented by Robin et al. (2003). All four stellar populations, thin disc, thick disc, halo and bulge (the latter not of relevance here) are included. Extinction by the gas and dust layer in the Galactic plane is taken into account using a simple analytical model. The impact of extinction on the white dwarf simulations presented here is not very large. Density laws, but not the zero points, and velocity dispersions were taken from Robin et al. (2003). Different from most white dwarf population synthesis models the age dependent scale height and velocity dispersion of the thin disc is taken into account. Stars are created at random positions according to the density laws, a random age and a random mass drawn from a Salpeter initial–mass–function. Short lived star bursts early in the lifetime of our Galaxy are assumed for halo and thick disc. The thin disc is modelled with a constant star formation rate.

The white dwarf progenitor lifetime is computed from the grids of evolutionary models created by the Padova group (Girardi et al. 2000). The simulations presented here used one single metallicity for every population. More realistic metallicity distributions are implemented now. The standard model uses the initial–final–mass relation (IFMR) from Weidemann (2000) and the cooling tracks are a blend of the Blöcker (1995) post-AGB tracks for hot white dwarfs and white dwarf sequences computed by Chabrier et al. (2000). Other IFMRs (e.g., that proposed by Dobbie et al. 2006) and Fontaine et al. (2001) cooling tracks are currently implemented and their effect will be tested.

A direct census of white dwarfs in the Galactic halo is impossible so far. All substantial samples of field white dwarfs are essentially collected from the local volume. Even at $V = 20$ (roughly corresponding to the spectroscopic white dwarf sample extracted from the SDSS archives) more than 98% of the observed white dwarfs are closer than 1 kpc in a region still dominated by the thin disc. However, classification of white dwarf membership in the Galactic populations is possible based on kinematic criteria. Pauli et al. (2003, 2006) combined accurate radial velocity measurements of white dwarfs from the SPY sample (Napiwotzki et al. 2001, 2003) with proper motion data from recent catalogues (UCAC2, USNO-B, SuperCosmos; see Pauli et al. 2006 for details) supplemented by own measurements. Distances were estimated from the spectroscopic analysis of the SPY spectra (Koester et al. 2001, Voss & Koester 2005). This work resulted in a sample of 400 white dwarfs brighter than $V = 16.5$ for which the complete 6-dimensional set of parameters in phase space (spatial and kinematical) is available.

Pauli et al. (2003, 2006) developed criteria to assign population memberships based on kinematical properties. An important tool for this analysis was the position in the $U - V$ plane (examples are shown in Fig. 1). U, V and W are the velocity components in the Galactic reference system. U is the component in direction of the Galactic centre, V the tangential component in direction of the orbit around the centre and W the velocity perpendicular to the Galactic plane. The classification was further refined using the position in the angular momentum–eccentricity plane and a check of the Galactic orbit as computed with the Odenkirchen & Brosche (1992) code. While thin disc stars have low eccentricity orbits confined to small distances from the Galactic plane, most thick disc stars have orbits of somewhat larger eccentricity reaching larger distances. Halo stars do not take part in the rotation of the Galactic disc around the centre and can have orbits which reach very large distances from the plane, sometimes with extreme eccentricities. Pauli et al. (2003, 2006) calibrated the classification criteria with main sequence stars, for which abundance information was available to assess population memberships. The resulting limits in the $U - V$ plane are shown in Fig. 1.

3. Galactic white dwarf populations

Pauli et al. (2006) classified a total of 398 white dwarfs using the criteria outlined above. The results are given in Table 1. We have to keep in mind that the numbers apply to a *brightness limited* sample ($V \leq 16.5$) of local white dwarfs. Three more steps are necessary before the white dwarf contributions to the Galactic mass budget can be estimated.

Table 1. Population classification of 398 white dwarfs from the SPY sample by Pauli et al. (2006).

population	$N_{\rm WD}$	
thin disc	368	92%
thick disc	23	6%
halo	7	2%

3.1. Correction for contamination

It is inevitable that, say, some thick disc stars have kinematical parameters which are typical for thin disc white dwarfs. This is not due to errors in the classification criteria, but an intrinsic property of the populations. This is demonstrated with simulated white dwarf samples in the $U-V$ plane in Fig. 1.

The first plot shows a simulated thin disc sample. The vast majority of stars lies within the thin disc boundaries defined by Pauli et al. (2006). Only a few white dwarfs are scattered into the thick disc or halo regions. Note that the kinematic properties of the white dwarfs in the simulation are independent of the Pauli et al. calibration, providing an independent verification of the chosen boundaries. The second plot shows the simulated thick disc sample. Again the outer boundaries are well selected, but it is obvious that quite a few thick disc white dwarfs will be misclassified as members of the thin disc. The third plot shows the result for the halo simulation. As expected the halo white dwarfs are distributed all over the $U-V$ plane, including the regions defined for the thin and thick disc.

The Monte Carlo model of the white dwarf populations can be used to correct for contamination or deficit. Since the corrections depend on the adopted space densities and the SPY results listed in Table 1 are used to calibrate the space densities in the population model, a few iterations are necessary until self-consistency of the corrected numbers is achieved. The final results are shown in Table 2. The main change to the uncorrected numbers is a substantial increase of the thick disc contribution. Corrections for the halo population cancel out.

3.2. Contributions in a volume limited sample

The corrected sample is still brightness limited. Many older and cooler white dwarfs are too faint to be observed. Since the thick disc and the halo populations are very old, they contain many old white dwarfs of very low luminosity. The luminosity functions in Fig. 2 show the much larger fraction of intrinsically faint white dwarfs compared to the thin disc. This results in an overrepresentation of the thin disc in any brightness selected sample of white dwarfs. The simulated white dwarf sample allows a straightforward correction of this bias (Table 2). We can use these results to compute the local space densities of the three populations. The local space density of all white dwarfs was estimated by Holberg et al. (2008) to be $\rho = 4.8 \times 10^{-3}\,{\rm pc}^{-3}$ based on a local sample of white dwarfs ($d \leq 13\,{\rm pc}$). Some objects were added or removed compared to an earlier version of the 13 pc sample (Holberg et al. 2002) and some low proper motion cool white dwarfs are probably still missing (Schröder et al. 2004), but overall this estimate of the

Journal of Physics: Conference Series **172** (2009) 012004 doi:10.1088/1742-6596/172/1/012004

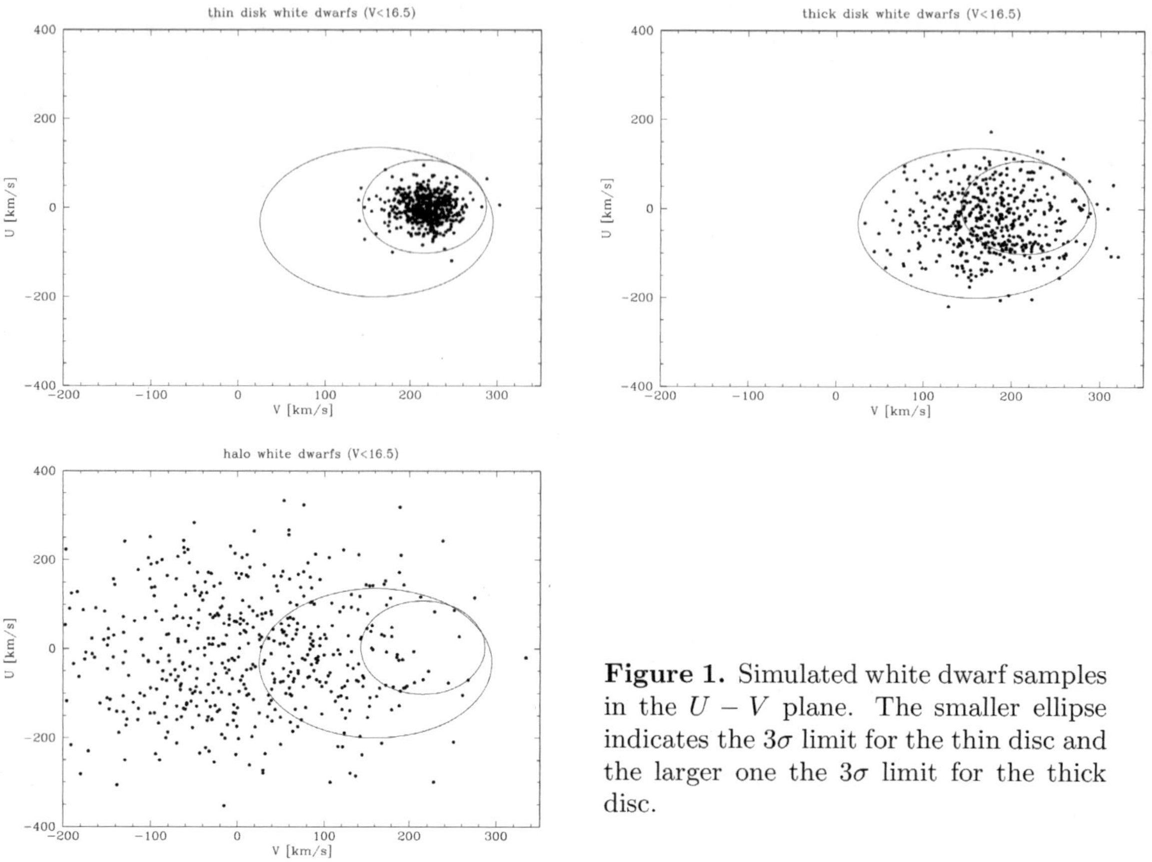

Figure 1. Simulated white dwarf samples in the $U - V$ plane. The smaller ellipse indicates the 3σ limit for the thin disc and the larger one the 3σ limit for the thick disc.

Table 2. Corrected results for white dwarf population membership. The column labelled SPY repeats the uncorrected fractions in the SPY sample given in Table 1. The next one (corr.) shows the fractions after correction for misclassifications. Relative fractions of the volume limited sample and local space densities ρ are given in the following columns. The last two columns give the relative fraction and total number extrapolated for the complete Galaxy.

population	SPY	corr.	vol. limited	ρ [pc^{-3}]	Galaxy	$N/10^9$
thin disc	92%	86%	59%	2.9×10^{-3}	17%	2.0
thick disc	6%	12%	35%	1.7×10^{-3}	34%	3.9
halo	2%	2%	6%	2.7×10^{-4}	49%	5.6

space density appears fairly robust and dramatic revisions are not expected for the future. The local white dwarf space densities of the three populations in Table 2 are computed combining the simulation results with the Holberg et al. (2008) white dwarf space density.

3.3. Galactic white dwarf populations

The final step is the extrapolation of results to the whole Galaxy. This can be easily done from the Monte Carlo simulations which produce a complete census of all white dwarfs in the

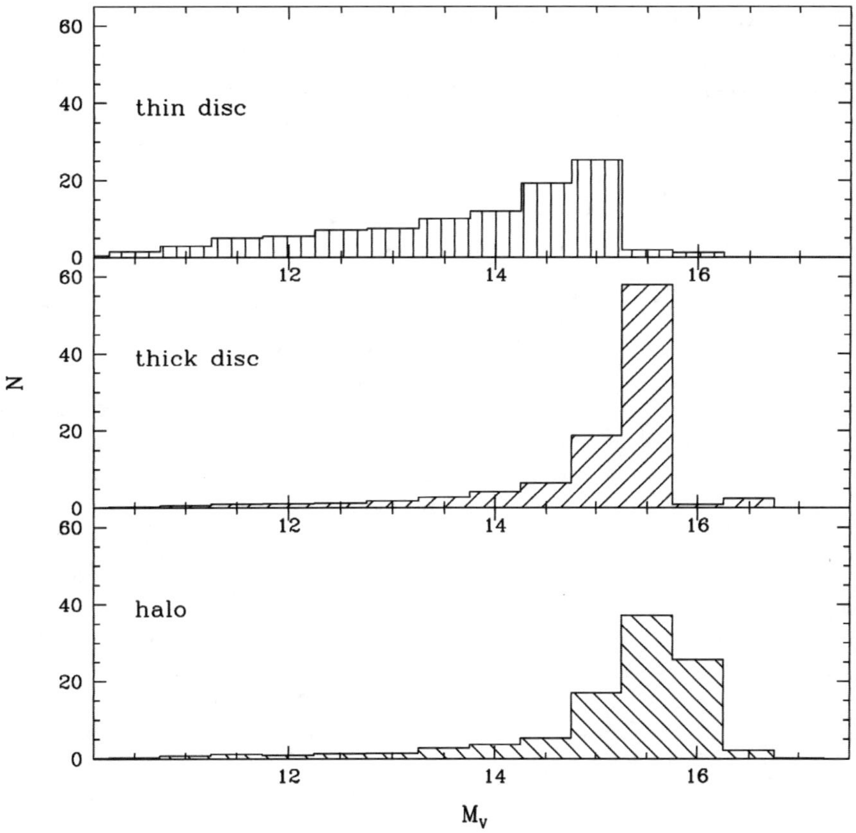

Figure 2. Simulated luminosity functions of thin disc, thick disc and halo white dwarfs. Each luminosity function is normalised to a total of 100 white dwarfs, although the real numbers of simulated white dwarfs are much higher.

Galaxy. Most of them are discarded in subsequent steps, because they are too distant or too faint, but the numbers are available after a run. The estimates of total white dwarf numbers and relative fraction of the Galactic population are given in Table 2. The results show that the Galactic population of white dwarfs is dominated by thick disc and halo. However, the white dwarf contribution to the total stellar mass of the Galaxy amounts not not more than $\approx$10%, far less than expected from the results of Alcock et al. (2000) and Oppenheimer et al. (2001).

A possible explanation for this apparent discrepancy will be discussed in the next section. However, let me first point out the limitations of these estimates. Obviously, the extrapolation relies on the validity of the model of the Galactic structure by Robin et al. (2003) implemented in the population model. Possible future modifications of state–of–the–art modelling of our Galaxy could result in significant changes of the white dwarf numbers. It can be expected that the relative numbers of white dwarfs and the relative contribution to the Galactic stellar mass are more robust than the total numbers. Another limitation comes from the number statistics. While the thin disc and thick disc white dwarf samples have a reasonable size, the size of the observed halo white dwarf sample is only seven. This causes a considerably statistical uncertainty. Improvement can only be expected from larger observed samples.

4. The Oppenheimer sample of high proper motion white dwarfs

Using the Monte Carlo model of the Galactic white dwarf population I have shown that the local, brightness limited sample observed by SPY implies a modest contribution of white dwarfs to the Galactic mass budget. This is in contrast to the claims by Alcock et al. (2000) and Oppenheimer et al. (2001) of large contributions of white dwarfs in the Galactic halo to the total mass of our Galaxy. Can the different results be reconciled? The interpretation of the MACHO results reported by Alcock et al. (2000) experienced some revision over the years. The reader is referred to Torres et al. (2008) and references therein for a discussion. Here I will have a look at the Oppenheimer et al. (2001) survey.

Oppenheimer et al. performed a survey of 4900 square degrees near the South Galactic Cap using digitised photographic plates. They selected high proper motion objects with $\mu > 0.33''/\mathrm{yr}$. These were plotted in a reduced proper motion diagram and a list of 126 candidates with a position consistent with halo membership was produced. Follow-up spectroscopy confirmed that most of them were white dwarfs. Oppenheimer et al. identified 38 candidates, which they claimed were cool white dwarfs in the Galactic halo.

I simulated a sample of white dwarfs using the local densities given in Table 2 and selection criteria similar to Oppenheimer et al.'s (2001) survey. The major difference being that the simulations constructed the reduced proper motion diagram using Johnson B and V filters instead of the photographic B_J and R measurements used in their investigation. I tried to define cuts similar to those of Oppenheimer et al. in the reduced proper motion diagram, but the match will not be perfect. Another effect not modelled in the simulation is caused by the large scatter associated with photographic brightnesses and colours. This can result in objects scattered into and out of the selection regions. However, the expected impact of both effects are relatively minor and will not affect the qualitative interpretation of the results.

The simulated sample consists of 140 candidates, which is consistent with the Oppenheimer et al. (2001) numbers. On the observational side it has to be expected that some candidates are missed, e.g. because of blending problems. Moreover, a certain error range has to be assumed for the SPY based calibration of white dwarf densities, as discussed in Sect. 3.3. Fig. 3 displays the temperature distributions of the simulated high proper motion sample. This plot verifies Oppenheimer et al.'s (2001) assumption that the thin disc contribution is small. However, the vast majority of cool-ish high white dwarfs in this sample belongs to the thick disc, not the halo population. This was already suspected by Reid et al. (2001). The temperature distributions in Fig. 3 are somewhat counter intuitive. As Fig. 2 shows, the (simulated) halo luminosity function contains even more faint and cool white dwarfs than the thick disc sample and typical space velocities are higher. However, space densities of the halo population are lower and typical distances of white dwarfs larger. Thus, while cool halo white dwarfs would easily make it through the proper motion criteria, many of them are not included in the survey because they are fainter than the brightness limit.

The misclassification of thick disc white dwarfs as halo objects is the crucial factor causing the very high estimate for the mass of the population of halo white dwarfs by Oppenheimer et al. (2001). They overestimated the local density of halo white dwarfs. Since the volume filled by the Galactic halo is much larger than the volume of the thick disc, the extrapolation to the whole Galaxy led to an dramatic overestimate of the mass. With correct classification of the white dwarfs the discrepancy disappears and SPY and Oppenheimer et al. sample are in good agreement.

5. Discussion and future work

A Monte Carlo simulation of the white dwarf population in our Galaxy is constructed. It was calibrated using population classifications derived by Pauli et al. (2006) from an analysis of the brightness limited SPY sample. It is straightforward to model selection criteria for observational

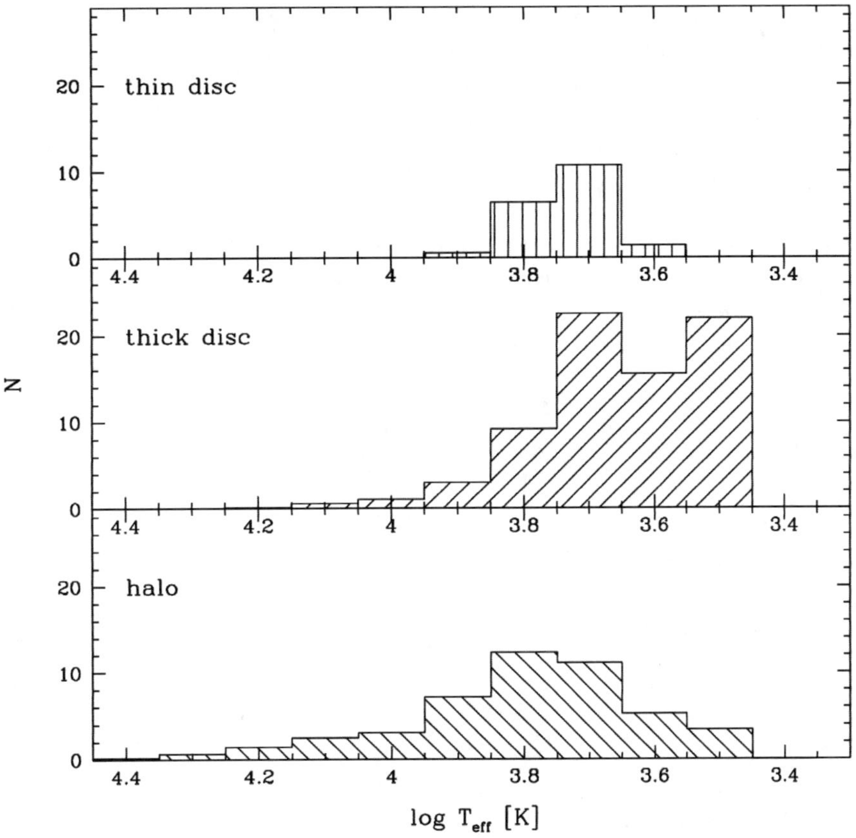

Figure 3. Simulated sample of high proper motion white dwarfs analogous to the selection criteria used by Oppenheimer et al. (2001).

samples in the Monte Carlo simulation. This was used to apply a statistical correction to the SPY results and compute the contribution of the Galactic populations to the local white dwarf space density and the Galactic number count. The estimated fraction of white dwarfs in the halo is ≈50% of the Galactic white dwarfs, but the overall contribution of white dwarfs to the Galactic mass budget is only modest. An important result of this study is that the often neglected thick disc white dwarfs contribute a third of the local white dwarf space density.

Neglect of the thick disc was also responsible for a gross overestimate by Oppenheimer et al. (2001) of the Galactic white dwarf mass. These authors aimed at classifying halo white dwarfs, but their sample contained a large fraction of thick disc white dwarfs. Extrapolation of their local numbers to the vast volume of the Galactic halo resulted in a very large estimate of the mass. The study presented here suggests a different interpretation with a more modest contribution of white dwarfs.

Open questions remain. Two scenarios were proposed to explain local white dwarfs with kinematical properties of the Galactic halo as thin disc stars, which were accelerated to high velocities after the companion exploded in a supernova type II (Davies et al. 2002) or in a single-degenerate supernova type Ia (Hansen 2003). Could this be true for the kinematically identified thick disc and halo white dwarfs in the SPY sample? The thin disc scenarios make two predictions: 1) the high-velocity white dwarfs should be single and 2) the masses should be similar to typical masses of thin disc white dwarfs. Although SPY is a radial velocity survey for binary white dwarfs, the first check can not be applied to the Pauli et al. (2006) sample, because suspected binary white dwarfs were intentionally excluded from the analysis. However, white dwarf masses are known from model atmosphere analysis of the SPY spectra (Koester et al. 2001, Voss & Koester 2005). Since star formation in the thick disc and halo ceased more than 10 Gyr ago, young white dwarfs of these populations have low mass progenitors and are thus expected to have low masses. Thick disc and halo candidates from the Pauli et al. (2006) sample have average masses of $0.48M_\odot$ and $0.49M_\odot$, respectively, well below the SPY average of $0.57M_\odot$ (Voss & Koester 2005). Thus there is little sign for a thin disc contamination of this sample. We are in the process of starting a deep photometric survey, which will trace the white dwarf populations to large distances from the Galactic plane and will allow direct measurements of space densities and scale heights.

Acknowledgments
The author gratefully acknowledges support by a PPARC/STFC Advanced Fellowship.

References
Alcock C Allsman R A Alves D R et al. 2000 *ApJ* **542**, 281–307.
Baugh C M Lacey C G Frenk C S Granato G L Silva L Bressan A Benson A J & Cole S 2005 *MNRAS* **356** 1191–1200.
Blöcker T 1995 *A&A* **299** 755–769.
Chabrier G Baraffe I Allard F & Hauschildt P 2000 *ApJ* **542** 464–472.
Davies M B King A & Ritter H 2002 *MNRAS* **333** 463–468.
Dobbie P D Napiwotzki R Burleigh M R Barstow M A Boyce D D Casewell S L Jameson R F Hubeny I & Fontaine G 2006 *MNRAS* **369** 383–389.
Fontaine G Brassard P & Bergeron P 2001 *PASP* **113** 409–435.
Girardi L Bressan A Bertelli G & Chiosi C 2000 *A&AS* **141** 371–383.
Hansen B M S 2003 *ApJ* **582** 915–918.
Holberg J B Oswalt T D & Sion E M 2002 *ApJ* **571**, 512–518.
Holberg J B Sion E M Oswalt T McCook G P Foran S & Subasavage J P 2008 *AJ* **135** 1225–1238.
Koester D Napiwotzki R Christlieb N et al. 2001 *A&A* **378** 556–568.
Napiwotzki R Christlieb N Drechsel H et al. 2001 *Astronomische Nachrichten* **322** 411–418.
Napiwotzki R Christlieb N Drechsel H et al. 2003 *The Messenger* **112** 25–30.
Odenkirchen M & Brosche P 1992 *Astron. Nachrichten* **313** 69–81.
Oppenheimer B R Hambly N C Digby A P Hodgkin S T & Saumon D 2001 *Science* **292** 698–702.
Pauli E M Napiwotzki R Altmann M Heber U Odenkirchen M & Kerber F 2003 *A&A* **400** 877–890.
Pauli E M Napiwotzki R Heber U Altmann M & Odenkirchen M 2006 *A&A* **447** 173–184.
Reid I N Sahu K C & Hawley S L 2001 *ApJ* **559** 942–947.
Robin A C Reylé C Derrière S & Picaud S 2003 *A&A* **409** 523–540.
Schröder K P Pauli E M & Napiwotzki R 2004 *MNRAS* **354** 727–736.
Torres S Camacho J Isern J & García-Berro E 2008 *A&A* **486** 427–435.
Voss B & Koester D 2005 *in* D Koester & S Moehler eds '14th European Workshop on White Dwarfs' Vol. **334** of *Astronomical Society of the Pacific Conference Series* pp. 655–658.
Weidemann V 2000 *A&A* **363** 647–656.

Axions and the white dwarf luminosity function

J Isern[1,2], S Catalán[1,2], E García–Berro[2,3] and S Torres[2,3]

[1]Institut de Ciències de l'Espai (CSIC), Facultat de Ciències, Campus UAB, 08193 Bellaterra, Spain

[3]Institut d'Estudis Espacials de Catalunya (IEEC), c/ Gran Capità 2–4, 08034 Barcelona, Spain

[3]Departament de Física Aplicada, Escola Politècnica Superior de Castelldefels, Universitat Politècnica de Catalunya, Avda. del Canal Olímpic s/n, 08860 Castelldefels, Spain

E-mail: isern@ieec.cat, catalan@ieec.cat, garcia@fa.upc.edu, santi@fa.upc.edu

Abstract. The evolution of white dwarfs can be described as a simple cooling process. Recently, it has been possible to determine with an unprecedented precision their luminosity function, that is, the number of stars per unit volume and luminosity interval. Since the shape of the bright branch of this function is only sensitive to the average cooling rate, we use this property to check the possible existence of axions, a proposed but not yet detected weakly interacting particle. We show here that the inclusion of the axion emissivity in the evolutionary models of white dwarfs noticeably improves the agreement between the theoretical calculations and the observational white dwarf luminosity function, thus providing the first positive indication that axions could exist. Our results indicate that the best fit is obtained for $m_{\rm a} cos^2\beta \simeq 2-6$ meV, where $m_{\rm a}$ is the mass of the axion and $cos^2\beta$ is a free parameter, and that values larger than 10 meV are clearly excluded.

1. Introduction

White dwarfs are the final evolutionary stage of low- and intermediate-mass stars ($M \leq 10 \pm 2M_\odot$). Since they are degenerate objects, they cannot obtain energy from thermonuclear reactions and their evolution is just a gravothermal process of cooling. They have a relatively simple structure composed by a degenerate core that contains the bulk of mass and acts as a reservoir of energy, and a partially degenerate envelope that controls the energy outflow. White dwarfs masses are in the range of $0.3 \leq M/M_\odot \leq 1.4$ (the Chandrasekhar's mass). Those with $M \leq 0.4 M_\odot$ have a core made of He, while those with a $M \geq 1.05 M_\odot$ have a core made of O and Ne. The remaining ones, the vast majority, have a core made of a mixture of C and O. All of them are surrounded by a thin helium layer with a mass ranging from 10^{-2} to $10^{-4}\, M_\odot$, which, in turn, is surrounded by an even thinner layer of hydrogen with a mass between 10^{-4} and 10^{-15} $M_\odot$, although about 25% of white dwarfs do not have such hydrogen envelopes. White dwarfs displaying hydrogen in their spectra are known as DA and the remainning ones as non-DA. Because of the different opacities involved, DA white dwarfs cool down much more slowly than non-DA white dwarfs.

Because of the simplicity of their structures, it has been proposed to use white dwarfs as laboratories to test new physics. The reasons for such a proposal are the following ones. Firstly, the evolution of white dwarfs is just a simple process of cooling. Secondly, the basic physical ingredients necessary to predict their evolution are well identified, although not necessarily well

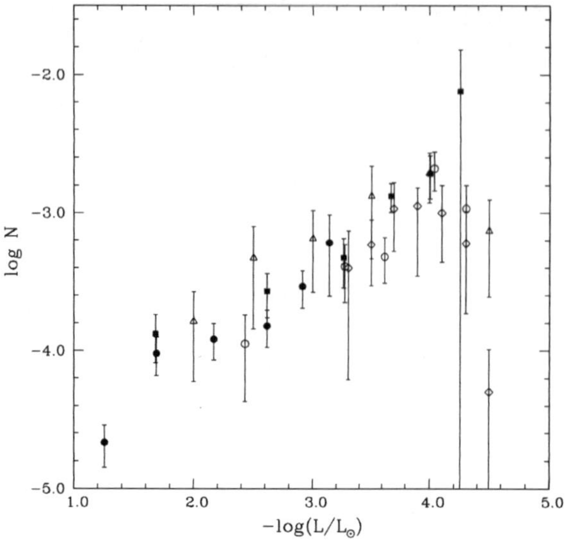

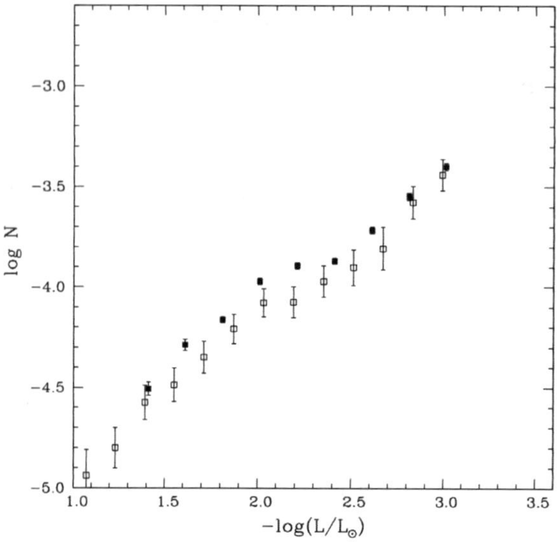

Figure 1. Left panel: luminosity functions obtained before the era of large surveys. The different symbols represent different determinations: Liebert, Dahn & Monet (1985), ● ; Evans (1992), ■; Oswalt, Smith, Wood & Hintzen (1996), △; Leggett, Ruiz & Bergeron (1998), ◇; Knox, Hawkins & Hambly (1999), ○ . Right panel: luminosity functions derived from the SDSS. One, ■, is composed by DA and non-DA white dwarfs identified from their photometry and proper motion (Harris et al. 2008). The other, □, is only composed by spectroscopically identified DA white dwarfs (DeGennaro et al. 2008)

understood, and, finally, there is a solid and continuously growing observational background that allows to test the different theories. Therefore, if an additional source or sink of energy is added, the characteristic cooling time is modified and these changes can be detected through the anomalies introduced in the luminosity function or in the secular drift of the pulsation period of degenerate variables — see Isern & García–Berro (2008) for details. In this paper we will limit ourselves to study the changes introduced by the introduction of axions in the white dwarf luminosity function.

2. The white dwarf luminosity function

The white dwarf luminosity function is defined as the number density of white dwarfs of a given luminosity per unit magnitude interval:

$$n(l) = \int_{M_{\rm i}}^{M_{\rm s}} \Phi(M)\,\Psi(\tau)\tau_{\rm cool}(l,M)\,dM \qquad (1)$$

where

$$\tau = T - t_{\rm cool}(l,M) - t_{\rm PS}(M) \qquad (2)$$

and $l = \log(L/L_\odot)$, M is the mass of the parent star (for convenience all white dwarfs are labeled with the mass of the main sequence progenitor), $t_{\rm cool}$ is the cooling time down to luminosity l, $\tau_{\rm cool} = dt/dM_{\rm bol}$ is the characteristic cooling time, $M_{\rm s}$ and $M_{\rm i}$ are the maximum and the minimum masses of the main sequence stars able to produce a white dwarf of luminosity l, $t_{\rm PS}$ is the lifetime of the progenitor of the white dwarf, and T is the age of the population under study. The remaining quantities, the initial mass function, $\Phi(M)$, and the star formation rate, $\Psi(t)$, are not known a priori and depend on the properties of the stellar population under study.

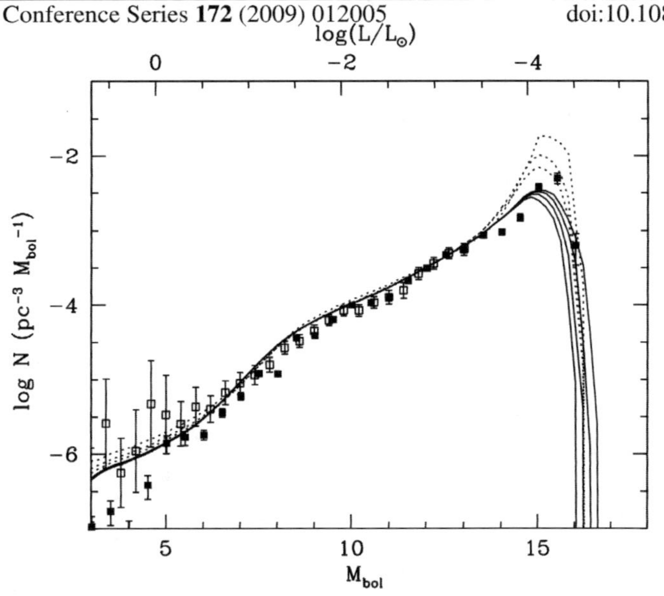

Figure 2. Theoretical luminosity functions of white dwarfs. The observational data are the same represented in the right panel of figure 1. Solid lines were obtained for different ages of the Galaxy — from left to right: 10, 11, 12 and 13 Gyr — and a constant star formation rate. Dotted lines were obtained using an age of 11 Gyr but exponentially decreasing star formation rates, with $\tau = 0.5$, 3 and 5 Gyr. All the luminosity functions have been normalized to the same observational data point.

The computed luminosity function is usually normalized to the bin with the smallest error bar, traditionally the one with $l = 3$, in order to compare theory with observations. Equation 1 shows that in order to use the luminosity function as a physical laboratory it is necessary to have good enough observational data and to know the galactic properties that are used in this equation (the star formation rate, the initial mass function and the age of the Galaxy).

The first luminosity function was derived four decades ago (Weidemann 1968) and since then it has been noticeably improved, see the left panel of Fig. 1. The monotonic behavior of this function clearly proves that the evolution of white dwarfs is just a cooling process. The sharp cut-off at low luminosities is a consequence of the finite age of the Galaxy. The availability of data from the Sloan Digital Sky Survey (SDSS) is noticeably improving the accuracy of the new luminosity functions. The one by Harris (2006) was built from a sample of 6000 DA and non-DA white dwarfs with accurate photometry and proper motions culled from the SDSS Data Release 3 and the USNO-B catalogue, whereas the one obtained by DeGennaro et al. (2008) was constructed from a sample of 3528 spectroscopically identified DA white dwarfs from the SDSS Data Release 4 (see the right panel of Figure 1). The quality of the data, specially in the central part of the bright branch, is so good that they allow to start testing the inclusion of new physics in the cooling process of white dwarfs.

It is evident that before introducing new ingredients it is necessary to have a good model of cooling able to reproduce as accurately as possible the observations. It is worthwhile here to remember that when the luminosity is large, $M_{\rm bol} < 8$, the evolution is dominated by neutrino emission. In this phase the main uncertainties come from our poor knowledge of the initial conditions. Fortunately, it has been shown that all the initial thermal structures converge towards a unique one. For smaller luminosities, $8 \leq M_{\rm bol} \leq 12$, the main source of energy is of gravothermal origin. In this phase, the Coulomb plasma coupling parameter is not large and the cooling can be accurately described. Furthermore, the energy flux through the envelope is controlled by a thick non-degenerate or partially degenerate layer with an opacity dominated

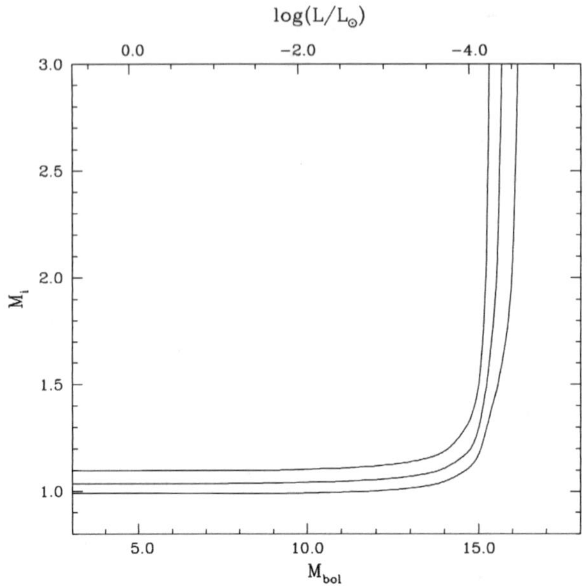

Figure 3. Minimum mass of the parent star able to produce a white dwarf of luminosity L for different ages of the Galaxy (9, 11 and 13 Gyr from top to bottom, respectively).

by hydrogen, when present, and helium, and it is weakly dependent on the metal content since metals sink towards the base of the envelope by gravitationally-induced diffusion. Below these luminosities, white dwarfs evolve into a region of densities and temperatures where the plasma crystallizes. When this happens, two additional sources of energy appear: the release of latent heat during crystallization and the release of gravitational energy induced by phase separation of the different chemical species (García–Berro et al. 1988). When the bulk of the star is solid, the white dwarf enters into the Debye cooling phase and the only important source of energy comes from the compression of the outer layers — see Isern et al (1998) for a detailed discussion. These late phases of cooling are not yet well understood (Isern et al 2000). Figure 2 shows that it is possible to construct a good standard model with the cooling sequences of Salaris et al. (2000), the initial-final mass relationship of Catalan et al. (2008), a constant star formation rate (solid line) and an age of the disk of 10.5 Gyr. The cooling models also assume a non-homogeneous distribution of carbon and oxygen in the core (Salaris et al. 1997), a pure helium layer of $10^{-2}\,M_*$ and on top of it a pure hydrogen layer of $10^{-4}\,M_*$, where M_* is the mass of the white dwarf.

The third condition to be fulfilled is the reliability of the Galactic data: the age of the Galaxy, the initial mass function and the star formation rate. An interesting feature of figure 2 is that the bright part of the white dwarf luminosity function — that with bolometric magnitude $M_{\rm bol} < 13$ — is almost independent of the assumed star formation rate. This can be explained with simple arguments. Since the characteristic cooling time is not strongly dependent on the mass of the white dwarf, Eq. (1) can be written as:

$$n = \langle \tau_{\rm cool} \rangle \int_{M_{\rm l}}^{M_{\rm u}} \phi(M)\psi(T - t_{\rm cool} - t_{\rm ps})\, dM. \tag{3}$$

Restricting ourselves to bright white dwarfs — namely, those for which $t_{\rm cool}$ is small — the lower limit of the integral is satisfied by low-mass stars and, as a consequence of the strong dependence of the main sequence lifetimes with mass, it takes a value that is almost independent of the luminosity under consideration (see figure 3). Therefore, if ψ is a well behaved function

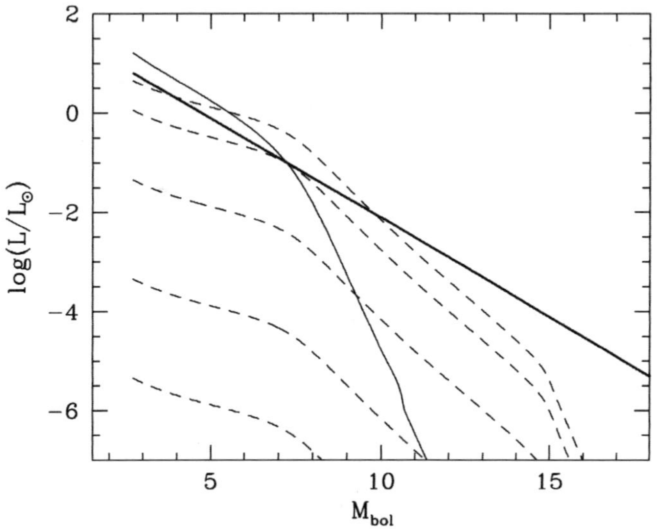

Figure 4. Energy losses for a typical $0.61\,M_{\odot}$ white dwarf as a function of the bolometric magnitude. The dashed lines represent the axion luminosity for different values of $m_{\rm a}\cos^2\beta$, where $m_{\rm a}$ is the mass of the axion, and $\cos^2\beta$ is a free parameter (from top to bottom: $m_{\rm a}\cos^2\beta = 10, 5, 1, 0.1, 0.01\,{\rm meV}$). The thick solid line represents the photon luminosity, while the thin solid line represents the neutrino luminosity.

and $T_{\rm G}$ is large enough, the integral is not sensitive to the luminosity, its value is absorbed by the normalization procedure and the shape of the luminosity function only depends on the averaged physical properties of white dwarfs. It is important to mention here that the initial-final mass relationship enters as a weight into the calculation of this average. Neverheless, since only those functions able to provide a good fit to the mass distribution of white dwarfs are acceptable, its influence on the shape of the bright branch of the luminosity function is minor (Isern et al 2008).

3. Axions

One solution to the strong CP problem of quantum chromodynamics is the Peccei-Quinn symmetry (Peccei & Quinn 1977a, b). This symmetry is spontaneously broken at an energy scale that gives rise to the formation of a light pseudo-scalar particle named axion (Weinberg, 1978; Wilczek, 1978). This scale of energies is not defined by the theory but it has to be well above the electroweak scale to ensure that the coupling between axions and matter is weak enough to account for the lack of positive detection up to now. The mass of axions and the energy scale are related by: $m_{\rm a} = 0.6(10^7\,{\rm GeV}/f_{\rm a})$ eV. Astrophysical and cosmological arguments have been used to constrain this mass to the range $10^{-4}{\rm eV} \leq m_{\rm a} \leq 10^{-4}{\rm eV}$. For this mass range, axions can escape from stars and act as a sink of energy. Therefore, if they exist, they can noticeably modify the cooling of white dwarf stars.

Axions can couple to photons, electrons and nucleons with a strength that depends on the specific implementation of the Peccei-Quinn mechanism. The two most common implementations are the KSVZ (Kim, 1979; Shifman et al., 1979) and the DFSZ models (Dine et al., 1981; Zhitniskii, 1980). In the first case, axions couple to hadrons and photons, while in the second they also couple to charged leptons. For the temperatures and densities of the white dwarfs under consideration, only DFSZ axions are relevant and in this case they can be emitted by Compton, pair annihilation and bremsstrahlung processes, but only the last mechanism turns out to be important. Figure 4 shows the energy losses for a typical white dwarf as a function of the bolometric magnitude. The axion emission rate (in erg/g/s) has been computed (Nakagawa

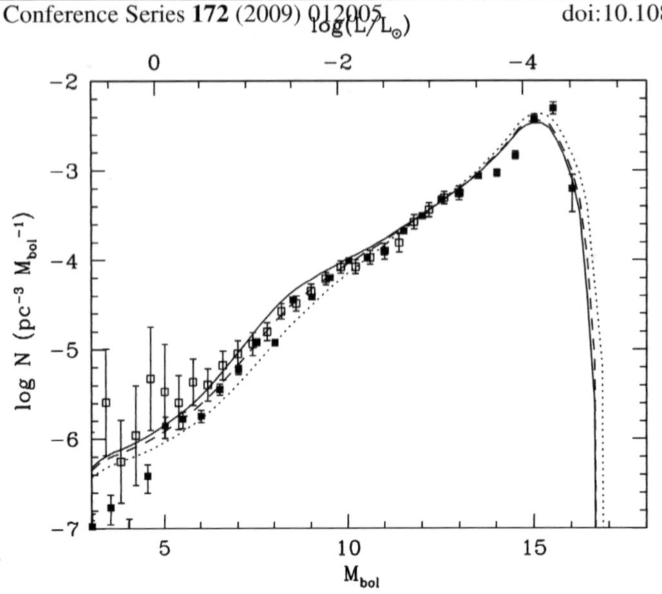

Figure 5. White dwarf luminosity functions for different axion masses: $m_{\rm a}\cos^2\beta = 0$ (solid line), 5 (dashed line) and 10 (dotted line) meV.

1987, 1988) as $e_{\rm a} = 1.08 \times 10^{23}\alpha Z^2/AT_7^4F$, where F is a function of the temperature and the density which takes into account the properties of the plasma, $\alpha = g_{\rm ae}^2/4\pi$ is related to the axion-electron coupling constant $g_{\rm ae} = 2.8\times10^{-11}m_{\rm a}\cos^2\beta/1\,{\rm eV}$. Since the core is almost isothermal, $L_{\rm a} \propto T^4$ in the region in which axions are the dominant sink of energy. The thick solid line represents the photon luminosity (Salaris et al 2000). For the region of interest $L_\gamma \propto T^\alpha$, with $\alpha \propto 2.6$, although this value changes as the white dwarf cools down. The thin solid line represents the neutrino luminosity (Salaris et al 2000) — which scales as $L_\nu \propto T^8$ — and is also dominated by the plasma and bremsstrahlung processes. Therefore, since the temperature dependence of the different energy-loss mechanisms is not the same, the luminosity function allows to disentangle the different contributions.

Figure 4 shows that in the region $M_{\rm bol} \sim 10$ the axion luminosity is not negligible when compared with the photon and neutrino ones. It also shows that the region around $M_{\rm bol} \sim 12$ provides a solid anchor point to normalize the luminosity function because there the observational data have reasonably small error bars, models are reliable, neutrinos are not relevant and axions, if they exist, are not dominant.

4. Results and conclusions

Figure 5 displays the luminosity function for different axion masses, a constant star formation rate and an age of the Galactic disk of 11 Gyr. As already mentioned, it is important to remember that the bright branch of the luminosity function is not sensitive to these last assumptions. All the luminosity functions have been normalized to the luminosity bin at $\log(L/L_\odot) \simeq -3$ or, equivalently, $M_{\rm bol} \simeq 12.2$. The best fit model — namely that which minimizes the χ^2 test in the region $-1 > \log(L/L_\odot) > -3$ (that is, $7.2 < M_{\rm bol} < 12.2$), which is the region where both the observational data and the theoretical models are reliable — is obtained for $m_{\rm a}\cos^2\beta \approx 5.5$ meV and solutions with $m_{\rm a}\cos^2\beta > 10$ meV are clearly excluded (Isern et al 2008). Figure 6 displays the behavior of χ^2 as a function of the mass of the axion for different normalization points. In all cases, χ^2 displays a pronounced minimum for masses in the range of roughly 2 to 6 meV except for $l = 2.51$ and 2.67 that are compatible with $m_{\rm a} = 0$. However, as it can be seen from figure 6, both points deviate from the behavior of the neighbouring values.

Journal of Physics: Conference Series **172** (2009) 012005 doi:10.1088/1742-6596/172/1/012005

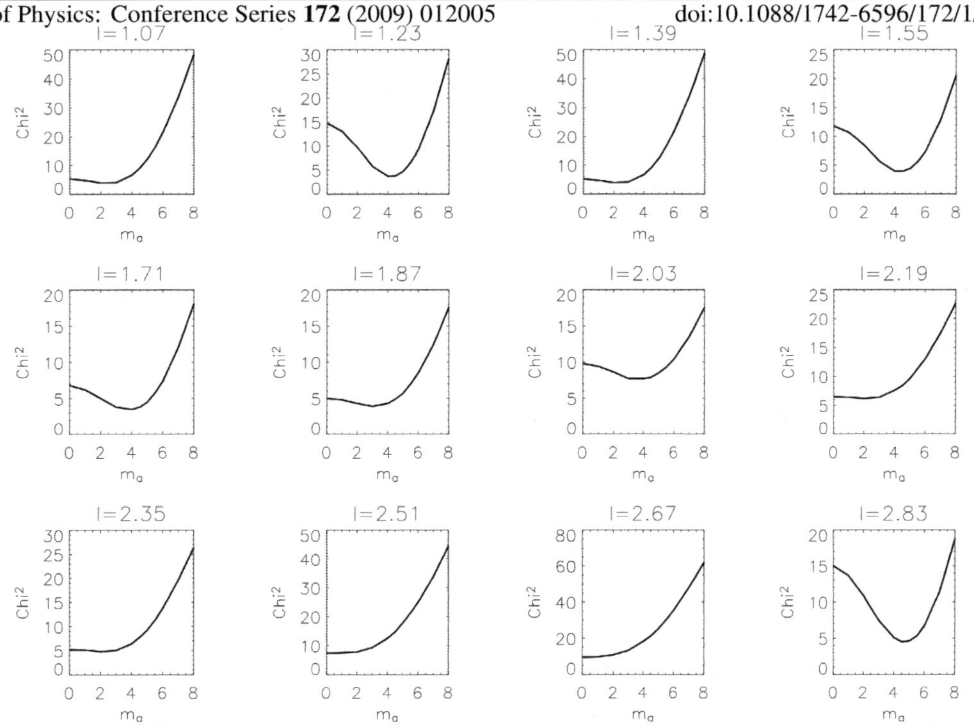

Figure 6. Value of χ^2 as a function of the mass of the axion adopting different normalization values.

Taken at face value, these results not only provide a strong constraint to the allowed mass of axions, but also a first evidence of their existence and a rough estimation of their mass. This is of course a strong statement and in order to be accepted it has to fulfill the following conditions: it has to be confirmed independently, it has to resist the introduction of any conventional effect not previously included, it has to provide testable predictions and it can not enter in contradiction with well established facts. A detailed discussion of the existing uncertainties is out of the scope of the present paper but it is worthwhile to enumerate some of them: influence of the metallicity in the age and core composition of the progenitor, transparency of the envelope and conversion of DA into non-DA white dwarfs, IMF, intial-final mass relationship, pathological SFRs or uncertainties in the observational luminosity function, especially on the role of He white dwarfs and the detailed shape of the brightest part of this function and others.

The results found here are completely compatible with the presently known constraints (Raffelt 2007), but since the predicted mass is near the allowed upper bound, it is natural to expect they will introduce some subtle changes in the late stages of stellar evolution as it is indeed the case of white dwarfs. A detailed discussion of this point is also out of the scope of this paper and we will just mention that these values are compatible with the bounds imposed by the drift of the pulsational period of the ZZ Ceti star G117−B15A (Isern et al. 1992; Córsico et al. 2001; Bischoff-Kim et al. 2008). It is also worthwhile to mention here that axions with $m_{\rm a}\cos^2\beta \approx 5$ meV would change the expected period drift of variable DB white dwarfs — which have values between $\dot{P} \sim 10^{-13}$ and 10^{-14} s s^{-1} (Córsico & Althaus 2004) — by a factor of 2, the exact value depending on the adopted temperature of the stellar core. The structure of the Sun would not be modified by the emission of axions of this masses since they would represent $L_{\rm a} \sim 10^{-6} L_{\odot}$. However, an instrument like CAST, able to operate in the region of masses $m_{\rm a} \sim$ meV and coupling constant $g_{{\rm a}\gamma} \sim 10^{-12}\,{\rm GeV}^{-1}$ could be able to detect them and confirm or rule out the masses predicted here. Cosmology could also provide some insight to this problem. Axions have been proposed as dark matter candidates but if we assume $\cos^2\beta = 1$, their contribution would be of the order of $\Omega_a h^2 \sim 10^{-4}$, where h is the Hubble constant normalized to 100 km/s/Mpc (Raffelt 2007), which rules out them as the main dark matter candidate. On the contrary, if dark energy is interpreted as the energy density of vacuum, axions could have an important role. The vacuum energy is obtained from the sum of the zero point energy of all the quantum fields (where bosons contribute positively and fermions negatively) and if we try to reproduce the observed value, that is positive, with just one quantum field we need a boson with a mass of the order or smaller than 10 meV (Friemann, Turner & Huterer, 2008), that is just what we have found here. This is indeed a remarkable coincidence.

To conclude, it is evident that our claim about the existence of axions has to be taken with caution, but it clearly shows that the white dwarf luminosity function can provide strong arguments to solve the long-lasting problem of the CP violation. Future work should be devoted to improving the observational data, especially for the brightest part of the white dwarf luminosity function. The theoretical models could also be improved and also additional efforts should be devoted to find new independent methods to detect axions. Finally, it is worthwhile to say that even in the case of a negative result, an accurate and precise luminosity function could be used to provide insight to many other problems like the hypothetical drift of the gravitation constant or the magnetic momentum of neutrinos just to cite two cases.

Acknowledgments

Part of this work was supported by the MEC grants AYA05-08013-C03-01 and 02, and ESP2007-61593, by the European Union FEDER funds and by the AGAUR.

References

Bischoff-Kim A Montgomery M H & Winget D E 2008 *ApJ* **675** 1512
Catalán S Isern J García–Berro E & Ribas I 2008 *MNRAS* **387** 1693
Córsico A H & Althaus L G 2004 *A&A* **428** 159
Córsico A H Benvenuto O G Althaus L G Isern J & García–Berro E 2001 *New Astron.* **6** 197
DeGennaro S von Hippel T Winget D E Kepler S O Nitta A Koester D & Althaus L 2008 *ApJ* **135** 1
Dine M Fishler W & Srednicki M 1981 *Phys. Lett. B* **104** 199
Evans D W 1992 *MNRAS* **255** 521
Friemann J A Turner M S & Huterer D 2008 *ARAA* **46** 385
García–Berro E Hernanz M Isern J & Mochkovitch R 1988 *Nature* **333** 642
Harris H C, et al. 2006 *AJ* **131** 571
Isern J Mochkovitch R García–Berro E & Hernanz M 1998 *J. Phys.: Cond. Matt.* **10** 11263
Isern J García–Berro E Hernanz M & Chabrier G 2000 *ApJ* **528** 397
Isern J & García–Berro E 2008 *Mem. SAI* **79** 545
Isern J García–Berro E Torres S & Catalán S 2008 *ApJ* **682** L109
Isern J Hernanz M & García–Berro E 1992 *ApJ* **392** L23
Kim J 1979 *Phys. Rev. Lett.* **43** 103
Knox R A Hawkins M R S & Hambly N C 1999 *MNRAS* **306** 736
Legget S K Ruiz M T & Bergeron P 1998 *ApJ* **497** 294
Liebert J Dahn C C & Monet D G 1988 *ApJ* **332** 891
Nakagawa M Kohyama Y & Itoh N 1987 *ApJ* **322** 291
Nakagawa M Adachi T Kohyama Y & Itoh N 1988 *ApJ* **326** 241
Oswalt T D Smith J A Wood M A & Hintzen P 1996 *Nature* **382** 692
Peccei R D & Quinn H R 1977a *Phys. Rev. D* **16** 1791
Peccei R D & Quinn H R 1977b *Phys. Rev. Lett.* **38** 1440
Raffelt G G 2007 *J. Phys. A: Math. Theor.* **40** 6607
Shifman M A Vainshtein A I & Zakharov V I 1980 *Nucl. Phys. B* **166** 43
Salaris M Domínguez I García–Berro E Hernanz M Isern J & Mochkovitch R 1997 *ApJ* **486** 413
Salaris M García–Berro E Hernanz M Isern J & Saumon D 2000 *ApJ* **544** 1036
Weidemann V 1968 *ARAA* **6** 351
Weimberg S 1978 *Phys. Rev. Lett.* **40** 223
Wilczek F 1978 *Phys. Rev. Lett.* **40** 279
Zhitnitskii A P 1980 *Sov. J. Nucl.* **31** 260

Journal of Physics: Conference Series **172** (2009) 012006 doi:10.1088/1742-6596/172/1/012006

SDSS White Dwarf mass distribution at low effective temperatures

D Koester[1], S O Kepler[2], S J Kleinman[3] and A Nitta[3]

[1] Institut für Theoretische Physik und Astrophysik, Universität Kiel, 24098 Kiel, Germany
[2] Instituto de Fisica, Universidade Federal do Rio Grande do Sul, 91501-970 Porto Alegre, RS Brazil
[3,4] Gemini Observatory, 670 N A'ohoku Pl., Hilo HI 96720 USA

E-mail: `koester@astrophysik.uni-kiel.de`

Abstract. The DA white dwarfs in the Sloan Digital Sky Survey, as analyzed in the papers for Data Releases 1 and 4, show an increase in surface gravity towards lower effective temperatures below 11500 K. We study the various possible explanations of this effect, from a real increase of the masses to uncertainties or deficiencies of the atmospheric models. No definite answer is found but the tentative conclusion is that it is most likely the current description of convection in the framework of the mixing-length approximation, which leads to this effect.

1. Introduction

Almost every contribution at this white dwarf meeting mentions the Sloan Digital Sky Survey (SDSS) or is even based on its data. Because of the huge impact the survey has on this field, it is very important to critically analyze all aspects of the data collection and reduction process, to understand possible uncertainties and biases. One problem related to white dwarfs, which was obvious already in the first white dwarf catalogue publication (Kleinman et al. 2004), was the apparent systematic increase of the mean surface gravity of DAs below effective temperatures of approximately 12000 K, determined from spectral fits, was again demonstrated in the data of the Data Release 4 (Eisenstein et al. 2006) and confirmed by Kepler et al. (2007) and DeGennaro et al. (2008).

While the huge number of objects and the homogeneity of the survey make this feature very obvious, it is by no means a new discovery. A similar trend was first recognized with data from the Palomar-Green Survey (Bergeron et al. 1990; Bergeron 1992, and several other papers of the Montreal group) and more recently by Liebert et al. (2005). A similar effect was also found in the data from the ESO Supernova Ia Progenitor Survey (SPY) by Voss (2006). A first conclusion therefore is that this effect is not a peculiarity of the SDSS data collection or reduction processes. In this paper we study the remaining possibilities

- the effect could be real, i.e. the DAs show a real increase of the average mass towards lower temperatures
- it could be caused by the fitting between observations and theoretical models used to determine the atmospheric parameters
- uncertainties in input physics or a flaw in the models used

For our study we use the data of DR4 as provided from the SDSS database, with slight changes in reduction software compared to Eisenstein et al. (2006). We also use very nearly the same models as used in that work for our "standard analysis" (see below). The major difference are the fitting algorithms: whereas the SDSS uses the whole spectra, adjusted with a polynomial using information from the photometry, we here use only the spectral lines, with the continuum adjusted on both sides of the line center. This technique minimizes the influence of uncertainties in the flux calibration. The SDSS photometry gives an independent estimate of parameters, which is used to distinguish between the two solutions, which exist for most objects between 8000 and 18000 K.

In the main part of this study we vary some of the parameters in the input physics, in the hope of finding some hints for the origin of the effect. This concerns the treatment of convection, of the non-ideal effects in the occupation numbers, and a possible admixture of invisible helium. All these possible explanations have been proposed and studied in various papers of the Montreal group. We partially repeat this work here with a much larger and very homogeneous sample.

2. Statement of the problem

Figure 1 shows a subsample of the brighter, cool (< 16000 K) DAs in the DR4 results. The parameters were determined by SJK; the only significant difference to Eisenstein et al. (2006) was a larger model grid. While the scatter of the distribution and trend towards large $\log g$ is similar for the two subsets, it is much smaller than in the total sample, which includes many fainter stars (see DR4 for a comparison). A S/N of at least 20, and preferably > 30 is a minimum requirement for reliable spectral analysis; we will therefore in the remaining tests use the bright sample with $g < 19$. The trend in $\log g$ becomes even clearer, if we bin the results in $T_{\rm eff}$ intervals of 500 to 1000 K (Fig. 2). The spectroscopic fits show a clear increase below $T_{\rm eff} = 11500$ K, which is somewhat more pronounced in the DR4 results than in our own fits. Since everything else is very similar, we attribute this difference to the different fitting methods. A tentative explanation could be that the SDSS polynomial correction to the flux calibration is problematic for the very extended wings of the broad Balmer lines, especially in this temperature range.

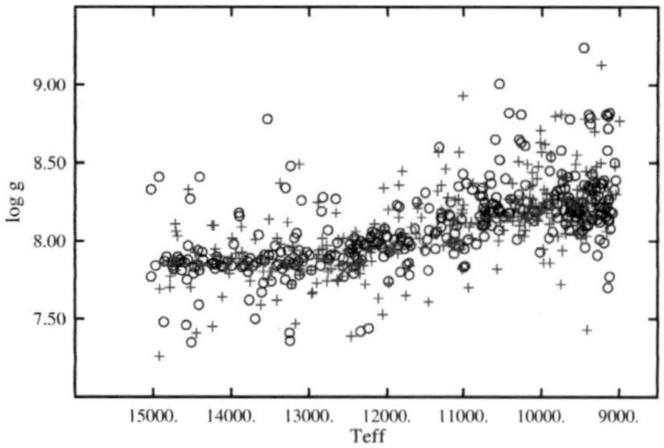

Figure 1. Distribution of $\log g$ with $T_{\rm eff}$ for a subsample of cool DAs from DR4. Crosses (red): 347 DAs with magnitude $g < 18$; circles (black): 360 DAs with g between 18 and 19.

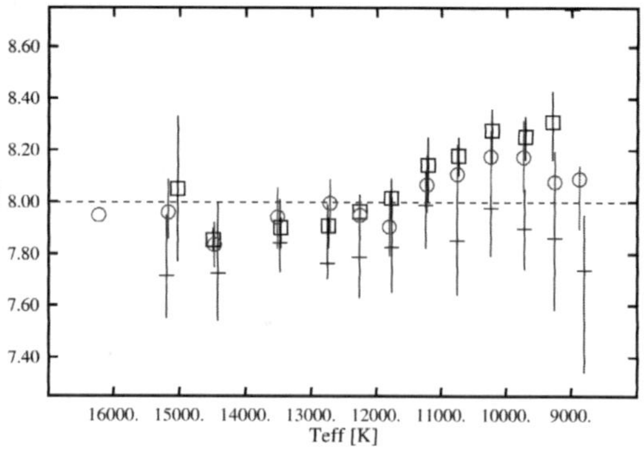

Figure 2. Average $\log g$ values and distribution in bins of $T_{\rm eff}$. Squares (black): original results from DR4; circles (red): our new fits to the spectra; horizontal bars (blue): results from photometry for the same objects. The vertical lines in all cases are the intervals containing the "inner" 50% of the distribution, cutting 25% on either side.

The surface gravity derived from fitting the 5 SDSS magnitudes simultaneously are slightly lower than the spectroscopic results. They also show a small increase of $\log g$ for lower $T_{\rm eff}$ but this is much less pronounced than in the spectroscopic data. We would like to make a few remarks regarding the SDSS photometry, which is not used as much for the analysis of white dwarfs as it deserves. In the case of the DAs the discrimination for $T_{\rm eff}$ and $\log g$ is very good in the range from 16000 K to 8000 K, and a determination using e.g. the two-color diagram $u - g$ vs. $g - r$ can in principle give results as accurate as from spectroscopy, without the problem of having two possible solutions above and below the Balmer line maximum. However, there is still some small uncertainty in the calibration of the magnitude zero-points. As a test, we have increased the observed g magnitude by 0.03, which results in a perfectly constant average $\log g$ sequence close to 8.0. While we do not imply that the g uncertainty is as large as this, we believe that because of the calibration uncertainties the photometry should at this time not have very high weight in a study of *systematic* effects in the stellar parameters. We will, however, in the following sections also show the effect of model changes on the photometric solutions, which in an ideal case would agree with the spectroscopic solution.

3. Could the mass change with $T_{\rm eff}$ be real?

The input physics for our "standard" models is the same as used for the DR4 fits and will be described in more detail below. Our spectroscopic fits with these models to the $\approx$ 700 DAs brighter than g = 19 and $T_{\rm eff}$ between 8500 and 16000 K is our reference set for this study. There are 61 DAs with multiple spectra in this set; from the differences in the determined parameters we estimate a typical uncertainty of 2.4% for $T_{\rm eff}$, and 0.15 in $\log g$. Since the instrumentation, reduction, models, and analysis methods were the same, this is certainly still an underestimate of the real errors. Nevertheless, the number of objects are large, and the $\log g$ increase so systematic that it certainly cannot be explained by uncertainties of the parameter determinations.

With our reference fits the mean mass of the DAs is 0.573 $M_\odot$ for the objects above 11500 K and 0.658 $M_\odot$ below. Could this increase of the white dwarf masses be a real effect? Possible explanations might be different cooling times for high versus low mass white dwarfs, or higher mass progenitors (and consequently high mass wds) for older stars, which originated from older progenitors with lower metal abundances. The latter argument is, however, contradicted by the findings of Hansen et al. (2007), that old white dwarfs in globular clusters have masses close to 0.5 $M_\odot$. All these arguments have been discussed in Kepler et al. (2007) and found not to be convincing. It is also very difficult to explain the rather sudden increase around 11500 K with these explanations, which all should lead to a more gradual transition if any.

4. Non-ideal effects in equation of state and occupation numbers

The standard treatment of these effects in white dwarf atmosphere models is the Hummer-Mihalas-Däppen occupation probability mechanism (Hummer & Mihalas 1988, Mihalas et al. 1988, Dappen et al. 1988). For any effect above 10000 K only the interactions with charged particles are of interest. Bergeron (1993) has demonstrated that a more consistent fit between low and higher Balmer lines can be achieved by increasing the parameter $\beta_{\rm crit}$ in the HMD formalism from 1 to 2. The effect is to decrease the absorption in the wings of the Balmer lines, compared to the standard treatment. We have experimented with this parameter and show two extreme cases in Figs. 3 and 4. For the smaller value (stronger line wings) the spectroscopic solution results in smaller $\log g$, as expected, and the values agree better with the photometric solution. The opposite is true for larger $\beta_{\rm crit}$. Neither of the changes offers any real improvements, since the jump at 11500 K remains. This could also be expected, because the effect of changing the strength of the charged particle interaction should be continuous around this critical region. We therefore exclude this change as a possible solution.

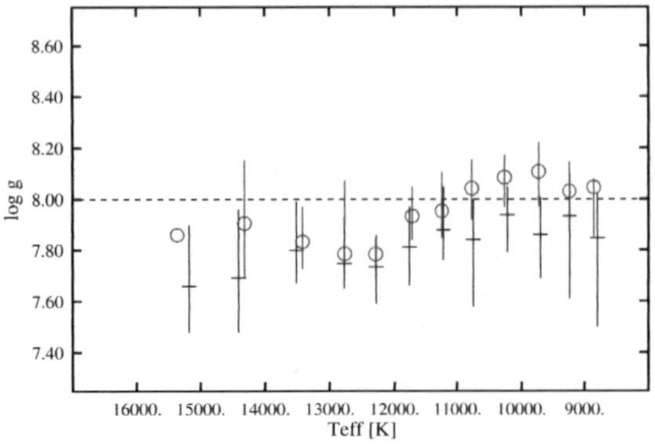

Figure 3. Average of $\log g$ as a function of T_{eff} using models with β_{crit} decreased to 1. Circles (red): spectroscopic solution; horizontal bars (blue): photometric solution

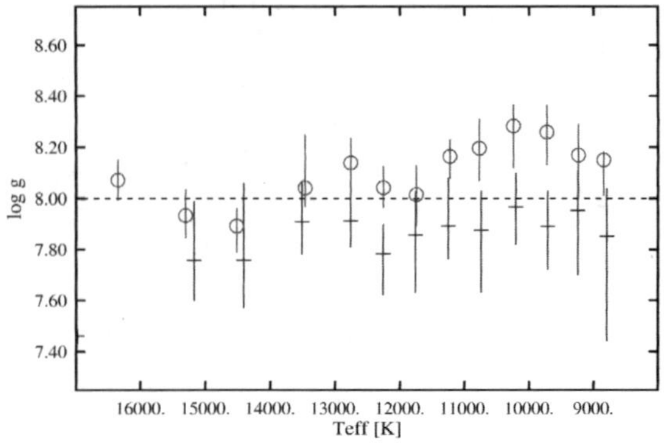

Figure 4. Similar to Fig. 3, but β_{crit} increased to 5.

5. Invisible helium in the atmosphere

This possible explanation was also first proposed by the Montreal group (Bergeron et al. 1990). Why could this solve the problem of increasing $\log g$? The surface gravity, which along with T_{eff} are the stellar parameters determined in an analysis, is only a proxy. The emission from a hot plasma is — in addition to the elemental abundances — determined by the thermodynamic variables, e.g. pressure and temperature. These variables are of course not constant throughout the visible atmosphere, but they can be described by characteristic values. For the temperature this is the effective temperature. The pressure is determined by the hydrostatic equation $dP/d\tau = g/\kappa$ Assuming a constant absorption coefficient (per mass) κ this can be integrated to $\tau = 1$ to give $P \sim g/\kappa$, showing that the pressure is inversely proportional to the absorption coefficient per mass. The latter is much smaller for helium than for hydrogen around 10000 – 12000 K, and therefore a significant helium pollution leads to higher pressure, mimicking a higher surface gravity.

A possible source of helium could be helium dredge-up by the convection zone (CVZ), which (with the standard description of convection in white dwarfs) increases significantly in depth for effective temperatures below 11500 K. If the total amount of hydrogen in the outer layer

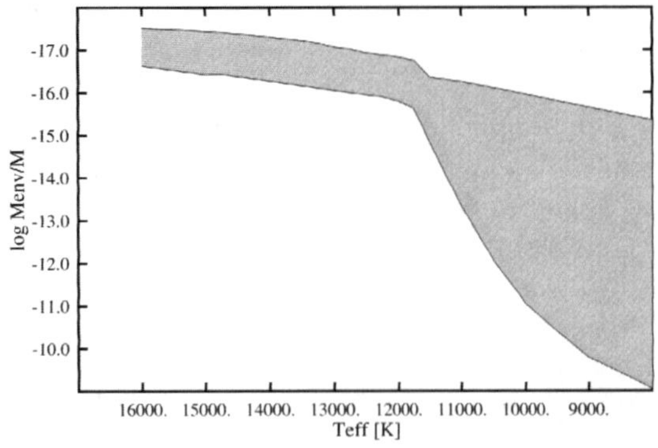

Figure 5. Size of the convection zone in DA white dwarfs as a function of T_{eff}. The two continuous (red) lines are the upper and lower boundary, the gray area is the convection zone. The vertical axis is a mass scale, giving the logarithm of the fractional mass counted from the surface of the star. Note the sharp increase in depth around 11500 K. The mixing-length parameters for this calculation are ML2/0.6 (see text)

is less than about 10^{-10} of the total mass, such a mixing could happen near the interesting temperature range.

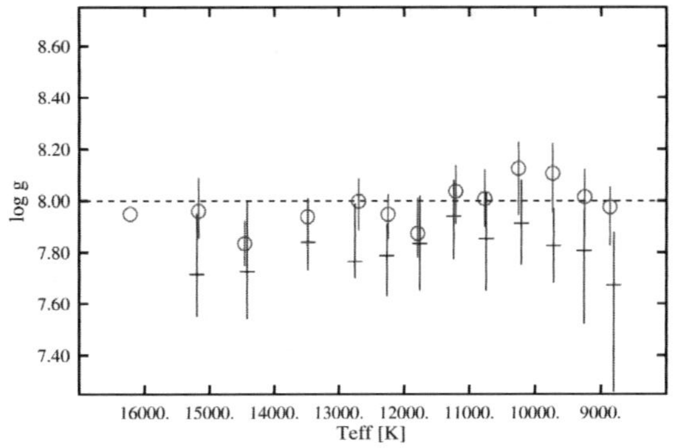

Figure 6. Effect of adding helium to the atmosphere with a number ratio of He:H = 0.1. Symbols and colors are the same as in the previous figures.

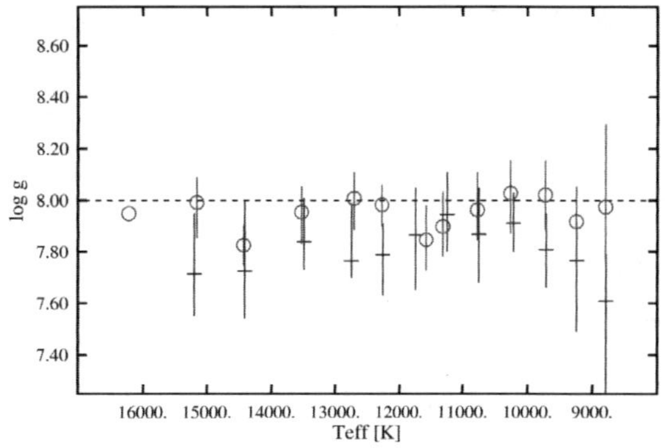

Figure 7. Adding helium with an abundance of He:H = 1 by numbers.

For simplicity we show in Figs. 6 and 7 the effect of adding a helium contamination with He:H = 0.1 and 1 by numbers in all models with $T_{\rm eff} \leq 11500$ K. Obviously this goes in the direction of a constant mass for all temperatures. In reality we would expect a helium pollution, which increases with decreasing $T_{\rm eff}$, with an even more convincing result (removing the dip between 11000 and 12000 K).

Could such a helium contamination be detected directly? We remind the reader of two recently discovered objects, which provide some guidance - HS 0146+1847 and GD362. Both stars were at first classified as massive DAZ, with very strong Balmer lines and CaII (Koester et al. 2005; Gianninas et al. 2004; Zuckerman et al. 2007). Yet, at closer inspection, they turned out to be in fact helium-rich, with normal masses. In both cases high resolution and high S/N spectra were necessary to discover the weak helium lines, although the helium content is far larger than necessary to explain the average $\log g$ increase. In order to find more similar objects we have systematically searched ≈ 1600 spectra from the SPY survey for weak helium lines, with no new discoveries. However, an abundance ratio of He:H = 1 or smaller would in general not be detectable at the typical S/N of that survey. We also note, that both helium-rich objects also have metal traces, indicating that some kind of accretion has happened and the stars are peculiar. Nevertheless, from the observational point alone, helium pollution could be the solution.

¿From the theoretical side this would be very difficult to understand. The latest asteroseismological study of ZZ Cetis (Castanheira & Kepler 2008) finds a few objects with a hydrogen layer $\log M_{\rm H}/M < -8$ (none < -10), but the vast majority have much thicker layers where no mixing is expected in the ZZ Ceti range.

6. The treatment of convection

The description of the energy flux by convection is one of the few remaining problems in stellar astrophysics. In the field of white dwarfs it has become customary to increase the freedom in the mixing-length approximation with 3 more parameters in addition to the mixing-length itself. The different versions are distinguished as ML1, ML2, ML3, with the mixing length added as in ML1/$\alpha = 2.0$ or short ML1/2.0 — see e.g. Fontaine et al. (1981) and Koester et al. (1994) for the nomenclature. For our current study the main point is that these versions differ in the efficiency of energy transport: the more efficient a version is the smaller can the temperature gradient be which transports the necessary flux. This temperature gradient in turn has a strong influence on the line profiles, and to a lesser degree also on the continuum energy distribution. We have experimented with various mixing-lengths in the framework of the ML1

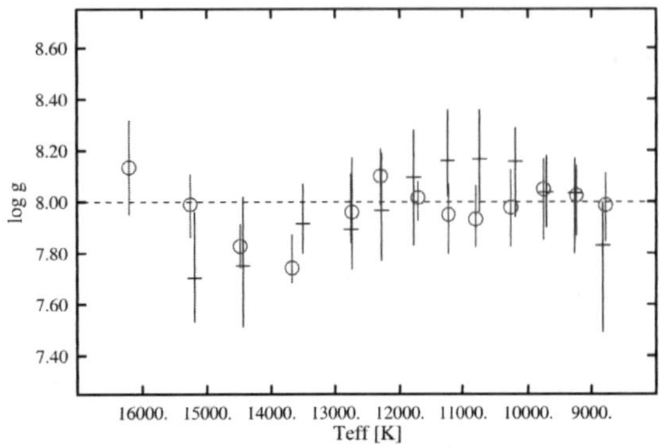

Figure 8. Convection completely suppressed, energy transport by radiation only.

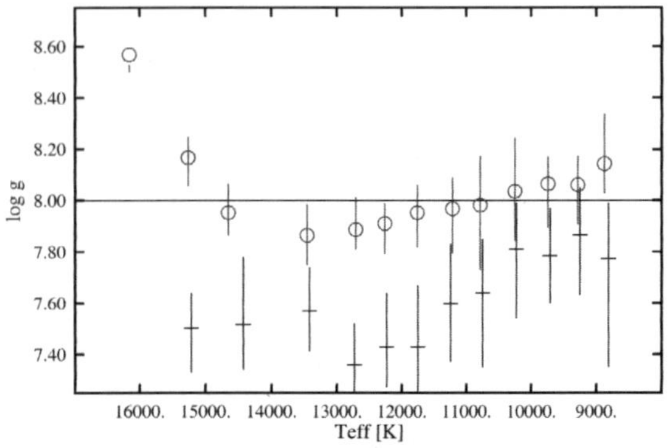

Figure 9. Temperature gradient forced to adiabatic in convective regions.

and ML2 versions. The changes always bring changes to the sequence of average $\log g$, but none solves the problem in the sense that a constant average results. As two extreme examples (which are outside the ML framework) we present in Figs. 8 and 9 a model set with convection totally suppressed (extremely inefficient) and another one with the convective gradient forced to the adiabatic gradient (infinite efficiency). In the second case this has an influence even at rather high temperatures, because there are thin convection zones, which are, however, normally inefficient and don't change the radiative gradient.

While this result is not encouraging, we nevertheless want to argue, that incorrect description of the convective gradient is still the most likely and in our opinion the only possibility to account for the observed effect.

First, there is a very striking coincidence near ≈ 11500 K between the steep increase in the thickness of the convection zone and the increase in the average $\log g$. Of the other possible solutions only the helium contamination would be able to explain this coincidence.

Second, and even more important: we definitely know that our current description of convection is incorrect. The current choice of parameters in the mixing-length approximation (ML2/0.6, or very similarly ML1/1.75) was introduced by Koester et al. (1994) and Bergeron et al. (1995) to allow a consistent parameter determination from optical and ultraviolet spectra. The effective temperatures found with this version for ZZ Cetis are in many cases so hot, that the thermal time scale of the CVZ is at most a few seconds, much shorter than the pulsation periods. On theoretical grounds, these timescales are expected to be of similar magnitude, and consequently Bergeron et al. (1995) concluded that no single choice of ML parameters can simultaneously reproduce optical and UV spectra *and* the location of the instability strip. Convection must be more efficient in deeper layers than in the atmosphere, at least when studied within the ML framework.

This same conclusion was reached completely independently by Ludwig et al. (1994). When comparing the temperature profile from a two-dimensional hydrodynamic simulation of a white dwarf with $T_{\rm eff} = 12600$ K, $\log g = 8$ with mixing-length models of the atmosphere and envelope they found that the mixing-length must increase with depth.

7. Conclusions

The frustrating result of this study is that we cannot propose any change of approximations in the input physics, which would clearly lead to the desired result — a constant average mass for all white dwarfs for all effective temperatures. Variations of parameters in the Hummer-Mihalas-Däppen occupation probabilities can be excluded. Helium pollution cannot be definitely excluded, but could be tested with high resolution, high S/N spectra of (apparent) DAs in the range 10000 – 12000 K. Because of the asteroseismology results discussed above we consider this solution unlikely.

That leaves by default convection as the only suspect. Given the current amount of uncertainty about the correct description, and the contradictions in the current applications, we consider it quite possible that a more physically sound description of the convective flux and gradient might solve the problem.

At the end of this study a final word of caution: fitting observed spectra with theoretical models is as much an art as an objective science. One should be very sceptical of results from spectra with $S/N < 30$, as well as DA spectra that do not include the higher Balmer lines with high S/N. Never believe the errors so readily calculated by a χ^2 routine; these are errors which consider only statistical errors in the observed data. With modern CCD spectra from large telescopes systematic effects of reduction, the fitting process, and model deficiencies result in much larger errors. These can only be estimated in a meaningful way from comparing results for the same objects from different observations, and if possible, from different authors, with different fitting routines and model grids.

Acknowledgments

D.K. wishes to thank the German DAAD and Brazilian CAPES for a grant to visit his colleagues in Brazil, as well as for their kind hospitality in Florianopolis and Porto Alegre.

References

Bergeron P 1992 *JRASC* **86** 309
Bergeron P 1993 in *"White Dwarfs: Advances in Observation and Theory"* Ed.: M A Barstow NATO ASIC Proc. **403** 267
Bergeron P Wesemael F Fontaine G & Liebert J 1990 *ApJ* **351** L21
Bergeron P Wesemael F Lamontagne R et al. 1995 *ApJ* **449** 258
Castanheira B Kepler S O 2008 *MNRAS* in press
Däppen W Mihalas D Hummer D G & Mihalas B W 1988 *ApJ* **332** 261
DeGennaro S von Hippel T Winget D E et al. 2008 *AJ* **135** 1
Eisenstein D J Liebert J Koester D et al. 2006 *AJ* **132** 676
Fontaine G Villeneuve B & Wilson J 1981 *ApJ* **243** 550
Gianninas A Dufour P & Bergeron P 2004 *ApJ* **617** L57
Hansen B M S Anderson J Brewer J et al. 2007 *ApJ* **671** 380
Hummer D G & Mihalas D 1988 *ApJ* **331** 794
Kepler S O Kleinman S J Nitta A et al. 2007 *MNRAS* **375** 1315
Kleinman S J Harris H C Eisenstein D J et al. 2004 *ApJ* **607** 426
Koester D Allard N F & Vauclair G 1994 *A&A* **291** L9
Koester D Napiwotzki R Voss B Homeier D & Reimers D 2005 *A&A* **439** 317
Liebert J Bergeron P & Holberg J B 2005 *ApJS* **156** 47
Ludwig H-G Jordan S & Steffen M 1994 *A&A* **284** 105
Mihalas D Dappen W & Hummer D G 1988 *ApJ* **331** 815
Voss B 2006 PhD thesis Universität Kiel
Zuckerman B Koester D Melis C Hansen B M & Jura M 2007 *ApJ* **671** 872

Testing the initial-final mass relationship of white dwarfs

S Catalán[1,2], J Isern[1,2], E García–Berro[1,3] and I Ribas[1,2]

[1]Institut d'Estudis Espacials de Catalunya, c/ Gran Capità 2–4, 08034 Barcelona, Spain
[2]Institut de Ciències de l'Espai, CSIC, Facultat de Ciències, UAB, 08193 Bellaterra, Spain
[3]Departament de Física Aplicada, Escola Politècnica Superior de Castelldefels, Universitat Politècnica de Catalunya, Avda. del Canal Olímpic s/n, 08860 Castelldefels, Spain

E-mail: catalan@ieec.uab.es

Abstract. In this contribution we revisit the initial-final mass relationship of white dwarfs, which links the mass of a white dwarf with that of its progenitor in the main-sequence. Although this function is of paramount importance to several fields in modern astrophysics, it is still not well constrained either from the theoretical or the observational points of view. We present here a revision of the present semi-empirical initial-final mass relationship using all the available data and including our recent results obtained from studying white dwarfs in common proper motion pairs. We have also analyzed the results obtained so far to provide some clues on the dependence of this relationship on metallicity. Finally, we have also performed an indirect test of the initial-final mass relationship by studying its effect on the luminosity function and on the mass distribution of white dwarfs.

1. Introduction

The initial-final mass relationship of white dwarfs connects the mass of a white dwarf with that of its progenitor in the main-sequence. This function is of fundamental importance for several aspects of modern astrophysiscs since it is an input required for studying the chemical evolution of galaxies, the determination of the ages of globular clusters and their distances, and also to understand the properties of the Galactic population of white dwarfs. Despite its importance, an accurate measurement of this relationship is still not available and, thus, more efforts are needed from both the theoretical and the observational sides to improve it.

The first attempt to empirically map this relationship was carried out by Weidemann (1977), who has also provided a recent revision (Weidemann 2000). Although many improvements have been achieved in these 30 years, there are still some pieces missing in the puzzle. For instance, the dependence of this function on the different stellar parameters — like, for instance, the metallicity, magnetic field and angular momentum of the progenitor star — is still not clear. On the other hand, numerous works have dealt with the calculation of a theoretical initial-final mass relationship (Domínguez et al. 1999; Marigo 2001) but the differences in the evolutionary codes, such as the treatment of convection, the value of the assumed critical mass — which is the maximum mass of a white dwarf progenitor — or the mass loss prescriptions used lead to very different results.

¿From an observational perspective, most efforts up to now have focused on the observation of white dwarfs in open clusters, since this allows to infer the total age and the original metallicity of white dwarfs belonging to the cluster. Open clusters have made possible the derivation of a semi-empirical initial-final mass relationship using more than 50 white dwarfs, although only covering the initial mass range between 2.5 and 7.0 $M_{\odot}$ because stellar clusters are relatively young and, hence, the white dwarf progenitors in these clusters are generally massive. The extension of this mass range to smaller masses was done recently by Catalán et al. (2008a) studying white dwarfs belonging to common proper motion pairs. This was the first study that used these pairs with the purpose of improving the initial-final mass relationship. The stars studied in Catalán et al. (2008a) are at shorter distances in comparison with star clusters and this allows a better spectroscopic study of both members of the pair, obtaining their stellar parameters with accuracy. At the same time, the study of these pairs enables a wide age and metallicity coverage of the initial-final mass relationship. A parallel attempt to cover the low-mass domain of this relationship was done recently by Kalirai et al. (2008) studying white dwarfs in old open clusters.

In this contribution we obtain the initial-final mass relationship grouping all the available results in order to derive a semi-empirical relation, which has also been tested indirectly by computing the white dwarf luminosity function and mass distribution.

2. The initial-final mass relationship

In order to derive a new semi-empirical initial-final mass relationship we have considered our results obtained studying white dwarfs in common proper motion pairs (Catalán et al. 2008a) and completed the sample with the available data in the literature, which is mainly based on open clusters. We have carried out a re-analysis of the open cluster data starting from the white dwarf atmospheric parameters ($T_{\rm eff}$ and $\log g$) given in the literature, as well as the ages and metallicities of the clusters used by the corresponding authors — see Catalán et al. (2008b) and references therein. The procedure followed consists basically in deriving the final mass ($M_{\rm f}$) and the cooling time of each white dwarf from the atmospheric parameters and the cooling sequences of Salaris et al. (2000). These cooling tracks consider a CO core white dwarf with a larger abundance of O the center of the core and with a H thick envelope ontop of a He buffer. Their masses are $q({\rm H}) = M_{\rm H}/M = 10^{-4}$ and $q({\rm He}) = M_{\rm He}/M = 10^{-2}$, respectively. Since from the age of the cluster we know the total ages of these white dwarfs we derived the main-sequence lifetimes of the progenitors, and from these, their initial masses using the stellar tracks of Domínguez et al. (1999).

In Fig. 1 we present the final masses versus the initial masses obtained for white dwarfs in common proper motion pairs and open clusters. The observational data that can be used to define the semi-empirical initial-final mass relationship contains now 62 white dwarfs. It is important to emphasize that all the values below 2.5 $M_{\odot}$ correspond to the data obtained from common proper motion pairs and the recent data based in old open clusters (Kalirai et al. 2008). Before these studies no data for the low-mass domain were available, since white dwarfs in stellar clusters are usually more massive, especially if the clusters are young. The coverage of the low-mass end of the initial-final mass relationship is specially important since it is the most populated bin according to standard initial mass functions and at the same time it guarantees, according to the theory of stellar evolution, the study of white dwarfs with masses near the most common value, $M \sim 0.57\,M_{\odot}$, which represent about 90% of the white dwarf population (Kepler et al. 2007). Thus, these new data increase considerably the statistical significance of the semi-empirical initial-final mass relationship. The results plotted in Fig. 1 reveal a clear dependence of the white dwarf masses on the masses of their progenitors, although a considerable scatter is present. The theoretical initial-final mass relationship of Domínguez et al. (1999) for solar composition has been also plotted in order to be consistent with the stellar tracks used to derive the initial masses. Although the distribution presents a large dispersion, which is larger than the error bars, a comparison of the observational data with these theoretical

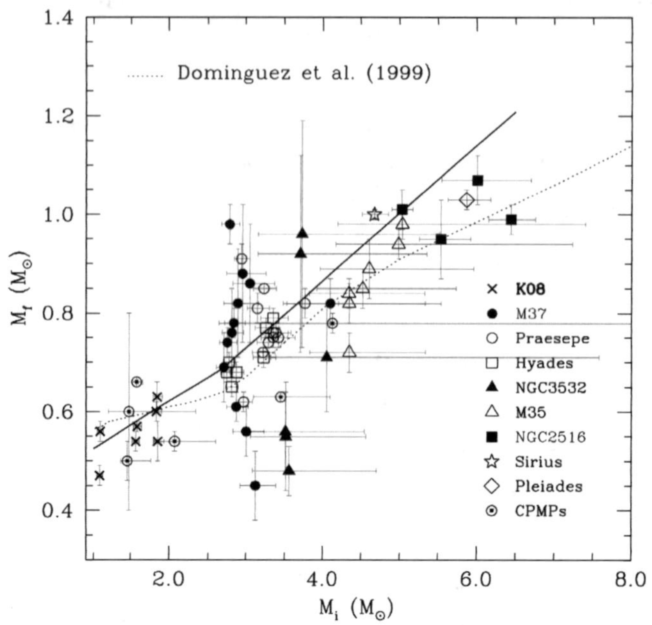

Figure 1. Final masses versus initial masses of the available data. The solid lines correspond to the analytical expressions derived from a weighted least squares linear fit. The dotted line represents the theoretical relation for solar composition of Domínguez et al. (1999).

relationships shows that they share the same trend. However, it should be noted that for each cluster the data presents an intrinsic spread in mass, which can be the result of a combination of different factors, which are discussed below.

Following closely recent works on this subject (Ferrario et al. 2005, Williams 2007), we assume that the initial-final mass relationship can be described by a linear function. As can be seen in Fig. 1 the semi-empirical data and the theoretical initial-final mass relationship (dotted line) can be divided in two different regions, both showing a linear trend. The threshold initial mass separating both regions is 2.7 $M_\odot$. It is worth noticing that the low-mass region shows a shallower slope probably due to the smaller efficiency of mass loss. Taking this into account we have performed a weighted least-squares linear fit for each region, obtaining for $M_{\rm i} < 2.7\ M_\odot$:

$$M_{\rm f} = 0.096 M_{\rm i} + 0.429 \qquad \Delta M_{\rm f} = 0.05\ M_\odot \tag{1}$$

whereas for $M_{\rm i} > 2.7\ M_\odot$ we obtain:

$$M_{\rm f} = 0.137 M_{\rm i} + 0.318 \qquad \Delta M_{\rm f} = 0.12\ M_\odot \tag{2}$$

These expressions have been overplotted in Fig. 1 as two solid lines. Taking into account the scatter of the data and the values of the reduced χ^2 of these fits (7.1 and 4.4, respectively) we have computed the dispersion of the derived final masses, obtaining 0.05 $M_\odot$ and 0.12 $M_\odot$, respectively. In past works, it was necessary to include in the fit a ficticious anchor point to represent the typical white dwarf mass, $M_{\rm f} \sim 0.57\ M_\odot$ (Kepler et al. 2007), since no data were available at the low-mass domain. In our case, this is not necessary since we are including new data that cover this typical mass value of white dwarfs.

The sample of white dwarfs studied here covers a range of metallicities from $Z = 0.006$ to 0.040. From a theoretical point of view it is well established that progenitors with large metallicity produce less massive white dwarfs. Thus, one should expect to see a dependence of the semi-empirical data on metallicity. We have performed a quantitative study of the correlation between the final masses and metallicity. We have computed the differences between the observed final masses and the final masses obtained using Eqs. (1) and (2). In order to quantify this correlation we have calculated the Spearman rank correlation coefficient obtaining -0.002 ± 0.128. The error has been derived from a bootstrapping. Thus, we can conclude that the final masses and metallicities of this sample are not correlated, and that the scatter in the distribution in Fig. 1 is not due, at least not exclusively, to the effect of metallicity.

3. Main systematic uncertainties

The results obtained in this work are dependent on different assumptions and approaches that we have considered during the procedure followed to derive the final and initial masses, which we discuss in the following.

3.1. Thicknesses of the envelopes

As previously pointed out, the observed white dwarf masses in clusters scatter considerably in the same region of initial masses — see Fig. 1. One of the reasons that may explain this, besides the possible stochastic character of mass-loss processes, is the assumption of a given internal composition and of the outer layer stratification of the white dwarfs under study. The thicknesses of the H and He layers is a key factor in the evolution of white dwarfs, since they control the rate at which white dwarfs cool down. To derive the final masses we have used cooling sequences with fixed thicknesses of the envelopes, which might be more appropriate in some cases than in others. In fact, the exact masses that the layers of H and He may have is currently a matter of debate, being the subject of several studies. For instance, models reducing the thickness of the H envelope to $q(\mathrm{H}) = 2.32 \times 10^{-6}$ were computed by Prada Moroni & Straniero (2002), obtaining cooling times shorter than those obtained in this work assuming a thick envelope H, $q(\mathrm{H}) = 10^{-4}$. This stems from the fact that H has a larger opacity than He. In the case of an even thinner H envelope, $q(\mathrm{H}) = 10^{-10}$, the cooling age could be reduced in 10% (1 Gyr) at $\log(L/L_{\odot}) = -5.5$. Thus, the uncertainty in the cooling times could be relevant in some cases, which would affect the estimates of the progenitor lifetimes and, in turn, the initial masses derived.

In order to estimate the effect that the thicknesses of H and He envelopes may have in the initial masses derived here, we have repeated the calculations using the cooling sequences of Fontaine et al. (2001) for a 50/50 CO core white dwarf with a standard He envelope, $q(\mathrm{He}) = 10^{-2}$, and two different thicknesses for the H envelope, a thick one, $q(\mathrm{H}) = 10^{-4}$, and a thin one, $q(\mathrm{H}) = 10^{-10}$. We have verified that the initial masses are indeed sensitive to the cooling sequences used, as expected. We have obtained larger initial masses when considering a thin envelope, due to the longer cooling times obtained in this case. As previously pointed out, in principle it can be expected that the cooling time should be smaller for a thin H envelope model, but this assumption is only true at low enough luminosities (Prada Moroni & Straniero 2002). In fact, at intermediate luminosities a white dwarf with a thinner H envelope evolves slower than the thicker counterpart because it has an excess of energy to radiate away (Tassoul et al. 1990). The maximum difference in the initial masses ($\sim 1\ M_{\odot}$) has been found to occur for high-mass progenitors ($M > 5\ M_{\odot}$), while this value is one order of magnitude smaller for smaller masses ($\sim 0.1\ M_{\odot}$). However, it should be noted that many other combinations are possible, for instance, with different thicknesses of the He envelope, which in this case has been kept fixed. However, since it is impossible to know which is the real chemical stratification of the outer layers of each individual white dwarf we have not formally introduced this error in the calculations, since in some cases we would be overestimating the error of the initial masses derived here.

3.2. Composition of the core

The most commonly assumed composition for a typical white dwarf is a mixture of carbon and oxygen. However it should be taken into account that white dwarfs can have also other internal compositions. Those white dwarfs more massive than 1.05 $M_{\odot}$ are thought to have a core made of ONe, while those with masses below 0.4 $M_{\odot}$ have an He-core. ONe white dwarfs cool faster than CO or He white dwarfs because the heat capacity of O and Ne is smaller than that of C or He. On the contrary, He white dwarfs are the ones that cool slower. Thus, those white dwarfs studied here with masses near the limits between different core populations would be introducing an uncertainty in the cooling times obtained, since their cooling times are different depending on the adopted composition. For instance, a 1 $M_{\odot}$ ONe white dwarf cools 1.5 times faster than a CO white dwarf of the same mass (Althaus et al. 2007). Thus, if an observed white dwarf has indeed a ONe core instead of the typical CO one, its progenitor lifetime would be underestimated in our analysis and, consequently, the initial mass derived would be larger than the real one. However, the exact impact of this depends also on the total age of the white dwarf. The smaller the total age, the higher the effect of considering a wrong internal composition.

3.3. Mass determinations when $T_{\rm eff} \leq 12000$ K

Spectroscopic masses can be determined precisely if high signal-to-noise spectra is used, thus, accurate atmospheric parameters are obtained. However, it should be taken into account that the atmospheres of DA white dwarfs below 12 000 K could be enriched in He while preserving their DA spectral type (Bergeron et al. 1992). This He is thought to be brought to the surface as a consequence of the development of a H convection zone. Depending on the efficiency of convection the star could still show Balmer lines, instead of being considered as a non-DA white dwarf. This will affect considerably the determination of the spectroscopic mass, since a large abundance of He mimics the pressence of a large surface gravity. As a consequence, the assumption of an unrealistic chemical composition could have a large impact on the cooling times estimated, although mainly at low effective temperatures. Nevertheless, a very large fraction of the stars in our sample ($\sim$95%) have temperatures well above this limit.

3.4. Total ages

The derived initial masses depend on the cooling times, the total ages, the metallicity and finally, on the stellar tracks used. Among these parameters the largest source of error is due to the uncertainty in the total ages of the white dwarfs. For white dwarfs in open clusters, the age can be usually derived with high accuracy from model fits to the turn-off location in a colour-magnitude diagram. The uncertainty on the age of a cluster is a systematic effect for stars belonging to the same cluster, since all the initial masses will be shifted together to larger or smaller masses in the final versus initial masses diagram. On the contrary, in the case of white dwarfs in common proper motion pairs, the accuracy in the total age depends on the evolutionary stage of the companion. The accuracy of the age using isochrone fitting could be large if the star is relatively evolved and located far away from the ZAMS. Using the X-ray luminosity method, the ages derived could also be quite precise — 8 to 20% per cent, see Catalán et al. (2008a) — if the star is relatively young ($t \leq 1$ Gyr).

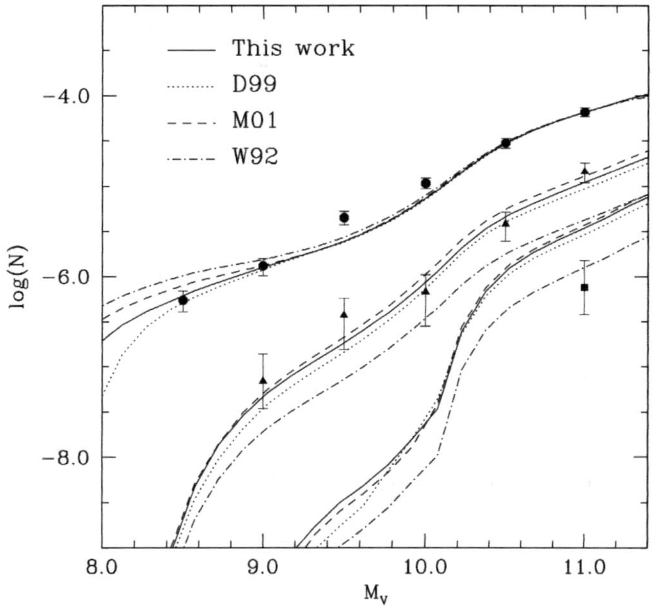

Figure 2. White dwarf luminosity functions versus visual magnitude using different initial-final mass relationships. Circles, triangles and squares correspond to the observational data of the PG Survey.

4. Luminosity function and mass distribution

We have computed a set of white dwarf luminosity functions considering an age of 11 Gyr for the Galactic disk and using bins of visual magnitude. In Fig. 2 we show from top to bottom the total luminosity function and the luminosity functions of white dwarfs with masses larger than 0.7 $M_{\odot}$ and 1.0 $M_{\odot}$, respectively. The total luminosity function (that is, considering the whole range of masses) was normalized to the bin corresponding to $M_{\rm V} = 11$, and then this normalization factor was used for the luminosity functions of white dwarfs more massive than 0.7 $M_{\odot}$ and 1.0 $M_{\odot}$. We have used the stellar evolutionary inputs of Domínguez et al. (1999) — D99 — Marigo (2001) — M01 — Wood (1992) — W92 — and the semi-empirical initial-final mass relationship presented in Sect. 2. Circles, triangles and squares correspond to the observational data of the Palomar Green (PG) survey (Liebert et al. 2005). Comparing the different theoretical luminosity functions, it can be noted that the predicted number of massive white dwarfs is larger when using the inputs of Domínguez et al. (1999) and Marigo (2001) with our analytical initial-final mass relationship that in the case in which the expressions of Wood (1992) are used. This was expected since the latter favors the production of low-mass white dwarfs. If we consider only the first three results it can be noted that it is not possible to evaluate which initial-final mass relationship produces a theoretical luminosity function that best fits the observational data, since the error bars of the observational data are larger than the differences between the theoretical results. In any case, what it can be clearly seen is that all the theoretical relations predict more massive white dwarfs than the observations when a mass cut of 1.0 $M_{\odot}$ is adopted, except in the case of Wood (1992). Nevertheless, conclusive results cannot be derived given the scarcity of observational data.

Following a similar procedure we have computed theoretical white dwarf mass distributions and compared our results with the recent data from the SDSS (DeGennaro et al. 2008). Since

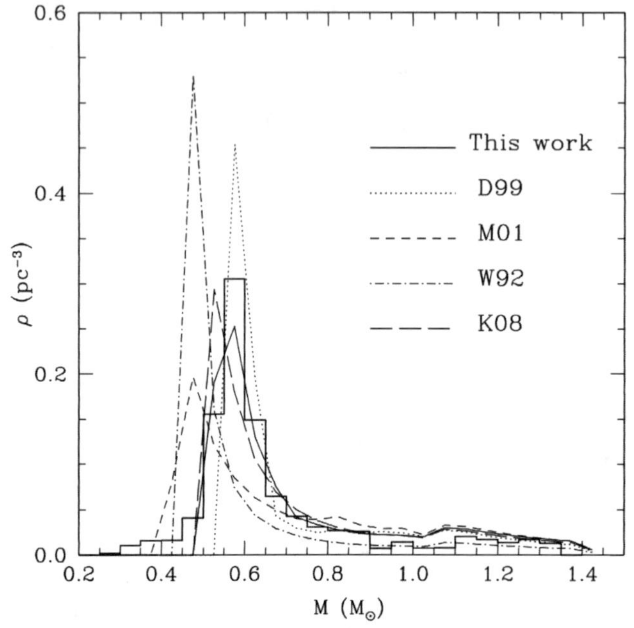

Figure 3. Mass distributions for white dwarfs with $T_{\rm eff} \geq 12\,000$ K considering different evolutive stellar models and initial-final mass relationships. The histogram corresponds to observational data from the SDSS.

the accuracy on the mass determinations decreases considerable when white dwarfs are cooler than 12 000 K, both the theoretical and the observational mass distributions only take into account white dwarfs with $T_{\rm eff} \geq 12\,000$ K. We have considered the same initial-final mass relationships as in the previous calculation, and also the recent semi-empirical relationship derived by Kalirai et al. (2008) — K08. All the white dwarf distributions have been normalized to the total density obtained in each case. As it can be noted, there is a well defined peak in all the mass distributions, the location of which is defined mainly by the initial-final mass relationship considered. On the contrary, the height of the peak depends also on the lifetime of the progenitors. The central peak is shifted to larger masses when the initial-final mass relationship considered favors the production of more massive white dwarfs for the low mass progenitors. If we compare the theoretical mass distributions with the SDSS observational data it can be noted how the location and height of the central peak is best fitted by the predictions obtained when using our semi-empirical relationship.

5. Conclusions

In this work we have derived a new semi-empirical initial-final mass relationship by considering our results obtained studying white dwarfs in common proper motion pairs and re-evaluating the available data in the literature, mainly based on open clusters. With this study we have extended the observational data to the low-mass domain ($M < 2.5\ M_\odot$), which was poorly studied until recently. The coverage of the low-mass end is of great importance since we are now studying the most populated region of initial masses according to the initial mass function, including in this way the well-established peak of the field white dwarf mass distribution in our study.

We have performed a weighted least-squares linear fit of the results, obtaining analytical expressions that connect the final and initial masses of progenitors of white dwarfs with massess ranging from 1 to 6.5 $M_{\odot}$. Since the sample of data collected for this study covers a wide range of ages and metallicites, we have performed an analysis to evaluate the dependence of the initial-final mass relationship on metallicity. The correlation factor obtained indicates that such dependence does not exist, or at least, it cannot be identified in our results. However, it should be taken into account that there are some other parameters that may play an important role (angular momentum, magnetism...) and that it is very difficult to study the effect of each of them separately.

We have also tested the initial-final mass relationship from an indirect point of view by computing the luminosity function and mass distribution of white dwarfs. For this purpose we have considered different stellar evolutionary inputs (stellar tracks and initial-final mass relationships). We have noted some differences between the theoretical luminosity functions, obtaining a clear dependence on the considered initial-final mass relationship. We have compared our results with the observational data from the PG Survey and we have always obtained a reasonable fit when the range of masses was constrained to $M > 0.7\ M_{\odot}$, except when using the initial-final mass relationship of Wood (1992). However, given the presently available observational data, any attempt to discern which initial-final mass relationship better fits the data is not feasible. Nevertheless, our results favor an initial-final mass relationship that favors the production of more massive white dwarfs (Catalán et al. 2008b). In the case of the white dwarf mass distribution we have obtained more conclusive results. From a comparison of our predictions with the observational data of the SDSS we have noted that the theoretical mass distribution obtained when considering the semi-empirical initial-final mass relationship derived in this work is the one that best fits the central peak of the observational mass distribution, being in this way the one which is more representative of the characteristics of the white dwarf population.

Acknowledgments

This research was supported by MEC grants AYA05–08013–C03–01 and 02, by the European Union FEDER funds and by the AGAUR.

References

Althaus L G García-Berro E Isern J et al. 2007 *A&A* **465** 249A
Bergeron P Wesemael F Fontaine G 1992 *ApJ* **387** 288
Catalán S Isern J García–Berro E et al. 2008a *A&A* **477** 213
Catalán S Isern J García–Berro E et al. 2008b *MNRAS* **387** 1693
DeGennaro S von Hippel T Winget D E et al. 2008 *AJ* **135** 1
Domínguez I Chieffi A Limongi M et al. 1999 *ApJ* **524** 226
Ferrario L Wickramasinghe D T Liebert J et al. 2005 *MNRAS* **361** 1131
Fontaine G Brassard P Bergeron P 2001 *PASP* **113** 409
Kalirai J S Hansen B M S Kelson D D et al. 2008 *ApJ* **676** 594
Kepler S O Kleinman S J Nitta A et al. 2007 *MNRAS* **375** 1315
Liebert J Bergeron P Holberg J B 2005 *ApJS* **156** 47
Marigo P 2001 *A&A* **370** 194
Prada Moroni P G Straniero O 2002 *ApJ* **581** 585
Salaris M García–Berro E Hernanz M et al. 2000 *ApJ* **544** 1036
Tassoul M Fontaine G Winget D E 1990 *ApJS* **72** 335
Weidemann V 1977 *A&A* **59** 418
Weidemann V 2000 *A&A* **363** 647
Williams K A 2007 in Proc. of the 15th European Workshop on White Dwarfs Eds.: R Napiwotzki and M Burleigh *ASP Conf. Ser.* **372** 85 (San Francisco)
Wood M A 1992 *ApJ* **386** 539

Journal of Physics: Conference Series **172** (2009) 012008 doi:10.1088/1742-6596/172/1/012008

Hot subdwarf binaries – Masses and nature of their heavy compact companions

Stephan Geier[1], Uli Heber[1], Heinz Edelmann[1], Thomas Kupfer[1], Ralf Napiwotzki[2] and Philipp Podsiadlowski[3]

[1] Dr. Remeis–Sternwarte, Institute for Astronomy, University Erlangen-Nürnberg, Sternwartstr. 7, 96049 Bamberg, Germany

[2] Centre of Astrophysics Research, University of Hertfordshire, College Lane, Hatfield AL10 9AB, UK

[3] Department of Astrophysics, University of Oxford, Keble Road, Oxford OX1 3RH, UK

E-mail: geier@sternwarte.uni-erlangen.de

Abstract. Neutron stars and stellar-mass black holes are the remnants of massive stars, which ended their lives in supernova explosions. These exotic objects can only be studied in relatively rare cases. If they are interacting with close companions they become bright X-ray sources. If they are neutron stars, they may be detected as pulsars. Only a few hundred such systems are presently known in the Galaxy. However, there should be many more binaries with basically invisible compact objects in non-interacting binaries.

Here we report the discovery of unseen compact companions to hot subdwarfs in close binary systems. Hot subdwarfs are evolved helium-core-burning stars that have lost most of their hydrogen envelopes, often due to binary interactions. Using high-resolution spectra and assuming tidal synchronisation of the subdwarfs, we were able to constrain the companion masses of 31 binaries. While most hot subdwarf binaries have white-dwarf or late-type main sequence companions, as predicted by binary evolution models, at least 5 % of the observed subdwarfs must have very massive companions: unusually heavy white dwarfs, neutron stars and, in some cases, even black holes. We present evolutionary models which show that such binaries can indeed form if the system has evolved through two common-envelope phases. This new connection between hot subdwarfs, which are numerous in the Galaxy, and massive compact objects may lead to a tremendous increase in the number of known neutron stars and black holes and shed some light on this dark population and its evolutionary link to the X-ray binary population.

1. Introduction

Neutron stars and stellar-mass black holes are the remnants of massive stars ending their lifes in supernova explosions. Detecting these exotic objects is possible when they are in a close orbit with another star. If matter is transferred from the companion star to the compact object, bright X-rays are emitted. Without ongoing mass transfer the companion remains invisible, but can be detected indirectly from the reflex motion of the visible star, which causes periodic variations of its radial velocity (RV). These variations are measureable via the Doppler effect from spectral line shifts. Stellar evolution models predict the existence of a hidden population of such compact objects. Subdwarf B or sdB stars are helium core burning stars of about half a solar mass with very thin hydrogen envelopes (Heber 1986). A large fraction of sdB stars resides in close binaries (Maxted et al. 2001; Napiwotzki et al. 2004).

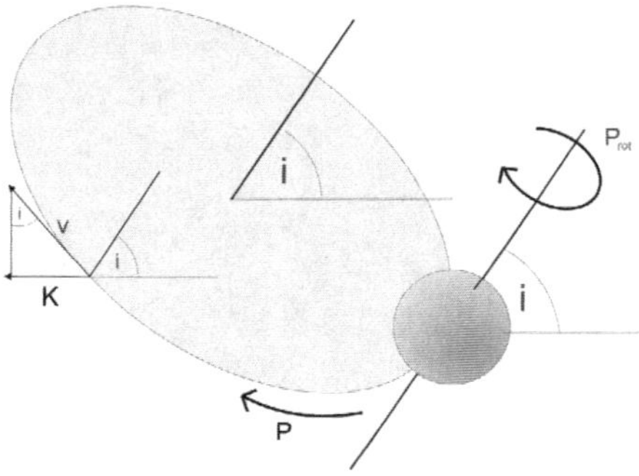

Figure 1. Schematic view of a single-lined binary system.

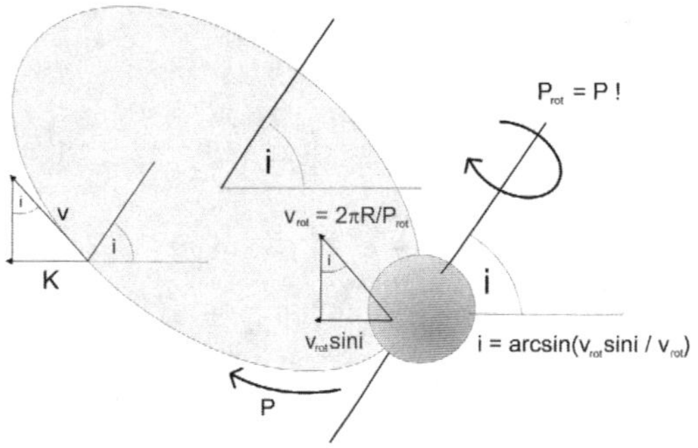

Figure 2. Schematic view of a single-lined binary system with synchronised rotation.

Because the components' separation in these systems is much less than the size of the subdwarf progenitor in its red-giant phase, these systems must have experienced a common-envelope and spiral-in phase (Han et al. 2002, 2003). In such a scenario, two main-sequence stars of different masses evolve in a binary system. If the primary reaches the red-giant phase and fills its Roche lobe, mass is transferred to the companion star. When mass transfer is unstable, a common envelope is formed. Due to friction with the envelope, the two stellar cores spiral towards each other until enough orbital energy has been deposited within the envelope to eject it. The end product is a much closer system containing the core of the giant, which then may become an sdB star, and a main-sequence companion. When the latter reaches the red-giant branch, another common-envelope phase is possible and can lead to a close binary consisting of a white dwarf and an sdB star. In all known cases the companions are either white dwarfs or late-type main-sequence stars.

2. Analysis method

Since the spectra of close binary subdwarfs are mostly single-lined, they reveal no information about the orbital motion of the sdB stars' companions. However, in very close systems, the rotation of the sdB star is expected to be tidally locked to the orbital motion. This allows us to constrain the inclination angle of the system and to derive the companion mass (Figs. 1, 4). In order to derive the parameters, the mass of the sdB primary has to be known. Binary population synthesis models constrain the mass range for the sdB binaries in question. The mass distribution shows a sharp peak at about $0.47\,M_{\odot}$, which we adopted to derive the most probable companion masses.

From the primaries orbital solution only the period P and the projected RV semi-amplitude K can be derived. The mass function $f_{\rm m} = \frac{M_{\rm comp}^3 \sin^3 i}{(M_{\rm comp}+M_{\rm sdB})^2} = \frac{PK^3}{2\pi G}$ then provides a lower limit for the companion mass $M_{\rm comp}$ for a given sdB mass $M_{\rm sdB}$ (Fig. 1). If the primary is synchronised (Fig. 2) the orbital period P equals the rotation period $P_{\rm rot}$.

- The stellar radius $R_{\rm sdB}$ is given by the mass radius relation $R_{\rm sdB} = \sqrt{M_{\rm sdB}G/g}$.
- The surface gravity $\log g$ can be obtained by a quantitative spectral analysis (for details see (Geier et al. 2007)).
- The rotational velocity $v_{\rm rot} = 2\pi R_{\rm sdB}/P$ can then be calculated and the projected rotational velocity $v_{\rm rot}\sin i$ can be measured from the spectral line broadening of weak metal lines in

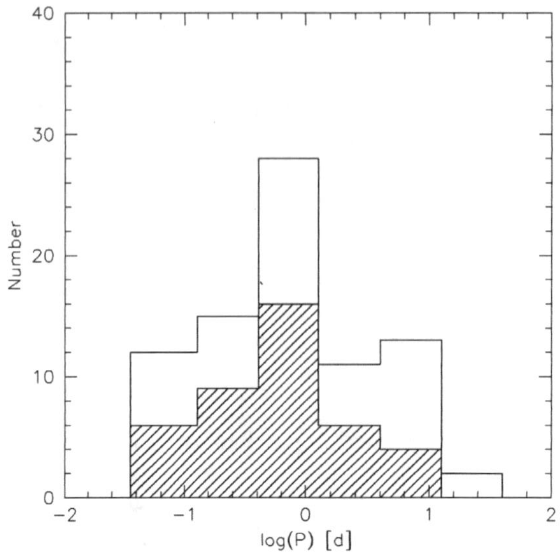

Figure 3. Number of binaries plotted against the logarithm of their orbital periods. The solid blank histogram marks all known sdB binaries from the catalogue of Ritter & Kolb (1998). The shaded histogram marks the studied sample of 41 binaries with a focus on systems with rather short periods.

the case of slow rotators (Geier et al. 2008) and of Balmer and helium lines for fast rotators (Geier et al. 2007).

Now the inclination angle i can easily be derived and the mass function can be solved for reasonable values of $M_{\rm sdB}$.

The stars in our sample were observed with the high-resolution ($R = 20\,000 - 48\,000$) spectrographs UVES at the ESO VLT, HIRES at the Keck telescope, HRS at the Hobby Eberly Telescope (HET), FEROS at the ESO 2.2 m telescope and FOCES at the CAHA 1.5 m telescope.

3. Nature of the unseen companions

Today, 81 sdB stars in close binary systems have been studied and their orbital parameters have been derived (see the catalogue of Ritter & Kolb 1998). From 51 radial velocity variable sdBs in our sample, we selected 41 binaries with known orbital parameters or half of the known sample (see Fig. 3). From these, 31 could be solved consistently under the assumption of synchronisation. In 10 cases the sdB primaries spin faster than synchronised. There are no spectral signatures of companions visible. Main sequence stars with masses higher than $0.45\,M_{\odot}$ could therefore be excluded because of their high luminosities in comparison to the sdB stars. In this case spectral features of the cool secondary (e.g. Mg I lines at $\approx 5\,160$ Å) get visible in the spectra and a flux excess in the infrared appears, which can be detected using 2MASS photometry (Lisker et al. 2005).

Another possibility to detect M dwarf companions are reflection effects in the binary light curves. Some of our programme stars have already been checked for modulations in their light curves. Unfortunately, this method only works, if the binary inclination is high and the orbital period short enough. Synthetic light curve modelling of sdB+M systems shows that the expected amplitude of a reflection effect drops below 1 mmag for orbital periods longer than 0.5 d. This

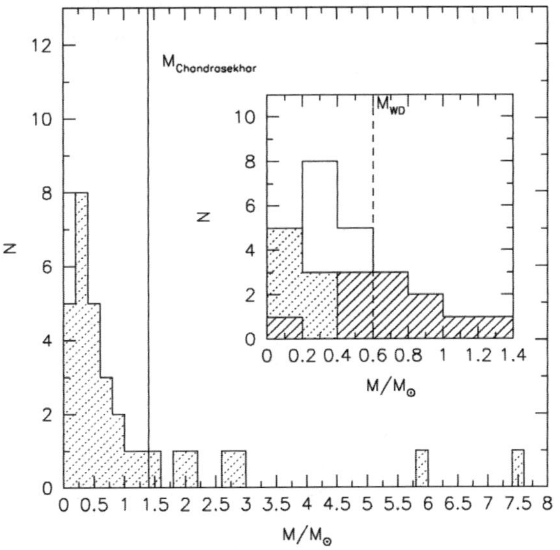

Figure 4. Mass distribution of the unseen companion stars. The companion mass is plotted against the total number of binaries under the assumption of canonical sdB mass $0.47\,M_{\odot}$. The solid vertical line marks the Chandrasekhar limit. The last bin at $7.5\,M_{\odot}$ is a lower limit. A detail of the mass distribution is shown in the inlet. The shaded solid histogram shows the fraction of subdwarfs with confirmed compact companions, the shaded dashed one the detected M dwarf companions. The dashed vertical line marks the average WD mass.

period therefore provides an upper limit for the detectability of reflection effects in sdB binaries from the ground.

- In seven sdB binaries the companion has been identified as late M star. In addition to the derived companion masses, reflection effects are visible in their lightcurves.
- Nine companions have to be white dwarfs, because of the derived masses and the absence of reflection in their lightcurves or flux contributions in their spectra. One sdB has a massive white dwarf companion and short orbital period. It qualifies as candidate for SN Ia.
- In seven cases this distinction cannot be made. The companions may be either late M stars or white dwarfs.
- Eight systems have massive compact companions near or even exceeding the Chandrasekhar limit. Their companions therefore may be either unusually heavy white dwarfs, neutron stars or black holes (see Fig. 3).

4. Orbital synchronisation of sdB binaries

The assumption of orbital synchronisation in close binary sdBs is essential for our analysis. Theoretical models predict synchronisation to be established up to orbital periods of about half a day to two days (Zahn 1977; Tassoul & Tassoul 1992). But it has to be pointed out that tidal dissipation in radiative stellar envelopes is still poorly understood and the timescales of synchronisation differ by orders of magnitude depending on the models. Empirical evidence is therefore needed to constrain such models and to answer the question which close binary sdBs are synchronised.

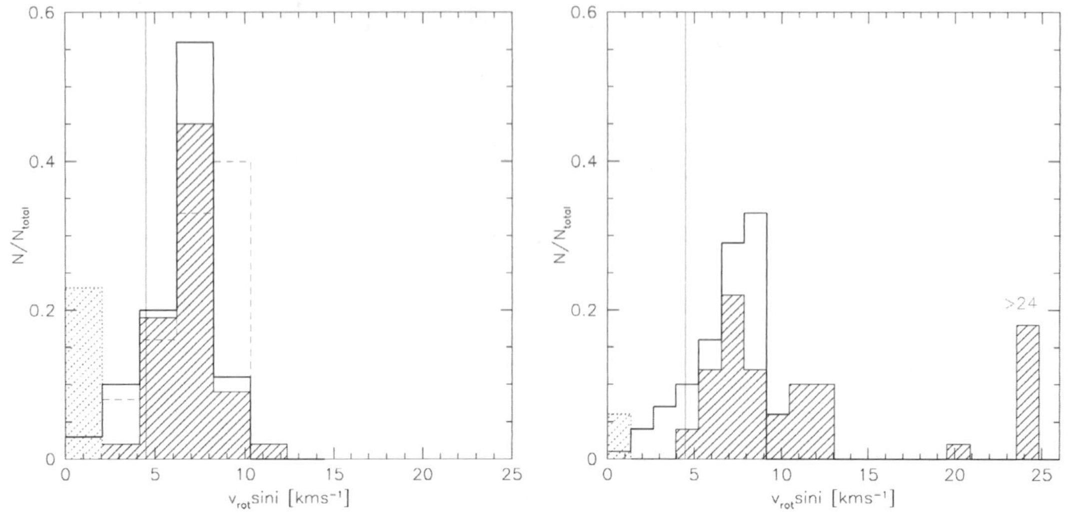

Figure 5. *Left panel* The measured $v_{\mathrm{rot}} \sin i$ of 49 single sdBs is plotted against relative fraction of stars as shaded histogram. The size fo the bins is given by the average error of the measurements. The blank histogram marks the expected uniform distribution of $v_{\mathrm{rot}} \sin i$ under the assumption of randomly oriented polar axes and the same rotational velocity $v_{\mathrm{rot}} = 8.3\,\mathrm{kms}^{-1}$ for all stars. The dashed histogram shows this distribution for $v_{\mathrm{rot}} = 9.0\,\mathrm{kms}^{-1}$. The solid vertical line at $v_{\mathrm{rot}} \sin i \approx 4.5\,\mathrm{kms}^{-1}$ marks the detection limit. All sdBs with lower $v_{\mathrm{rot}} \sin i$ are stacked into the first bin (dotted histogram). *Right panel* The measured $v_{\mathrm{rot}} \sin i$ of 51 RV variable sdBs is plotted in the same way. The size of the bins is given by the average error of the measurements and is therefore slightly different than in the left panel. The blank histogram marks the uniform distribution for $v_{\mathrm{rot}} = 8.3\,\mathrm{kms}^{-1}$. All sdBs with $v_{\mathrm{rot}} \sin i$ higher than $24\,\mathrm{kms}^{-1}$ are stacked into the last bin.

The timescale of the synchronisation process is highly dependent on the tidal force exerted by the companion. If the companion is very close and the orbital period therefore very short, synchronisation is established much faster than in binaries with longer orbital periods. If the sdB in a binary with given orbital period is proven to be synchronised, all other sdBs in binaries with shorter orbital periods should be synchronised as well. Although the timescales also scale with sdB radius and companion mass, the orbital period is dominating at first order. Most of the non-synchronised systems in our sample have orbital periods exceeding 1.3 d.

To study the influence of close companions on the rotational properties of sdBs we measured the projected rotational velocities of 49 single sdBs from the SPY survey (Lisker et al. 2005). We found that single sdBs are slow rotators spinning with an almost uniform rotational velocity of about $8\,\mathrm{kms}^{-1}$. A comparison with the $v_{\mathrm{rot}} \sin i$-distribution of our close binary sample revealed that the rotation of sdBs in close binary systems is clearly affected by the tidal influence of the invisible companions (see Fig. 5). This alone does not prove that all sdBs in these binaries are synchronised, but it shows that the tidal influence of the companions is strong enough to significantly change the rotational properties of the sdBs. Although these stars may be not yet be synchronised, the synchronisation process is already at work.

The parameters of the eclipsing binaries PG 1336−018 (Vučković et al. 2007), HS 0705+6700 (Drechsel et al. 2001) and HW Vir (Edelmann 2008) derived from light curve analyses are consistent with the parameters derived with our method assuming synchronised orbits. This essentially means that the calculated $v_{\rm rot}\sin i$ for synchronous rotation is consistent with the measured value. In eclipsing systems, all these parameters can be measured. This provides clear empirical evidence that at least the upper layers of the stellar envelope are synchronised to the orbital motion of the eclipsing sdB binaries in our sample. We therefore conclude that all sdBs in close binaries with orbital periods up to 0.12 d should be synchronised as well.

Two known sdBs clearly show ellipsoidal variations in their light curves with periods exactly half the orbital periods (KPD 0422+5421 (Orosz & Wade 1999); KPD 1930+2752 (Maxted, Marsh & North 2000; Geier et al. 2007). This alone is only an indication for tidal synchronisation. To really prove synchronisation it is necessary that the stellar parameters determined independently from the light curve analysis are consistent with a synchronised orbit. This is the case for KPD 0422+5421 as well as KPD 1930+2752. Both ellipsoidal variable systems have very short periods of ≈ 0.1 d and high inclinations. Otherwise ellipsoidal variations are very hard to detect.

This was possible in the case of the sdB+WD binary PG 0101+039 using a high precision light curve obtained with the MOST satellite (Geier et al. 2008). We found a strong indication that the surface rotation of the sdB star PG 0101+039 is tidally locked to its orbit. Hence, other sdB stars in close binaries should also be synchronised if their orbital period is less than that of PG 0101+039 (P = 0.567 d). We conclude that tidally locked surface rotation is at least established in sdB binaries with orbital periods of less than half a day.

An independent method to proof orbital synchronisation is provided by asteroseismology. Van Grootel et al. (2008) were able to reproduce the main pulsation modes of the short period pulsating sdB Feige 48 ($P \approx 0.38$ d), derived the surface rotation from the splitting of the modes and concluded that the subdwarf rotates synchronously. Charpinet et al. (2008) reach a similar conclusion for the short period eclipsing binary PG 1336−018 ($P \approx 0.10$ d). Asteroseismic analyses revealed that sdB binaries up to orbital periods of about 0.4 d are synchronised. We therefore conclude that all sdBs in close binaries with shorter periods should be synchronised as well.

From the eight candidate sdB+NS/BH binaries found in our sample four have orbital periods shorter than 0.4 d where synchronisation could be proven by asteroseismology. Another two binaries have periods shorter than 0.6 d, the lower limit for synchronised rotation derived by binary light curve analysis. And the longest binary period is only 0.8 d. We therefore conclude that all these binaries should be synchronised and the derived parameters are therefore correct.

5. Formation of sdB+NS/BH binaries

The evolution that leads to such systems requires an initial binary, consisting of a primary star massive enough to produce a neutron star or black hole, and a companion of typically several solar masses. These systems experience two mass-transfer phases and one supernova explosion (see Fig. 4). The second mass-transfer phase had to be unstable, leading to a common-envelope and spiral-in phase. Population synthesis estimates suggest that of order 1 % of all hot subdwarfs should have neutron-star or black-hole companions (Podsiadlowski, Rappaport & Pfahl 2002; Pfahl, Rappaport & Podsiadlowski 2003).

6. The fraction of NS/BH companions and selection effects

This high fraction of massive compact companions (20 %) is in part caused by an interplay of several different selection effects (e.g. RV variability selection, bias on short period binaries, see Fig. 3). Especially the fact that most of the binaries with the highest derived companion masses have low inclination seems suspicious. It can be explained by the limitations of our analysis method. Nevertheless, high inclination systems should be more numerous than binaries seen at low inclination. The fact that no such system was found up to now (81 binaries with solved

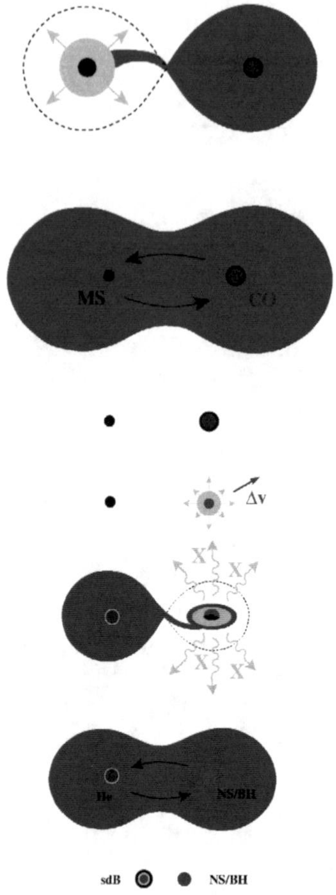

Figure 6. Schematic diagram of formation scenarios leading to hot subdwarf binaries with neutron-star (right hand panel) or black-hole (left hand panel) companions.

orbits are known), puts an upper limit of a few percent to the number fraction of sdB+NS/BH binaries.

Both observational constraints and binary population synthesis models suggest a fraction of a few percent for sdBs with NS/BH companions. The fraction we found in our sample is too high to be consistent with these results even if selection effects are taken into account. It has to be pointed out that all of our candidate binaries have short orbital periods ($0.2 - 0.8\,$d) and should therefore be most likely synchronised. Adopting the lowest possible mass for the sdBs ($0.3\,M_\odot$, Han et al. 2002, 2003) only two of our candidate binaries ($\approx 5\,\%$) have companion masses exceeding the Chandrasekhar limit. This number is consistent with theoretical and observational constraints. We therefore conclude that most sdBs in our candidate systems should have masses lower than canonical.

7. HYPERMUCHFUSS survey

The results presented here provide strong indications for the existence of sdB+NS/BH binaries, but rely on the assumption of orbital synchronisation and are seriously affected by selection effects. Most of the binaries with massive unseen companion are seen at low inclination. High inclination systems must exist as well and should be even more numerous. In this case a determination of the orbital parameters would be sufficient to prove their existence, because in binaries with such high RV amplitudes the lower limit for the companion mass (see Fig. 1) should already exceed the Chandrasekhar mass. The HYPERMUCHFUSS-survey has been launched in 2008 to search for sdBs with high RV variability and massive compact companions as well as the recently discovered class of hypervelocity stars (see Tillich et al. these proceedings).

References

Charpinet S van Grootel V Reese D et al. 2008 *A&A* **489** 377
Edelmann H 2008 *ASPCS* **392** 187–194
Drechsel H Heber U Napiwotzki R et al. 2001 *A&A* **379** 893–904
Geier S Nesslinger S Heber U et al. 2007 *A&A* **464** 299–307
Geier S Nesslinger S Heber U et al. 2008 *A&A* **477** L13–L16
Han Z Podsiadlowski P Maxted P F L Marsh T R & Ivanova N 2002 *M.N.R.A.S.* **336** 449–466
Han Z Podsiadlowski P Maxted P F L & Marsh T R 2003 *M.N.R.A.S.* **341** 669–691
Heber U 1986 *A&A* **155** 33–45
Lisker T Heber U Napiwotzki R et al. 2005 *A&A* 430 223–243
Maxted P F L Heber U Marsh T R & North R C 2001 *M.N.R.A.S.* **326** 1391–1402
Maxted P F L Marsh T R & North R C 2000 *M.N.R.A.S.* **317** L41–L44
Napiwotzki R Karl C A Lisker T et al. 2004 *Ap&SS* **291** 321–324
Orosz J A & Wade R A 1999 *M.N.R.A.S.* **310** 773–783
Pfahl E Rappaport S & Podsiadlowski Ph 2003 *ApJ* **597** 1036–1048
Podsiadlowski Ph Rappaport S & Pfahl E D 2002 *ApJ* **565** 1107–1133
Ritter H & Kolb U 1998 *A&AS* **129** 83–85
Tassoul J -L & Tassoul M 1992 *ApJ* **395** 259–267
Van Grootel V Charpinet S Fontaine G & Brassard P 2008 *A&A* **483** 875
Vučković M Aerts C Østensen R et al. 2007 *A&A* **471** 605–615
Zahn J -P 1977 *A&A* **57** 383–394

16th European White Dwarfs Workshop IOP Publishing
Journal of Physics: Conference Series **172** (2009) 012009 doi:10.1088/1742-6596/172/1/012009

The HYPERMUCHFUSS Campaign - an undiscovered high velocity population

Alfred Tillich[1], Stephan Geier[1], Uli Heber[1], Heiko Hirsch[1], Pierre Maxted[2], Boris Gaensicke[3], Tom Marsh[3], Ralf Napiwotzki[4], Roy Østensen[5] and Chris Copperwheat[3]

[1] Dr. Remeis-Sternwarte , Sternwartstr.7, 96049 Bamberg, Germany
[2] Astrophysics Group, School of Physics and Geographical Sciences, Lennard-Jones Laboratories, Keele University, ST5 5BG, United Kingdom
[3] Department of Physics, University of Warwick, Coventry CV4 7AL, United Kingdom
[4] Centre of Astrophysics Research, University of Hertfordshire, College Lane, Hatfield AL10 9AB, UK
[5] Instituut voor Sterrenkunde, Katholieke Universiteit Leuven, Celestijnenlaan 200D, 3001 Leuven, Belgium

E-mail: `Alfred.Tillich@sternwarte.uni-erlangen.de`

Abstract. We present an overview and a status report of HYPERMUCHFUSS (HYPER velocity or Massive Unseen Companions of Hot Faint Underluminious Stars Survey) aiming at the detection of a population of high velocity subluminous B stars and white dwarfs. The first class of targets consists of hot subdwarf binaries with massive compact companions, which are expected to show huge radial velocity variations. The second class is formed by the recently discovered Hyper-velocity stars, which are moving so fast that the dynamical ejection by a supermassive black hole seems to be the only explanation for their origin. Until now only one old HVS has been found, but we expect a larger population. We applied an effecient selection technique for hot subdwarfs and white dwarfs with high galactic restframe velocities from the *SDSS* spectral data base, which serve as first epoch observations for our campaign with the ESO VLT and NTT in Chile, the 3.5 m telescope at DSAZ observatory (Calar Alto) in Spain and the WHT on La Palma. The survey is nearing completion and provides us with promising candidates which will be followed up to measure their RV-curves to uncover massive companions or prove their nature as HVS.

1. Introduction

Our HYPERMUCHFUSS campaign aims basically at two different fast populations amongst hot subdwarf stars. Hot subdwarf stars (sdB/sdOs) are considered to be helium core burning stars with very thin hydrogen envelopes and masses around $0.5\,M_{\odot}$. A large fraction $(40-70\%)$ of the sdBs are members of short period binaries ($P = 0.1 - 10\,\mathrm{d}$) (Maxted et al. 2001). They are the result of one or two common-envelope ejection phases in which the companion to a giant star is engulfed in the outer layers. Due to dynamical friction the companion spirals in towards the core of the red giant and the envelope is ejected. About 10% of these companions are low-mass M stars, the remainder are compact stars, mostly white dwarfs. In addition 20% of sdB stars are in wide binaries with G-K type companions.

- Geier et al. (this conference) presented the discovery of a hidden population of heavy companions (including stellar black holes) in low inclination hot subdwarf systems. The lack of high inclination systems remains mysterious and calls for a dedicated short-period binary survey. Therefore the first group of our fast targets are short period binaries. Geier et al. studied in detail 50 sdB binaries and derived companion masses by measuring the projected rotational velocities and surface gravities from high resolution UVES spectra. For short period systems (P=0.1 to 0.6 d) we assume that the stars are rotating synchronously with the orbit. This assumption has been verified from ellipsoidal light variations and asteroseismology. Hence the method is applicable for orbital periods of up to at least 0.6d, which is the case for more than half of the sample. While most masses are compatible with normal WDs, a couple of systems have unusually heavy companions. About 4-8% of massive companions were found to exceed the Chandrasekhar mass limit. A massive main sequence companion can be excluded as it would outshine the sdB. The companions are therefore most likely neutron stars or black holes. All of these systems are of low inclination and have short orbital periods. Obviously the inclination does not have a preferred direction and therefore also a population of high inclination systems has to exist. These systems would then have radial velocity (RV) variations as high as 500 kms^{-1} at orbital periods of less than 0.5 days. Considering the fraction of $4-8\%$ together with the very low number of known neutron star or black hole binaries which are detected as X-ray sources, a confirmation would mean a tremendous rise of the total number of these heavy compact objects in the galaxy.
- The second group of our fast targets consists of single stars. Hyper-velocity stars (HVS) have been found among young late B-type stars (Brown et al. 2005) as well as among old population hot subdwarf stars (Hirsch et al. 2005). The first systematic survey targeted late B-type HVS and has been performed with great success (Brown et al. 2006). Therefore we initiated the first intensive HVS survey dedicated to old population stars. HVS are predicted to have been formed through the tidal disruption of a binary by the supermassive blck hole (SMBH) in the galactic center (GC)(Hills 1988). Other formation scenarious like e.g. the scattering of multiple stars (or binary stars) by SMBH+IMBH systems and the multiple star scattering in dense clusters have been proposed recently. This rises the number of potential HVSs significantly and justifies extended searches for these interesting objects. Especially as their trajectories contain information on the important parts of the mass distribution in our galaxy: the GC and the dark matter Galactic potential. We focus particularly on the old population HVS stars i.e. hot subdwarf stars as we discovered and successfully analysed the first one of this kind (Hirsch et al. 2005).
- By combining the two parts of the project we might even find a *HVS binary*. Our campaign is a very promising approach for finding such systems, as we target subdwarfs and especially short period binaries. These systems more likely might remain intact after interacting with a SMBH/IMBH pair than others (Sesana, Madau & Haardt 2008). As pointed out by Lu, Yu & Lin (2007), the detection of just one such object would provide strong evidence that the massive black hole in the GC is a binary.

2. Hyper-velocity stars

HVSs have attracted great interest as they allow implication on a wide field of astrophysics, which we intend to outline here.

HVS move so fast that dynamical ejection by massive black holes appears to be the only viable mechanism for their origin. It is well established that the GC hosts a SMBH (Schödel et al. 2002; Ghez et al. 2005) with a mass of about $3\times10^6 M_\odot$. Already in the late eighties it was predicted from numerical experiments that a star can be ejected from the GC with velocities exceeding the escape velocity of the Galaxy by the disruption of a binary through tidal interaction with

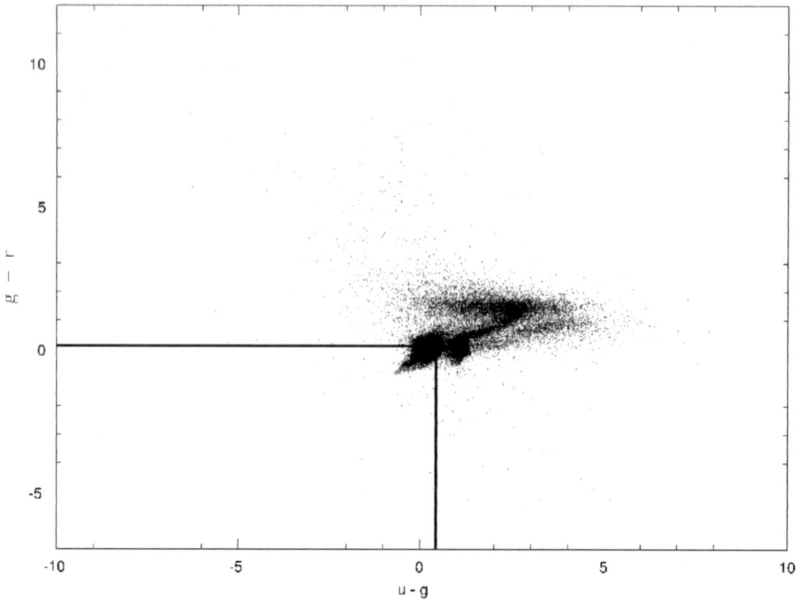

Figure 1. Colour indices of all stars in *SDSS* DR6 with the colour criteria used.

a central massive black hole (Hills 1988). The detection of one such HVS would be evidence for the existence of a massive black hole in the GC since none of the alternative scenarios put forward to explain the phenomenon of run-away stars with such high velocities (Hills 1988).

Due to the exceptional origin of HVS, they must urgently be separated from the already well known *runaway stars*, which are high velocity O and B type stars at high Galactic latitudes with galactic restframe (GRF) velocities in the range of $100 - 200\,\mathrm{kms}^{-1}$. For *runaway stars* we know two capable ejection mechanisms. (i) In the binary supernova scenario a massive star undergoes a core collapse explosion and its companion star is released at about its orbital velocity (up to $200\,\mathrm{kms}^{-1}$). (ii) Close single star/binary (or binary/binary) encounters in open clusters can lead to ejection velocities up to $300\,\mathrm{kms}^{-1}$.

These scenarios have met with considerable success to explain most of the observed run-away stars known (Conlon et al. 1992). Recently, the unbound star HD 271791 was found to be an extreme SN binary run-away B-star and termed hyper-run-away star (Przbilla et al. 2008a). Therefore we suggest a clear classification of the fastest stars.

The first HVS was found by Brown et al. (2005), a faint B-type star with a heliocentric RV of 853 km/s. Soon thereafter two members of our group (Hirsch et al. 2005; Edelmann et al. 2005) discovered HVSs No. 2 and 3 (a sdO star US 708, $RV = 708\,\mathrm{kms}^{-1}$ and an early B-main sequence star HE 0437−5439, $RV = 723\,\mathrm{kms}^{-1}$), both unbound to the Galaxy. Dedicated systematic surveys for late B-type stars have resulted in the discovery of additional 13 HVSs (Brown, Geller & Kenyon 2008). Brown et al. (2006) estimated that the Galactic halo holds $\approx 2\,000$ HVS in a sphere of 120 kpc radius. Additionally they introduced the new class of bound HVS (Brown et al. 2007) with GRF velocities of more than 275 kms^{-1}, which have most likely been ejected by the SMBH in the GC but obviously do not exceed the local Galactic escape velocity. The most important attribute of HVSs is, that they provide insight into the environment of the SMBH and its nature as well as on the shape and density distribution of the Galactic dark matter halo (Gnedin et al. 2005, Bromley et al. 2006).

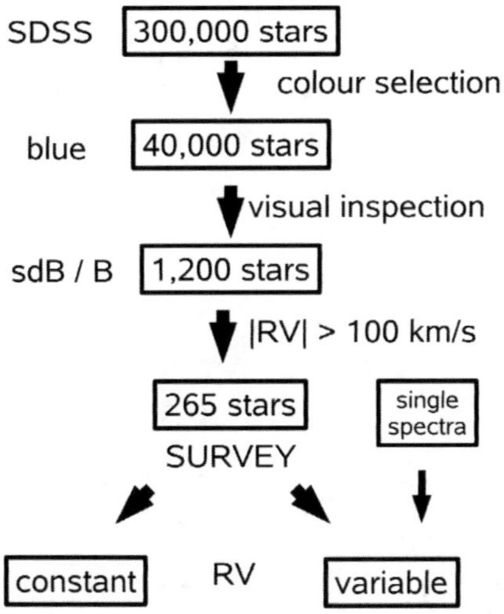

Figure 2. Target selection method based on Sloan Digital Sky Survey.

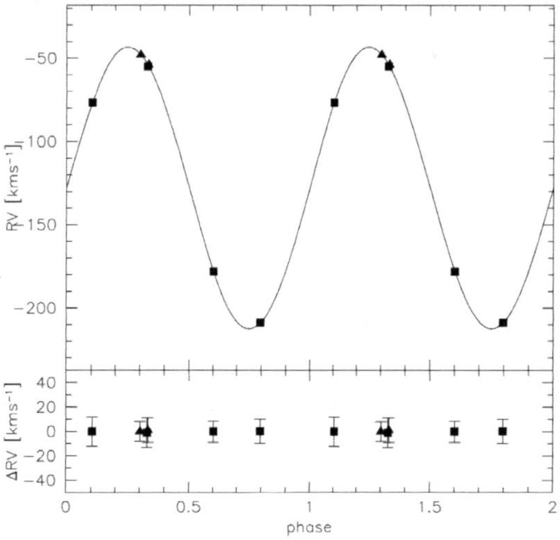

Figure 3. Radial velocity curve of an sdB binary based on *SDSS* individual spectra only. Period and amplitude are well determined.

There are several independent observational results concerning the SMBH environment, which fit in with the existence of HVS: e.g the reason for the high eccentricity of orbits of stars orbiting the Galactic centre may be that they are former companions of HVSs in binaries disrupted by tidal interaction with the SMBH (Ginsburg & Loeb 2006).

A first extrapolation by Yu & Tremaine (2003) showed a HVS formation rate of 10^{-5}/yr. If the GC hosts a tight *binary* of two massive black holes the formation rate of HVSs would be 10 times larger than predicted for a single massive black hole. Therefore once we have found enough HVS to use a statistical approach we could deduce on the structure of the GC.

Strikingly, 13 out of 16 HVS are of late B-type. However a discrimination between a main sequence and a horizontal branch nature remains dificult, as both branches overlap in the HRD. Only with the help of quantitative spectroscopy one can analyse these stars in great detail and draw conclusions as to their nature. Recently one of these stars was discovered to be a chemically peculiar (CP) main sequence star (Przybilla et al. 2008b). Another star seems to be originated in the LMC (Przybilla et al. 2008c), due to its abundance pattern and its location.

Due to this various implications it is obvious that there is an urgent need to find new HVS and to analyse already know ones to the highest available precision.

3. Target selection

Unfortunately until now, surveys have only targeted late B-type stars. The only known hot subdwarf was found serendipitously and no systematic search has been undertaken yet. This was one reason for us to start the HYPERMUCHFUSS campaign.

The enormous *SDSS* database with its current data release 6 is a perfect starting point for us as the survey is directed towards high Galactic latitudes. The best sky coverage is reached at northern hemisphere at Galactic longitudes close to the GC and the anticenter (Fig. 4). To date *SDSS* is a rich source for stellar astronomy, led to various exciting discoveries and has therefore been well evaluated in terms of systematic errors and accuracy.

Our target selection method is shown in Fig. 2. After a colour selection (g-r<0.1 and u-

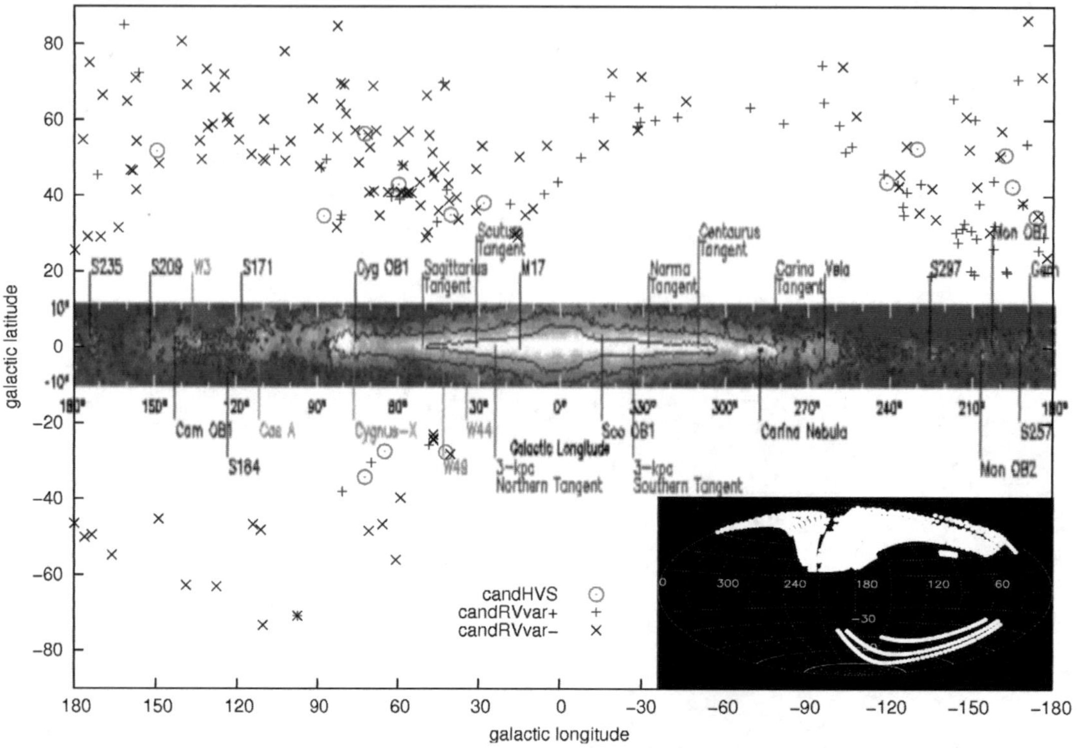

Figure 4. Target distribution in galactic coordinates ("x":RV<0, "+":RV>0) and the complete *SDSS* DR6 target distribution (lower right corner).

g<0.4, see Fig.1) and a visual inspection of spectra we select only stars with heliocentric radial velocities higher than 100 kms^{-1}. This reduces the number of interesting stars from 1200 to 265. Only those stars can reach sufficiently high galactocentric velocities to leave the galaxy.

As all *SDSS* spectra are a composite of at least three consecutive 15 min exposures, we also extracted the individual spectra for each target, brighter than 18 mag. If the S/N ratio is sufficiently good we are already able to detect rapid RV variability without any additional observations. For some of the stars with more images ($\gg 3$ spectra) at different epochs one can even find orbital solutions just from these individual spectra (see Fig 3 for an example).

Our target list consists of a total number of 265 target stars. Fig. 4 shows the target distribution in Galactic coordinates. Although we selected only the fastest stars, the spatial distribution shows remarkable similarities with the general *SDSS* target sample. Among the targets on the left side in Fig. 4 we find more blue-shifted spectra as the sun approaches them on its disc orbit around the GC, while among the targets on the right side there are more red-shifted spectra for the same obvious reason.

Naturally some spectral misclassification occured, as some of these stars turned out to be other B-type stars rather than sdB stars. But those objects were of interest in their own right.

Finally we calculated the Galactic rest-frame (GRF) velocity and extracted the stars with absolute GRF velocity values of more than 275 kms^{-1} as primary targets for the HVS search. This observational cut is motivated by Brown et al. (2007) to distinguish the bound HVS and runaway halo stars (see above). Henceforth the GRF velocity is used to rank our targets for our survey. Thus we eliminate stars of the disc population and binaries of small RV amplitude,

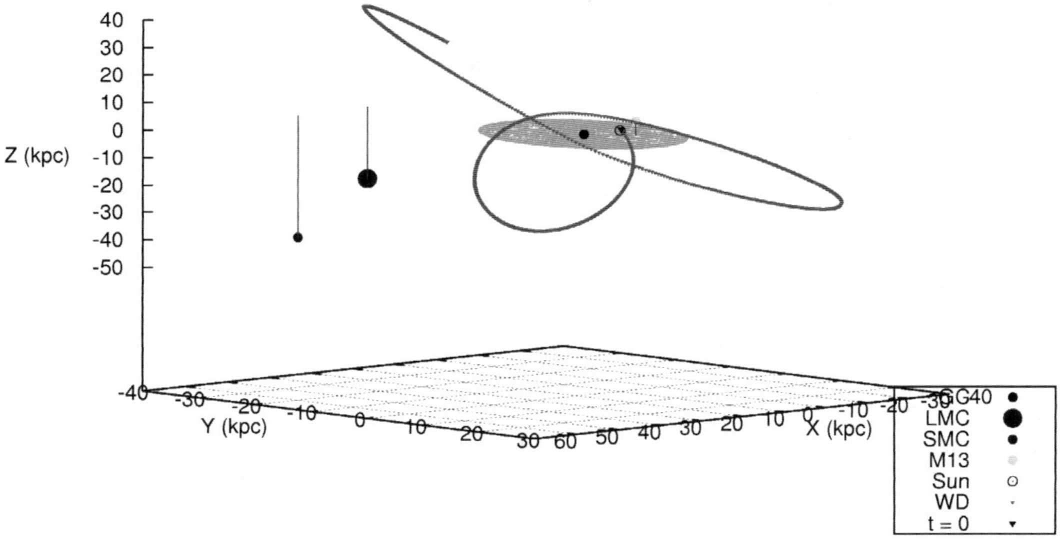

Figure 5. Trajectory of a halo white dwarf from our sample (known proper motion).

which will mostly harbour companions of relatively low mass.

4. Aims and techniques

In the primary step our campaign aims at obtaining a second epoch spectra for all our targets (survey), as we use the *SDSS* spectra as a first epoch measurement. For each star in our campaign there are obviously only two possible results: either the star shows constant RV or we find RV variability. The second step (follow-up) of the campaign then depends highly on what we reveal: in the first case the star is a candidate HVS and a spectroscopic distance determination from high S/N spectra is required to constrain the kinematics of the star. In the second case we simply need to get more spectra in order to derive the period and the RV amplitude.

A χ^2-fitting of synthetic spectra is used to determine the velocity, $T_{\rm eff}$ and $\log g$. To avoid misclassification problems we use our synthetic model spectra for the appropriate stellar types even for the RV fitting. Using the stellar parameters we can infer the stellar masses by applying different evolutionary tracks. With the help of the atmospheric parameters, stellar mass and brightness we calculate the spectroscopic distances using the astrophysical fluxes. After that we search the archives for a proper motion measurement of the star or derive a proper motion. It is worth noting that proper motion measurements are not available for any of Browns HVSs, making it impossible to determine their trajectories. But with this final step we would be able to calculate a complete trajectory for the star in the galactic gravitational potential (Dehnen & Binney 1998) and probe the galaxy. The future ESO GAIA satellite mission will provide complete proper motion for stars all over the sky as faint as 20 mag.

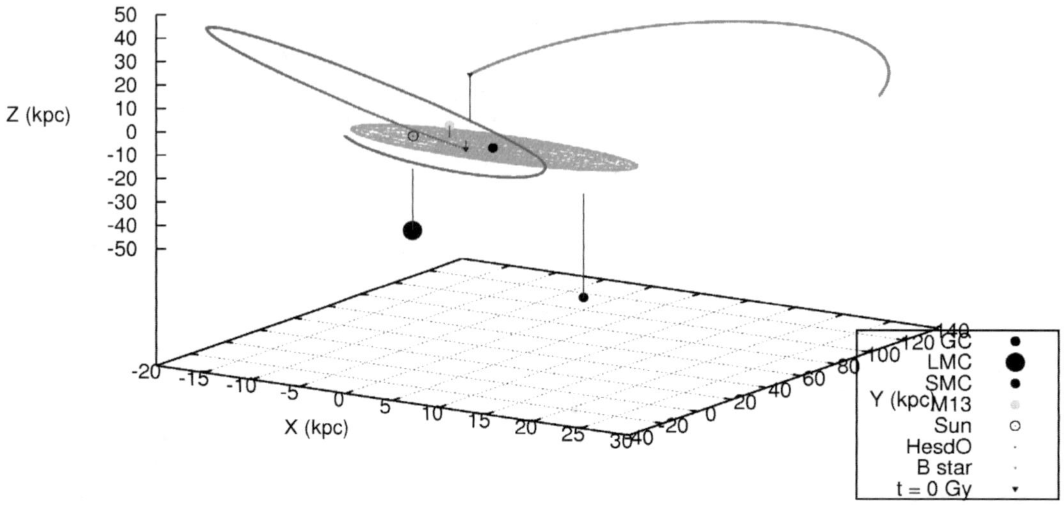

Figure 6. 3D plot of the trajectory of a He-sdO star(dark grey) and a main sequence B star(light grey) from our sample (unknown proper motion).

5. Preliminary results

Our campaign started in early 2007 and until now we obtained data from the 3.5m telescope at the DSAZ observatory (Calar Alto) in Spain, the ESO NTT and VLT in Chile and the WHT on La Palma. The first step of the campaign consisting of second epoch observations for all possible targets (survey) is currently running and the main part of the data analysis is in progress.

Until now we observed and analysed 65 out of our 265 targets, i.e. $\approx 25\%$. Accordingly 51 of these stars have constant RV within our detection limits and at most 17 stars show indication for binarity, leading to a lower limit for the binary fraction of 27%. Selecting only the hot subdwarfs we derived an even lower limit of only 15%, which is less than previous estimates (40-70%) but may be expected if many of our targets belong to population II as the observed binary fraction amongst globular cluster sdBs is very low (Moni Bidin, Catelan & Altmann 2008).

In Fig. 5 we show the trajectory for a halo white dwarf from our sample, with $T_{\rm eff}$=16,871 K and $\log g = 7.38$dex. Due to the intrinsic faintness of white dwarfs they are much closer to us than the HVSs, which is confirmed by our method ($d_{\rm WD} \approx 400$ pc). Nevertheless its high GRF velocity combined with a high proper motion value drives the star far out into the halo and indicates population II membership.

The trajectories shown in Fig. 6 belong to a He-rich subdwarf O star and a late-type B-type star from our sample. According to the atmospheric parameters ($T_{\rm eff} \approx 14800$ K, $\log g = 4.0$dex) the B-type star is a main sequence star and hence cooler and bigger than we had expected.

The GRF at their present locations are $-373\,{\rm kms}^{-1}$ for the subdwarf O star and $+303\,{\rm kms}^{-1}$ for the B-type star. If massive, the B-type star is very far away (30 kpc) and the Galactic gravitational potential at its present location is fairly small, which cooresponds to a Galacic escape speed of only about $350\,\mathrm{kms}^{-1}$. Accordingly this star is also probing far out in the halo. Unfortunately we still do not know its transversal motion. At this location a tiny proper motion already will set this star unbound and turn it into a HVS.

6. The Search Project

The HYPERMUCHFUSS campaign targets two different populations, which probe both the GC with its SMBH(s?) and the dark halo as well as a hidden population of black holes. Due to our sophisticated target selection the success of the campaign is guaranteed. Using various state-of-the-art techniques to analyse and interpret the spectra we will derive orbital solutions for the short-period binaries and peform kinematical studies of the HVS.

We already carried observing runs at DSAZ observatory (Calar Alto) in Spain, ESO NTT & VLT in Chile and the WHT on La Palma. During the second half of 2008 we plan to complete the first step of the campaign (survey). But the second step of the campaign (follow-up), in which we shall obtain high S/N spectra of the found HVSs and measure precisely the binary orbits, has still to come.

References

Bromley B C Kenyon S J Geller M J Barcikowski E Brown W R and Kurtz M J 2006 *ApJ* **653** 1194–1202 (*Preprint* arXiv:astro-ph/0608159)
Brown W R Geller M J Kenyon S J & Kurtz M J 2005 *ApJ* **622** L33–L36 (*Preprint* arXiv:astro-ph/0501177)
Brown W R Geller M J Kenyon S J & Kurtz M J 2006 *ApJ* **640** L35–L38 (*Preprint* arXiv:astro-ph/0601580)
Brown W R Geller M J Kenyon S J Kurtz M J and Bromley B C 2007 *ApJ* **660** 311–318 (*Preprint* arXiv:astro-ph/0701600)
Brown W R Geller M J and Kenyon S J 2008 *ArXiv e-prints* (*Preprint* 0808.2469)
Conlon E S Dufton P L Keenan F P McCausland R J H and Holmgren D 1992 *ApJ* **400** 273–279
Dehnen W and Binney J 1998 *MNRAS* **294** 429–+ (*Preprint* arXiv:astro-ph/9612059)
Edelmann H Napiwotzki R Heber U Christlieb N and Reimers D 2005 *ApJ* **634** L181–L184 (*Preprint* arXiv:astro-ph/0511321)
Ghez A M Salim S Hornstein S D Tanner A Lu J R Morris M Becklin E E and Duchêne G 2005 *ApJ* **620** 744–757 (*Preprint* arXiv:astro-ph/0306130)
Ginsburg I and Loeb A 2006 *MNRAS* **368** 221–225 (*Preprint* arXiv:astro-ph/0510574)
Gnedin O Y Gould A Miralda-Escudé J and Zentner A R 2005 *ApJ* **634** 344–350 (*Preprint* arXiv:astro-ph/0506739)
Hills J G 1988 *Nature* **331** 687–689
Hirsch H A Heber U O'Toole S J & Bresolin F 2005 *A&A* **444** L61–L64 (*Preprint* arXiv:astro-ph/0511323)
Lu Y Yu Q & Lin D N C 2007 *ApJ* **666** L89–L92 (*Preprint* 0707.1872)
Maxted P f L Heber U Marsh T R and North R C 2001 *MNRAS* **326** 1391–1402 (*Preprint* arXiv:astro-ph/0103342)
Moni Bidin C Catelan M and Altmann M 2008 *A&A* **480** L1–L4 (*Preprint* 0801.4795)
Przybilla N Nieva M F Heber U and Butler K 2008a *ApJ* **684** L103–L106
Przybilla N Nieva M F Tillich A Heber U Butler K and Brown W R 2008b *A&A* **488** L51–L54
Przybilla N Nieva M F Heber U Firnstein M Butler K Napiwotzki R and Edelmann H 2008c *A&A* **480** L37–L40 (*Preprint* 0801.4456)
Schödel R Ott T Genzel R Hofmann R Lehnert M Eckart A Mouawad N Alexander T Reid M J Lenzen R Hartung M Lacombe F Rouan D Gendron E Rousset G Lagrange A M Brandner W Ageorges N Lidman C Moorwood A F M Spyromilio J Hubin N and Menten K M 2002 *Nature* **419** 694–696 (*Preprint* arXiv:astro-ph/0210426)
Sesana A Madau P and Haardt F 2008 *ArXiv e-prints* (*Preprint* 0810.3848)
Yu Q and Tremaine S 2003 *ApJ* **599** 1129–1138 (*Preprint* arXiv:astro-ph/0309084)

Journal of Physics: Conference Series **172** (2009) 012010 doi:10.1088/1742-6596/172/1/012010

A determination of the luminosity function of DA white dwarfs from the Kiso Survey

Marie-Michèle Limoges and Pierre Bergeron

Département de Physique, Université de Montréal, C.P. 6128, Succ. Centre-Ville, Montréal, Québec H3C 3J7, Canada

E-mail: limoges@astro.umontreal.ca, bergeron@astro.umontreal.ca

Abstract. We rederive the luminosity function for the sample of DA white dwarfs from the Kiso Schmidt ultraviolet excess survey (KUV stars) using the spectroscopic method where the atmospheric parameters ($T_{\rm eff}$ and $\log g$) and absolute visual magnitudes for each star are obtained from detailed model atmosphere fits to optical spectroscopic data. We compare the result of our determination with that obtained by Darling (1994) based on empirical photometric calibrations. Our luminosity function is also compared with that derived spectroscopically from the PG survey. Misclassified objects are also briefly discussed.

1. Introduction

The Palomar Green (PG) survey has identified 1874 ultraviolet excess objects and it is estimated to be 84% complete (Green et al. 1986). The luminosity function of DA white dwarfs from the PG survey was determined from photometric calibrations by Fleming et al. (1986) and improved by Liebert et al. (2005) using the spectroscopic method for measuring effective temperatures, surface gravitites, and absolute visual magnitudes (see, e.g., Bergeron et al. 1992). Similarly, Darling (1994) derived a luminosity function of all white dwarfs from the Kiso Schmidt ultraviolet excess survey (KUV) — all spectral types included — using M_V values obtained from photometric calibrations. However, the $B-V$ versus (photographic) color index relation used to derive the absolute visual magnitudes showed a considerable dispersion, and it is believed that the estimated M_V values probably reflect a similar dispersion (Liebert et al. 2005). These authors (see their Fig. 10) made a detailed comparison of their luminosity function with those of Fleming et al. (1986) and Darling (1994). They evaluate the completeness of the PG survey to 75%, while Darling (1994) found a lower value of 58%. In this paper, we present preliminary results from our study aimed at improving the luminosity function by applying the spectroscopic method to the DA stars in the KUV sample. One of our goals is to improve the comparison of the DA luminosity functions derived from the PG and KUV samples.

2. White Dwarfs from the KUV Survey

The Kiso Ultraviolet Excess Survey (KUV) is a photometric search for UV-excess objects, performed with the 105-cm Schmidt telescope at Kiso Observatory. The 1186 objects were found in 44 fields in a belt from the northern to the southern galactic pole at a galactic longitude of 180°. The fields of the survey have a mean limiting magnitude of 17.7 and cover a total area of 1400 square degrees (Noguchi et al. 1980; Kondo et al. 1984).

We have secured optical spectroscopic observations for all white dwarfs in the KUV survey. Spectra have been obtained with the Steward Observatory 2.3-m telescope equipped with the Boller & Chivens spectrograph and a Loral CCD detector. The 4.5 arcsec slit together with the 600 l/mm grating in first order provided a spectral coverage of $\lambda\lambda$3200–5300 at an intermediate resolution of $\sim$ 6 Å FWHM and a signal-to-noise ratio of at least 50. We have reclassified several objects (see Table 1): 30 objects from Darling's sample turned out to be main sequence or subdwarf stars. Our analysis also includes seven new white dwarfs that have been identified in the KUV survey since 1994. Only DA stars are analyzed here, for a total of 172 white dwarfs (including the 5 magnetic DAs). Sample spectra are displayed in Figure 1.

Table 1. Spectroscopic content of the KUV survey.

	DA	DB	DQ/DZ/DC	DO	Mag DA	Total
This work	167	23	15	1	5	211
Darling	192	25	15	0	2	234

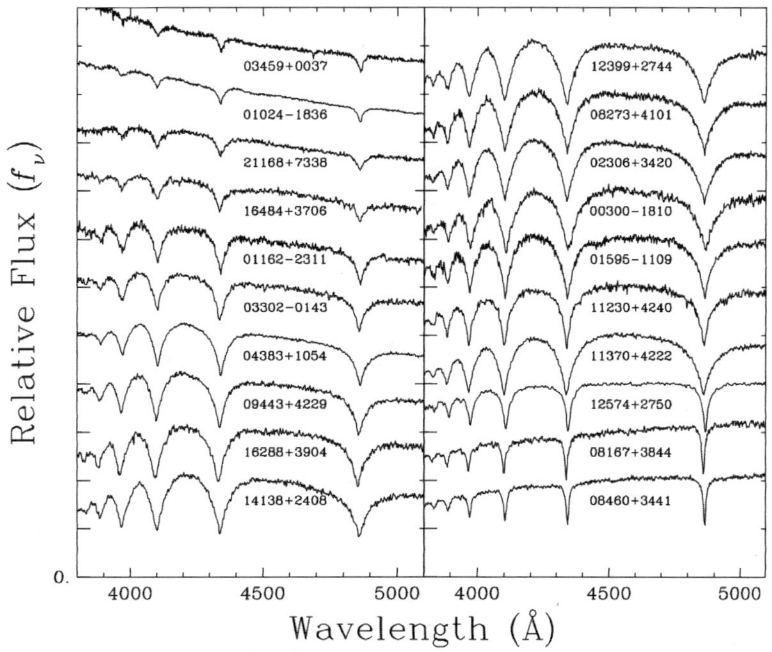

Figure 1. Optical spectra for a subset of DA stars from the KUV sample. The effective temperature decreases from upper left to bottom right.

With our sample properly defined, we proceed to measure the atmospheric parameters ($T_{\rm eff}$ and $\log g$) of each star using the spectroscopic fitting technique and model atmospheres described at length in Liebert et al. (2005) and references therein. Our results for the KUV sample are summarized in Figure 2 in a $\log g$ vs $\log T_{\rm eff}$ diagram. Absolute visual magnitudes are then calculated from evolutionary models following the prescription of Holberg and Bergeron (2006). Our spectroscopic M_V values are compared in Figure 3 with those derived by Darling (1994), which are based on empirical photometric calibrations. The differences observed are quite significant, with the empirical estimates of Darling being generally fainter than the M_V

values determined spectroscopically. It is therefore expected that the luminosity function based on these absolute magnitudes will be affected as well.

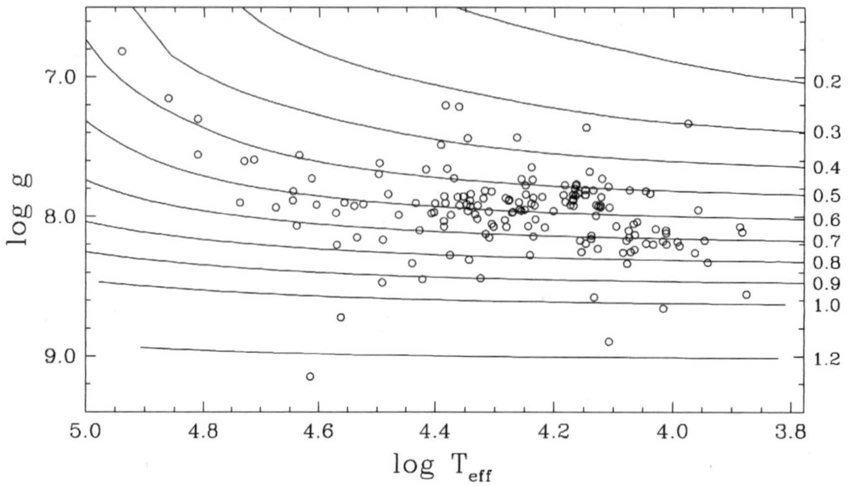

Figure 2. $T_{\rm eff}$ and $\log g$ values for all DA stars from the complete KUV sample. The solid lines represent carbon-core evolutionary models with thick hydrogen layers — see Liebert et al. (2005) and references therein. Numbers on the right hand side of the figure indicate the mass of each model in solar masses.

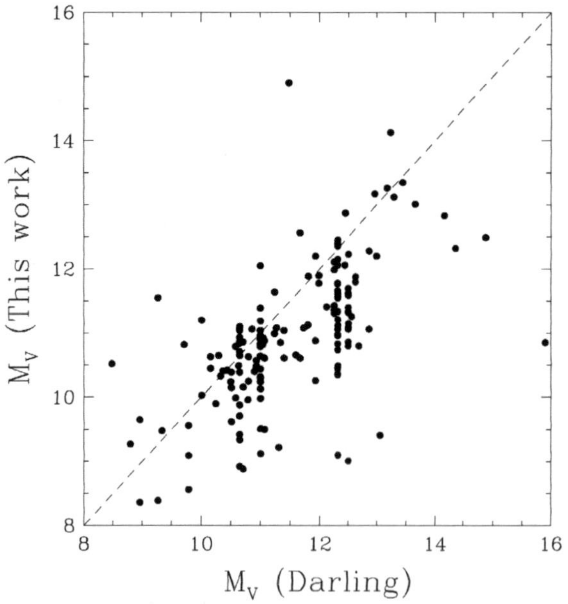

Figure 3. Comparison of the absolute visual magnitudes obtained from the empirical photometric calibration of Darling (1994) and from the spectroscopic method.

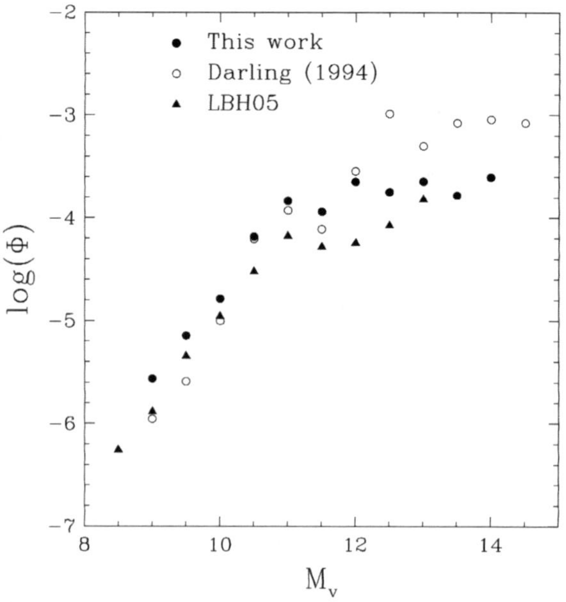

Figure 4. Comparison of the luminosity functions of DA stars from the KUV survey (this work and Darling 1994) and from the PG survey (Liebert et al. 2005).

3. The Luminosity Function

The luminosity function of white dwarf stars is a measure of the local density of this type of object by interval of absolute magnitude. It contains information about the history of the local galactic disk and stellar evolution. It represents also a direct measure of the space density of white dwarfs and of the stellar death rate in the local galactic disk. Compared to theoretical luminosity functions, it can lead to an estimate of the age of the local galactic disk, as discussed in Fontaine et al. (2001) for instance. We derive the luminosity function for the DA stars from the KUV survey by adding the inverse of the contribution of each star to the space density for each magnitude bin, following the well established $1/V_{\rm max}$ method (Schmidt 1968). The completeness of the sample is evaluated from the $\langle V/V_{\rm max}\rangle$ method described in Green (1980). Using this technique, we obtain a limiting magnitude of 17.31 for our sample of DA stars.

Our improved luminosity function for the KUV sample based on spectroscopic determinations of M_V is compared in Figure 4 (filled circles) with the results of Darling (1994) based on empirical photometric calibrations (open circles). Also shown is the luminosity function for the DA stars from the PG survey derived by Liebert et al. (2005) using the same spectroscopic technique as that used in our analysis (filled triangles). Our results for the KUV sample are systematically larger than the PG sample in each magnitude bin, which suggests that the KUV survey is more complete than the PG survey. This is particularly true at the faint end of the luminosity function where the PG sample is known to be fairly incomplete (Liebert et al. 2005). We note, however, that the differences do not appear as large as previously estimated from the comparison with the results of Darling — see also Liebert et al. (2005), Fig. 10. Indeed, above $M_V \sim 12$, our spectroscopic determination of the luminosity function is systematically lower than that of Darling. This is explained by the fact that a lot of the objects in the fainter bins in Darling's analysis have been shifted to brighter magnitude bins as a result of his overestimates of the M_V values (see Fig. 3). We must also mention that because of our spectral reclassification (see above), our sample contains fewer DA white dwarfs than the original sample analyzed by Darling. According to Darling, the completeness of the PG survey is 58%, while the KUV survey is 75% complete. The total space density of DA stars based on our analysis of the KUV sample is 2.10×10^{-3} pc^{-3}. Liebert et al. (2005) obtained for the PG survey a value of 0.5×10^{-4} pc^{-3} for DA white dwarfs with $M_V < 12.75$. For the same range of M_V, we derive a value of 7.58×10^{-4} pc^{-3}, i.e. a space density ~ 15 times larger. The next step in our analysis will be to include the DB stars observed in our survey to improve our determination of the complete luminosity function of white dwarf stars.

Acknowledgments

We would like to thank the director and staff of Steward Observatory for the use of their facilities. This work was supported in part by the NSERC Canada and by the Fund FQRNT (Québec). P.B. is a Cottrell Scholar of the Research Corporation for Science Advancement.

4. References

Bergeron P Saffer R & Liebert J 1992 *ApJ* **394** 247
Green R F 1980 *ApJ* **238** 685
Green R F Schmidt M & Liebert J 1986 *ApJS* **61** 305
Darling G W 1994 *PhD Thesis* Dartmouth College Hanover NH
Fleming T A Liebert J & Green R F 1986 *ApJ* **308** 176
Fontaine G Brassard P & Bergeron P 2001 *PASP* **113** 409
Holberg J B & Bergeron P 2006 *ApJ* **132** 1221
Kondo M Noguchi T & Maehara H 1984 *Ann. Tokyo Astron. Obs.* **20** 130
Liebert J Bergeron P & Holberg J B 2005 *ApJ* **156** 47
Noguchi T Maehara H & Kondo M 1980 *Ann. Tokyo Astron. Obs.* **18** 55
Schmidt M 1968 *ApJ* **151** 393

Journal of Physics: Conference Series **172** (2009) 012011 doi:10.1088/1742-6596/172/1/012011

C/O white dwarfs of very low mass: 0.33-0.5 $M_\odot$

P G Prada Moroni[1,2] **and O Straniero**[3]

[1] Physics Department "E. Fermi", University of Pisa, largo B. Pontecorvo 3, I-56127, Pisa, Italy

[2] INFN, largo B. Pontecorvo 3, I-56127, Pisa, Italy

[3] INAF – Osservatorio Astronomico di Collurania, via Maggini, I-64100, Teramo, Italy

E-mail: prada@df.unipi.it

Abstract. The standard lower limit for the mass of white dwarfs (WDs) with a C/O core is roughly 0.5 $M_\odot$. In the present work we investigate the possibility to form C/O WDs with mass as low as 0.33 $M_\odot$. Both the pre-WD and the cooling evolution of such nonstandard models are described.

1. Introduction

It is by now a well-known and commonly accepted result of the stellar evolution theory that stars of low and intermediate mass will end their lives as white dwarfs (WDs). Since the pioneering papers by Chandrasekhar (1931) and Stoner (1932), it has been known that there must be an upper limit of the mass of WDs. The current value for such a limit, now universally known as Chandrasekhar's mass, is M_{Ch}= 1.456 $(2/\mu_e)^2$ $M_\odot$(where μ_e is the electronic mean molecular weight, Hansen, Kawaler & Trimble 2004). Below this critical mass there are essentially three different families of WDs concerning the chemical composition of their cores: the O/Ne/Mg, C/O and He. The upper mass limit for C/O WDs, which is still subject to some debate, since it is the aftermath of the evolution of a star with initial mass slightly lower than M_{up}, the lowest mass for star that succeeds in igniting carbon burning before the asymptotic giant branch (AGB), it is around 1.05 $M_\odot$(Dominguez et al. 1999, Weideman 2000, Prada Moroni & Straniero 2007, Catalan et al. 2008, Meng, Cheng & Han 2008).

On the other hand, the value for the minimum mass of WDs with a C/O core is commonly believed to be around 0.5 $M_\odot$ (Weideman 2000, Meng et al. 2008). The theory of stellar evolution predicts that an isolated single star with mass lower than about 0.5 $M_\odot$, the exact value depending on the chemical composition (0.46 and 0.50 for Z=0.04 and Z=0.0001, respectively, Dominguez et al. 1999, Girardi 1999), do not ignite the helium-burning and die as WDs with an helium rich core. For these stars, during the red giant phase the electronic thermal conduction and the emission of neutrinos, which cool the core, prevail on the heating due to the contraction caused by the accretion of fresh helium processed by the hydrogen-burning shell. This is the main reason why WDs of mass less than about 0.5 $M_\odot$ are commonly believed to have a He-core.

On the other hand, as early as 1985 Iben & Tutukov, in a very beautiful and instructive paper, showed that a star with an initial mass of 3 $M_\odot$ evolving in a close binary can produce a remnant of 0.4 $M_\odot$ with a sizeable C/O core. A similar result was obtained 15 years later by Han, Tout & Eggleton (2000), who computed the evolution of close binary systems with several different parameters. They showed that in a binary system with the primary star of mass M_1= 2.51 $M_\odot$, the mass ratio q=M_1/M_2=2 and the initial period P=2.559 d, the primary star succeeded to ignite the helium-burning. They did not followed the evolution of the remnant C/O WD, since their code met numerical problems when the primary star was as low as 0.33 $M_\odot$ with a C/O core of 0.11 $M_\odot$.

In this paper we investigate both a possible evolutionary scenario able to produce C/O WDs of low mass, that is $M < 0.5$ $M_\odot$, and their cooling evolution. We will show the results of detailed evolutionary computations performed with a full Henyey code able to follow consistently the evolution of stars from the initial pre-main sequence to the final cooling phase of WDs.

2. The red giant phase transition

The stellar models showed in the present work have been computed with an updated version of FRANEC (Prada Moroni & Straniero 2002, Degl'Innocenti et al. 2008), a full Henyey evolutionary code. We adopted a metallicity Z=0.04 and an initial helium abundance Y=0.32 suitable for the very metal-rich stars belonging to some open galactic cluster, such as NGC6791.

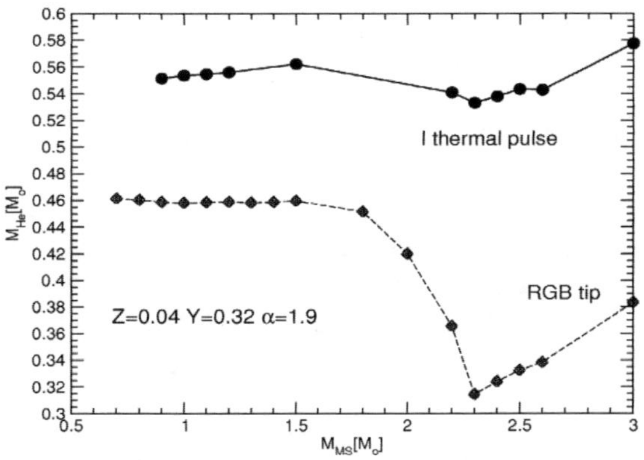

Figure 1. Mass of the He-core at the tip of the RGB (diamond and dashed line) and mass of the hydrogen exhausted core at the first thermal pulse on the AGB (circles and solid line) as a function of the initial mass for stars with Z=0.04 Y=0.32.

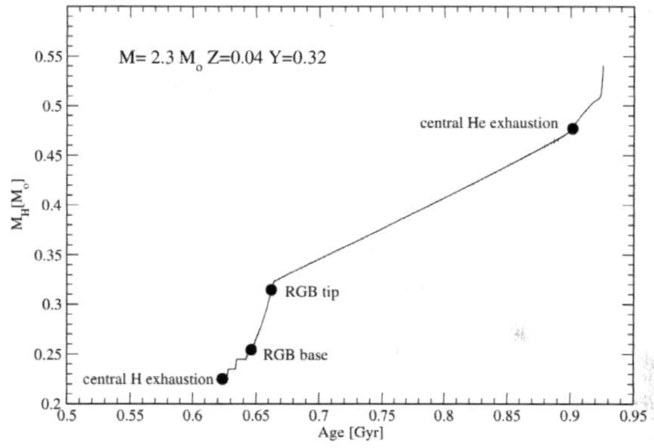

Figure 2. Mass of the hydrogen exhausted core as a function of the age for a star of M=2.3 $M_\odot$, Z=0.04 Y=0.32. The circles mark: the central hydrogen exhaustion, the base and the tip of the RGB and the central helium exhaustion.

The figure 1 shows the mass of the hydrogen exhausted core M_H at the first thermal pulse on the AGB (upper curve) and the mass of the helium core M_{He} at the tip of the red giant branch (RGB tip, lower curve) as a function of the initial mass. As it is well known, M_H at the first thermal pulse is nearly constant around 0.55 $M_\odot$, the exact value depending on the chemical composition, for initial masses lower than 3 $M_\odot$. On the other hand, the behavior of the M_{He} at the tip of RGB as a function of the initial mass is much less smooth than the previous one, in particular a deep and sharp minimum is present around 2.3 $M_\odot$ for this chemical composition (Castellani, Chieffi, Straniero 1992). Such a behavior is the consequence of the physical conditions present in the helium-core at the onset of the 3α nuclear reaction.

For initial mass lower than 1.5-1.7 $M_\odot$, an electron-degenerate core develops at the beginning of the red giant phase, and the onset of the 3α occurs in a nuclear runaway, the so-called helium-flash (Hoyle & Schwarzschild 1955). For these stars, the M_{He} at the tip of the RGB is almost constant, about 0.46 $M_\odot$ for metal-rich stars and 0.50 $M_\odot$ for metal-poor ones, and is essentially produced during the hydrogen-shell burning phase.

From this point on, the larger the initial mass, the weaker the degeneracy of the core in the red giant phase, thus the weaker the He-flash and the lower the mass of the helium-core required for the 3α burning. When the initial mass is as high as about 2.3 $M_\odot$, the helium-burning ignites quiescently, without a flash. This explains the steep decrease of the M_{He} at the tip of the RGB between 1.8 and 2.3 $M_\odot$. For higher masses, the core in the red giant phase is not degenerate any more. From this point on, the larger the initial mass, the larger the M_{He} at the tip of the RGB, as the larger the convective core during the previous main sequence phase. Thus, the minimum mass of the helium-core M_{He} at the onset of the 3α occurs in the transition between these two regimes, around 2.3 $M_\odot$, reaching a value of 0.315 $M_\odot$.

Notice that, as previously stated, this deep minimum is almost canceled at the first thermal pulse. In fact, the hydrogen burning shell continues to process hydrogen and move outward, then increasing M_H, during the central helium-burning phase. Moreover, it is a well established result of the theory of stellar evolution that the lower the mass of the He-core at the onset of the 3α, the fainter the star and the longer the central helium-burning phase. Thus, as the initial mass of the helium-rich core decreases the hydrogen-burning shell will work for a longer time resulting in an larger ΔM_H. This is the reason why star with initial mass lower than 3 $M_\odot$ enter the AGB phase with nearly the same M_H (Dominguez et al. 1999).

The first paper devoted to the study of the evolutionary characteristics of stars belonging to this sharp and deep minimum in the mass of the He-core at the tip of the red giant phase was Iben (1967). Since then, many studies have been focused on this transition, named by Renzini & Buzzoni (1983) red giant phase transition (hereafter RGB phase transition), but a detailed discussion of such an important feature is well beyond the aim of the present paper. We will dwell only upon a feature of the evolution of a star with M= 2.3 $M_\odot$ important to our purpose. Figure 2 shows the evolution of the mass of the hydrogen-exhausted core M_H for this transition mass. The circles mark the following evolutionary phases: the central-hydrogen exhaustion, the base and the tip of the RGB, respectively, and finally the central-helium exhaustion. As one can easily see, when the star exhausts its central hydrogen, the mass of the He-core is already 0.225 $M_\odot$, which is more than the 70% of the value at the RGB tip. In fact, a sizeable convective core was developed during the preavious main sequence phase. When the star reaches the base of the RGB, the mass of the helium core is 0.254 $M_\odot$, which is about the 80% of the mass required to onset He-burning. This means that, since a large fraction of the helium-core required for the ignition of the He-burning is already developed at the beginning of the red giant evolution, a star in the RGB phase transition, that is around 2.3 $M_\odot$, that undergoes a very strong mass loss during the red giant phase can produce a C/O WD with a mass significative lower than 0.5 $M_\odot$.

3. C/O white dwarfs of very low mass

In order to check such an idea and provide models of C/O WDs with mass lower than 0.5 $M_\odot$, we computed the evolution at constant mass of a star with M=2.3 $M_\odot$ until the red giant phase. Then at about $\log L/L_\odot$=1.34, when the He-core was M_{He}= 0.2569 $M_\odot$, we turned on the mass loss, with a constant rate of the order of 10^{-7} $M_\odot$ yr^{-1}. We stopped the mass loss once the desired final mass was obtained and then we followed the next evolution until the final cooling phase.

The figures 3 - 7 show the evolutionary tracks in the HR-diagram for different values of the final mass, from 0.48 $M_\odot$ to 0.33 $M_\odot$, from the red giant to the final WD phase. The model showed in figure 3, with a total mass of 0.48 $M_\odot$, succeded to ignite He-burning and become a WD with a C/O core whose mass fraction is about 89 %, where we defined the core edge to be the point where the helium abundance became larger than 50%. A very similar evolution is followed by the model with 0.461 $M_\odot$, as can be seen in figure 4. In this case, the resulting WD has a C/O core of 85% of the total mass.

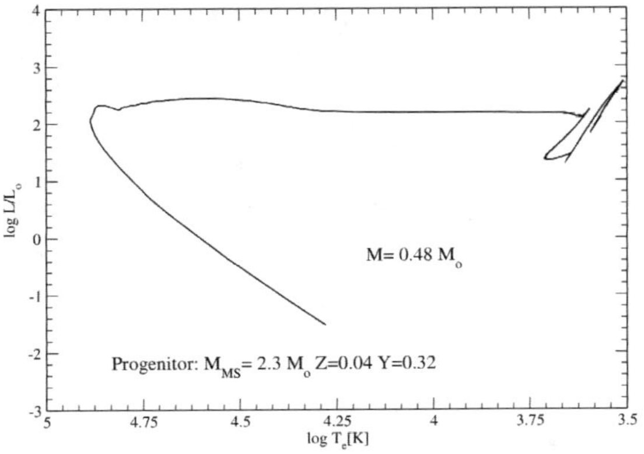

Figure 3. Evolutionary track of a star of M=0.48 $M_\odot$ whose porgenitor of M= 2.3 $M_\odot$ underwent a strong mass loss episode during the red giant phase.

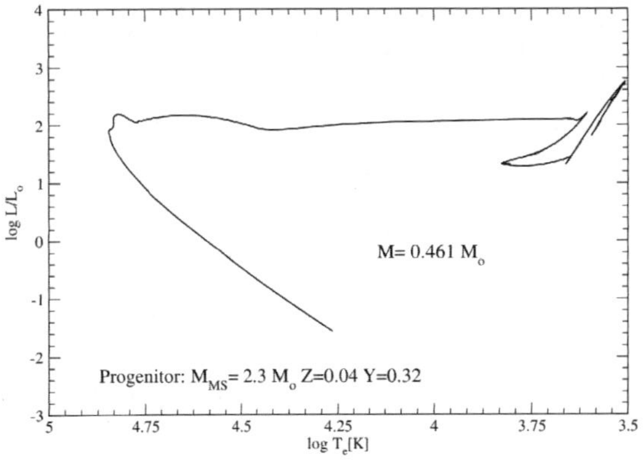

Figure 4. Same as in figure 3 but for a remnant of mass M=0.461 $M_\odot$.

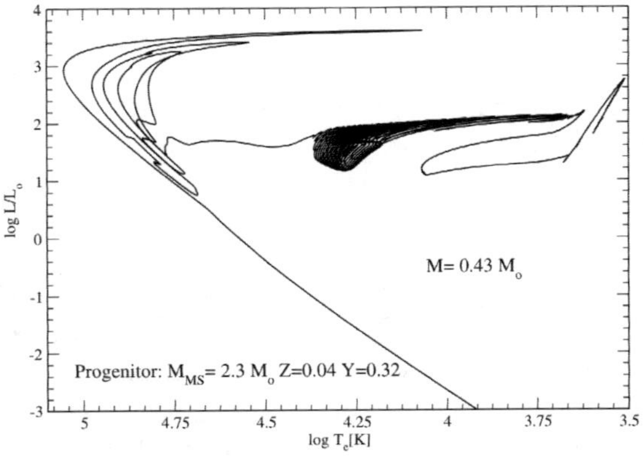

Figure 5. Same as in figure 3 but for a remnant of mass M=0.43 $M_\odot$.

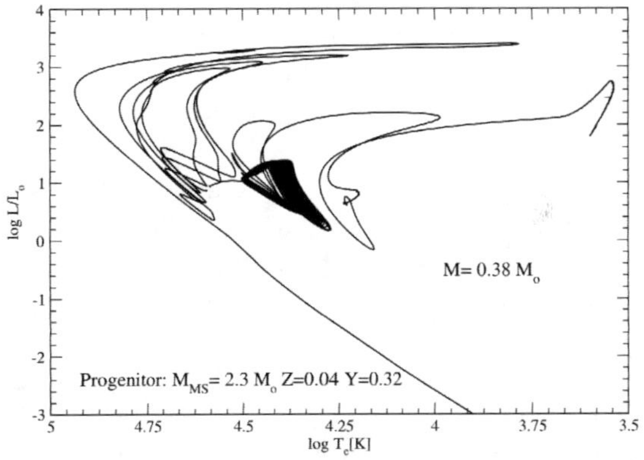

Figure 6. Same as in figure 3 but for a remnant of mass M=0.38 $M_\odot$.

The evolution becomes more complex and difficult to compute for lower masses. The figure 5 shows the evolution of the model with 0.43 $M_\odot$. The star onsets the 3α nuclear reaction and after the central He-burning phase it experiences a series of He-shell thermal pulses: during each flash the star describes a loop in the HR diagram. After the end of the thermal pulses the helium-burning shell continues quiescently and the model moves toward higher effective tempearures. Then the star, approaching the cooling track, experiences a few strong hydrogen-shell flashes, more specifically CNO-flashes. During these episodes, the hydrogen rich outermost layer is progressively eroded, until the star can cool down as a WD with a C/O core of 79% of the total mass. The evolution of the model with 0.38 $M_\odot$, shown in figure 6, presents again both the He-thermal pulses and the CNO-flashes. In addition, after the end of themal pulses, when the model moves toward the blue it experiences a late thermal pulse. The resulting WD has a C/O core of 50% of the total mass.

Notice that the lower the mass of the remnant, the lower the mass fraction of the C/O core, thus the larger the mass of the helium-rich buffer. In standard C/O WD, the thickness of the He-rich layer is of the order of a few percent (1-2%), while in the present models of very low

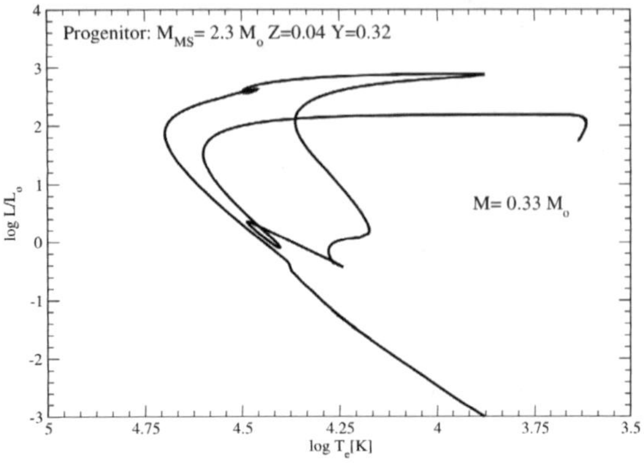

Figure 7. Same as in figure 3 but for a remnant of mass M=0.33 $M_\odot$. The model does not succeed to ignite 3α and eventually cools as a He-core WD.

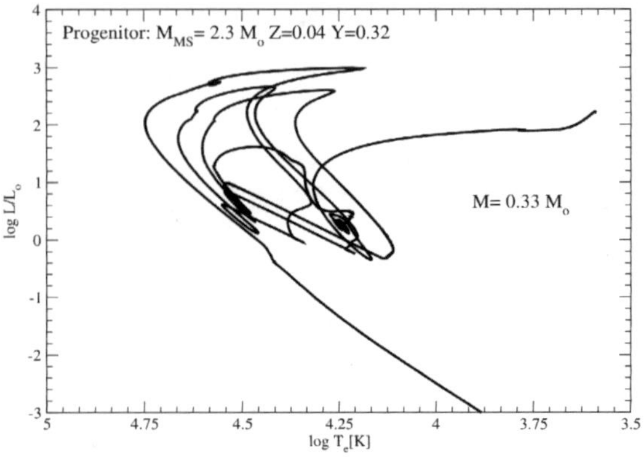

Figure 8. Same as in figure 3 but for a remnant of mass M=0.33 $M_\odot$. In this case the model ignites He-burning and eventually cools as a C/O-core WD.

mass C/O WDs it can reach the 50% of the total mass.

Notice that the helium-shell thermal pulses we showed above are not the same that occur in AGB stars. In fact, in that case the thermal instability is the consequence of the accumulation of a critical mass of helium accreted by the quiescent hydrogen-burning shell. While in this case, they are self-excited relaxation oscillations, where the compression is due to the relaxation of the star after a previous contraction. At variance with thermal pulses in AGB, these pulses affect the entire star, with large oscillations in central temperature and density, as early shown in the pioneering paper by Iben et al. (1986).

The evolution of the remntant with mass M= 0.33 $M_\odot$ deserves a detailed discussion. In fact, adopting the same mass loss rate along the red giant phase of the progenitor star of 2.3 $M_\odot$ as in the previous cases, the 0.33 $M_\odot$ does not experience 3α burning and becomes a WD with a helium core. Figure 7 shows the related evolutionary track on the HR diagram. On the other hand, if the mass loss is turned on later on the RGB phase of the progenitor star, when the mass of the helium core is larger ($\Delta M_{\rm He} \approx 1\%$), the final model with 0.33 $M_\odot$ succeeds to ignite He-burning and eventually evolves to the final WD stage with a C/O core of about the 53% of the total mass. Figure 8 shows its evolutionary track. This is the lowest C/O WD we managed to produce for this chemical composition, in fact the models with mass smaller than 0.33 $M_\odot$ do not ignite the 3α and finish their evolution as He-core WDs. The present evolutionary computations prove that it is possible to have C/O WDs with mass as low as 0.33 $M_\odot$, significantly lower than 0.5 $M_\odot$, the classical and commonly accepted lower limit. This means that in the mass range 0.33 - 0.5 $M_\odot$ both He and C/O core WDs can exist. Thus it is interesting to compare the main characteristics of their structure and evolution. A detailed description of the cooling evolution of low mass WDs has been recently published by Panei et al. (2007). Figure 9 shows the comparison between the core chemical profiles of the two remnants of 0.33 $M_\odot$. While in one case the core is constituted by nearly pure helium, in the other a mixture of carbon and oxygen is present. Notice that in the center there is a significant abundance of carbon, X_{12C}= 0.285, as in standard C/O WDs, thus these very low mass WDs should not be called simply oxygen WDs. As previously stated, a peculiar feature of these very low mass C/O WDs is that the helium-rich buffer is very thick, up to about 50% of the total mass, while the value of typical C/O WDs is of the order of 1-2 %. The thickness of the hydrogen rich outermost layer is almost the same in the two models, that is M= 0.0014 $M_{\rm WD}$ for the WD with a helium

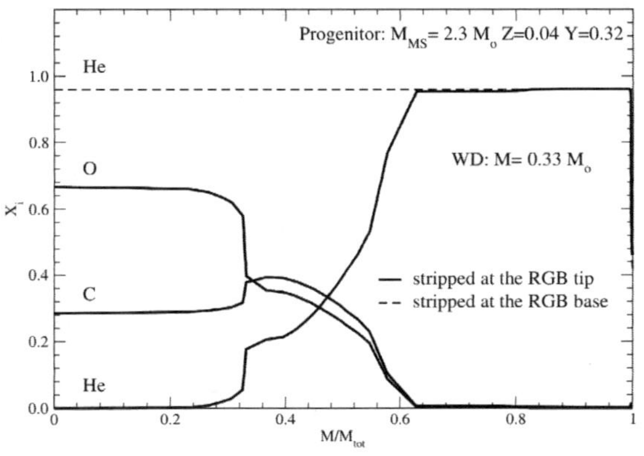

Figure 9. Abundances in mass of helium, carbon and oxygen as a function of the mass coordinate for the two WDs with M= 0.33 $M_\odot$, the He-core (dashed line) and the C/O core (solid lines).

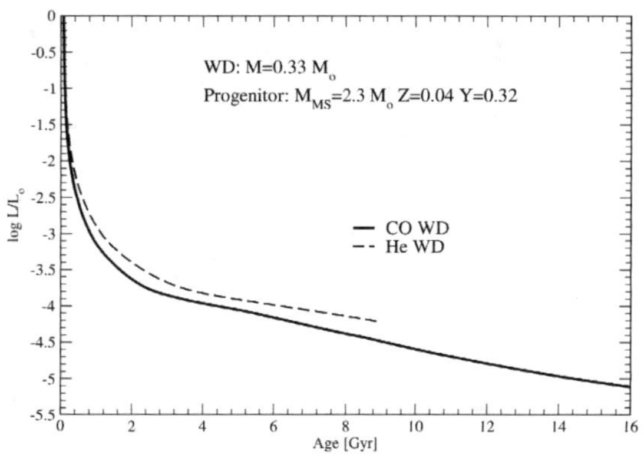

Figure 10. Comparison between the cooling curves, logL/$L_\odot$ vs. time, of the two WDs of 0.33 $M_\odot$, that with the He-core (dashed line) and with C/O core (solid line).

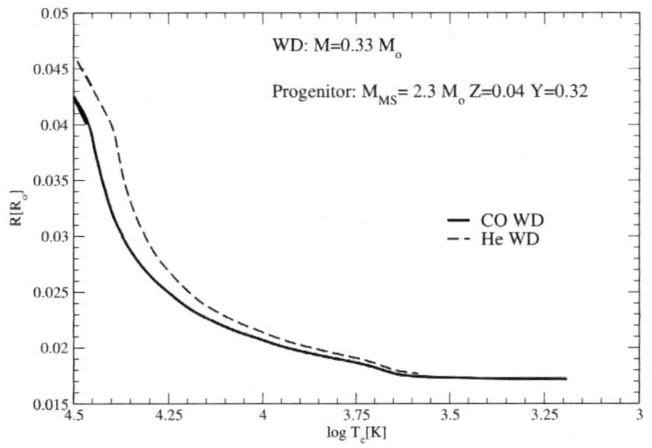

Figure 11. Radius vs. effective temperature of the two WDs of 0.33 $M_\odot$, that with the He-core (dashed line) and with C/O core (solid line).

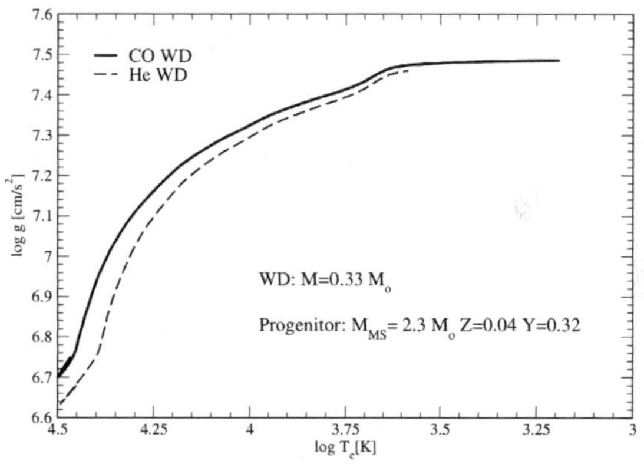

Figure 12. Surface gravity vs. effective temperature of the two WDs of 0.33 $M_\odot$, that with the He-core (dashed line) and with C/O core (solid line).

rich core and M= 0.0015 M_{WD} for that with C/O one.

Figure 10 shows the comparison between the cooling curves of the He-core WD (dashed line) and the C/O core (solid line). As expected, the remnant with the helium rich core cools slower than that with a C/O one, as the specific heat per gram of helium is larger than that of carbon and oxygen; thus the thermal content of the He WD is larger than that of a C/O one with the same total mass. Such a difference in the cooling times is slightely counterbalanced at $\log L/L_\odot \approx$ -4 as the C/O core crystallizes, with the concomitant energy release, while the He core does not, in the presently computed range of temperatures and densities.

Figures 11 and 12 show the comparison between the radii and surface gravities, respectively, as a function of the effective temperature for the two WDs of 0.33 $M_\odot$. For a given effective temperature, the He-core WD is more expanded than the other one, difference not negligible at the beginning of the cooling ($\sim$ 8.5%). Figure 13 shows the comparison between the central

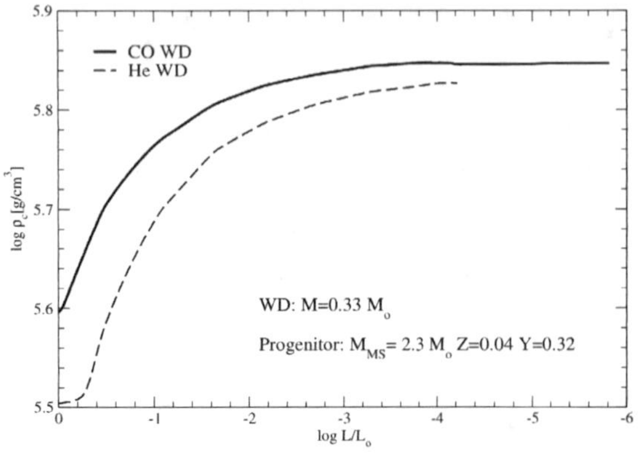

Figure 13. Central density ρ_c vs. luminosity of the two WDs of 0.33 M$_\odot$, that with the He-core (dashed line) and with C/O core (solid line).

density ρ_c as a function of the luminosity for the two WDs of 0.33 M$_\odot$, with the He-core (dashed line) and C/O one (solid line). We did not show the plot with the central temperatures T_c versus luminosity relations for the two WDs because they are nearly identical.

4. Conclusion

In conclusion, as already anticipated by Iben & Tutukov (1985), we have proved, by means of fully and consistent evolutionary computations, that the minimum mass for a C/O WD is about 0.33 M$_\odot$, near the minimum possible mass of the helium-core at the onset of He-burning, which occurs at the RGB phase transition (M$\approx$ 2.3 M$_\odot$).

As a consequence, in the mass range 0.33-0.5 M$_\odot$ both He and C/O core WDs can exist. As expected and already shown by Panei et al. (2007), the cooling times of these two classes of WDs are quite different, being the He-core remnants significantly slower than the C/O ones. In addition, we have also shown that He WDs are more expanded that their C/O counterparts at a given effective temperature.

Acknowledgments

It's a pleasure to thank Giuseppe Bono, who kindly read the paper, for the many useful and pleasant discussions.

References

Castellani V Chieffi S & Straniero O 1992 *ApJSS* **78** 517
Catalan S Isern J García-Berro E & Ribas I 2008 *MNRAS* **387** 1693
Chandrasekhar S 1931 *ApJ* **74** 81
Degl'Innocenti S Prada Moroni P G Marconi M & Ruoppo A 2008 *Ap&SS* **316** 25
Dominguez I Chieffi A Limongi M & Straniero O 1999 *ApJ* **524** 226
Girardi L 1999 *MNRAS* **308** 818
Hansen C J Kawaler S D & Trimble V 2004 *Stellar Interiors* (Springer)
Han Z Tout C A & Eggleton P P 2000 *MNRAS* **319** 215
Hoyle F & Schwarzschild M 1955 *ApJS* **2** 1
Iben I Jr 1967 *ApJ* **147** 650
Iben I Jr Fujimoto M Y Sugimoto D & Miyaji S 1986 *ApJ* **304** 217
Iben I Jr & Tutukov A V 1985 *ApJSS* **58** 661
Meng X Cheng X & Han Z 2008 *A&A* **487** 625
Panei J A Althaus L G Chen X & Han Z 2007 *MNRAS* **382** 779
Prada Moroni P G & Straniero O 2002 *ApJ* **581** 585
Prada Moroni P G & Straniero O 2007 *A&A* **466** 1043
Renzini A & Buzzoni A 1983 *memSAIT* **54** 739
Stoner E C 1932 *MNRAS* **92** 662
Weidemann V 2000 *A&A* **363** 647

Journal of Physics: Conference Series **172** (2009) 012012 doi:10.1088/1742-6596/172/1/012012

A Progress Report on the Carbon Dominated Atmosphere White Dwarfs

P Dufour[1], J Liebert[1], B Swift[1], G Fontaine[2] and T Sukhbold[1]

[1] Steward Observatory, University of Arizona, 933 North Cherry Avenue, Tucson, AZ 85721
[2] Département de Physique, Université de Montréal, C.P.6128, Succ. Centre-Ville, Montréal, Québec, Canada H3C 3J7

E-mail: dufourpa@as.arizona.edu

Abstract. Recently, Dufour et al. (2007) reported the unexpected discovery that a few white dwarfs found in the Sloan Digital Sky Survey had an atmosphere dominated by carbon with little or no trace of hydrogen and helium. Here we present a progress report on these new objects based on new high signal-to-noise follow-up spectroscopic observations obtained at the 6.5m MMT telescope on Mount Hopkins, Arizona.

1. Introduction

White dwarfs have traditionally been divided into two main categories depending on the composition of the primary constituent of their atmosphere. Stars with a hydrogen rich surface composition are classified as DA white dwarfs while those with a helium dominated surface composition belong to the non-DA category (DO, DB, DC, DZ and DQ spectral type). A third category was recently uncovered by Dufour et al. (2007) when an analysis of Hot DQ white dwarfs from the Sloan Digital Sky Survey (SDSS) revealed that their surface composition was dominated by carbon and not helium as was initially assumed by Liebert et al. (2003).

These strange objects are quite rare. Only 9 were found in the SDSS DR4 catalog of spectroscopically identified white dwarfs which contained nearly 10,000 stars (Eisenstein et al. 2006). A detailed analysis of the spectroscopic and photometric data by Dufour et al. (2008) showed that the hot DQ white dwarfs are found in a rather narrow range of effective temperature centered around $\sim$ 20,000 K.

No carbon dominated atmosphere white dwarfs with an effective temperature higher than $\sim$ 24,000 K have been found from a careful inspection of thousands of spectra in SDSS DR6 with $u - g < -0.2$ and $g - r < -0.3$. The absence of hotter counterparts for carbon dominated atmosphere white dwarfs suggest that they probably cool as helium atmosphere stars until a convection zone develops in the underlying carbon-rich mantle due to the recombination of that element. If this scenario is correct, a subphotospheric carbon convection zone would develop around 24,000 K and dilute the surface helium layer in the much more massive carbon envelope. This would transform a helium dominated DB star with a very thin surface helium layer into a carbon dominated Hot DQ white dwarf (see Dufour et al. 2008 for more on this scenario).

Why would some helium rich white dwarfs have thin helium layers is still not entirely understood yet. One possibility is that these stars are the result of the evolution of massive stars that have burned carbon and have an oxygen-magnesium-neon core (see Garcia-Berro et al. 1997 and references therein). Alternatively, they could have experienced a late thermal pulse that eliminated most of the helium, a phenomenon similar to the one that is generally believed to explain the existence of other hydrogen deficient stars (Werner & Herwig 2006). There might also be other ways to eliminate the superficial helium layers that still need to be explored. However, a scenario involving close binary interaction, as proposed to explain the variability of J1426+5752 (Montgomery et al. 2008), now appears very unlikely since i) this star has been found to show multi-periodic variations (Green et al. 2009, see also Fontaine et al. in these proceedings) ii) no radial velocity variations have been detected in J1426+5752 (see again Green et al. 2009) and iii) more importantly, it doesn't explain why *all* Hot DQ's are found within the same temperature range and *none* at higher or lower effective temperatures. We thus believe that the convective mixing scenario presented in Dufour et al. (2008) is the most appealing.

One essential parameter that would give important clues about the origin and evolution of these objects is the mass. The mass of a white dwarf is generally obtained from a determination of the surface gravity by matching the observed line profiles with those of model atmosphere calculations. However, in the case of the Hot DQ white dwarfs, the surface gravity is poorly known due to several factors.

First, the atmospheric parameter determinations of Dufour et al. (2008) were based on the analysis of rather noisy optical SDSS spectra. Years of experience from analyzing DA and DB white dwarfs has taught us that high signal-to-noise ratio spectroscopic observations are required for a precise measurement of the surface gravity (and effective temperature). Rough estimates of log g were obtained by Dufour et al. (2008) but the quality of the SDSS spectra is clearly insufficient to provide the needed accuracy.

Second, most of the flux emitted by these stars is in the ultraviolet part of the electromagnetic spectrum. This region of the spectrum contains many more absorption features than there is in the optical. All these absorption lines have an important influence on the thermodynamic structure (temperature and pressure as a function of optical depth) of the atmosphere since the flux absorbed at short wavelength tends to be redistributed at longer wavelength (see Fig. 4 of Dufour et al. 2008). Hence, small errors in our modeling of the UV part of the spectrum will have an important effect on the line profiles in the optical. Unfortunately, an assessment of the quality of our modeling of the UV will only be possible by comparing UV observations (which we should obtain, if the HST repair mission goes as planed, sometime next year) with our model calculations.

Finally, any determination of the surface gravity from the line profiles must rely on a state-of-the-art theory of the broadening of the spectral lines. At the time of the Dufour et al. (2008) analysis, detailed calculations for the Stark broadening of CII lines were not available and the standard scaled classical approximation was used. Therefore, surface gravities (and effective temperatures) obtained in Dufour et al. (2008) can only be considered as preliminary approximations until new tables for the Stark broadening of CII (which are currently being calculated) are incorporated in our next generation of model atmosphere.

In summary, atmospheric parameters will only be truly reliable when each of the three issues above have been properly addressed. This paper reports our development on the first of these issues.

2. New MMT Observations

As mentioned above, in order to obtain a precise determination of the atmospheric parameters, it is of utmost importance to have high signal-to-noise ratio spectroscopic observations. Due to the nature of the Sloan Digital Sky Survey acquisition procedure, spectra of faint objects such as the Hot DQ are of rather poor quality. We have thus recently undertaken a program to reobserve all the known carbon dominated atmosphere white dwarfs with the Mt. Hopkins 6.5m MMT telescope. We have obtained 7 nights of MMT time and succeeded in securing high signal-

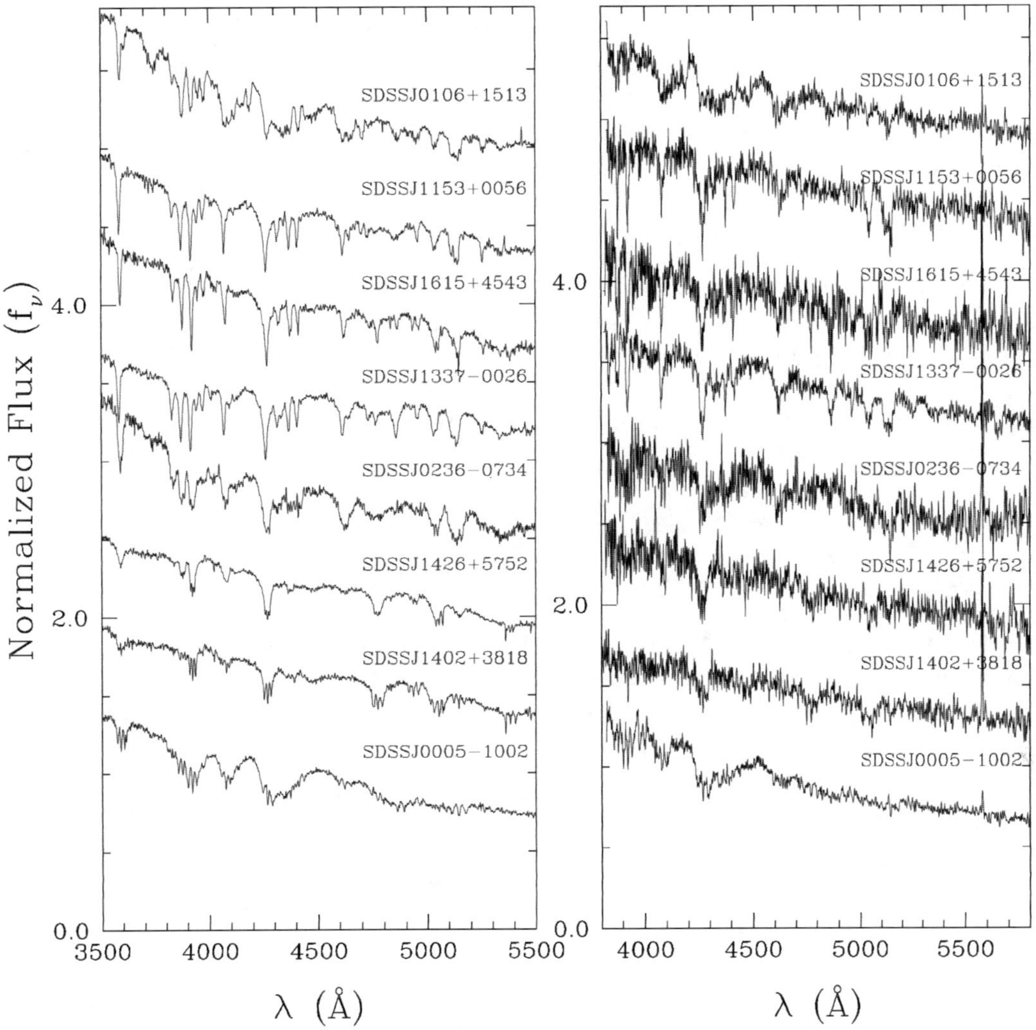

Figure 1. Left: New spectroscopic observations from the 6.5m MMT telescope for 8 of the 10 known Hot DQ white dwarfs. Spectra are normalized to unity at 4500 Å and offset from each other for clarity. The stars are approximately ordered by increasing effective temperature from bottom to top. Right: Corresponding SDSS spectra for each of the Hot DQ that have been reobserved with MMT. Note that we have applied for clarity a three-point average window smoothing in the display of the SDSS spectroscopic data.

to-noise ratio spectra (typically S/N of 60 or more) for 8 of the 10 (9 from the DR4 WD catalog and a 10th one recently found in DR6) known Hot DQ white dwarfs (see Figure 1). We used the Blue Chanel with the 500 line mm^{-1} grating with a 1" slit, resulting in a $\sim$ 3.6 Å FWHM spectral resolution at 5500 Å. The spectra were reduced with standard IRAF packages. The setup used allowed a coverage of a few hundred angstroms bluer than that of SDSS although at the expense of an equally good coverage of the redder part of the electromagnetic spectrum (our spectra cover a wavelength range of $\sim$3400-6300 Å vs $\sim$3800-9000 Å for SDSS).

In Figure 1, we present the new high signal-to-noise ratio spectrum for all the Hot DQ that have been reobserved with MMT so far. We also show, for comparison, the corresponding SDSS spectra that were used for the preliminary analysis of Dufour et al. (2008). The noise level and sharpness of the observed spectral lines are now suitable for an eventual precise spectroscopic analysis. However, it would be meaningless at this point to present updated atmospheric parameters based on these new observations since our grid does not incorporate the new tables for the Stark broadening of CII yet. We prefer to wait for the next generation of models, and also the HST observations, before sharing the results of our reanalysis for these stars. Nevertheless, even without proceeding with a full detailed analysis, several interesting observations are worth discussing.

The most remarkable concerns the fraction of Hot DQ stars that are found to be magnetic. In Figure 1, we clearly see magnetic splitting of the spectral lines for at least 4 objects. These are the 4 objects starting from the the bottom in Figure 1. The separation between the σ and π components indicates that the fields are of the order of 1 MG. The splitting was not observable in the noisier SDSS spectra except for the brightest Hot DQ SDSS J0005-1002. Spectropolarimetric measurement has also confirmed the magnetic nature for SDSS J0005-1002 (see Dufour et al. 2008). This is in contrast with the next 3 stars in Figure 1 that show instead very sharp CII lines where no trace of magnetic splitting is detectable. Given the spectral resolution of our observations, a limit of about 200 kG can be put on the strength of the magnetic field for these objects. The hottest stars in our sample, SDSS J0106+1513, do not show sharp and well defined lines which suggest perhaps the presence of a weak magnetic field but more precise spectropolarimetric measurements will be needed to confirm magnetism. Note also that one of the two stars that remains to be observed with MMT, SDSS J2200-0741, seems also to have lines slightly broadened by a weak magnetic field.

It thus seems that at least 4 (and perhaps 6) of the 10 known Hot DQ are magnetic. This is an incredibly high fraction considering that the fraction of magnetism of nearby white dwarfs is around 10% (Liebert, Bergeron & Holberg 2003). Is this an indication that Hot DQ have higher mass than average since magnetic white dwarfs tend to be more massive in general? The preliminary analysis of Dufour et al. (2008) indicated that most of these stars have surface gravities around log g $\sim$ 8.0 although this could possibly be revised if it is found that our approximation for the Stark broadening considerably overestimated the broadening of the lines.

This unusual high fraction could indicate that magnetism plays a significant role in the evolution of these objects. But if this is the case, why do we detect magnetism in only about 50% and not 100% of these stars? Could it be that they are indeed all magnetic but that in some cases the strength of the field is not strong enough to be detected from line splitting? More sensitive spectropolarimetric measurements, which we plan to obtain soon, should elucidate those questions.

We also detect the HeI 4471 line, indicating the presence of a significant amount of helium (C/He $\sim$0.5), in two of the coolest Hot DQ (SDSS J1426+5752 and SDSS J1402+3818). The presence of helium in SDSS J1426+5752 was also needed to explain the pulsational properties recently discovered by Montgomery et al. (2008) since pure carbon atmosphere white dwarfs are not expected to pulsate at this temperature (see Fontaine et al. 2008 and Fontaine et al., these proceedings).

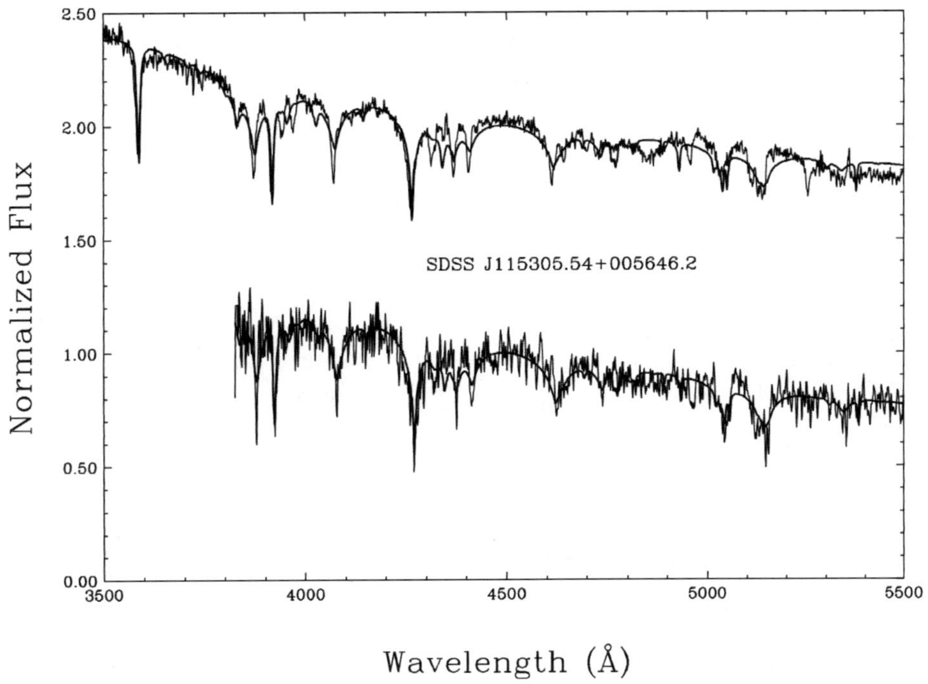

Figure 2. Comparison of the fits of the SDSS (bottom) and MMT (top) spectra for SDSS J1153+0056. The solution is the same as that presented in Dufour et al. (2008), that is log =8 and $T_{\rm eff}$ = 21,650 K.

We will conclude this discussion with some remarks on our models. The solutions presented in Dufour et al. (2008) were not bad considering the level of noise of SDSS spectra. However, with the new MMT observations, it is now obvious that improvement in our models will be needed before we can claim that we know the atmospheric parameters with great accuracy. This is illustrated in Figure 2 where we show the Dufour et al. (2008) solution and a very similar fit to the MMT data for SDSS J1153+0056. While the fit to the SDSS data looks quite acceptable, there are significant discrepancies that appear when observing in detail the MMT fit. It seems that our models do not reproduce correctly the strength of many lines. This is most evident when looking at the group of lines near 4400-4500 Å(and it can also be observed at many other wavelength). Some of the lines are quite well reproduced with our models while others seem to require a different log g or $T_{\rm eff}$ (see Figure 1 of Dufour et al. 2008 for example). This is usually symptomatic of a wrong thermodynamic structure. This could be due, as already mentioned above, to an improper modeling of the UV part of the spectrum. The treatment of the Stark broadening of CII lines might also cause such a discrepancy. The exact reason is probably a combination of both, as well as other parameters such as the abundance of other trace elements, treatment of the convective efficiency or perhaps even uncertainties in the atomic data used. Carbon dominated atmosphere modeling is relatively new and has obviously not reached yet the level of accuracy that is attained for DA and DB stars. We hope to achieve this in a near future.

3. Conclusions

White dwarfs with a carbon dominated atmosphere represent a new challenge to stellar evolution. Their origin and evolution are very mysterious and as such, they deserve all our attention. Progress on our understanding of these objects will necessitate i) new models with state-of-the-art physics (Stark broadening) ii) observation in the UV part of the electromagnetic spectrum (HST observations are coming soon, assuming no problems with the next repair mission), and iii) high signal-to-noise ratio spectroscopic data. In this short report, we presented new optical spectra taken with MMT. Their analysis will eventually lead to a more precise measurement of the atmospheric parameters. One of the most outstanding characteristics that emerges from these new observations is the unusually high fraction of Hot DQ with a large magnetic field. Whether or not the magnetic field is intimately related to their evolution and origin remains unclear. Undoubtedly, further study of Hot DQ white dwarfs will increase our knowledge of stellar evolution. The new challenges that they bring to our field will be of the most interesting to solve.

Acknowledgments

P.D. acknowledges the financial support of NSERC. This work was also supported by the National Science Foundation through grant AST 03-07321. Support from the Reardon Foundation is also gratefully acknowledged.

References

Dufour P Fontaine G Liebert J Schmidt G D & Behara N 2008 *ApJ* **683** 978
Dufour P Liebert J Fontaine G & Behara N 2007 *Nature* **450** 522
Eisenstein D J et al. 2006 *ApJ* **167** 40
Fontaine G, Brassard P & Dufour P 2008 *A&A* **483** L1
Garcia-Berro E Ritossa C & Iben I Jr 1997 *ApJ* **485** 765
Green E M Dufour P & Fontaine G in preparation
Liebert J et al. 2003 *AJ* **126** 2521
Liebert J Bergeron P & Holberg J B 2003 *AJ* **125** 348
Montgomery M H Williams K A Winget D E Dufour P DeGennaro S & Liebert J 2008 *ApJL* **678** L51
Werner K & Herwig F 2006 *PASP* **118** 183

Journal of Physics: Conference Series **172** (2009) 012013 doi:10.1088/1742-6596/172/1/012013

Measuring the Surface Inhomogeneity of Metals on Accreting White Dwarfs

M H Montgomery[1,3], S E Thompson[2,3] and T von Hippel[1,4]

[1]Department of Astronomy, University of Texas at Austin, Austin, TX, USA
[2]Department of Physics and Astronomy, University of Delaware, Newark, DE
[3]Delaware Asteroseismic Research Center, Mt. Cuba Observatory, Greenville, DE
[4]Department of Physics, Siena College, Loudonville, NY

E-mail: mikemon@astro.as.utexas.edu

Abstract. Due to the short settling times of metals in DA white dwarf atmospheres, any white dwarfs with photospheric metals must be actively accreting. It is therefore natural to expect that the metals may not be deposited uniformly on the surface of the star. We present calculations showing how the temperature variations associated with white dwarf pulsations lead to an observable diagnostic of the surface metal distribution, and we show what constraints current data sets are able to provide.

1. Astrophysical context

There are two main classes of white dwarf stars: those with hydrogen-rich atmospheres (spectral type DA) and those with helium-rich atmospheres (non-DA spectral types). The reason for this is that the high surface gravities of white dwarfs lead to efficient gravitational settling, with the lightest elements rising to the surface. In addition, some DA white dwarfs have spectra showing metal lines of elements such as Ca and Mg (Zuckerman et al. 2003), and these stars are referred to as DAZs; about 20% of all DAs fall into this category. Recently, Dufour et al. (2007) have announced a new class of white dwarf with carbon-dominated atmospheres, the "hot DQ" stars, several examples of which have been found in the Sloan Digital Sky Survey (Liebert et al. 2003).

The presence of metals in the DAZs is intriguing since the settling time scale for the metals may be many orders of magnitude shorter than the evolutionary age of these objects. Indeed, for DAZs with $T_{\rm eff} \sim 12{,}000$ K, the settling time scale can be on the order of days or weeks, meaning that these objects are experiencing ongoing accretion (Koester & Wilken 2006). This ongoing accretion is consistent with the fact that nearly a dozen of these objects have detected dust disks ("debris disks," see Tokunaga, Becklin & Zuckerman 1990; Becklin et al. 2005; Kilic et al. 2005; Reach et al. 2005; Kilic et al. 2006; von Hippel et al. 2007; Jura, Farihi & Zuckerman 2007a, Jura et al. 2007b; Farihi, Zuckerman & Becklin 2008) and these disks are assumed to be the sources of the metal lines seen in these white dwarf atmospheres. The best studied object of this DAZ class with an observed disk, G29-38, is also a multi-periodic variable white dwarf (DAV), pulsating in non-radial g-modes with periods of a few hundred to one thousand seconds.

The technique of asteroseismology uses the observed pulsation modes of a star to infer and constrain the interior structure of the star, thus obtaining information on the star's structure as a function of radius (Bradley & Winget 1994; Kawaler & Bradley 1994; Metcalfe, Salaris &

Winget 2002; Montgomery, Metcalfe & Winget 2003). In contrast to this, our approach in this paper is to use the different *angular* dependence of the pulsations to constrain the accretion process. Since the accretion is most likely occurring through a disk, it is natural to suppose that the metals may not be uniformly distributed across the star's surface. We present calculations showing how we can place constraints on the non-uniformity of the accretion process.

2. Gravitational settling and horizontal diffusion

For all our calculations, we use the DAZ G29-38 as our template, since, given its brightness, it has the greatest potential for successful measurement of a surface inhomogeneity. In addition, we have archival data on this star appropriate to this application. In this section, we will therefore attempt to quantify the importance of gravitational settling and horizontal diffusion assuming a model with parameters similar to those of G29-38.

Bergeron et al. (2004) find $T_{\rm eff}$ = 11,820 K and $\log g$ = 8.14 for this star, while Koester, Provençal & Shipman (1997) find $T_{\rm eff}$ = 11,600 K and $\log g$ = 8.05. Interpolating in the tables of Koester & Wilken for Ca yields a settling time of $\sim$ 13 days for the first set of parameters[1] and a settling time of $\sim$ 23 days for the second. We note that while unseen helium in the atmosphere could lengthen these settling times considerably (e.g., García-Berro et al. 2007), its presence is inconsistent with the pulsation results for this star: such an amount would imply a *much* deeper surface convection zone, in conflict with that found by Montgomery (2005).

Since these stars should have surface convection zones, the dominant form of horizontal transport of Ca will be due to the turbulent viscosity. We can estimate the size of this diffusion coefficient as $D \sim v_C l_h$, where v_C is a typical convective velocity and l_h is the assumed "mixing length" for convection. From our white dwarf evolution code (e.g., see Montgomery et al. 1999) we find for both sets of stellar parameters that $D \approx 1.5 \cdot 10^{10} \rm{cm}^2/\rm{sec}$.

For these simple exploratory calculations we assume azimuthal symmetry for both the accretion and the surface metal distribution, i.e., $Z = Z(\theta, t)$ and $S = S(\theta, t)$, where $Z(\theta, t)$ is the metal abundance, $S(\theta, t)$ is the source function of metals accreting onto the white dwarf, θ is the co-latitude of a point on the star's surface, and t is time.

Since convection will uniformly mix material vertically, we can treat the convective region as a single zone and write an equation for the time evolution of Z as a function of θ and t:

$$\frac{\partial Z(\theta,t)}{\partial t} = -\gamma\, Z(\theta,t) + D\, \nabla_{\rm h}^2 Z(\theta,t) + S(\theta,t), \tag{1}$$

where $\gamma \equiv 1/(\text{settling time})$ is the settling rate, $\nabla_{\rm h}^2$ is the horizontal part of the Laplacian operator, and the other variables are as defined above. An estimate of the relative importance of sinking to spreading is $\eta \equiv \gamma\, R_\star^2/D$, where $R_\star$ is the radius of the white dwarf: $\eta \sim 50$ using the Bergeron et al. (2004) values whereas $\eta \sim 30$ for those of Koester et al. (1997). When $\eta \gg 1$ the metals will sink before they have a chance to diffuse horizontally, while for $\eta \ll 1$ the metals will have a chance to mix thoroughly horizontally before sinking, producing a nearly uniform surface distribution.

In Figure 1 we show the metal distributions which arise from solutions of equation 1. The dashed curve is the equilibrium distribution which results from constant accretion centered at the poles and the solid curve is that which results from constant accretion centered on the equator.

3. The diagnostic

The flux variations observed in pulsating white dwarfs are due almost entirely to temperature changes on the surface of the stars (Robinson, Kepler & Nather 1982). These same temperature

[1] Due to a mis-reading of Table 2 in Koester & Wilken (2006), von Hippel & Thompson (2007) calculated a settling time of $\sim$ 7 days.

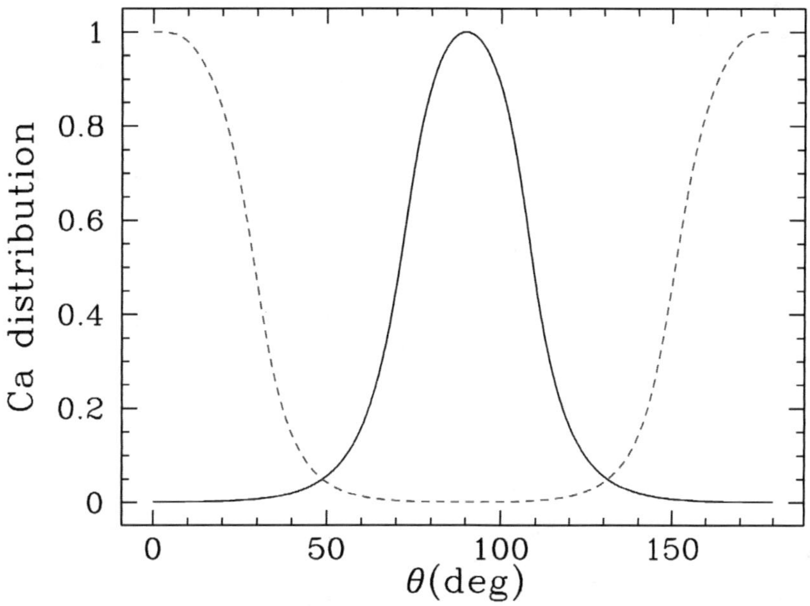

Figure 1. Possible surface metal distributions for accretion centered on the poles (dashed curve) or the equator (solid curve) as a function of the polar angle (co-latitude).

changes will also affect the equivalent widths (EWs) of any spectral lines, and in particular the EWs of metal lines. These metals may not be uniformly distributed across the star's surface and since the temperature variations are also non-uniform, we hope to be able to constrain the surface metal distribution.

The relevant diagnostic we have developed, denoted by R, is the ratio of the fractional EW amplitude to the fractional flux amplitude:

$$R \equiv \frac{\delta\langle EW\rangle/\langle EW\rangle}{\delta F_X/F_X} = \frac{A_{\rm EW}}{A_{\rm Flux}}, \tag{2}$$

where $A_{\rm EW}$ is the fractional amplitude of EW variations, and $A_{\rm Flux}$ is the amplitude of the fractional flux variations observed in the given passband X. In Figure 2 we show separately the amplitudes of the photometric (top panels) and equivalent width (lower panels) variations as a function of the inclination angle.

To compute R, we assume a particular surface temperature perturbation of the form $\delta T/T \propto Y_{\ell m}(\theta,\phi)$. We then use model atmospheres to turn this into EW and flux variations on the surface of the star, taking into account the fact that the EW of the Ca lines will be directly proportional to the local abundance of Ca (e.g., the curves in Figure 1). For the passband X we assume a wavelength response appropriate to the Argos CCD with a BG40 filter on the 2.1m telescope at McDonald Observatory (Nather & Mukadam 2004); the wavelength range is taken to be 3000 Å to 7000 Å, with a peak response at 5400 Å. Finally, we integrate the result across the visible surface of the star. Limb darkening is automatically taken into account by this procedure.

We note for a uniform distribution of metals that $R = 2.71$ for $\ell = 1$ and $R = 2.59$ for $\ell = 2$. Thus, even by measuring R for a single mode it may be possible to tell whether metals are uniformly distributed on the star's surface. To obtain further constraints, we need R determinations from other modes and/or additional information such as the inclination angle of the star.

Journal of Physics: Conference Series **172** (2009) 012013 doi:10.1088/1742-6596/172/1/012013

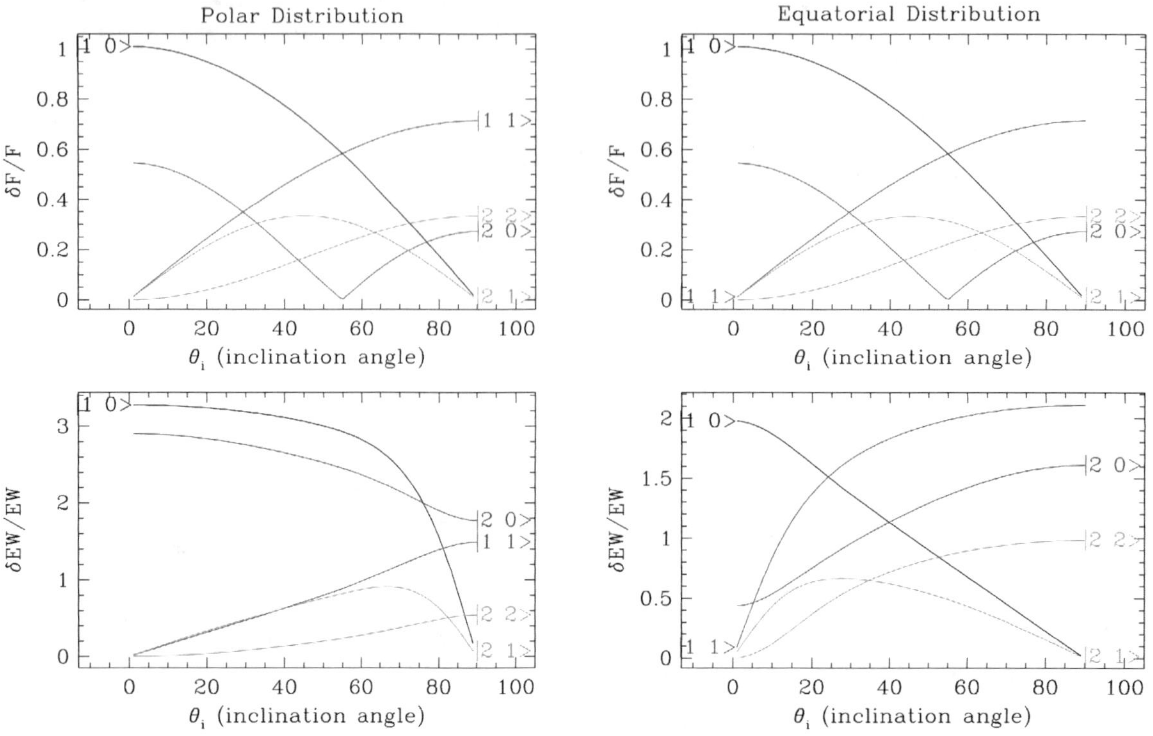

Figure 2. Left panels: The flux amplitude (top panel) and the EW amplitude (lower panel) as a function of θ_i for the polar distribution of metals shown in Figure 1. Right panels: the same as the left panels but for the case of the equatorial distribution of metals in Figure 1.

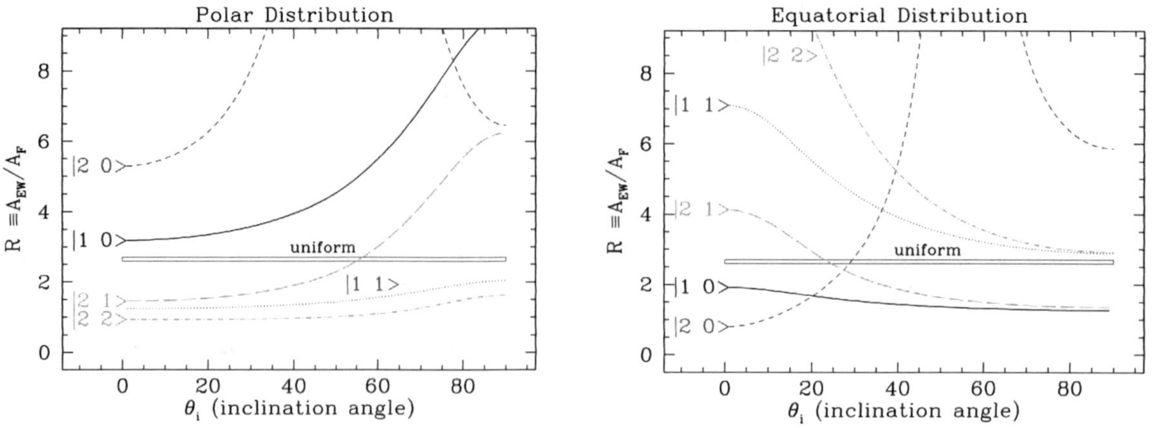

Figure 3. The diagnostic R as a function of inclination angle θ_i. The left panel shows the results for the polar distribution of metals shown in Figure 1 and the right panel shows the results for the equatorial distribution. The different curves are labeled by the $|\,\ell\, m\,\rangle$ values of the relevant pulsation modes.

In Figure 3 we show the R diagnostic as a function of inclination angle θ_i. The different curves are labeled according to their ℓ and m values as $|\,\ell\, m\,\rangle$. The left-hand plot is for the polar distribution of metals given in Figure 1 and the right-hand plot is for the equatorial

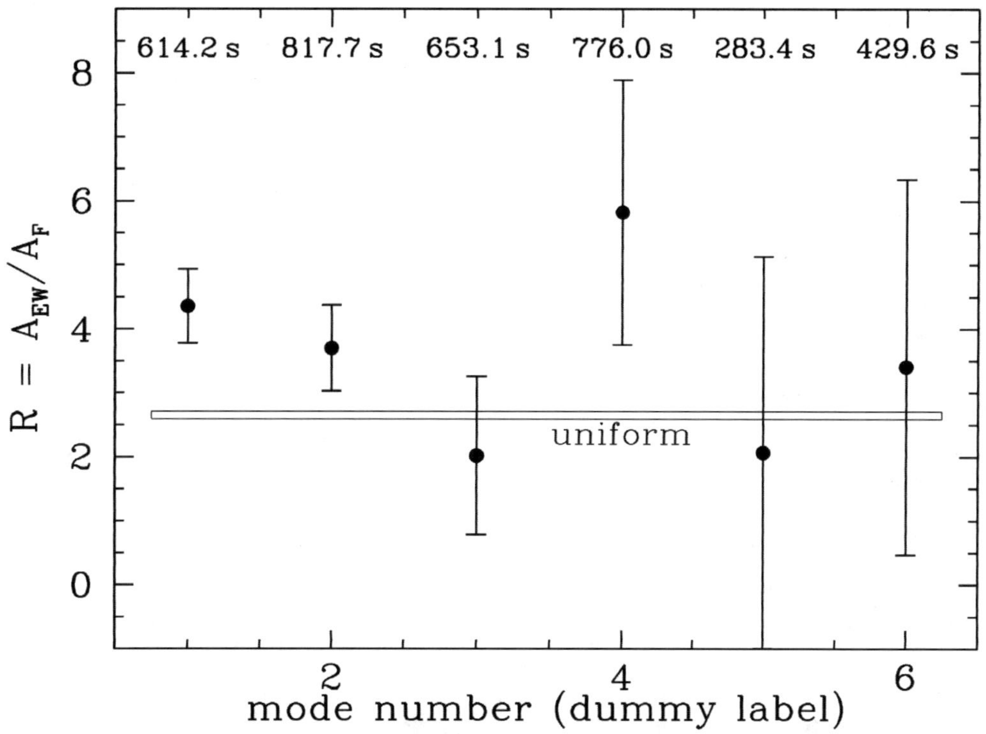

Figure 4. Values of R for modes in the DAV G29-38 as obtained from 4 hours of Keck data in 1999. The values for the first 6 highest amplitude modes are shown, with the period in seconds indicated above each point. Only the first two modes have a signal-to-noise ratio large enough to be useful.

distribution. The thin horizontal boxed region shows the value of R expected if the metal distribution is uniform; we see that a non-uniform distribution is very unlikely to produce a value of R in this range.

4. Comparison with observations

In 1996 Clemens & van Kerkwijk obtained over 4 hours of time-resolved spectroscopy of the DAV G29-38 (Clemens, van Kerkwijk & Wu 2000; van Kerkwijk, Clemens & Wu 2000; Clemens et al. 1999). While not originally intended for this purpose, we can use this as an example data set for the technique proposed in the previous section. We take the amplitudes for the EW variations from the analysis of von Hippel & Thompson (2007) and and we use the amplitudes of the broadband (5200–5500 Å) flux variations as determined by van Kerkwijk et al. (2000)[2]. We calculate $R = A_{EW}/A_{\rm Flux}$ for each of the modes, taking into account the errors on all quantities.

We show the results of this procedure in Figure 4. Of the 6 modes we identified, only the first two have R values with small enough error bars to provide any constraint on the Ca distribution. The 614 s mode provides the most convincing evidence for a non-uniform Ca distribution, since its R value is a full 3σ above the range produced by a uniform distribution. The 818 s mode may also provide some evidence, although it is only about 1.3σ above the value expected from a uniform distribution.

[2] Even though this wavelength range is much narrower than that used in the previous section for Figure 3, the central wavelengths of the two passbands are nearly the same and the results obtained are virtually indistinguishable from one another.

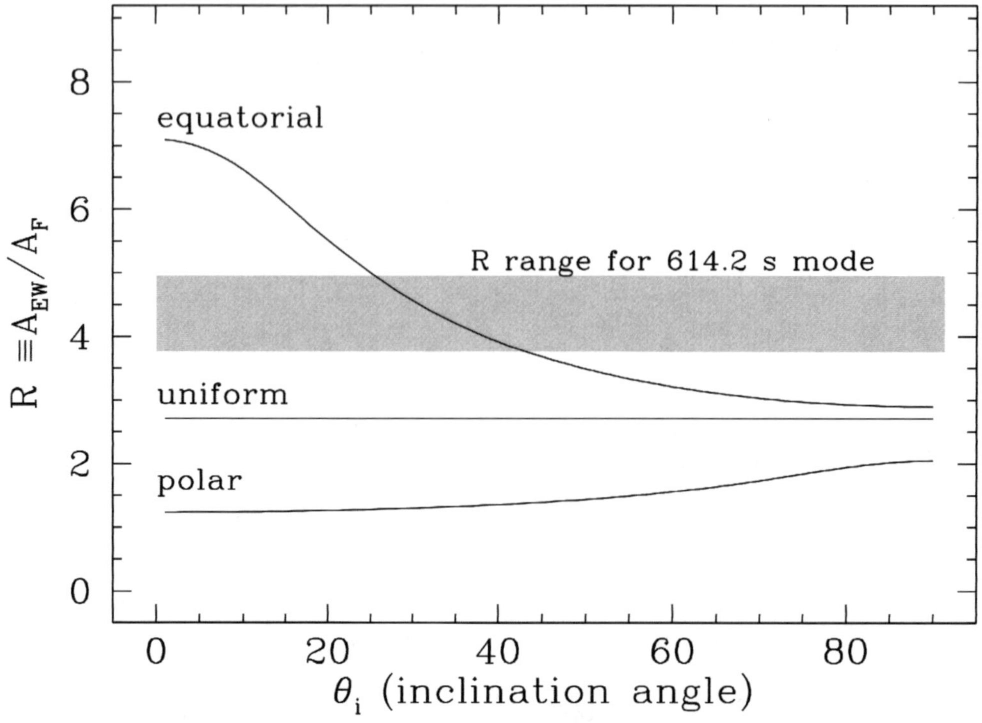

Figure 5. Comparison of the observed R value for the 614 s mode in G29-38 (shaded region) to the value expected for a polar, an equatorial, and a uniform distribution of Ca, as a function of inclination angle θ_i. The data are only consistent with the equatorial case.

With only one statistically significant value of R, we cannot hope to infer anything further about the Ca distribution. Fortunately, we do have additional information. Clemens et al. (2000) used time-resolved spectroscopy to determine that this mode has $\ell = 1$. In addition, Montgomery (2005), by modeling the nonlinear pulse shape of this mode, was able to further specify that it is an $\ell = 1$, $m = 1$ mode. With just this one additional constraint, we are somewhat surprisingly able to constrain our models of the Ca distribution. Figure 5 shows the R values derived using an equatorial, a polar, and a uniform distribution, for a variety of inclination angles. The 614 s measurement is inconsistent with a polar distribution of metals, and is most consistent with an equatorial distribution.

5. Conclusions

We have shown how the temperature variations due to stellar pulsation can be used to constrain the metal distribution on the surface of a white dwarf, and we have shown that data currently in hand for the star G29-38 suggests that the metal distribution, and therefore the accretion, may be equatorial. If further studies support an equatorial distribution for Ca in G29-38, this argues against magnetic accretion onto spots near the poles. Non-magnetic equatorial accretion would further imply that the inner edge of the disk is much thinner than the white dwarf's radius. A physically thin disk is consistent with observations to date (e.g., Jura et al. 2007a; von Hippel et al. 2007). Thus, the pulsations allow us to probe the physics of accretion and gravitational settling in these systems.

Acknowledgments

M.H.M. and T.v.H are grateful for the financial support of the National Science Foundation, under awards AST-0507639 and AST-0607480, respectively. M.H.M. and S.E.T. gratefully acknowledge the support of the Delaware Asteroseismic Research Center.

References

Becklin E E Farihi J Jura M Song I Weinberger A J and Zuckerman B 2005 *ApJ* **632** L119
Bergeron P Fontaine G Billères M Boudreault S and Green E M 2004 *ApJ* **600** 404
Bradley P A and Winget D E 1994 *ApJ* **430** 850
Clemens J C van Kerkwijk M H Wu Y and Kleinman S J 1999 *ASP Conf. Ser. 169: 11th European Workshop on White Dwarfs* p 122
Clemens J C van Kerkwijk M H and Wu Y 2000 *MNRAS* **314** 220
Dufour P Liebert J Fontaine G and Behara N 2007 *Nature* **450** 522
Farihi J Zuckerman B and Becklin E E 2008 *ApJ* **674** 431–446
García-Berro E Lorén-Aguilar P Pedemonte A G Isern J Bergeron P Dufour P and Brassard P 2007 *ApJ* **661** L179
Jura M Farihi J and Zuckerman B 2007a *ApJ* **663** 1285
Jura M Farihi J Zuckerman B and Becklin E E 2007b *AJ* **133** 1927
Kawaler S D and Bradley P A 1994 *ApJ* **427** 415
Kilic M von Hippel T Leggett S K and Winget D E 2005 *ApJ* **632** L115
Kilic M von Hippel T Leggett S K and Winget D E 2006 *ApJ* **646** 474
Koester D and Wilken D 2006 *A&A* **453** 1051
Koester D Provençal J and Shipman H L 1997 *A&A* **320** L57
Liebert J et al. 2003 *AJ* **126** 2521
Metcalfe T S Salaris M and Winget D E 2002 *ApJ* **573** 803
Montgomery M H Metcalfe T S and Winget D E 2003 *MNRAS* **344** 657
Montgomery M H 2005 *ApJ* **633** 1142
Montgomery M H Klumpe E W Winget D E and Wood M A 1999 *ApJ* **525** 482
Nather R E and Mukadam A S 2004 *ApJ* **605** 846
Reach W T Kuchner M J von Hippel T Burrows A Mullally F Kilic M and Winget D E 2005 *ApJ* **635** L161
Robinson E L Kepler S O and Nather R E 1982 *ApJ* **259** 219
Tokunaga A T Becklin E E and Zuckerman B 1990 *ApJ* **358** L21
van Kerkwijk M H Clemens J C and Wu Y 2000 *MNRAS* **314** 209
von Hippel T and Thompson S E 2007 *ApJ* **661** 477
von Hippel T Kuchner M J Kilic M Mullally F and Reach W T 2007 *ApJ* **662** 544
Zuckerman B Koester D Reid I N and Hünsch M 2003 *ApJ* **596** 477

Journal of Physics: Conference Series **172** (2009) 012014 doi:10.1088/1742-6596/172/1/012014

Modeling He-rich subdwarfs through the hot-flasher scenario

M M Miller Bertolami[1,2], L G Althaus[1,2], K Unglaub[3] and A Weiss[4]
[1]Facultad de Ciencias Astronómicas y Geofísicas, Universidad Nacional de La Plata, Paseo del Bosque s/n, (1900) La Plata, Argentina.
[2]Instituto de Astrofísica La Plata, UNLP-CONICET
[3]Dr. Remeis-Stenwarte Bamberg, Sterwartstr. 7, D96049 Bamberg, Germany
[4]Max-Planck-Institut für Astrophysik, Karl-Schwarzschild-Str. 1, 85748, Garching, Germany.

E-mail: `mmiller@fcaglp.unlp.edu.ar`

Abstract. We present 1D numerical simulations aimed at studying the hot-flasher scenario for the formation of He-rich subdwarf stars. Sequences were calculated for a wide range of metallicities and with the He core flash at different points of the post-RGB evolution (i.e. different remnant masses). We followed the complete evolution from the ZAMS, through the hot-flasher event, and to the subdwarf stage for all kinds of hot-flashers. This allows us to present a homogeneous set of abundances for different metallicities and all flavors of hot-flashers. We extend the scope of our work by analyzing the effects in the predicted surface abundances of some standard assumptions in convective mixing and the effects of element diffusion. We find that the hot-flasher scenario is a viable explanation for the formation of He-sdO stars. Our results also show that element diffusion may produce the transformation of (post hot-flasher) He-rich atmospheres into He-deficient ones. If this is so, then the hot-flasher scenario is able to reproduce both the observed properties and distribution of He-sdO stars.

1. Introduction

Hot subdwarf stars can roughly be grouped into the cooler sdB stars, whose spectra display typically no or only weak helium lines, and the hotter sdO stars, which have higher helium abundances on average and can even be dominated by helium. The diversity of spectra observed among sdO stars and the helium-enhanced surface abundances observed in many of these stars (Lemke et al. 1997) pose a challenge to stellar evolution theory. As a consequence, some non-canonical evolutionary scenarios were proposed to explain the formation of He-rich subdwarf stars (i.e. He-sdB, He-sdO stars). Among them, the merger of two white dwarfs (Saio & Jeffery 2000) and a helium core flash (HeCF) after departure from the red giant branch (i.e. "hot-flasher scenario"; D'Cruz et al. 1996, Sweigart 1997) offer the most promising explanations of their formation (Ströer et al. 2007). The hot-flasher scenario was also proposed to explain the existence and characteristics of blue hook stars in some globular clusters (see Moehler et al. 2007 and references therein). Following Lanz et al. (2004) hot-flashers can be classified into *early* hot-flashers, which experience the HeCF during the evolution at constant luminosity (after departure from the RGB) and become hot subdwarfs with standard H/He envelopes; and *late* hot-flashers, which end as He-enriched hot subdwarfs either by burning or dilution of the remaining H-rich envelope. The latter case is the object of the present article.

Table 1. Initial (ZAMS) values of our standard sequences and range of remnant masses in which *late* hot-flashers are obtained.

M_{ZAMS} ($M_\odot$)	Initial abundances ($X_0/Y_0/Z_0$)	Remnant mass range for *late* hot-flashers ($M_\odot$)
0.88	0.769/0.230/0.001	0.4916 – 0.4810
0.98	0.736/0.254/0.010	0.4780 – 0.4662
1.03	0.702/0.278/0.020	0.4740 – 0.4611
1.04	0.668/0.302/0.030	0.4658 – 0.4522

2. Description of the present work

The core feature of this work is to present a homogeneous set of simulations of the *l*ate hot-flasher scenario (i.e. those events that lead to He-enhanced stellar surfaces) for which the chemical abundances are consistently calculated by simultaneously solving the mixing and burning equations within a diffusive convection picture. We then extend the scope of this work by analyzing how these results can be altered by some non-standard assumptions, such as deviations from the standard mixing length theory (MLT) or element diffusion. Particular emphasis is placed on on the predicted surface properties of the models.

The sequences presented in this work were calculated with `LPCODE`, a numerical code for solving the equations of stellar structure which is extensively described in Althaus et al. (2005). In our simulations we have followed the evolution from the ZAMS, through the hot-flasher event and then to the He-core burning stage. Simulations of the hot-flasher scenario were performed for 4 different metallicities and with the HeCF taking place at different points of the post-RGB evolution (which corresponds to different remnant masses, see Table 1). Initial masses and abundances are shown on Table 1 together with the range of post-RGB remnant masses for which He-enriched subdwarfs are obtained. For a more extensive discussion on the numerical and physical aspects of the models, as well as on the results of the present study, we refer the reader to Miller Bertolami et al. (2008).

3. Results and discussion

Standard sequences: Convection in our standard sequences is described by the standard MLT with convective boundaries given by a bare Schwarzschild criterium. H-deficiency in our standard sequences is attained by dilution and/or burning of the remaining H-rich envelope, depending on the exact moment of the post-RGB evolution at which the HeCF takes place. If the star is already at the white dwarf cooling track when the HeCF develops (*deep mixing* events; DM) then the convective zone generated by the HeCF is able to penetrate into the H-rich envelope and H is violently burned (a H-flash), before the remaining H-envelope is diluted by a very shallow convective zone. On the other hand, if the H-burning shell is still active when the HeCF develops then, the convective zone generated during the HeCF does not reach the H-rich envelope and no H-burning happens. In these cases (*shallow mixing* events; SM) a mild H-deficiency can be attained when one of the convective zones generated by the HeCF moves towards the H-He transition and merges with a inwardly growing convective envelope. In our simulations there are some transition cases (labelled SM* in Table 2) in which the HeCF driven convective zone partially penetrates into the H-rich envelope, and some H-burning takes place, but no H-flash finally developes. Then H-deficiency is attained by a combination of burning and dilution of the H-rich envelope.

In Table 2 we present the surface abundances for selected sequences calculated in this work. As it is shown in Table 2, DM episodes predict *immediately after the flash* an extremely H

Table 2. Surface abundances (by mass fractions) for sequences calculated in this work.

Final Mass ($M_\odot$)	Z_0	H	He	^{12}C	^{13}C	N	O
0.49145 (SM*)	0.001	0.2137	0.7475	0.0365	1.57×10^{-6}	1.22×10^{-4}	1.52×10^{-3}
0.49104 (DM)	0.001	2.42×10^{-4}	0.9450	0.0267	7.47×10^{-3}	0.0106	4.77×10^{-5}
0.48545 (DM)	0.001	4.60×10^{-6}	0.9524	0.0264	7.73×10^{-3}	0.0134	4.53×10^{-5}
0.48150 (DM)	0.001	2.41×10^{-6}	0.9666	0.0107	5.79×10^{-3}	0.0185	3.79×10^{-5}
0.47770 (SM)	0.01	0.0780	0.8682	0.0426	6.26×10^{-6}	1.30×10^{-3}	6.84×10^{-4}
0.47744 (SM*)	0.01	0.0261	0.8996	0.0642	2.36×10^{-6}	6.59×10^{-4}	7.43×10^{-4}
0.47725 (DM)	0.01	2.27×10^{-5}	0.9464	0.0382	6.12×10^{-3}	4.35×10^{-3}	2.94×10^{-4}
0.46921 (DM)	0.01	1.02×10^{-5}	0.9560	0.0211	6.13×10^{-3}	0.0128	2.92×10^{-4}
0.46644 (DM)	0.01	4.61×10^{-6}	0.9592	0.0193	5.52×10^{-3}	0.0122	3.16×10^{-4}
0.47378 (SM)	0.02	0.3822	0.5978	1.23×10^{-3}	8.96×10^{-5}	7.57×10^{-3}	4.69×10^{-3}
0.47250 (SM*)	0.02	9.56×10^{-3}	0.9273	0.0425	1.93×10^{-5}	2.94×10^{-3}	8.64×10^{-4}
0.47112 (DM)	0.02	9.05×10^{-6}	0.9389	0.0420	4.27×10^{-3}	4.88×10^{-3}	6.96×10^{-4}
0.46410 (DM)	0.02	1.05×10^{-5}	0.9460	0.0273	6.41×10^{-3}	0.0120	7.680×10^{-4}
0.46150 (DM)	0.02	4.66×10^{-6}	0.9480	0.0263	5.67×10^{-3}	0.0116	1.01×10^{-3}
0.46521 (SM)	0.03	0.2227	0.7271	0.0197	7.79×10^{-5}	8.09×10^{-3}	5.18×10^{-3}
0.46470 (SM)	0.03	4.71×10^{-3}	0.9239	0.0380	4.12×10^{-4}	6.10×10^{-3}	1.58×10^{-3}
0.46367 (SM*)	0.03	1.63×10^{-3}	0.9249	0.0390	3.25×10^{-3}	6.48×10^{-3}	1.62×10^{-3}
0.46282 (DM)	0.03	7.48×10^{-6}	0.9333	0.0420	1.74×10^{-3}	7.04×10^{-3}	1.55×10^{-3}
0.45660 (DM)	0.03	9.22×10^{-6}	0.9382	0.0341	4.09×10^{-3}	0.0103	1.67×10^{-3}
0.45234 (DM)	0.03	1.94×10^{-5}	0.9411	0.0273	4.76×10^{-3}	0.0134	2.14×10^{-3}

deficient surface with H-abundances preferentially between 10^{-5} and 10^{-6} (by mass fraction), while SM/SM* predict a more wide range of H-abundances between almost solar and 10^{-3}.

Although evolutionary tracks pass close to the location of He-sdO stars (see Figure 1), their distribution in the $\log g - \log T_{\rm eff}$ diagram is not correctly reproduced. In fact, according to (standard) theoretical predictions, He-sdO should cluster around the He-core burning stage where evolutionary timescales are longer. However, this inconsistency can be solved when taking into account the effects of element diffusion in the outer layers (see next paragraphs).

Effects of departures from the standard MLT: we have performed additional simulations including the effects of extramixing at convective boundaries (as in Herwig et al. 1997) and/or the effects of chemical gradients (as in Grossman & Taam 1996). Our results show that the role of chemical gradients as an extra barrier to the penetration of the He-core flash driven convective zone into the H-rich envelope, is significant and has to be studied in detail (probably by means of hydrodynamical simulations of the event). In fact, in our simulations, the stabilizing effect of chemical gradients can even prevent the penetration when no extramixing is considered at convective boundaries. On the other hand, extramixing processes alone does not produce significant departures from standard results.

Effects of element diffusion on the outer layers: Recently, Unglaub (2008) showed that chemical homogeneous winds in these stars are not possible at $\log g > 5.5$ and, consequently, H and He can hardly be expelled from the stars. In view of these results we have performed simulations of element diffusion in the absence of mass loss by means of the method described in Unglaub & Bues (1998). Figure 1 shows the evolution of the surface abundaces predicted for two of our sequences due to the effects of element diffusion. As it is shown in Figure 1 in the absence of mass loss the SM* sequence becomes H-dominated almost instantly, compared to the evolutionary timescale, and its surface is expected to become H-dominated as soon as

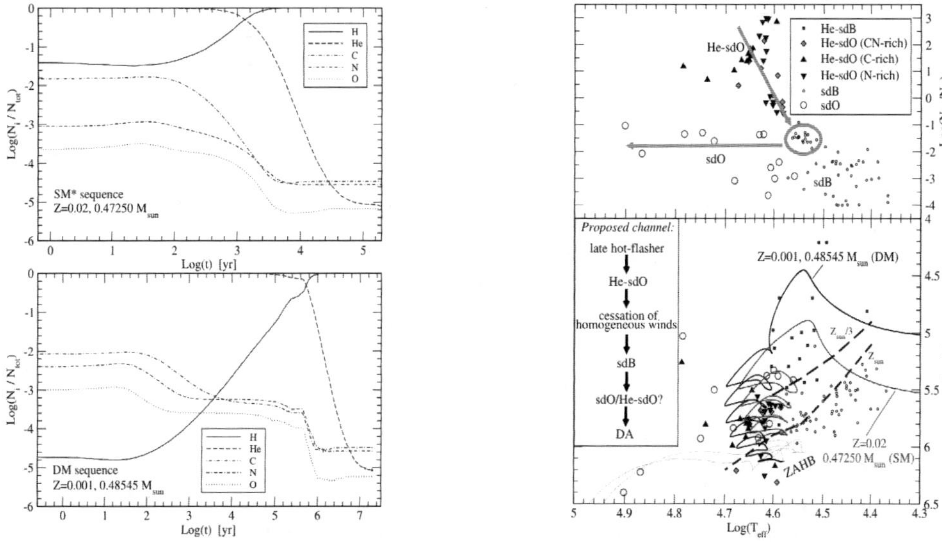

Figure 1. *Left:* Evolution of the surface ($\tau_{\mathrm{Rosseland}} = 2/3$) abundances predicted in two of our hot-flashers due to the effects of element diffusion in the absence of mass loss and assuming $T_{\mathrm{eff}} = 40\,000$ K and $\log g = 6$. *Bottom Right:* Evolution of the same two hot-flasher sequences in the $\log g - \log T_{\mathrm{eff}}$ diagram. Thick dashed lines mark the wind limit, for two different metallicities, below which homogeneous winds are not possible according to Unglaub (2008). Dotted lines denoted the portion of the evolutionary track where the star should have already became H-deficient according to element diffusion simulations. *Top Right:* Sketch of the possible evolution in the $\log N_{\mathrm{He}}/N_{\mathrm{H}} - \log T_{\mathrm{eff}}$ diagram after a deep mixing event.

homogeneous winds halt. On the other hand for the DM sequence the timescale in which the star becomes H-dominated is similar to the evolutionary timescale from the primary HeCF to the settlement on the He-core burning stage (~ 1 Myr). In this case, the surface is expected to become H-dominated as soon as the He-core burning stage begins and evolutionary timescales become $\sim$ 40 times longer. Note that, with the inclusion of element diffusion, the surface properties and distribution of He-sdO stars predicted by the hot-flasher scenario correctly reproduces the observed ones (see Figure 1).

Acknowledgments

Part of this work was supported by PIP 6521 grant from CONICET. Also the European Association for Research in Astronomy is acknowledged for an EARA-EST fellowship under which this work was started.

References

Althaus L G Serenelli A M Panei J A et al. 2005 *A&A* **435** 631
Cassisi S Schlattl H Salaris M Weiss A 2003 *ApJL* **582** L43
D'Cruz N Dorman B Rood R O'Connell R 1996 *ApJ* **466** 359
de Boer K Drilling J Jeffery C & Sion E 1997 in *"Third Conference on Faint Blue Stars"* L. Davis Press 515
Grossman S & Taam R 1996 *MNRAS* **283** 1165
Herwig F Blöcker T Schönberner D & El Eid M 1997 *A&A* **324** L81
Lanz T Brown T Sweigart A Hubeny I Landsman W 2004 *ApJ* **602** 342
Lemke M Heber U Napiwotzki R et al. 1997 in *"Third Conference on Faint Blue Stars"* L. Davis Press 375
Miller Bertolami M M Althaus L G Unglaub K & Weiss A 2008 *A&A* DOI: 10.1051/0004-6361:200810373
Moehler S Dreizler S Lanz T et al. 2007 *A&A*, **475** L5
Saio H & Jeffery S 2000 *MNRAS* **313** 671
Ströer A Heber U Lisker T et al. 2007 *A&A* **462** 269
Sweigart A 1997 in *"Third Conference on Faint Blue Stars"* L. Davis Press 3
Unglaub K 2008 *A&A* **486** 923
Unglaub K & Bues I 1998 *A&A* **338** 75

16th European White Dwarfs Workshop IOP Publishing
Journal of Physics: Conference Series **172** (2009) 012015 doi:10.1088/1742-6596/172/1/012015

Carbon abundances of sdO stars from SPY

Heiko Hirsch and Uli Heber

Dr. Remeis–Sternwarte, Institute for Astronomy, University Erlangen-Nürnberg, Sternwartstr. 7, 96049 Bamberg, Germany

E-mail: heiko.hirsch@sternwarte.uni-erlangen.de

Abstract. Ströer et al. (2007) recently suggested a classification of sdOs according to supersolar and subsolar helium abundances, with only the helium-enriched stars showing signes of carbon and/or nitrogen in their optical spectra. We aim to derive reliable carbon and nitrogen abundances by fitting synthetic spectra to data obtained with the UVES spectrograph at ESO. Here we present our first results of the analysis of carbon abundances in hot subdwarf O stars. By constructing a grid of model atmospheres consisting of hydrogen, helium and carbon we were able to derive atmospheric parameters of nine carbon rich sdOs. We find $\log(N_{\mathrm{C}}/N_{\mathrm{total}})$ up to ten times higher than the solar value, while the mean value for the effective temperature and the surface gravity is slightly lower than derived by helium-hydrogen models only. Surprisingly, we also find three fast rotators among our program stars.

1. Introduction

Subdwarf O stars (sdO) and their cooler siblings, the sdBs, form what is called the "Extended Horizontal Branch (EHB)" in the HRD. They are core helium burning stars with virtually no hydrogen envelope. While the sdBs are quite homogeneous, showing low helium abundances and their metal abundances are explainable by diffusion, the sdO stars show a wide variance in T_{eff}, $\log(g)$ and $\log(N_{\mathrm{He}}/N_{\mathrm{H}})$. Ströer et al. (2007) divide them into helium-enriched and helium-deficient stars with the dividing line being the solar helium abundance, $\log(N_{\mathrm{He}}/N_{\mathrm{H}})_{\odot} = -1$. The helium-enriched sdOs are clustered in a narrow region in the $T_{\mathrm{eff}} - \log(g)$-plane at $T_{\mathrm{eff}} \approx 45\,000\,\mathrm{K}$ and $\log(g) \approx 5.6$.

2. Evolution

Comparison between observed parameters of sdOs and theoretical evolutionary tracks (see Fig. 1) show that the helium deficient stars (open symbols) most likely are the progeny of the sdB stars. The helium-enriched stars (filled symbols) however are not so easily explained as their overdensity strongly disagrees with the timescales of the post-EHB stracks.

3. Beyond hydrogen and helium

Ströer at al. (2007) find that the helium-enriched sdOs can further be divided into C-, N-, and CN-types, depending on their spectra showing nitrogen or carbon or both. Two scenarios for the origin of He-enriched sdOs have been proposed. *a) Late hot flashers* (Sweigart et al., 1997; Lanz et al., 2004) are stars that undergo a delayed helium core flash when they are already on the WD cooling curve. This scenario comes in two different flavours, "deep" and "shallow" mixing.

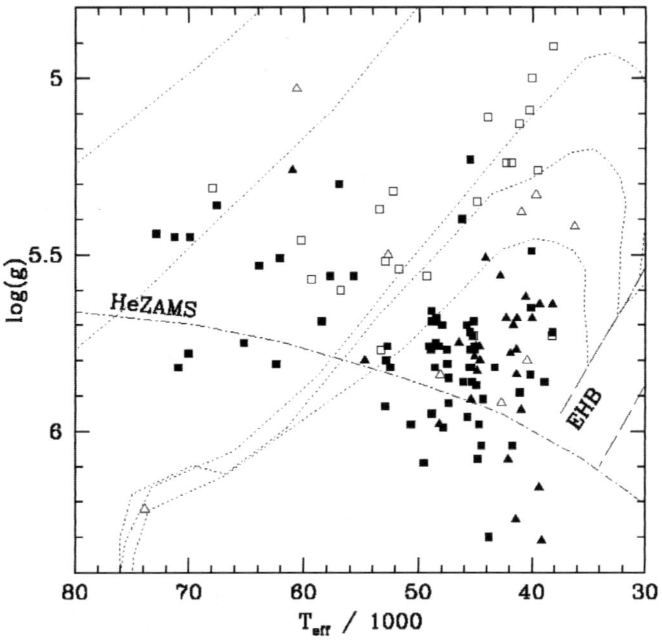

Figure 1. Data from SPY (triangles) and SDSS (squares). Filled symbols are helium-enriched stars, open symbols are helium-deficient. Also shown are the locations of the EHB and the main sequence for helium burning stars. The two evolutionary tracks in the upper left are post-AGB stars (Schönberner et al., 1983). Evolutions from the EHB towards the White Dwarf graveyard are shown by three tracks on the right (Dorman et al., 1993).

In the former, the hydrogen envelope will be mixed into the star's interior and will be burned completely, while in the latter case the mixing will take place without the hydrogen being burnt.

b) He-WD mergers (Han et al., 2002; Saio & Jeffery, 2002) result from the coalescence and eventual merger of He-WDs due to radiation of gravitational waves.

These scenarios may become distinguishable by the predicted abundances.

4. Spectral analysis

To further investigate this issue, we are doing an analysis of sdO stars from the SPY survey (Napiwotzki et al. 2001), which provides the nessecary resolution and SNR for this analysis. The spectra were observed with the UVES spectrograph at ESO and cover the wavelength range from 3,300 Å to 6,600 Å at high resolution.

As a first step in this direction we identified all elements present in SPY sdO stars. While the helium-deficient sdOs spectra are virtually void of any element other than helium and hydrogen, we find the C-type helium enriched ones to show traces of nitrogen, silicon and oxygen besides carbon. In the N-type spectra we find neon, sulfur, titanium, silicon, oxygen and carbon besides the dominant nitrogen.

Using model atoms of H, HeI, HeII, CIII and CIV we constructed an extensive model grid of $\log(N_{\rm He}/N_{\rm H}) = -1 \cdots + 2$, $T_{\rm eff} = 40000 \cdots 60000\,{\rm K}$, $\log(g) = 5.00 \cdots 6.40$ and $\log(N_{\rm C}/N_{\rm total}) = -2 \cdots - 6$. The atmospheric structure was calculated using the PRO2 code by Werner & Dreizler (1999) and Dreizler (2003) under consideration of lineblanketing of the present model atoms. The model spectra were folded to the instrument profile taking the seeing

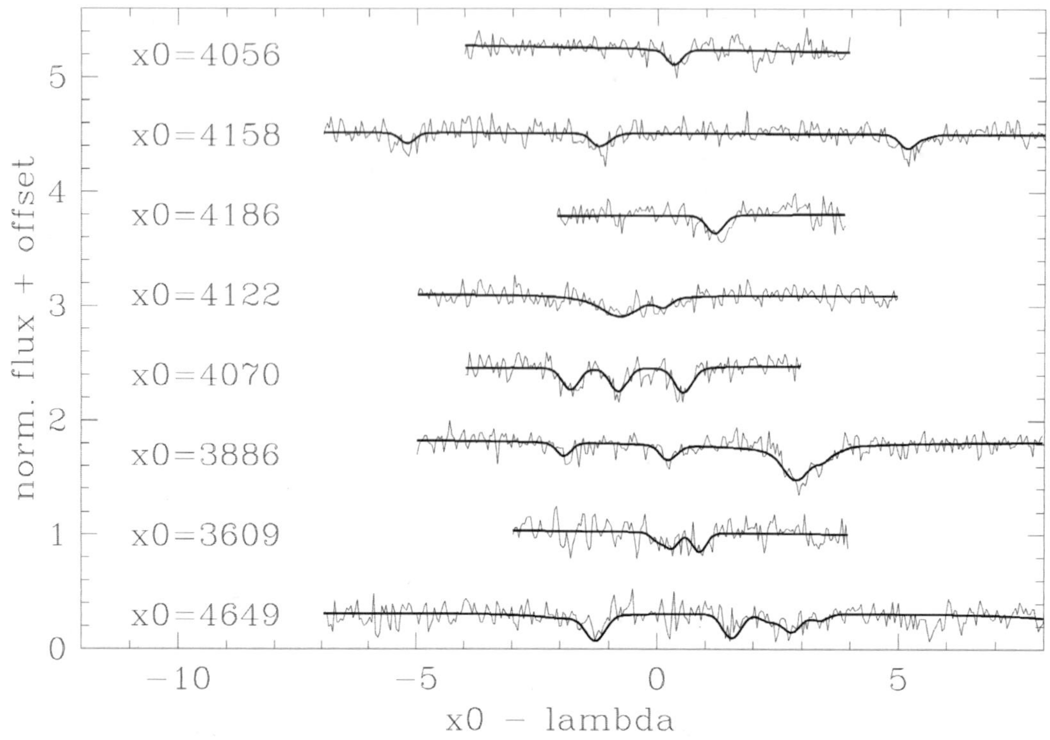

Figure 2. An example of a fit of $\log(N_C/N_{total})$ and $v_{rot} \sin i$. The best fit model here is $T_{eff} = 44200$ K, $\log(g) = 5.64$, $\log(N_C/N_{total}) = -2.8$ and $v_{rot} \sin i = 12\,\mathrm{km\,s^{-1}}$. All lines shown are CIII with the exception of their blend with HeII 3889 Å and HeI 4121 Å. All fitted ranges shown are shiftet to $\lambda = 0$ with their original central wavelength given on their left side.

condition into account.

Spectral line fitting was performed by interpolating in $\log(N_{He}/N_H)$ to the value obtained by Ströer et al. (2007), then simultaniously fitting T_{eff} and $\log(g)$ using the H- and He-lines. Care was taken that no contanimation of carbon lines did occur in the broad line wings. Fixing T_{eff} and $\log(g)$, a new fit using only CIII-lines was done to determine $\log(N_C/N_{total})$ and $v_{rot} \sin i$, see Fig 2.

With one exeption, for every star two or more spectra were available. We use these multiple results to get a crude error estimate of $\Delta T_{eff} = 500$ K, $\Delta \log(g) = 0.1$ and $\Delta \log(N_C/N_{total}) = 0.3$.

5. Results

Atmospheric parameters: For all nine C-type helium-enriched sdOs in the SPY-sample we determined new T_{eff} and $\log(g)$. The results are plotted in Fig. 3. The new model atmospheres lead to some changes of parameters compared to the values by Ströer et al. 2007. The mean T_{eff} is lowered by 1 300 K, the mean $\log(g)$ by 0.14 dex. Whether these differences are due to additional lineblanketing of the carbon or to different analysing methods is still open for investigation.

Carbon abundances: We find the carbon abundances to lie between $\log(N_C/N_{total}) =$

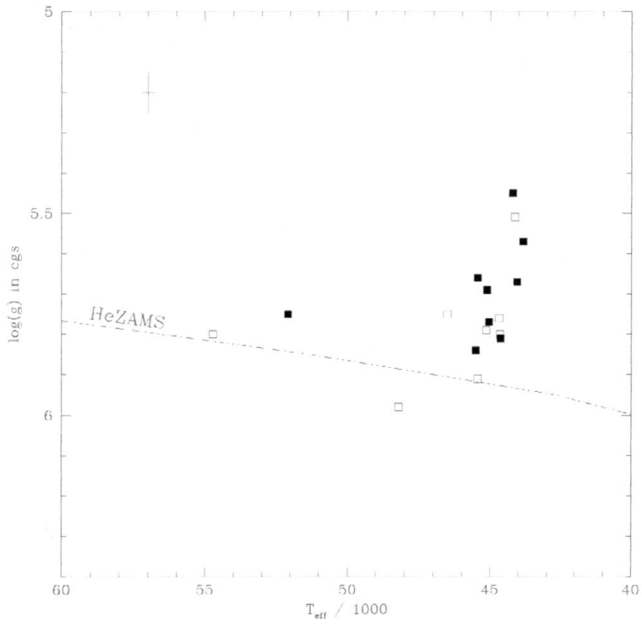

Figure 3. Filled symbols are the new values for $T_{\rm eff}$ and $\log(g)$, the values from Ströer et al. (2007) are open symbols. The origin of this difference still has to be investigated.

$-2.3 \cdots -3.1$, that is up to ten times to the solar value ($\log(N_{\rm C}/N_{\rm total}) = -3.5$), but no trends in $T_{\rm eff} - \log(N_{\rm C}/N_{\rm total})$ or $\log(N_{\rm He}/N_{\rm H}) - \log(N_{\rm C}/N_{\rm total})$ can be seen. These first measured abundances enable us to compare the observations with theoretical late hot flasher calculations, Miller Bertolami, M., this proceedings.

Rotational velocity: Surprisingly, we find three out of nine sdOs to be "fast" rotators with $v_{\rm rot} \sin i = 23$, 23 and $28\,{\rm km\,s^{-1}}$ — fast compared to sdBs which mostly show $v_{\rm rot} \sin i < 10\,{\rm km\,s^{-1}}$ (Geier et al., in prep.). This discovery seems to favour the WD-merger theory.

Acknowledgments
We like to thank Thomas Rauch for providing us with the carbon model atoms. This work was supported by DFG through grant HE1356/44-1.

References
Dorman B Rood R T and O'Connell R W 1993 *ApJ* **419** 596-614
Dreizler S 2003 *ASP Conf. Ser.* **288** 69-82
Han Z Podsiadlowski Ph Maxted P F L Marsh T R and Ivanova N 2002 *MNRAS* **336** 449-66
Lanz T Brown T Sweigart A Hubeny I Landsman W 2004 *ApJ* **602** 342-55
Miller Bertolami M M Althaus L G 2007 *Preprint* ArXiv:0711.0140
Napiwotzki R et al. 2001 *Astronomische Nachrichten* **322** 411-18
Saio C H and Jeffery C S 2002 *MNRAS* **333** 121-132
Schönberner D 1983 *ApJ* **272** 708-14
Ströer A Heber U Lisker T Napiwotzki R Dreizler S Christlieb N and Reimers D 2007 *A&A* **462** 269-80
Sweigart A 1997 *Third Conference on Faint Blue Stars, L. Davies Press* 3
Werner K and Dreizler S 1999 *Preprint* ArXiv: astro-ph/9906130

Journal of Physics: Conference Series **172** (2009) 012016 doi:10.1088/1742-6596/172/1/012016

New results on the sdB star PG 0014+067

P Brassard and G Fontaine

Département de Physique, Université de Montréal, C.P. 6128, Succ. Centre-Ville, Montréal, Québec, Canada H3C 3J7

E-mail: brassard@astro.umontreal.ca, fontaine@astro.umontreal.ca

Abstract. We present here our third generation of sdB star models geared for asteroseismology. As a test of these new models, we return to a thorough analysis of the pulsating sdB PG 0014+067. We also provide a new treatment of the observed data. Many possible working hypotheses for the mode identification have been tested. In an outcome that may seem surprising to some, we found that almost all of them seem to produce very similar seismic models.

1. Observations

From the discovery of its variability in 1998, PG 0014+067 has been the subject of four important observational missions to decipher its period structure. The first one, that established the variability of the star, provided 9.6 hours of observations at the CFH Telescope (Brassard et al. 2001, hereafter referred to as CFTH1) in June 1998. A follow-up mission, again at the CFHT in August 1999, provided us with another 16.2 hours of observations (Charpinet et al. 2005, hereafter referred to as CFHT2). After that, 29.5 hours where obtained at the WHT in August 2004 with the Ultracam photometer (Jeffery et al. 2005, hereafter referred to as the WHT run) and $\sim$ 180 hours with the WET in October 2004 (Vučković et al. 2006, hereafter referred to as the WET run).

For our experiments, we have used the periods extracted from the combined run CFHT1 and CFHT2. In spite of the rather large time gap between to the two runs, we were able to extract 19 periodicities from the joint data, thanks to the very low noise level obtained at the CFHT. We recovered all of the 10 periodicities found in the WET data. We also recovered 13 of the 19 from the WHT. For the asteroseismic analysis, we have kept only the central component of the multiplets. This left us with 10 observed periods, one more (154.72 s) and one less (130.37 s) than those used by CFTH2 and Brassard et al. (2008). Values are given in Table 1.

2. Models

Along with our previous generation of envelope models which incorporate iron profiles that take into account the effects of radiative levitation, we are now able to complete the structure of our models with a nucleus that include the description of the nuclear burning and neutrinos effects. We have used the updated nuclear reaction rates of Caughlan & Fowler (1988) available at http://www.phy.ornl.gov/astrophysics/data/cf88/ and the neutrinos rates from N. Itoh and collaborators. These are static models built under the assumption of mechanical and thermal equilibrium.

Table 1. Observed Periods in PG0014+067.

P (s)	A (%)	P (s)	A (%)	P (s)	A (%)
100.293	0.104	141.065	0.252	154.947	0.064
116.928	0.030	146.495	0.207	154.974	0.068
126.308	0.039	150.744	0.054	161.421	0.027
139.167	0.071	150.762	0.128	161.467	0.048
139.193	0.033	150.784	0.087	168.830	0.057
141.016	0.166	154.552	0.065	168.926	0.047
141.033	0.109				

The basic input parameters used to build our models are the mass of the star ($M_\star/M_\odot$), the mass of the envelope ($M_{\rm env}/M_\star$), the mass of the convective core ($M_{\rm core}/M_\star$) along with its He/C/O mixture composition. The non-convective part of the core is assumed to be pure He. The envelope is the same than in our previous generation of models, i.e., almost pure H with traces of Fe in radiative levitation.

One can note here that the surface gravity ($\log g$) and the effective temperature ($T_{\rm eff}$) of a given model are obtained after the computation of the model. This can complicate the direct comparison with observations and does explain the approach used in the next section.

3. Methodology

Beforehand, we have built a huge grid of nearly 3,750,000 stellar models and computed their pulsation periods. Such a grid is useful in many ways. First, we can select models with specified $\log g$ and $T_{\rm eff}$ (within any given ranges). Then, we can rapidly test various hypotheses on the mode identifications (all the hard work having already been done). Also, the grid resolution is sufficient to ensure that we have a global minimum instead of a local one. Finally, the grid can be reused to analyze other stars.

Our analysis is made in two steps. First we select all models within 2σ in the observed $T_{\rm eff}$ and $\log g$ of the star in our pre-built grid. Then we minimize the quantity $S^2 = \sum(P_{\rm obs} - P_{\rm th})^2$ using a matching algorithm. The matching can be done under different hypotheses for the modes identification. In a second step, best-matching models are refined using a finer resolution in the basic parameters.

4. Observational Constraints

A first constraint arises from the analysis of the multi-color data of the WHT run made by Tremblay et al. (2006). They show that the modes at 141.033 s and 146.496 s can only be identified with ℓ = 0, 1 or 2. If we look at Table 1, only 4 periods can be candidate for a $\ell = 0$ mode. Also, if we restrict ourselves to high amplitude modes, we are left with only two possibilities: 100.29 s and 146.496 s. If we look at the CFHT1 and CFHT2 runs individually, we find that the 100.29 s is very stable in frequency and amplitude. This is not the case with the mode at 146.496 s. Indeed, there is a possibility for a mode at 146.475 s. This mode was not selected in Table 1 because it falls short on our selection criterion (amplitude slightly under the 3σ noise level). Another constraint arises from the paper of Randall et al. (2005) in which the authors clearly show that $\ell = 3$ modes are difficult to observe in sdB stars.

Table 2. Solutions near 34,130 K

Mode ID		Hypothesis 1		Hypothesis 2	
ℓ	k	$P_{\rm th}$ (s)	$P_{\rm obs}$ (s)	$P_{\rm th}$ (s)	$P_{\rm obs}$ (s)
0	5	99.333		98.666	
0	4	113.313		113.173	
0	3	123.572		124.141	
0	2	150.008		149.068	**146.50**
0	1	162.845		163.318	
0	0	188.940		187.511	
1	6	95.376		95.078	
1	5	104.612		104.475	
1	4	119.980		119.159	
1	3	141.743	141.04	141.809	141.04
1	2	154.802	154.72	154.111	154.72
1	1	188.343		186.892	
2	6	89.119		89.065	
2	5	100.096	100.29	99.464	100.29
2	4	115.939	116.93	115.453	116.93
2	3	126.772	126.31	126.724	126.31
2	2	151.061	150.77	149.981	150.77
2	1	168.863	168.88	169.238	168.88
2	0	187.872		186.482	
4	5	95.179		94.802	
4	4	105.146		105.062	
4	3	119.694		118.932	
4	2	139.736	139.18	139.077	139.18
4	1	146.583	**146.50**	145.662	
4	0	160.394	161.45	160.366	161.45
4	1	186.440		185.196	

Table 3. Properties of PG 0014+067

Quantity	Spectroscopy	Hypothesis 1	Hypothesis 2
$\log g$	5.77	5.751	5.754
$T_{\rm eff}$ (K)	34,130	34,124	34,172
$M_\star/M_\odot$		0.450	0.451
$\log(M_{\rm env}/M_\star)$		−3.941	−3.975
$\log(M_{\rm core}/M_\star)$		−0.328	−0.310
X(C)		0.469	0.467
X(O)		0.468	0.466
S^2 (Hyp. 1)		3.254	6.599
S^2 (Hyp. 2)		15.687	12.628

Journal of Physics: Conference Series **172** (2009) 012016 doi:10.1088/1742-6596/172/1/012016

5. Results

We have tested many hypotheses for the mode identification for PG 0014+067. Each of these hypotheses were tested with the 10 pulsation modes extracted from Table 1 and also with the subset of the 5 modes with the highest observed amplitudes. Detailed results will be produced elsewhere, but we summarize the main results here.

First, there is only one solution assuming that all the modes have $\ell \leq 2$. This solution have the modes at 139.27 s and 150.76 s with a $\ell = 0$ index. This is clearly in contradiction with the observations. It follows that we reassert the result of CFHT1 that the light curve of PG0014+067 can only be deciphered if we allow the presence of modes with $\ell \geq 3$. Second, with the notable exception of the hypothesis that the modes can have any values with $\ell \leq 4$, we find that all our hypotheses produce very similar models with $-3.9 \leq \log(M_{\rm env}/M_\star) \leq -4.0$ (for the general case with $\ell \leq 4$, numerous solutions are possible in the range $-3.75 \leq \log(M_{\rm env}/M_\star) \leq -4.45$). Third, we do not find unique solutions, but instead a family of solutions that produces essentially the same fit but at various masses for the star. Fortunately the observed quantity $T_{\rm eff}$ can be used to pinpoint a solution fully consistent with the observations.

For the rest of the paper we take a closer look at two interesting possibilities consistent with the observational constraints.

Hypothesis 1: We exclude modes with $\ell = 0$ and try to obtain solutions with $\ell = 1$, 2 and 4 only. The member of the family of solution with $T_{\rm eff}$ near the observed effective temperature of 34,130 K that produce the best fit is given in Table 2 and Table 3. This is a very interesting solution from a strict theoretical point of view since all the modes have consecutives k values. This solution can also accommodate the possibility that the 100.3 s observed period is a $\ell = 0$ mode because there is a nearby $\ell = 0$ mode with 99.3 s in the model. The only problem with this model is that the condition that the 146.50 s must have $\ell = 0$, 1 or 2 is not respected.

Hypothesis 2: The mode at 146.50 s may have $\ell = 0$, 1 or 2, the mode at 141.04 s may have $\ell = 1$ or 2, and the other observed modes may have $\ell = 1$, 2 or 4. In this case, we have a solution fully consistent with observational constraints. What is interesting here, is that this new solution is essentially the same than the previous one! The mode at 146.50 s as simply moved from $\ell = 3$ to $\ell = 0$.

6. Conclusions

We thus found another case for which a priori partial mode identification seems to be immaterial for defining the optimal seismic model in sdB asteroseismology. This may seem surprising, even "heretic" to many, but it can be readily explained by the particular period structure found in sdB pulsators. Modes with the same radial order but different values of the degree index tend to have very similar periods. At the present level of accuracy with which observed periods can be matched with our models of sdB stars, a priori mode identification does not seem necessary. This is a good thing because pulsating sdB stars are relatively faint, and constraints on the values of the modal degree from multicolor photometry and/or time-resolved spectroscopy are likely to be limited to the brighter members of the class.

References

Brassard P Fontaine G 2008 *ASP Conference Series* **392** 261
Brassard P Fontaine G Billères M et al. 2001 *ApJ* **563** 1013
Caughlan G R Fowler W A 1988 *Atomic Data and Nuclear Data Tables* **40** 283
Charpinet C Fontaine G Brassard P et al. 2005 *ASP Conference Series* **334** 619
Jeffery C S Aerts C Dhillon V S et al. 2005 *MNRAS* **362** 66
Randall S K Fontaine G Brassard P et al. 2005 *ApJS* **161** 456
Tremblay P-E Fontaine G Brassard P 2006 *ApJ* **165** 551
Vučković M Kawaler S D O'Toole S et al. 2006 *ApJ* **646** 1230

16th European White Dwarfs Workshop IOP Publishing
Journal of Physics: Conference Series **172** (2009) 012017 doi:10.1088/1742-6596/172/1/012017

Calibrating cosmological chronometers: white dwarf masses via astrometry

John P. Subasavage[1], Todd J. Henry[1], Wei-Chun Jao[1], Edmund P. Nelan[2], Hugh C. Harris[3] and Conard C. Dahn[3]

[1] Department of Physics and Astronomy, Georgia State University, Atlanta, GA 30302-4106, USA [2] Space Telescope Science Institute, 3700 San Martin Drive, Baltimore, MD 21218, USA [3] US Naval Observatory, Flagstaff Station, P.O. Box 1149, Flagstaff, AZ 86002-1149, USA

E-mail: subasavage@chara.gsu.edu, thenry@chara.gsu.edu, jao@chara.gsu.edu, nelan@stsci.edu, hch@nofs.navy.mil, dahn@nofs.navy.mil

Abstract.
In an effort to increase the number of accurate dynamical masses for white dwarfs (WDs), we have begun an initiative using *Hubble Space Telescope*'s Fine Guidance Sensors (FGS) to resolve suspected binary WDs. With the increasing number of WD trigonometric parallaxes becoming available via CTIO's and the USNO's ongoing parallax programs, we have targeted objects that are overluminous at V magnitude and are presumably unresolved multiple systems. A few targets were selected because of spectral anomalies or possible perturbations evident in the residuals of the trigonometric parallax solutions. A total of 16 HST orbits were designated to this program and 12 are completed. Of the eleven WDs observed thus far (one object was observed twice), all but one were unresolved. Analysis of a recent orbit's data indicate a pair was resolved with a separation of 70 mas and a delta V magnitude of $\sim$1.4. Coupled with astrometric data from the USNO parallax program, we have obtained preliminary constraints on component masses.

1. Introduction

White Dwarfs (WDs) are, in general, rather well understood physically and thus, serve as proxies to astrophysically compelling questions regarding ages. For instance, WDs have been used to gauge the ages of the nearest globular clusters (Hansen et al. 2004, 2007) as well as the age of the Milky Way's thick disk (Reid 2005 and references therein). The two components needed to model WD age are effective temperature and mass. The effective temperature can be obtained by spectral energy distribution fitting to the photometry and/or spectroscopy and the mass is usually obtained by invoking the theoretical mass-radius (M-R) relation (using the radius obtained either by spectral line fitting or trigonometric parallax). The M-R relation is dependent on the internal composition of the WD (either pure or a mixed combination of He, C, N, Ne, Mg, or Fe). Thus, WD masses need to be determined accurately and independently to stress test the theoretical M-R relation and reveal the true chemical makeup of WDs. The most direct method of determining accurate masses is to astrometrically map orbits of binary systems.

To date, only three WDs have dynamical masses determinations known to better than 5% (Sirius B, Procyon B, and 40 Eri B, Provencal et al. 2002). While these three cover a fairly broad range of masses (on the low end, 0.50 $M_\odot$ for 40 Eri B, and on the high end, 1.02 $M_\odot$ for

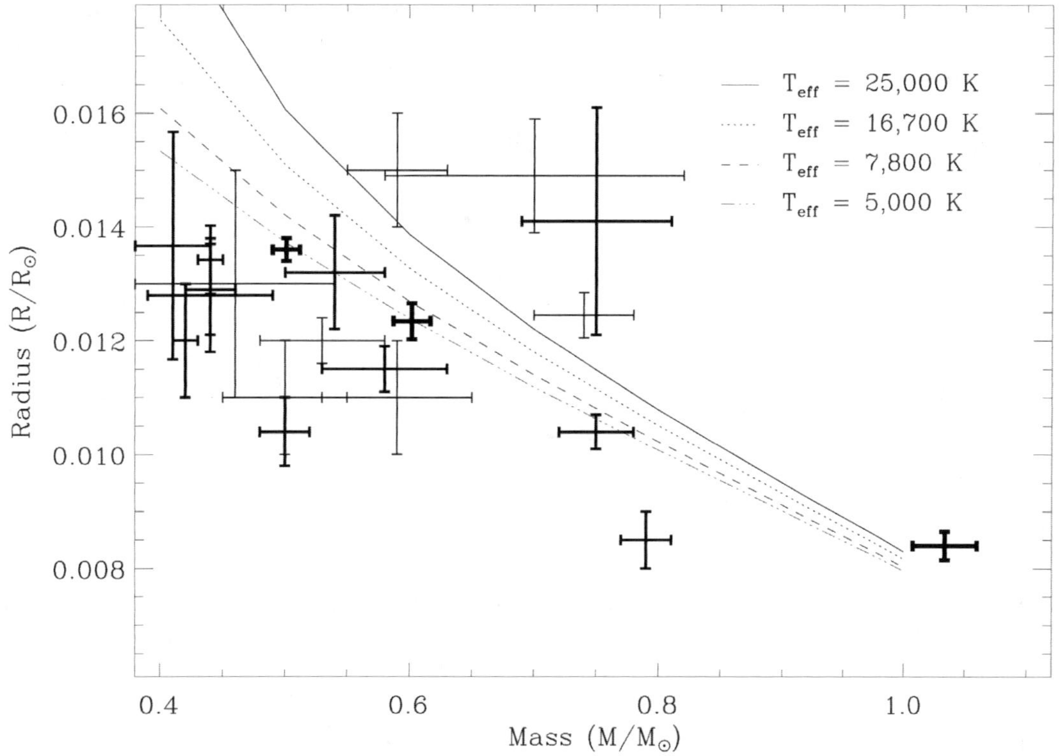

Figure 1. Current empirical confirmations of the theoretical M-R relation reproduced from Provencal et al. (2002). The objects included are the three mentioned in the text that have dynamical mass determinations (*extra-thick bars*), eleven are field WDs with masses derived from spectral modeling (*thick bars*), and seven are members of wide common proper motion binaries whose masses were determined via gravitational redshift (*thin bars*). Theoretical curves for the four temperatures listed are derived from the Wood (1995) C/O-core models with 10^{-4} fractional H and 10^{-2} fractional He layers.

Sirius B), they hardly provide a robust empirical confirmation of the theoretical M-R relation. To scale projected dimensions to true dimensions (necessary to constrain the sum of component masses), an accurate distance determination via trigonometric parallax is necessary. Thus, it is not surprising that these three systems rank among the 50 nearest systems to the Sun (Sirius AB = fifth, Procyon AB = thirteenth, and 40 Eri ABC = fiftieth, www.recons.org). In fact, the largest component of the error budgets of Sirius B's and 40 Eri B's mass determinations are the uncertainties of the parallaxes (Provencal et al. 1998). It is necessary to probe nearby WDs for companions and ultimately obtain dynamical masses and independent checks on the fundamental and widely used theoretical M-R relation.

2. Current mass-radius relation

There currently exists 21 empirical checks to the theoretical M-R relation with varying degrees of quality (see Figure 1). However, as mentioned previously, only 3 have masses known to better than 5%. Eleven WDs have masses determined based on spectroscopic analyses coupled with trigonometric parallaxes. In effect, observed spectra are compared to synthetic spectra derived from model atmospheres to obtain estimates of T_{eff} and log g, the latter being constrained by the parallax. This method is limited to those WDs with accurate trigonometric parallaxes and suffers drawbacks. Uncertainties include pressure effects that lead to overestimated values of log g (i.e., the presence of trace amounts of He), the importance of and ability to model temperature-dependant convective properties, the reliability of broadening theory, and the inherent uncertainties of the observationally constrained parameters (Provencal et al. 1998).

Seven WDs are members of wide common proper motion systems. In these cases, the radial velocities are measured for both components (the non-WD components are usually red dwarfs) with the assumption that any orbital motion is negligible. Thus, any difference between the measured radial velocities of the components is attributed to the gravitational redshift of the WD, hence a mass determination is possible (when coupled with a radius determined via trigonometric parallax). The drawback with these systems is their distances. They are, on average, more distant than the visual binaries and as such, the fractional parallax errors (i.e., parallax error relative to parallax) are larger and correspond to larger mass errors. As is evident in Figure 1, all but the three dynamical mass determinations are of limited use as empirical checks to the theoretical M-R relation.

3. Identifying new candidate double degenerate systems

A double degenerate (DD) system is one that contains two WDs in orbit around each other. Such systems can be extremely useful to test the theoretical M-R relation because, when the orbit is mapped (and the distance is determined), dynamical masses can be obtained for two WDs at once. However, as a consequence of binary evolution, DDs are often found in short period systems because their progenitors filled their Roche lobes as they evolved off of the main sequence. Thus, they underwent a common envelope phase where friction inside the envelope drastically decreased the binary separation. The orbital energy and angular momentum removed from the system during the process then expelled the envelope revealing the two components. Additional angular momentum loss via gravitational radiation further degraded the orbit such that periods range from hours to days (Rebassa-Mansergas et al. 2008). Thus, there is a bimodal distribution of DDs with those that underwent a common envelope phase having very short orbital periods and those that essentially evolved as single stars having periods of decades to centuries.

Post-common envelope binary WDs are rather common ($\sim$100 with known orbital periods, Morales-Rueda et al. 2005) yet because of their highly interactive pasts (the details of which are not fully understood), they are not good candidates for empirical checks of the theoretical M-R relation. Instead, those systems whose components evolved as single stars are more relevant to test the M-R relation given that this relation is invoked to estimate the masses of single field WDs. However, finding such systems with orbital periods that are not prohibitively long such that the system's orbital parameters can be sufficiently constrained is non-trivial.

In general, one must exclude any system that is resolved by standard ground-based imaging[1]. As previously mentioned, proximity to the Sun is important to minimize the fractional parallax error as this is often the largest contribution to the mass error budget. We have targeted nearby WDs for a more thorough evaluation in search of DD systems.

3.1. Overluminous white dwarfs

As part of the southern hemisphere parallax program, CTIOPI (Cerro Tololo Inter-American Observatory Parallax Investigation), we have added 35 new WD systems to the 25 pc sample (horizons of the Catalog of Nearby Stars and the NStars database) thus far in addition to

[1] Although 40 Eri B is resolved, the orbital period is $\sim$250 years and we have yet to observe one complete orbit. Its proximity to the Sun (5.03 $\pm$ 0.02 pc, www.recons.org) and the fact that it has been observed since 1851 allows for an accurate mass determination.

measuring accurate parallaxes to a number of WD systems beyond 25 pc. With accurate trigonometric parallaxes available, we generated a Hertzsprung-Russell (H-R) diagram, including the known sample of WDs with parallaxes from Bergeron et al. (2001). In some cases, there are a few WDs that lie above the locus of the rest, thus implying that they are overluminous. There are two obvious scenarios that would lead to a WD being overluminous, (1) the object is an unresolved DD with each component contributing to the total flux, or (2) the object is a very low mass WD with a large radius and hence a relatively large luminosity. The second scenario is limited by the age of our Galaxy such that the lowest-mass WD that could be formed via single star evolution is $\sim$0.47 $M_{\odot}$ (Iben & Renzini 1984)[2]. Given that the first scenario is the more likely, the apparently overlumious WDs are potentially resolvable DD systems.

3.2. White dwarfs with spectral anomalies

During co-author Subasavage's spectroscopic search for new WD discoveries in the southern hemisphere, two spectra show anomalies that hint at the possibility that the spectra are convolutions of two different WD components. The first (WD 0121−429) is a cool hydrogen-rich DA WD that exhibits Zeeman splitting of Hα and Hβ indicative of a magnetic field on order of 9.5 MG (Subasavage et al. 2007). The trigonometric parallax obtained via CTIOPI implies that if this object is a single star, its mass is 0.43 $\pm$ 0.03 $M_{\odot}$ which is unlikely to form via single star evolution as discussed earlier. Also, the spectral fit to the central line of Hα (the strongest and most important to model correctly) requires a 50% dilution factor (see Figure 2). This can be interpreted as a DD system with one component being a magnetic DA WD and the other component being a featureless DC component.

A second object (WD 0622−329) is classified spectroscopically as a DAB WD (prominent hydrogen lines with weaker He lines). However, it is not possible to model the spectrum with a single temperature. By assuming the spectrum is a convolution of two WDs of different temperatures, the fit is much better (see Figure 3). Thus, it is likely that this system is comprised of a hotter DB WD and a cooler DA WD (Subasavage et al. 2007).

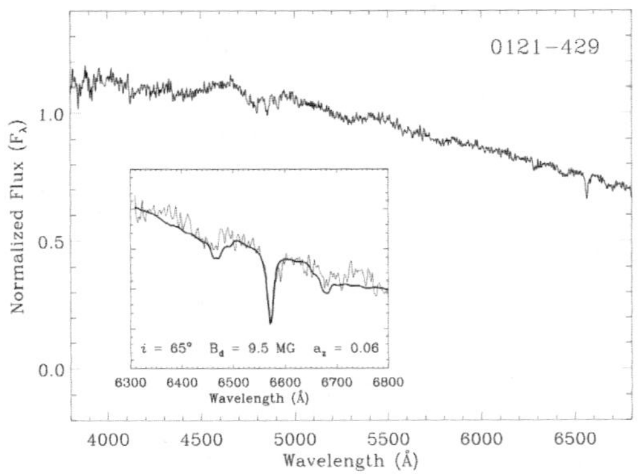

Figure 2. Spectral fit for WD 0121-429 (*inset plot dark line*) including a 50% dilution factor.

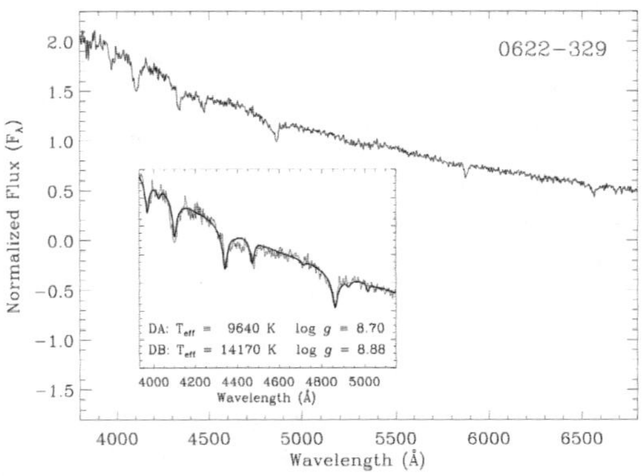

Figure 3. Spectral fit for WD 0622-329 (*inset plot dark line*) assuming an unresolved double degenerate.

[2] We note that recent work by Kilic et al. (2008) suggests that WD masses lower than this can be formed via single star evolution in the solar neighborhood. It is proposed that a relatively metal-rich progenitor may undergo significant mass loss while on the red giant branch such that it will never ignite He and thus continue its evolution as a lower mass He core WD. This scenario would allow for a single star's minimum mass to be $\sim$0.40 $M_{\odot}$.

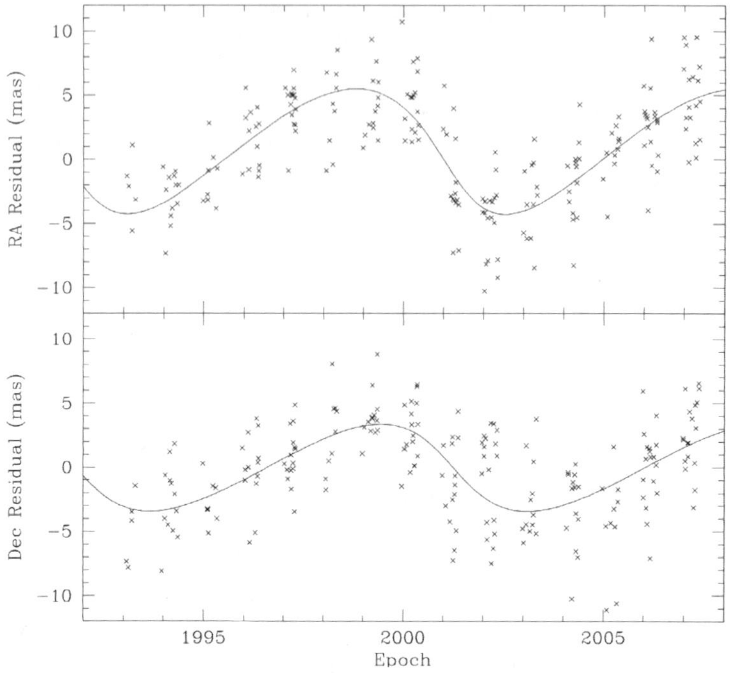

Figure 4. Plot of the residuals for WD 1132+470 once proper motion and parallax are fit and removed as well as a best-fit orbit (*solid line*).

3.3. White dwarfs with apparent astrometric perturbations

Ongoing trigonometric parallax programs both at CTIO and USNO have collected data on numerous nearby WDs that span several years (in some cases on the USNO program, more than a decade). On a rare occasion, once proper motion and parallax have been fit to the astrometric data and removed, there is an indication of a periodic perturbation caused by an unseen companion. On the CTIOPI program, it was thought that three such WD systems were identified; however, it was recently determined that they were erroneous signatures. The Johnson *V* filter used at CTIO to obtain parallax data since 1999 cracked in early 2005. Thus, a different (though similar in terms of transmission) Johnson *V* filter replaced the cracked filter. This caused a slight shift in the centroids that resulted in a "dip" in the residuals by ~ 5 mas for mid-2005 data. As more data were collected, the residuals began to resemble sinusoidal arcs as a consequence of the parallax fitting routine attempting to fit this dip. This was not noticed to be a systematic effect until years later when there was sufficient post-filter-switch data that the dip was clearly evident in numerous *V* filter targets. Once all targets with possible perturbations that were observed in *V* are eliminated, there are no obvious perturbations with periods less than $\sim$5 years.

However, one target (WD 1132+470) on the USNO program that has been observed for $\sim$15 years, shows a clear periodic trend in the both right ascension and the declination residuals[3]. These data indicate a period of $\sim$9 years and a photocentric semi-major axis of $\sim$6 mas (see Figure 4). There is an ambiguity as to the nature of the companion based solely on these signatures. It could be either (1) a nearly equal mass, equal luminosity object, or (2) a substantially lower mass, lower luminosity object. If the first scenario is correct, this DD system is a prime target for measuring WD masses.

[3] A second USNO WD target shows a similar trend but was not analyzed in this work.

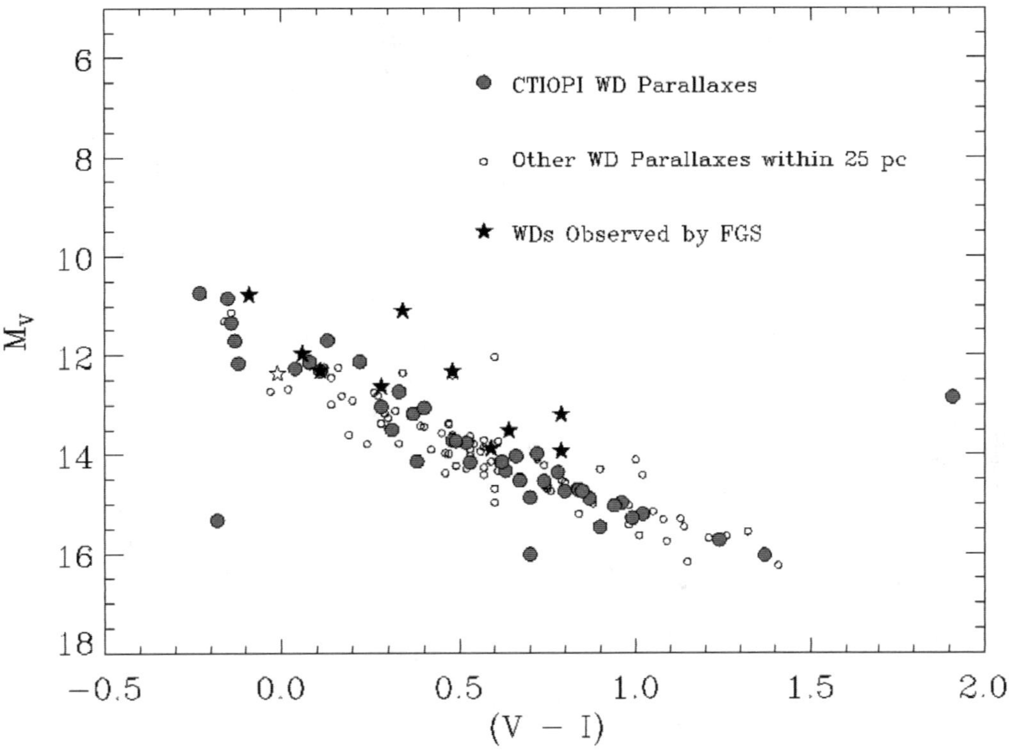

Figure 5. WD region of the H-R diagram. Plotted are WDs from the CTIOPI program (*filled circles*), known WDs within 25 pc from Bergeron et al. 2001 (*open circles*), and WDs observed with FGS via this effort (*filled stars*). WD 1132+470 is plotted as an open star.

4. Astrometric measurements using *HST*'s Fine Guidance Sensors

The Fine Guidance Sensors (FGS) on board *Hubble Space Telescope* (*HST*) are responsible for precision guiding of the telescope during observations that utilize other science instruments. Thus, it is a high precision astrometric instrument. It is also capable of stand-alone astrometric observations because there are three individual sensors, one that can be positioned on the target of interest while the other two provide guiding. It can aquire targets as faint as $V = 16.6$ and can resolve companions with separations as small as 10 mas (assuming modest Δmag). Translating these values into physical units, FGS would resolve an equal magnitude DD system at 25 pc with separations larger than 0.25 AU.

A total of 16 orbits spread over *HST* cycles 15 and 16 were awarded to this effort to resolve probable DD systems. To date, 12 are completed with 11 objects observed (one was observed twice) and are plotted in the H-R diagram in Figure 5. Of the 11 objects observed, only one was definitively resolved (WD 1132+470), and another was marginally resolved (WD 0121−429). In the second case, the signature may due to the evolution of the Y-axis fringe so a single star calibrator taken close in time to the target observation is needed to determine the degree of evolution. Another orbit has been assigned to this target as a safe measure. It is possible that the apparently overluminous targets that were unresolved are short period post-common envelope binaries. High resolution radial velocity measurements are necessary for confirmation.

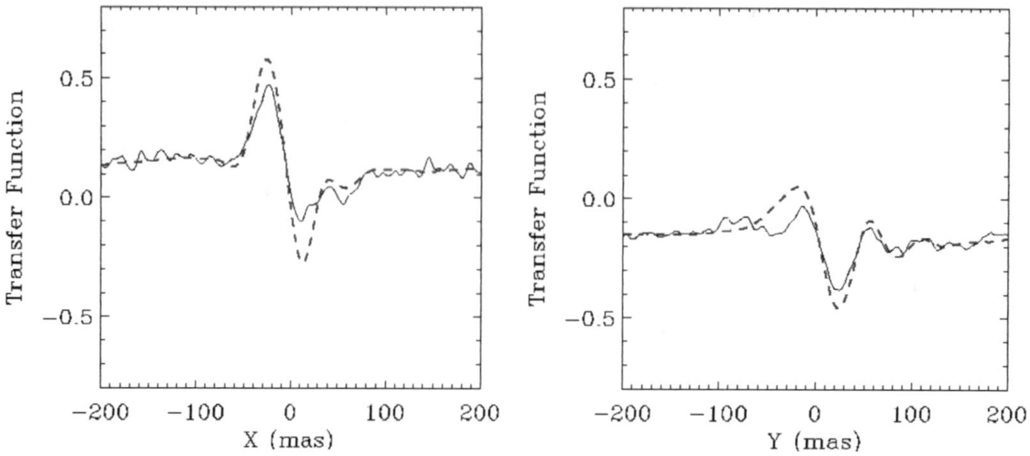

Figure 6. Fringes in both axes for WD 1132+470 (*solid line*) vs. those of a single star calibrator scaled to the same magnitude (*dashed line*).

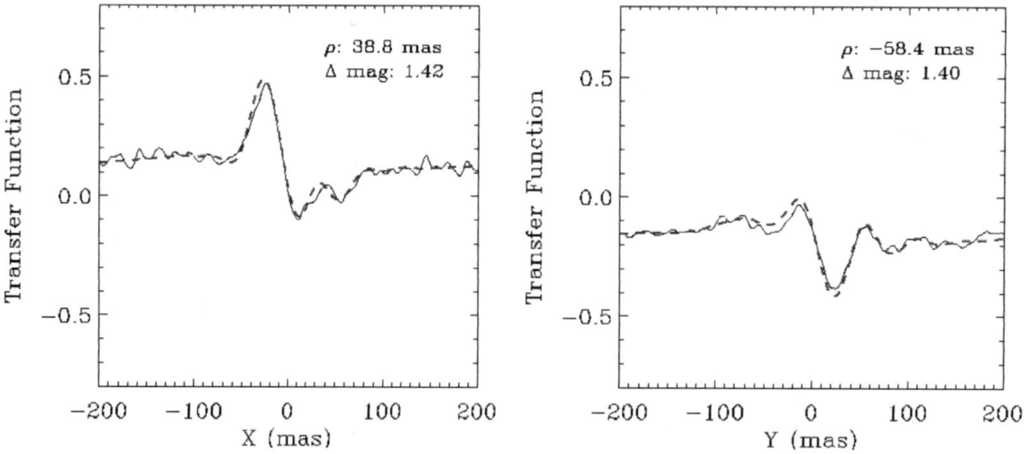

Figure 7. Fringes in both axes for WD 1132+470 (*solid line*) as well as the binary model fit (*dashed line*).

4.1. WD 1132+470 results

By comparing the fringes in both axes of WD 1132+470 to that of a single star calibrator of the same magnitude (see Figure 6), it is clear that a second luminous source is distorting the profile of the fringes. Once a binary model is fit to the fringes (see Figure 7), a companion with a Δmag = 1.4 and a separation of 70 mas is identified.

Combining this data with the USNO astrometric data, we find that each of these WD components is massive ($\sim$1.1 - 1.4 $M_\odot$). There is some discrepancy when combining the two datasets in that the Δmag determined by FGS at V is too large to produce the small photocentric shift seen in the USNO data. This can be partially rectified if the two WDs are of different temperatures such that the Δmag at R (the filter used to detect the astrometric perturbation) is smaller. Another visit using FGS is warranted to confirm the initial Δmag that was determined and to obtain another data point on the orbit. Further evaluation should allow for a pair of well constrained WD dynamical masses that will anchor the high-mass end of the M-R diagram.

5. Conclusions

With only three WDs having dynamical mass determinations known to better than 5%, it is vital that additional WD systems are identified with orbital parameters conducive to accurate orbit determinations. In addition, to remove any uncertain variables associated with common envelope phase systems, the WD systems should have evolved essentially as single stars. Co-author Nelan has identified two DD systems (WD 1639+153 and WD 1818+126) with ~20 year periods as well as another DD system (WD 0727+483) with a ~18 year period. In a relatively short time, including the resolved system presented here (WD 1132+470), these objects will provide eight new empirical points on the M-R diagram. Also, co-author Nelan has identified an additional four DD systems that were only marginally resolved with FGS such that no accurate orbital determinations were possible. Should astrometric resolving capabilities improve in the future (e.g., via *Space Interferometry Mission*), these systems would be resolvable and accurate orbits could be mapped. Thus, we can be confident that the number of accurate dynamical WD masses will substantially increase. We plan to continue FGS observations of new nearby WDs to further populate the M-R Diagram and better scrutinize the theoretical M-R relation.

References

Bergeron P Leggett S K & Ruiz M T 2001 *ApJS* **133** 413

Hansen B M S et al. 2004 *ApJS* **155**, 551

Hansen B M S et al. 2007 *ApJ* **671** 380

Iben I & Renzini A 1984 *Phys. Rep.* **105** 329

Kilic M Stanek K Z & Pinsonneault M H 2007 *ApJ* **671** 761

Morales-Rueda L Marsh T R Maxted P F L Nelemans G Karl C Napiwotzki R & Moran C K J 2005 *MNRAS* **359** 648

Provencal J L Shipman H L Hog E & Thejll P 1998 *ApJ* **494** 759

Provencal J L Shipman H L Koester D Wesemael F & Bergeron P 2002 *ApJ* **568** 324

Rebassa-Mansergas A et al. 2008 *MNRAS* **1112**

Reid I N 2005 *ARA&A* **43** 247

Wood M A 1995 *White Dwarfs, Proceedings of the 9th European Workshop on White Dwarfs Held at Kiel, Germany, 29 August - 1 September 1994. Lecture Notes in Physics* vol. 443 eds D. Koester and K. Werner (Heidelberg: Springer-Verlag) p 41

Journal of Physics: Conference Series **172** (2009) 012018 doi:10.1088/1742-6596/172/1/012018

White Dwarfs and the Ages of Open Clusters

Elizabeth J. Jeffery[1], Steven DeGennaro[1], Ted von Hippel[1,2], David van Dyk[3], William H. Jefferys[1,4] and Nathan Stein[1]

[1] Department of Astronomy, University of Texas at Austin, Austin, TX 78712, USA
[2]Department of Physics, Siena College, Loudonville, NY 12211, USA
[3]Department of Mathematics and Statistics, University of Vermont, Burlington, VT 05405, USA
[4]Department of Statistics, University of California at Irvine, Irvine, CA 92697 78712, USA

E-mail: ejeffery@astro.as.utexas.edu

Abstract.
Open clusters provide the ideal environment for the calibration of ages determined from main sequence evolutionary theory (via cluster isochrones) and ages determined from white dwarf cooling theory. In an effort to measure more precise cluster ages, our group has developed a new technique using Bayesian statistics. Here we will discuss new capabilities of the technique, as well as the first application to real data, using the Hyades as a test case. Because the faintest white dwarfs have likely evaporated from the Hyades, we also demonstrate the first successful application of the bright white dwarf technique for deriving ages from the bright cluster white dwarfs alone.

1. Introduction

Accurate age measurements are fundamental to many areas of astrophysics. Currently there are two main techniques for independently determining the ages of stellar populations: main sequence (MS) evolutionary theory (via cluster isochrones) and white dwarf (WD) cooling theory. Ages determined from the MS turnoff (MSTO) of globular clusters provide the most reliable age of the Galactic halo (Chaboyer, Demarque & Sarajedini 1996), while WD cooling ages provide the most reliable age of the Galactic disk (Winget et al. 1987). Before ages determined by these two techniques can be meaningfully compared, they must be calibrated to the same absolute scale. Additionally, such a calibrations provides insights to thoroughly understanding the physics of each.

The best way to do this calibration is to measure and compare WD ages and MSTO ages in several open clusters with a wide range of ages and metallicities. Additionally, being able to compare WD theory against MSTO theory provides an excellent opportunity to refine our understanding of both theories. Recent studies in calibrating MSTO and WD ages in open clusters have shown good agreement for cluster ages up to 4 Gyr. A summary of recent studies was done by von Hippel (2005). To aid us in the endeavor to determine more precise cluster ages, our group has developed a powerful new Bayesian statistical approach (von Hippel et al. 2006, Jeffery et al. 2007)

This paper is organized as follows: in Section 2 we briefly summarize the Bayesian technique as well as discuss its new capabilities. We review the bright WD technique as a new tool to measure cluster ages in Section 3. In Section 4 we discuss the first application of the Bayesian

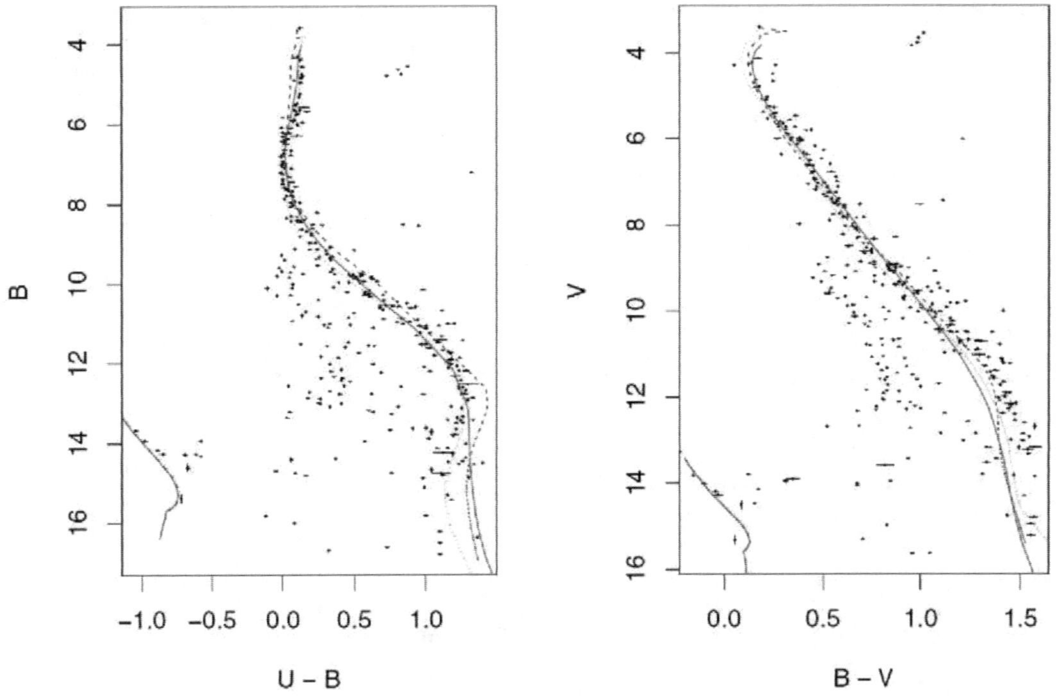

Figure 1. Isochrones for the three different MS models, set at the nominal age, distance, and metallicity of the Hyades. Solid lines are the Girardi models, dotted lines are the Yale-Yonsei models, and the dashed lines are the DSED models.

technique to real data, using the Hyades as a test case. Our result also demonstrates the power of the bright WD technique. We end with concluding remarks in Section 5.

2. The Bayesian technique

Our goal is to improve the age precision in both WD and MSTO age techniques, to about 5%. Despite many high quality datasets having been collected on open clusters, this 5% age precision is still generally out of reach. We therefore believe that the greatest gains we can currently make in age precision will require improved modeling techniques (see also work done by Tosi et al. 1991; Tosi, Bragaglia & Cignoni 2007; Hernandez & Valls-Gabaud 2008).

Briefly, the Bayesian technique we have developed derives a posterior probability distribution for a cluster's age, metallicity, distance, and line-of-sight absorption by objectively incorporating our prior knowledge of stellar evolution, star cluster properties, and data quality estimates.

Since first being presented by von Hippel et al. (2006), we have expanded the capabilities of the technique to included binary stars, field stars, as well as including additional MS models. Each of these points is expounded below.

Originally, the technique modeled each point on the color-magnitude diagram (CMD) as an individual stellar system with a single mass. We are now able to treat each system as if it were a binary, and parameterize it in terms of the primary mass, and by the ratio of the two masses. Stars without binary companions can also be modeled, by setting the mass ratio equal to 0. We are currently only considering MS-MS binaries.

During any given iteration of the technique, each object in the CMD is assigned to be a cluster member or a field star. In the final analysis, the membership probability of each star in

Journal of Physics: Conference Series **172** (2009) 012018 doi:10.1088/1742-6596/172/1/012018

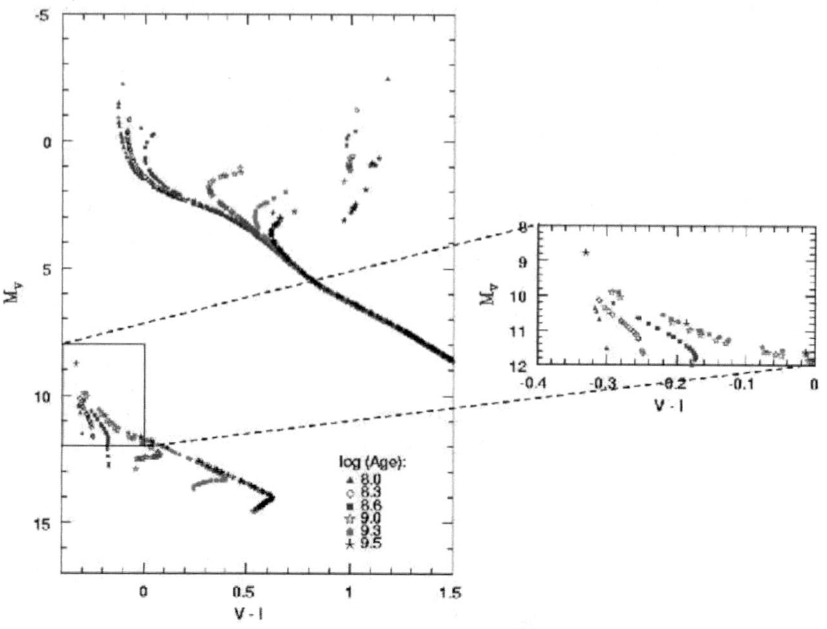

Figure 2. Simulated clusters for several different ages. The expanded region shows the regime of the brighter WDs, clearly showing the subtle differences in the slopes and positions of the WD cooling sequences relative to the MS for clusters of different ages. This makes it possible to extract age information without observing the faintest WDs.

the cluster can be determined by dividing the number of total iterations during which it was a cluster member by the total number of iterations in the run.

In von Hippel et al. (2006), we incorporated a Miller & Scalo (1979) initial mass function, MS and giant branch stellar evolution time scales of Girardi et al. (2000), the initial-final mass relation from Weidemann (2000), WD cooling timescales of Wood (1992), and WD atmosphere colors from Bergeron, Wesemael & Beauchamp (1995). We now include two additional sets of MS models – Yale-Yonsei (Yi et al. 2001), and a finer grid of models from the Dartmouth Stellar Evolution Database (DSED, Dotter et al. 2008), as well as updated versions of Bergeron WD atmosphere models. Figure 1 shows CMDs for the Hyades with the three MS models overlaid. Solid lines are the Girardi models, dotted lines are the Yale-Yonsei models, while the dashed lines are the DSED models.

3. The bright white dwarf technique

In Jeffery et al. (2007) we demonstrated the theoretical feasibility of determining WD ages of clusters using the brighter WDs alone. Briefly, this technique relies on the subtle differences in slope and position of the WD cooling sequence relative to the MS for clusters of different ages (as illustrated in Figure 2).

The bright WD technique assumes that the initial-final mass relation (IFMR) is universal and single-valued, which is the general consensus among researchers (Weidemann 2000). Because of this dependence on the IFMR, this technique is a relative age indicator and requires calibration. As will be discussed in the next section, this work is the first step in such a calibration.

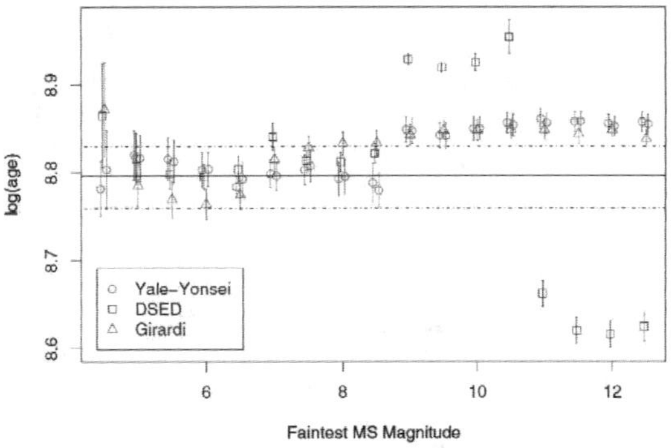

Figure 3. Our derived age for the WD age of the Hyades as a function of the faintest MS magnitude star included in the analysis. The solid lines are the mean and $\pm$ 1σ deviations of the most reliable age estimates for the age of the Hyades.

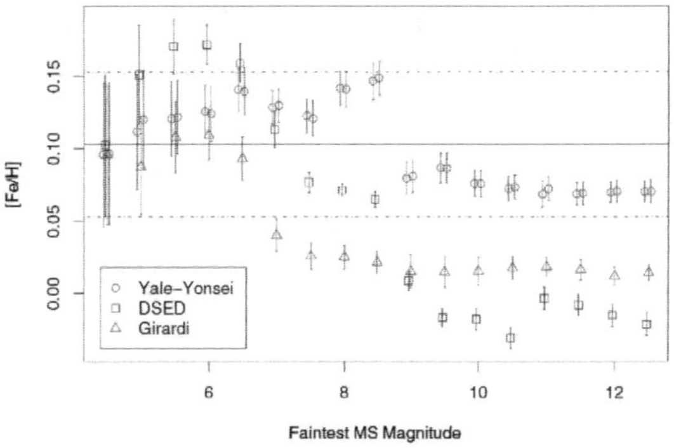

Figure 4. Our derived metallicity for the Hyades as a function of the faintest MS magnitude star included in the analysis.

4. The Hyades

The Hyades is one of the most well-studied open clusters. Perryman et al. (1998) report a MSTO age for this cluster of 625 $\pm$ 50 Myr and a distance to the center of the cluster (based on trigonometric parallaxes from Hipparcos) of $m - M = 3.33 \pm 0.01$. High resolution spectroscopy has been used to determine the metallicity to high accuracy, [Fe/H] = +0.103 $\pm$ 0.008 (Taylor & Joner 2005), based on their re-analysis of Paulson, Sneden & Cochran (2003). In order to apply our Bayesian technique to real data for the first time, this cluster was the logical choice.

Data used here for our analysis of the Hyades is taken from The General Catalog of Photometric Data (GCPD – http://obswww.unige.ch/gcpd/gcpd.html – Mermilliod, Mermilliod & Hauck 1997). They determine a weighted mean and dispersions in the V band photometry in addition to the B - V and U - B color indices.

4.1. Results

Because we are presently interested in measuring the WD age of the cluster, we removed all stars brighter than V = 4.5 (i.e., the MSTO). The divergence among MS models in the lower MS (see Figure 1) was also a point of concern, as our ability to fit the best model to the data relies on the models being accurate. In order to test the sensitivity of our results and errors on the poorly modeled lower MS, we performed the Hyades analysis with varying amounts of MS. Each data set contains all the WDs and the MS between V = 4.5 and some low cutoff, in half magnitude intervals.

The results of this analysis are shown in Figures 3 through 6. In each of these figures, triangles represent the Girardi models, squares are the DSED models, and the circles are the Yale-Yonsei models. The horizontal lines are the most recently accepted values for the parameters and their standard errors. As demonstrated by these figures, all three models give very reasonable results for all the parameters.

These figures demonstrate how the choice of model and amount of MS included affect the results. We see there is a clear difference between V $\leq$ 8.5 and V $>$ 8.5, for all models in each parameters as the MS diverges from the models. Assuming the models are accurate down to V = 8.5, we take a weighted average of the three models for each parameter for the runs which include MS stars down to V = 8.5. Errors were determined rather conservatively, by adding

Journal of Physics: Conference Series **172** (2009) 012018 doi:10.1088/1742-6596/172/1/012018

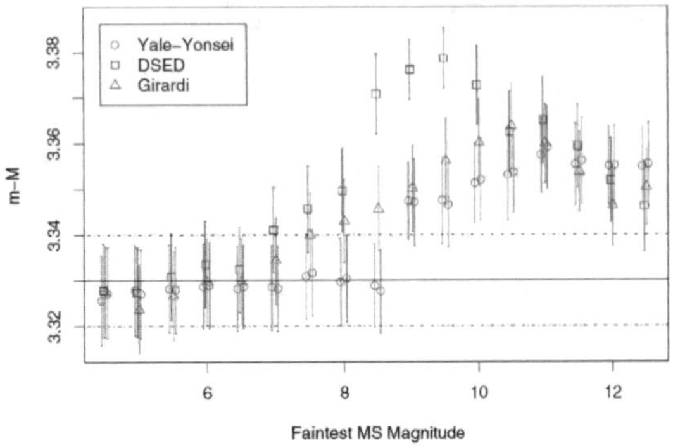

Figure 5. Our derived distance modulus for the Hyades as function of the faintest MS magnitude star included in the analysis.

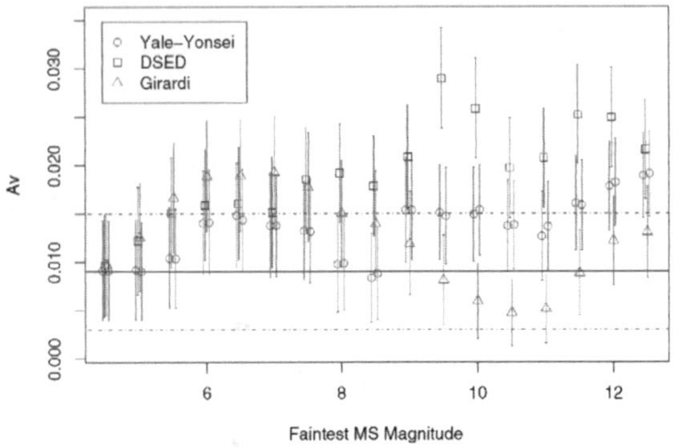

Figure 6. Our derived A_V for the Hyades as a function of faintest MS magnitude star included in the analysis.

the standard deviations for the three determinations in quadrature to the average error of the individual determinations.

Therefore, we measure a WD age for te Hyades of 648 $\pm$ 45 My; an [Fe/H] of 0.078 $\pm$ 0.065; m-M of 3.35 $\pm$ 0.02, and A_V of 0.014 $\pm$ 0.007.

4.2. The bright white dwarf age of the Hyades

Previous studies to determine the WD age of the Hyades cluster have produced a result (300 Myr; Weidemann et al. 1992) that is about half the measured MSTO age (625 Myr; Perryman et al. 1998). Weidemann et al. (1992) suggested that this discrepancy is due to the dynamical evaporation of stars from this cluster; the coolest WDs are no longer present. In the absence of any data on these missing faint WDs, traditional techniques to determine WD ages can provide at best a lower limit to the WD age.

As summarized earlier, in Jeffery et al. (2007) we showed the possibility of determining cluster WD ages from just the bright WDs, when the coolest WDs are not observed. Because the coolest WDs are missing from the Hyades, we require the bright WD technique to measure the true WD age (rather than a lower limit, as was done previously). Our Hyades results provide empirical evidence that the Bright WD technique yields reasonable and precise ages for real data, as well as providing an important step in calibrating the technique.

Figure 7 is an updated version of Figure 1 from von Hippel (2005), plotting WD age vs. MSTO age for open clusters up to 4 Gyr. Our measurement of the bright WD age of the Hyades brings the WD age into agreement with the MSTO age for the first time. The gray point shows the previously most reliable WD age for the Hyades (Weidemann et al. 1992). The solid line is the one-to-one correspondence between WD and MSTO ages.

5. Conclusions

We are continuing in an effort to calibrate WD and MSTO ages using open clusters. In order to improve accuracy of the measured ages, our group has developed a technique using Bayesian statistics. We have presented current progress on our Bayesian technique, including an expansion of the code to include more MS models as well as treatment of binaries and field stars.

We have demonstrated the technique by applying it on the Hyades as a test case. Results for cluster parameters, namely WD age, metallicity, distance, and reddening, have yielded results consistent with previous measurements. Of particular interest is the determination of the bright WD age of the Hyades, required due to the missing cool WDs in this cluster. Our results bring

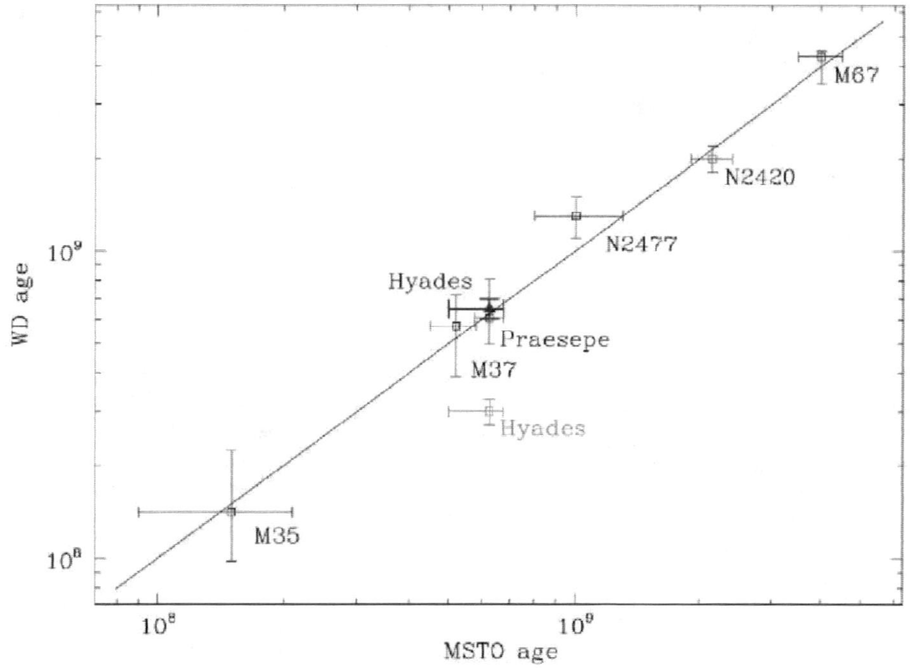

Figure 7. WD versus MSTO age for seven clusters, adapted and updated from von Hippel (2005). The age we derive from the WDs in the Hyades using our bright WD technique brings the WD age of the Hyades into agreement with the MSTO age for the first time (solid triangle). The solid line shows the one-to-one correspondence between the WD and MSTO ages, and the gray point shows the most reliable WD age of the Hyades prior to this work (Weidemann et al. 1992).

the WD age of this cluster in agreement with the MSTO age for the first time, as well as provide a first step in calibrating the bright WD technique.

Acknowledgments

This material is based upon work supported by the National Aeronautics and Space Administration under Grant No. NAG5-13070 issued through the Office of Space Science.

References

Chaboyer B Demarque P & Sarajedini A 1996 *ApJ* **459** 558
Winget D E Hansen C J Liebert J van Horn H M Fontaine G Nather R E Kepler S O & Lamb D Q 1987 *ApJ* **315** L77
von Hippel T 2005 *ApJ* **622** 565
von Hippel T Jefferys W H Scott J Stein N Winget D E DeGennaro S Dam A & Jeffery E J 2006 *ApJ* **645** 1436
Jeffery E J von Hippel T Jefferys W H Winget D E Stein N & DeGennaro S 2007 *ApJ* **658** 391
Tosi M Greggio L Marconi G & Focardi P 1991 *AJ* **102** 951
Tosi M Bragaglia A & Cignoni M 2007 *MNRAS* **378** 730
Hernandez X & Valls-Gabaud D 2008 *MNRAS* **383** 1603
Miller G E & Scalo J M 1979 *ApJS* **41** 513
Girardi L Bressan A Bertelli G & Chiosi C 2000 *A&AS* **141** 371
Weidemann V 2000 *A&A* **363** 647
Wood M A 1992 *ApJ* **386** 539
Bergeron P Wesemael F & Beauchamp A 1995 *pasp* **107** 1047

Yi S Demarque P Kim Y-C Lee Y-W Ree C H Lejeune T Barnes S 2001 *ApJS* **136** 417
Dotter A Chaboyer B Jevremovic D Kostov V Baron E & Ferguson J W 2008 *ApJS* in press arXiv:0804.4473
Perryman M A C et al. 1998 *A&A* **331** 81
Taylor B J & Joner M D 2005 *ApJS* **159** 100
Paulson D B Sneden C & Cochran W D 2003 *AJ* **125** 3185
Mermilliod J-C Mermilliod M & Hauck B 1997 *A&AS* **124** 349
Weidemann V Jordan S Iben I J & Casertano S 1992 *AJ* **104** 1876

The Age of White Dwarf Companions

S Weston and R Napiwotzki

Centre for Astrophysics Research, University of Hertfordshire, College Lane, Hatfield, Herts, AL10 9AB, UK

E-mail: s.1.weston@herts.ac.uk

Abstract. We carried out a spectroscopic investigation of single lined white dwarfs (WDs) in double degenerate (DD) systems and discuss their binary evolution. Simulated spectra of the Hα region are used to derive upper limits on the temperature of the invisible component and thus lower limits on the cooling age. This is done for a range of hypothetical secondary masses and a minimum cooling age deduced. Results are compared with the well known parameters of the visible primary, which allows us to determine a lower limit for the cooling age difference of both WDs. Most of the ten systems in our sample have a minimum age difference of not larger than 0.5 Gyr and their small orbital separation is highly suggestive of at least one unstable mass transfer phase. However, a stable first mass transfer phase is feasible as the age difference is less then 1 Gyr. The results imply that unstable mass transfer is the most likely final contact binary scenario to have occurred in DD systems but the first mass transfer phase is not constrained.

1. Introduction

Between 5 and 10% of white dwarfs (WD) reside in close binary systems with a WD companion (Napiwotzki et al. 2003, Marsh 2000), known as double degenerates (DDs). DDs must have come into contact and fallen towards each other as it is not possible for the stars to have evolved so close during giant branch phases. Orbital periods range from hours to a few days corresponding to separations of 2-3 $R_\odot$ at most. This implies that the binary must have gone through phases of intensive interaction affecting the evolution of the WD progenitors. One of the binary components evolving up the giant branch (RGB or AGB) will fill its Roche lobe and start to transfer mass to its companion. This can result in three possible outcomes (Nelemans et al. 2001):

- Conservative/stable mass transfer – The transferred mass is accreted by the companion and no mass is lost to the surroundings. The size of the orbit remains relatively large and can even increase.
- Standard common envelope – The rate of mass transfer is so high that not all of it can be accreted. Most of it is lost and forms a common envelope around the system. Friction causes a large reduction in the orbital separation.
- Envelope eject or double spiral-in – During the ejection of the common envelope both stars lose their envelope. Orbital radius greatly reduced and age difference very small (within 0.1Myr).

Different formation time scales can be expected for the three scenarios. Only very small age differences are allowed for DDs resulting from the double spiral-in channel. Stable mass transfer

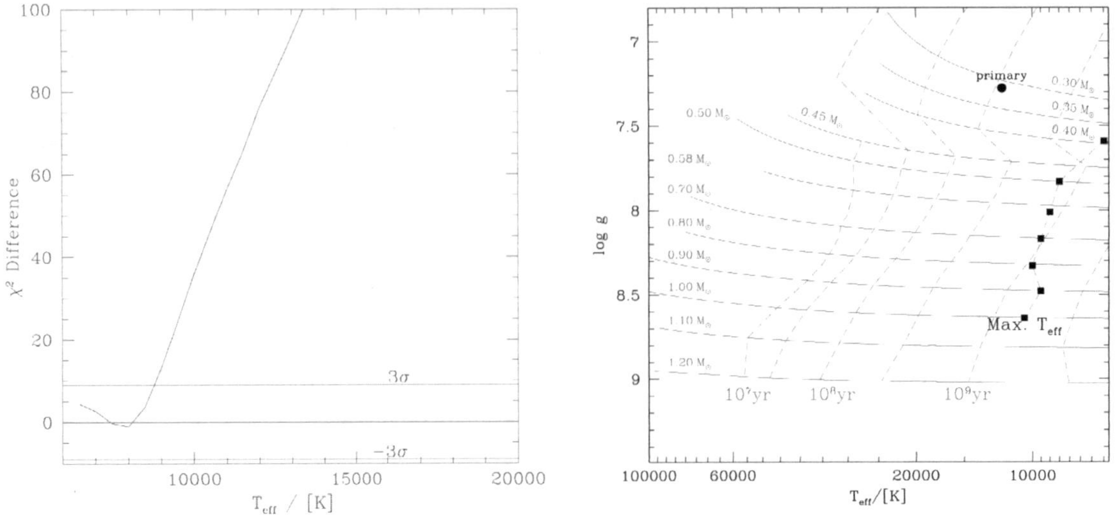

Figure 1. *Left:* Fit quality as a function of temperature for a $0.5\,M_\odot$ companion of HE0320−1917. The maximum possible temperature of the secondary is 8500 K. *Right:* H-R diagram with maximum $T_{\rm eff}$ for hypothetical seconday masses of HE0320−1917. Evolutionary tracks of Benvenuto & Althaus (1999) for masses (solid lines) are shown along with cooling ages (dashed lines).

results in a substantial mass increase of the companion, which will speed up its subsequent evolution. Since it is a very short lived phase, no such effect is expected during the common envelope. We derive lower limits on the age of the secondary WD and therefore an age difference estimate of the primary and secondary's formation. This may be used with the theoretical predictions to rule out possible evolutionary scenarios for a binary system. The aim of this investigation is to compare the resulting limits with predicted distributions for the different channels.

2. Determining Maximum Secondary Temperatures

The DDs investigated have been classified SB1 in the Supernova type Ia Progenitor surveY (SPY; Napiwotzki et al., 2003), meaning that no obvious contribution from the secondary is present in the spectra. We use SPY spectra of the Hα region to constrain the secondary's contribution in a quantitative way. We consider just the Hα line because its core is the most prominent feature in the spectra of DA WDs. SPY is a high resolution, high signal-noise spectroscopic study which uses a large input sample of WDs combined from many surveys. Systems are confirmed DD due to radial velocity (RV) fluctuations. The data set contains six binary systems for which orbital solutions and parameters for the brighter primary star are already known and published in Nelemans et al. (2005). Four other non-published systems were used with well established primary parameters. For the latter four systems the RV and thus position of the Hα core is not constrained. In these cases a worst case was assumed with the secondary line core close to the primary's. Using a fitting program, FITSB2 (Napiwotzki et al. 2004), synthetic spectra are created and compared to the observed, and the χ^2 calculated. The parameters which produce fits outside 3σ confidence limits can be excluded with our attention focused on the possible solutions. The last good fit defines a maximum temperature for a given mass component of the secondary star. This was done for secondary masses of 0.4–1.0 $M_\odot$ at a 0.1 $M_\odot$ increment. Fig. 1 shows HE0320−1917 as an example.

Table 1. Overview of the primary parameters and secondary limits. The columns give primary mass (M_1), effective temperature of the primary ($T_{\rm eff,1}$), cooling age of the primary ($\tau_{\rm cool,1}$), assumed secondary mass (M_2), maximum secondary temperature (Max. $T_{\rm eff,2}$) and minimum secondary cooling age (Min. $\tau_{\rm cool,2}$). The last column indicates if the fit quality increased by more than 3σ with a secondary component, therefore making the system a SB2 candidate. We show minimum ages for all hypothetical masses of HE0320−1917 and then the youngest calculated age with its corresponding mass for all other systems.

Object	M_1 [$M_\odot$]	$T_{\rm eff,1}$ [K]	$\tau_{\rm cool,1}$ [Gyr]	M_2 [$M_\odot$]	Max. $T_{\rm eff,2}$ [K]	Min. $\tau_{\rm cool,2}$ [Gyr]	SB2?
HE0320−1917	0.29	12 000	0.3327	0.4	6 500	2.337	Y
				0.5	8 500	0.7505	
				0.6	9 000	0.8344	
				0.7	9 500	0.9409	
				0.8	10 000	1.116	
				0.9	9 500	1.685	
				1.0	10 500	1.561	
HE1511−0448	0.48	50 000	0.0001	0.4	41 000	0.0026	Y
WD0326−273	0.51	9 300	0.6883	0.5	6 500	1.474	-
WD1013−010	0.44	8 000	0.8449	0.5	6 500	1.474	-
WD1210+140	0.29	32 000	0.0232	0.6	9 000	0.834	-
WD0101+048	0.48	8 600	1.517	0.5	6 500	1.474	-
HE0131+0149	0.33	14 500	0.2541	0.8	26 000	0.0646	Y
WD0216+143	0.34	27 000	0.0433	0.5	30 000	0.0088	-
WD0341+021	0.30	21 500	0.0747	1.0	26 000	0.1419	Y
WD1824+040	0.31	14 000	0.2817	0.5	10 000	0.4492	Y

3. Discussion on Companion Ages and Possible SB2 Systems

The results are summarised in Table 1 and suggest nearly all of the systems companions have a larger cooling time than the primary. Only in three systems the secondary could be younger but it must be emphasised that we only determine lower age limits. In addition, for one of these cases (and some other) we are constrained by our cool model grid boundary of 6500 K. In the cases where the secondary is definitely older, all of the secondaries are older by more than 0.1 Myr therefore ruling out a double spiral-in scenario. Nelemans et al. (2001) consider 1Gyr an approximate maximum age difference for a secondary to have undergone an initial stable mass transfer before a common envelope. In all but one of our sample this is a feasible evolutionary path.

Some of our χ^2 results indicate possible detections of secondary components in the spectra; HE0320−1917, HE1511−0448, HE0131+0149, WD0341+021 and WD1824+040 are all candidate SB2s and further analysis of the spectra may confirm the features of the secondary. Fig 2 shows the spectra and χ^2 plots of WD1824+040, a SB2 candidate. These systems may be useful as they are close to the SB1/SB2 borderline and will show at what mass and temperature one might expect to see features appear in the combined spectra.

4. Conclusion

We test possible evolutionary scenarios of DDs by computing the cooling age of ten systems and comparing the primary with hypothetical secondary masses. The secondary age in nearly

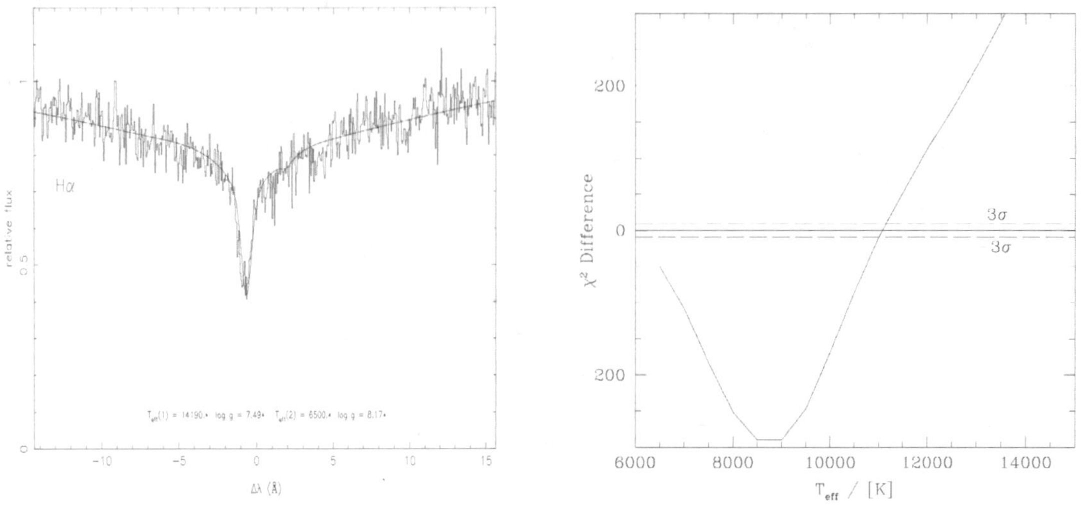

Figure 2. *Left:* Fit to one of the spectra of WD1824+040 with an apparent detection of the seconqdary. The primary (1) and secondary (2) atmospheric parameters are displayed in the plot. *Right:* Summary of fit quality for various different temperature secondaries. The maximum temperature of the secondary is 11,000 K.

every case is larger than the primary as expected. None of the older secondaries have a small enough cooling age difference for a double spiral-in to have occured. In all systems the minimum age difference is below 1Gyr, therefore the first mass transfer phase may have been stable or unstable. The results agree with current models which suggest, if the hypothetical mass is close to the primary's the cooling ages are similar. Additionally, the results of the systems without orbital solutions show that limits can be obtained without knowing orbital parameters of the system and with a conservative estimate a meaningful limit can be deduced.

References

Benvenuto O G & Althaus L G 1999 *MNRAS* **303** 30–38.
Marsh T R 2000 *New Astronomy Review* **44** 119–124.
Napiwotzki R Christlieb N Drechsel H et al. 2003 *The Messenger* **112** 25–30.
Napiwotzki R Yungelson L Nelemans G et al. 2004 *in* R W Hilditch H Hensberge & K Pavlovski eds 'Spectroscopically and Spatially Resolving the Components of the Close Binary Stars' Vol. 318 of *Astronomical Society of the Pacific Conference Series* 402–410.
Nelemans G Napiwotzki R Karl C et al. 2005 *A&A* **440** 1087–1095.
Nelemans G Yungelson L R Portegies Zwart S F & Verbunt F 2001 *A&A* **365** 491–507.

Journal of Physics: Conference Series **172** (2009) 012020 doi:10.1088/1742-6596/172/1/012020

The next SDSS white dwarf catalog

S J Kleinman[1], A Nitta[1] and D Koester[2]

[1]Gemini Observatory, 670 N. A'ohoku Place, Hilo HI 96720

[2]Institut für Theoretische Physik und Astrophysik, University of Kiel, D-24098 Kiel, Germany

E-mail: `skleinman@gemini.edu`

Abstract. We have begin work on the next Sloan Digital Sky Survey (SDSS) spectroscopically-identified white dwarf catalog. It will feature white dwarf stars from Data Release 7, the final data release from the SDSS-I and SDSS-II surveys. Besides incorporating the latest spectrophotometry improvements from the SDSS, we have improved our spectroscopic models and expanded our grid. We expect on the order of 5000–6000 additional white dwarf stars not included in the earlier Data Release 4 catalog.

1. Introduction

As many of the presentations at this conference will attest, the Sloan Digital Sky Survey (SDSS: York et al. 2000) has had a huge impact on white dwarf astronomy. Its rich, homogeneous data have produced multiple, large catalogs of white dwarf stars. Using the Early Data Release, Harris et al. (2003) first presented 269 white dwarf stars from 190 square degrees of SDSS data. Using data from the first SDSS Data Release (DR1), data, Kleinman et al. (2004) reported on 2551 white dwarf stars from 2100 sq. degrees of the survey. Whereas these previous catalogs were based on the SDSS spectroscopic data, Harris et al. (2006) presented a luminosity function based on 6000 white dwarf stars from the SDSS DR3, using observed SDSS colors and proper motions to form their sample. Eisenstein et al. (2006) followed the DR1 catalog with another spectroscopic catalog, this time based on DR4 data and including 9316 white dwarf stars from 4783 square degrees of data.

We are now preparing the latest spectroscopic white dwarf catalog based on SDSS Data Release 7 (Abazajian et al. 2008), the final data release from SDSS-I and SDSS-II. DR7 includes all data from the original SDSS and SEGUE-I footprints. Future SDSS data releases will provide even more white dwarf stars from SEGUE-II and possibly other survey components. DR7 covers 8250 sq. degrees on the sky. Spectrophotometric improvements in reductions were made for DR6 (Adelman-McCarthy et al. 2008), with further improvements included in DR7 as well.

Our new catalog will include spectroscopic fits based on our latest atmospheric models from D. Koester and the latest SDSS DR7 reductions. It will be include all new and old objects found in the previous spectroscopic catalogs. That is, we will re-analyze all the white dwarfs presented in the EDR, DR1, and DR4 spectroscopic catalogs and put them on a uniform basis in this new catalog.

2. Sample Selection

Each of the SDSS white dwarf catalogs mentioned above used a different set of sample selection criteria. The DR1 catalog (Kleinman et al., 2004) used a series of different criteria to form a

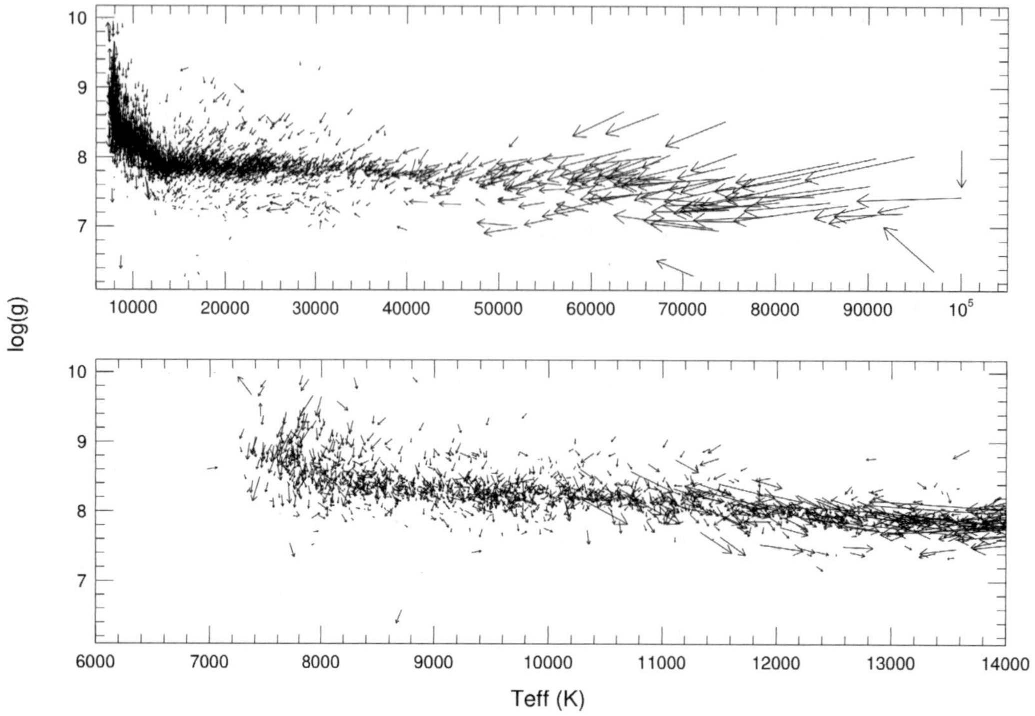

Figure 1. DR4 (base) vs. DR6 (arrow) reductions for a sample of $g < 19$ DAs. The results show the effects of the new DR6 spectrophotometry on our fits. The hotter DAs had their temperatures significantly reduced, making them more in agreement with other fits.

catalog of some 9000 candidate white dwarf stars. The spectra of these candidates were then all looked at by eye and classified as a white dwarf or not. The resulting white dwarf sample was then fit by our spectroscopic fitting code, *autofit*. Eisenstein et al. (2006) used a smaller set of criteria to select candidates from two different SDSS spectroscopic reduction pipelines. They then used *autofit* to fit the spectra of the candidates, automatically triggering human inspection only for objects with uncertain fits, the sample size simply being too cumbersome for easy human inspection of the entire sample. However, by using a second set of reductions not available via the standard SDSS data servers (http://cas.sdss.org), their candidate selection is difficult to reproduce.

For the new catalog, we have chosen to follow an approach similar to that of Eisenstein et al. (2006), but using new outputs now available from the SDSS Catalog Archive Server to replace the second set of data reductions used by Eisenstein et al. (2006). We have coded their main selection criteria in SQL and optionally added constraints requiring the object either to have been targeted for observing based on it being a possible white dwarf star or to have had a reduced spectrum best matched by one of the white dwarf templates now used in the SDSS/SEGUE spectroscopic pipeline. When both sets of criteria were insisted upon, the success rate was very high with nearly 75% of the candidates being well-fit as a white dwarf star by *autofit*. Even more white dwarf stars will eventually be extracted by finishing the human inspections on the uncertain fits. Our candidate selection is still ongoing, but we currently have 3683 white dwarfs stars not included in the DR4 or DR1 catalogs. This sample, however, is based only on DR6 data and does not include any human inspections. Combined with the previous spectroscopic

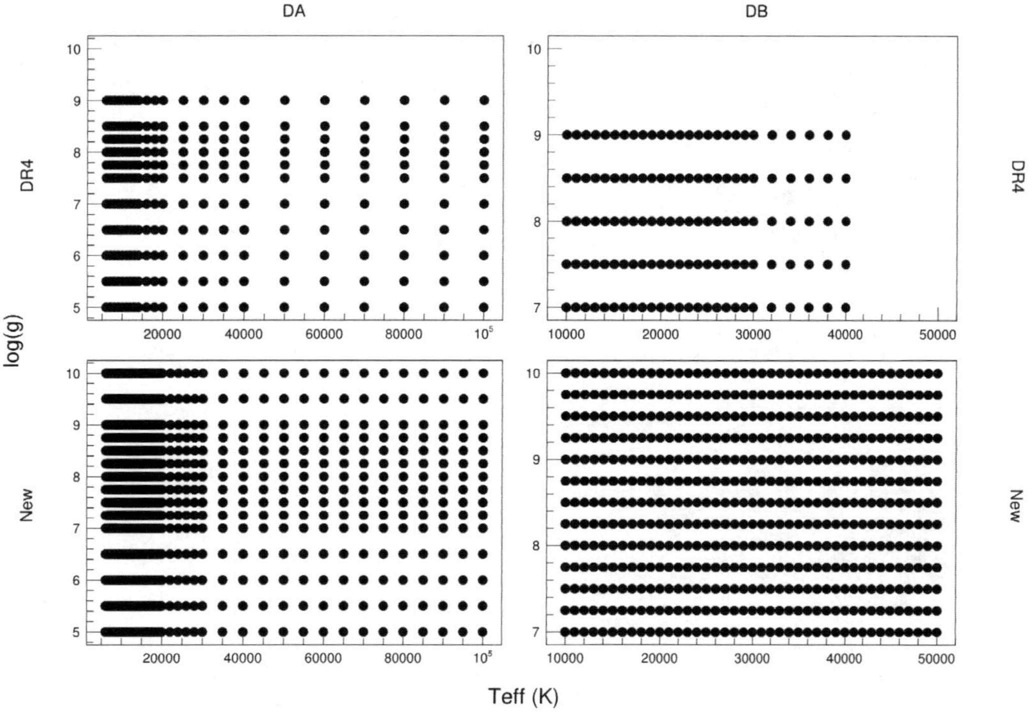

Figure 2. A schematic showing the new vs. the DR4 catalog model grids.

catalogs, we should eventually have of order 15,000 white dwarf stars, with the latest reductions of their spectra fit with the latest version of our fitting code.

3. Spectrophotometry

For the first time since DR2, the SDSS changed its spectroscopic reductions for DR6 and again for DR7. In particular, wavelengths near the H absorption lines now have more reliable spectrophotometry. Using the new DR6 reductions, most hot DAs moved to lower $\log g$ and $T_{\rm eff}$ as seen in Figure 1. This change should go a long way in solving the trend reported in the DR1 and DR4 catalogs of our hot DA fits being systematically hotter than those from other fits of the same objects. The DB fits changed much less systematically, as expected, as a result of these SDSS reduction changes. DR7 reductions have even more improved spectrophotometry corrections included, but we have not yet fit these reductions. We are confident, though, that these reductions will at least partially improve the comparison between our fits and those in the literature for hot DAs (as reported in both the DR1 and DR4 catalogs as well as Kleinman, 2007).

4. New models

Our fitting code, *autofit*, has changed very little since DR4, but we have made many improvements to the model grid it uses. First of all, our grid has expanded. We now have a denser sampling of $\log g$, $T_{\rm eff}$. We have also extended the grid to $\log g = 10.0$ in both the DA and DB grids. Figure 2 illustrates the old DR4 and new catalog grids.

Additional changes have been made to the models themselves. The most significant changes include the consistent use of neutral line broadening in the models. For DBs, the results are

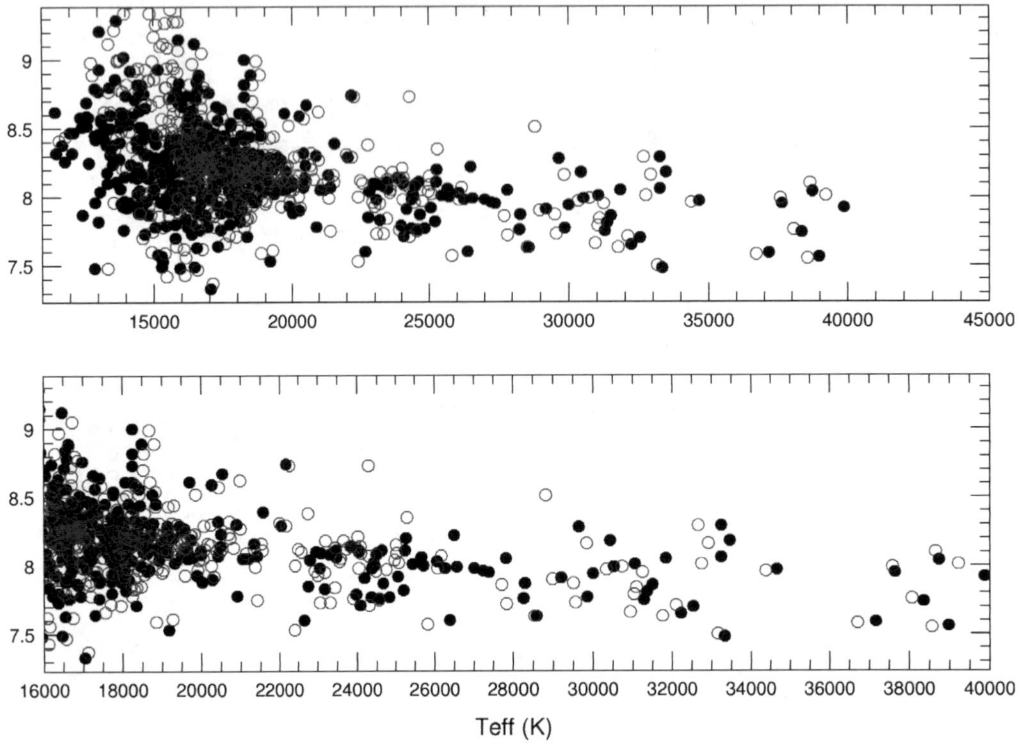

Figure 3. Sample new (solid circles) vs. old (hollow circles) DB fits showing the result of our new DB models. The $\log g$ values below $T_{\rm eff} = 16,000$ K are now much more consistent with those above 16,000 K.

clearly seen at $T_{\rm eff} < 16000$ K, where $\log g$ is now much more consistent across this temperature range, as shown in Figure 3.

We also now have an updated grid of DB models with an ML2/$\alpha = 1.25$ convection treatment which seems to better represent convection in DBs (see for example, Beauchamp et al., 1999 and Montgomery, 2007) than do the previously used ML2/$\alpha = 0.6$ models. We have not yet run our fits with this grid, but will do so for the final catalog.

5. Conclusions

We are still in the process of preparing the latest SDSS white dwarf catalog. It will contain unified fits of all the new and previous SDSS white dwarf stars and use the latest DR7 reductions. Our model grid has improved both in its density and its physics. We will explore fitting the DBs with our new ML2/$\alpha = 1.25$ grid. The improved spectrophotometry has already been shown to help our DA temperature determinations in the high $T_{\rm eff}$ range and the improved model physics should help our fits in the low $T_{\rm eff}$ regime. The new catalog should feature uniform fits on some 15,000 or so spectroscopically-verified white dwarf stars.

Acknowledgments

SJK thanks Daniel Eisenstein for his development of our *autofit* code and Jim Liebert, Hugh Harris, Jurek Krzesinski, Don Winget, and S.O. Kepler for helpful discussions in producing this work and the upcoming catalog.

References

Abazajian K et al. 2008 *ApJ Supp* submitted
Adelman-McCarthy J et al. 2008 *ApJ Supp* **175** 297
Beauchamp A Wesemael F Bergeron P Fontaine G Saffer R A Liebert J & Brassard P 1999 *ApJ* **516** 887
Eisenstein D E et al. 2006 *ApJ Supp* **167** 40
Harris H et al. 2003 *AJ* **126** 1023
Harris H et al. 2006 *AJ* **131** 571
Kleinman S J et al. 2004 *ApJ* **607** 426
Kleinman S J Eisenstein D J Liebert J & Harris H C 2007 in *"Proc. of the 15th European Workshop on White Dwarfs"* ASP Conf. Ser. **372** 121
Montgomery M H 2005 *Communications in Asteroseismology* **150** 253
York D G et al. 2000 *AJ* **120** 1579

Journal of Physics: Conference Series **172** (2009) 012021 doi:10.1088/1742-6596/172/1/012021

Spectroscopic analysis of DA white dwarfs from the McCook & Sion catalog[1]

A Gianninas[1], P Bergeron[1] and M T Ruiz[2]

[1] Département de Physique, Université de Montréal, C.P. 6128, Succ. Centre-Ville, Montréal, Québec H3C 3J7, Canada

[2] Departamento de Astronomía, Universidad de Chile, Casilla 36-D, Santiago, Chile

E-mail:
gianninas@astro.umontreal.ca,bergeron@astro.umontreal.ca,mtruiz@das.uchile.cl

Abstract. For some years now, we have been gathering optical spectra of DA white dwarfs in an effort to study and define the empirical ZZ Ceti instability strip. However, we have recently expanded this survey to include all the DA white dwarfs in the McCook & Sion catalog down to a limiting visual magnitude of V=17.5. We present here a spectroscopic analysis of over 1000 DA white dwarfs from this ongoing survey. We have several specific areas of interest most notably the hot DAO white dwarfs, the ZZ Ceti instability strip, and the DA+dM binary systems. Furthermore, we present a comparison of the ensemble properties of our sample with those of other large surveys of DA white dwarfs, paying particular attention to the distribution of mass as a function of effective temperature.

1. Introduction

Although the Sloan Digital Sky Survey (SDSS) has unearthed thousands of new white dwarfs, it is our belief that there is a significantly brighter sample of white dwarfs whose scientific potential has never truly been exploited. With this in mind, we have undertaken a systematic survey of DA white dwarfs based, in large part, on the last published version of the catalog of spectroscopically identified white dwarfs of McCook & Sion (1999). We will first describe our survey and how it compares to other large surveys of DA white dwarfs and examine some of the ensemble properties of our current sample. We will then take a closer look at a few choice subsamples of objects. In particular, we will look at the DAO white dwarfs, the ZZ Ceti instability strip, and the DA+dM binaries. Finally, we will look at one specific object, CBS 229, which we believe to be a binary system with a magnetic component.

2. Survey Overview

Over the last few years, we have been obtaining optical spectra for DA white dwarfs near the ZZ Ceti instability in an effort to constrain its empirical boundaries. More recently, we expanded this observational survey to include all the DA white dwarfs from the McCook & Sion catalog down to a limiting visual magnitude of V=17.5. Many of these stars have never been analyzed with modern CCD spectroscopy and the only information available is a spectral classification, which

[1] Based on observations made with ESO Telescopes at the La Silla or Paranal Observatories under program ID 078.D-0824(A) and with the Las Campanas 2.5 m Irénée du Pont telescope.

is often erroneous. The bulk of this project has been conducted using Steward Observatory's 2.3 m telescope at Kitt Peak. However, we were also able to obtain time on the ESO 3.6 m telescope at La Silla (Chile) as well as Carnegie Observatories' 2.5 m du Pont telescope at Las Campanas (Chile), allowing us to extend our survey into the southern hemisphere.

How does our work compare to other large surveys of DA white dwarfs? The SPY project (Koester et al. 2001), which obtained high resolution spectra for several hundred white dwarfs, had a limiting magnitude of $V = 16.5$ and was conducted in the southern hemisphere at the VLT. There is actually quite a large overlap with our own sample of stars. Indeed, close to 80% of the stars surveyed in SPY are included in this work. In contrast, we have only a very small overlap with the SDSS (see Figure 1) owing to the fact that the majority of their objects are quite faint due to the nature of the SDSS itself. The faintness of the objects in the SDSS also limits the quality of their spectra since a single exposure time is set for all the objects observed on a given plate. Therefore, the dimmer an object, the lower the signal-to-noise ratio (S/N) of the spectrum. This makes for an extremely inhomogeneous sample of data as far as S/N is concerned. In contrast, we observe one star at a time and adjust our exposure times to obtain spectra with a minimum S/N of approximately 50 (see Fig. 2). Thus our sample, although not as large as the SDSS, comprises data of much higher quality overall. This is key as we have shown in Gianninas, Bergeron & Fontaine (2005) the importance of high S/N for measuring the atmospheric parameters of DA white dwarfs.

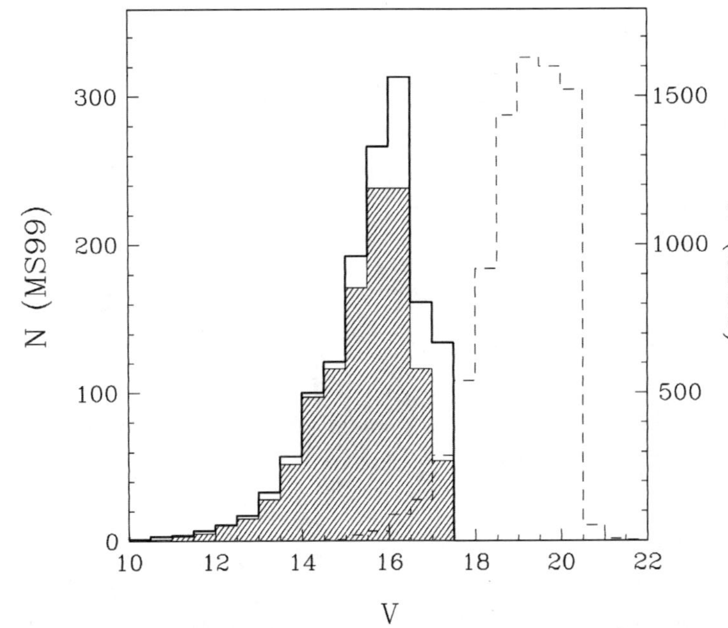

Figure 1. Distribution of visual magnitudes, V, for our sample selected from McCook & Sion (1999; *bold histogram*) and for the white dwarfs we have observed to date (*hatched histogram*). In comparison, the distribution for the SDSS sample as of Data Release 4 (Eisenstein et al. 2006) is also shown (*dashed histogram*). Note that the scale is different for the McCook & Sion sample (*left*) and the SDSS sample (*right*).

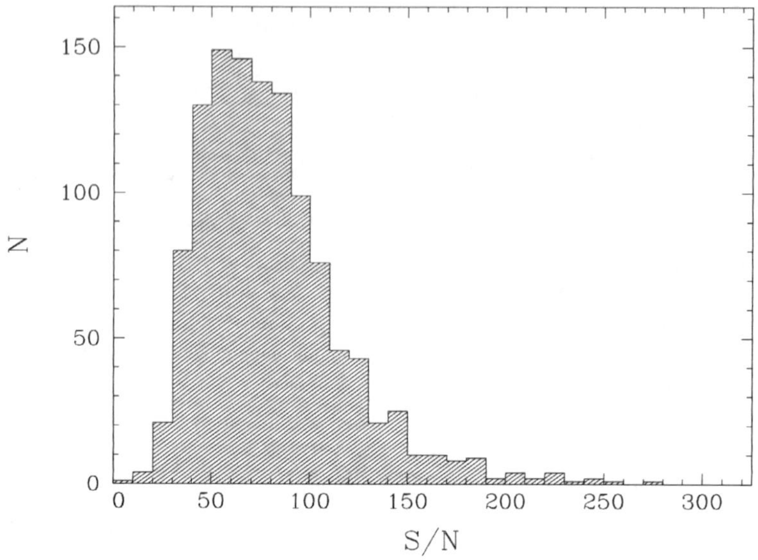

Figure 2. Distribution of S/N for the sample of DA white dwarfs for which we have obtained optical spectra. The majority of the spectra have a S/N of 50 or greater.

3. Mass distribution

We show in Figure 3 the mass distribution of our sample as a function of effective temperature. The atmospheric parameters, $T_{\rm eff}$ and $\log g$, are determined using the spectroscopic technique described in Bergeron, Saffer & Liebert (1992; see also Liebert, Bergeron & Holberg 2005) and the masses are derived from evolutionary models with carbon/oxygen cores and thick hydrogen layers (see references in Liebert et al. 2005). Our mass distribution shows the usual increase in mass at lower temperatures as seen in the PG sample (Liebert et al. 2005), for example. This phenomenon is still not understood although many possible solutions have been proposed over the years (Bergeron, Gianninas & Boudreault 2007). We also notice a certain number of white dwarfs with masses less than $\sim$0.45 $M_\odot$. This population of low mass white dwarfs is necessarily the product of binary evolution as a progenitor with the appropriate mass could not yet have evolved to the white dwarf stage. As such, these objects are important as they represent a separate evolutionary channel for white dwarf stars.

4. DAO white dwarfs

DAO stars are hydrogen atmosphere white dwarfs whose optical spectra also contain lines of ionized helium, usually He II λ4686. The mechanism which maintains the helium in the atmosphere is still unclear, although a weak stellar wind has already been proposed. Figure 4 shows the location of the DAO stars from our sample in the $T_{\rm eff}$-$\log g$ plane along with the regular DA white dwarfs that populate that same region of the diagram. First, we notice that the sequence of DAO stars seems to be best reproduced by the 0.5 $M_\odot$ cooling track in contrast with the DA stars that follow the 0.6 $M_\odot$ cooling track, which is consistent with the accepted mean mass for white dwarfs. This can be explained if we assume that these DAO stars are the product of post-EHB evolution whereby the progenitor was not massive enough to climb back up the asymptotic giant branch (AGB) and evolved to the white dwarf stage directly from the extreme horizontal branch (EHB). We note, however, that this scenario does not apply to all DAO white dwarfs. As shown in Napiwotzki (1999), there is a sequence of DAO stars that are

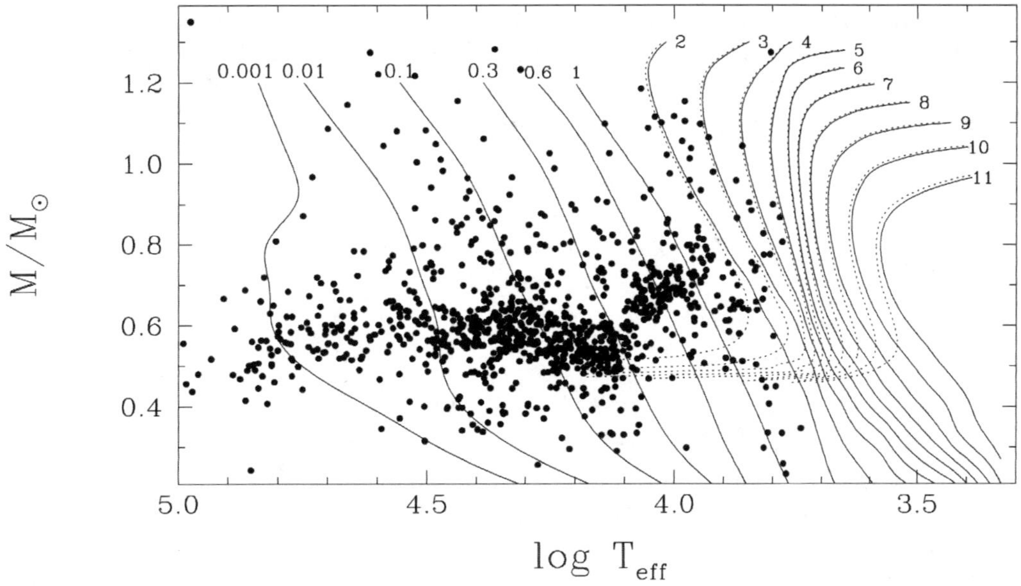

Figure 3. Mass distribution as a function of effective temperature for our entire sample. The solid lines represent isochrones which take into account only the white dwarf cooling time whereas the dotted lines include also the main sequence lifetime. Each isochrone is labeled by its age in Gyr.

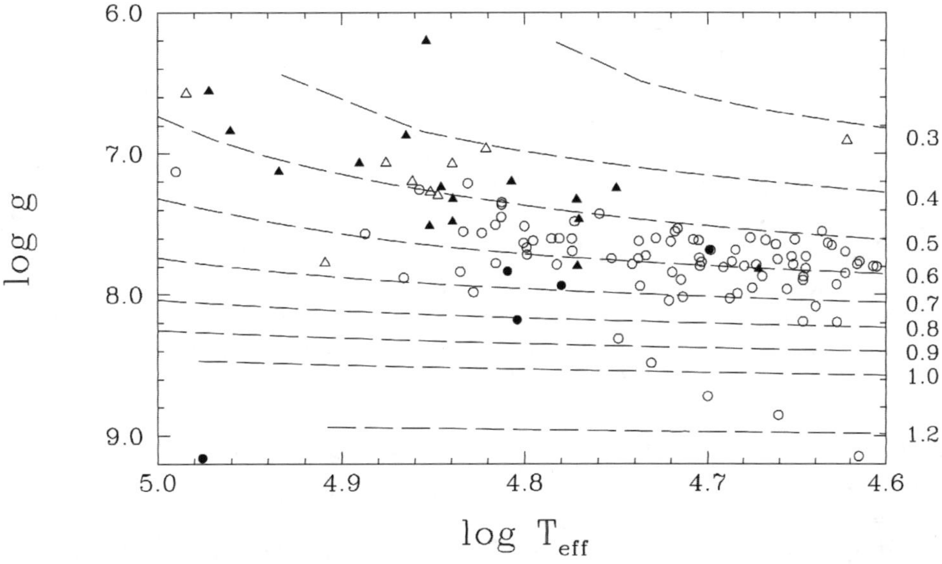

Figure 4. Section of the $\log g$ vs $\log T_{\rm eff}$ diagram showing the hot end of the DA white dwarf cooling sequence. Circles represent DA white dwarfs and triangles correspond to the DAO stars. The filled symbols indicate those stars that exhibit the Balmer line problem. The dashed lines are white dwarf cooling tracks with thick hydrogen layers and masses (in $M_\odot$) indicated to the right of the figure.

consistent with normal post-AGB evolution like the majority of white dwarfs.

We also notice in Figure 4 that most of the DAO stars exhibit the Balmer line problem as first reported by Napiwotzki (1992). The problem manifests itself as an inability to obtain consistent values of $T_{\rm eff}$ and $\log g$ from the spectroscopic fitting technique for the individual Balmer lines. Werner (1999) eventually showed that the problem could be solved by including C, N, and O in the models with proper Stark broadening of the metallic lines. We hope to include this solution within the next generation of our models in order to properly analyze these DAO white dwarfs as well as the other DA stars which exhibit the same phenomenon.

5. The ZZ Ceti instability strip

Figure 5 shows our most up to date vision of the ZZ instability strip. In particular, we have re-observed several of the variables in the strip and with the exception of one pulsator, we have spectra with S/N $\gtrsim$ 70 for all our ZZ Ceti stars. However, one also notices the presence of a photometrically constant star in the middle of the instability strip, HS 1612+5528. This object had been reported as NOV (not observed to vary) by Voss et al. (2006). We obtained our own spectrum of this star and confirmed its position within the instability strip. Consequently, we decided to conduct our own observations to determine whether this star was variable or not. We observed HS 1612+5528 at the Observatoire du mont Mégantic using the Montréal three-channel photometer LAPOUNE for several hours on the night of 2006 July 20 and we detected no variations down to a limit of 0.2%. However, there are known ZZ Ceti stars that pulsate with amplitudes as low as 0.05% so we hope to obtain new high-speed photometry to determine once and for all the status of this object. For the moment, it is the only photometrically constant star contaminating the strip, but it is entirely possible that we, and Voss et al., observed HS 1612+5528 during a period of destructive interference. Alternatively, HS 1612+5528 could also represent a ZZ Ceti star whose pulsations are not detectable due to our line of sight with respect to the pulsation modes of the star.

Finally, we see that our survey as once again uncovered several new objects which lie near or within the empirical boundaries of the strip and we are exploring the possibility of obtaining high-speed photometric measurements for these stars as well.

6. DA+dM binary systems

Although the majority of stars in our survey are isolated DA white dwarfs, there are several objects whose optical spectra contain the tell-tale signs pointing to the presence of a main sequence companion, usually an M dwarf. The spectrum of the M dwarf will contaminate one or several of the Balmer lines from the white dwarf spectrum which renders our normal technique of fitting the observed Balmer line profiles very difficult. To try and get around this problem, we either exclude certain spectral lines or certain portions of the spectral lines from our fitting procedure. However, this means that our determinations of the atmospheric parameters for these stars are quite uncertain. In an effort to determine more accurately the atmospheric parameters of these white dwarfs, we have begun gathering spectra which cover the entire Balmer series from Hα to H8 thus providing sufficient wavelength coverage to determine the spectral type of the M dwarf.

Our analysis will involve a 5-parameter fit to the entire spectrum: the effective temperature and surface gravity of the white dwarf primary, the spectral type of the M dwarf secondary and the relative intensities of the two energy distributions. We will essentially be co-adding our usual synthetic spectra of DA white dwarfs with the M dwarf spectral templates from Bochanski et al. (2007). We will then use the results of the fit to subtract an appropriate template, with the proper flux level, from our initial spectrum. We see in Figure 6 a sample of some of the spectra we have obtained so far. The kink we see in the red wing of Hβ is due to a TiO band. Some spectra also show a MgH band in the blue wing of Hβ as well as Ca I between Hγ and Hδ. We

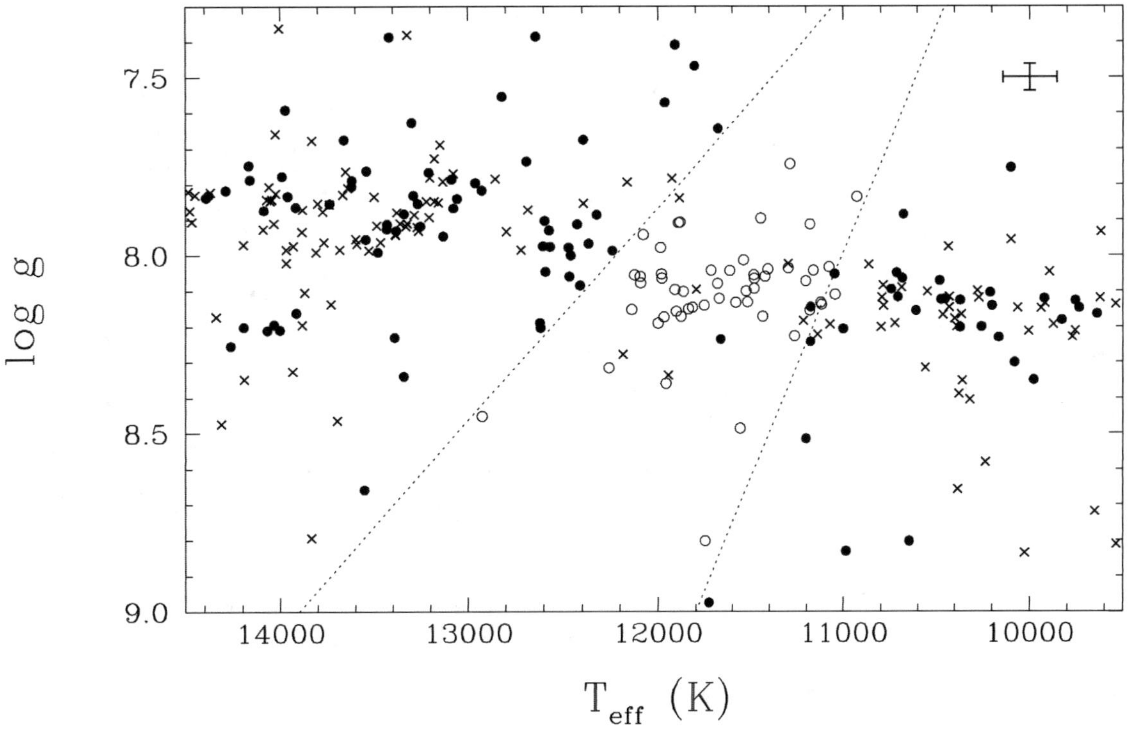

Figure 5. Section of the log g vs $T_{\rm eff}$ diagram showing the ZZ Ceti instability strip. Open circles correspond to ZZ Ceti stars while filled circles represent stars that are photometrically constant. The × symbols are stars from our survey which have never been observed for variability. The dashed lines correspond to the empirical boundaries of the instability strip and the typical error bars in this region of the $T_{\rm eff}$-log g plane are shown in the upper right corner.

also notice the prominent Na D line at 5895 Å. Furthermore, hydrogen-line emission from the M dwarf can contaminate the center of the Balmer lines of the white dwarf.

7. CBS 229

Large surveys often uncover unique and interesting objects. In the course of our survey of the DA+dM binaries described in the previous section, we came across CBS 229, an object which turned out to be rather interesting. We had initially thought that the feature in the red wing of Hβ observed in our blue spectrum (not shown here) was the usual TiO absorption band produced by the presence of an M dwarf. But our full optical spectrum, shown at the top of Figure 7, does not show any signs of an M dwarf companion. We became aware after the fact that CBS 229 had also been observed as part of SDSS, and classified as magnetic with an estimated polar magnetic field of $B_p \sim 20$ MG (Vanlandingham et al. 2005). Indeed, the SDSS spectrum (also displayed in Fig. 7) with a better spectral resolution shows what are clearly magnetic components near Hα. But the observed profile at Hβ, or even at Hα, is really at odds with the predictions of magnetic models. For instance, the models for KPD 0253+5052 shown in Figure 6 of Wickramasinghe & Ferrario (2000), with a comparable magnetic field, predict a much weaker Hα central Zeeman component with respect to the shifted components than observed in CBS 229. Furthermore, the predicted Hβ profile and the higher Balmer lines are totally smeared out, in sharp contrast with the strong and sharp Balmer lines observed in CBS 229.

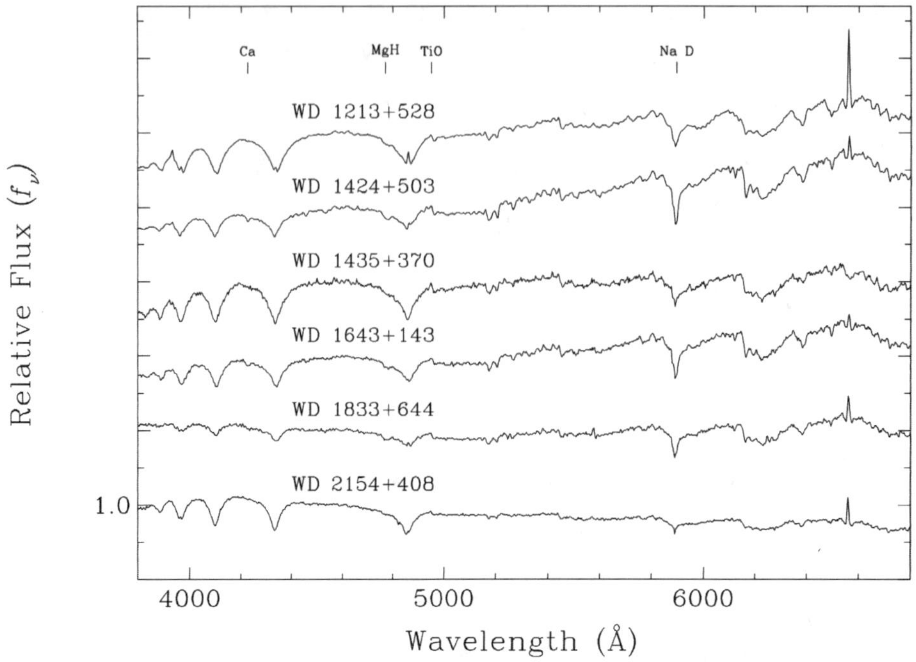

Figure 6. Optical spectra of DA+dM binary systems. Tick marks indicate metallic features in the spectrum of the M dwarf companion's spectrum.

Instead, we suggest that CBS 229 is an unresolved double degenerate binary composed of a magnetic DA star and a normal DA star. We show in Figure 7 a very preliminary attempt to deconvolve the spectrum into its two separate components. We subtracted from the SDSS data a synthetic spectrum with $T_{\rm eff} = 15,000$ K and $\log g = 8.5$. We assumed that the non-magnetic DA contributes 40% of the flux at 5500 Å; this sets the relative intensities of the spectra. The residual spectrum clearly shows magnetic features at Hα and Hβ that bear a strong resemblance with the 17 MG model shown in Figure 6 of Wickramasinghe & Ferrario (2000). To our knowledge, there are not many double degenerate systems with only one magnetic component that have been discovered and analyzed, with the exception of the LB 11146 system reported by Liebert et al. (1993), for which the companion to the non-magnetic DA white dwarf turned out to be a highly magnetic helium-atmosphere white dwarf. We plan a more detailed analysis of CBS 229 that can hopefully shed some more light on the evolution of such unique systems.

8. Future Outlook

We have several upcoming observing runs at Kitt Peak during which we hope to complete our survey of the McCook & Sion catalog in the northern hemisphere. In parallel, we will continue to obtain spectra for the remaining DA+dM binary systems that we have identified in our sample. We are also securing telescope time in Chile to complete our survey in the southern hemisphere. We hope that within a year's time our survey will essentially be complete.

Our analysis of these stars is ongoing on several fronts. We are working on adding the necessary physics in our models to analyze the DAO stars exhibiting the Balmer line problem. Our analysis of the DA+dM binaries, and in particular, our ability to extract the white dwarf spectrum will soon yield results as will our analysis of CBS 229.

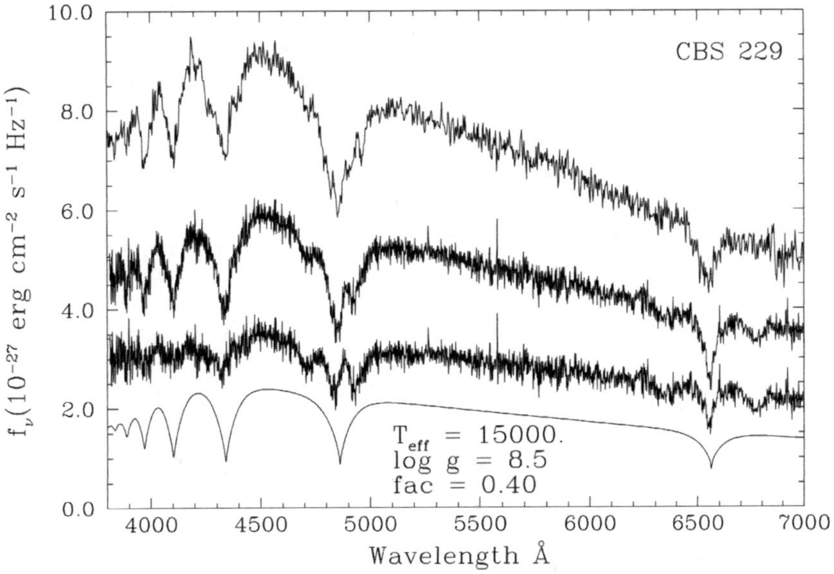

Figure 7. Our own spectrum (first from the top) and the SDSS spectrum (second from the top) of CBS 229. At the bottom, a synthetic spectrum of a DA white dwarf corresponding to the atmospheric parameters indicated in the figure. The quantity $fac = 0.40$ implies that the synthetic spectrum contributes 40% of the flux of the sytem at 5500 Å. The spectrum just above is the residual left over from subtracting the synthetic DA model from the SDSS spectrum.

Acknowledgments

We would like to thank the director and staff of Steward Observatory for the use of their facilities. We would also like to thank G. Fontaine for his assistance in conducting the high-speed photometry and analysis of HS 1612+5528. This work was supported in part by the NSERC Canada and by the Fund FQRNT (Québec). A. G. acknowledges the contribution of the Canadian Graduate Scholarships. P. B. is a Cottrell Scholar of Research Corporation for Science Advancement. M. T. R. received support from FONDAP 15010003 and PFB-06 (CONICYT).

References

Bergeron P Gianninas A and Boudreault S 2007 *ASP Conf. Series* vol 372, eds R Napiwotzki and M R Burleigh (San Francisco: ASP) p 29
Bergeron P Saffer R and Liebert J 1992 *ApJ* **394** 247
Bochanski J J, West A A, Hawley S L and Covey K R 2007 *AJ* **133** 531
Eisenstein et al. 2006 *ApJS* **167** 40
Gianninas A Bergeron P and Fontaine G 2005 *ApJ* **631** 1100
Koester D et al. 2001 *A&A* **378** 556
Liebert J Bergeron P and Holberg J B 2005 *ApJS* **156** 47
Liebert J Bergeron P Schmidt G D and Saffer R A 1993 *ApJ* **418** 426
McCook G P and Sion E M 1999 *ApJS* **121** 1
Napiwotzki R 1992 *Lecture Notes in Physics* vol 401, ed U Heber and C S Jeffery (Berlin: Springer) p 310
Napiwotzki R 1999 *A&A* **350** 101
Vanlandingham K M et al. 2005 *ApJ* **130** 734
Voss B Koester D Ostensen R Kepler S O Napiwotzki R Homeier D and Reimers D 2006 *A&A* **450** 1061
Werner K 1999 *ApJ* **457** L39
Wickramasinghe D T and Ferrario L 2000 *PASP* **112** 873

Journal of Physics: Conference Series **172** (2009) 012022 doi:10.1088/1742-6596/172/1/012022

What Fraction of White Dwarfs are Members of Binary Systems?

J B Holberg[1]

[1] Lunar and Planetary Laboratory, University of Arizona, 1541 E. University Blvd., Sonett Space Sciences Bld., Tucson AZ 85721 USA

holberg@argus.lpl.arizona.edu

Abstract. White dwarfs were originally discovered as the subordinate faint companions of bright nearby stars (i.e. Sirius B and 40 Eri B). Several general categories of binary systems involving white dwarfs are recognized: Sirius-like systems, where the white dwarf may be difficult to detect, binary systems containing white dwarfs and low mass stars, where the white dwarf is often readily discerned; and double degenerate systems. Different modes of white dwarf discovery influence our perception of both the overall binary fraction and the nature of these systems; proper motion surveys emphasize resolved systems, while photometric surveys emphasize unresolved systems containing relatively hot white dwarfs. Recent studies of the local white dwarf population offer some hope of achieving realistic estimates of the relative number of binary systems containing white dwarfs. A sample of 132 white dwarfs within 20 pc indicates that an individual white dwarf has a probability of 32 ± 8% of occurring within a binary or multiple star system.

1. Binary Systems

Binary or multiple star systems containing white dwarfs are important for a number of reasons. For example, in systems such as Sirius and Procyon where accurate binary orbits can be determined it is possible to establish dynamic white dwarf masses which can be compared with the degenerate mass-radius relations. Additionally resolved systems often allow the measurements of gravitational redshifts and in some cases lead to the determination of individual points on the white dwarf initial-final mass relation (Catalán et al. 2007). It is also of interest to understand the frequency of binary systems involving white dwarfs as well as the distribution of companion masses and separations. In close unresolved systems mass transfer may occur which leads to various forms of binary star evolution.

2. The Local Sample

A good place to examine the frequency of white dwarf binary systems is the local sample of degenerate stars within 20 pc of the sun (Holberg et al. 2008). This sample is by definition volume-limited, and can be demonstrated to have a high degree of completeness (~80%). It is relatively unbiased with respect to white dwarf luminosity and most of the sample members have been studied extensively over the years.

Since publication in April of 2008, the number of known white dwarfs within 20 pc has grown from 126 individual degenerates to 132. Four additional individual WDs were discovered by Subasavage et al. (2008) and two Sirius-like Systems: GJ86 (WD0210-508) (Mugrauer and Neuhauser 2005) and HD27442 (WD0416-593) (Mugrauer et al. 2007) were subsequently found in the exoplanet literature, where they turned up as close but resolved companions in near IR images of the primary stars containing extra solar planets. The white dwarf around GJ86 is significant since the distance of the system is 10.8 pc, placing it well within the 13 pc subsample from which the local space density of white dwarfs is established. The revised white dwarf space density, including GJ86B, is now $4.9 \pm 0.5 \times 10^{-3}$ pc^{-3}. The statistics of the local sample reported here are based on the additional stars included as of July 2008.

Of the 132 WDs in the local sample 92 are single stars and the remainder are members of binary or multiple star systems of various descriptions. This is summarized in Table 1.

Table 1. The Local White Dwarf Sample

White Dwarfs	Number	Percent	Comments
Single WDs	92	70	
WD +dM	16	12	
Sirius-Like	10	8	3 contain exoplanets
WD + WD	7	5+5	3 resolved, 4 unresolved
Total WDs	132	100	

Thus, on this basis white dwarfs have a 32 ± 8% probability of occurring in a binary or multiple star system, including double degenerate systems. Note that this should be regarded as a lower limit since the full 20 pc sample is only 80% complete, which implies that possibly 33 ± 4 white dwarfs remain to be discovered and a significant fraction of these will be in binary systems. However, even if 50% of the remaining white dwarfs are ultimately found in binary systems it is difficult to increase the binary fraction above 35%. It is also of interest that three white dwarfs (WD0210-508, WD0416-593 and WD1620-391) are members of Sirius-Like Systems that include extrasolar planets, and one white dwarf (WD2326+049 = G29-38) is encircled by a dusty debris disk. Finally, one system, 40 Eri B, is a member of a hierarchical triple system.

3. Sirius-Like Systems

One interesting and very under represented type of binary system are Sirius-like systems. In these systems the white dwarf orbits a main sequence star of spectral type K or earlier. In close binary systems the distinction between Sirius-like systems and those composed of a white dwarf and an M dwarf is primarily observational. M dwarf systems are much easier to identify since the white dwarf is generally not dominated by the M dwarf companion. One of the large unknowns about Sirius-like systems is their true frequency.

Historically most Sirius-like systems were discovered as resolved proper motion pairs. Later, with the advent of UV spectroscopy, several unresolved systems were discovered through the presence of UV excesses in the spectra of G and K stars. The largest number, however, were discovered as part of the Extreme Ultraviolet All Sky Surveys of the *ROSAT* Wide Field Camera and the *Extreme Ultraviolet Explorer* (Barstow and Holberg 2003). The author maintains a list of some 75 known Sirius-like systems with companions ranging from B5 to K7. The mean distance for these systems is 101 pc. This would seem to indicate that these are relatively rare systems. However, there is strong evidence that they are quite common. In Table 1 the local sample contains eleven Sirius-like systems, all discovered either as common proper motion companions or recently resolved by near IR imaging. Indeed, one such system (LHS 2231 + WD1009-184) was discovered during the analysis of the 20 pc sample (Holberg et al. 2008).

In Table 2 the 11 known Sirius-like systems within 20 pc are listed along with the composition of the system, the angular separation, estimated period and distance. If the 20 pc sample is representative of the larger solar neighbourhood then we might expect a similar number in the distance interval between 20 pc and 25 pc, which has virtually the same volume. In fact there is only one identified system in this interval. The possible addition to this interval is the recent discovery of an unseen radial velocity companion to Regulus (α Leo) a B8IV star at a distance of 24.3 pc. Gies et al. (2008) argue that this companion is a white dwarf. The dearth of Sirius-like systems beyond 20 pc persists at all distances, as shown in Figure 1, where the number of systems for successive shells of equal volume is plotted as a function of distance. Unless the local sample contains an unusual number of Sirius-like systems then the cumulative number for all bins in Figure1 should be on the order of 65.

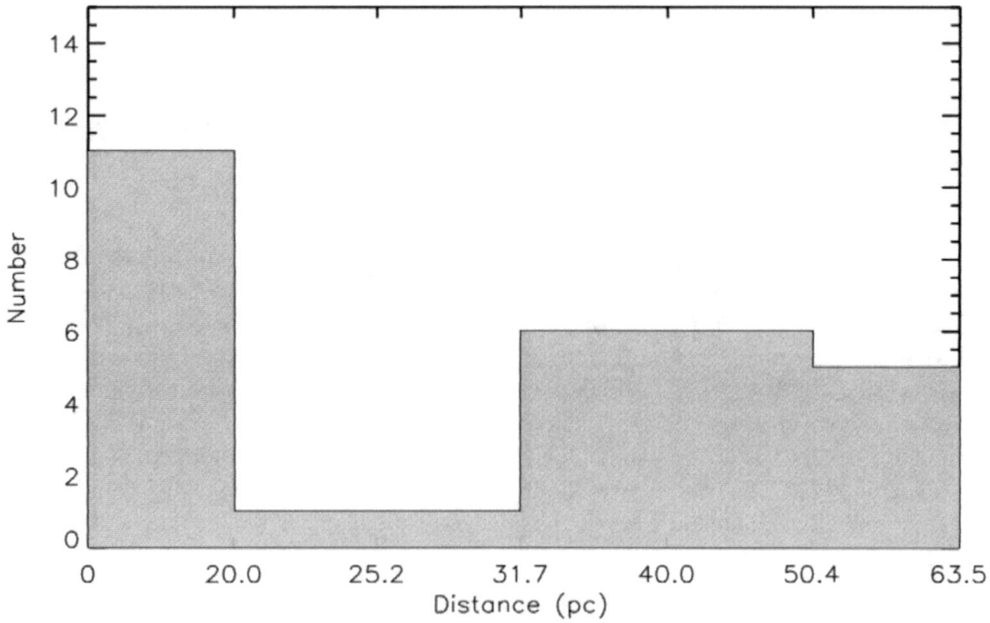

Figure 1 The observed numbers of Sirius-like systems as a function of distance. Each bin represents a shell containing a volume equal to that of the 20 pc sample.

Clearly many relatively nearby Sirius-like systems are being overlooked. Where might they be found? Considering the separations in Table 1, searches for proper motion companions of nearby bright stars could find over 50 % of them. Other potential sources of new systems come from the scrutiny that exoplanet searches are applying to nearby main sequence stars. In terms of radial velocity studies, it is clear that at least three of the systems in Table 2, with orbital periods less than 100 years, could be detected as low level secular drifts in the velocities of the primary star. Another source of potential detections is near IR or adaptive imaging of nearby stars, indeed this is how two such systems were recently discovered. Finally, there may be some systems lurking among the suspected binaries in the *Hipparcos* catalogue. Ultimately it is probably *Gaia* that will find most of the Sirius-like systems, a decade from now, as resolved proper motion pairs or as unresolved proper motion variables.

Table 2. Sirius-Like Systems Within 20 pc

WD Num.	Alt. ID	System	Separation (arc sec)	Period (yrs)	Dist. (pc)
WD0642–166	Sirius B	A0V+DA2	7.5	50.1	2.63
WD0736+053	Procyon B	F5V+DQZ6.5	4.27	40.8	3.51
WD0413–077	40 Eri B	K1V+DA3	83.42	7700	5.04
WD1132–325	VB04B	K0V+DC?	4.94	270	9.56
WD0210–508*	GJ86B	K1V+DA10	1.93	80	10.8
WD1620–391*	CD-38° 10980	G2V+DA2	345.05	$2.3x10^5$	12.87
WD0743–336	VB03 B	G0V+DC9	1351.3	$2.3x10^6$	15.21
WD1544–377	HR 5864 B	G6V+ DA4.7	15.32	3033	15.85
WD1009–184	LHS 2231 B	K7V+DZ7.8	399.96	$5.2x10^5$	17.18
WD0433+270	HD283750	K2V+DA9	124.05	$8.8\ x10^4$	17.85
WD0416–593*	HD 27442 B	K2IV+DA3.8	12.90	3056	18.24

Acknowledgements
The author acknowledges support form NSF Grant AST 0507797.

References

Barstow M A and Holberg J B 2003 in *Extreme Ultraviolet Astronomy*, (Cambridge University Press: Cambridge pp 87-90
Catalán S Isern J García-Berro E Ribas I Allende Prieto C Bonanos A Z 2007 *A&A* **477** 213
Gies R et al. 2008 *ApJ* **682** L117
Holberg J B Sion E M Oswalt T McCook G P Foran S and Subasavage J P 2008 *AJ* **135** 1225
Mugrauer M and Neuhauser R 2005 *MNRAS* **361** L15
Mugrauer M Neuhauser R and Mezeh T 2007 *A&A* **469** 775
Subasavage J P Henry T J Bergeron P Dufour P and Hambly N C 2008 *AJ* **136** 899

Journal of Physics: Conference Series **172** (2009) 012023 doi:10.1088/1742-6596/172/1/012023

Spectral analysis of 636 WD - M star binaries from the Sloan Digital Sky Survey (Data Release 6)

R Heller[1], D Homeier[2], S Dreizler[3] and R Østensen [4]

[1]Hamburger Sternwarte, Gojenbergsweg 112, 21029 Hamburg, Germany
[2],[3]Institut für Astrophysik, Georg-August-Universität Göttingen, Friedrich-Hund-Platz 1. 37077 Göttingen, Germany
[4]Institute of Astronomy, Katholieke Universiteit Leuven, Celestijnenlaan 200D, 3001 Leuven, Belgium

E-mail: [1] rheller@hs.uni-hamburg.de, [2] derek@astro.physik.uni-goettingen.de, [3] dreizler@astro.physik.uni-goettingen.de, [4] roy@ster.kuleuven.be

Abstract. We present a catalog of 857 WD (WD)-M star binaries from the sixth data release of the Sloan Digital Sky Survey (SDSS), most of them known before. For 636 of them, we give a spectral analysis and derive the basic parameters of the stellar constituents and their distances from Earth. Our purpose is to test the modeled spectra applied in the fitting procedure on the one hand and to confine the respective evolutionary scenarios on the other hand. We use a χ^2 minimization technique in order to decompose each combined spectrum and yield independent parameter estimates for the components. 41 of the stellar duets in our spectroscopic sample are optically resolved on the respective SDSS images. Despite various selection effects, the fraction of 6.4 % of WD-M star binaries with orbital separation around 500 AU and orbital perdiods of $\sim 10^4$ yr is a criterion for evolutionary models of stellar binary systems. We find 20 out of 636 WDs being fitted as DOs, with 16 of them indicating $T_{\rm eff}^{\rm WD}$ around 40 000 K. This excess of cool DOs is most likely due to additional WDs in the DB-DO $T_{\rm eff}$ range, for which no detailed fitting was done. Furthermore, we identify 70 very low-mass objects, meaning secondaries with masses smaller than about 0.1 $M_\odot$, as candidate substellar companions. A trend of WD masses towards higher values in a binary constellation compared to those of field WDs is compatible with our results.

1. Introduction

The Sloan Digital Sky Survey (SDSS) offers a considerable improvement of observational contribution to hat subject. Raymond (2003) threw a first glance at WD-M dwarf (dM) pairs in the SDSS and identified 109 of such objects. 99 of them are in our master sample but we discarded 32 of them for our purpose by reasons given at the end of Sect. 2. Other cataloges and analyses on WDs and WD-main-sequence (MS) binaries have been presented by (Greenstein 1986, Luyten 1997, Luyten 1999, Wachter et al. 2003, Kleinman et al. 2004, Eisenstein et al. 2006, Hügelmeyer 2006, Silvestri 2007, Rebassa-Mansergas et al. 2007). 153 systems from our sample were not included in any of the mentioned studies.

The present work introduces a novel method of analyzing WD-dM binaries by simultaneously fitting model spectra for both the WD and the MS star to the composite spectrum to derive the atmospheric parameters of both stars. Unlike the aforementioned studies, we thus obviate the need to calibrate empirical relations between physical parameters and spectral types. Although current cool star atmosphere models may still have some shortcomings in fitting dM spectra, this approach avoids possible biases due to systematic differences between single dMs and those in close binaries.

2. Observations

A detailed desription of the SDSS' instrumental setup can be found in Adelman-McCarthy et al. (2008). The spectroscopic dataset used for our study consists of 857 objects that were labeled as WD-M dwarf binaries during a search for hot subdwarfs in the SDSS dataset. Our selection procedure resembles that of the SDSS' own in that we investigated a large fraction of the color-color space in the EDR release, and refined our search area in the following releases. Of particular importance for the WD-M sample is the cutoff in $g - r$. For the EDR we imposed no cutoff, but for later releases we imposed $g - r < 0.2$ in addition to the $u - g < 0.8$ criterion used to select stars with UV excess. A more detailed explanation of the color-color loci of WD-M star systems within the SDSS is given in Smolčić et al. (2004).

Our sample is made up of 857 objects: this is our master sample. From that, we discard noisy spectra and those, in which either the WD or the red companion shows rather weak features. Only spectra with either clear DA or DO features are respected. Furthermore, we reject spectra with pollution of nearby light sources, which we verified with the aid of the photometric data. WD-M pairs with cool WDs and dominant red components are preferentially dismissed from our sample by the selection method – those systems with significant light contribution due to mass overflow, too. The buildup of that subsample of 636 systems is consequently subject to severe selection effects.

We assume a mass-radius relation for unevolved main sequence stars by using the evolutionary models of Chabrier & Baraffe (1997) for a fixed M star age of 10^{10} yr to deduce the radius ($R_{\rm M}$) and mass ($M_{\rm M}$) of the M star from its fitted effective temperature ($T_{\rm eff}^{\rm M}$) and metallicity $[{\rm Fe/H}]_{\rm M}$. For the mass range considered here, evolution effects on the main sequence are negligible within a Hubble time, thus only pre-main sequence stars with ages $\lesssim 5 \cdot 10^8$ yr might introduce larger errors in the radius estimate. The errors in surface gravity of the WD ($g_{\rm WD}$) from our fits are in most cases inadequately high for a meaningful deduction of the WD radii. But, as shown by Hu, Wu & Wu (2007), the major part of the SDSS DAs has masses clustering closely around the peak of the field WD mass distribution at $0.58\,M_\odot$. There are observational indications for higher WD masses in close binaries compared to the masses of field WDs (Ramsay 2000, Ritter & Kolb 1998, Cropper, Ramsay & Wu 1998). We therefore use a fixed WD mass ($M_{\rm WD}$) of $0.6\,M_\odot$ together with the WD temperature ($T_{\rm eff}^{\rm WD}$) from the fit to derive the radius of the WD ($R_{\rm WD}$) from the evolutionary tracks calculated by Wood (1994). These models are available for 0.4, 0.5, 0.6, 0.7, 0.8 and $1.0\,M_\odot$.

WD-dM alignment by chance is found to be negligible. The probability $P_{\rm M}$ to find at least one M star within a circle with the $3'$ diameter of the SDSS fiber on the celestial plane by chance is given by the Poisson probability

$$P_{\rm M}(\nu \geq 1) = 1 - e^{-\mu} \quad , \qquad \mu = A \cdot \rho_{\rm M} \tag{1}$$

where $A = \pi\,(1.5')^2$ is the probed area and $\rho_{\rm M}$ is the area density of M stars on the celestial plane. To estimate $\rho_{\rm M}$, we refer to private communication with J. Bochanski who measured the field luminosity function of stars in the DR6. We get $\rho_{\rm M} = 9.55 \cdot 10^4\,(')^{-2}$ and $P_{\rm M} = 6.73 \cdot 10^{-3}$. The expectation value is $P_{\rm M} \cdot 636 \approx 4.3$ for the number of WD-M binaries aligned by chance within our sample.

3. Models

Models for the full parameter range of interest were pre-computed, providing two grids for the spaces of WD parameters $T_{\rm eff}^{\rm WD}$, $g_{\rm WD}$ and M dwarf parameters $T_{\rm eff}^{\rm M}$, $g_{\rm M}$ and $[{\rm Fe/H}]_{\rm M}$, respectively. White dwarf spectra have been calculated for pure hydrogen atmospheres covering surface gravities of $7 \leq \log(g_{\rm WD}) \leq 9$ with a step size of 0.5 dex in a temperature range from 6 to 90 kK. We had 1 kK steps between 6 and 30 kK, 2 kK steps between 30 and 50 kK and 5 kK steps for temperatures above 50 kK. This core model grid was supplemented with existing models for extremely hot ($T_{\rm eff}^{\rm WD} > 90$ kK) or lower-gravity ($\log(g_{\rm WD}) \in \{5, 6\}$) atmospheres and for hot helium-rich atmospheres. The latter were available for $\log(g_{\rm WD}) = 7.5$ in the range of $40\,{\rm kK} \leq T_{\rm eff}^{\rm WD} \leq 80$ kK every 5 kK. The grid of the MS spectra was almost complete for $2\,600\,{\rm K} \leq T_{\rm eff}^{\rm M}$ every 200 K, $[{\rm Fe/H}]_{\rm M} \in \{-1.5, -1, -0.5, 0, 0.3\}$ and $\log(g_{\rm WD}) \in \{2, 3, 4, 4.5, 5, 5.5\}$.

The WD models have been computed by D. Koester using his codes developed for static, plane-parallel stellar atmospheres in radiative, hydrostatic and local thermodynamic equilibrium (LTE), as described in detail in Finley, Koester & Basri (1997). The main body of this grid covers fully-blanketed pure hydrogen (DA) atmospheres. These models take convective flux into account according to the mixing length approximation (MLT), using a variation of the standard formulation of Mihalas (1978) designated as ML2 in the notation of Finley, Koester & Basri (1997) and references therein, and adopting a mixing length of $\alpha = 0.6$ (in units of pressure scale height). This setup has been demonstrated to be one of the best available configurations for reproducing DA spectra at high resolution and S/N (Homeier et al. 1998, Koester et al. 2001). This grid was extended by an existing set of hydrogen atmosphere models based on the same input physics to cover the more extreme parts of the WD parameter space in high $T_{\rm eff}^{\rm WD}$ and lower gravity (sdB-like atmospheres). In addition for those hot WDs that were not adequately reproduced by DA spectra the sequence of helium-rich (DO) models was used. These models assume a helium-to-hydrogen mixing ratio of 100:1 and the ML1 MLT version with $\alpha = 1.0$ as described in Jordan et al. (1997). These latter models only cover the spectrum up to $\lambda \leq 8\,000$ Å, and thus an important part of the dM flux had to be masked out in the analysis. However among our final results none of the primaries ended up in the low-gravity or ultra-hot domain of the DAs/sdBs; and a quantitative analysis of the helium-rich WDs was beyond the scope of this work, as detailed below.

The secondary spectra were calculated with version 14.2 of the multi-purpose stellar atmosphere code `PHOENIX` (Hauschildt & Baron 1999) for 1D spherically symmetric, static atmospheres in LTE, also considering convective instability in the framework of MLT according to Mihalas (1978) with a mixing length parameter $\alpha = 2.0$. Our models follow the general setup used for the first GAIA grid (Brott & Hauschildt 2005), which includes a number of updates from the NextGen grid (Hauschildt et al. 1999) and the microphysics described by (Allard et al. 2001), but ignores effects of condensate formation. We have therefore restricted the range of validity of these models to $T_{\rm eff} \geq 2800$ K.

4. Mathematical Treatment

With our χ^2 fitting method we move within a 5D parameter space, spanned by $T_{\rm eff}^{\rm WD}$, $T_{\rm eff}^{\rm M}$, $g_{\rm WD}$, $g_{\rm M}$ and $[{\rm Fe/H}]_{\rm M}$. Each observed flux data point $F_i^{\rm obs}$ in a binary spectrum with a total number of m observed data points shall be reproduced by a combination of two single-star models x_i and y_i. These have to be weighted with scaling factors a and b respectively, depending on the stars' distances. χ^2 is defined a

$$\chi^2 := \sum_i^m \frac{(F_i^{\rm obs} - F_i^{\rm mod})^2}{\sigma_i^2} = \sum_i^m \frac{(F_i^{\rm obs} - a\,x_i - b\,y_i)^2}{\sigma_i^2} =: [(F_i^{\rm obs} - a\,x_i - b\,y_i)^2] \tag{2}$$

with σ_i as the observational error given in the SDSS .fits file of the object. We use $[w_i]$ as an abbreviation for $\sum_i^m w_i/\sigma_i^2$. With

$$\nabla\,\chi^2(a,b) = -2\left([x_i\,(F_i^{\rm obs} - a\,x_i - b\,y_i)]\,,\,[y_i\,(F_i^{\rm obs} - a\,x_i - b\,y_i)]\right) \stackrel{!}{=} 0 \tag{3}$$

we obtain

$$\begin{pmatrix}[F_i^{\rm obs}\,x_i]\\ [F_i^{\rm obs}\,y_i]\end{pmatrix} = \underbrace{\begin{pmatrix}[x_i^2]\,[x_i\,y_i]\\ [y_i\,x_i]\,[y_i^2]\end{pmatrix}}_{Z}\begin{pmatrix}a\\ b\end{pmatrix} \quad , \tag{4}$$

which is equivalent to

$$\begin{pmatrix}a\\ b\end{pmatrix} = Z^{-1}\begin{pmatrix}[F_i^{\rm obs}\,x_i]\\ [F_i^{\rm obs}\,y_i]\end{pmatrix} = \begin{pmatrix}[F_i^{\rm obs}\,x_i]\,[y_i^2] - [x_i\,y_i]\,[F_i^{\rm obs}\,y_i]\\ [F_i^{\rm obs}\,y_i]\,[x_i^2] - [y_i\,x_i]\,[F_i^{\rm obs}\,x_i]\end{pmatrix}\frac{1}{[x_i^2]\,[y_i^2] - [x_i\,y_i]^2} \quad . \tag{5}$$

(5) gives us the scaling factors for both the WD and the M star model, which we then use to compute χ^2 with (2). Using that procedure, we avoid a mutual dependence of a and b since the system of equations is uniquely solvable; we also avoid the possibility of running into a local χ^2 minimum.

4.1. Distances

Once R_{WD} is drawn from the fixed M_{WD}, the fitted $T_{\mathrm{eff}}^{\mathrm{WD}}$ and using models from (Wood 1994), the only thing we still need to obtain the distance d_{WD} of the WD from Earth is the scaling factor $a = (R_{\mathrm{WD}}/d_{\mathrm{WD}})^2$ from our fitting method which scales the model flux down to the observed flux (and analog for the dM). To make the estimates of the respective distances comparable among all the spectra, we introduce the coefficient

$$C := \frac{\frac{\sqrt{2}}{2}\,(d_{\mathrm{WD}} - d_{\mathrm{M}})}{\frac{d_{\mathrm{WD}}+d_{\mathrm{M}}}{2}} = \sqrt{2}\,\frac{d_{\mathrm{WD}} - d_{\mathrm{M}}}{d_{\mathrm{WD}} + d_{\mathrm{M}}} \tag{6}$$

which weighs the displacement of a certain binary system from the diagonal in the $(d_{\mathrm{WD}} - d_{\mathrm{M}})$ plane with the average distance $\overline{d} = (d_{\mathrm{WD}} + d_{\mathrm{M}})/2$ of the system from Earth.

We use the averaged distance $\overline{d}$ of the system from Earth, the displacement D of the stars on the SDSS images in units of pixels and the given resolution ϱ of the images in $''$/pixel to calculate the angular separation α in units of radian via $\alpha = \varrho\, D/3600 \cdot \pi/180$ and finally, simply using geometry, the projected mutual distance d_{proj}

$$d_{\mathrm{proj}} = 2\,\overline{d}\,\tan\left(\frac{\alpha}{2}\right) \quad . \tag{7}$$

We may benefit from that visible separation and estimate the period of the system via

$$P \gtrsim 2\pi\,\sqrt{\frac{d_{\mathrm{proj}}^3}{G\,(M_{\mathrm{WD}} + M_{\mathrm{M}})}} \quad . \tag{8}$$

5. Results

The temperature function for our WDs can be seen in Fig. 1. Since most of the WDs in our sample have a surface gravity between 7 and 8 dex, the shape of the WD-distance function in Fig. 2 is quite similar to that of the temperature function.

The standard deviations for the deduced parameters are quite weak in terms of physical significance, which shows that the systematic errors dominate the mathematical ones. For a conservative estimate of the errors resulting from the the incomplete molecular data for the M star models, SDSS flux calibration errors and interstellar reddening, we refer to Hügelmeyer (2006) and assume an uncertainty of 2 000 K for the WDs with temperatures smaller than 50 000 K and a 1-σ interval of 100 K for the MS stars. For $T_{\mathrm{eff}}^{\mathrm{WD}} > 50\,000$ K the absolute value of our accuracy is given by the half of the model step width of 5 000 K. Our accuracy in the surface gravities is limited by the low resolution of the SDSS spectra and the step size of our model grid and thus we have $\sigma_{\log(g_{\mathrm{WD}})} \approx 0.5$ dex $\approx \sigma_{\log(g_{\mathrm{M}})}$. The metallicity determination is merely accurate to about 0.3 dex.

5.1. White Dwarfs Showing He Lines

616 spectra from our original input sample can be fitted well using a DA primary, but among the remaining systems, a small number shows He features in the primary spectrum. These tend to be found at relatively high $T_{\mathrm{eff}}^{\mathrm{WD}}$, since spectra showing clear DB features have been efficiently removed in the initial selection. Among the remainder, which are mostly dominated by the He II Bracket-equivalent series, we found 20 to be better fitted by spectra from our DO model library. These stars cluster predominantly at the cool end of the DO sequence: 16 of 20 are fitted with the lowest effective temperature in the DO grid

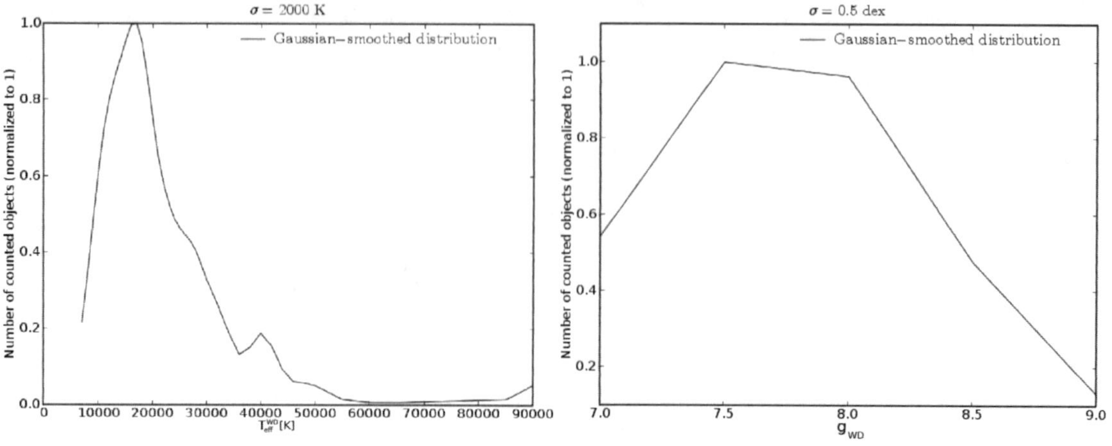

Figure 1. *Left:* Temperature function for the WDs; While the major peak at 17 000 K belongs to the DAs, the bump at 40 000 K is due to the preferential selection of cool DOs for our sample. This plot looks very similar to that shown in Silvestri et al. (2006), except for the DO feature and that it is smoothed. *Right:* Distribution of the WD surface gravities.

of 40 000 K. Many of the latter show stronger He I lines in addition to He II, and would thus probably be better classified as DBO according to the nomenclature of Wesemael et al. (1993). We suspect that this distribution does not reflect the true luminosity function of Helium-rich WDs, but is rather biased by a contribution of WDs below the T_{eff} limit of our DO models.

From the 167 spectra with H emission features, we substract 8 known PCEB candidates and 4 active CVs, which yields 155 WD-M star binaries, corresponding to a fraction of 24.4 %, that probably harbor an active M dwarf. This value is conform with the fraction of 24.4 % for active field M dwarfs in the SDSS as found by West et al. (2004). However, Silvestri et al. (2006) found a significantly larger fraction of active M dwarfs in WD-M binaries. Considering only unresolved binaries, as they did, our fraction of binary systems with H_α emission that probably comes from the M component even decreases to 23.2 % and exacerbates the discord between our results and theirs.

5.2. Optically Resolved Binaries

Of our 636 binaries, 41 are chosen for follow-up studies because, firstly, the red star is found to be located within an area around the WD that was covered by the SDSS fiber and, secondly, these stars have a separation wide enough to distinguish between the two components. For photometric binaries with a separation clearly larger than the SDSS fiber radius of 1.5′ we act on the assumption that the M star on the SDSS image is not the one represented in the spectrum. The respective fitting results are marked and have to be taken with a pinch of salt since we may not state clearly if a significant fraction of the light contribution from the red star was caught by the fiber anyway, despite the separation larger than 1.5′. The results for the resolved binaries can be found in the supplementary data to this acticle. The typical projected distances are large: of the order of some hundred AU with a mean value of roughly 650 AU. The widest separation we find is 1 700 AU for J1006+5633, where the real spatial separation should be even larger. The orbital periods we derive are typically in the order of some 10^4 yr with J1006+5633 indicating a period of $\gtrsim$ 70 000 yr.

6. Discussion

The M star temperatures we fitted are in good accordance with the spectral types derived by Rebassa-Mansergas et al. (2007). In Fig. 3 we compare our fits for $T_{\text{eff}}^{\text{M}}$ with their spectral types for 41 objects (left panel) and with the spectral classification from (Silvestri et al. 2006) (right panel) for 446 objects,

Journal of Physics: Conference Series **172** (2009) 012023 doi:10.1088/1742-6596/172/1/012023

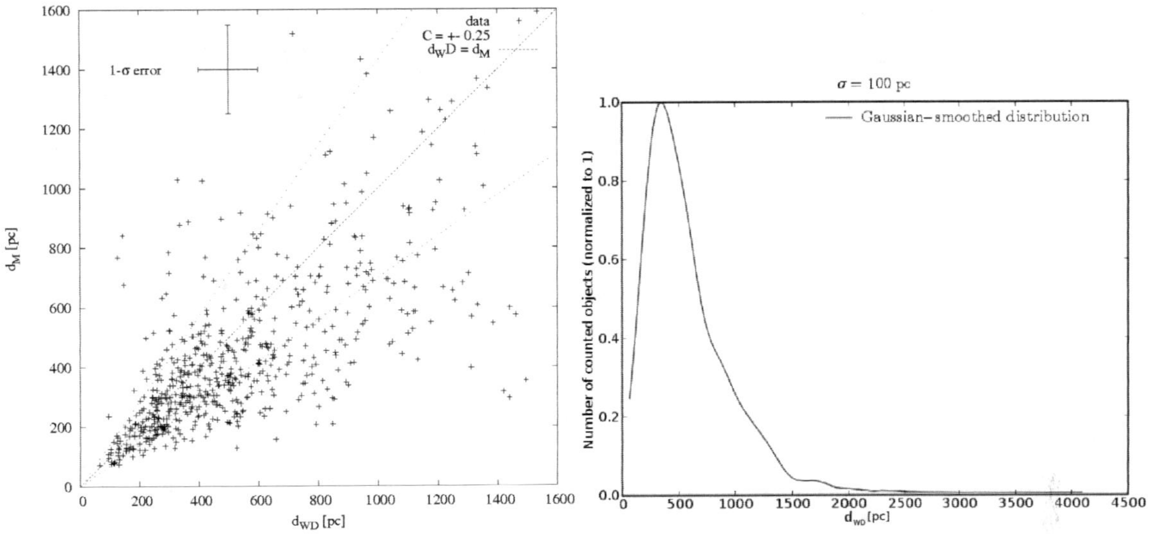

Figure 2. *Left*: Distribution of the derived distances of the stellar components of our systems; The diagonal is the ideal for physical binaries. The dotted curves span a tolerance fan for $C = 0.25$. *Right*: The distribution of the WD distances in our sample is a consequence of the restricted magnitude range of the SDSS. It has its maximum at 354 pc.

which were both in ours and their sample. The authors of the latter publication derived their dM types on the basis of template spectra and color indices as described in Hawley et al. (2002). In both comparisons the maximum is reached at (3 200 K , M4) and the plot shows a monotone decrease of the spectral type with increasing temperature. These counts probably do not mirror the true $T_{\rm eff}^{\rm M}$ or spectral type function due to selection effects like the growing number of M dwarfs towards later types on the one hand and decreasing visibility of the secondary component on the other hand.
In our study, we see a trend towards $d_{\rm M}$ being smaller than $d_{\rm WD}$, and we consider three possible effects:

i. Reiners (2005) has shown the absorption in the TiO ϵ-band to be systematically underestimated by `PHOENIX` M star spectra. Since this band system as a primary temperature indicator deepens with decreasing $T_{\rm eff}^{\rm M}$, model spectra fits to this feature risk to systematically underrate $T_{\rm eff}^{\rm M}$, too. This effect, if present in our models as well, would cause $d_{\rm M}$ to be systematically underestimated.

ii. As mentioned in Sect. 2, there is observational evidence that the masses of WDs in close binaries cluster rather between $0.7\,M_\odot$ and $0.8\,M_\odot$ than around $0.6\,M_\odot$ as for field WDs. Only 12 out of the 636 objects presented here are known to be close systems, i.e. CVs or PCEBs, while 41 are pretty widely separated as inferred from the optical detachment. We cannot give an assessment of the remaining orbits from our data and we may not say precisely if the WD masses are rather close to 0.7 or even $0.8\,M_\odot$ but a potential trend towards higher WD masses would be compatible with the bias towards $d_{\rm WD} > d_{\rm M}$. An increase in mass leads to a decrease in radius for a WD and consequently a smaller distance to Earth is required in order to reproduce the observed flux. Our possible underestimation of $M_{\rm WD}$ could thus contribute to a systematic overestimation of $d_{\rm WD}$.

iii. The M star radii that we deduced from the Chabrier & Baraffe (1997) model tracks, based on the effective temperature and the metallicity from our spectral fits, are systematically underestimated for low-mass objects ($M_{\rm M} \lesssim 0.3\,M_\odot$, see Fig. 2). This general discrepancy between theory and observations is known in the community of low-mass stars science and emerged, hitherto, particularly in observations of eclipsing binary systems with a low-mass component (Ribas et al. 2007). With our – under statistical aspects – large sample we support this claim.

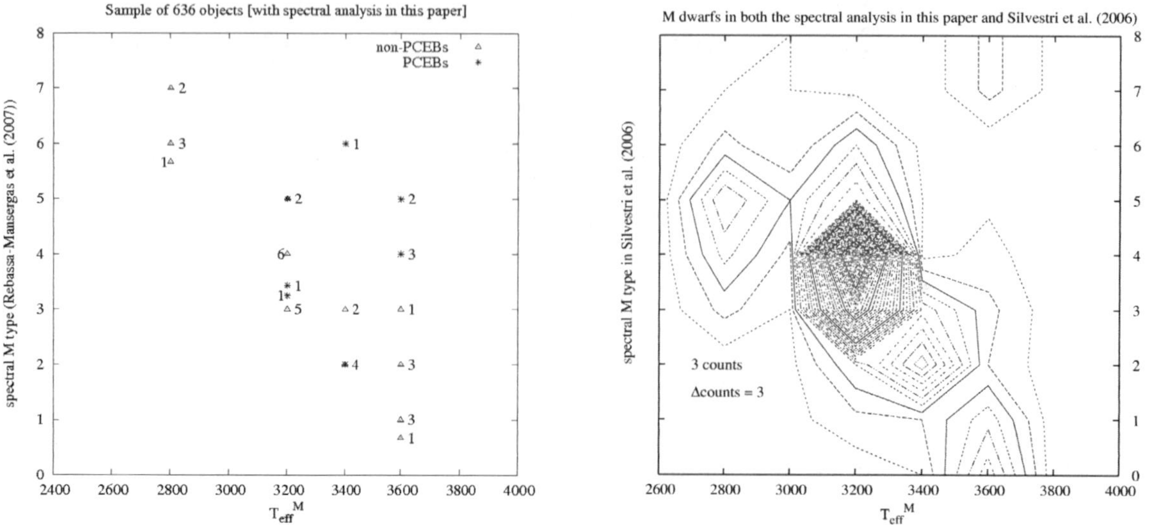

Figure 3. *Left*: Comparison of the spectral types from Rebassa-Mansergas et al. (2007) with our $T_{\rm eff}^{\rm M}$ for 41 M stars, that match both their sample and ours; 11 of them are marked as PCEBs. Labels indicate the counts per grid point. *Right*: Comparison of the spectral types given in Silvestri et al. (2006) with our $T_{\rm eff}^{\rm M}$ for 446 M stars, that match both their sample and ours; The number of dMs per grid point is projected onto the ($T_{\rm M}^{\rm eff}$-SpT) plane. The outer contour marks the path of 3 counts, each subsequent contour symbolizes an increase of 3 counts. The maximum is at (3 200 K , M4) with 115 counts. Their average spectral mismatch was ±1, while our 1-σ accuracy for $T_{\rm eff}^{\rm M}$ is 100 K.

There is a relatively large number of widely-separated binaries with orbital distances $\gtrsim$ 250 AU and up to 1 700 AU: 41/636 = 6.4 %. This is, however, not a stable assessment; we see five effects that may smear the optical-binary fraction to either higher or lower values.

i. If two objects are so close together that they appear as one elongated object with off colors, the SDSS classification procedure may preselect it as a possible galaxy. That would mean that we have underestimated the true optical-binary fraction.

ii. A small number of physical pairs aligned along the line of sight could also drive the true optical-binary fraction to a higher value. We would have missed them just because of their unfavorable geometrical constellation.

iii. Some of the stellar duets on the SDSS images with angular distances > 1.5′ could also be physical pairs (but the secondary on the image would not be the one represented in the respective SDSS spectrum).

iv. A significant contingent of binaries with mass overflow from the MS companion to the WD are supposed to be located out of the color-color region that we studied in order to make up our sample. These stellar duets are close systems with no optical separation on the SDSS images, which pushes the optical-binary fraction to lower values.

v. The statistical considerations presented in Sect. 2 show that about 4 of the binaries in our sample should be aligned by chance – without a common evolutionary background. Probably none, or at most one, of the optical binaries is one of these mavericks.

Asked if stellar duets with such large orbital separations are stable over long timescales, we refer to the paper of Weinberg et al. (1987) and Fig. 2 therein. Their calculations, including both stars passing by and encounters with subclumps within giant molecular clouds, show that binaries with a total mass of

1 $M_\odot$ and initial orbital separations around 650 AU have a typical lifetime of more than the age of the Universe.

References

Adelman-McCarthy J K *et al.* 2008 *ApJs* **175** 297–313 (*Preprint* arXiv:0707.3413)
Allard F Hauschildt P H Alexander D R Tamanai A and Schweitzer A 2001 *ApJ* **556** 357–372 (*Preprint* arXiv:astro-ph/0104256)
Brott I and Hauschildt P H 2005 *ESA SP-576: The Three-Dimensional Universe with Gaia* pp 565–+
Cropper M Ramsay G and Wu K 1998 *MNRAS* **293** 222–+
Chabrier G and Baraffe I 1997 *A&A* **327** 1039–1053 (*Preprint* arXiv:astro-ph/9704118)
Eisenstein D J *et al.* 2006 *ApJs* **167** 40–58 (*Preprint* arXiv:astro-ph/0606700)
Finley D S Koester D and Basri G 1997 *ApJ* **488** 375–+
Greenstein J L 1986 *AJ* **92** 867–+
Hauschildt P H Allard F Ferguson J Baron E and Alexander D R 1999 *ApJ* **525** 871–880 (*Preprint* arXiv:astro-ph/9907194)
Hauschildt P H and Baron E 1999 *Journal of Computational and Applied Mathematics* **102** 41–63
Hawley S L *et al.* 2002 *AJ* **123** 3409–3427 (*Preprint* arXiv:astro-ph/0204065)
Homeier D Koester D Hagen H J Jordan S Heber U Engels D Reimers D and Dreizler S 1998 *A&A* **338** 563–575
Hügelmeyer S D Dreizler S Homeier D Krzesiński J Werner K Nitta A and Kleinman S J 2006 *A&A* **454** 617–624 (*Preprint* arXiv:astro-ph/0605551)
Hu Q Wu C and Wu X B 2007 *A&A* **466** 627–639 (*Preprint* arXiv:astro-ph/0701197)
Jordan S Napiwotzki R Koester D and Rauch T 1997 *A&A* **318** 461–465 (*Preprint* arXiv:astro-ph/9605007)
Kleinman S J *et al.* 2004 *ApJ* **607** 426–444 (*Preprint* arXiv:astro-ph/0402209)
Koester D *et al.* 2001 *A&A* **378** 556–568
Luyten W J 1997 *VizieR Online Data Catalog* **1130** 0–+
Luyten W J 1999 *VizieR Online Data Catalog* **3070** 0–+
Mihalas D 1978 *Stellar atmospheres /2nd edition/* (San Francisco, W H Freeman and Co., 1978 650 p.)
Ramsay G 2000 *MNRAS* **314** 403–408 (*Preprint* arXiv:astro-ph/9912420)
Raymond S N *et al.* 2003 *AJ* **125** 2621–2629 (*Preprint* arXiv:astro-ph/0302405)
Rebassa-Mansergas A Gänsicke B T Rodríguez-Gil P Schreiber M R and Koester D 2007 *MNRAS* **382** 1377–1393 (*Preprint* arXiv:0707.4107)
Reiners A 2005 *Astronomische Nachrichten* **326** 930–933 (*Preprint* arXiv:astro-ph/0510502)
Ribas I Morales J Jordi C Baraffe I Chabrier G and Gallardo J 2007 *ArXiv e-prints* **711** (*Preprint* 0711.4451)
Ritter H and Kolb U 1998 *A&As* **129** 83–85
Silvestri N M *et al.* 2007 *AJ* **134** 741–748 (*Preprint* arXiv:0704.0789)
Silvestri N M *et al.* 2006 *AJ* **131** 1674–1686
Smolčić V *et al.* 2004 *ApJl* **615** L141–L144 (*Preprint* arXiv:astro-ph/0403218)
Wachter S Hoard D W Hansen K H Wilcox R E Taylor H M and Finkelstein S L 2003 *ApJ* **586** 1356–1363 (*Preprint* arXiv:astro-ph/0212174)
Weinberg M D Shapiro S L and Wasserman I 1987 *ApJ* **312** 367–389
Wesemael F Greenstein J L Liebert J Lamontagne R Fontaine G Bergeron P and Glaspey J W 1993 *PASP* **105** 761–778
West A A *et al.* 2004 *AJ* **128** 426–436 (*Preprint* arXiv:astro-ph/0403486)
Wood M 1994 *IAU Colloq. 147: The Equation of State in Astrophysics* pp 612–+

Journal of Physics: Conference Series **172** (2009) 012024 doi:10.1088/1742-6596/172/1/012024

White dwarf post common envelope binaries from the SDSS

M R Schreiber[1], B T Gaensicke[2], M Zorotovic[3,4], A Rebassa-Mansergas[1,2], A Nebot Gomez-Moran[5], J Southworth[2], A D Schwope [5], S Pyrzas[2], C Tappert [4] and L Schmidtobreick[3]

[1] Departamento de Física y Astronomía, Facultad de Ciencias, Universidad de Valparaíso, Valparaíso, Chile

[2] Department of Physics, University of Warwick, Coventry CV4 9BU, UK

[3] European Southern Observatory, Casilla 19001, Santiago 19, Chile), AB(European Southern Observatory, Casilla 19001, Santiago 19, Chile

[4] Departamento de Astronomia y Astrofisica, Pontificia Universidad Catolica, Casilla 306, Santiago 22, Chile

[5] Astrophysikalisches Institut Potsdam, An der Sternwarte 16, D-14482 Potsdam, Germany

E-mail: matthias@dfa.uv.cl

Abstract. Post common envelope binaries (PCEBs) consisting of a white dwarf and a main sequence star are ideal systems to calibrate current theories of angular momentum loss in close compact binary stars. The potential held by PCEBs for further development of close binary evolution could so far not be exploited due to the small number of known systems and the inhomogeneity of the sample. The Sloan Digital Sky Survey is changing this scene dramatically, as it is very efficient in identifying white dwarf/main sequence (WDMS) binaries, including both wide systems whose stellar components evolve like single stars and - more interesting in the context of close binary evolution - PCEBs. We pursue a large-scale follow-up survey to identify and characterise the PCEBs among the WDMS binaries that have been found with SDSS. We here present preliminary results of our survey and derive first constraints on current models of white dwarf binary evolution.

1. Introduction

Close white dwarf binary stars span a wide range of interesting objects and are important in many astrophysical contexts. Type Ia supernovae, white dwarfs in close binaries driven over their Chandrasehkar limit, are routinely used as “standard candles” to probe the nature of dark energy. AM CVn stars, very short orbital period binaries with a helium rich donor star, are expected to significantly contribute to the gravitational wave background which will be probed by missions such as LISA (Laser Interferometer Space Antenna). Furthermore, Cataclysmic Variables (CVs), close binaries in which a white dwarf accretes gas from its Roche-lobe filling companion, are excellent laboratories to study the physics of accretion.

The evolution of initially wide main sequence binaries with long orbital periods (>years) into short orbital period (<days) binaries containing white dwarfs depends primarily on the rate at which angular momentum is extracted from the binary orbit. If we wish to understand the populations and relative numbers of close white dwarf binaries in general, we need to understand the angular momentum loss mechanisms driving their formation and evolution. It is thought

that a significant fraction of the angular momentum is extracted from the binary orbit during the short-lived ($< 1 - 100$ years) but crucial common envelope phase (Webbink 2007): once the progenitor of the white dwarf leaves the main sequence, dynamically unstable mass transfer may lead to the formation of a gaseous envelope around the entire binary. Friction within this envelope is significantly reducing the binary separation as the core of the giant and the main sequence star spiral in towards the centre of mass. Once the common envelope has been ejected, the evolution of the remaining post common envelope binary towards shorter orbital periods is driven by orbital angular momentum loss through gravitational radiation and – much more efficient – magnetic wind braking.

The general concepts of both, the common envelope phase as well as magnetic braking in close binaries, have been developed more than 25 years ago. Meanwhile, quite detailed binary population synthesis (BPS) studies have been carried out and are a commonly adopted method to predict the outcome of binary star evolution, and to investigate e.g. the nature of SN Ia progenitors, or the properties of the galactic white dwarf plus main sequence (WDMS) binary population. However, in current BPS models the crucial common envelope phase is usually simply approximated by a parameterized energy equation and the efficiency to "use" the orbital energy to expel the envelope is very uncertain (see Willems & Kolb 2004 for a recent review). Similarly, most of the results obtained by BPS calculations sensitively depend on the prescriptions used for magnetic braking. Present models of magnetic braking differ by up to two orders of magnitude, and, even worse, it is not clear whether magnetic braking is present in all PCEBs, or whether it breaks down in systems where the secondary star is fully convective. In order to explain the orbital period gap observed in the period distribution of cataclysmic variables, one in fact needs to assume the latter "disrupted magnetic braking" (Rappaport et al. 1983), while observations of single low mass stars do not show any evidence for such a discontinuity (Sills et al. 2000). To sum up, the diagnostic potential and predictive power of BPS studies is severely limited by the lack of understanding the detailed physics of the common envelope phase and subsequent orbital angular momentum loss.

On the other hand, it is fair to say that there is also a lack of observational tests on the outcome of close binary evolution, that could be provided by measuring e.g. the fraction of binaries that underwent a common envelope, or the orbital period distribution of post common envelope (PCEB) binaries. We here review the first results obtained from a large observational project especially designed to fill this gap and to provide the much needed observational constraints on models of compact binary formation and evolution. The article is organized as follows. In Section 2 and 3 we review in more detail our current understanding of angular momentum loss in close white dwarf binaries and emphasise the potential of observational population studies to constrain angular momentum loss. In Section 4 we describe our survey and present preliminary survey results in Section 5 and 6.

2. The common envelope efficiency

Since Paczyński (1976) outlined the first ideas of common envelope evolution, it is thought that this phase of binary evolution is producing virtually all the known types of close white dwarf binaries. Common envelope evolution results from dynamically unstable mass transfer from the evolving more massive star to the less massive main sequence star (Paczyński 1976; Webbink 1984; Hjellming 1989). This situation occurs especially if the evolving more massive star fills its Roche-lobe when having a deep convective envelope (usually on the giant or asymptotic giant branch). Then, as a response to mass transfer, the radius of the mass donor may increase while its Roche-radius is decreasing. The resulting runaway mass transfer drives the mass gainer out of thermal equilibrium as it accretes on a time scale faster than its thermal time scale. Consequently, the lower mass star also expands until filling its Roche-lobe, which then leads to a common envelope configuration.

In current binary population models the energy equation relating the total energy of the binary before and after the common envelope phase is scaled with two uncertain dimensionless parameters, called the structural binding energy parameter (λ) and the common envelope efficiency (α). Both parameters can be combined to obtain one uncertain parameter $\alpha\,\lambda$:

$$\frac{GM_gM_e}{R_g} = \lambda\alpha\left(\frac{GM_cM_2}{2a_f} - \frac{GM_gM_2}{2a_i}\right). \tag{1}$$

Here, M_g, M_c, R_g, M_e, and M_2 are the masses and radii of the giant, its core, its envelope, and the accreting secondary star respectively, while a_f and a_i are the final and initial binary separation.

Nelemans et al. (2000) and Nelemans & Tout (2005) developed an algorithm to reconstruct the common envelope phase for observed white dwarf binaries. They derive the possible masses and radii of the progenitors of the white dwarfs in these binaries from fits to detailed stellar evolution models Hurley et al. (2000). This information can then be used to reconstruct the mass-transfer phase in which the white dwarf was formed. However, to constrain the common envelope efficiency a much larger and representative sample of white dwarf post common envelope binaries than the one available to Nnelemans & Tout (2005) needs to be analyzed.

3. Disrupted magnetic braking

Once the common envelope has been ejected, the evolution of the post common envelope binary is driven by angular momentum loss through gravitational radiation and magnetic wind braking. While the first is well understood, the latter represents an additional big uncertainty in white dwarf binary formation.

Based on Skumanich (1972) who analyzed the spin down rates of G-type stars, Verbunt & Zwaan (1981) derived a prescription for angular momentum loss due to magnetic braking of the form $\dot{J} \propto {P_{\mathrm{orb}}}^{-3}$. In order to explain the orbital period gap observed in the period distribution of cataclysmic variables (CVs), Rappaport et al. (1983) proposed the disrupted magnetic braking scenario by assuming that magnetic braking ceases when the secondary star becomes fully convective at $M_{\mathrm{sec}} \sim 0.3\mathrm{M}_{\odot}$ (which corresponds to $P_{\mathrm{orb}} \sim 3\mathrm{h}$). The disrupted magnetic braking idea is based on the result that the transition region between radiative core and convective envelope plays a crucial role for the stellar dynamo, i.e. for producing stellar magnetic fields (McDermott & Taam 1989). As the donor stars in CVs above the gap are driven out of thermal equilibrium by strong mass transfer (caused by efficient magnetic braking), disruption of magnetic braking for fully convective secondary stars can indeed cause the donor star to relax to thermal equilibrium, to shrink within its Roche-lobe, and to become a detached white dwarf main/sequence binary.

The most important observational support for the disrupted magnetic braking hypothesis has been mentioned already, the famous orbital period gap, i.e. the significant deficit of CVs in the orbital period range of $2-3\,\mathrm{h}$. Additional support comes from the fact that (a) the current mass transfer rates derived from observations of CVs above the gap are significantly higher than those of CVs below the gap, (b) the mean accretion rates derived from accretion induced compressional heating are systematically higher above than below the gap (Townsley & Bildsten 2003; Townsley & Gaensicke 2008), and (c) the donor stars in CVs above the gap seem to be slightly expanded compared to main sequence stars which is consistent with the donor stars driven out of thermal equilibrium (Patterson et al. 2005).

In spite of this success, there also exist serious doubts on the reliability of the disrupted magnetic braking scenario. Most importantly, observations of the spin down rates of single stars do drastically disagree with both the strength of magnetic braking according to Verbunt & Zwaan (1981) as well as the disrupted magnetic braking scenario proposed by Rappaport et al. (1983). From observations of young open clusters, Sills et al. (2000) derived slow spin down rates

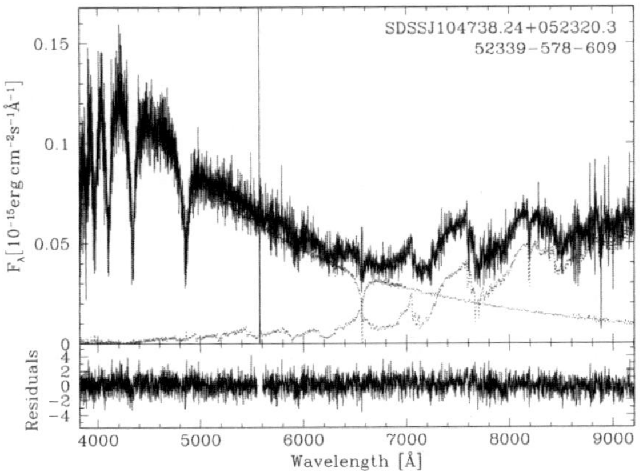

Figure 1. Spectral decomposition of the WDMS binary SDSSJ104738.24+052320.3.

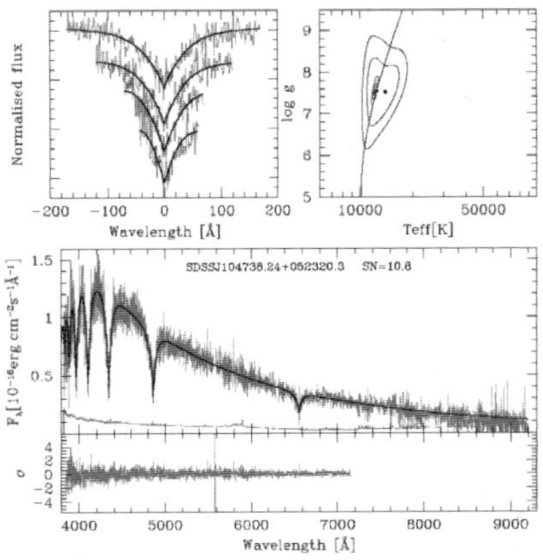

Figure 2. Spectral model fits to the white dwarf in SDSSJ104738.24+052320.3. For details see e.g. Schreiber et al. (2008)

and did not find any signs for a discontinuity in the braking law of rapidly rotating low-mass stars. Recently, Briggs et al. (2007) measured the X-ray activity of secondary stars in PCEBs and find that there seems to be no discontinuity in the magnetic activity of the secondary stars at the fully convective boundary.

Apparently, independent tests of disrupted magnetic braking are required. Fortunately, Schreiber & Gaensicke (2003) and, more recently, Politano & Weiler (2006) and Davis et al. (2008) demonstrated that observational population studies of post common envelope binaries consisting of a white dwarf and a main sequence star can provide such tests of disrupted magnetic braking. However, performing these tests requires immense observational efforts in order to establish a large and unbiased sample of PCEBs.

4. WDMS binaries from the SDSS

Schreiber & Gaensicke (2003) carried out a detailed theoretical study of the 30 well-observed PCEBs known at that time and were the first to realize the full potential of observational population studies of PCEBs in the context of close compact binary evolution. However, their sample was not only small but also heavily biased. Hence, no clear constraints on white dwarf binary evolution and formation could be derived.

Since the SDSS turned out to be extremely efficient in identifying detached WDMS binaries, this situation changed drastically. The published DR5 SDSS-WDMS sample (Silvestri et al. 2007) consists of more than 1200 mostly young systems containing hot white dwarfs. We find $\sim$ 1600 WDMS in DR6 and add $\sim$ 300 systems containing mostly cold white dwarfs from a dedicated survey of WDMS that has been implemented in SEGUE (see also Schreiber et al. 2007). The population of these $\sim$ 1900 WDMS binaries consists of (a) wide binary systems whose stellar components never interact and evolve like single stars and (b) systems with short orbital periods that went through a common envelope phase – post common envelope binaries (PCEBs). Given the large number of WDMS binaries identified, the number of SDSS PCEBs among them should be sufficient to provide clear new constraints on white dwarf binary formation.

In 2006 we started an extended survey performing follow-up observations at several 4-8m telescopes to identify and characterize the PCEBs among the SDSS WDMS binaries. The project follows a three-step approach. First we derive the stellar parameters of the systems (i.e. the white dwarf mass, the white dwarf temperature, the secondary spectral type, and the distance) using

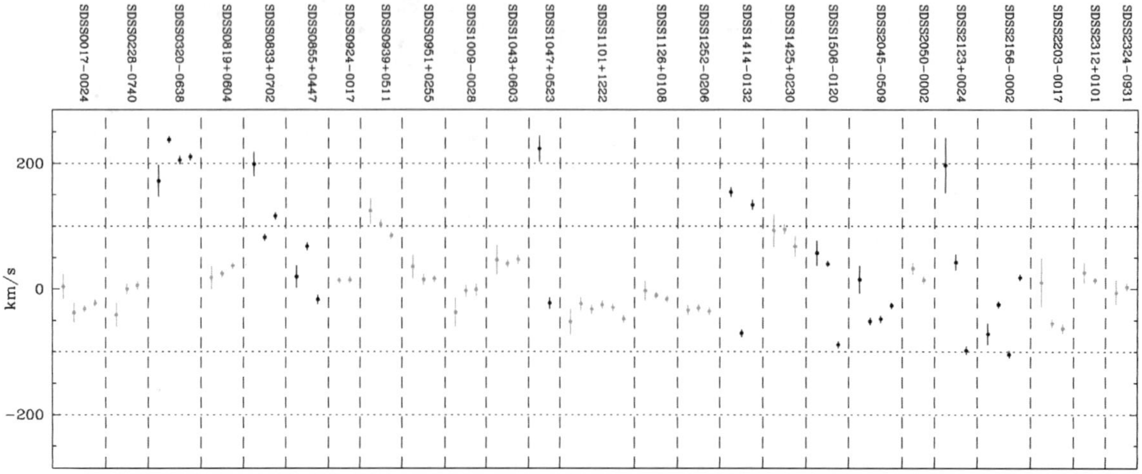

Figure 3. Radial velocity measurements using VLT/FORS2 for 25 white dwarf/main sequence binaries. The systems in black show 3σ radial velocity variations and we consider them to be strong post common envelope binary candidates.

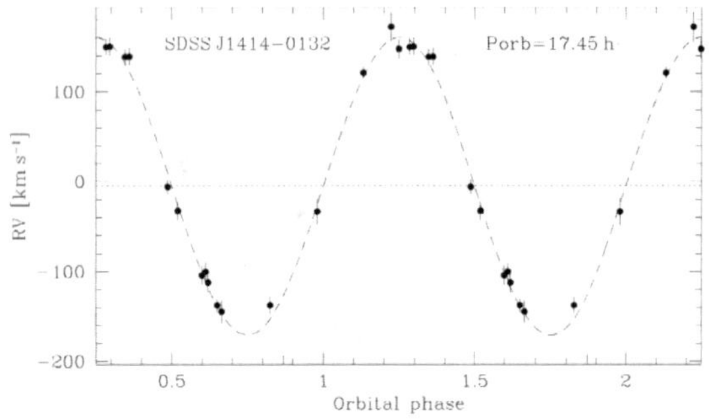

Figure 4. Radial velocity curve of the secondary star in SDSSJ141451.73-013242.7. The orbital period is 17.45 hr.

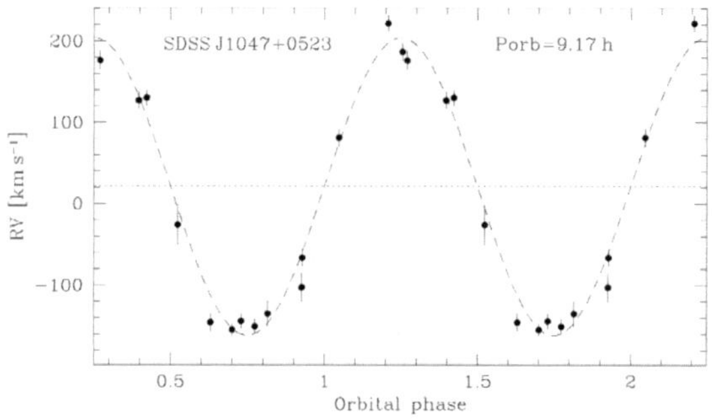

Figure 5. Radial velocity curve of the secondary star in SDSSJ104738.24+052320.3. The orbital period is 9.17 hr.

a spectral decomposition method as shown in Fig. 1 and 2 and as explained in detail in Rebassa-Mansergas et al. (2007, 2008). The second step is represented by the identification of PCEBs as close binaries from at least two spectra taken in different nights. Strong radial velocity variations unambiguously identify the PCEBs among the WDMS population and qualify the system for stage three, i.e. the determination of the orbital period from time-series spectroscopy. We obtained at least two spectra of 143 WDMS and identified 55 PCEBs due to the detection of significant radial velocity variations. It appears that the fraction of PCEBs among the SDSS WDMS is about 35%, i.e. slightly higher than predicted by theoretical population synthesis studies. As an example, Fig. 3 shows radial velocity measurements of 25 WDMS with VLT/FORS2 (see also Schreiber et al. 2008). Once a strong PCEB candidate has been identified, we determine the orbital period of the binary by measuring the radial velocity curve of the secondary star. Two examples are given in Fig. 4 and Fig. 5. By the time of writing we have characterized a sample of 43 PCEBs with measured orbital period this way. Our final goal is to establish a sample of $\sim$ 200 PCEBs until 2010/2011.

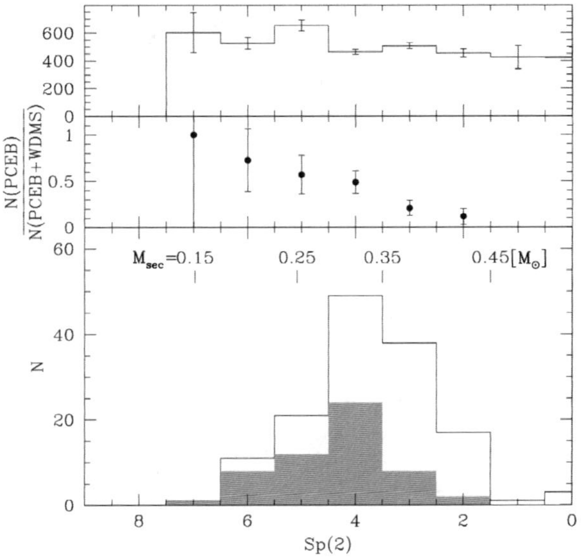

Figure 6. Summary of our results of WDMS radial velocity surveys carried out at the ESO/VLT and the WHT over the past semesters. The bottom panel shows the distribution of systems with radial velocity information in white, and those identified as PCEBs in gray. The middle panel gives the PCEB-fraction which shows an increase at $\sim 0.35 M_\odot$ as predicted by disrupted magnetic braking. The top panel shows the mean distance to the observed stars.

5. Testing disrupted magnetic braking

As mentioned previously, the disrupted magnetic braking model postulates that magnetic braking vanishes for fully convective secondary stars. As shown by Politano & Weiler (2006), disrupted magnetic braking implies much shorter evolutionary time scales for non-fully convective secondary stars than for PCEBs having a fully convective companion. Hence, the fraction of PCEBs among WDMS binaries should be significantly higher for the latter. The measurements described in the previous section allow to test this prediction directly. Figure 6 shows the fraction of PCEBs among WDMS binaries as a function of spectral type. Indeed, there seems to be a steep increase around the fully convective boundary, in perfect agreement with the predictions of disrupted magnetic braking. This finding can be considered as a strong argument in favour of disrupted magnetic braking. However, the error bars in Fig. 6 are still rather large, especially for low mass secondary stars, and further measurements are required to derive definite conclusions.

6. Reconstructing the history of the binaries

In Rebassa-Mansergas et al. (2008) we discussed a preliminary orbital period distribution and showed that the measured orbital period distribution seems to imply a low value of α when compared with those predicted by binary population synthesis models. However, the orbital period distributions predicted by BPS models depend not only on the common envelope efficiency but also on the assumed initial mass function, the initial distribution of binary separations, and the modelling of mass transfer.

Here we follow a different approach. We reconstruct the history of the observed population of PCEBs and derive all possible initial parameters for a given system. First, we follow Schreiber & Gaensicke (2003) and reconstruct the post common envelope evolution of the PCEBs for a given angular momentum loss prescription (we here assume classical disrupted magnetic braking according to Rappaport et al. 1983). This requires knowing the cooling ages of the white dwarfs which we derive using the cooling tracks of Wood (1995). As most of the observed PCEBs are relatively young and as most of our PCEBs have low mass secondary stars where gravitational radiation is assumed to be the only angular momentum loss mechanism, the reconstructed zero-age post-CE orbital period distribution is not dramatically different to the observed distribution but peaks at a slightly longer period ($\log(P_{\rm orb}[days]) \sim 0.5$). Using the calculated orbital periods the PCEBs had at the end of the CE phase we can now reconstruct the CE phase following Nelemans & Tout (2005). We assume that the white dwarf mass is equal to the core mass of the giant at the onset of mass transfer and that the mass of the secondary star does not change during common envelope evolution. This allows to reconstruct all possible histories for a given PCEB. Hereby each assumed progenitor mass corresponds to one value of $\alpha\lambda$. The derived values of $\alpha\lambda$ are given in Fig. 7. The different vertical lines for a given system correspond to the uncertainty in the derived white dwarf masses. Apparently, all our systems have solutions

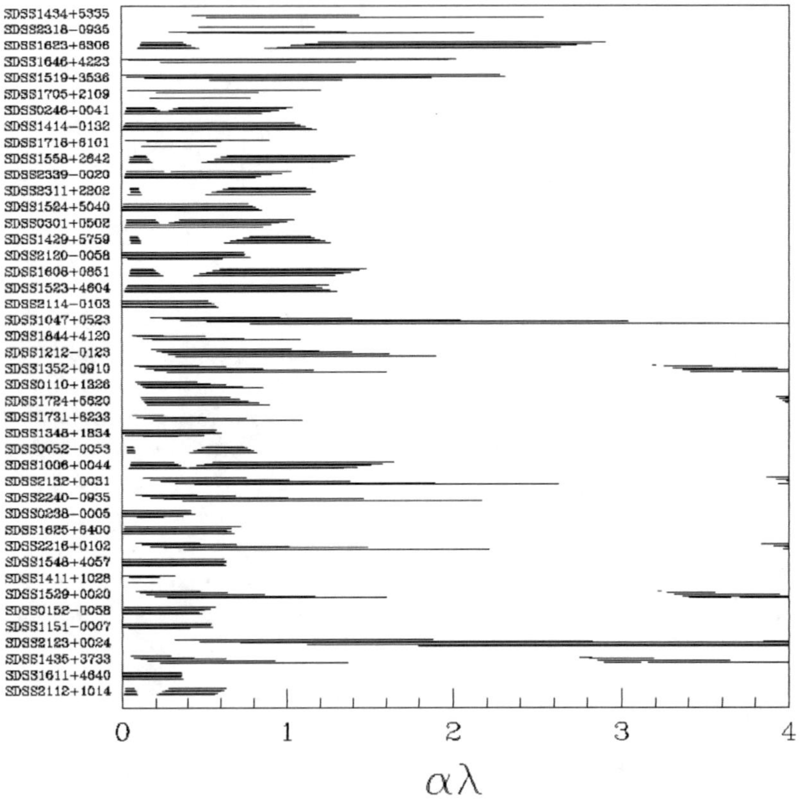

Figure 7. Reconstruction of the common envelope phase of SDSS PCEBs. Assuming the core of the progenitor of the WD to be equal to the white dwarf mass and assuming that the mass of the secondary does not change during the CE, we get one value of $\alpha\lambda$ for a given mass of the Roche-lobe filling giant. Solutions with progenitors on the giant branch are in black while solutions in blue and green correspond to progenitors on the early or late AGB.

for small values of $\alpha\lambda$. This indicates that the efficiency of using orbital energy to expel the envelope is rather small. In other words, if there exits a universal value of $\alpha\lambda$ it should be $\lesssim$0.5. However, it has been shown that the binding energy parameter λ sensitively depends on the evolutionary stage of the giant when the binary enters the common envelope phase. The envelopes of giants on the asymptotic giant branch are less bound than those of stars on the first giant branch. This effect has to be included before final conclusions on the common envelope efficiency can be drawn.

7. Summary

The currently available large sample of white dwarf/main sequence (WDMS) binaries provided by the SDSS offer a unique possibility to progress with our understanding of close compact binary evolution. We have performed a large follow-up study of these WDMS which is especially designed to identify a large sample of white dwarf post common envelope binaries. So far we identified 55 PCEBs and measured the orbital period of 43 systems. The preliminary results of the project can be summarized as follows.

- The fraction of PCEBs among WDMS binaries seems to depend sensitively on the mass of the secondary star. While the fraction of PCEBs is $\gtrsim 40\%$ for late spectral type secondary stars, it is only $\lesssim 20\%$ for more massive secondary stars. This result is in agreement with the predictions of the disrupted magnetic braking scenario.
- The large majority of PCEBs from the SDSS has orbital periods shorter than a day. This indicates that the common envelope efficiency is probably smaller than generally assumed.

In the near future we will continue our observational campaign to further verify the preliminary conclusions listed above. We will also incorporate an appropriate prescription of the binding energy parameter in the algorithm we use to reconstruct the common envelope phase. Then, we expect to give more definite answers on the two big questions in close white dwarf binary formation and evolution.

References

Briggs K R Napiwotzki R Maxted P F L and Wheatley P J 2007 *15th European Workshop on White Dwarfs* (*Astronomical Society of the Pacific Conference Series* vol 372) ed Napiwotzki R and Burleigh M R p 497

Davis P J Kolb U Willems B and Gänsicke B T 2008 *MNRAS* **398** 1563-1576

Hjellming M S 1989 *Rapid mass transfer in binary systems* Ph.D. thesis AA(Illinois Univ. at Urbana-Champaign, Savoy.)

Hurley J R Pols O R and Tout C A 2000 *MNRAS* **315** 543–569

McDermott P N and Taam R E 1989 **ApJ 342** 1019–1027

Nelemans G and Tout C A 2005 *MNRAS***356** 753–764

Nelemans G, Verbunt F, Yungelson L R and Portegies Zwart S F 2000 *A&A 360* 1011–1018

Paczyński B 1976 *IAU Symp. 73: Structure and Evolution of Close Binary Systems* p 75

Patterson J Kemp J Harvey D A Fried R E Rea R Monard B Cook L M Skillman D R Vanmunster T Bolt G Armstrong E McCormick J Krajci T Jensen L Gunn J Butterworth N Foote J Bos M Masi G and Warhurst P 2005 *PASP* **117** 1204–1222

Politano M and Weiler K P 2006 *ApJ* **641** L137–L140

Rappaport S Joss P C and Verbunt F 1983 *ApJL* **275** 713–731

Rebassa-Mansergas A Gänsicke B T Rodríguez-Gil P Schreiber M R and Koester D *MNRAS* **382** 1377-1393

Rebassa-Mansergas A Gänsicke B T Schreiber M R Southworth J Schwope A D Gomez-Moran A N Aungwerojwit A Rodríguez-Gil P Karamanavis V Krumpe M Tremou E Schwarz R Staude A and Vogel J *MNRAS* **390** 1635-1646

Schreiber M R and Gänsicke B T 2003 *A&A* **406** 304

Schreiber M R Gänsicke B T Southworth J Schwope A D and Koester D 2008 *A&A* **484** 441–450

Schreiber M R Nebot Gomez-Moran A and Schwope A D 2007 *Astronomical Society of the Pacific Conference Series* (*Astronomical Society of the Pacific Conference Series* vol 372) ed Napiwotzki A and Burleigh M R pp 459–+

Sills A, Pinsonneault M H and Terndrup D M 2000 *ApJ* **534** 335–347

Silvestri N M Lemagie M P Hawley S L West A A Schmidt G D Liebert J Szkody P Mannikko L Wolfe M A Barentine J C Brewington H J Harvanek M Krzesinski J Long D Schneider D P and Snedden S A 2007 *AJ* **134** 741–748

Skumanich A 1972 *ApJ* **171** 565

Townsley D M and Bildsten L 2003 *ApJ* **596** L227–L230

Townsley D M and Gaensicke B T 2008 (*Preprint* ArXiv e-prints/0811.2447)

Verbunt F and Zwaan C 1981 *A&A* **100** L7–L9

Webbink R F 1984 *ApJ* **277** 355–360

Webbink R F 2007 **704** (*Preprint* ArXiv e-prints/0704.0280)

Willems B and Kolb U 2004 *A&A* **419** 1057–1076

Wood M A 1995 *White Dwarfs* (*LNP* no 443) ed Koester D and Werner K (Heidelberg: Springer) pp 41–45

Journal of Physics: Conference Series **172** (2009) 012025 doi:10.1088/1742-6596/172/1/012025

The physical properties of white dwarf-main sequence binaries from SDSS

A Rebassa-Mansergas[1], B T Gänsicke[1] and D Koester[2]
[1]University of Warwick, Coventry CV4 7AL, UK
[2] Institut für Theoretische Physik und Astrophysik, University of Kiel, 24098 Kiel, Germany
E-mail: A.Rebassa-Mansergas@warwick.ac.uk

Abstract. We present a catalogue of 1596 white dwarf-main sequence binaries (WDMS) from the spectroscopic Sloan Digital Sky Survey Data Release 6 (SDSS DR6). We select WDMS binary candidates from template fitting all 1.27 million DR6 spectra, using combined constraints in both χ^2 and signal-to-noise ratio (S/N). In addition, we use Galaxy Evolution Explorer (GALEX) and UKIRT Infrared Sky Survey (UKIDSS) magnitudes to search for objects in which one of the two components dominates the SDSS spectrum. We use a decomposition/fitting technique to measure the effective temperatures, surface gravities, masses and distances to the white dwarfs, as well as the spectral types and distances to the companions in our catalogue.

1. Introduction

Binaries containing a white dwarf primary plus a main sequence companion were initially main sequence binaries in which the more massive star evolved through the giant phase and became a white dwarf. In the majority of cases the initial separation of the main sequence binary is wide enough to allow the evolution of both stars as if they were single. Nevertheless, a small fraction are believed to undergo mass transfer interactions (Webbink 1984, de Kool 1992, Willems & Kolb 2004). The more massive star will engulf its companion while evolving into a red giant, and the system enters a common envelope (CE, Iben & Livio 1993, Webbink 2007). Friction inside this envelope causes a rapid decrease of the binary separation. Henceforth orbital energy and angular momentum are extracted from the binary orbit and lead to the ejection of the envelope, exposing a post-common-envelope binary (PCEB). After the ejection of the envelope, close WDMS binaries evolve to shorter orbital periods through angular momentum loss via magnetic braking.

Population synthesis models have been developed for the understanding of both CE evolution and magnetic braking (e.g. Politano & Weiler 2007, Davis et al. 2008). Nonetheless the results provided by these works clearly lack observational constraints, and the advance in our understanding of CE evolution and magnetic braking is consequently most likely to arise from the analysis of PCEBs that are both numerous and well-understood in terms of their stellar components. It becomes then necessary to dispose of a large sample of observable WDMS binaries to constrain the evolution of close binaries. The SDSS (York et al. 2000, Stoughton et al. 2002, Adelman-McCarthy et al. 2008) has offered the opportunity of increasing the number of WDMS binaries available for follow-up studies. Here we make use of the SDSS spectroscopic DR6 to create a catalogue of 1596 WDMS binaries and candidates. In a parallel work, within

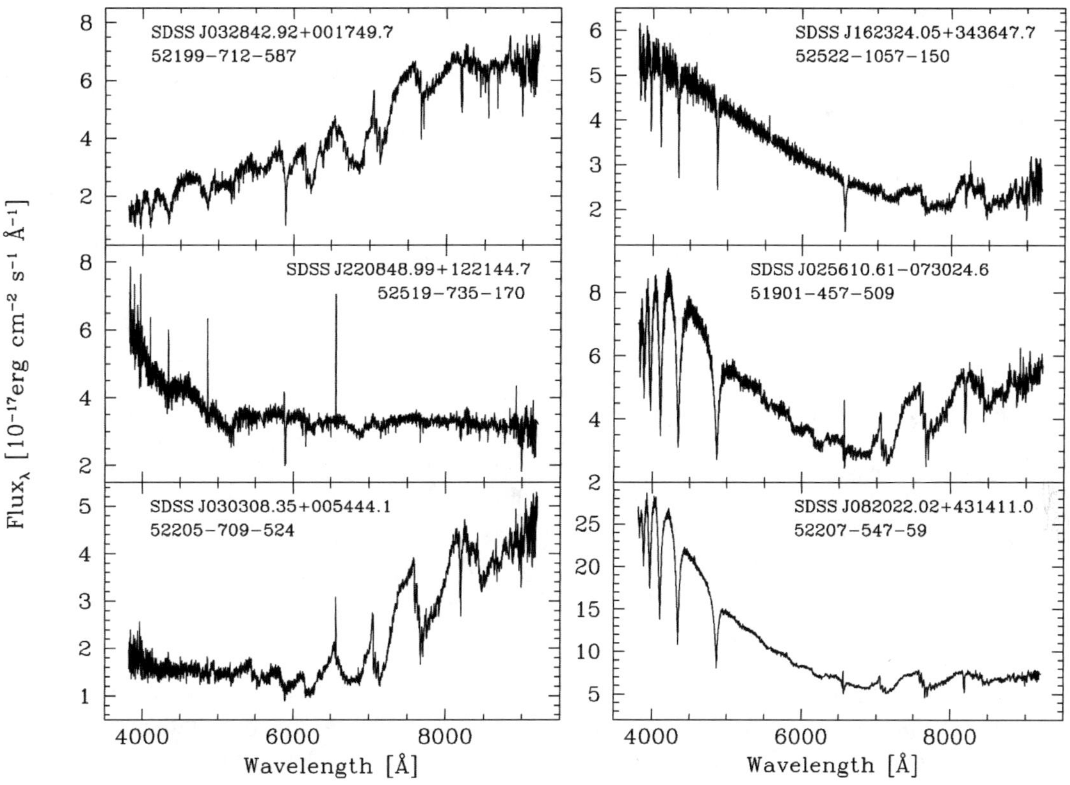

Figure 1. Six examples of previously known WDMS binaries used in this work as WDMS binary templates. SDSS names and MJD-PLT-FIB identifiers are indicated for each of them.

SEGUE we are also running a dedicated program to identify WDMS binaries containing cold white dwarfs (Schreiber et al. 2007). This sample of more than 1500 WDMS binaries is an excellent database for future follow-up observational studies of WDMS binaries (Schreiber et al. 2008, Rebassa-Mansergas et al. 2008, Pyrzas et al. 2008, Nebot Gomez-Moran et al. 2008).

2. Identification of WDMS binaries in SDSS

2.1. Computational method

We have developed a procedure based on χ^2 template fitting in order to automatically identify WDMS binary candidates from the SDSS DR6 spectroscopic data base. Our initial template set consisted of several dozen SDSS spectra of confirmed WDMS binaries from Eisenstein et al. (2006) and Silvestri et al. (2007). These spectra were chosen to sample a wide variety of white dwarfs and companion stars, and to be of high S/N. A set of representative templates is shown in Fig. 1. In addition, we compiled a set of 17 single DA white dwarf template spectra from Eisenstein et al. (2006)'s list, covering the entire observed range of $T_{\rm eff}$ and $\log g$, as well as the M0-M9 Bochanski et al. (2007) M-dwarf templates.

Each of these WDMS binary, white dwarf, and M-dwarf templates were then fitted to the full 1.27 million spectra in DR6. In this process, the template spectrum was normalised to the SDSS spectrum under scrutiny, and a reduced χ^2 was calculated using the errors of the two spectra added in quadrature. In practice, our fitting procedure produced for each of the WDMS

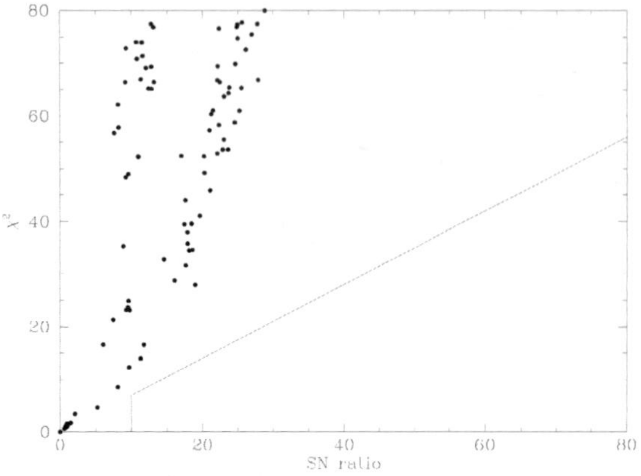

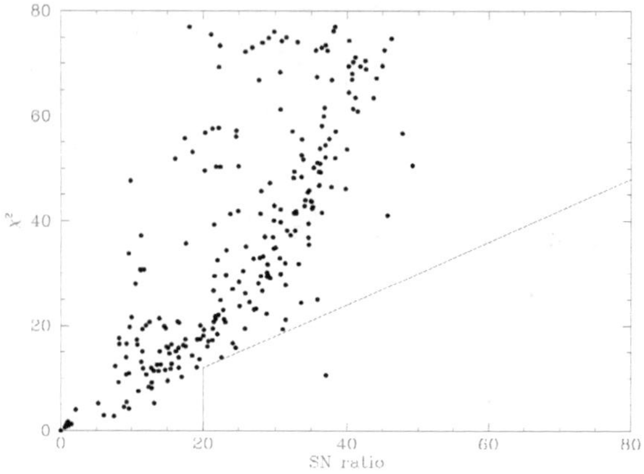

Figure 2. χ^2 – S/N plane obtained fitting two of our WDMS binary templates (SDSS J103121.97+202315.1, left and SDSS J204431.44-061440.2, right) to the 640 objects contained in the SDSS spectral plate PLT = 2295. Objects falling in the area defined by $\chi^2_{\rm spec} < a \times {\rm S/N}$ and ${\rm S/N_{spec}} > {\rm S/N_{min}}$ were considered WDMS binary candidates. Left panel: ${\rm S/N_{min}} = 10$ and $\chi^2_{\rm max} = 0.7 \times {\rm S/N}$, no WDMS binary candidate is identified. Right panel: ${\rm S/N_{min}} = 20$ and $\chi^2_{\rm max} = 0.6 \times {\rm S/N}$, one WDMS binary candidate is found.

binary, white dwarf, and M-dwarf templates a list of spectrum identifier (MJD-PLT-FIB), S/N of the spectrum, and χ^2 for all SDSS DR6 spectra. For each of the templates, we plotted χ^2 as a function of S/N, and defined a minimum value of S/N, ${\rm S/N_{min}}$, and a linear relation $\chi^2_{\rm max} = a \times {\rm S/N}$. We considered any spectrum with

$$\chi^2_{\rm spec} < \chi^2_{\rm max} \quad \text{and} \quad {\rm S/N_{spec}} > {\rm S/N_{min}} \tag{1}$$

as a WDMS binary, white dwarf, or M-dwarf (depending on the current template). ${\rm S/N_{min}}$ had to be set to avoid being swamped by spectra that are too noisy for any definite identification, and the S/N-dependent form of $\chi^2_{\rm max}$ accounts for the increase of χ^2 for higher values of S/N. Both constraints were defined individually for each of the templates, as the different spectral shapes resulted in a large spread of χ^2 distributions. The (χ^2, S/N) planes obtained from fitting the spectra of a single spectral plate (640 spectra) are shown for two different WDMS binary templates in Fig. 2.

After a first run through the DR6 spectra, we complemented the template set with the spectra of a number of newly identified WDMS binaries, and re-run the fitting for those new templates again. That process was repeated until no new WDMS binary candidates were found – at which point we had used a total of 163 different WDMS binary template spectra.

2.2. Red and blue excess in SDSS spectra: help from GALEX and UKIDSS

While the template fitting proved to be a robust method to find WDMS binaries in which both stellar components contribute clearly visible amounts of flux, the procedure is prone to misclassify white-dwarf dominated WDMS binaries as single white dwarfs, and M-dwarf dominated WDMS binaries as single M-dwarfs. We therefore decided to probe more specifically for the presence of excess flux at the red (blue) end of the SDSS spectra in objects classified initially as single white dwarfs (M-dwarfs).

For the search of red flux excess in single white dwarf candidates, we fitted synthetic white dwarf spectra computed with the code described by Koester et al. (2005) to the SDSS spectra, and then calculated the reduced χ^2 over the wavelength ranges 4000 – 7000 Å ($\chi^2_{\rm b}$)

and 7000 – 9000 Å ($\chi^2_{\rm r}$). Objects with $\chi^2_{\rm r}/\chi^2_{\rm b} > 1.5$ were "promoted" from single white dwarf candidates to WDMS binary candidates (see top and middle left panels of Fig. 3).

The search for blue flux excess proceeded in an analogous fashion for the single M-dwarf candidates, only that we used the set of high S/N M-dwarf templates from Rebassa-Mansergas et al. (2007) instead of model spectra, and calculated the reduced χ^2 over the wavelength ranges 4000 – 5000 Å and 7000 – 9000 Å. Objects with $\chi^2_{\rm b}/\chi^2_{\rm r} > 1.5$ were "promoted" from single M-dwarf candidates to WDMS binary candidates (see top and middle right panels of Fig. 3). However, in several cases, the detection of blue or red flux excess is rather marginal.

As a final step in our search for WDMS binaries, we have cross-correlated our entire list of WDMS binaries candidates identified above with the GALEX DR4, providing near- and far-ultraviolet magnitudes, and with the DR4 of UKIDSS, providing infrared $yJHK$ magnitudes. We then inspected the observed ultraviolet-optical spectral energy distribution of all secondary star dominated WDMS binary candidates with available GALEX magnitudes, and the optical-infrared spectral energy distribution of all white dwarf dominated WDMS binary candidates with available UKIDSS magnitudes. Objects where a clear ultraviolet or infrared excess was detected were then included in our WDMS binary sample (see bottom panels of Fig. 3).

2.3. SDSS images

As a final check on the nature of the WDMS binary candidates, we inspected their SDSS DR6 images for morphological problems, and found primarily two types of issues.

Occasionally, single white dwarfs (M-dwarfs) may be located close to very bright M-dwarfs (white dwarfs or A-stars), which will cause scattered light to enter the spectroscopic fibre, and result in an (apparent) two-component spectrum. A spectacular example is SDSSJ 073531.86+315015.2 (left panels of Fig. 4), where the SDSS spectrum clearly exhibits an M-dwarf at red wavelengths, and a blue component with strong Balmer lines in the blue – the SDSS image reveals that this is a single M-dwarf at a distance of 12 arcmin of Castor A/B – two $V = 2 - 3$ A-stars. The SDSS magnitudes (black dots) are superimposed to the SDSS spectrum (gray) and are consistent with those of a single red star. In addition, single M-dwarfs are also likely to be found superimposed with single early-type stars in the same image. An example is SDSSJ 005827.24+005642.6 (see middle top panel of Fig. 4). At first glance one could be tempted to consider a resolved WDMS binary pair. Nevertheless, the SDSS spectrum (middle bottom panel of the same figure) shows the typical Balmer lines of an A star in the blue, while at redder wavelengths the typical spectral features of a low-mass star can be detected. This implies that these two systems are superimposed stars in the same image rather than being a resolved WDMS binary pair (see below).

SDSS images can also help identifying WDMS binaries among our sample that are spatially resolved in the SDSS images, but close enough that flux from both stars will enter into the spectroscopic fibre. In such cases, the SDSS magnitudes are often discrepant with the flux-calibrated SDSS spectrum, and/or have large errors as consequence of the deblending applied by the photometric pipeline. Fig. 4 (top right panel) shows the SDSS image of SDSSJ 025306.37+001329.2, which clearly reveals a spatially resolved pair of stars. The SDSS spectrum of SDSSJ 025306.37+001329.2 contains the typical signatures of a WDMS binary, i.e. broad Balmer lines from the white dwarf and TiO absorption bands from the M-dwarf, however, the errors on the SDSS magnitudes are untypically large, and do not match well the flux calibrated SDSS spectrum (bottom left panel of Fig. 4).

3. Stellar parameters

The spectroscopic data provided by the SDSS project are of sufficient quality to estimate the stellar parameters of the WDMS binaries in our catalogue. For this purpose, we have developed a procedure which decomposes the WDMS binary spectrum into its white dwarf and main

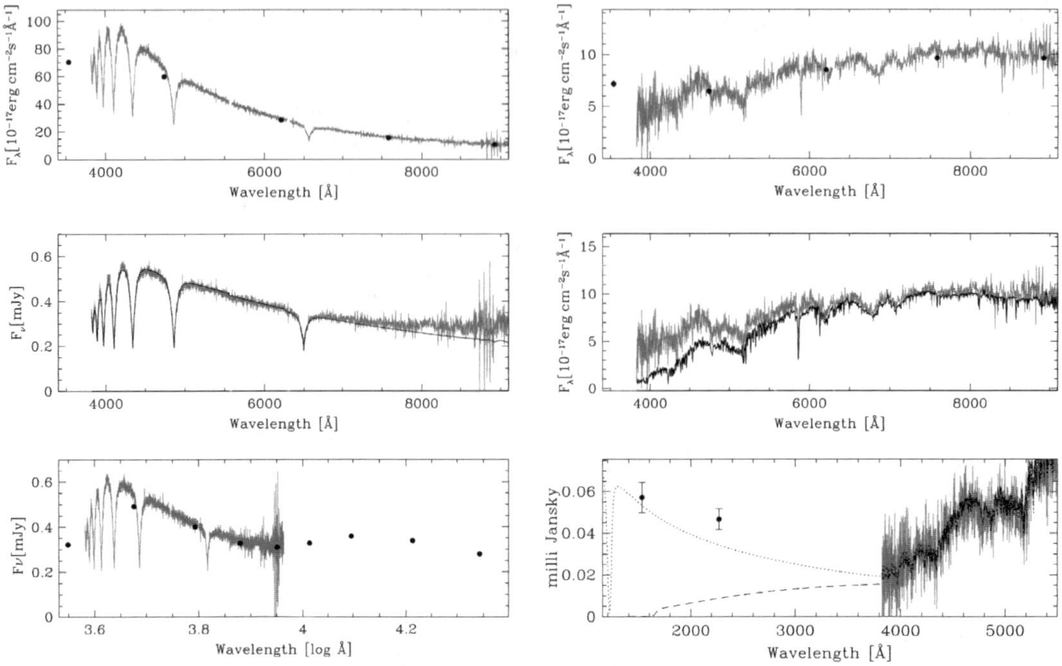

Figure 3. Top left: SDSS spectrum (gray) of SDSSJ 132925.21+123025.5, a WDMS candidate initially catalogued as single DA white dwarf. The black dots represent the SDSS magnitudes. Middle left panel: the best white dwarf model fit is superimposed in black, unambiguously identifying the red excess of the secondary star in the binary. Bottom left panel: SDSS and UKIDSS magnitudes (black dots) superimposed to the SDSS spectrum (gray). Again, the UKIDSS magnitudes clearly show the presence of a low-mass companion. Top right panel: the same for SDSSJ 131928.80+580634.2, an initially catalogued early M-type star. Middle and bottom right panels: the best M-type fit and the near- and far-ultraviolet GALEX magnitudes (black dots) clearly confirm the presence of a white dwarf primary, respectively. The dotted and dashed lines represent the white dwarf solutions (dotted for the hot, dashed for the cold) obtained from decomposing/fitting the spectrum (see Sect 3).

sequence star components, determines the spectral type of the companion by means of template fitting, and derives the white dwarf effective temperature and surface gravity from spectral model fitting (Rebassa-Mansergas et al. 2007). Assuming an empirical spectral type-radius relation for the secondary star and a mass-radius relation for the white dwarf from an updated list of Bergeron et al. (1995)'s tables, two independent distance estimates are calculated from the flux scaling factors of the template/model spectra.

In the course of decomposing/fitting the WDMS binary observations, we make use of a grid of observed M-dwarf templates, a grid of observed white dwarf templates, and a grid of white dwarf model spectra. High S/N ratio M-dwarf templates matching the spectral coverage and resolution of the WDMS binary data were produced from a few hundred late-type SDSS spectra from DR4. These spectra were classified using the M-dwarf templates of Beuermann et al. (1998). We also compiled a library of 490 high S/N DA white dwarf spectra from DR4 covering the entire observed range of $T_{\rm eff}$ and $\log g$. Finally, we computed a grid of synthetic DA white dwarf spectra using the model atmosphere code described by Koester et al. (2005), covering

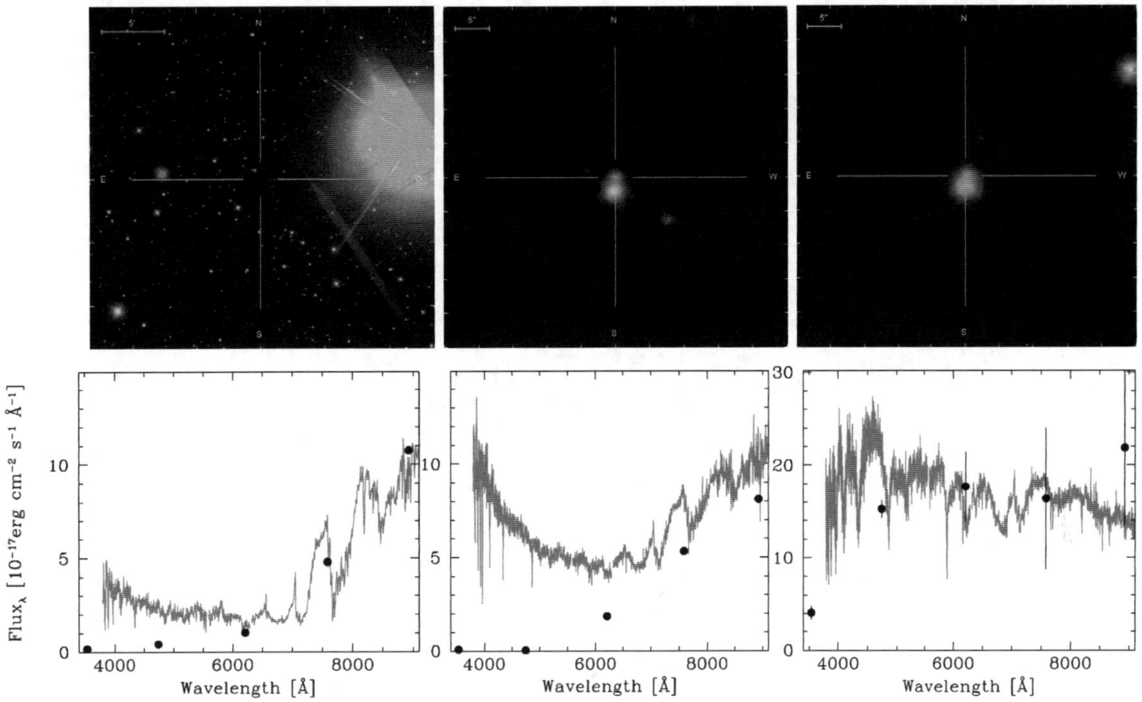

Figure 4. Top left panel: SDSS image of SDSSJ 073531.86+315015.2 (5 arcmin scale), a single red star initially considered as a WDMS binary candidate. Bottom left panel: SDSS magnitudes (black dots) and spectrum (gray) of the same system. The light from the saturated bright star (Castor A/B) is also dispersed in the spectrum. The magnitudes are consistent with those of a single red star. Top middle panel: SDSS image of SDSSJ 005827.24+005642.6 (5 arcsec scale). The image suggests a resolved WDMS binary pair. Middle bottom panel: the detection of the Balmer lines typical of an A star in the blue, together with the typical spectral features of a low-mass main sequence star in the red (gray), indicate that these are two single stars superimposed in the same image rather than a resolved WDMS binary pair. SDSS magnitudes are indicated with black dots, and are consistent with those of a low-mass star. Top right panel: SDSS image of SDSSJ 025306.37+001329.2 (5 arcsec scale), a resolved WDMS binary in our sample. Bottom right panel: SDSS magnitudes (black dots) and spectrum (gray) of the same system. Whilst the SDSS spectrum clearly shows both components, the SDSS magnitude errors are untypically large (see text for details).

$\log g = 5.0 - 9.5$ in steps of 0.25 and $T_{\rm eff} = 6000 - 100000$ K in 37 steps nearly equidistant in $\log(T_{\rm eff})$.

Our approach is a two-step procedure. In a first step, we fitted the WDMS binary spectra with a two-component model and determined the spectral type of the M-dwarf. Once the best-fit M-dwarf template has been determined and scaled appropriately in flux, it is subtracted from the WDMS binary spectrum. The residual white dwarf spectrum is then fitted with the grid of DA models described above. We then use the $T_{\rm eff}$ and $\log g$ from the fits to the whole spectrum, continuum plus lines, to select the "hot" or "cold" solution from the line profile fits (see Rebassa-Mansergas et al. 2007 for further details).

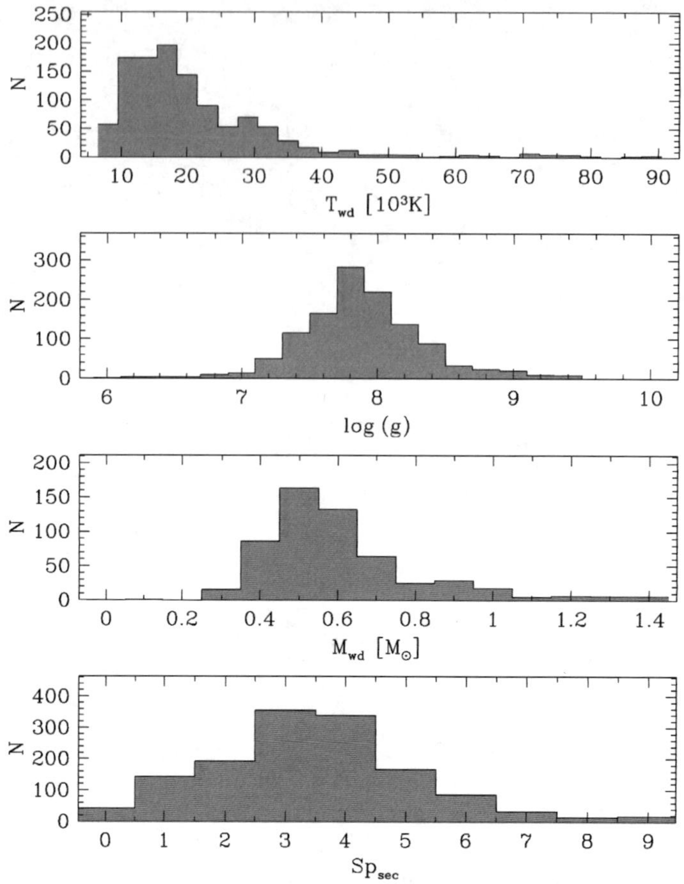

Figure 5. White dwarf $T_{\rm eff}$, $\log g$, mass, and spectral type of the secondary histograms obtained from the SDSS WDMS binaries in our catalogue. Excluded are those systems with individual white dwarf masses, $T_{\rm eff}$, and $\log g$ associated to relative errors larger than 25 per cent.

3.1. Distribution of the stellar parameters

Having determined stellar parameters for each individual system in Sect. 3, we now look at their global distribution within our sample of WDMS binaries. Figure. 5 shows histograms of the white dwarf effective temperatures, masses, $\log g$, and the spectral types of the main sequence companions. The most frequent white dwarf temperatures are between 10 000–20 000 K, white dwarf masses cluster around $M_{\rm wd} \simeq 0.5\,{\rm M}_{\odot}$, and the companion stars have most typically a spectral type M3–4, with spectral types later than M7 or earlier than M1 being very rare.

At closer inspection, the distribution of white dwarf masses has a pronounced tail towards lower masses. A tail of lower-mass white dwarfs, peaking around $0.4\,{\rm M}_{\odot}$ is observed also in well-studied samples of single white dwarfs (Liebert et al. 2005), and is interpreted as He-core white dwarfs descending from evolution in a binary star. In a sample of WDMS, a significant fraction of systems will have undergone a CE phase, and hence the fraction of He-core white dwarfs among WDMS is expected to be larger than in a sample of single white dwarfs.

The cut-off at early spectral types in the bottom panel of Fig. 5 is due to the fact that WDMS with K-type companions can only be identified from their spectra/colours if the white dwarf is very hot – and hence, very young, and correspondingly only few of such systems are in the total

Journal of Physics: Conference Series **172** (2009) 012025 doi:10.1088/1742-6596/172/1/012025

SDSS WDMS sample. The cut-off seen for low-mass companions is not so trivial to interpret. Obviously, very late-type stars are dim and will be harder to be detected against a moderately hot white dwarf. Such a bias was discussed by Schreiber & Gänsicke (2003) for a sample of ~ 30 well-studied WDMS which predominantly originated from blue-colour (= hot white dwarf) surveys. However, old WDMS with cool white dwarfs should be much more common (Schreiber & Gänsicke 2003), and SDSS, sampling a much broader colour space than previous surveys, should be able to identirfy WDMS containing cool white dwarfs plus very late type companions. The relatively low frequency of such systems in the SDSS spectroscopic data base suggests that either SDSS is not efficiently targeting those systems for spectroscopic follow-up, or that they are rare in the first place, or a combination of both.

References

Adelman-McCarthy J K et al. 2008 *ApJS* **175** 297-313
Bergeron P Wesemael F and Beauchamp A 1995 *PASP* **107** 1047
Beuermann K Baraffe I Kolb U and Weichhold M 1998 *A&A* **339** 518-524
Bochanski J J West A A Hawley S L and Covey K R 2007 *AJ* **133** 531-544
Davis P J Kolb U Willems B and Gänsicke B T 2008 *MNRAS* **389** 1563-1576
de Kool M 1992 *A&A* **261** 188-202
Eisenstein D J et al. 2006 *ApJS* **167** 40-58
Iben I J and Livio 1993 *PASP* **105** 1373-1406
Koester D Napiwotzki R Voss B Homeier D and Reimers D 2005 *A&A* **439** 317-321
Liebert J Bergeron P and Holberg J B 2005 *ApJS* **156** 47-68
Nebot Gomez-Moran A et al. 2008, submitted
Politano M and Weiler K P 2007 *ApJ* **665** 663-679
Pyrzas S et al. 2008 submitted
Rebassa-Mansergas A Gänsicke B T Rodríguez-Gil P Schreiber M R and Koester D 2007 *MNRAS* **382** 1377-1393
Rebassa-Mansergas A et al. 2008 *MNRAS* **390** 1635-1646
Schreiber M R and Gänsicke B T 2003 *A&A* **375** 937
Schreiber M R Nebot Gomez-Moran A and Schwope A D 2007 *Astronomical Society of the Pacific* **372** 459
Schreiber M Gänsicke B T Southworth J Schwope A D and Koester D 2008 *A&A* **484** 441-450
Silvestri M N et al. 2007 *AJ* **134** 741-748
Stoughton C et al. 2002 *AJ* **123** 485-548
Webbink R F 19884 *ApJ* **277** 355-360
Webbink R F 2007 *Preprint* ArXiv e-prints/0704.0280
Willems B and Kolb U 2004 *A&A* **419** 1057-1076
York D G et al. 2000 *AJ* **120** 1579-1587

Journal of Physics: Conference Series **172** (2009) 012026 doi:10.1088/1742-6596/172/1/012026

Searching for white dwarfs candidates in Sloan Digital Sky Survey Data

Mirosław Nalężyty[1], **Agnieszka Majczyna**[1,2], **Anna Ciechanowska**[1] **and Jerzy Madej**[1]

[1]University of Warsaw Astronomical Observatory, Al. Ujazdowskie 4, 00-478 Warsaw, Poland
[2]The Andrzej Sołtan Institute for Nuclear Studies, Hoża 69, 00-681 Warsaw, Poland

E-mail: `nalezyty@astrouw.edu.pl`

Abstract. Large amount of observational spectroscopic data are recently available from different observational projects, like Sloan Digital Sky Survey. It's become more urgent to identify white dwarfs stars based on data itself i.e. without modelling white dwarf atmospheres. In particular, existing methods of white dwarfs identification presented in Kleinman et al. (2004) and in Eisenstein et al. (2006) did not allow to find all the white dwarfs in examined data. We intend to test various criteria of searching for white dwarf candidates, based on photometric and spectral features.

1. Introduction

Because of their physical properties, white dwarfs stars are difficult to observe and identification. The first version of the McCook & Sion catalogue (1987) contains 1279 white dwarfs only. Recently, observational techniques was considerable improve, so the amount of photometric and spectral data rapidly grows. For this reason also the number of identified white dwarfs grows, but it is very needed to elaborate efficient identification and classification methods.

Until now, many of the white dwarf samples were selected, based on different selection criteria. For example Fleming et al. (1996) published a white dwarfs catalogue based on ROSAT data (Trumper 1983, 1992). A classification was made by searching for position coincidence between ROSAT sources and hot white dwarfs from McCook & Sion catalogue. Sources with no optical counterparts were marked as white dwarf if was visible only in the softest X-ray energy range. White dwarfs in EUV, selected also from the ROSAT all-sky survey by Marsh et al. (1997a,b), 129 white dwarf stars from Bergeron, Saffer & Liebert (1992), 200 white dwarfs from the Palomar Green survey (Green, Schmidt & Liebert (1986) analyzed by Liebert & Bergeron (1995) - it is only a few samples of these objects.

The Sloan Digital Sky Survey project (York et al. 2000) provide us large amount of very interesting data. Based on Data Release 1 and 4 (DR1 and DR4, respectively) there were made two catalogues by Kleinman et al. (2004) and Eisenstein et al. (2006), which contain 2551 and 9316 objects, respectively. Used identification methods did not allow to find all white dwarfs with whole range of effective temperatures and types in data sample. Kleinman et al. (2004) chose white dwarf candidates based on dereddened colour indexes $u - g$ and $g - r$ and a magnitude in u filter. Selected white dwarf candidates was next verified by visual inspection and prescribed its type - DA, DB, DZ etc. After this procedure, spectra of selected objects were fitted by theoretical DA and DB spectra.

Eisenstein et al. (2006) disposed larger amount of data so they used more automatic selection procedure. They select stars fulfilled applicable criteria (see details in Eisenstein et al. 2006). In this manner they obtain sample of blue stars, which observational spectra were fitted by theoretical spectra. Stars with parameters characteristic for white dwarfs were next verified and classified during different tests including visual inspection. This method has some limitations, for example does not work for stars cooler than 8000 K.

We intend to find a method of the white dwarf candidates selection - or, if it is possible, white dwarfs itself - based only on photometric and spectral features, chosen to obtain the best distinguishing between white dwarf stars and other type objects. Before we start fitting theoretical model atmospheres to observing spectra.

Not for the first time we have used data from the Sloan Digital Sky Survey (York et al. 2000). We chose Data Release 5 (DR5) and we selected spectral data for 10 % randomly choosing objects. In this manner we obtained very imposing amount of nearly 16000 different type objects, including galaxies, quasars, stars, and of course white dwarfs. Additionally we also used white dwarfs identified in SDSS DR4 by Eisenstein et al. (2006), treating these stars on the one hand as a potential help to improve our white dwarfs selection method, on the other hand as a control data.

2. Balmer lines and their parameters

The most characteristic feature of many DA white dwarf spectra is a presence of wide, hydrogen absorption Balmer lines (except for the hottest objects). It would appear that it should be a good selection criteria of degenerated stars with high surface gravity, but not every white dwarf have wide Balmer lines. Nevertheless we decided to check this possibility. Because of large amount

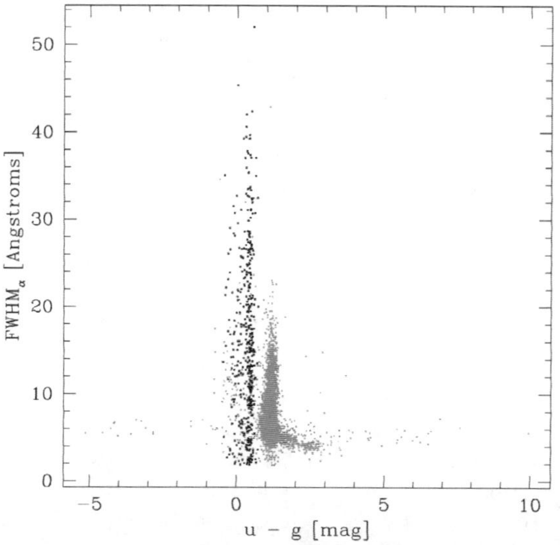

Figure 1. Colour index $u - g$ versus FWHM of the H_α Balmer line objects with H_α, H_β, and H_γ lines present in their spectra (dark gray dots). White dwarfs from SDSS DR4 belonging to our sample with detected Balmer lines was marked with black dots.

(15998) of objects in our sample we had to create proper software, which automatically finds hydrogen lines from Balmer series, and then calculates its various parameters like line widths at given depth, fluxes in lines etc. Fig. 1 shows one of the obtained results, a dependency of FWHM of Balmer H_α line on colour index $u - g$ for objects with H_α, H_β, and H_γ lines present in their spectra. Clearly seeing bimodal structure is not a by-product of our Balmer lines searching, but it is a result of a bimodal distribution of the colour index $u - g$. A location of the white dwarfs identified in SDSS DR4 (Kleinman et al. 2004), and belonged to our sample with detected Balmer lines, denoted by black dots, suggests, that all of the white dwarfs lies in the left part of this diagram, with colour indexes $u - g < 0.7$. In general, this is not a truth. Visual inspection shows, that some amount of white dwarf stars also lies in the right part of the diagram, and have larger values of colour indexes $u - g$. However, our requirement for the presence of hydrogen Balmer lines in spectra reduces our sample to practically star-like type objects only. In fact, it is decreases a number of white dwarf candidates from 15998 to 4613. Unfortunately, reduced in this way sample does not contain the hottest white dwarf stars, because there is no Balmer lines visible in their spectra.

3. Fluxes in given ranges

Colour indexes based on *ugriz* photometry are not very useful for us, generally because the filters are too wide. So we decide to define own artificial filters, i.e. own wavelength ranges, in which we calculated fluxes. Choosing our filters we intended to not contain any Balmer lines.

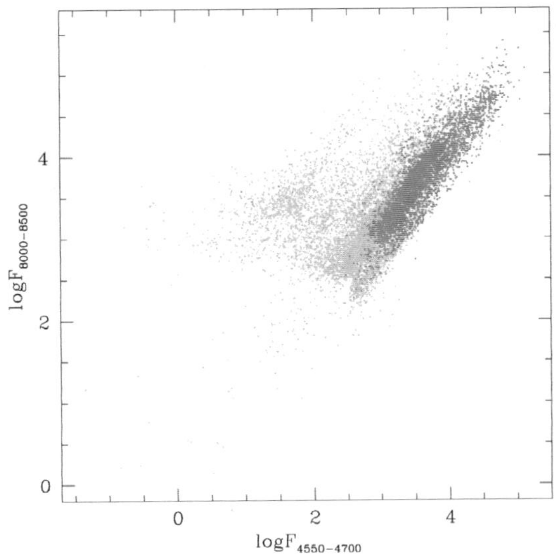

Figure 2. Logarithms of fluxes relation, calculated in two wavelength ranges: $4550 - 4700$Å and $8000 - 8500$Å for all of 15998 objects in our sample (light gray dots). Objects with detected H_α, H_β, and H_γ lines was marked by dark gray dots.

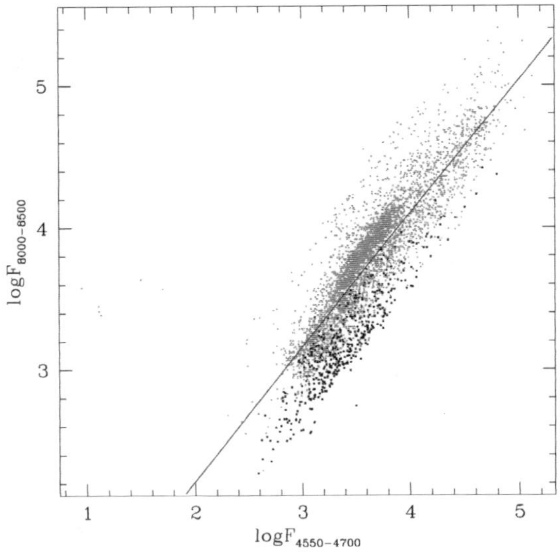

Figure 3. The same as in Fig. 2, but for 4613 objects with Balmer lines only (dark gray dots). Black dots denote white dwarfs from SDSS DR4. Linear function $y = 0.94118x + 0.33235$ is an approximate boundary between two regions visible in this diagram. Preliminary results of the visual inspection shows, that almost all of the 2202 objects located in the lower-right region looks like white dwarf stars.

In Fig. 2 we present a dependency of two flux logarithms, calculated in two arbitrarily chosen wavelength ranges: $4550 - 4700$Å and $8000 - 8500$Å for all of 15998 objects in our sample. Objects with detected H_α, H_β, and H_γ lines (dark gray dots) occupy well defined area, so we decided to watch them closer. Fig. 3 shows logarithms of fluxes in the same wavelength ranges as in Fig. 2, but for 4613 objects with Balmer lines only. It is not very difficult to see some structure in this diagram. At least part of the objects divides to, not so bad, separate regions. This impression is additionally magnified by black dots showing positions of the white dwarfs from DR4, which occupy one of the mentioned above regions only. We decided to approximate a boundary between these two areas by the linear function $y = 0.94118x + 0.33235$, with empirically chosen constants. Preliminary results of the visual inspection shows, that spectra of almost all of the 2202 objects located in the lower-right region looks like white dwarf spectra. Because of relatively poor separation of these two areas, between 2411 objects belonging to the upper-left region we still could find relatively small amount of white dwarf stars, located close to the borderline.

4. Conclusions

Our method of selecting white dwarf candidates is based on searching for objects with hydrogen Balmer lines visible in their spectra, and on flux calculations in well selected wavelength ranges. Preliminary results show that our method allows to select quite complete sample of white dwarf stars, under the above assumptions. Of course this method should be tested in detail, and it needs some improvements (for example increasing separation between two areas in Fig. 3, probably by choosing better wavelength ranges). It is quite possible, that using this method we shall be able to find also the hottest white dwarfs, which are too hot to show Balmer lines in their spectra, although we will need the other criteria to select star-like type objects. We plan to do this in future.

Acknowledgments

This work has been supported by the Polish Ministry of Science and Higher Education grant No. N N203 4061 33. We also thank Institut d'Estudis Espacials de Catalunya (IEEC) for financial support.

Funding for creation and distribution of the SDSS Archive has been provided by the Alfred P. Sloan Foundation, the Participating Institutions, the National Aeronautics and Space Administration, the National Science Foundation, the US Department of Energy, the Japanese Monbukagakusho, and Max Planck Society. The SDSS Web site is http://www.sdss.org/ .

The SDSS is managed by the Astrophysical Research Consortium (ARC) for the Participanting Institutions. The Participanting Institutions are The University of Chicago, Femilab, the Institute for Advanced Study, the Japan Participation Group, the Johns Hopkins University, Los Alamos National Laboratory, the Max-Planck-Institute for Astronomy (MPIA), the Max-Planck-Institute for Astrophysics (MPA), New Mexico State University of Pittsburgh, Princeton University, the United States Naval Observatory, and the University of Washington.

References

Bergeron P Saffer R A and Liebert J 1992 *ApJ* **477** 313
Eisenstein D J Liebert J Harris H C et al. 2006 *ApJS* **167** 40
Fleming T A et al. 1996 *A&A* **316** 147
Green R F Schmidt M and Liebert J 1986 *ApJS* **61** 305
Kleinman S J Harris H C Eisenstein D J et al. 2004 *ApJ* **607** 426
Liebert J and Bergeron P 1995 *White Dwarfs* eds D Koester and K Werner (Berlin: Springer) p12
Marsh M C Barstow M A Buckley D A et al. 1997a *MNRAS* **286** 369
Marsh M C Barstow M A Buckley D A et al. 1997b *MNRAS* **287** 705
McCook G P and Sion E M 1987 *ApJS* **65** 603
Trumper J 1983 *Adv.Sp.Res.* **2** 241
Trumper J 1992 *QJRAS* **33** 165
York D G Adelman J Anderson J E Jr et al. 2000 *AJ* **120** 1579

SDSS121258.25–012310.1: a new eclipsing post common envelope binary

A Nebot Gómez-Morán[1] , A D Schwope[1] , M R Schreiber[2] and B T Gänsicke[3]

[1] Astrophysikalisches Institut Potsdam, An der Sternwarte 16, D-14482 Potsdam
[2] Departamento de Fisica y Astronomia, Universidad de Valparaiso , Avenida Gran Bretana 1111, Valparaiso, Chile
[3] Department of Physics, University of Warwick, Coventry, CV4 7AL, UK

E-mail: angm@aip.de

Abstract. From optical photometry we show that SDSS121258.25–012310.1 is a new eclipsing post common envelope binary with an orbital period of $\sim$ 8.06 hours and eclipse length of 23 minutes. From spectroscopic follow up observations we measured the semi-amplitude of the radial velocity for the secondary star, $K_2 = 181 \pm 3$ km s^{-1}. By fitting the optical SDSS spectrum we constrained the stellar parameters. The white dwarf temperature is 17700 ± 300 K, its surface gravity $\log g = 7.53 \pm 0.2$, the spectral type of the secondary star is M4 $\pm$ 1 and the distance to the system $d = 230 \pm 20$ parsec. Masses are estimated to be in the range $0.25 - 0.38$ M$_\odot$ for the red dwarf and $0.33 - 0.48$ M$_\odot$ for the white dwarf. We finally discuss the evolutionary state of the system.

1. Introduction

Most of the stars in the Galaxy are born in binary or multiple star systems, and some of them will interact at some point of their lives. If the orbital separation of the initial binary is small enough as the more massive star evolves mass transfer occurs and, if the rate at which it happens is high, it may engulf its companion, entering a common envelope (CE) phase. The stars start a spiral-in process, the envelope is not co-rotating with the binary and friction within the common envelope leads to a shrinkage of the binary separation and angular momentum and energy are extracted from the orbit, eventually expelling the CE. Once the CE is expelled the binary can be formed by a remnant of the more massive star, i.e. a white dwarf, and a main sequence star, perhaps sorrounded by the ejected material. These are the so called post common envelope binaries (PCEB) and further evolution, mainly driven by angular momentum loss due to gravitacional radiation and magnetic braking, can bring the system into semi-contact, where the secondary star fills in its Roche lobe and mass transfer occurs again but in a stable way, forming a cataclysmic variable (Paczyński 1976; Iben & Livio 1993).

There are theories for the CE phase, but this phase is still poorly understood and even more, there is a lack of observational constraints (Schreiber & Gänsicke 2003). Eclipsing binaries are of most relevance as we can derive stellar parameters with highest accuracy. Recently the number of known eclipsing PCEBs is increasing (Nebot et al. 2008; Pyrzas et al. 2008). Our serendipitous discovery of a new eclipsing PCEB, SDSSJ121258.25–012310.1 (henceforth SDSS1212–0123), in

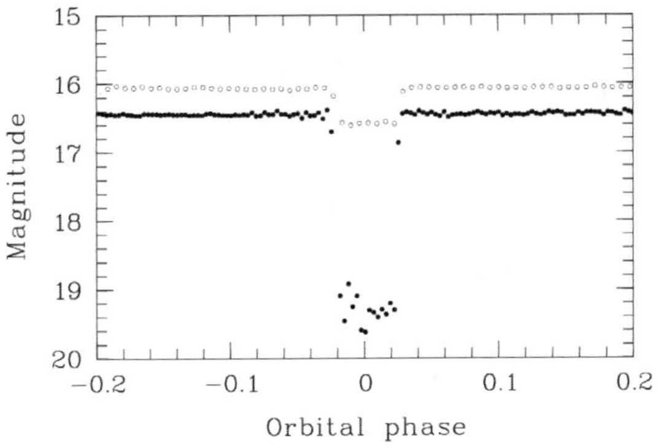

Figure 1. Optical lightcurve of SDSS1212–0123 in the V (black circles) and I (open circles) bands from bottom to top.

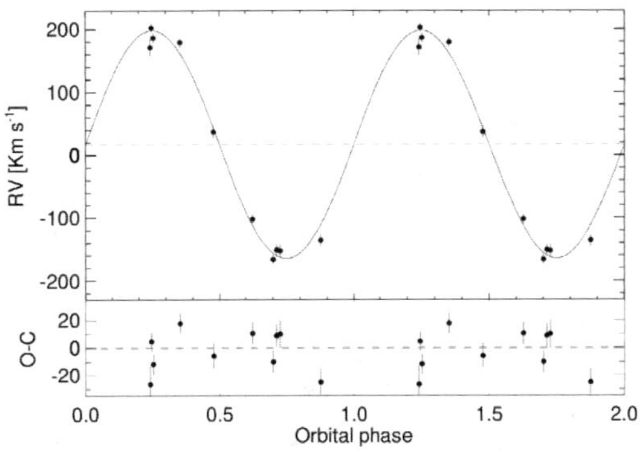

Figure 2. Radial velocity curve, sine fit and residuals.

our ongoing search for PCEBs (Schreiber et al. 2008; Rebassa et al. 2007; 2008), triggered further observations.

2. Eclipse ephemeris from optical photometry

We observed SDSS1212–0123 over 11 nights using two different telescopes, the IAC80 (OT, Spain) and the 70 cm telescope of the Astrophysikalisches Institut Potsdam (Babelsberg, Germany) in three different bands. Differential magnitudes were obtained with repect to the star SDSSJ121302.39–012343.5. The light curve is presented in Fig. 1, the deeper eclipse in the V band represents the fact that the white dwarf is being occulted by its companion. We measured the eclipse length, 23 ± 1 minutes, and determined the ephemeris of eclipse center to $\mathrm{HJD_{mid}} = 2454104.7086(2) + 0.3358706(5) \times \mathrm{E}$, where numbers in parenthesis indicate the 1σ uncertainty in the last digits.

3. Spectroscopic follow up observations

Spectroscopic follow up observations were obtained for SDSS1212–0123 in the period May 16-19, 2007, with the LDSS3 imaging spectrophotograph at the Magellan Clay telescope. We measured the radial velocities from the NaI absorption doublet (8183.27Å, 8194.81Å) present in the spectrum of SDSS1212–0123 and originating from the secondary star. The radial velocity semaimplitude was obtained from a least-squares fit of a sine function, $v_r = \gamma_2 + K_2 \sin\left[\frac{2\pi(t-t_0)}{P}\right]$, to the observed radial velocity measurements. Orbital period, P, and epoch of mid eclipse, t_0, were determined photometrically and were kept fixed. We find the systemic velocity $\gamma_2 = 17{\pm}3\,\mathrm{km\,s^{-1}}$ and $K_2 = 181 \pm 3\,\mathrm{km\,s^{-1}}$ (see Fig.2)

4. Decomposition of the SDSS spectrum

From the optical SDSS spectrum we determined the stellar parameters of SDSS1212–0123 following the procedure described in Rebassa et al. (2007). We determine the spectral type of the secondary star by fitting the spectrum (Fig. 3, black line) to composite spectra made of M-dwarfs and white dwarfs templates (Fig. 3, dotted line). We subtract the best dM and we fit the white dwarf to modeled atmospheres of white dwarfs on a $T_{\rm eff}$–$\log g$ grid, getting two results, a 'cold' and a 'hot' solution (Fig. 4, upper panel). To discern we fit the overall continuum plus lines (Fig. 4, bottom panel).

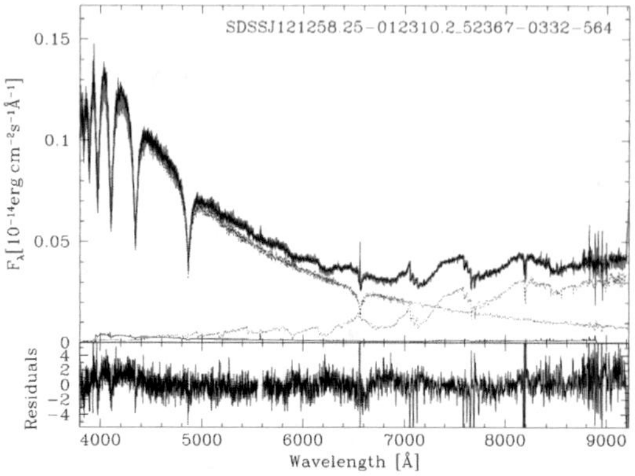

Figure 3. Best fit (grey line) of the SDSS optical spectrum of SDSS1212–0123 (black line) to composite spectra made of white dwarf and M-dwarfs templates (dotted lines). Residuals are shown in the bottom.

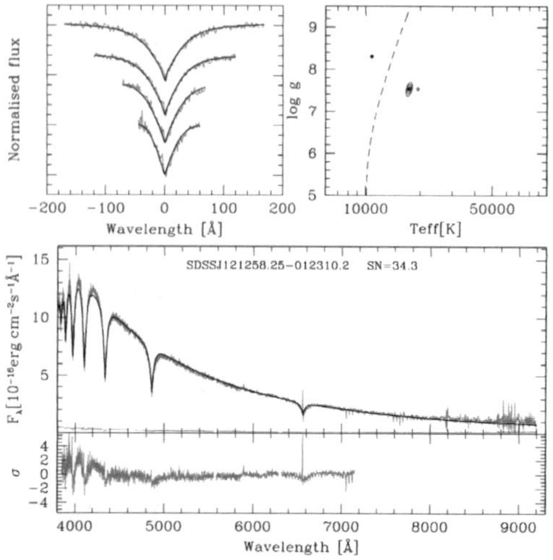

Figure 4. *Hα–Hϵ* lines are fitted to model atmospheres on a $T_{\rm eff}$–$\log g$ grid (left upper panel) giving two solutions, a 'hot' and a 'cold' one (right upper panel). The overall spectrum is fitted (bottom panel) braking the degeneracy (dark dot, upper right panel).

The spectral type of the secondary was determined to be M4 $\pm$ 1 implying a distance $d_{\rm sec} = 320 \pm 95\,$pc, mass range of the secondary $M_{\rm sec} = 0.255 - 0.380$ $\rm M_{\odot}$ and radius range $R_{\rm sec} = 0.258 - 0.391$ $\rm R_{\odot}$, using Rebassa et al. (2007) spectral type-mass and spectral type-radius empirical relations respectively. The best fit was found for $T_{\rm eff} = 17700 \pm 300$ K and $\log g = 7.53 \pm 0.05$ (implying a white dwarf mass $M_{\rm wd} = 0.39 \pm 0.02$ $\rm M_{\odot}$, and $R_{\rm wd} = 0.018 \pm 0.001$ $\rm R_{\odot}$). All the quoted errors are purely statistical, the true uncertainty of the white dwarf spectral parameters is clearly larger than suggested by the derived numbers. We estimate the systematic uncertainty of our $\log g$ determination to be of the order of 0.2 dex, which results in rather large ranges of possible values for the mass and the radius of the primary, i.e. $M_{\rm wd} = 0.33 - 0.48$ $\rm M_{\odot}$ and $R_{\rm wd} = 0.015 - 0.021$ $\rm R_{\odot}$. The derived distance to the white dwarf is $d_{\rm wd} = 226 \pm 8\,$pc (assuming the statistical error only). The two distance estimates differ, $d_{\rm sec}$ being larger than $d_{\rm wd}$, but in agreement within the erros, taking into account systematic errors we obtain $d_{\rm wd} = 230 \pm 20\,$pc as the distance to the system.

5. Evolutionaty state

Following Schreiber & Gänsicke (2003) we predicted the post common envelope evolution of SDSS1212–0123. From the mass and the temperature of the primary we estimated the cooling age to be 6.8×10^7 years, using the cooling tracks from Wood (1995) (see Fig. 5). From the derived mass of the secondary we conclude that the only mechanism reducing the orbit of the binary is gravitational radiation (magnetic braking gets disrupted when the secondary star becomes completely convective, ~ 0.3 $\rm M_{\odot}$). SDSS1212–0123 left the CE phase with an orbital period of about 8 hours and, according to its cooling age, SDSS1212–0123 has passed only a small fraction of its PCEB lifetime. The system appears to be different to the progenitors of the current CV population as the calculated PCEB lifetime exceeds the Hubble time. SDSS1212–0123 will become a CV within or slightly above the orbital period gap.

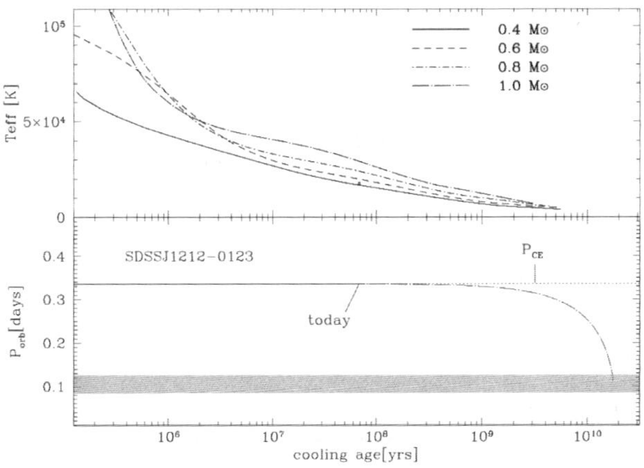

Figure 5. Upper panel: We derived a cooling age for SDSS1212–0123 of $\sim 7 \times 10^7$ years by interpolating the cooling tracks of Wood (1995). Bottom panel: We reconstructed the evolution of SDSS1212–0123 finding that it has only lived a small fraction of its PCEB lifetime, it will take $\sim 1.8 \times 10^{10}$ years until SDSS1212–0123 will become a CV.

6. Conclusions

From optical photometry we conclude SDSS1212–0123 is a eclipsing PCEB with an orbital period of 0.336 days and an eclipse length of 23 min. From spectroscopic follow-up observations we have derived a systemic velocity of 17 ± 3 km/s and a semi-amplitude of the radial velocity of 181 ± 3 km/s. From the SDSS spectrum we derived $T_{\rm eff} = 17700 \pm 300$ K, $\log g = 7.53 \pm 0.2$ and a secondary spectral type M4$\pm$1 and a distance to the system of 230 ± 20 parsecs. We estimated $M_{\rm sec} = 0.25 - 0.38$ M$_\odot$ and $M_{\rm wd} = 0.33 - 0.48$ M$_\odot$ from the eclipse length, using the mass function and an empirical mass radius relation for main sequence stars. We have reconstructed and predicted the post CE evolution of SDSS1212–0123, finding that SDSS1212–0123 at the end of the CE phase had a very similar orbital period. The only mechanism involved in shrinking the orbital period is and has been gravitational radiation. As the PCEB lifetime of SDSS1212–0123 exceeds the Hubble time we conclude that it is not representative of the progenitors of the current CV population.

References

Iben I J & Livio M 1993 *PASP* **105** 1373–1406
Nebot Gomez-Moran A Schwope A D Schreiber M R Gänsicke B T & Southworth J 2008 (*Preprint* submitted)
Paczyński B 1976 *IAU Symp. 73: Structure and Evolution of Close Binary Systems* p 75
Pyrzas S Gänsicke B T Marsh T R Aungwerojwit A Rebassa-Mansergas A Rodríguez-Gil P Southworth J Schreiber M R Nebot Gomez-Moran A & Koester D 2008 (*Preprint* arXiv:0812.2510v1)
Rebassa-Mansergas A Gänsicke B T Rodríguez-Gil P Schreiber M R & Koester D 2007 *M.N.R.A.S.* **382** 1377–1393
Rebassa-Mansergas A Gänsicke B T Schreiber M R Southworth J Schwope A D Gomez-Moran A N Aungwerojwit A Rodríguez-Gil P Karamanavis V Krumpe M Tremou E Schwarz R Staude A & Vogel J 2008 *M.N.R.A.S.* **390** 1635–1646
Schreiber M R & Gänsicke B T 2003 *A&A* **406** 305–321
Schreiber M R Gänsicke B T Southworth J Schwope A D & Koester D 2008 *ApJ* **484** 441–450
Wood M A 1995 *White Dwarfs* (*LNP* no 443) ed Koester D and Werner K (Heidelberg: Springer) pp 41–45

Journal of Physics: Conference Series **172** (2009) 012028 doi:10.1088/1742-6596/172/1/012028

Three eclipsing white dwarf plus main sequence binaries from SDSS.

S Pyrzas[1,2], B T Gänsicke[1], T R Marsh[1], A Aungwerojwit[1,3], A Rebassa-Mansergas[1], J Southworth[1], P Rodríguez-Gil[2,4], M R Schreiber[5] and D Koester[6]

[1] Department of Physics, University of Warwick, Coventry, CV4 7AL, UK
[2] Isaac Newton Group of Telescopes, Apartado de correos 321, S/C de la Palma, E-38700, Canary Islands, Spain
[3] Department of Physics, Faculty of Science, Naresuan University, Phitsanulok 65000, Thailand
[4] Instituto de Astrofísica de Canarias, Vía Láctea, s/n, La Laguna, E-38205, Tenerife, Spain
[5] Departamento de Fisica y Astronomia, Universidad de Valparaiso, Avenida Gran Bretana 1111, Valparaiso, Chile
[6] Institut für Theoretische Physik und Astrophysik, University of Kiel, 24098 Kiel, Germany

E-mail: S.Pyrzas@warwick.ac.uk

Abstract. We identify SDSS 0110+1326, SDSS 0303+0054 and SDSS 1435+3733 as three eclipsing white dwarf plus main sequence binaries from the Sloan Digital Sky Survey, and report on their follow-up observations. Orbital periods for the three systems are established through multi-season photometry. Time-resolved spectroscopic observations lead to the determination of the radial velocities of the secondary stars. A decomposition technique of the SDSS spectra is used to estimate the surface gravities and effective temperatures of the white dwarfs, as well as the spectral types of the secondaries. By combining the constraints from the spectral decomposition, the radial velocity data and the modeling of the systems' light curves, we determine the physical parameters of the stellar components. Two of the white dwarfs are of low mass ($M_{\rm WD} \sim 0.4\,M_{\odot}$), while the third white dwarf is unusually massive ($M_{\rm WD} \sim 0.8-0.9\,M_{\odot}$) for a post-common envelope system.

1. Introduction

We have selected possibly eclipsing white dwarf plus main sequence (WDMS) binaries from the Sloan Digital Sky Survey (SDSS) based on the available information on the radial velocities of their companion stars, and/or evidence of a strong reflection effect.

Initially, we used SDSS spectroscopy to measure the radial velocity of the companion star either from the Na I $\lambda\lambda$ 8183.27,8194.81 absorption doublet, or from the Hα emission line. SDSS J030308.35+005444.1 (henceforth SDSS 0303+0054) and SDSS J143547.87+373338.6 (henceforth SDSS 1435+3733) exhibited the largest secondary star radial velocities among $\sim$ 1150 WDMS binaries which have SDSS spectra of sufficiently good quality. For SDSS J011009.09+132616.1 (henceforth SDSS 0110+1326) two SDSS spectra are available, which differ substantially in the strength of the emission lines from the heated companion star. Photometric time series observations revealed white dwarf eclipses in all three objects. Table 1 lists the full coordinates and SDSS u, g, r, i, z magnitudes of the three systems.

Table 1. Full coordinates and u, g, r, i, z magnitudes of the three systems.

SDSS J	u	g	r	i	z
011009.09+132616.1	16.51	16.53	16.86	17.02	16.94
030308.35+005444.1	19.14	18.60	18.06	16.89	16.04
143547.87+373338.5	17.65	17.14	17.25	16.98	16.66

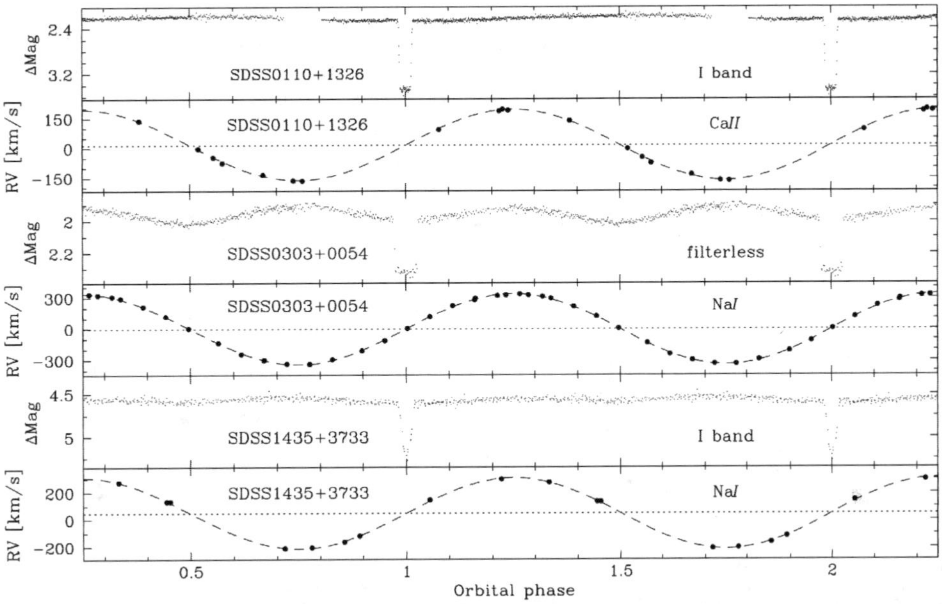

Figure 1. Sample phase folded light curves and radial velocity curves of the three systems. The system's name, the filter used in the photometric observations and the spectral line used to measure the respective radial velocity are indicated in each panel. A full orbit has been duplicated for clarity.

2. Orbital periods and radial velocities

Follow-up photometric and spectroscopic observations were obtained at the 4.2 m WHT, the 2.2 m Calar Alto, the 1.2 m Mercator and the 80 cm IAC telescopes, for a total of 32 nights for all systems. Figure 1 show sample light curves and radial velocity curves of the three systems.

For each system, we measured the eclipse mid-points in the light curves and established the orbital periods and the orbital ephemerides. These are: $T_0(\mathrm{HJD}) = 2453994.447876(95) + 0.3326873265(62)E$ for SDSS 0110+1326, that is, $P_{\mathrm{orb}} = 7.984495836(89)\mathrm{h}$, $T_0(\mathrm{HJD}) = 2453991.616419(26) + 0.1344377259(15)E$ for SDSS 0303+0054, that is, $P_{\mathrm{orb}} = 3.226505422(80)\mathrm{h}$ and $T_0(\mathrm{HJD}) = 2454148.703615(62) + 0.125631039(04)E$ for SDSS 1435+3733, that is, $P_{\mathrm{orb}} = 3.015144936(10)\mathrm{h}$. The ephemerides were then used to fold the available photometric and spectroscopic data over phase.

From our spectroscopic observations we were able to measure the radial velocities of the secondary stars for all three systems. For SDSS 0110+1326 we used the Ca II $\lambda\lambda$ 8498.02,8542.09,8662.14 emission triplet, whereas for SDSS 0303+0054 and SDSS 1435+3733 the Na I $\lambda\lambda$ 8183.27,8194.81 absorption doublet. The results, regarding the radial velocity and the systemic velocity of each system, are: $K_{\mathrm{sec}} = 178.8 \pm 2.4\,\mathrm{km\,s^{-1}}$ and

$\gamma = 15.2 \pm 2.4\,\mathrm{km\,s^{-1}}$ for SDSS 0110+1326, $K_{\rm sec} = 339.7 \pm 1.9\,\mathrm{km\,s^{-1}}$ and $\gamma = -4.0 \pm 1.4\,\mathrm{km\,s^{-1}}$ for SDSS 0303+0054 and $K_{\rm sec} = 260.9 \pm 2.9\,\mathrm{km\,s^{-1}}$ and $\gamma = 47.4 \pm 2.2\,\mathrm{km\,s^{-1}}$ for SDSS 1435+3733.

3. Spectral decomposition

A decomposition technique, based on template spectra, was used to isolate the white dwarf and the M-dwarf contributions. M-dwarf templates were used in the first step, to determine the spectral type of the secondary and its flux contribution. After the best fit was found (see also Figure 2), it was subtracted from the total spectrum, leaving only a residual white dwarf spectrum. The Balmer line profiles of the DA white dwarfs in SDSS0110+1326 and SDSS1435+3733 were then fitted with model white dwarf spectra (see also Figure 3). In this fashion, we also determined the white dwarf effective temperatures and surface gravities, as well as the distances to the systems. Using these values, one can obtain an initial estimate for the mass of the white dwarf. We should note that SDSS0303+0054 contains a DC white dwarf, hence only the spectral type of the M-dwarf was determined.

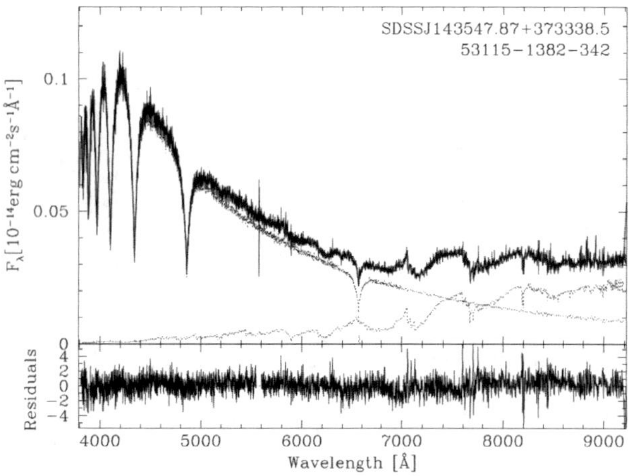

Figure 2. Two-component fit to the spectrum of SDSS 1435+3733.

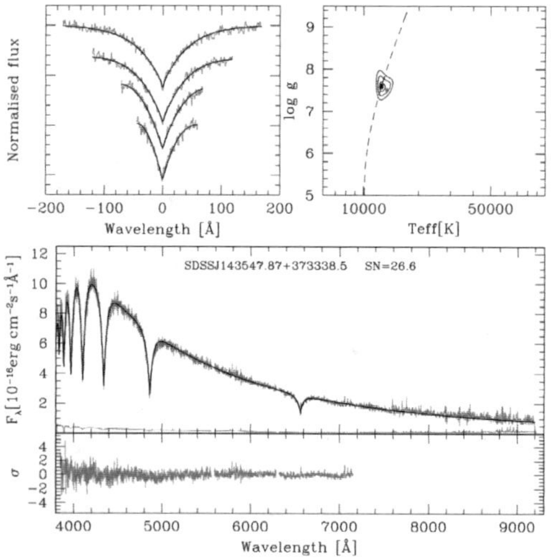

Figure 3. Spectral model fit to the white dwarf in SDSS 1435+3733, obtained after subtracting the best-fit M-dwarf template.

4. Light curve model fitting

We used TRM's close binary systems fitting code to model the light curves of the three systems and obtain additional constrains for the stellar parameters, mainly the masses and radii of the stellar components and the inclination angles of the three systems. The code uses a file of supplied parameters (the mass ratio q, the inclination angle i, the two radii scaled by binary separation $r_1 = R_{\rm WD}/a$ and $r_2 = R_{\rm sec}/a$ and the effective stellar temperatures $T_{\rm WD}$ and $T_{\rm sec}$) to compute the model, which is afterwards fitted to the data using Levenberg-Marquardt minimisation. To illustrate the quality achieved in the fits, Figures 4 and 5 show sample fits for SDSS 0110+1326 and SDSS 0303+0054 respectively.

5. Results

Table 2 contains the values for $\log g$ and $T_{\rm eff,WD}$, as well as the spectral type of the secondary, obtained from the decomposition of the spectra, as described in Section 3. These values, along

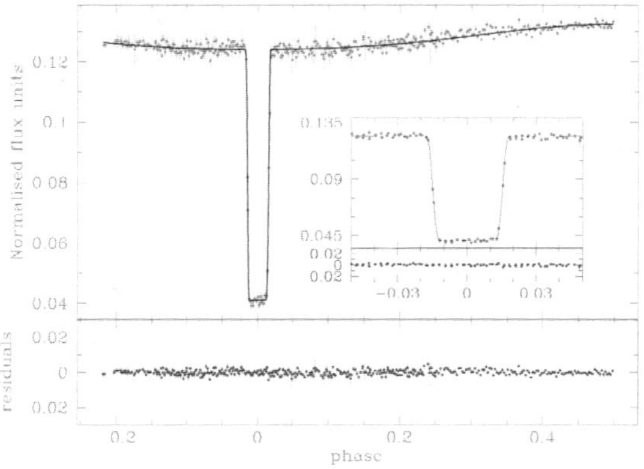

Figure 4. Sample model fit to the light curve of SDSS 0110+1326, for $M_{\rm WD} = 0.48\,M_\odot$ and $M_{\rm sec} = 0.22\,M_\odot$.

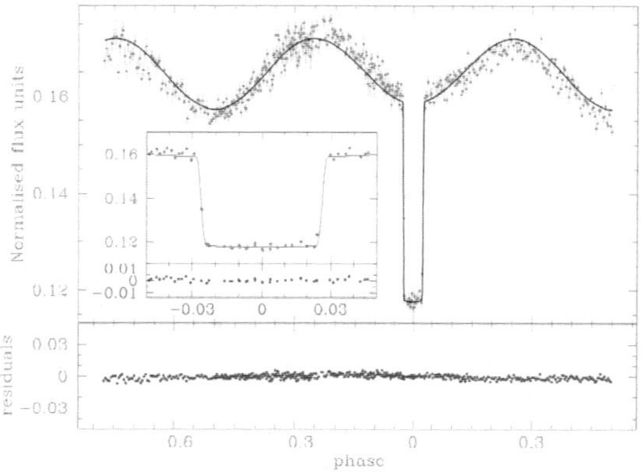

Figure 5. Sample model fit to the light curve of SDSS 0303+0054, for $M_{\rm WD} = 0.91\,M_\odot$ and $M_{\rm sec} = 0.25\,M_\odot$.

Table 2. Mass and radius values for the stellar components, white dwarf surface gravities and effective temperatures and secondary spectral types for all three systems, after combining the constrains from the available photometric and spectroscopic data sets.

SDSS J	$M_{\rm WD}$ $[M_\odot]$	$R_{\rm WD}$ $[R_\odot]$	$\log g$	$T_{\rm eff,WD}$ [K]	$M_{\rm sec}$ $[M_\odot]$	$R_{\rm sec}$ $[R_\odot]$	Sp(2)
0110+1326	0.45-0.49	0.016-0.017	7.65	∼26000	0.20-0.30	0.21-0.29	M4 ± 1
0303+0054	0.88-0.95	0.008-0.009	8.45-8.56	<8000	0.22-0.28	0.25-0.27	M4.5 ± 0.5
1435+3733	0.48-0.53	0.014-0.015	7.58	∼12500	0.19-0.25	0.22-0.25	M4.5 ± 0.5

with the results of the light curve modeling, presented in Section 4, lead to a range of possible masses and radii for the stellar components. Both the former and the latter are also shown in Table 2.

6. Conclusion

The three systems presented here bring the total number of known eclipsing WDMS binaries to 14 systems. SDSS 1435+3733 has the shortest orbital period among them all. SDSS 0110+1326 and SDSS 1435+3733 have masses in line with the expectations from close binary evolution. SDSS 0303+0054 seems to be unusually massive. All three systems have late M-type secondaries. Higher temporal resolution light curves, leading to more accurate determination of masses and radii, will provide valuable data for the calibration of M-R relations for low-mass stars and testing the M-R relation of white dwarfs.

References

Adelman-McCarthy J K et al. 2006 *ApJS* **162** 38
Bergeron P et al. 1995 *PASP* **107** 1047
Koester D et al. 2005 *A&A* **439** 317
Pyrzas S et al. 2008 *MNRAS* submitted
Rebassa-Mansergas A et al. 2007 *MNRAS* **382** 1377
Silvestri N M et al. 2006 *AJ* **131** 1674
Steinfadt J D R et al. 2008 *ApJ Lett.* **677** L113
Stoughton C et al. 2002 *AJ* **123** 485

Journal of Physics: Conference Series **172** (2009) 012029 doi:10.1088/1742-6596/172/1/012029

Hubble Space Telescope Observations of White Dwarfs in Detached Binaries

Howard E. Bond

Space Telescope Science Institute, 3700 San Martin Dr., Baltimore, MD 21218 USA

E-mail: bond@stsci.edu

Abstract. I describe three programs of *Hubble Space Telescope* (*HST*) observations of white dwarfs (WDs) in detached binaries. (1) I summarize the *HST* UV spectroscopy of the Hyades eclipsing binary V471 Tauri, containing a dK main-sequence star and a hot WD. By phasing time-resolved STIS spectra on the 9.25-min rotation period of the WD, we have shown that there are two opposite spots on the WD where there is magnetic accretion from the K star's wind. (2) Several programs of direct imaging and FGS astrometry are yielding high-precision visual orbits and dynamical masses, including the famous systems of Sirius and Procyon, as well as three fainter binaries in which both components are DC WDs. (3) Finally I discuss the bizarre nucleus of the low-surface-brightness planetary nebula EGB 6. The central star is associated with a compact emission-line nebula, which *HST* imaging shows is associated with a resolved dM companion. Why there should be a compact nebula around the cool star is puzzling; we speculate that it could be an accretion disk of material captured from the outflow that produced the surrounding faint PN. In a recent development, *Spitzer* has detected a 24 μm excess from EGB 6, indicating a dust component at a temperature of $\sim$260 K; the location of this dust relative to the hot and cool stars and the compact nebula remains uncertain at present.

1. Introduction

This paper reports on several observational programs with the *Hubble Space Telescope* (*HST*) that focus on spectroscopy, direct imaging, and astrometry of white dwarfs (WDs) in detached binary systems and in planetary nebulae (PNe). I will present progress reports on the following topics:

- Time-resolved UV spectroscopy of the WD in V471 Tauri;
- Dynamical masses of WDs in visual binaries from *HST* astrometry;
- *HST* imaging of the bizarre WD+dM binary nucleus of the PN EGB 6.

But first I would like to start by thanking a large number of collaborators on these projects, as well as several astronomers who kindly provided me with results in advance of publication:

- V471 Tau: Ed Sion, Don Lindler, Sean O'Brien, Jean Dupuis, Patrick Godon, Fred Walter
- Visual binaries: Ed Nelan, Gail Schaefer, Martin Barstow, Matt Burleigh, Jay Holberg, Terry Girard, Ron Gilliland, Russ Makidon
- EGB 6: Jim Liebert, Kate Su, Jana Bilikova, Alvio Renzini, Mike Jura, You-Hua Chu

2. UV Spectroscopy of the Post-Common Envelope Hyades Binary V471 Tauri

V471 Tau is an eclipsing binary containing a hot DA WD and a dK main-sequence companion. With an orbital period of only 12.5 hr, it is the prototype of binaries that must have gone through a common-envelope (CE) interaction. It is of special importance because it is a member of the Hyades cluster. The WD rotates with a period of only 9.25 min. Fig. 1 shows a scale model of the V471 Tau system. Because the K star is constrained to rotate at the short orbital period, it is very active and spotted, and has a significant stellar wind. A small part of this wind is accreted onto the WD companion.

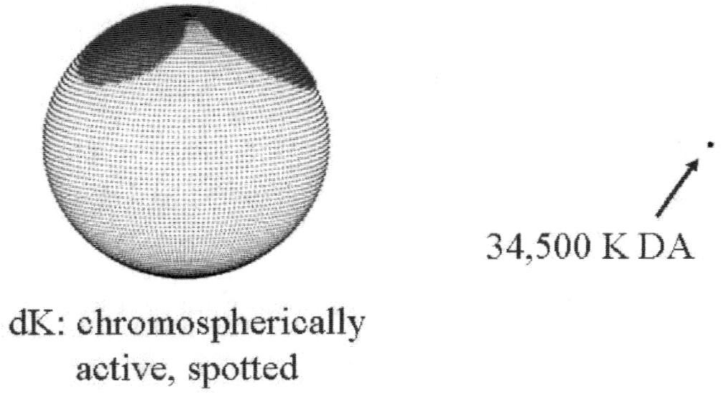

Figure 1. Scale model of the V471 Tauri system. The chromospherically active dK star is orbited every 12.5 hr by its hot DA companion. Figure prepared using D. Bradstreet's BinaryMaker 3.0 program.

HST has obtained UV spectroscopy of V471 Tau with both the Goddard High Resolution Spectrograph (GHRS) and, more recently, the Space Telescope Imaging Spectrograph (STIS).

2.1. GHRS Results: the Paradox of V471 Tauri

The GHRS programs on V471 Tau were the following:

- 1992: PI H. Shipman, G160M grating, resolution 20000, exploratory observations
- 1993-4: PI H.E. Bond, G160M, taken at quadrature of the binary
- 1995: PI F.M. Walter, G140M & G140L, resolution 25000 & 2200, taken near times of eclipse

The main published results from these observations are: (1) the discovery of coronal mass ejections from the K star, producing transient absorption lines as they are silhouetted against the WD (Bond et al. 2001; Walter 2004); (2) direct demonstration of a spot of accreted metals at one magnetic pole of the WD through detection of metallic absorption lines varying on the 9.25-min rotation period (Sion et al. 1998); and (3) determinations of the radial-velocity (RV) curve of the WD, from which we derived the radii and dynamical masses of both stellar components (O'Brien et al. 2001).

In the O'Brien et al. study, we obtained dynamical masses of $M_{\rm WD} = 0.84\,M_\odot$ and $M_{\rm dK} = 0.93\,M_\odot$. These results give rise to what we called the evolutionary paradox of V471 Tau: at $T_{\rm eff} = 34,500$ K, the WD is the hottest one in the Hyades cluster, with a cooling age of only $\sim 10^7$ yr. However, at the mass we derived, it is the most massive WD in the cluster. This is the opposite of expectation, since the hottest, and therefore youngest, WD in a cluster should be the *least* massive, as it would be descended from a lower-mass progenitor than the older WDs. O'Brien et al. proposed that the progenitor system of V471 Tau was originally a triple, containing a close pair of A stars orbited by a more-distant dK companion. The pair of A stars was close enough that when the more massive one began to evolve and increase its radius, after an interval of a few hundred million years, it merged with its companion to make a single blue straggler of $\sim 5\,M_\odot$. The straggler then evolved to the AGB, and only then engulfed the dK star in a CE. The orbit then spiralled down to a 12.5-hr period, ejecting the CE in the process and forming a PN that would have been visible for a few thousand years. We are now about 10^7 yr after the CE event, and observe a hot DA orbiting the dK star, with the PN now long gone.

2.2. STIS Observations and Preliminary Results

During the 1997 astronaut Servicing Mission, the GHRS spectrograph was removed from *HST* and replaced with STIS. STIS was operational until its failure in 2004. It is hoped that it will be brought back into operation during the 2009 Servicing Mission 4.

STIS observed V471 Tau in three programs:

- 1999 & 2002: PI F.M. Walter, E140M, resolution 46000 covering 1120–1710 Å, taken near eclipse
- 2000: PI H.E. Bond, E140M, taken at quadrature

Apart from the paper by Walter (2004) on signatures of mass loss from the dK star, details of these spectra remain unpublished. In collaboration with Ed Sion and Don Lindler, we are now systematically analyzing these observations, and I give a progress report here.

Lindler has assembled all of the STIS data taken between 1999 and 2002. He has phased the spectra on the 9.25-min WD rotation period, resulting in time-resolved UV spectra as well as continuum UV light curves. Fig. 2 shows a "photographic" representation of the time-resolved spectra. Absorption lines of various metallic species are marked, and we clearly see variations in these lines that are synchronized with the WD rotation. This image is shown normalized to the mean of all of the spectra; thus, although the metallic lines are always either in absorption or absent, there is an offset in the image stretch such that when a line is absent there is an apparent "emission" feature at its location.

Fig. 2 demonstrates that there are *two* absorption-line dips per WD rotation period. Thus, we see that the WD in V471 Tau has two magnetic caps on opposite sides of the star, both of them accreting material from the dK star's wind. The right-hand side of the figure shows the total UV flux phased on the WD rotation period. The light curve has two peaks per WD rotation, corresponding to the two polar caps. It can be seen that the continuum is *brightest* when the metallic absorption is strongest, i.e., the polar caps are *fainter* in the UV than their surroundings on the surface of the WD.

The UV light variations are thus *in phase* with the 9.25-min "pulsations" detected in time-resolved optical photometry of the system (e.g., Clemens et al. 1992). From UV to optical wavelengths the magnetic poles are brightened by flux redistribution from short wavelengths. In the soft X-ray and EUV range (and in the UV absorption lines), the continuum variations are 180° out of phase, because the poles are darkened by metallic absorption.

In our work underway on the collection of STIS spectra, we will do the following: (1) refine the RV measurements using the sharp metallic absorption lines, and thus refine the masses and radii of the two stars; (2) determine chemical abundances at the two poles; (3) estimate the accretion rates at the two poles; and (4) attempt to detect Zeeman splitting of the lines, in order to determine the magnetic fields.

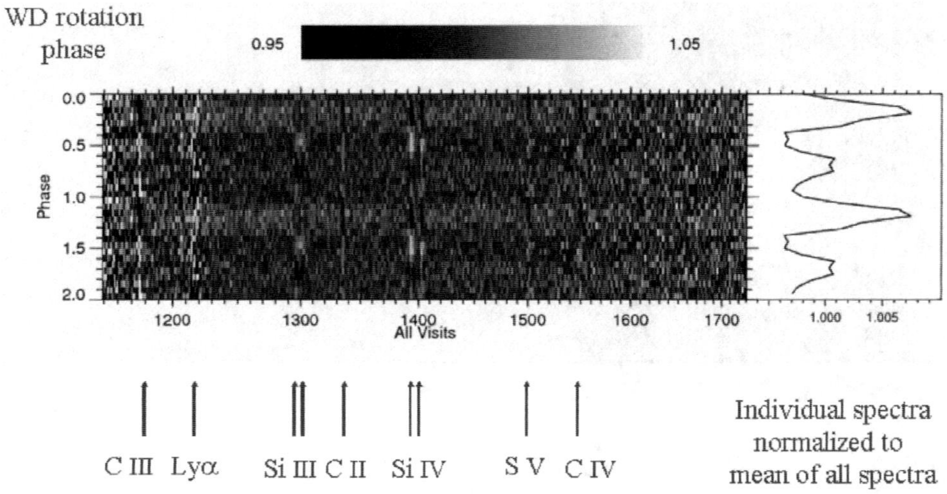

Figure 2. Photographic representation of the STIS time-resolved UV spectra of V471 Tau, phased to the 9.25-min rotation period of the WD. As described in the text, the apparent "emission" features correspond to absence of the absorption lines, and the dark features correspond to deep absorption. The absorption lines of the metallic species marked at the bottom all vary at the WD rotation period; since there are two absorption dips per rotation, these data demonstrate the presence of *two* opposite polar caps of accreted metals. The curve on the right shows the time variation of the integrated continuum flux, i.e., the UV light curve phased on the WD rotation period.

3. HST Astrometry of Visual Binaries Containing White Dwarfs

In separate *HST* programs, we have been making astrometric measurements of WDs in visual binaries. The ultimate aims of these programs are the direct determination of dynamical masses for WDs, and a search for perturbations due to third bodies (including those of planetary mass).

3.1. Direct Imaging

For several years, we have been regularly imaging the two famous bright and nearby stars that have faint, resolved WD companions: Sirius and Procyon.

Although notoriously difficult to measure from the ground, because of the enormous magnitude differences, the companions of Sirius and Procyon are easily detected in *HST* images. Fig. 3 shows a WFPC2 image of Sirius, with Sirius B clearly visible on the lower left. In the case of Sirius, the primary star is saturated even at the shortest possible WFPC2 exposure time of 0.11 sec; hence we do the centroiding of Sirius A by using the diffraction spikes. For Procyon, we take a series of short and long exposures at the same telescope pointing.

In Fig. 4 (top) we show a plot of the historical ground-based measurements of the orbit of Sirius B, from data assembled by Jay Holberg. Note the large scatter, due to the huge magnitude difference ($\sim$10) between the stars, as well as the paucity of points near closest approach (which limits determination of the semi-major axis and thus of the masses). The two bottom panels in Fig. 4 show magnified views of the portions of the orbit for which *HST* measurements from WFPC2 images are available, in the years 1997 and 2001 through 2008. The small black dots show the measurements, and the surrounding circles show the positions predicted by an orbit calculated by Gail Schaefer. The errors of the *HST* measures are less than 2 mas. Careful inspection leads us to suspect a small amount of jitter around the solution, which might suggest

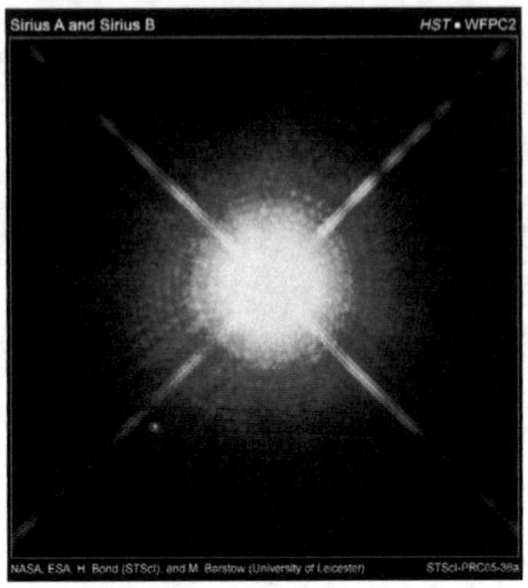

Figure 3. *HST* WFPC2 image of the Sirius system. Sirius B is easily detected; it is the faint star adjacent to the diffraction spikes of Sirius A at the lower left.

presence of a low-mass third body. Note that a 1 Jupiter-mass companion in a 6-year orbit around Sirius B would produce a deflection of 2 mas, and we are seeing apparent residuals several times that value. However, we still need to verify our method of centroiding Sirius A using the diffraction spikes, based on a set of calibration observations we obtained recently on a fainter A-type star where we have both overexposed images and unsaturated ones taken at the same telescope pointing.

Schaefer's orbit solution, based on a weighted combination of the *HST* measures, the ground-based historical data, and the astrometric motion of Sirius A (Gatewood & Gatewood 1978), yields dynamical masses of $M_A = 2.06 \pm 0.04\,M_\odot$ and $M_B = 1.02 \pm 0.01\,M_\odot$. Liebert et al. (2005) used masses of 2.02 and $1.00\,M_\odot$ to infer an initial mass of the WD progenitor of $5.1\,M_\odot$. This value, adjusted slightly for our new results, provides an important point on the WD initial/final-mass relation.

For Procyon we have *HST* data from 1995 to the present. By combining the *HST* measures, ground-based visual observations (which have large errors but cover a long time base), the single-lined RV orbit of Procyon A, and the *Hipparcos* parallax, Schaefer has obtained a solution implying masses of $M_A = 1.50 \pm 0.04\,M_\odot$ and $M_B = 0.55 \pm 0.02\,M_\odot$.

3.2. FGS Measurements

We also have monitoring programs underway using the *HST* Fine Guidance Sensors (FGS). The FGS system provides excellent astrometry of visual binaries when the magnitude differences are not too large and the primary stars are not too bright. Our FGS programs are concentrating on three binaries in which both components are WDs: G 107-70, WD 1639+153, and WD 1818+126. All three systems are classified DC in the Villanova WD Catalog. The binarity of G 107-70 was discovered by Harrington et al. (1981) at the USNO, and that of the other two systems by Ed Nelan during an earlier *HST* FGS snapshot survey of cool WDs.

Fig. 5 shows an example of the orbit quality that results from the high-precision FGS measurements. We show the *HST* FGS measurements of G 107-70 as filled circles, and a set of earlier observations of the system with WFPC2 as open circles. The observations, which now cover 1995 to 2008, extend over 70% of the 18.55 yr orbital period. The solid ellipse shows Schaefer's orbital solution. For comparison, the ellipse shown as a dashed line is Harrington's 1981 ground-based orbit. Schaefer's solution implies a total system mass of $1.19 \pm 0.06\,M_\odot$. This may be compared with an analysis of the system by Bergeron et al. (2001), who used model atmospheres to fit the combined spectrum and energy distribution and the observed delta magnitude, and found $\log g$ values of 7.92 and 8.12. The implied individual masses are 0.53 and $0.66\,M_\odot$, for a total of $1.19\,M_\odot$, in precise agreement with the dynamical mass.

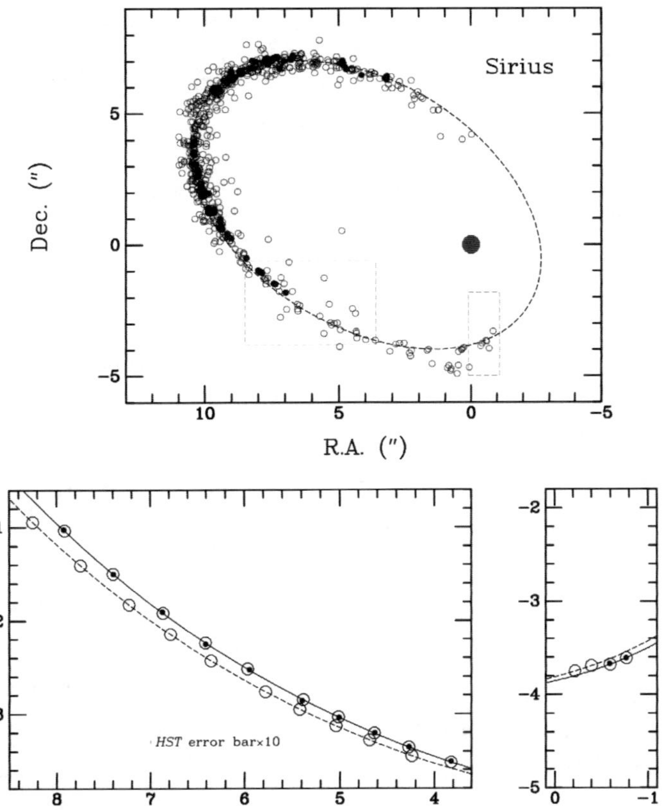

Figure 4. Top: historical measurements of the orbit of Sirius B, assembled by J. Holberg. *Filled circles:* photographic observations, 1956–1986; *open circles:* visual observations, 1862–1981. *Dashed ellipse:* best-fit ground-based orbit (Holberg in prep.). *Dashed boxes:* locations of the portions of the orbit magnified in the two bottom panels. **Bottom panels:** Zoom-ins showing *HST* measurements (*filled circles*) of the position of Sirius B in 2001–2008 (*left panel*) and in 1997 (*right panel*). The *dashed line and open circles* show the ground-based orbit and predicted positions, and the *solid line and open circles* show a new orbit and predicted positions based on a solution by G. Schaefer including the *HST* points with high weight. The small cross in the lower left panel shows an *HST* error bar of about ±2 mas *multiplied by a factor of 10.* Note the very large departures of the observations from the ground-based predicted positions ($\sim 0\farcs4$ approximately west; i.e., well to the right of the predictions) in both panels, and much smaller, but still significant, departures from the *HST*-based orbit, implying noticeable "jitter."

Our data are less extensive for the other two systems. Preliminary orbits give periods of 3.87 and 11.7 yr for WD 1639+153 and WD 1818+126, respectively.

4. The Bizarre Central Star of EGB 6

EGB 6 is a large (diameter 13′), very faint PN that I discovered serendipitously on a Palomar Sky Survey print in 1978 (see Ellis, Grayson, & Bond 1984). My subsequent spectroscopy of the blue central star showed strong [O III] emission, far too strong to arise in the faint PN. When I obtained a second spectrum at a location 10″ off the star, it showed no emission (see Liebert et al. 1989). We initially interpreted this result as indicating that the nucleus was surrounded by a compact, unresolved nebula.

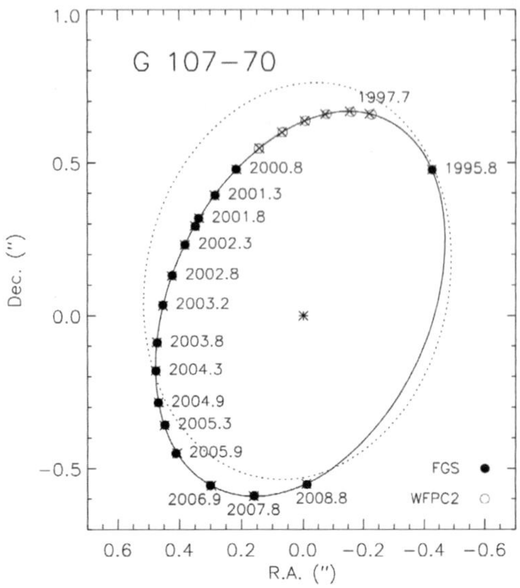

Figure 5. *HST* FGS and WFPC2 measures of the orbit of the double WD G 107-70. We have now observed this system over 70% of its 18.55 yr orbit. FGS observations are shown as filled circles, and WFPC2 measures as open circles. The solid ellipse is G. Schaefer's orbital fit, and the crosses show the predicted positions from this fit. The dashed ellipse is the solution of Harrington et al. (1981) from ground-based data.

In order to investigate the nature of this compact nebula, we used *HST* to observe it in the early 1990's using both direct and grism imaging. To our surprise, these observations showed that the compact emission-line source is not the central star, but is a point source located 0.″18 away from the nucleus.

Fig. 6 illustrates the *HST* imaging. At the upper left is a Faint Object Camera (FOC) image taken in a narrow-band [O III] filter. The companion source is clearly resolved. At the bottom I show a grism image, taken with the old WF/PC-1 camera. At the left is the zero-order image, and at the right is the bright, blue continuum of the hot nucleus. At 5007 Å, the emission-line companion is clearly resolved, with a brightness consistent with *all* of the [O III] emission coming from the companion.

The next development was that Fulbright & Liebert (1993) discovered that the nucleus shows an IR excess in *JHK* photometry. They interpreted this as arising from a dM companion of the central star. We obtained *HST* images in the *I* band with both WF/PC-1 and later with WFPC2, which clearly resolve this cool companion, and show that it is *coincident with the [O III] emission source.* The upper right panel in Fig. 6 shows the resolution of the cool companion in a WF/PC-1 *I*-band image.

A spectroscopic analysis of the hot nucleus by Liebert, Bergeron, & Holberg (2005) yields $T_{\rm eff} = 108,000$ K, $\log g = 7.4$, and an estimated distance of 645 pc. The projected separation of the cool, emission-line companion from the central star is 120 AU, and its nominal orbital period is $\sim$1400 yr. Mike Jura has proposed that the nebula around the dM companion is an accretion disk captured by the star during the outflow that produced the large, faint PN, and is photoionized by radiation from the central star. If this scenario is correct, the disk has survived for the cooling age of the central star, $\sim$135,000 yr.

The most recent development concerning this already remarkable system is the *Spitzer* detection of a 24 μm excess (Kate Su & Jim Liebert, priv. comm.). Thus there are now three components known in the system: the hot nucleus, the dM companion, and dust at a temperature of $\sim$260 K.

The questions raised by EGB 6 are: what is the origin of the compact nebula around the cool companion; how is this nebula confined (or replenished); where is the cool dust found by

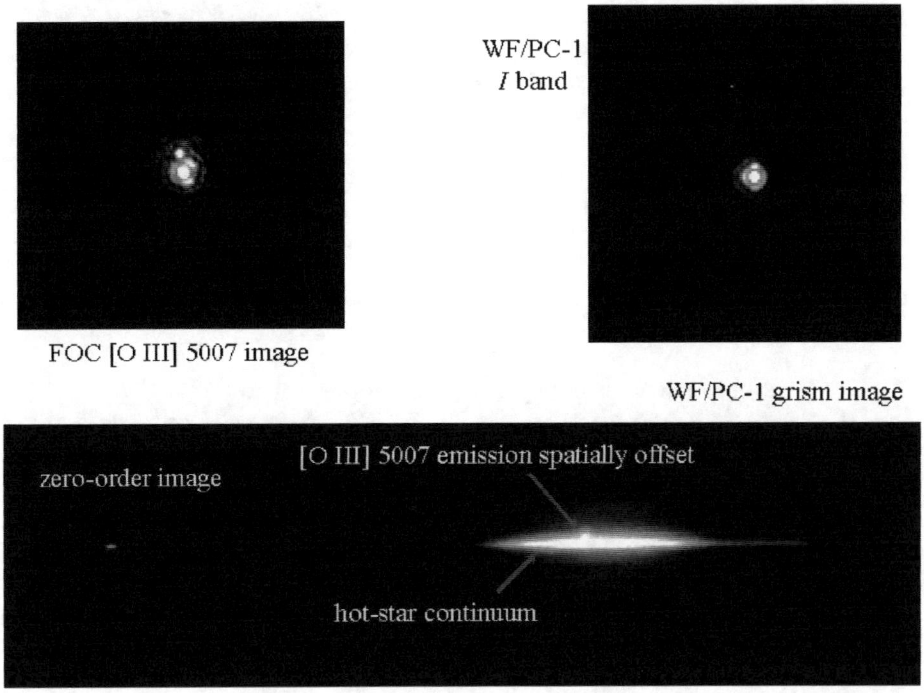

Figure 6. *HST* direct and grism images of the central star of EGB 6. **Upper left:** FOC image taken in [O III] 5007 Å, showing the resolved emission-line source (at about "11:30") separated by 0″.18 from the central star. **Bottom:** A WF/PC-1 grism image of the central star, described in the text. **Upper right:** WF/PC-1 *I*-band image, showing that the continuum of the resolved cool companion is detected in the near-IR.

Spitzer located in the system; and what, if anything, does EGB 6 have to do with the 24 μm excesses found in other PN nuclei and WDs, as discussed in other papers at this workshop?

Acknowledgments

I acknowledge several STScI grants that have supported the work described above over many years.

References

Bergeron P Leggett S K & Ruiz M T 2001 *ApJS* **133** 413
Bond H E Mullan D J O'Brien M S & Sion E M 2001 *ApJ* **560** 919
Clemens J C et al. 1992 *ApJ* **391** 773
Ellis G L Grayson E T & Bond H E 1984 *PASP* **96** 283
Fulbright M S & Liebert J 1993 *ApJ* **410** 275
Gatewood G D & Gatewood C V 1978 *ApJ* **225** 191
Harrington R S Christy J W & Strand K A 1981 *AJ* **86** 909
Liebert J Bergeron P & Holberg J B 2005 *ApJS* **156** 47
Liebert J et al. 1989 *ApJ* **346** 251
Liebert J Young P A Arnett D Holberg J B & Williams K A 2005 *ApJ* **630** L69
O'Brien M S Bond H E & Sion E M 2001 *ApJ* **563** 971
Sion E M Schaefer K G Bond H E Saffer R A & Cheng F H 1998 *ApJ* **496** L29
Walter F M 2004 *AN* **325** 241

Journal of Physics: Conference Series **172** (2009) 012030 doi:10.1088/1742-6596/172/1/012030

Monte Carlo simulations of the binary white dwarf population: a progress report

J Camacho[1], **S Torres**[1,2] **and E García–Berro**[1,2]

[1] Departament de Física Aplicada, Escola Politècnica Superior de Castelldefels, Universitat Politècnica de Catalunya, Avda. del Canal Olímpic 15, 08860 Castelldefels, Spain

[2] Institute for Space Studies of Catalonia, c/ Gran Capità 2–4, Edif. Nexus 104, 08034 Barcelona, Spain

E-mail: judit@fa.upc.edu

Abstract. We present a detailed Monte Carlo simulator of the population of binary stars within the solar neighborhood. We have used the most updated models for stellar evolution, a complete treatment of the Roche lobe overflow episodes, as well as a full implementation of the orbital evolution. Special emphasis has been placed on processes leading to the formation of binary systems in which one of the members is a white dwarf.

1. Introduction

A wide variety of interesting astrophysical problems, with relevance to many current and planned terrestrial and space-borne observatories like Chandra and XMM-Newton — which are devoted to study X-ray binaries — LIGO and LISA — which will study gravitational radiation sources — or Swift — whose main aim is to study gamma-ray burst progenitors — require extensive population synthesis studies. Additionally, several other interesting problems require modeling the Galactic populations of binary stars. Among these we find the study of Type Ia supernova progenitors, of binary systems containing a white dwarf and a neutron star and of double white dwarf binary systems. Including the selection and observational biases in these kind of studies is of the largest importance and can only be reliably done using a Monte Carlo simulator. Here we describe the preliminary results of such a simulator.

2. Building the sample

A detailed description of our code will be presented in a forthcoming publication. Consequently, here we will only summarize its most important inputs. Our model is based on a Monte Carlo simulator of the disk population of single white dwarfs (García–Berro et al. 1999). The code is based in a random number generator (James 1990) which provides an uniform probability density within the interval $(0, 1)$ and ensures a repetition period of 10^{18}, which is virtually infinite for our purposes. The masses of each of the components of the binary system were obtained using a standard initial mass function (Scalo 1988). We only considered stellar masses smaller than $20\,M_{\odot}$. Also, a constant star formation rate and a disk age of 11 Gyr were adopted. In addition, orbital separations were randomly drawn according to a logarithmic probability distribution (Nelemans & Tout 2005). The eccentricities were also randomly drawn according to a thermal distribution (Heggie 1975), $f(e) = 2e$ for $0 \leq e \leq 0.9$. Finally, we computed the

Table 1. Summary of the results obtained for the frequency of binary systems that have suffered none or one common envelope episode.

	Relative to the entire binary population	Binary systems with white dwarfs	White dwarf type for each formation channel		
			He	CO	ONe
Case A	15%	0%	0%	0%	0%
Case B	7%	23%	100%	0%	0%
Case C	6%	0%	0%	0%	0%
Case TPAGB	1%	11%	1%	88%	11%
Detached binaries	71%	66%	0%	96%	4%

orbital evolution of the binary taking into account circularization and synchronization (Zahn 1977, 1989; Hut 1981). The stellar ingredients adopted for each of the components of the binary system are the analytical fits to detailed stellar evolutionary tracks of Hurley et al. (2000). These tracks cover the evolutionary stages until the EAGB. Mass loss through stellar winds was not taken into account until the beginning of the thermally pulsing phase TPAGB. In order to model the TPAGB phase we used the radius-luminosity relation of Hurley et al. (2000) but incorporating the luminosity supplied by Marigo et al. (1996) for masses smaller than $5\,M_{\odot}$, and from Poelarends et al. (2008) for masses larger than this. For the initial-final mass (which links the mass of the white dwarf with that of its progenitor) we used that of Catalán et al. (2008), complemented with the theoretical fit of Iben et al. (1997) for those progenitors with masses larger than $6\,M_{\odot}$.

The Roche lobe radius was modeled according to the prescription of Eggleton (1983), and during the overflow episodes both rejuvenation and ageing were taken into account. Likewise, the overflow treatment was performed following the treatment of Webbink (1985), except when the two stars have a common envelope phase with convective envelopes. In this case we used the double common envelope formalism of Belczynski et al. (2008). During the first common envelope episode gravitational radiation losses and magnetic braking were disregarded. Within these formalisms the mass transfer episodes can be modeled using the radius-mass exponent, $\zeta \equiv d\ln R/d\ln M$. For this we computed $\zeta_{\rm L}$, which represents the response of the Roche lobe itself, $\zeta_{\rm ad}$ which is the adiabatic hydrostatic response of the star, and $\zeta_{\rm eq}$, which is the thermal equilibrium stellar response. It follows that when $\zeta_{\rm ad} < \zeta_{\rm L}$ a dynamical mass transfer ensues, when $\zeta_{\rm eq} < \zeta_{\rm L} < \zeta_{\rm ad}$ a thermal mass transfer takes place, whereas nuclear evolution only occurs when $\zeta_{\rm L} > \zeta_{\rm ad}, \zeta_{\rm eq}$. Additionally, the fraction of mass accreted was modeled taking into account the Kelvin-Helmholtz timescales of both the donor and the accretor.

3. Results

In Table 1 we present a summary of our results. These results are preliminary, as we have only focused on binary systems experiencing at most one common envelope episode. Additionally, in this work we pay special attention to those systems in which at least one of the components is a white dwarf. Systems in which both components are white dwarfs have been counted twice to obtain the frequencies of occurrence. A total of $\sim 9.0 \times 10^4$ binary systems were generated. Of these binary systems, 71% remained detached, 15% underwent a Roche-lobe overflow during the main sequence phase (case A), 7% binaries overflowed its Roche lobe before He ignition (case B), 6% did it before C ignition (case C) and $\sim$ 1% did it during the TPAGB. As a result, 66% of the binary systems containing a white dwarf are detached systems. The most representative channel of formation of white dwarfs when a mass-transfer episode takes place corresponds to a case B

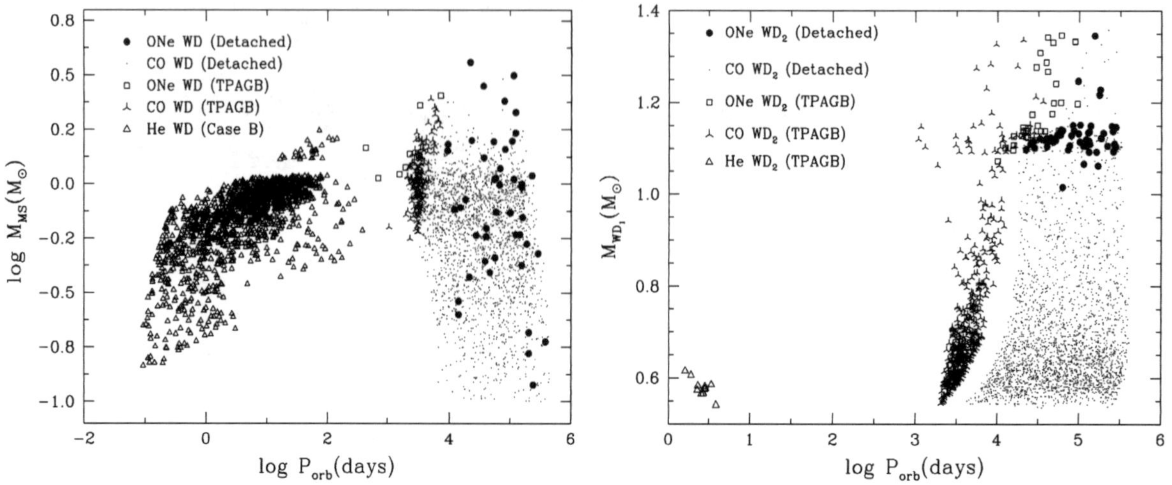

Figure 1. White dwarfs in detached systems or systems which have evolved through one common envelope episode. Left panel: main sequence mass versus orbital period for binary systems composed by a main sequence star and a white dwarf. Right panel: double white dwarf binaries.

Roche lobe overflow (23%), followed by a common envelope episode during the TPAGB phase (11%), whereas no case A or case C Roche lobe overflow results in binary systems containing a white dwarf. Detached binaries in which one of the components is a white dwarf amount to 66%.

If we classify the systems according to the type of the companion we find that He white dwarfs formed in a case B Roche lobe overflow have always main-sequence companions. Here we do not consider those cases in which the resulting system is composed of two He white dwarfs, since these systems are expected to merge due to gravitational wave radiation in a time smaller than the age of the Galaxy. When a case C Roche-lobe overflow is considered the outcome is in all the cases a He white dwarf with a He giant star companion, although the companion star is expected to evolve through a possible second common envelope episode, which we have not yet modeled. For systems resulting from a Roche lobe overflow during the TPAGB phase we obtained than the most favored outcome is that in which a binary system consisting of two CO white dwarfs is formed. This occurs in 59% of the cases. In 25% of the cases the companion of the CO white dwarf is a main sequence star. In 7% of the cases we obtain a double white dwarf binary system composed of a CO white dwarf and an ONe white dwarf. Double ONe white dwarfs occur in 6% of the cases, while systems containing a He white dwarf and a CO white dwarf represent a 2%. Finally, systems in which the primary is an ONe white dwarf and the secondary is a main sequence star occur in only 1% of the cases. Regarding detached binaries, the most common systems are those in which one of the stars is a CO white dwarf while the companion is a main sequence star (65%), followed by those systems in which the two components are CO white dwarfs (30%), while systems consisting of an ONe and a CO white dwarf only amount to a modest 3%. The rest of the possible outcomes are quite infrequent. For instance, systems in which both member of the system are ONe white dwarfs amount to 1%, whereas systems in which the primary is an ONe white dwarf while the secondary is a main sequence star represent 1% of the cases.

The left panel of Fig. 1 shows the characteristics of those systems in which one of the components is a white dwarf while the companion is a main sequence star. We have chosen to show only detached systems, systems which have undergone a case B episode or systems which have evolved through one common envelope episode during the TPAGB phase. As can be seen, those systems in which the white dwarf has a He core have orbital periods shorter than 10^2 days while when the companion is a CO or an ONe white dwarf the orbital periods are larger than 10^3 days. Moreover, white dwarfs formed during a common envelope episode during the TPAGB present a narrow distribution of orbital periods and have main sequence companions with masses ranging from $\sim 0.6\,M_\odot$ to $3\,M_\odot$.

The right panel of Fig. 1 illustrates the properties of our simulated double white dwarf binary systems. Again, we have chosen to show only white dwarfs in detached systems or systems which have evolved through one common envelope episode during the TPAGB phase. For these systems we show the distribution of masses of the most massive white dwarf as a function of the orbital period. As can be seen, those systems in which the primary is a CO white dwarf while the secondary is a He white dwarf have orbital periods around $\sim$ 2 days, while the rest of white dwarf pairs have orbital periods larger than 10^3 days. Also, double CO white dwarf pairs formed from a Roche-lobe overflow during the TPAGB have orbital periods ranging from $\sim 10^3$ to $\sim 10^4$ days, while detached double white dwarfs present a wider range of orbital periods. In the same way, those systems in which both components are ONe white dwarfs formed during a common envelope episode during the TPAGB have shorter orbital periods than those formed in detached systems.

4. Summary and conclusions

We have presented the preliminary results obtained using a Monte Carlo simulator of the Galactic population of binary systems containing at least one white dwarf. Specifically, we have performed an study of those systems in which one of the components is a white dwarf and are either detached systems or have suffered only one common envelope phase. Special emphasis has been placed in studying the resulting orbital parameters (eccentricities and orbital separations) as a previous step to compare the theoretical results with observational data.

Acknowledgments

Part of this work was supported by MEC grant AYA05-08013-C03-01, by the European Union FEDER funds and by the AGAUR.

References

Belczynski K Kalogera V Rasio F A Taam R E Zezas A Bulik T Maccarone T J & Ivanova N 2008 *ApJS* **174** 223
Catalán S Isern J García-Berro E & Ribas I 2008 *M.N.R.A.S.* **387** 1693
Eggleton P P 1983 *ApJ* **268** 368
García–Berro E Torres S Isern J & Burkert A 1999 *M.N.R.A.S.* **302** 173
Heggie D C 1975 *M.N.R.A.S.* **173** 729
Hurley J R Pols O R & Tout C A 2000 *M.N.R.A.S.* **315** 543
Hut P 1981 *A&A* **99** 126
Iben I Ritossa C & García-Berro E 1997 *ApJ* **489** 772
James F 1990 *Comput. Phys. Commun.* **60** 329
Marigo P Bressan A & Chiosi C 1996 *A&A* **313** 545
Nelemans G & Tout C A 2005 *M.N.R.A.S.* **356** 753
Poelarends A J T Herwing F Langer N & Heger A 2008 *ApJ* **675** 614
Scalo J 1998 in *"The Stellar Initial Mass Function"* Eds.: G Gilmore & D Howell (San Francisco: PASP Conference Series), **142** 201
Webbink R F 1985 in *"Interacting binaries"*, Eds.: J E Pringle & R A Wade (Cambridge: Cambridge University Press) 39
Zahn J P 1977 *A&A* **57** 383
Zahn J P 1989 *A&A* **220** 112

Journal of Physics: Conference Series **172** (2009) 012031 doi:10.1088/1742-6596/172/1/012031

The WD perspective on the PN binary hypothesis

Orsola De Marco, J Farihi and J Nordhaus

American Museum of Natural History, Central Park West at 79th Street, New York, NY 10024, USA

Department of Physics & Astronomy, University of Leicester, Leicester LE1 7RH, UK

Dept. of Astrophysical Sciences, Princeton University, Princeton, NJ 08544, USA

E-mail: orsola@amnh.org

Abstract. There is no solid theoretical precept through which we can understand the amazing shapes of planetary nebulae (PN). The only plausible theoretical explanation of most PN shapes is that a companion interacts with the mass-losing progenitor of the central star of PN. Plausibility, however, is not sufficient to prove reality and efforts are ongoing to test the *PN binary hypothesis* observationally. Here, a short review is presented of the problem, but this time set within the context of binarity in the white dwarf (WD) population. We find that observationally the PN and WD binary populations agree. Most noticeably, both populations point to a deep flaw with current theories of the *common envelope* binary interaction. However, unexplained issues remain, several of which point to a sizable population of as yet undetected intermediate period binaries. Unfortunately, today the WD binaries do not constrain sufficiently the PN binary hypothesis.

1. Introduction

Text books tell us that a planetary nebula (PN) is produced by a mass-losing asymptotic giant branch (AGB) star and later ionized by the AGB star core, on its way to becoming a white dwarf (WD). The great majority of PNe are not spherical. They display axi- and point-symmetries, sometime with the addition of jet-like structures. These shapes have been attributed to the action of stellar rotation and/or magnetic fields (García-Segura et al. 1999, 2005), which result in an equatorially-concentrated AGB mass-loss. When the post-AGB star heats up, it blows a fast but tenuous wind, which is constrained at the equator, resulting in an elliptical or bipolar PN. In a minority of cases, the action of a companion is thought to be required in order to produce jets, point symmetries, or to explain off-center central stars (see Balick and Frank 2002 for a review).

For the last two decades the community debated whether stellar rotation and global magnetic fields can be sustained in a *single* AGB star (Soker and Livio 1989, Soker 1997, De Marco 2006, Zijlstra 2007). Recently, the debate has been rekindled (Soker 2006, Nordhaus et al. 2007) by the argument that in a large majority of cases, single AGB stars are unlikely be able to sustain large scale magnetic fields for long enough to affect shaping, because the field drains the star of angular momentum on short time scales and quenches itself. As the models have shown, magnetic fields can be an effective ingredient in shaping many PNe, but in those models the magnetic field strength was assumed constant and was not coupled to the negative feedback action of the stellar envelope. Without the action of an angular momentum source, the field would vanish.

Journal of Physics: Conference Series **172** (2009) 012031 doi:10.1088/1742-6596/172/1/012031

Such angular momentum source could effectively be provided by a companion. If this is what actually happens, then PNe would be, by and large, a binary-interaction phenomenon.

1.1. Constraints on the binary hypothesis from an observational point of view

Ideally, we would detect the central star binary fraction and put an end to the debate. We know that 10-15% of all central stars are short period binaries with P$\lesssim$3 days (Bond 2000). All these stars have gone through a common envelope (CE), whereby the companion is engulfed by the primary when it expands on the red giant branch (RGB) or, in the case of central binaries of PN, the AGB. The core of the giant and the companion therefore find themselves inside the envelope of the giant and gravitational drag forces reduce their orbital separation. Deposition of energy and angular momentum into the envelope eventually results in its ejection and the emergence of a close binary (Paczynski 1976). We also know that about 10% of all central stars are in wide (visual) binaries (Ciardullo et al. 1999), which are unlikely to have interacted.

Radial velocity surveys have been carried out (e.g., De Marco et al. 2004, Afšar and Bond 2005) to detect binaries with P$\gtrsim$3 days, but central stars have high luminosities and strong winds, resulting in strong spectral line variability which mask even the strongest binary signal (De Marco et al 2008a). Near IR excess techniques have, when applied to central stars of PN, their own problems, such as the fact that central stars are bright even in the near IR and their distances tend to be large, such that (the common) M-type main sequence companions are hard to detect. This said, Frew and Parker (2008) has analyzed a sample of 2MASS and DENIS-observed objects and deduced that a third to one half of them have an IR excess indicative of binarity. The caveat here is that companions fainter than M4-8V stars (depending on the luminosity of the primary) cannot be detected.

1.2. Constraints on the binary hypothesis from a population synthesis point of view.

Moe and De Marco (2006) determined from a population synthesis study that the number of PN in the Galaxy should be 6 times larger than is observed, at the 3σ level, if single stars and binaries within a certain mass range make PNe. Their prediction was based on the standard rules of stellar and galactic evolution. They therefore concluded that only some of these objects, maybe those which have gone through a binary interaction, can make a PN. Moe & De Marco (in preparation) have determined not only the fraction of all "classically-produced" PN that go through a binary interaction, but produce a binary period distribution for the central stars of PN and WDs that we will use later.

In this paper, we cast the information about central stars of PN in binaries in the context of the WD binary population and use the combined information to constrain the PN binary hypothesis.

2. Binarity on and after the AGB: consolidating period distributions

In Fig. 1 we show a cartoon Hertzsprung-Russell (HR) diagram portraying the evolution of binaries from the AGB, to the WD phase. Those binaries that enter a CE will quickly move to the hot side of the HR diagram, become close binary central stars of PN and eventually become close WD binaries, while those binaries that avoid a CE will become intermediate period binary post-AGB stars, intermediate period central star binaries and eventually intermediate period WD binaries. Below we review what we know of the periods of post-AGB binary populations and compare their distributions.

2.1. The WD binary periods distribution

All evolved populations should reflect the binary fraction and period distribution of their parent population with some exceptions. The main exceptions are that, binaries with periods shorter

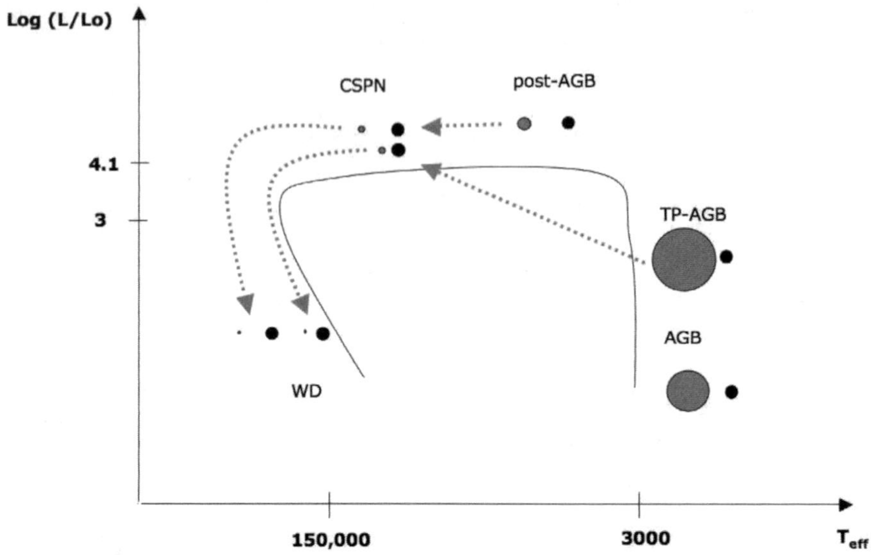

Figure 1. Cartoon HR diagram portraying life as a binary. The lighter star evolves from the lower AGB, to the thermally-pulsing (TP) AGB to the post-AGB phase and then to the WD phase. The darker star is an unevolved low mass main sequence star. Depending on the binary separation on the AGB the resulting binary will be close (post-CE) or intermediate/wide. The post-AGB phase can be skipped if a CE takes place.

than a certain threshold will have their period shortened further by a CE interaction. In addition, some periods will be altered by other interactions, for instance tidal captures. There will be then a small fraction of binaries that merge and, finally, there are those binaries that are too wide to interact and have their separations widened further when one or both of the components suffer mass-loss.

Farihi et al. (2006) determined the projected separation distribution of a sample of WDs known to have near IR excess indicative of a cool companion. We have plotted the 42 objects from Table 5 of Farihi et al. (2006) that are marked WD+RD (red dwarf) in Fig. 2 where we have de-projected all separations by dividing them by 0.67, the average of $\sin i$, and obtained periods by using a WD mass of 0.6 $M_\odot$ and a RD mass of 0.34 $M_\odot$ (corresponding to an M4V star). The group at lower periods are all upper limits. The gap between the two groups is the result of the CE interaction shortening orbital separations. The cluster of lower limit periods constitutes 30% of the total, possibly pointing to the fact that only 10-15% of all WD binaries are post-CE systems (assuming a total WD binary fraction of 30-40%; Holberg, these proceedings).

Schreiber et al. (2008) determined *directly* that 35% of all WD+RD binaries are post-CE systems. This is approximately the same figure one would deduce from the work of Farihi et al. (2006) and translates to about 10-15% of all WDs, assuming the same WD binary fraction of 30-40%. Schreiber et al. (2008) determined the periods of 7 post-CE WD+RD binaries, which we plot on a separate histogram (Fig. 3) alongside the work of Farihi et al. (2006). What is extremely surprising in their finding is that *all* post-CE binaries detected so far have periods smaller than a day, where one can confidently state that if binaries with periods longer than that limit and shorter than a few weeks had been present they would have been easily detected. This is in contradiction with *all* predictions of period distributions for post-CE populations (e.g.,

Yungelson et al. 1993, Han et al. 1995), no matter what values of the CE efficiency parameter is chosen. As we shall see below, this is also in line with the post-CE central stars of PN, as suggested already by De Marco et al. (2008b).

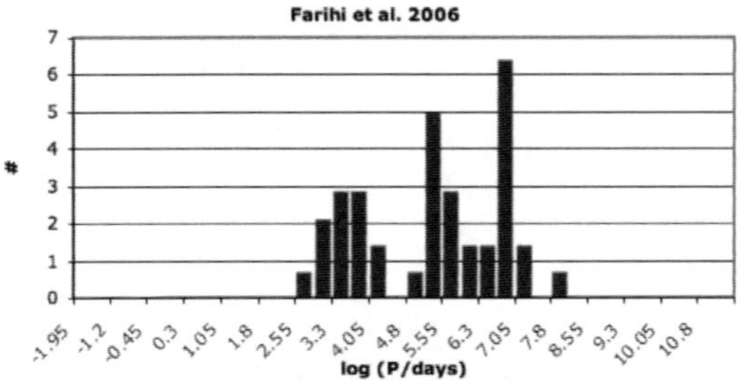

Figure 2. The WD+RD period distributions obtained by the HST imaging project of Farihi et al. (2006). The lower period group are upper limits, presumably corresponding to the distribution revealed by Schreiber et al. (2008) in Fig.3. The distributions is scaled to a total of 30, representing the ~30% binary fraction of WD.

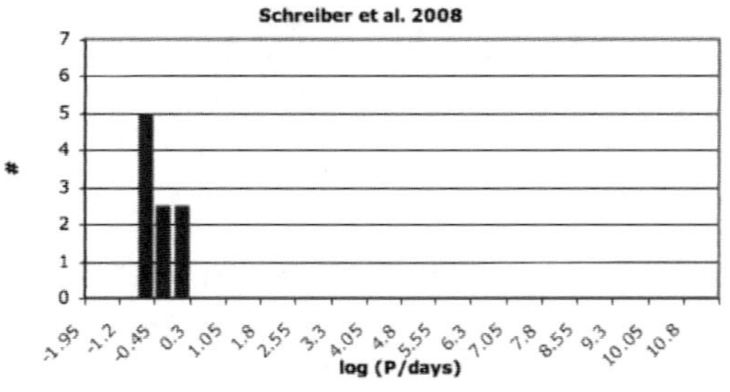

Figure 3. The WD+RD period distributions obtained by the radial velocity program of Schreiber et al. (2008). The distributions are scaled to a total of 10, representing the ~10% post-CE binary fraction for WD.

2.2. The predicted vs. observed WD period distribution

In Figs. 4 and 5, we have combined the WD period distribution from Farihi et al. (2006) but have substituted the short period group in that distribution with the post-CE WD period distribution of Schreiber et al. (2008). This is in accordance with what we concluded in § 2.1. In the same Figure we also present a preliminary prediction by Moe & De Marco (in preparation) which evolved a population of G-F stars (using the period distribution of Duquennoy and Mayor 1991) to the WD stage. This distribution is scaled to 55%, which accounts for a 60% binary fraction determined by Duquennoy and Mayor (1991) minus a small percentage of stars which have merged. The predicted distribution should only be taken as indicative, except for the following four characteristics: (i) the WD binary fraction, at 30-40% seems lower than what might be expected if they derive from main sequence stars with spectral type G-F (which should approximately represent the mean WD progenitors as can be judged by the mean WD mass as determined by Liebert et al. 2005 and the initial-final mass relation of Weidemann 2000). (ii) The predicted post-CE period distribution extends to longer periods compared to the observations. While these predictions are still preliminary they are similar to those of previous population synthesis work (Yungelson et al. 1993, Han et al. 1995). (iii) The prediction has some intermediate period binaries. While the exact period of these objects is very uncertain, it is likely that such objects exist, since several companions should be tidally captured by the AGB giant, but will avoid CE because the giant rotation and orbit of the companion become synchronized before capture takes place. (iv) Finally, the position of the upper edge of the CE gap is predicted well.

2.3. The binary central star of PN period distribution

As we have already described, binary central stars are hard to detect. 10-15% are known to have periods smaller than ~3 days. This has long been presumed to be a lower limit, due to the fact that the irradiation technique used to detect these object is less effective for larger orbital

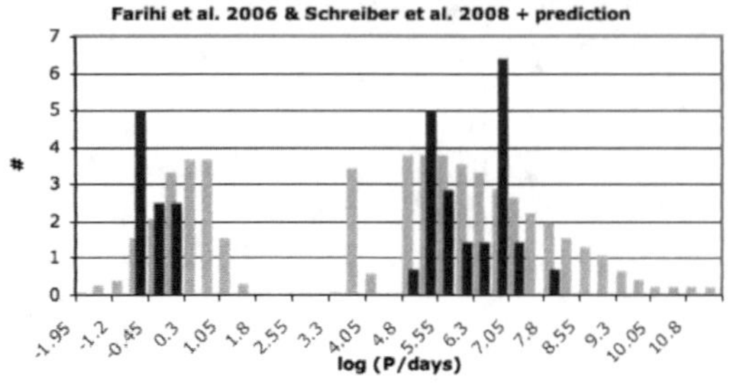

Figure 4. The WD period distribution (darker grey/green), scaled to 30 as in Fig. 2, compared with a preliminary prediction by Moe and De Marco (in preparation; light grey/blue), scaled to 55 as explained in the text.

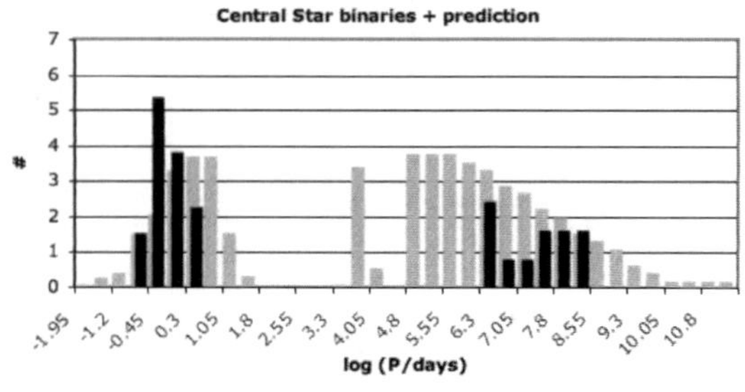

Figure 5. The central star of PN binary period distribution (dark grey/burgondy, scaled to 22) compared with a preliminary prediction by Moe and De Marco (in preparation; light grey/blue).

separations. However, De Marco et al. (2008b) showed that if systems with 3 days $\lesssim$ P $\lesssim$ 15 days were plentiful, the technique should have detected them. This led De Marco et al. (2008b) to suggest that either there is another bias that prevents the detection of systems with periods in the ~3-15 day range, or that these systems do not exist.

The wide central star binary period distribution was derived by using data from Ciardullo et al. (1999), de-projecting projected separations and using the same mass constraints to derive periods as used for the WD wide binaries in § 2.1. The binary fraction for this population is 9%.

The comparison of *short period* systems between the WDs and the central stars can be carried out by looking at Figs. 4 and 5 where we have used the predicted distribution to facilitate the comparison. The obvious lessons to be drawn from the comparison are: (i) post-CE central star binaries might extend to slightly longer periods (~3 days) than post-CE WD binaries (~1 day), although this needs to be checked with a larger sample. (ii) Both central stars and WD close binary distributions fall short of the maximum period predicted for post-CE systems. We emphasize here once again that such prediction relies critically on the value chosen for the CE efficiency, but that no matter what this choice is, *all* predictions suggest that CEs can produce a substantial number of binaries with periods $\gtrsim$1-3 days, contrary to these observations. (iii) for the wider binaries, the WD distribution is closer to the upper edge of the CE gap, while the central stars cluster at much larger periods. In fact, there are no wide central star binaries with separations near the HST resolution limit (0.05 arcsec or 50 AU at a typical PN distance of 1 kpc), while predictions suggest a rising number of binaries for decreasing periods down to the gap limit. (iv) We are not surprised at the dearth of WD and central star wide binaries with the very longest periods, since this is due to biases of the respective detection techniques.

We leave the discussion of the intermediate period binaries seen in the prediction to § 2.4.

2.4. The post-AGB binary period distribution.

We finally discuss the intermediate period binaries. First of all, we point out that detecting central star, or WD, binaries with periods from several months to several years is extremely difficult. For all intents and purposes we can therefore assume that the large period gap between short and long periods WD and central star binaries could well be partly populated.

As we explained in § 2.2, Moe and De Marco (in preparation) predict that there could be a population of binaries in that period gap where the secondary was tidally captured but avoided the CE. This statement is in line with the existence of a population of binaries where the primary is a post-AGB star and the secondary is, most likely, a main sequence star. Post-AGBs are stars that are still too cool to ionize the circumstellar material that was ejected during the AGB and so they cannot yet have a PN. It is however likely that these stars will in due course ionize the PN and become intermediate period central star of PN and eventually WD+RD intermediate period binaries. In Fig. 1 these are labeled as post-AGB binaries. We should also point out that close post-AGB binaries are not possible since post-AGB stars have radii that are multiple times the solar radius (5-50 $R_{\odot}$) so that a stable companion must be at a few stellar radii, implying periods of the order of 100 days and above.

It appears that ~30% of the entire post-AGB population has such intermediate period companions (P=100-1500 days; van Winckel 2003). If so we could expect to find a sizable population of central star and WD binaries in this period range. This would in turn raise the overall binary fraction to be more in line with what we expect from the main sequence binary fraction (to something of the order of 40-45%=10-15%+30%). (We note, however, that while this is plausible for the central star population, binary searches for WDs are more complete and it is likely that a population of intermediate period WD binaries would have been already discovered.) We plot these binaries along with the central stars and the prediction in Fig. 6, where we scaled the post-AGB binaries to a value of 30. As can be seen from the figure, their period range is not that predicted. However, predictions of these periods are extremely sensitive to assumptions and should be considered approximate.

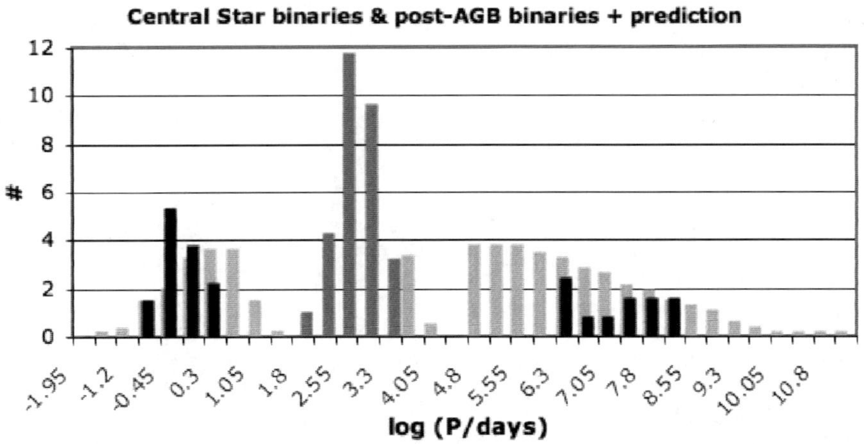

Figure 6. The central star of PN binary period distribution (black/burgondy), plotted along with the post-AGB binary period distribution (dark grey/dark orange) and compared with a preliminary prediction by Moe and De Marco (in preparation; light grey/blue).

3. Conclusion

The comparison of WD and central star period distributions has allowed us to draw a number of conclusions.

- There is a fundamental agreement between the period distributions of the shortest periods WD and central star binaries. This points to a deficit in our understanding of the CE interaction in that even using the whole range of plausible CE efficiency parameters, post-CE binaries are predicted with periods *longer* than we observe.

- There is an overall dearth of binaries in the WD population compared to the main sequence population which is plausibly its progenitor, but which could be reconciled with expectations if there were a population of intermediate period binaries that is as yet undetected.
- We must be aware that disks and tori present around a binary (e.g., all binary post-AGB objects have such disks) can alter their periods.
- The lack/dearth of wide central star binaries with separations in the 20-500 AU range is puzzling in that one would expect there to be more binaries at the short period end of the wide binary period/separation distribution; this is indeed the case for the WDs. Although more objects need to be detected to determine whether this is an effect of low number statistics, we wonder whether these wide binary central stars could be triple systems (one is known to be; A 63; Ciardullo et al. 1999, De Marco et al. 2008b). We know that $\sim$10% of all main sequence stars of type G-F are indeed triples (Duquennoy and Mayor 1991). If so, it is possible that the wider of the three stars needs to be farther away to allow a stable system.
- Finally, even if an additional 30% of intermediate period central star binaries is detected, the central star binary fraction with an orbital separation that allows a past interaction, would only be about 40-45%, too small to justify the entire PN phenomenon with binarity. This unfortunately does not yet disprove the PN binary hypothesis, although it might be already suggesting that a binary-only evolutionary channel for PN might not be correct. Difficult to observe interactions with very faint companions such as brown dwarfs and planets cannot however be excluded and may make up a sizable part of the PN population. For a full review see De Marco (2009).

Acknowledgments
I acknowledge conversations with Matthias Schreiber, Alberto Rebassa-Mansergas. I also acknowledge financial support from NSF grant AST-0607111.

References
Afşar M and Bond H E 2005 *Memorie della Societa Astronomica Italiana* **76** 608
Balick B and Frank A 2002 *ARAA* **40** 439–486
Bond H E 2000 *ASP Conf. Ser. 199: Asymmetrical Planetary Nebulae II: From Origins to Microstructures* p 115
Ciardullo R Bond H E Sipior M S Fullton L K Zhang C Y and Schaefer K G 1999 *AJ* **118** 488–508
De Marco O 2006 *Planetary Nebulae in our Galaxy and Beyond* (*IAU Symposium* vol 234) ed Barlow M J and Méndez R H pp 111–118
De Marco O 2009 *PASP*, in press
De Marco O Bond H E Harmer D and Fleming A J 2004 *ApJL* **602** L93–L96
De Marco O Wortel S Bond H E and Harmer D 2008a *Asymmetric Planetary Nebulae IV, in press* pp ArXiv e–prints 0709.1508
De Marco O Hillwig T C and Smith A J 2008b *AJ* **136** 323–336
Duquennoy A and Mayor M 1991 *A&A* **248** 485–524
Farihi J Hoard D W and Wachter S 2006 *ApJ* **646** 480–492 (*Preprint* arXiv:astro-ph/0603747)
Frew D J and Parker Q A 2008 *Asymmetric Planetary Nebulae IV, in press*
García-Segura G Langer N Różyczka M and Franco J 1999 *ApJ* **517** 767–781
García-Segura G López J A and Franco J 2005 *ApJ* **618** 919–925 (*Preprint* arXiv:astro-ph/0409595)
Han Z Podsiadlowski P and Eggleton P P 1995 *MNRAS* **272** 800–820
Liebert J Bergeron P and Holberg J B 2005 *ApJS* **156** 47–68
Moe M and De Marco O 2006 *ApJ* **650** 916–932 (*Preprint* arXiv:astro-ph/0606354)
Nordhaus J Blackman E G and Frank A 2007 *MNRAS* **376** 599–608 (*Preprint* arXiv:astro-ph/0609726)
Paczynski B 1976 *IAU Symp. 73: Structure and Evolution of Close Binary Systems* ed Eggleton P, Mitton S and Whelan J pp 75
Schreiber M R Gänsicke B T Southworth J Schwope A D and Koester D 2008 *AAP* **484** 441–450 (*Preprint* arXiv:0709.4545)
Soker N 1997 *ApJS* **112** 487

Soker N 2006 *PASP* **118** 260–269 (*Preprint* arXiv:astro-ph/0501647)
Soker N and Livio M 1989 *ApJ* **339** 268–278
van Winckel H 2003 *AARA* **41** 391–427
Weidemann V 2000 *A&A* **363** 647–656
Yungelson L R Tutukov A V and Livio M 1993 *ApJ* **418** 794
Zijlstra A A 2007 *Baltic Astronomy* **16** 79–86 (*Preprint* arXiv:astro-ph/0610558)

Journal of Physics: Conference Series **172** (2009) 012032 doi:10.1088/1742-6596/172/1/012032

(F)UV Spectroscopy of the hybrid PG 1159-type central stars of the planetary nebulae NGC 7094 and Abell 43

M Ziegler[1], T Rauch[1], K Werner[1], L Koesterke[2] and J W Kruk[3]

[1]Institute for Astronomy and Astrophysics, Kepler Center for Astro and Particle Physics, Eberhard Karls University, Tübingen, Germany

[2]Texas Advanced Computing Center, University of Texas, Austin, U.S.A.

[3]Department of Physics and Astronomy, Johns Hopkins University, Baltimore, U.S.A.

E-mail: ziegler@astro.uni-tuebingen.de

Abstract. Hydrogen-deficient post-AGB stars have experienced a late helium-shell flash that mixes the hydrogen-rich envelope and the helium-rich intershell (between hydrogen- and helium-burning shell). The amount of hydrogen remaining in the stellar envelope depends on the particular moment when this late thermal pulse occurs. Previous spectral analyses of hydrogen-deficient post-AGB stars, namely PG 1159−035 and the central stars of the planetary nebulae (CSPN) K 1−16, NGC 7094, and Abell 78, surprisingly revealed strong iron deficiencies of up to 1 dex. A possible explanation may be neutron captures due to an efficient s-process on the AGB that transformed iron into heavier elements. An increased abundance of these would be a strong indication for this scenario. Since reliable atomic data for highly ionized species heavier than iron is only available for cobalt and nickel, we can presently determine only the nickel abundance. We performed a detailed spectral analysis by means of NLTE model-atmosphere techniques based on high-resolution UV observations of the two PG 1159-type CSPN of NGC 7094 and Abell 43 which are spectroscopic twins, i.e., they exhibit very similar spectra. We confirmed a strong iron-deficiency of at least one dex in both stars. The search for nickel lines in their UV spectra was entirely negative. We find that both stars are even nickel deficient by about one dex.

1. Introduction

The hydrogen-deficient PG 1159 stars are the result of a late helium-shell flash (late thermal pulse, LTP; for details see Werner & Herwig 2006). The occurrence of such a flash (the re-ignition of helium-shell burning in a post-AGB star or white dwarf), was predicted in earlier investigations e.g. by Iben et al. (1983). For a schematic picture of the inner structure of an AGB star see Fig. 1. The time when the flash occurs determines the remaining photospheric hydrogen (Herwig 2001). Still on the AGB (AGB final thermal pulse, AFTP), flash-induced mixing of the hydrogen-rich envelope ($10^{-2}\,M_\odot$) with the helium-rich intershell layer ($10^{-2}\,M_\odot$) reduces the hydrogen abundance but Balmer lines remain detectable. After the departure from the AGB the hydrogen-rich envelope is less massive ($10^{-4}\,M_\odot$). If the nuclear burning is still "on", i.e. the star evolves at constant luminosity, the mixing due to a LTP reduces hydrogen below the detection limit (which is about 10 % at the high surface gravities of $\log g \approx 7$). After a post-AGB star has entered the white-dwarf cooling sequence and the nuclear burning is "off", a very late thermal pulse (VLTP) will cause convective mixing of the entire hydrogen-rich envelope down to the bottom of the helium-burning shell where hydrogen is completely burned. In that case, the star will become completely hydrogen free at that time. The hybrid PG 1159 stars, e.g. the CSPN of NGC 7094 and Abell 43, are AFTP stars.

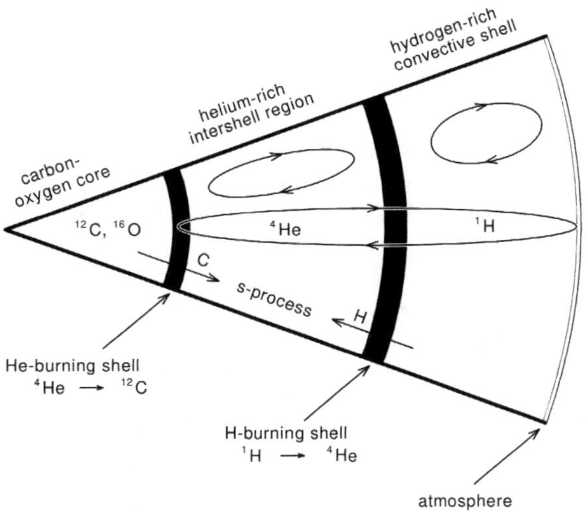

Figure 1. Inner structure of an AGB star. Note that the convection zone from the surface to the bottom of the helium-burning shell is established during a late thermal pulse only.

Previous spectral analysis of three PG 1159 stars (NGC 7094, K 1-16, PG 1159-035) as well as one [WC]-PG 1159 transition object (Abell 78) have revealed a strong, unexpected Fe-deficiency of at least 1 dex (Miksa et al. 2002, Werner et al. 2003a). A possible explanation for the Fe-deficiency is the transformation of Fe into heavier nuclei due to s-process. Our analysis aims to determine the Ni abundance in the CSPN of NGC 7094 and Abell 43. An increased nickel abundance would be a hint for the transformation of Fe nuclei into heavier elements due to n-capture.

NGC 7094 was discovered by Lewis Swift in 1888 and it was classified as a planetary nebula (PN) in 1963 (Kohoutek 1963, PN G066.7−28.2). NGC 7094 is located at a distance of 2.2 kpc (Napiwotzki 1999) and has an apparent size of $102''\!.5 \times 99''\!.4$ (Tylenda et al. 2003). The photospheric parameters of its central star (CS) were previously determined by Dreizler et al. (1997) ($T_{\rm eff}$ = 110 kK, $\log g$ = 5.7 (cm/sec^2), H:He:C=0.36,0.43,0.21 in mass fractions). Abell 43 was discovered in 1955 (Abell 1955, PN G036.0+17.6). Its apparent size is $80''\times80''$ (Napiwotzki & Schönberner 1995) and it is located at a distance of 2.2 kpc (Napiwotzki 1999). The CSPN of Abell 43 is a spectroscopic twin of the CSPN of NGC 7094 (Miksa et al. 2002).

We note that the CSPN of Abell 43 as well as of NGC 7094 are non-radial g-mode pulsators (Vauclair et al. 2005, Solheim et al. 2007). According to Solheim et al. (2007), the pulsation periods of Abell 43 lie in a range between 2 380 and 6 075 sec, which agrees reasonably well with the predicted theoretical periods (2 604 − 5 529 sec, Quirion et al. 2005). The theoretical pulsation periods of NGC 7094 (2 550 − 5 413 sec, Quirion et al. 2005) are also in good agreement with the observed periods (2 040 − 4 980 sec, Solheim et al. 2007).

In this analysis, we determine the photospheric parameters of the CSPN of NGC 7094 and Abell 43 precisely. In Sect. 2 we briefly describe the observations. Then, in a first step, we scrutinize the photospheric parameters determined by Dreizler et al. (1997) (Sect. 3). A detailed spectral analysis based of models that consider opacities of all elements from H to Ni is presented in Sect. 4.

2. Observations

A precise spectral analysis of hot stars requires high-resolution and high-S/N UV observations because most of the strategic lines of highly ionized metals are located in this wavelength range. E.g. in the Far Ultraviolet Spectroscopic Explorer (*FUSE*) spectra (905 – 1187 Å, resolving power $R \approx 20\,000$), strong Fe VII and Ni VI lines might be found.

For our analysis, we performed *HST/STIS* observations of the CSPN of NGC 7094 (Table 1). They cover the range between 1150 – 1730 Å with $R \approx 45\,800$. The spectra were retrieved from the MAST archive (standard pipeline reduction in version of 2006) and then co-added and subsequently smoothed with a Savitzky-Golay filter (Savitzky & Golay 1964). The achieved S/N is > 50. For Abell 43 no *HST/STIS* observation is available. *FUSE* observations were performed for both objects (Table 1), and processed in the standard pipeline with CalFUSE v3.1.

Table 1. *HST/STIS* and *FUSE* observation log for the CSPN of NGC 7094 and Abell 43.

Object	Instrument	Dataset	Start time (UT)	Aperture	Exp. time / sec
NGC 7094	HST/STIS	O8MU02010	2004-06-24 20:43:30	E140M	650
NGC 7094	HST/STIS	O8MU02020	2004-06-24 22:19:29	E140M	656
NGC 7094	HST/STIS	O8MU02030	2004-06-24 23:55:29	E140M	655
NGC 7094	FUSE	P1043701000	2000-11-13 08:53:28	LWRS	22754
Abell 43	FUSE	B0520202000	2001-08-03 22:18:20	LWRS	9528
Abell 43	FUSE	B0520201000	2001-07-29 20:41:47	LWRS	11438

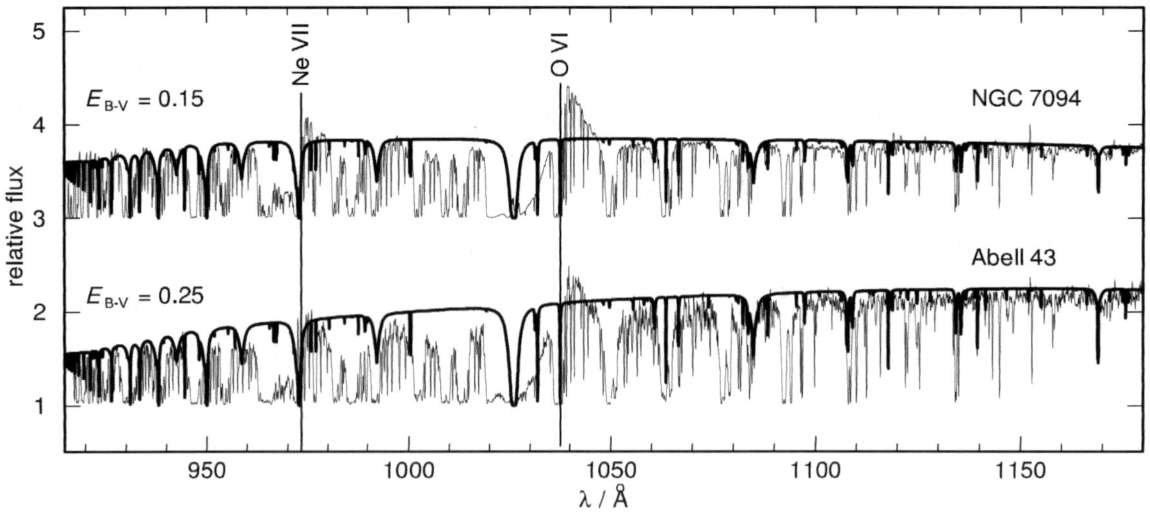

Figure 2. Comparison of the *FUSE* spectra of the CSPN of NGC 7094 (top) and Abell 43 with synthetic spectra of our final model.

The *FUSE* observations of the CSPN of NGC 7094 and Abell 43 are strongly contaminated by interstellar line absorption (Fig. 2). We determine a reddening of $E_{B-V} = 0.15$ and $E_{B-V} = 0.25$ for the CSPN of NGC 7094 and Abell 43, respectively. A prominent P Cygni profile of O VI $\lambda\lambda\,1031.9, 1037.6$ Å is visible in the *FUSE* spectra of both stars. Ne VII $\lambda\,973.3$ Å shows a weaker P Cygni profile (Fig. 5).

3. Model Atmospheres and Preliminary Analysis

Dreizler et al. (1997) used medium-resolution optical spectra of the CSPN of NGC 7094 for their analysis. Since then, our model atmospheres improved and high-resolution, high-S/N ($R = 18\,00$) spectra were obtained during the SPY campaign (Napiwotzki et al. 2001) with *VLT/UVES*. Before we started to calculate detailed model atmospheres with all elements from H to Ni, we wanted to check the previously determined parameters.

Spectral energy distributions (SEDs) from *TheoSSA*[1], a service provided by the German Astrophysical Virtual Observatory (*GAVO*[2]), showed that $T_{\rm eff} = 100\,\rm kK$ and $\log g = 5.5$ with a H/He abundance ratio of H:He=0.17:0.69 gives a much better fit (Fig. 3) compared to the "old" parameters given by Dreizler et al. (1997). Especially the hydrogen abundance is much less than previously thought. We adopt $T_{\rm eff} = 100\,\rm kK$, $\log g = 5.5$, and H:He=0.17:0.69 for our further analysis. The abundances of the other metals that are considered in our models are adjusted to fit the observation (Sect. 4).

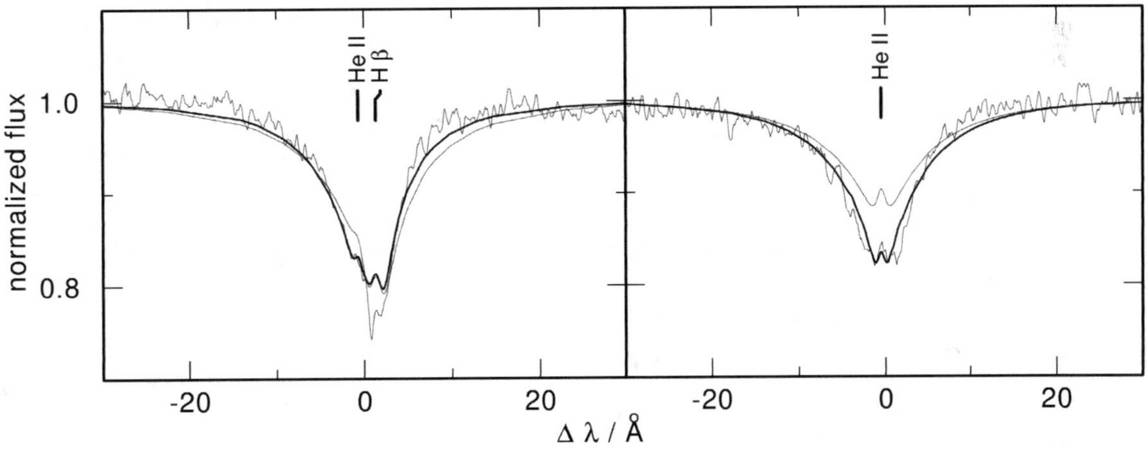

Figure 3. Synthetic line profiles of He II $\lambda\,4860.7\,\text{Å}$ / H β (left) and He II $\lambda\,5411.5\,\text{Å}$ (right) calculated from two H+He+C+N+O model atmospheres (thick: $T_{\rm eff} = 100\,\rm kK$, $\log g = 5.5$, H:He=0.17:0.69; thin: $T_{\rm eff} = 110\,\rm kK$, $\log g = 5.7$, H:He=0.36:0.43, Dreizler et al. 1997) compared with the *VLT/UVES* observation of the CSPN of NGC 7094.

4. Metal Abundances

We employed *TMAP*, the Tübingen NLTE Model Atmosphere Package (Rauch & Deetjen 2003, Werner et al. 2003b) for the calculation of plane-parallel, hydrostatic atmospheres in radiative equilibrium (for details on atomic data etc., see Rauch et al. 2007). Because the photospheric spectra of the two stars are so similar, we used a single set of models for both stars in our initial analysis. The calculated model atmospheres considered the opacities of 23 elements, including the iron-group (Ca–Ni). We calculated a small grid of model atmospheres with all elements and adjusted their abundances (Table 2).

The comparison of synthetic spectra calculated with different iron abundances shows that an upper limit of 0.1 times solar can be determined ($[\rm Fe] = -1$, Fig. 4). This confirms a similar result by Miksa et al. (2002). The strongest nickel lines are not visible in the *FUSE* spectra.

[1] http://vo.ari.uni-heidelberg.de/ssatr-0.01/TrSpectra.jsp?
[2] http://www.g-vo.org/www

Table 2. Photospheric abundances of the CSPN of NGC 7094 and Abell 43. [X] denotes log (mass fraction / solar mass fraction) of element X. Solar abundances are adopted from Asplund et al. (2005). The other iron-group elements are not detected but considered with solar abundance (observational upper limit).

X	H	He	C	N	O	F	Ne	Si	P	S	Fe	Ni
[X]	−0.62	0.45	1.76	−0.85	−1.81	0.34	0.00	−0.21	−1.15	0.16	< -1	< -1

The comparison with synthetic spectra gives an upper limit of the Ni abundance of 0.1 times solar ([Ni] = −1, Fig. 4).

Due to the similarity of their spectra, we used the same synthetic UV and FUV spectra for NGC 7094 and Abell 43. Despite a difference in the reddening, observation and model fit equally well in the UV wavelength range (Fig. 2). This is confirmed by comparison to optical spectra as well. We therefore adopt the same parameters for both stars.

The *FUSE* observations of the CSPN of NGC 7094 and Abell 43 are strongly contaminated by interstellar line absorption. To identify weak photospheric lines that are possibly hidden in this ISM absorption, it is necessary to model the ISM spectrum as well. This was done using *OWENS*, a code that considers a number of ISM clouds with different parameters such as radial and turbulent velocity, chemical composition, column densities of the included elements and the cloud temperature. We calculated a normalized ISM absorption spectrum which is then multiplied with the stellar model-atmosphere spectrum. An example of the quality of our fit to the *FUSE* observation of NGC 7094 can be seen in Fig. 5. However, we were not able to identify weak photospheric line hidden between ISM lines.

The FUSE observations of NGC 7094 and Abell 43 exhibit strong P Cygni profiles (Fig. 2, indicating ongoing mass loss driven by radiation pressure of the hot CS. The P Cygni profiles of Ne VII λ 973.3 Å, the O VI λ 1031.9, 1037.6 Å resonance doublet, F VI λ 1139.5 Å, and the C IV $\lambda\lambda$ 1548.2, 1550.8 Å resonance doublet are shown in Fig. 5. Previous analyses of P Cygni profiles of NGC 7094 determined a terminal wind velocity of $v_\infty = 3\,900$ km/sec (Kaler & Feibelman 1985). Koesterke et al. (1998) determined a mass-loss rate of $\log \dot{M}/\mathrm{M}_\odot/\mathrm{yr} = -7.3$ from C IV $\lambda\lambda$ 1548.2, 1550.8 Å using *HST/GHRS* and *IUE* observations. Koesterke & Werner (1998) used *ORFEUS* observations and analyzed O VI $\lambda\lambda$ 1031.9, 1037.6 Å. They derived a slightly different value of $\log \dot{M}/M_\odot/\mathrm{yr} = -7.7$.

For the calculation of the wind profiles (Fig. 5), we employed *HotBlast*, a newly developed NLTE code for spherically expanding model-atmospheres. It utilizes the static model-atmosphere structure provided by *TMAP* and calculates the expanding part of the atmosphere. A combination of a modeled ISM and wind spectrum agrees well with the observation. We can confirm the previously determined mass-loss rate for the CSPN of NGC 7094. Fig. 2 shows that the CSPN of Abell 43 has a significantly weaker wind. A detailed analysis of its properties is in progress.

5. Results and Conclusions

Our re-analysis of the basic photospheric parameters of NGC 7094 based on better observations and improved NLTE model-atmospheres revealed slightly different $T_{\mathrm{eff}} = 100 \pm 15$ kK and $\log g = 5.5 \pm 0.2$ (previous values were 110 and 5.7, respectively). The H:He=0.17:0.69 ratio is much lower than previously determined (0.36:0.43, Dreizler et al. 1997). Fine-tuning of the metal abundances (Table 2) yields a good agreement with the observation. We note that the CSPN of NGC 7094 and Abell 43 have a subsolar oxygen abundance. If they are AFTP stars, carbon and oxygen should be dredged up during the thermal pulse and therefore be enhanced.

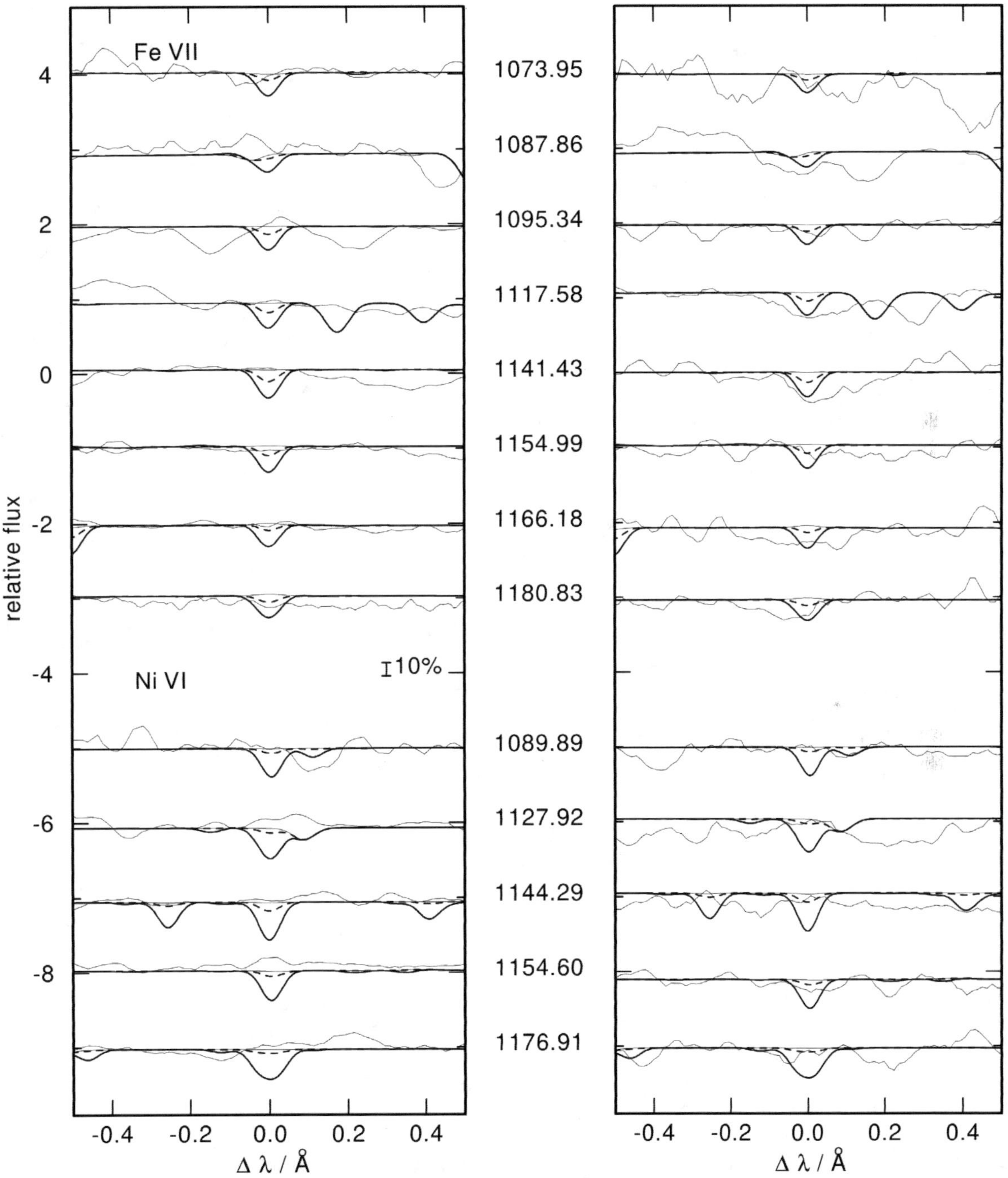

Figure 4. Comparison of the strongest Fe VII (top panels) and Ni VI (bottom panels) lines (wavelengths are given in the middle) in the *FUSE* spectra of NGC 7094 (left) and Abell 43 (right) with synthetic spectra ($T_{\rm eff} = 100\,\rm kK$, $\log g = 5.5$, abundances in Table 2) and different Fe and Ni abundances (thick: $[\rm Fe] = [\rm Ni] = 0$, dashed: $[\rm Fe] = [\rm Ni] = -1$, and thin: $[\rm Fe] = [\rm Ni] = -2$). 10 % of the continuum-flux level are indicated by a vertical bar.

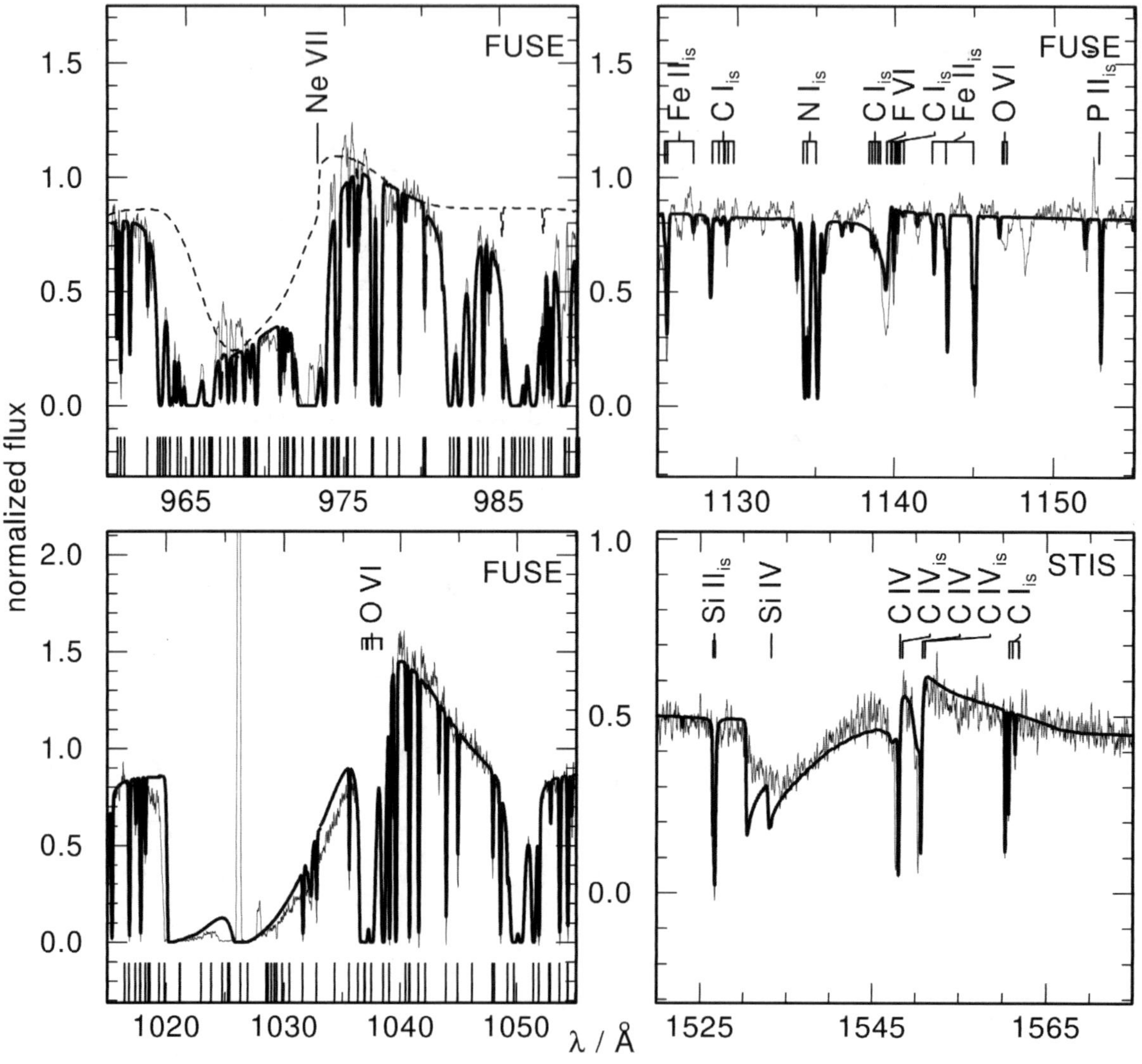

Figure 5. Sections of the *FUSE* and *HST/STIS* observations of NGC 7094 around the strongest P Cygni profiles compared with combined synthetic stellar and interstellar spectra. For Ne VII $\lambda\, 973.3\,\AA$ (top left) the pure synthetic stellar spectrum (dashed) is shown, too. Marks at the bottom (left panels only) denote interstellar H_2 lines.

This discrepancy poses a problem to AGB nucleosynthesis that is to be solved.

For the CSPN of Abell 43, a detailed spectral analysis based on individual model-atmospheres is ongoing. This will show whether it is still a spectroscopic twin of the CSPN of NGC 7094. The *FUSE* observations showed that at least their wind properties (Fig. 2) are different.

Our analysis of the two hybrid PG 1159 stars NGC 7094 and Abell 43 confirmed the previously found Fe-deficiency. Although we have derived new parameter values which improved the agreement between model and observation, we were not able to find any Ni enhancement. Instead, we found Ni to be deficient by at least one dex for a model atmosphere with $T_{\rm eff} = 100\,\rm kK$ and $\log g = 5.5$. It is therefore possible that s-process also converted Ni into even heavier trans-iron group elements. Unfortunately, we are presently not able to search for these elements because of the lack of reliable atomic data of the expected high ionization stages.

Modelling of the interstellar line absorptions improves the agreement with the observation. In the case of the CSPN of NGC 7094 and Abell 43, this did not help to identify weak photospheric lines. Prominent P Cygni profiles that are heavily blended by interstellar line absorption, show a good fit only if the ISM is properly modelled (Fig. 5). We conclude that without ISM modelling, the wind properties cannot be derived precisely.

Acknowledgments

T.R. is supported by the *German Astrophysical Virtual Observatory* (GAVO) project of the German Federal Ministry of Education and Research (BMBF) under grant 05 AC6VTB. J.W.K. is supported by the FUSE project, funded by NASA contract NAS5−32985. We used the profile fitting procedure OWENS developed by M. Lemoine and the FUSE French Team.

References

Abell G O 1955 *PASP* **67** 258

Asplund M Grevesse N & Sauval A J 2005 in *"Cosmic Abundances as Records of Stellar Evolution and Nucleosynthesis in honor of David L. Lambert"*, Eds.: T G Barnes III, & F N Bash, ASP Conference Series **336** 25

Dreizler S Werner K & Heber U 1997 IAU Symposium **180** 103

Herwig F 2001 *ApSS* **275** 15

Iben I Jr Kaler J B Truran J W & Renzini A 1983 *ApJ* **264** 605

Kaler J B & Feibelman W A 1985 *ApJ* **297** 724

Koesterke L & Werner K 1998 *ApJ* **500** 55

Koesterke L Dreizler S & Rauch T 1998 *A&A* **330** 1041

Kohoutek L 1963 *Bulletin of the Astronomical Institute of Czechoslovakia* **14** 70

Miksa S Deetjen J L Dreizler S et al. 2002 *A&A* **389** 953

Napiwotzki R 1999 *A&A* **350** 101

Napiwotzki R & Schönberner D 1995 *A&A* **301** 545

Napiwotzki R Christlieb N Drechsel H et al. 2001 *AN* **322** 411

Quirion P O Fontaine G & Brassard P 2005 *A&A* **441** 231

Rauch T & Deetjen J L 2003 in *"Stellar Atmosphere Modeling"* Eds.: I Hubeny D Mihalas & K Werner ASP Conference Series **288** 103

Rauch T Ziegler M Werner K et al. 2007 *A&A* **470** 317

Savitzky A & Golay M J E 1964 *Anal. Chem.* **36** 1627

Solheim J E Vauclair G Mukadam A S Janulis R & Dobrovolskas V 2007 *A&A* **468** 1057

Tylenda R Siódmiak N Górny S K Corradi R L M & Schwarz H E 2003 *A&A* **405** 627

Vauclair G Solheim J E & Østensen R H 2005 *A&A* **433** 1097

Werner K & Herwig F 2006 *PASP* **118** 183

Werner K Deetjen J L Dreizler S Rauch T Barstow M A & Kruk J W 2003a in *"White Dwarfs"* Eds.: de Martino D Silvotti R Solheim J E & Kalytis R NATO ASI Proc. 105

Werner K Deetjen J L Dreizler S et al. 2003b in *"Stellar Atmosphere Modeling"* Eds.: I Hubeny D Mihalas & K Werner ASP Conference Series **288** 31

16th European White Dwarfs Workshop IOP Publishing
Journal of Physics: Conference Series **172** (2009) 012033 doi:10.1088/1742-6596/172/1/012033

Central Stars of Planetary Nebulae in SDSS and IPHAS

S Weston[1], **R Napiwotzki**[1] **and S Sale**[2]
[1]Centre for Astrophysics Research, University of Hertfordshire, College Lane, Hatfield, Herts, AL10 9AB, UK
[2]Imperial College London, Blackett Laboratory, Exhibition Road, London, SW7 2AZ, UK
E-mail: s.l.weston@herts.ac.uk

Abstract. Space densities and birthrates of Planetary Nebulae (PNe) are highly uncertain. A large range of formation rates has been derived by different studies, which has led to contradicting ideas for the final evolutionary phases of low and intermediate mass stars. We started a project to deduce a birthrate using a sample of PNe within 2 kpc. The central stars will be identified in the PNe fields by their photometric colours and then used to establish improved distance estimates. To facilitate this we have created grids of photometric colours which are used to constrain stellar parameters. Our study has concentrated on PNe in SDSS and the INT Photometric Hα Survey (IPHAS) so far. IPHAS is a nearly complete northern galactic plane survey in Hα, r' and i' bands. Many previously unknown PNe have been discovered with IPHAS. We investigate implications of a more complete local sample on PN birthrate estimates.

1. Introduction

Planetary Nebulae (PNe) are thought to be the final evolutionary phase of most low/intermediate mass ($0.1M_{\odot} < M < 8M_{\odot}$) stars as they leave the asymptotic giant branch (AGB) and evolve onto the white dwarf (WD) cooling sequence. PNe are an important tool for an understanding of the final phases of stellar evolution.

A hot topic of discussion is whether PNe are formed mainly (or solely) by single or binary stellar systems. In the standard single star scenario the PN is formed when the star leaves the AGB after heavy mass loss. The ejected envelope lights up as the central star contracts and increases in temperature. Alternatively, De Marco & Moe (2005) and Moe & De Marco (2006) suggest that PN are largely created by binary stars. They argue that if the single star scenario is dominant then too many PNe will be formed in our galaxy compared to the number observed, even considering corrections for incompleteness. In the binary scenario, closely orbiting binaries will come into contact and orbital energy lost to the common envelope. The energy is used to eject the surrounding gas and drive the expansion of the nebula. Further support for a dominant binary channel comes from the non-spherical morphology of most PNe (Nordhaus et al. 2007).

A complete population census of PNe can be used to compare observations with theoretical predictions resulting from evolutionary time scales and birthrates. Current investigations predict a range of formation rates $(0.2 - 8.0) \times 10^{-12}$ pc^{-3} yr^{-1} (Soker 2006, Ishida & Weinberger 1987). As each PN forms a WD, the WD birthrate should be equal or greater than the PNe formation

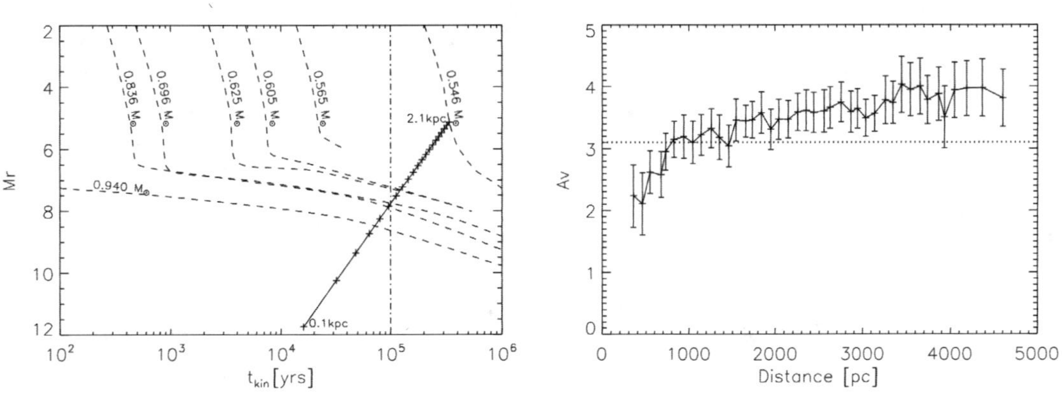

Figure 1. *Left:* PN G158.8+37.1 as an example $M_{r'} - t_{kin}$ plot with the evolutionary tracks (dashed lines) of Schönberner (1983) and Blöcker (1995). The dash-dot line indicates the time scale limit of 100,000 years. The resulting upper limit of the PN distance is 0.6 kpc. *Right:* Example of a PN (PN G126.6+01.3) in IPHAS and the projected line of sight extinction. This PN lies between 0.7 and 4.0 kpc away.

rate. The WD birthrate of $1.0 \pm 0.25 \times 10^{-12}\mathrm{pc}^{-3}\mathrm{yr}^{-1}$ (Liebert et al. 2005) poses a real problem for many PNe estimates which exceed this. Napiwotzki (2001) states that one reason for very high estimates of PNe space densities and birthrates are underestimates of PNe distances.

We intend to improve the local density estimate by collating known and newly discovered PNe within 2 kpc. An important task is to ensure we only include PNe within our volume limited sample which requires a distance estimate. We will obtain distances by using the central star of the PN (CSPN). Therefore, we must identify the CSPN within the PN field as not all are located in the centre. We discuss locating the central star in Section 2. Two methods for determining distance to CSPNe are explained in Section 3. Finally, we give details of model grids which will improve the locating the central star method previously mentioned as well as give stellar parameters using photometry in Section 4.

2. Locating the Central Star

Central stars have unique photometric colours which distinguish them from most other stellar objects, so a semi-empirical selection region was defined for the whole range of possible CSPNe parameters. The region was constructed using synthetic photometry of OB stars (Fitzpatrick & Massa 2005) and DA WDs (Holberg & Bergeron 2006) extended to hotter temperatures and adjusted to find known CSPNe in SDSS. Fitzpatrick & Massa (2005) produce Johnson photometry so we convert it to SDSS. The result positively showed that most CSPN fall within the defined region (with the remaining either saturated or their colours distorted by a companion) and no contaminate objects are detected as possible central stars. This gives us confidence, at least with no or little reddening, that CSPN are well defined with four colours and can be identified within a PN field.

3. Distance Estimates

Using the central stars we apply two ways to estimate distances. Candidate local PNe are rejected or confirmed from our sample based on the results, or inspected further if close to our distance limit.

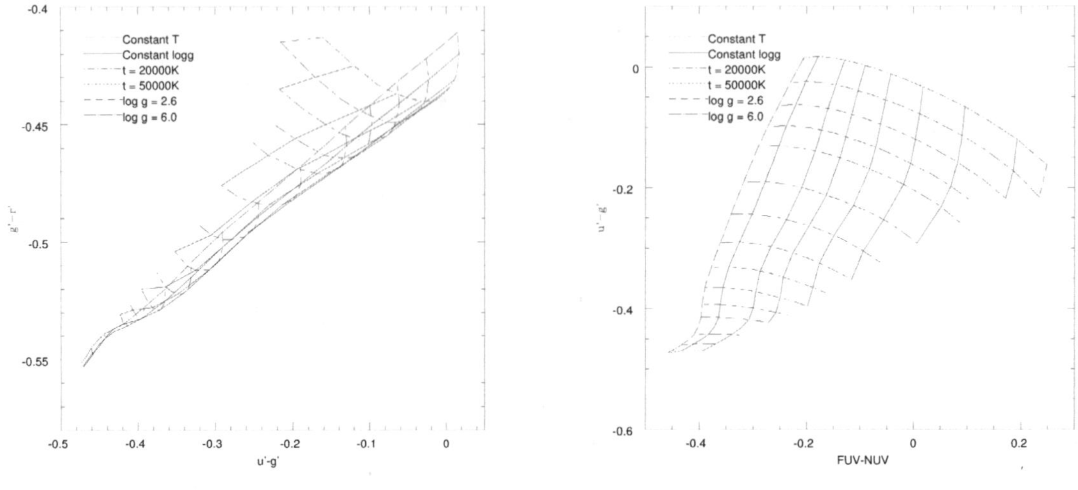

Figure 2. *Left:* PN G158.8+37.1 as an example $M_{r'} - t_{kin}$ plot with the evolutionary tracks (dashed lines) of Schönberner (1983) and Blöcker (1995). The dash-dot line indicates the time scale limit of 100,000 years. The resulting upper limit of the PN distance is 0.6 kpc. *Right:* Example of a PN (PN G126.6+01.3) in IPHAS and the projected line of sight extinction. This PN lies between 0.7 and 4.0 kpc away.

3.1. Distance Estimates from Evolutionary Tracks

For a PNe with the distance known, an absolute magnitude of the CSPN and a kinematic age, t_{kin}, can be computed from the nebula radius and expansion velocity. Fig. 1 (left) shows the post-AGB evolutionary tracks of Schönberner (1983) and Blöcker (1995) converted to the $M_{r'} - t_{\text{kin}}$ plane. This plot can be used to constrain the distance of PNe. For any hypothetical distance $M_{r'}$ and t_{kin} lie on a line, each point corresponding to a CSPN mass. We assume $0.6M_{\odot}$ as the standard value as 80% of all white dwarfs have $M = 0.6 \pm 0.1M_{\odot}$ (Liebert et al. 2005). The angular diameters and expansion velocities are taken from Acker et al. (1992). Where the expansion velocity is unknown a value of 20 kms^{-1} is assumed. Extinction is calculated using $E(r' - i')$ assuming an intrinsic $r' - i'$ value of -0.35.
We compared our results to distances from Cahn et al. (1992) and Acker et al. (1992). The results are mixed with some distances in agreement, however, most have large discrepancies. Although some uncertainty results from the the unknown CSPN mass, the statistical methods used by Cahn et al. (1992) and Acker et al. (1992) have large systematic errors (Napiwotzki 2001).

3.2. Distance Estimates from Extinction

Distances can be estimated using relations between interstellar absorption and distance. A good distance estimate can only be obtained with a detailed 3D dust map. The maps of Schlegel et al. (1998) give only the integrated extinction and thus can't be used for this purpose. Our future sample will include observations from the INT Photometric Hα survey (IPHAS, Drew et al., 2005). We will exploit the work of Sale et al. (2008) who present an algorithm which will produce a 3D extinction map across the entire Northern Galactic plane (Sale et al., in prep.). All CSPNe have a narrow range of intrinsic $r' - i'$ colours ($r' - i' = -0.35 \pm 0.04$) and so we can estimate an accurate $E(r' - i')$. Once converted to an r' band extinction, $A_{r'}$, the distance can be extracted from Sale et al.'s projection of the galactic dust. An example can be seen in Fig. 1 (right). Combining both methods, we will be able to compile a sample of good candidates for the local PN population.

4. Grids of Synthetic Photometry for CSPNe

We are in the process of refining the photometric selection and analysis by computing synthetic photometry in all filters required from model Spectral Energy Distributions (SEDs). This will be extended to include the whole parameter range for CSPNe and filters covering the UV to near-IR range. So far we have created photometric grids for OB type CSPNe from ATLAS9 models (Kurucz 1991) and hot DA WDs from Finley et al. (1997). If the star was observed with the UV satellite GALEX as well, then we can easily determine stellar parameters from photometry (Fig. 2). With the current model grids for SDSS, GALEX and IPHAS we will be able to locate the central star, and obtain an approximate distance (or range) and stellar parameters. Furthermore, with the SEDs many bands and systems (therefore surveys) can be added and the project begin to compile an accurate and more complete local sample.

References

Acker A, Marcout J, Ochsenbein F, Stenholm B & Tylenda R 1992 *Strasbourg-ESO catalogue of galactic planetary nebulae.* Garching: European Southern Observatory, 1992.
Blöcker T 1995 *A&A* **299**, 755.
Cahn J H, Kaler J B & Stanghellini L 1992 *A&AS* **94**, 399.
De Marco O & Moe M 2005 *in* R Szczerba, G Stasińska & S. K Gorny, eds, 'Planetary Nebulae as Astronomical Tools' Vol. 804 of *American Institute of Physics Conference Series* p. 169.
Drew J E, Greimel R, Irwin M J et al. 2005 *MNRAS* **362**, 753.
Finley D S, Köster D & Basri G 1997 *ApJ* **488**, 375.
Fitzpatrick E L & Massa D 2005 *AJ* **129**, 1642.
Holberg J B & Bergeron P 2006 *AJ* **132**, 1221.
Ishida K & Weinberger R 1987 *A&A* **178**, 227.
Kurucz R L 1991 *in* 'NATO ASIC Proc. 341: Stellar Atmospheres – Beyond Classical Models' p. 441.
Liebert J, Bergeron P & Holberg J B 2005 *ApJS* **156**, 47.
Moe M & De Marco O 2006 *ApJ* **650**, 916.
Napiwotzki R 2001 *A&A* **367**, 973–982.
Nordhaus J, Blackman E G & Frank A 2007 *MNRAS* **376**, 599.
Sale S E, Drew J E, Unruh Y C et al. 2008 *ArXiv e-prints* **0810.2547**.
Schlegel D J, Finkbeiner D P & Davis M 1998 *ApJ* **500**, 525.
Schönberner D 1983 *ApJ* **272**, 708.
Soker N 2006 *ApJ* **645**, L57.

Journal of Physics: Conference Series **172** (2009) 012034 doi:10.1088/1742-6596/172/1/012034

Mass transfer dynamics in double degenerate binary systems

M. Dan, S. Rosswog and M. Brüggen

School of Engineering and Science, Jacobs University Bremen, Campus Ring 1, 28759, Bremen, Germany

E-mail: m.dan@jacobs-university.de

Abstract. We present a numerical study of the mass transfer dynamics prior to the gravitational wave-driven merger of a double white dwarf system. Recently, there has been some discussion about the dynamics of these last stages, different methods seemed to provide qualitatively different results. While earlier SPH simulations indicated a very quick disruption of the binary on roughly the orbital time scale, more recent grid-based calculations find long-lived mass transfer for many orbital periods. Here we demonstrate how sensitive the dynamics of this last stage is to the exact initial conditions. We show that, after a careful preparation of the initial conditions, the reportedly short-lived systems undergo mass transfer for many dozens of orbits. The reported numbers of orbits are resolution-biased and therefore represent only lower limits to what is realized in nature. Nevertheless, the study shows convincingly the convergence of different methods to very similar results.

1. Introduction

White dwarfs that merge under the influence of gravitational waves have been suspected to be related to a variety of phenomena. Although not being the most popular model for type Ia supernovae, white dwarf mergers have despite several ups and downs defended their position as a serious, possible progenitor model (Iben & Tutukov 1984; Webbink 1984; Yoon et al. 2007). If ignited at the surface, merger remnants have been found to be transformed into an O-Ne-Mg white dwarf that finally forms a neutron star in an accretion-induced collapse (Saio & Nomoto 1985; Nomoto & Kondo 1991; Saio & Nomoto 2004). Merged low-mass white dwarf mergers are thought to produce extreme helium stars and the majority of R Corona Borealis stars, e.g. Saio & Jeffery (2002). A fraction of white dwarf binaries may evolve into systems that survive the onset of mass transfer and evolve to longer periods with ever decreasing mass transfer rate (e.g. Paczyński 1967; Nelemans et al. 2001), so-called AM CVn systems.

The merger process of two white dwarfs has been modeled by a number of groups, Benz et al. (1990), Rasio & Shapiro (1995), Segretain et al. (1997), Guerrero et al. (2004), Yoon et al. (2007), all of the mentioned approaches used the SPH method. Recently, grid-based simulations (D'Souza et al. 2006; Motl et al. 2007) were performed where the authors carefully tried to reduce the angular momentum non-conservation due to advection errors that often plagues grid-based codes, see e.g. New & Tohline (1997). Their results are very different from most previous SPH simulations in the sense that their donor stars are not destroyed on a dynamical time scale, but instead show long-lived mass transfer over many orbital periods. These results have sparked some discussion about the stability of mass transfer in double degenerate systems during these stages, see e.g. Fryer & Diehl (2008).

Journal of Physics: Conference Series **172** (2009) 012034 doi:10.1088/1742-6596/172/1/012034

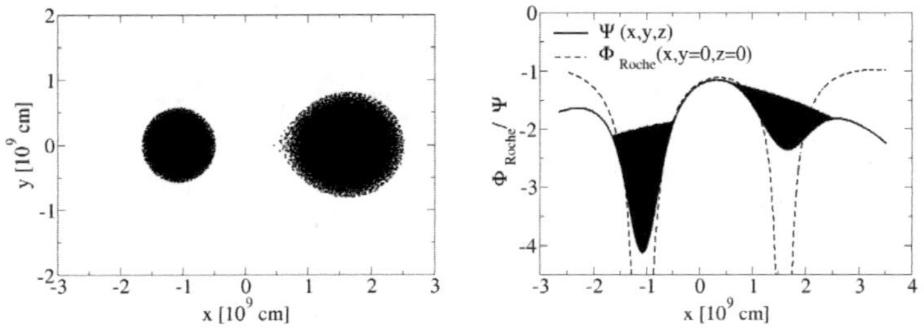

Figure 1. Equilibrium configuration (0.9 and 0.6 $M_\odot$) at onset of mass transfer. *Left*: projection of SPH particles onto the orbital plane, *right*: projection onto the $(x, \Phi_{\rm Roche}/\Psi)$ plane. The dashed line is the Roche potential as calculated for a point mass binary, $\Phi_{\rm Roche}(x, y = 0, z = 0)$. The solid black line is the corresponding quantity as derived from the real mass distribution, $\Psi(x, y = 0, z = 0)$, and is given by $\Psi(\boldsymbol{r}) = \phi(\boldsymbol{r}) - \frac{1}{2}(\boldsymbol{\omega} \times \boldsymbol{r})^2$, where $\phi(\boldsymbol{r})$ is the gravitational potential.

In the following, we will demonstrate the sensitivity of the simulation dynamics on the exact initial conditions and show that with accurately constructed starting configurations long-lived mass transfer ensues.

2. Model

2.1. Numerical model

We use the smoothed particle hydrodynamics method to solve the 3D equations of fluid dynamics. Our code is documented in Rosswog et al. (2008). The system of fluid equations is closed by the HELMHOLTZ equation of state (Timmes & Swesty 2000). It accepts an externally calculated nuclear composition and allows a convenient coupling to a nuclear reaction network. We use a minimal nuclear reaction network (Hix et al. 1998) to determine the evolution of the nuclear composition and to include the energetic feedback onto the gas from the nuclear reactions. A set of only seven abundance groups greatly reduces the computational burden, but still reproduces the energy generation of all burning stages from He burning to NSE accurately. We use a binary tree (Benz et al. 1990) to search for the neighbor particles and to calculate the gravitational forces.

2.2. Initial conditions

The dynamics of the binary systems is very sensitive to the initial conditions, therefore, it is vital to start the simulations from a setup that is as accurate as possible. Most of the previous SPH simulations (e.g. Benz et al. 1990; Segretain et al. 1997; Guerrero et al. 2004; Yoon et al. 2007) started from rather approximate initial conditions and consequently found a very large initial mass transfer rate and a subsequent disruption of the donor star within a few orbits.

In this work, we restrict ourselves to corotating initial configurations, but very carefully construct initial conditions along the strategy outlined by Rosswog et al. (2004). In a first step the (isolated) stars are "relaxed" individually into hydrostatic equilibrium by adding a velocity-proportional damping term to the momentum equation. Subsequently, we place the stars in the corotating frame at a separation a, that is large enough to avoid immediate mass transfer. This separation is then adiabatically reduced according to $\dot{a} = -\frac{a}{\eta \cdot \tau_{\rm dyn}}$, so that $\frac{a}{\dot{a}} \gg \tau_{\rm dyn}$. Here, $\tau_{\rm dyn}$ is the larger of the two white dwarf dynamical time scales and η takes values between 10 and 30. Once the first SPH particle crosses the L_1 point of the Roche potential, the binary is set on a circular orbit. This moment was adopted as the time origin of the simulation. An example of a corotating equilibrium configuration at this stage (masses of 0.6 $M_\odot$ and 0.9 $M_\odot$) is shown in Fig. 1, left panel. The right panel shows a comparison between (point mass) Roche potential and the corresponding values as calculated from the real matter distribution.

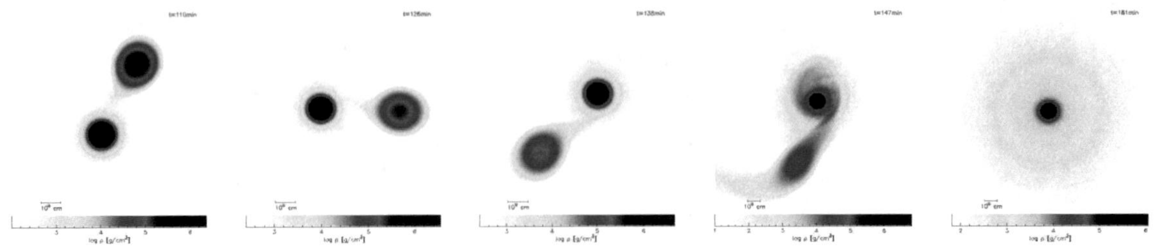

Figure 2. Dynamical evolution of a double white dwarf system with $q = 0.5$ (at 44, 50, 56, 59 and 67 times the initial binary period.)

3. From the onset of mass transfer to a possible disruption

Exemplary for a large set of simulations that will be presented elsewhere (Dan et al. 2009) we show the evolution of two binary systems: one with $q = M_{\rm donor}/M_{\rm accretor} = 0.5$ to be compared to Rasio & Shapiro (1995), D'Souza et al. (2006), and another one with $q = 0.27$. Rasio & Shapiro (1995) also used an SPH approach while D'Souza et al. (2006) used a grid-based code. Although we think that our investigated systems are close enough to Rasio & Shapiro (1995), D'Souza et al. (2006) to warrant a fair comparison, one has to keep in mind that the latter investigations use 5/3-polytropes while we use a physical EOS. Since the EOS governs, via the mass-radius relationship, how the star reacts on mass loss, it also influences the orbital evolution. Thus, the comparison should be taken with a grain of salt.

We carry out the simulation of the $q = 0.5$ case (0.3 and 0.6 $M_\odot$) at two different resolutions. First, we use $\approx 4 \times 10^4$ SPH particles to conform the simulations of Rasio & Shapiro (1995). We find an initial separation where (numerically resolvable) mass transfer sets in that is larger by 27 percent than in Rasio & Shapiro (1995). While in Rasio & Shapiro (1995) the binary was disrupted within five orbits, we count more than 18 full orbits before the disruption sets in.

In a second simulation of the same system we use 2×10^5 SPH particles, see Fig. 2, and consequently the initial separation increases slightly since we are now better resolving the outer stellar layers. The main purpose of this run is the comparison with D'Souza et al. (2006) who, in turn, ran their calculation to compare against Rasio & Shapiro (1995). D'Souza et al. (2006) found the orbital separation to be constant for about 20 orbits, then it increased and the mass transfer rate leveled off without any sign of a merger. After 32 orbits the simulation was stopped. We followed our simulation for an even longer time. Similar to D'Souza et al. (2006) the orbital separation increases slightly but in the end the donor is still disrupted, although only after as many as 60 orbital revolutions, see panel 4 in Fig. 2. Thus, for as long as we can compare, our simulations are in perfect agreement with D'Souza et al. (2006). We also explore the evolution of a binary with a mass ration of $q = 0.27$ (0.3 and 1.1 $M_\odot$). Contrary to our previous example where we saw a direct impact onto the accretor, see Fig. 2, panel 1, now the circularization radius is larger than the accretor radius and a disk forms. This configuration feeds back angular momentum into the orbit thereby increasing the orbital separation, Fig. 3, bottom left. This de-circularizes the orbits from the beginning. A fraction of the angular momentum is transferred to the accretor and used to spin it up, Fig. 3, bottom right. Even after 41 orbits there is no sign of a merger, instead the separation increases continuously and the mass transfer decreases for almost 20 full orbits (Fig. 3, center). Such a system could possibly evolve into a AM CVn system.

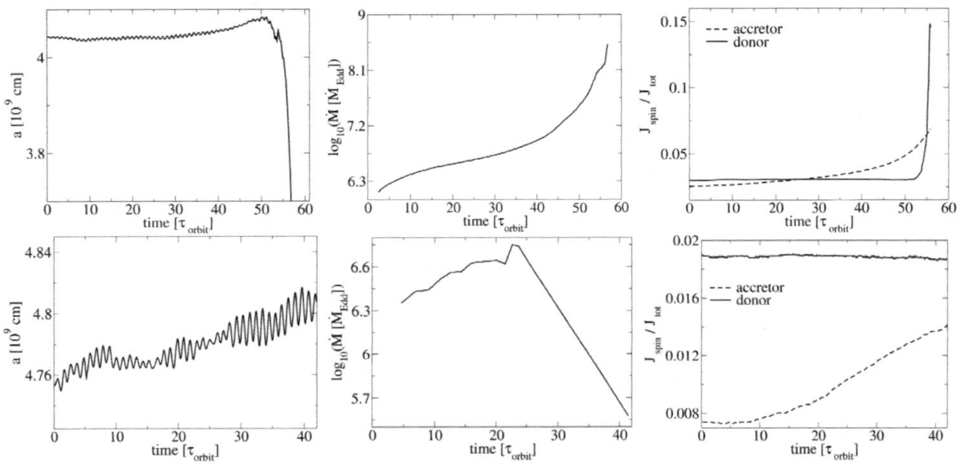

Figure 3. Evolution of the orbital separation (left), the mass ratio (center) and the spin angular momentum for both the accretor and the donor (right) for a system with $q = 0.5$ (upper panels) and $q = 0.27$ (lower panels).

4. Conclusions

We have demonstrated the sensitivity to the initial conditions used in double degenerate binary simulations and have put particular emphasis on the accuracy of our starting configurations. Contrary to earlier SPH simulations our binaries are *not* disrupted after a few orbital periods, but instead transfer mass for dozens of orbits. In as far as we can compare it, our simulations are in perfect agreement with the recent simulations of D'Souza et al. (2006).

Acknowledgments

This work was supported by DFG grant no. RO 3399/4-1.

References

Benz W Bowers R Cameron A Press W 1990 *ApJ* **348** 647
Dan M et al. 2009 *to be submitted*
D'Souza M C R Motl P M Tohline J E and Frank J 2006 *ApJ* **643** 381
Fryer C L and Diehl S 2008 *"Hydrogen-Deficient Stars ASP Conf. Series"* **391** 335
Guerrero J García-Berro E Isern J 2005 *A&A* **413** 257
Hix W R Khokhlov A M Wheeler J C and Thielemann F K 1998 *ApJ* **503** 332
Iben I Jr and Tutukov A 1984 *ApJS* **55** 335
Motl M P Frank J Tohline J E and D'Souza M C R 2007 *ApJ* **670** 1314
Nelemans G Portgeis Zwart S F Verbunt F and Yungelson L R 2001 *A&A* **368** 939
New K C B and Tohline J E 1997 *ApJ***490** 311
Nomoto K and Kondo Y 1991 *ApJ* **367** L19
Paczyński B 1967 *Acta Astron* **17** 287
Rasio F A and Shapiro S L 1995 *ApJ* **438** 887
Rosswog S Ramirez-Ruiz E Hix W R and Dan M 2008 *Comp. Phys. Comm.* **179** 184
Rosswog S Speith R and Wynn G A 2004 *M.N.R.A.S.* **351** 1121
Saio H and Jeffery C S 2002 *M.N.R.A.S.* **333** 121
Saio H and Nomoto K 1985 *A&A* **150** L21
Saio H and Nomoto K 2004 *ApJ* **615** 444
Segretain L Chabrier G and Mochkovitch R 1997 *ApJ* **481** 355
Timmes F X and Swesty F D 2000 *ApJS* **126** 501
Webbink R F 1984 *ApJ* **277** 355
Yoon S C Podsiadlowski Ph and Rosswog S 2007 *M.N.R.A.S.* **380** 933

Journal of Physics: Conference Series **172** (2009) 012035 doi:10.1088/1742-6596/172/1/012035

The gravitational waveforms of white dwarf collisions in globular clusters

P Lorén–Aguilar[1,2], J A Lobo[3,2], J Isern[3,2] and E García–Berro[1,2]

[1]Departament de Física Aplicada, Escola Politècnica Superior de Castelldefels, Universitat Politècnica de Catalunya, Av. del Canal Olímpic, s/n, 08860 Castelldefels, Spain

[2]Institute for Space Studies of Catalonia, c/Gran Capità 2–4, Edif. Nexus 104, 08034 Barcelona, Spain

[3]Institut de Ciències de l'Espai, CSIC, Campus UAB, Facultat de Ciències, Torre C-5, 08193 Bellaterra, Spain

E-mail: loren@fa.upc.edu

Abstract. In the dense central regions of globular clusters close encounters of two white dwarfs are relatively frequent. The estimated frequency is one or more strong encounters per star in the lifetime of the cluster. Such encounters should be then potential sources of gravitational wave radiation. Thus, it is foreseeable that these collisions could be either individually detected by LISA or they could contribute significantly to the background noise of the detector. We compute the pattern of gravitational wave emission from these encounters for a sufficiently broad range of system parameters, namely the masses, the relative velocities and the distances of the two white dwarfs involved in the encounter.

1. Introduction

The Galactic population of field binary white dwarf is one of the most promising sources of gravitational waves, since it was shown long ago that LISA could detect them individually (Bender et al. 1998). Field binary white dwarfs are the natural outcome of the evolution of many binary systems and it can be shown that due to tidal interactions their orbits are usually circular. Thus, the gravitational wave pattern of these systems is easy to compute since the two white dwarfs can be safely assumed to be point masses. Due to the ensuing emission of gravitational waves, the orbital separation of these systems decreases and the final fate of these systems is to coalesce. Consequently, the gravitational wave emission resulting form the merger of such systems has been the subject of some recent studies (Guerrero et al. 2004; Lorén–Aguilar et al. 2005). These studies show that the resulting pattern of gravitational wave emission is rather simple. In particular, it has been shown that these waveforms do not show a prominent peak and that the gravitational wave signal rapidly fades in a few orbital periods.

The typical stellar densities of globular clusters are $\sim 10^3$ pc^{-3} and, due to their long lifetimes, their central regions contain many collapsed and degenerate objects, such as white dwarfs. In such a dense environment, where stars can pass very close one to another, every star is expected to suffer one or more strong encounters during its lifetime. In fact, it has been predicted (Willems et al. 2007) that dynamical interactions in globular clusters can form double white dwarf systems which could be eventually detectable by LISA. Moreover, in globular clusters, dynamical interactions can form binary white dwarf systems at an enhanced rate compared to

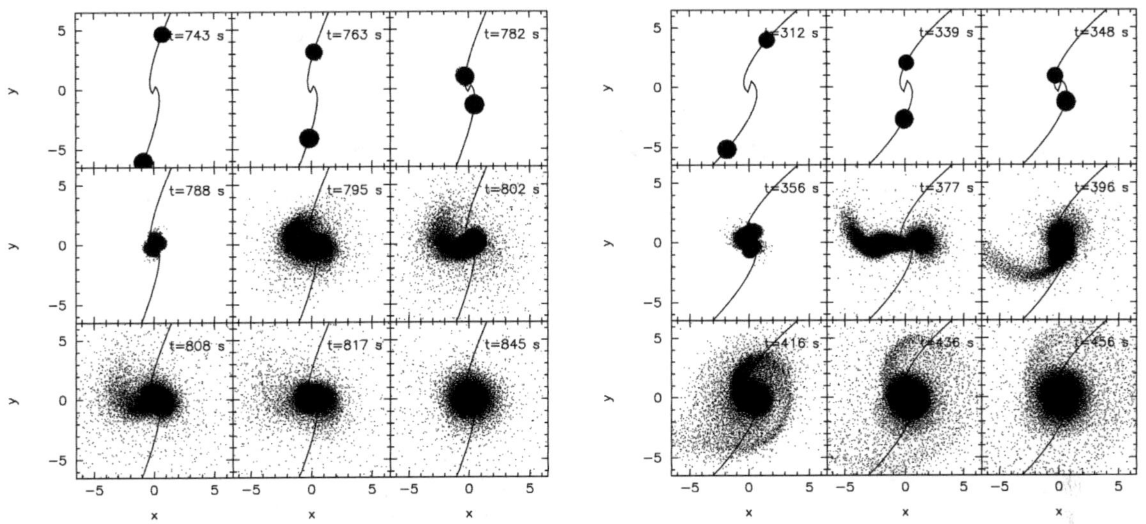

Figure 1. Temporal evolution of the close encounters of the double white dwarf systems discussed here. The left panels show the evolution of the positions of the particles for the 0.6+0.8 $M_{\odot}$ system, when an initial velocity $v_i = 50$ km/s and an impact parameter of $b = 0.6\,R_{\odot}$ are adopted. The right panels show the evolution of the same system when the initial conditions are $v_i = 150$ km/s and $b = 0.3\,R_{\odot}$. The positions of the particles have been projected onto the xy plane. The units of positions are 10^9 cm. Times are shown in the right upper corner of each panel. These figures have been done using the visualization tool SPLASH (Price 2007).

the field (Shara & Hurley 2002). Additionally, such systems are generally formed with non-zero eccentricities. This is an important issue because tidal interactions in eccentric systems substantially modify the gravitational wave emission pattern, thus providing us with a unique opportunity to use LISA to check our understanding of the interiors of white dwarfs (Willems et al. 2007). Here we study the dynamical interaction of white dwarfs in globular clusters and we compute the associated gravitational wave radiation. Due to the complicated dynamics of the process, numerical simulations of such encounters are essential to obtain realistic sets of waveforms. Consequently, we have performed a series of SPH simulations of strong encounters of white dwarfs in globular clusters, covering a broad enough range of system parameters, which include impact parameter and relative stellar velocities for two representative masses of the white dwarfs involved in the interaction.

2. Input physics and method of calculation

We follow the hydrodynamic evolution of the interacting white dwarfs using Smoothed Particle Hydrodynamics (SPH). This method is totally lagrangian and does not require a grid making it specially suitable for studying an intrinsically three-dimensional problem. We will not describe in detail the most basic equations of our numerical code, since this is a well-known technique. Instead, the reader is referred to the abundant literature for descriptions of this technique, and to Guerrero et al. (2004) where all the relevant input physics for the problem at hand and of our specific implementation of this technique are explained in depth. We compute the gravitational wave emission in the slow-motion, weak-field quadrupole approximation (Misner e tal. 1973). In order to compute a large enough set of gravitational wave patterns we have relaxed two initial white dwarf models with masses 0.6 $M_{\odot}$ and 0.8 $M_{\odot}$, respectively. The mass of the SPH

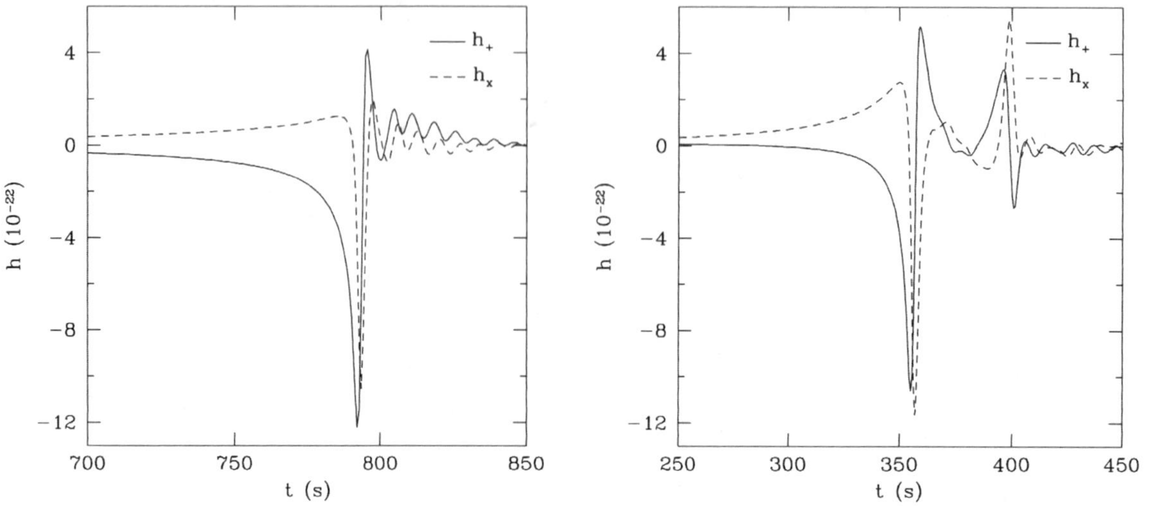

Figure 2. Gravitational wave emission for the cases shown in Fig. 1. The left panel corresponds to the case in which two white dwarfs of masses 0.6 and 0.8 $M_\odot$, respectively, interact to produce a head-on collision. The right panel corresponds to the case these two white dwarfs interact and a lateral impact occurs. The dimensionless strains h_+ and $h_\times$ are measured in units of 10^{-22}. The source is located at a distance of 10 kpc.

particles were the same in both cases in order to avoid numerical artifacts. Consequently, the $0.6\,M_\odot$ white dwarf was relaxed using 2.0×10^4 particles, whereas the $0.8\,M_\odot$ white dwarf needed 2.6×10^4 particles. For the interactions in which two equal-mass white dwarfs were involved we have used the otherwise typical $0.6\,M_\odot$ model, whereas for the other cases we have used the $0.6\,M_\odot$ and the $0.8\,M_\odot$ models. We have fixed the initial distance x between the stars and their spin (fixed to typical values in field white dwarfs) allowing the impact parameter b (distance between stars on the y direction) and the initial velocity $\vec{v}_i = (v_i, 0, 0)$ to be our free parameters. We have performed 11 simulations with impact parameters ranging from $0.3\,R_\odot$ to $0.9\,R_\odot$ and initial velocities from 50 to 150 km/s. These values of the initial relative velocities and impact parameters are typical of globular clusters.

3. Results

We have obtained three different kinds of behavior, depending on the initial parameters adopted for the close encounter. When the initial velocity v_i is relatively small and the impact parameters are not very large the gravitational forces bend the initial trajectories and the result of the interaction is a head-on collision. For increasing initial velocities there is a transition region in which the initial trajectories are less affected by the gravitational interaction and the final outcome turns out to be a lateral collision. Finally, for sufficiently large initial velocities and above a threshold impact parameter which depends on the initial relative velocity of the interacting white dwarfs, the stars do not collide but, instead, form an eccentric binary system.

In this work we pay special attention to those systems in which a collisions occurs. The cases in which an eccentric binary system is formed will be discussed at length elsewhere. Fig. 1 shows two typical examples of the temporal evolution of some of the close encounters studied here. The left panels of Fig. 1 show a case in which a head-on impact occurs. The masses of the colliding white dwarfs are, respectively, 0.6 and 0.8 $M_\odot$. The initial conditions adopted in this case were $v_i = 50$ km/s and $b = 0.6\,R_\odot$. The right panels of Fig. 1 show the case in which a lateral collision of the two intervening white dwarfs occurs. In this case the initial conditions were $v_i = 150$ km/s and $b = 0.3\,R_\odot$. The trajectories of the centers of mass of each individual white dwarf are shown using solid lines. These lines can be used as a visual aid to follow the trajectories of the intervening white dwarfs. As can be seen in these panels, once mass transfer starts it occurs on a dynamical timescale and the coalescence occurs in a few seconds. The final outcome of the merger process is in both cases a central object with ellipsoidal shape. The characteristics of these central objects will be discussed in detail in a forthcoming publication and here we concentrate in describing the gravitational wave pattern radiated by these systems. In all the cases studied here, at early times the system does not radiate gravitational waves because the accelerations are very small. Once the stars approach sufficiently each other, the gravitational wave signal rapidly grows. In the case of a head-on collision the signal first grows and then suddenly decreases — see the left panel of Fig. 2. This peak is followed by a series of peaks of smaller amplitude, the so-called ring-down phase. These peaks are caused by the circularization of the orbits of a spiral arm which forms during the merging process. In the case of a lateral collision the gravitational wave emission is more irregular and does present several amplitude peaks — see the right panel of Fig. 2. This occurs because in a lateral collision before the final merger, the two interacting white dwarfs describe orbits of decreasing separation, in which sizeable mass transfer occurs between the two components of the system. Finally, in those cases in which an eccentric binary forms the gravitational wave patterns present a regular, periodical behavior, which we do not show here for the sake of conciseness.

4. Conclusions

We have computed the emission of gravitational waves arising from the dynamical interaction of two white dwarfs in the dense central regions of a globular cluster. These interactions are thought to be quite frequent and thus it is important to characterize the resulting gravitational waveforms. We have done so for a sufficiently broad range of initial relative velocities and impact parameters and for two different component masses. We have presented here a set of calculations corresponding to close encounters in which a typical $0.6\,M_\odot$ white dwarf and a rather massive $0.8\,M_\odot$ companion are involved. We have found that there are three possible different outcomes, depending on the initial relative velocity and impact parameter of the system, which lead to distinctive patterns of gravitational wave radiation. Specifically, we have characterized the range of parameters for which head-on mergers, lateral collisions and formation of an eccentric binary system occur. For the cases in which the outcome is a collision (either a head-on one or a lateral one) we find that signal is a strong burst of gravitational waves, followed in some cases by a series of smaller peaks. Such events will not be detectable by LISA. On the contrary if the initial conditions are such that the dynamical interaction leads to the formation of an eccentric binary system, the signal is periodical and will probably be detectable with LISA.

Acknowledgments

Part of this work was supported by MEC grants AYA05–08013–C03–01 and 02, ESP2007–61593, by the European Union FEDER funds and by the AGAUR.

References

Bender P L et al. 1998 *"LISA: Laser Interferometer Space Antenna for the detection and observation of GW"*, Pre-fase A Report (Garching: Max Planck-Institut für Quantenoptik)
Guerrero J García–Berro E & Isern J 2004 *A&A* **413** 257
Lorén–Aguilar P Guerrero J Isern J Lobo J A & García-Berro E 2005 *MNRAS* **356** 627-636
Misner C W Thorne K S & Wheeler J A 1973 *"Gravitation"* (New York: W.H. Freeman)
Price D J 2007 *P.A.S.A.* **24** 159
Shara M M & Hurley J R 2002 *ApJ* **571** 830
Willems et al. 2007 *ApJ* **665** L59

Tidally-induced thermonuclear Supernovae

Stephan Rosswog[1], Enrico Ramirez-Ruiz[2] and W Raphael Hix[3]

[1] School of Engineering & Science, Jacobs University Bremen, 28759 Bremen, Germany
[2] Dept. of Astronomy & Astrophysics, University of California, Santa Cruz, CA 95064, USA
[3] Physics Division, Oak Ridge National Laboratory, Oak Ridge, TN37831-6374, USA

E-mail: srosswog@jacobs-university.de

Abstract. We discuss the results of 3D simulations of tidal disruptions of white dwarfs by moderate-mass black holes as they may exist in the cores of globular clusters or dwarf galaxies. Our simulations follow self-consistently the hydrodynamic and nuclear evolution from the initial parabolic orbit over the disruption to the build-up of an accretion disk around the black hole. For strong enough encounters (pericentre distances smaller than about 1/3 of the tidal radius) the tidal compression is reversed by a shock and finally results in a thermonuclear explosion. These explosions are not restricted to progenitor masses close to the Chandrasekhar limit, we find exploding examples throughout the whole white dwarf mass range. There is, however, a restriction on the masses of the involved black holes: black holes more massive than $2 \times 10^5\,M_\odot$ swallow a typical $0.6\,M_\odot$ white dwarf before their tidal forces can overwhelm the star's self-gravity. Therefore, this mechanism is characteristic for black holes of moderate masses. The material that remains bound to the black hole settles into an accretion disk and produces an X-ray flare close to the Eddington limit of $L_{\rm Edd} \simeq 10^{41}$erg/s $(M_{\rm bh}/1000\,M_\odot)$, typically lasting for a few months. The combination of a peculiar thermonuclear supernova together with an X-ray flare thus whistle-blows the existence of such moderate-mass black holes. The next generation of wide field space-based instruments should be able to detect such events.

1. Introduction

Two classes of black holes are established beyond reasonable doubt: black holes of a few solar masses ("stellar-mass") make themselves known by interacting with a companion star (McClintock & Remillard 2003) and "supermassive" black holes with masses of $10^6 - 10^{10}\,M_\odot$ seem to lurk in the centres of most, if not all galaxies, see e.g. (Kormendy & Richstone 1995; Ferrarese & Merritt 2000). A third class, so-called "intermediate-mass black holes" (IMBH) with masses of $\sim 1000\,M_\odot$, represents a plausible, yet to date still missing link. The evidence for their existence is mounting, but to date it is not entirely conclusive. Based on kinematical studies Gerssen et al. (Gerssen et al 2002; 2003) have suggested the presence of a $3.9 \times 10^3\,M_\odot$ black hole in the core of the globular cluster M15. More recently, there were further reports on IMBHs in the globular clusters NGC 2808 (Maccarone & Servillat 2008) and omegaCen (Noyola et al. 2008). Additional evidence comes from ultraluminous, compact X-ray sources in young star clusters (Zezas et al. 2002; Pooley & Rappaport 2006) and from n-body simulations (Portegies et al. 2004) that indicate that runaway collisions in dense young star clusters can produce rapidly growing black holes. None of the above arguments is strong enough to close the case, but their different nature gives this hypothesis credibility.

At the very large number densities in the environments that may habor such black holes stars are disrupted at a rate of $\sim 10^{-7}\text{yr}^{-1}\ M_{\text{bh},3}^{4/3}\ n_{*,6}\ \sigma_{10}^{-1}\ (R_{\text{per}}/R_{\text{t}})$, where $M_{\text{bh},3}$ is the black hole mass in units of 1000 $\text{M}_\odot$, $n_{*,6}$ is the central stellar density in units of 10^6 pc^{-3}, σ_{10} the velocity dispersion in 10 kms^{-1} and R_{per} is the distance of closest approach. The quantity $R_{\text{t}} = (M_{\text{bh}}/M_*)^{1/3} R_*$ is the tidal radius, the approximate separation where the hole's tidal field wins the battle against the star's self-gravity. It is proportional to the stellar radius R_*, but only grows $\propto M_{\text{bh}}^{1/3}$, i.e. substantially slower than the gravitational radius, $R_{\text{S}} = 2GM_{\text{bh}}/c^2$, of the black hole. As a consequence, too massive black holes, $M_{\text{bh}} > M_{\text{bh,lim}} \simeq 2.5 \times 10^5\, M_\odot\ R_{\text{wd},9}^{3/2} M_{\text{wd},0.6}^{-1/2}$ will swallow stars before disrupting them. Here, $R_{\text{wd},9}$ is the white dwarf radius in 10^9 cm and $M_{\text{wd},0.6}$ its mass in units of 0.6 $\text{M}_\odot$. Thus, white dwarfs with their small radii and their (at least in principle) available nuclear fuel can only be tidally squeezed and disrupted by black holes of moderate masses. Therefore they are precious tools to probe the existence of this type of black hole.

Luminet & Pichon (1989) were the first to discuss the tidal disruption of white dwarfs by black holes in the context of the affine star model (Carter & Luminet 1983; 1985). Wilson and Mathews (2004) believe that general relativistic effects may lead to a tidal ignition even for supermassive black holes and at separations of many gravitational radii. A hydrodynamic simulation with a Newtonian Adaptive Lagrangian Eulerian (ALE) code was performed recently (Dearborn et al. 2005) where the gravitational constant was adjusted to mimic general relativistic effects. We have recently performed a large set of 3D hydrodynamic simulations (Rosswog et al. 2008a; 2008d) that incorporate the nuclear energy generation self-consistently with the hydrodynamics.

2. Ingredients for the simulations

Here we briefly summarize the ingredients of our simulations, both in terms of physics and numerical methods. We described the numerical aspects in Rosswog et al. (2008b) and in Rosswog et al. (2008d) we concentrated on the physics and the astrophysical implications of such tidal disruptions.

We use the smoothed particle hydrodynamics method (SPH) to follow the hydrodynamic evolution of the white dwarf matter. Being completely Lagrangian and conserving mass, energy, momentum and angular momentum by construction the method is ideal for the study of such a violent stellar disruption. General reviews on the method can be found in Benz (1990a), Monaghan (1992) and Monaghan (2005), our particular implementation is documented in Rosswog et al. (2008b). We calculate the forces from self-gravity via a binary tree (Benz et al. 1990b) and those from the central black hole via a relativistic pseudo potential (Paczyński & Wiita 1980), details of how the singularity is treated numerically are laid out in Rosswog (2005a). We close the system of hydrodynamics equations via the HELMHOLTZ equation of state (EOS) (Timmes & Swesty 2000). It accepts an externally calculated chemical composition, facilitating the coupling to nuclear reaction networks. The electron-/positron equation of state makes no assumptions about the degree of degeneracy or relativity, the exact expressions are integrated numerically and are subsequently tabulated. The interpolation in this table enforces thermodynamic consistency by construction (Timmes & Swesty 2000). The nuclei in the gas are treated as a Maxwell-Boltzmann gas, the photons as blackbody radiation. Most importantly in this context, we couple a minimal nuclear reaction network (Hix et al. 1998) to the hydrodyamics. Throughout this simulation set we assume a uniform nuclear composition accross the white dwarfs. Stars with masses < 0.6 $\text{M}_\odot$ are instantiated as pure helium, more massive stars are modeled with a mass fraction of 50% carbon and 50% oxygen. All simulations are started from parabolic orbits, the strength of an encounter is parametrized by the penetration factor $\beta = R_{\text{t}}/R_{\text{per}}$. For more details we refer the interested reader to Rosswog et al. (2008b, 2008d).

Journal of Physics: Conference Series **172** (2009) 012036 doi:10.1088/1742-6596/172/1/012036

3. Results

3.1. Explosion mechanism

The tidal gravitational field is, to lowest order, determined by the second-order tidal tensor and the third-order deviation tensor. The latter determines the deviation of an extended body from a point particle trajectory of the same mass. The tidal tensor determines how the fluid will be shaped. It possesses one positive and two negative eigenvalues, see e.g. Brassart & Luminet (2008), and, as a consequence, the tidal field stretches the star in one eigen-direction and compresses it along the other two. Most severely, the star is compressed perpendicular to the orbital plane, this process halts when a shock forms (Kobayashi et al. 2004a; Brassart & Luminet 2008; Rosswog et al. 2008d) that reverts the compression into an expansion. For deep enough penetrations matter reaches nuclear statistical equilibrium, see Fig. 1, upper right panel. The star is squeezed through a point of maximum compression on a time scale of $\tau_{\rm comp} \sim R_{\rm wd}/v_{\rm p} \simeq 0.2{\rm s}\ (M_{\rm wd,0.6})^{-1/6}\,(R_{\rm wd,9})^{3/2}\,(M_{\rm bh,3})^{-1/3}$, where $v_{\rm p} \sim (R_{\rm g}/R_\tau)^{1/2}c \simeq 5\times10^9{\rm cm\ s}^{-1}\ M_{\rm wd,0.6}^{1/6}\ R_{\rm wd,9}^{-1/2} M_{\rm bh,3}^{1/3}$ is the peri-centre passage velocity. This velocity exceeds thermonuclear flame speeds by orders of magnitude, therefore, flame propagation effects can be safely neglected. The compression time scale needs to be compared to the dynamical time scale of the star, $\tau_{\rm dyn} = (G\bar\rho)^{-1/2} \simeq 7.2\ {\rm s}\ M_{\rm wd,0.6}^{-1/2}\ R_{\rm wd,9}^{3/2}$, with $\bar\rho$ being the average stellar density and to the nuclear reaction time scale, $\tau_{\rm nuc}$. Only for $\tau_{\rm nuc} \ll \tau_{\rm comp}, \tau_{\rm dyn}$ can a substantial nuclear energy release be expected. Our large simulation set (Rosswog et al. 2008d) shows that white dwarfs of all masses can be thermonuclearly exploded provided that the penetration factor exceeds $\beta \simeq 3$. For definiteness, we illustrate here the results at the example of a typical 0.6 $M_\odot$ carbon-oxygen (CO) white dwarf that passes a 500 $M_\odot$ black hole with a penetration factor of $\beta = 5$, see Fig. 1.

3.2. Nucleosynthesis

Nucleosynthesis is triggered predominantly in the point of maximum compression, see Fig. 1, upper right panel, the concomitant nuclear energy release inflates a hot bubble in the debris centre, see Fig. 1, lower panels. The high-density centre of the star produces an iron-group core of 0.18 $M_\odot$[1], if mainly composed of ^{56}Ni this amount is comparable, but at the lower end of what is deduced for standard type Ia supernovae (Mazzali et al. 2007). The iron core is surrounded by a 0.21 $M_\odot$ shell of silicon-group elements which in turn is covered by a sheath of unburned carbon-oxygen material, see Fig. 2. In reality, such a disruption would result in even stronger carbon-enhanced outer layers than shown in Fig. 2. For simplicity, we have instantiated our initial carbon oxygen white dwarf models ($M_{\rm WD} \geq 0.6\,M_\odot$) as homogeneously mixed stars with a 50% mass fraction of each nucleus. While such internal chemical profiles are likely accurately realized in nature in very massive white dwarfs ($\sim 1\ M_\odot$) (Mazzitelli & Dantona 1986), for lower masses the gravothermal adjustment of the interior during the cooling phase produces oxygen-enhanced stellar cores surrounded by very carbon-rich mantles ($X_C \sim 0.8$). The exact radial distribution depends on the exact value of $^{12}C(\alpha,\gamma)^{16}O$ rate and the details of how convection proceeds (Mazzitelli & Dantona 1986; Salaris et al. 1997; Straniero et al. 2003), but this general stratification tendency is well-established. Thus, the disruption of a standard 0.6 $M_\odot$ white dwarf should produce a highly carbon-enriched remnant atmosphere.

We found exploding examples throughout the whole white dwarf mass range. An extreme case is a low-mass white dwarf that can (due to its large stellar and correspondingly large tidal radius) be sent very deeply into the tidal radius of a black hole. A 0.2 $M_\odot$, pure helium white dwarf disrupted by a 1000 $M_\odot$ black hole ($\beta = 12$) produced an explosion with about 0.03 $M_\odot$ of iron-group elements. At the other extreme, a 1.2 $M_\odot$ CO white dwarf exploded with 0.66 $M_\odot$

[1] Note that our network uses element groups, individual isotopes would need to be recovered in a post-processing step with larger network.

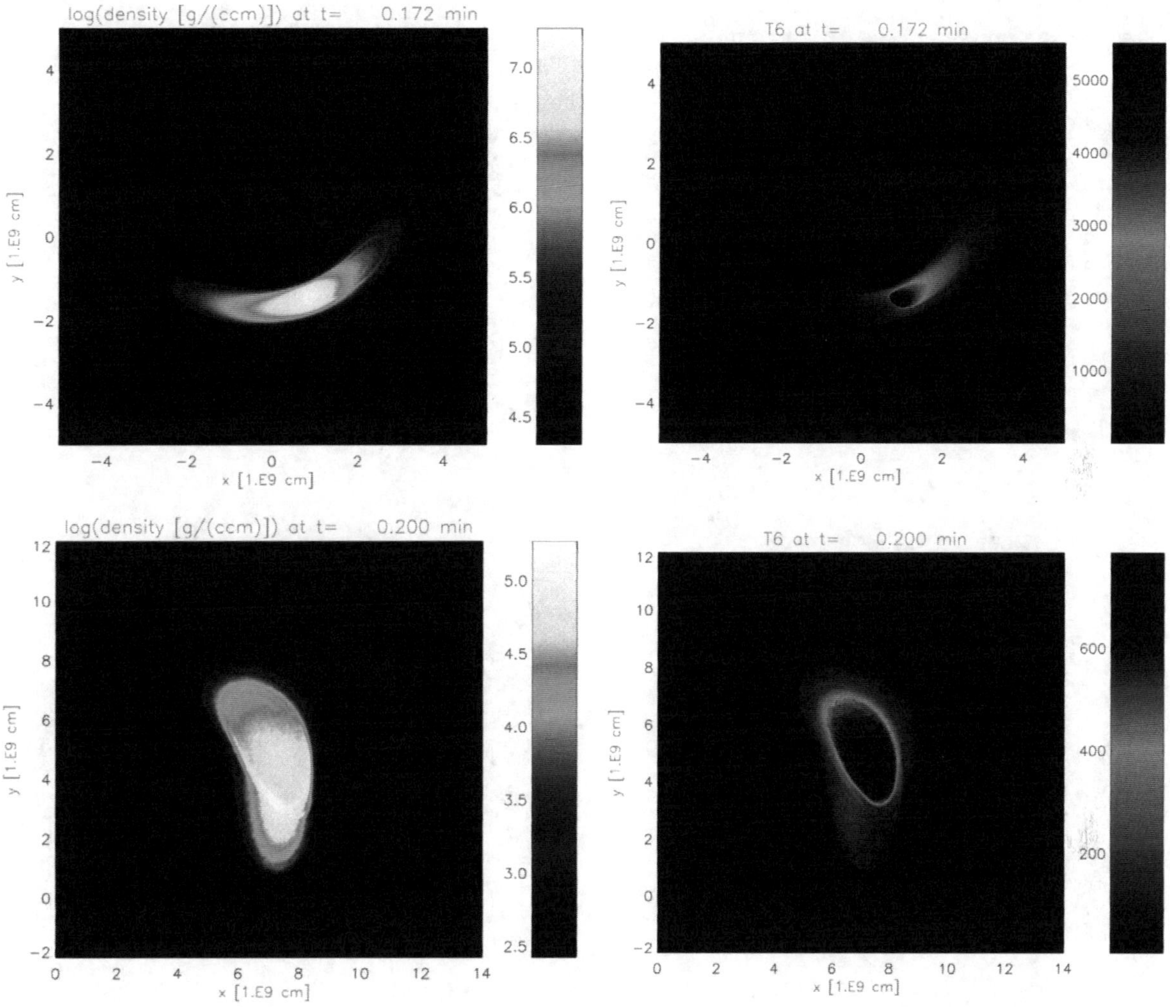

Figure 1. Density (left column) and temperature (right column) evolution during the disruption of 0.6 $M_\odot$ CO white dwarf by a 500 $M_\odot$ black hole. The distance of closest approach was 1/5 of the tidal radius.

of iron group elements after passing a 500 $M_\odot$ black hole with $\beta = 2.6$. For more details and examples, the interested reader should consult Rosswog et al. (2008d). A detailed comparison of the lightcurves of tidally-induced supernovae and "normal" type Ia supernovae is currently being prepared (Kasen et al. 2008).

3.3. Accompanying signatures

If the release of nuclear binding energy is disregarded, about half the stellar mass is ejected, the rest is still bound to the black hole and, in principle, available to be accreted. Nuclear energy release during pericentre passage can substantially increase the amount of unbound mass to about $\sim 65\%$ (Rosswog et al. 2008a). Before the energy contained in the bound matter can be released, an accretion disk has to form. On returning to the black hole, the orbits of the bound matter become focussed once more towards peri-center. The spread of specific energies across the accretion stream is very large (Rees 1988) so that matter that passes on the inside

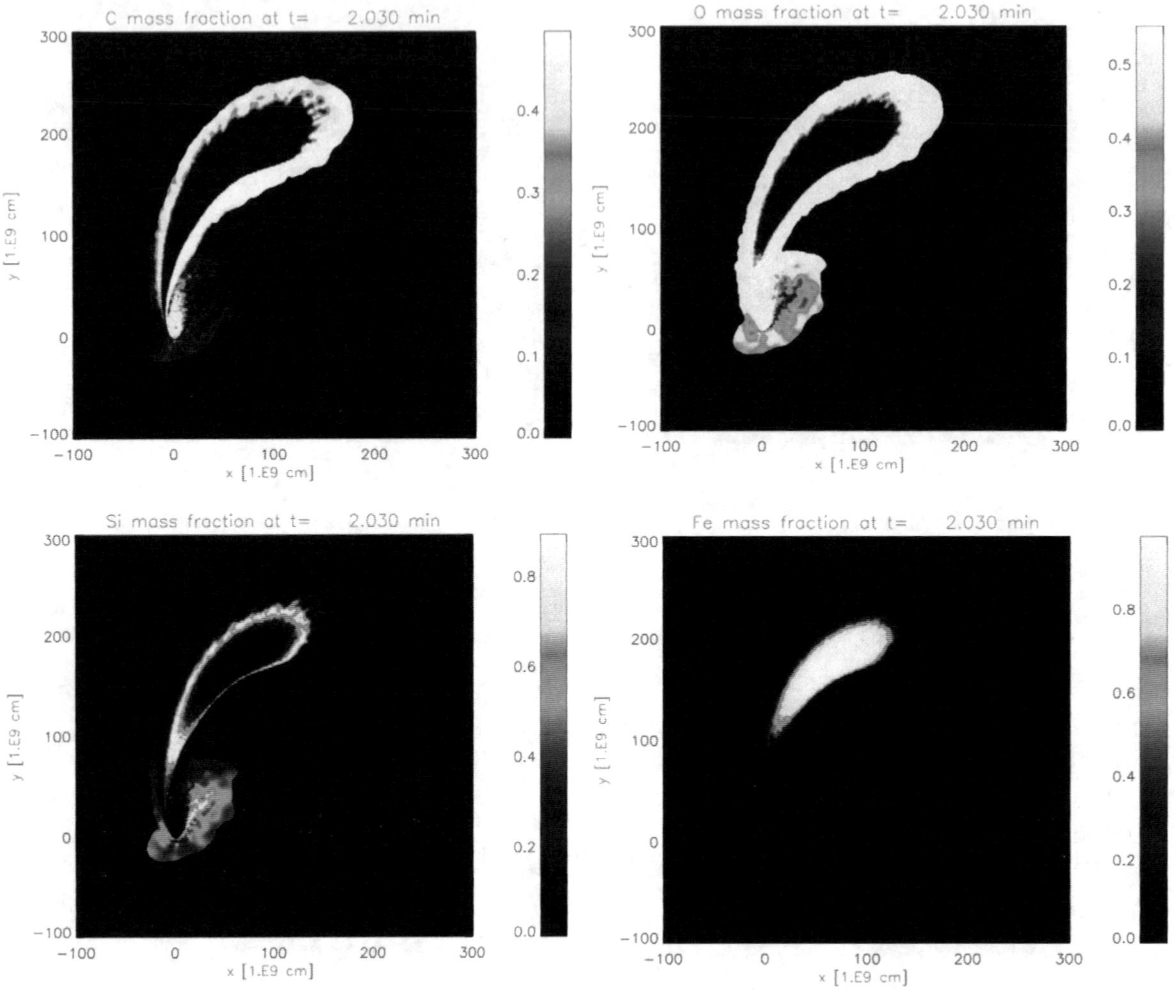

Figure 2. Mass fractions resulting from the disruption of 0.6 $M_\odot$ CO white dwarf by a 500 $M_\odot$ black hole. The distance of closest approach was 1/5 of the tidal radius.

of the stream, i.e. closer to the black hole, is substantially stronger bound and subsequently only reaches moderate apocentre distances while matter on the outside track is relaunched into a substantially wider orbit. Thus, after having passed the hole for the second time, the matter stream spreads in a fan-like manner (Rees 1988; Rosswog et al. 2008d). It subsequently collides with the still infalling material and this self-interaction produces an angular momentum redistribution shock that circularizes the forming accretion disk, see e.g. Fig. 4 in (Rosswog et al. 2008a). In the early stages, mass is fed towards the black hole at a rate that carries the imprint of the internal structure of the star (Lodato et al. 2008; Ramirez-Ruiz & Rosswog 2008), at later stages the rate falls off $\propto t^{-5/3}$ (Rees 1988, Phinney 1989). Once a disk has formed, it evolves under the influence of viscosity, radiatively driven winds and the fallback rate by which it is fed. We expect a luminosity that is comparable to the Eddington value, $L_{\rm Edd} \simeq 10^{41}$erg/s $(M_{\rm bh}/1000\, M_\odot)$, typically lasting for a few months.

4. Summary and discussion

We have explored in detail the fate of white dwarfs that approach a moderately-massive black hole close enough to become tidally disrupted. Above a limiting mass of $\sim 10^5\,M_\odot$, the detailed value depending on the white dwarf, the black hole swallows the star as a whole before disrupting it. Thus white dwarfs represent a precious tool to probe the existence of moderate-mass black holes. White dwarfs that penetrate the tidal radius deeply enough are heated by the tidal compression and ensuing shock to the temperatures of nuclear statistical equilibrium. In favorable cases more than the gravitational self-binding energy of the star can be released via nuclear reactions thus triggering a thermonuclear explosion of the white dwarf. The amount of iron-goup nuclei produced is rather sensitive to the densities at which nuclear burning occurs, and therefore to the white dwarf mass. The amount produced in the exploding cases of our simulation set (Rosswog et al. 2008d) ranges from 0.03 $M_\odot$ for a $0.2\,M_\odot$ pure helium white dwarf to $0.66\,M_\odot$ for a $1.2\,M_\odot$ carbon-oxygen white dwarf, i.e. if composed of mainly ^{56}Ni the favorable cases produce amounts comparable to standard type Ia supernovae (Mazzali et al. 2007).

These tidally-induced supernovae are, however, in several respects different from what is considered a "normal" type Ia supernova. First, they are not restricted to progenitor masses close to the Chandrasekhar limit. We find exploding examples throughout the full explored mass range from 0.2 to $1.2\,M_\odot$. Typical examples of these explosions are close to peak of the white dwarf mass distribution, i.e. near $0.6\,M_\odot$, or slightly above this value, since in the dense stellar environments around such black holes close encounters between stars may lead to mass segregation (Binney & Tremaine 2008). Thus the average progenitor will be less massive than for a normal type Ia. Second, and closely related to the first point, the progenitor composition is not restricted to carbon-oxygen. Also pure helium white dwarfs can explode via this mechanism, but since white dwarfs near and slightly beyond the peak of the mass distribution are thought to consist of carbon and oxygen this is the most common progenitor composition. Due to the initial chemical layering of the white dwarfs, the outer shells of the most common explosion remnants consist of unburned, highly carbon-enriched ($X_C \sim 0.8$) material. Third, the remnant geometry is most likely much less spherical than a standard type Ia, therefore the optical lightcurve should be rather unique as a result of the radiating material being highly squeezed into the orbital plane. The large velocities of the ejected debris ($> 10^4$ km/s) should produce large Doppler shifts. And last but not least, this peculiar type of thermonuclear supernova is accompanied by an X-ray flare close to the Eddington luminosity that lasts for a few months.

The estimated event rates for this type of transient are $\sim 10^{-3}$ of the type Ia supernova rate. Though being substantially less frequent, they occur often enough to warrant a search for this new class of optical transient. Upcoming supernova searches hope to discover several thousand to several hundred thousand type Ia-like events (Riess & Livio 2006; Aldering et al. 2007) per year. Chances are good to find examples of tidally-induced thermonuclear supernovae among them. Maybe, the recently detected carbon-rich transient SCP 06F6 accompanied by an X-ray signal (Gaensicke et al. 2008) is already the first example for this class of object.

Acknowledgments

We thank Holger Baumgardt, Peter Goldreich, Jim Gunn, Piet Hut, Dan Kasen, Bronson Messer and Martin Rees for very useful discussions. E. R. acknowledges support from the DOE Program for Scientific Discovery through Advanced Computing (SciDAC; DE-FC02-01ER41176). The simulations presented in this paper were performed on the JUMP computer of the Höchstleistungsrechenzentrum Jülich. Oak Ridge National Laboratory is managed by UT-Battelle, LLC, for the U.S. Department of Energy under contract DE-AC05-00OR22725.

References

Aldering G Kim A G Kowalski M Linder E V Perlmutter S 2007 Snapping supernovae at z > 1 7 *Astroparticle Physics* **27** 213–225

Benz W 1990a Smooth particle hydrodynamics: A review in: J Buchler (Ed) *Numerical Modeling of Stellar Pulsations* Kluwer Academic Publishers Dordrecht p 269

Benz W Bowers R Cameron A Press W 1990b Dynamic mass exchange in doubly degenerate binaries i - 0 9 and 1 2 solar mass stars *ApJ* **348** 647

Binney J Tremaine S 2008 *Galactic Dynamics: Second Edition* ISBN 978-0-691-13026-2 (HB) Published by Princeton University Press Princeton NJ USA

Brassart M Luminet J 2008 Shock waves in tidally compressed stars by massive black holes *A&A* **481** 259–277

Carter B Luminet J P 1983 Tidal compression of a star by a large black hole I Mechanical evolution and nuclear energy release by proton capture *A&A* **121** 97–113

Carter B Luminet J P 1985 Mechanics of the affine star model *M.N.R.A.S.* **212** 23–55

Dearborn D S P Wilson J R Mathews G J 2005 Relativistically Compressed Exploding White Dwarf Model for Sagittarius A East *ApJ* **630** 309–320

Ferrarese L Merritt D 2000 A Fundamental Relation between Supermassive Black Holes and Their Host Galaxies *ApJL* **539** L9–L12

Gaensicke B T Levan A J Marsh T R Wheatley P J 2008 SCP06F6: A carbon-rich extragalactic transient at redshift z˜0 14 ArXiv e-prints

Gerssen J van der Marel R P Gebhardt K Guhathakurta P Peterson R C Pryor C 2002 Hubble Space Telescope Evidence for an Intermediate-Mass Black Hole in the Globular Cluster M15 II Kinematic Analysis and Dynamical Modeling *AJ* **124** 3270–3288

Gerssen J van der Marel R P Gebhardt K Guhathakurta P Peterson R C Pryor C 2003 Addendum: Hubble Space Telescope Evidence for an Intermediate-Mass Black Hole in the Globular Cluster M15 II Kinematic Analysis and Dynamical Modeling *AJ* **125** 376–377

Hix W R Khokhlov A M Wheeler J C Thielemann F -K 1998 The Quasi-Equilibrium-reduced alpha -Network *ApJ* **503** 332

Kasen D Ramirez-Ruiz E Rosswog S 2008 in preparation

Kobayashi S Laguna P Phinney E S Mészáros P 2004 Gravitational Waves and X-Ray Signals from Stellar Disruption by a Massive Black Hole *ApJ* **615** 855–865

Kormendy J Richstone D 1995 Inward Bound–The Search For Supermassive Black Holes In Galactic Nuclei *Ann Rev Astron Astrophys* **33** 581

Lodato G King A R Pringle J E 2008 Stellar disruption by a supermassive black hole: is the light curve really proportional to $t^{-5/3}$? ArXiv e-prints

Luminet J P Pichon B 1989 Tidally-detonated nuclear reactions in main sequence stars passing near a large black hole *A&A* **209** 85–102

Maccarone T J Servillat M 2008 Radio observations of NGC 2808 and other globular clusters: constraints on intermediate-mass black holes *M.N.R.A.S.* **389** 379–384

McClintock J E Remillard R A 2003 Black Hole Binaries ArXiv Astrophysics e-prints

Mazzali P A Röpke F K Benetti S Hillebrandt W 2007 A Common Explosion Mechanism for Type Ia Supernovae *Science* **315** 825–

Mazzitelli I Dantona F 1986 The relation between initial and minimum final white dwarf mass for Population I stars *ApJ* **311** 762–773

Monaghan J J 1992 Smoothed particle hydrodynamics *Ann Rev Astron Astrophys* **30** 543

Monaghan J J 2005 Smoothed particle hydrodynamics *Reports on Progress in Physics* **68** 1703–1759

Noyola E Gebhardt K Bergmann M 2008 Gemini and Hubble Space Telescope Evidence for an Intermediate-Mass Black Hole in ω Centauri *ApJ* **676** 1008–1015

Paczyński B Wiita B 1980 Thick accretion disks and superficial luminosities **88** 23

Phinney E S 1989 Manifestations of a Massive Black Hole in the Galactic Center in: M Morris (Ed) *IAU Symp 136: The Center of the Galaxy* p 543

Pooley D Rappaport S 2006 X-Rays from the Globular Cluster G1: Intermediate-Mass Black Hole or Low-Mass X-Ray Binary? *ApJL* **644** L45–L48

Portegies Zwart S F Baumgardt H Hut P Makino J McMillan S L W 2004 Formation of massive black holes through runaway collisions in dense young star clusters *Nature* **428** 724–726

Ramirez-Ruiz E Rosswog S 2008 The star ingesting luminosity of intermediate mass black holes in globular clusters ArXiv e-prints

Rees M J 1988 Tidal disruption of stars by black holes of 10 to the 6th-10 to the 8th solar masses in nearby galaxies *Nature* **333** 523–528

Riess A G Livio M 2006 The First Type Ia Supernovae: An Empirical Approach to Taming Evolutionary Effects

in Dark Energy Surveys from SNe Ia at z>2 *ApJ* **648** 884–889
Rosswog S 2005 Mergers of Neutron Star-Black Hole Binaries with Small Mass Ratios: Nucleosynthesis Gamma-Ray Bursts and Electromagnetic Transients *ApJ* **634** 1202–1213
Rosswog S Ramirez-Ruiz E Hix R 2008a Atypical thermonuclear supernovae from tidally crushed white dwarfs *ApJ* **679** 1385
Rosswog S Ramirez-Ruiz E Hix R 2008d Tidal disruption and ignition of white dwarfs by moderately massive black holes submitted to *ApJ*
Rosswog S Ramirez-Ruiz E Hix R 2008 Simulating black hole white dwarf encounters *Comp Phys Comm*
Salaris M Dominguez I Garcia-Berro E Hernanz M Isern J Mochkovitch R 1997 The Cooling of CO White Dwarfs: Influence of the Internal Chemical Distribution *ApJ* **486** 413–+
Straniero O Domínguez I Imbriani G Piersanti L 2003 The Chemical Composition of White Dwarfs as a Test of Convective Efficiency during Core Helium Burning *ApJ* **583** 878–884
Timmes F X Swesty F D 2000 The Accuracy Consistency and Speed of an Electron-Positron Equation of State Based on Table Interpolation of the Helmholtz Free Energy *ApJS* **126** 501–516
Wilson J R Mathews G J 2004 White Dwarfs near Black Holes: A New Paradigm for Type I Supernovae *ApJ* **610** 368–377
Zezas A Fabbiano G Rots A H Murray S S 2002 Chandra Observations of "The Antennae" Galaxies (NGC 4038/4039) III X-Ray Properties and Multiwavelength Associations of the X-Ray Source Population *ApJ* **577** 710–725

Journal of Physics: Conference Series **172** (2009) 012037 doi:10.1088/1742-6596/172/1/012037

Accretion induced collapse of white dwarfs in binary systems and their observational properties

D T Wickramasinghe[1], Jarrod R Hurley[2], Lilia Ferrario[1], Christopher A Tout[3] and Paul D Kiel[2]

[1] Mathematical Sciences Institute, The Australian National University, Canberra, Australia
[2] Centre for Astrophysics and Supercomputing, Swinburne University of Technology, P.O. Box 218, VIC 3122, Australia
[3] Institute of Astronomy, Cambridge, UK

E-mail: jhurley@swin.edu.au

Abstract. Neutron stars can form through core collapse (CC) following a supernova explosion of a massive star, or from a white dwarf that first forms in a binary system and then collapses into a neutron star via accretion induced collapse (AIC). So far, there have been no unambiguous identifications of AIC neutron stars in our Galaxy, although it has been suspected that they may exist among the population of binary millisecond pulsars (BMSPs). We summarise results of new calculations on the expected birth rates of binary systems with AIC neutron stars and discuss the predicted orbital period of systems with He, CO and O-Ne-Mg white dwarfs and main sequence companions. We argue that AIC systems may make an important contribution to the observed population of BMSPs.

1. Introduction

The BMSPs have neutron stars with magnetic fields of $10^8 - 10^9$ G, much smaller than the fields of $10^{11} - 10^{13}$ G found in the bulk of isolated radio pulsars. The origin of the low fields have been the subject of much debate.

One view is that the neutron stars in the BMSPs have formed from the core collapse (CC) of a massive star with fields of $\sim 10^{11} - 10^{13}$ G that are essentially of crustal origin. The crustal field decays by 5 orders of magnitude in response to accretion (Konar and Bhattacharya 1997), and the neutron star spins-up maintaining spin equilibrium with the inner edge of the disc which recedes towards the neutron star. During these phases, the system is seen as a low mass or an intermediate mass X-ray binary (LMXB/IMXB). At the end of the accretion disc phase, the rapidly spinning and low field neutron star emits radio emission and is seen as a BMSPS.

Another often discussed channel for the production of BMSPs involves the accretion induced collapse (AIC) of a white dwarf (Michel 1987). Here, during the course of mass transfer, a white dwarf reaches the Chandrasekhar limit, and collapses to form a neutron star (Bhattacharya and Van den Heuvel 1991). An AIC is the expected outcome of thermal time scale mass transfer in a binary system with orbital periods ($P_{\rm orb}$) of a few days if accretion occurs onto an O-Ne-Mg white dwarf (e.g. Ivanova and Taam 2004).

In the AIC scenario, we may expect the magnetic field distribution of the BMSPs to reflect in some way the magnetic field distribution of their progenitor white dwarfs. Here, the results of recent studies of the magnetic field distribution in white dwarfs may be relevant. There is now good evidence that the white dwarf field distribution is bi-modal with a high field component (HFMWD: $B \sim 10^6 - 10^9$ G) that peaks at 3×10^7 G comprising some 12 – 15% of the white dwarf population, and a low field component (LFMWD: $B \sim 10^3 - 10^5$ G) that extends down to at least the current limit of detectability ($\sim 10^3$ G) comprising $\sim$ 15% of the white dwarf population (Jordan et al. 2007). The implication for the AIC scenario is that magnetic white dwarfs that are massive enough to undergo collapse by mass accretion could potentially lead to a group of rapidly spinning neutron stars with fields amplified to neutron star values.

The role played by accretion in modifying field strength and structure is still under debate. The competition between the advection of the field due to accretion of un-magnetised material, and the decay of the field due to Ohmic dissipation has been discussed by many investigators both in the context of neutron stars (Bisnovatyi-Kogan and Komberg 1974; Urpin and Geppert 1995; Cumming, Zweibel and Bilsden 2001), and white dwarfs (Cumming 2002) on the assumption of spherical accretion. The complications introduced by non-spherical accretion have been discussed more recently by Zhang and Kojima (2006) and Payne and Melatos (2007).

In the field evolutionary scenario discussed by Cumming (2002) for the magnetic cataclysmic variables, the fields in intermediate polars are screened during the accretion disc phase, and re-emerge to pre-screened AM Her type fields when accretion ceases in the period gap. The field decay is minimal. In the CC scenario for the BMSPs the fields in the neutron stars are assumed to decay by five orders of magnitude, and reach an asymptotic value as the current carrying regions in the crust are pushed towards regions of higher conductivity by the accretion flow (Konar and Bhattacharya 1997). In the AIC scenario, a decay of the field may or may not be necessary depending on whether the white dwarf that collapses belongs to the high field or low field group of magnetic white dwarfs.

In this paper, we present a comparative study of the population properties of binary systems that contain neutron stars that form from CC and AIC extending previous similar studies that have focussed exclusively on the CC route. We show that the different routes predict different relative populations of neutron stars with main sequence, He, CO, O Ne Mg white dwarfs, different orbital period distributions, and total birth rates which can potentially be used to distinguish between the different scenarios. We show that it may be possible to have a scenario in which the bulk of the observed BMSPs come form the accretion induced collapse of white dwarfs with different implications for field evolution and decay in accreting compact stars.

2. Models and Results

We have carried out a population synthesis study that allows for the formation of neutron stars by both the core collapse (CC) and the accretion induced collapse (AIC) routes. The model is based on the the rapid binary star evolution (BSE) code constructed by Hurley et al. (2002). The model allows for tidal interactions (circularization and synchronization), supernova kicks, angular momentum loss by gravitational radiation and magnetic braking, mass transfer on nuclear, thermal and dynamical timescales and via winds and covers all aspects of the evolution of stars from the main-sequence up to and including the remnant stages. The details of our calculations will be presented elsewhere (Hurley et al. 2009). Here, we present results for our standard model which has small kicks for AIC neutron stars (dispersion of $20\,\mathrm{km\,s^{-1}}$), imposes the Eddington limit for accretion, and uses the reduced helium star wind from Kiel and Hurley (2006).

The evolution of CC systems has been discussed by Podsiadlowski, Rappaport and Pfahl (2002). The period evolution bifurcates into two branches at an orbital period of $\sim$ 1 d (the bifurcation period). If the secondary star is less massive that $\sim 2\,\mathrm{M}_\odot$, the system evolves towards shorter orbital periods after the bifurcation period driven by magnetic braking and gravitational radiation. On the other hand, systems with more massive secondaries evolve to longer orbital periods driven by thermal time scale evolution. The systems in the former group

Table 1. Birth rates of the different types of binary system.

		BMSP	BMSP(observable)	LMXB (observable)
AIC NS	all	3.5548E-04	1.4757E-04	2.3968E-04
	MS	9.7418E-07	0.0000E+00	9.7418E-07
	WD	2.0631E-04	1.4399E-04	6.2320E-05
CC NS	all	5.2864E-05	1.2813E-05	5.2864E-05
	MS	8.8235E-06	0.0000E+00	8.8235E-06
	WD	2.2716E-05	1.1884E-05	2.2716E-05

have short periods and low mass He WD, CO WD or O-Ne-Mg WD companions at the end of their orbital evolution. Those in the latter group have long orbital periods at the end of their evolution and have He WD companions.

In the AIC route, the primary star is less massive and therefore first evolves into a white dwarf and then subsequently collapses into a neutron star due to accretion. The orbital evolution bifurcates into two routes as with the CC route so both routes predict binary systems of the same type, but with different orbital period distributions and in different proportions. The model also predicts a significant class of extremely short period ($P_{\rm orb} \sim 2 \times 10^{-4} - 10^{-2}$ days binaries with low mass He WD,CO WD and O-Ne-Mg WD companions). These are systems that have been through two phases of common envelope evolution (see Hurley et al. 2009).

We show in Figure 1 the period distributions predicted by our calculations of systems that result from the CC and AIC routes that are detached and observable as BMSPs. Both routes predict a broad peak centered at $P_{\rm orb} \sim 10$ d roughly as observed. These are systems that have evolved towards longer periods and with He WD companions. The secondary peak at $P_{\rm orb} \sim 100$ d is only weakly present in the AIC route and even less so in the CC route. A similar discrepancy was found by Podsiadlowski et al. (2002). This suggests inadequacies in current treatments of binary stellar evolution that determine the bifurcation period. The birth rates of the different types of binary system arising from the two routes are summarised in Table 1, where we have also given the rates of systems potentially observable as BMSPs. The AIC route has a birth rate that is $\sim$ 10 times higher for systems with WD companions suggesting that this route makes the dominant contribution to the observed BMSP population. On the other hand, the birth rate of systems with MS companions is significantly higher for the CC route compared to the AIC route. The CC route is therefore likely to be the major contributor to LMXBs.

3. Discussion and Conclusions

We have explored the possibility that neutron stars that result from the accretion induced collapse (AIC) of a white dwarf may play a role in explaining the BMSPs. Our calculations have shown that both the CC and AIC routes result in populations of binary systems that can be identified with LMXB/IMXBs while transferring matter, and BMSPs at the end of an accretion phase. Furthermore, both routes predict binary systems with MS, He WD, CO WD, O-Ne-Mg WD companions, albeit in different relative proportions.

If we focus on BMSPs with WD companions, which is the dominant group in the observed sample, our results indicate birth rates that are $\sim$ 10 times higher for BMSPs that come from the AIC route. On the other hand, if we focus on LMXBs with main sequence companions, we find birth rates that are 10 times larger for the CC route. Both routes make a contribution to the observed population of ultra compact LMXBs.

For BMSPs that come from the AIC route, the fields would primarily be a consequence of the white dwarf ancestry of the neutron star. The field in such a neutron star at birth is

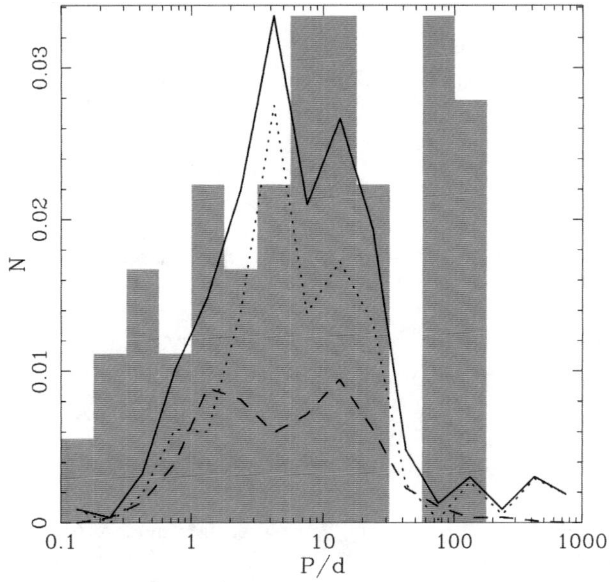

Figure 1. Period distribution of binary MSPs formed in the model from the AIC route (upper dotted line) and the CC route (lower dashed line) compared with observations (gray histogram). The solid line gives the combined distribution for AIC and CC routes. The observations are normalized to the combined model distribution.

essentially the field that would have resided in the core of the white dwarf that is compressed by the collapse. White dwarfs massive enough for AIC in the low and high field groups will lead to neutron stars with fields of $10^8 - 10^{10}$ G and $10^{12} - 10^{15}$ G respectively. Depending on which group dominates, field decay may or may not be necessary to explain the origin of the BMSPs.

References

Bhattacharya D and van den Heuvel E P J 1991 *Physical Review*, **203** 1
Bisnovatyi-Kogan G S and Komberg B V 1974 *Soviet Astronomy* **18** 217
Cumming A 2002 *MNRAS* **333** 589
Cumming A Zweibel E and Bilsden L 2001 *ApJ* **557** 958
Ferrario L and Wickramasinghe D T 2007 *MNRAS* **375** 1009
Hurley J R Tout C A and Pols O R 2002 *MNRAS* **329** 897
Hurley J R Wickramasinghe D T Ferrario L Tout C A and Kiel P D 2009 *MNRAS* to be submitted
Ivanova N Taam R E 2004 *ApJ* **601** 1058
Jordan S Aznar Cuadrado R Napiwotzki R Schmid H M and Solanki S K 2007 *A&A* **462** 1097
Kiel P D and Hurley J R 2006 *MNRAS* **369** 1152
Konar S and Bhattacharya D 1997 *MNRAS* **84** 311
Manchester R N Hobbs G B Teoh A and Hobbs M 2005 *ApJ* **129** 1993
Michel F C 1987 *Nature* **329** 310
Payne D J B and Melatos A 2007 *MNRAS* **376** 609
Pfahl E Rappaport S and Podsiadlowsky Ph 2003 *ApJ* **597** 1036
Urpin V and Geppert U 1995 *ApJ* **275** 1117
Zhang C M and Kojima Y 2006 *MNRAS* **366** 137

Journal of Physics: Conference Series **172** (2009) 012038 doi:10.1088/1742-6596/172/1/012038

The white dwarf in V842 Cen (Nova Cen 1986)

Brian Warner[1,2] and Patrick A. Woudt[1]

[1]Department of Astronomy, University of Cape Town, Rondebosch 7700, South Africa
[2]School of Physics and Astronomy, University of Southampton, Highfield, Southampton S017 1BJ, UK

E-mail: Brian.Warner@uct.ac.za

Abstract. High speed photometric measurements made in February and March 2008 of the remnant of Nova Cen 1986 show that in the interval since 2000 a strong 56.83 s signal has appeared. We interpret this as the rotation period of the white dwarf (the third shortest known), now visible since the mass transfer rate has dropped, allowing the primary's magnetosphere to expand and produce an intermediate polar structure. Optical sidebands to the 56.83 s signal support the model and enable us to measure an orbital period of 3.94 h; there is also a strong negative superhump in the light curve, with a period of 3.78 h. An additional signal at 2.886 h is of unknown origin, but is probably an example of the mysterious GW Lib phenomenon.

1. Introduction

V842 Cen is the remnant of Nova Centauri 1986, a moderately fast nova with a pre-eruption magnitude ~ 18.5 (McNaught 1986). Our current measurements give V ~ 16.3, so 22 years after maximum it is still ~ 2 magnitudes above its expected quiescent brightness. Schmidtobreick et al. (2005) report a spectrum obtained in 2003 which shows a strong blue continuum with weak high ionization lines (e.g. C IV), probably arising in the nova ejecta, but accompanied by strong Balmer emission lines running to high series members and by He I emission, yet with only moderate He II and 4650 Å Bowen fluorescence lines, which they describe as more characteristic of dwarf novae in quiescence than of a nova.

Our previous high speed photometry of V842 Cen, runs S6096 and S6097 obtained in 2000 (Woudt & Warner 2003: hereafter WW03), showed rapid activity with flares up to 0.25 mag on time scales ~ 5 mins, but no coherent short period brightness modulations (though quasi-periodic oscillations in the range 750 – 1300 s were observed).

To check on the current status of the nova remnant we re-observed it in February and March 2008 – and found that it had developed a number of optical modulations since we last observed it, some of which are of relevance to the state of the cooling white dwarf.

2. Observations and Analysis

Our observations were made on the 74-in Radcliffe telescope at the Sutherland site of the South African Astronomical Observatory, using the University of Cape Town's frame transfer CCD photometer with no optical filter. Integration times were 6 s. This 'white light' photometry was placed on an approximate magnitude scale using a white dwarf standard star. A list of our observations is given in Table 1.

Table 1. Observing log of photometric observations

Run No.	Date of obs. (start of night)	HJD of first obs. (+2450000.0)	Length (h)	V (mag)
S7791	2008 Feb 17	4514.54882	2.14	16.2
S7795	2008 Feb 18	4515.45960	3.18	16.3
S7798	2008 Feb 19	4516.45845	4.34	16.3
S7803	2008 Mar 13	4539.41017	5.83	16.3
S7806	2008 Mar 14	4540.39963	3.67	16.3
S7809	2008 Mar 16	4542.39838	6.25	16.3
S7811	2008 Mar 17	4543.43785	3.05	16.3
S7813	2008 Mar 18	4544.40490	6.12	16.2

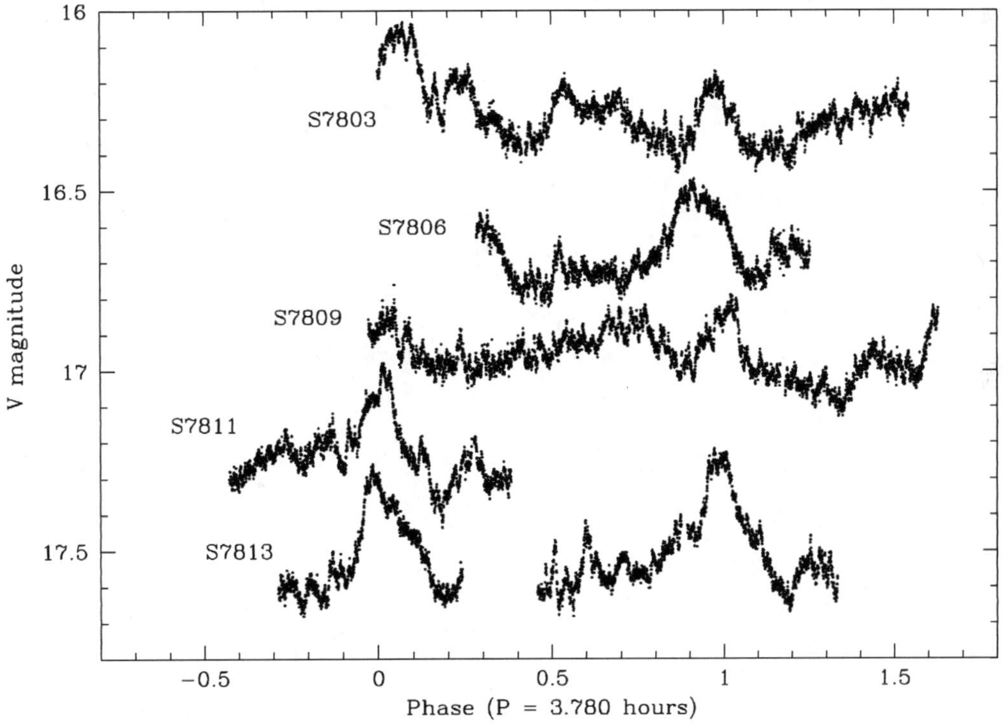

Figure 1. The 2008 March light curves phase folded on the 3.780 h superhump period. The light curves are labelled with the run numbers (see Table 1) and offset vertically for display purposes.

A selection of the light curves is shown in Figure 1, where we have phased them on a period of 3.780 h, which illustrates the periodically recurring narrow peak that we attribute to a superhump, arising from a precessing disc, which is not of direct relevance here. These light curves show also the presence of rapid low amplitude variations which we find have periodic and quasi-periodic components, but mostly they are flickering caused by the mass transfer process.

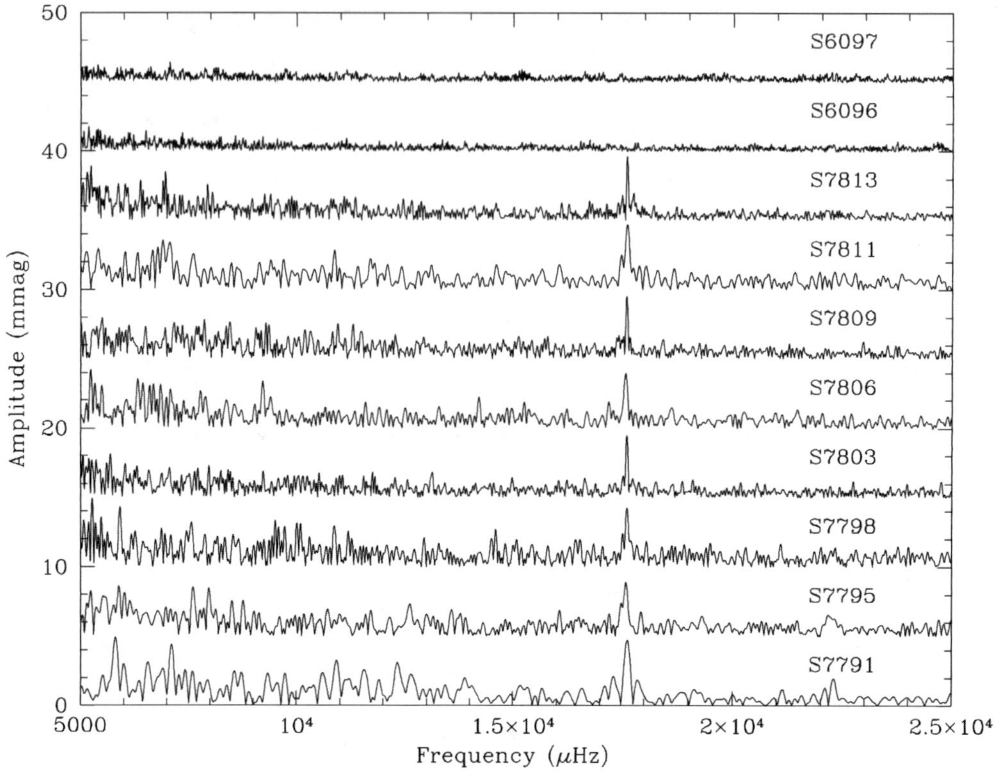

Figure 2. Fourier transforms in the range of 5000 – 25000 μHz for individual observing runs on V842 Cen.

In Figure 2 we compare the Fourier transform (FT) of our two nights of observation made in 2000 (WW03) with that of our observations in 2008. The most immediately obvious difference is the appearance of a strong coherent periodicity at ~ 57 s in the latter data set. We begin our analysis by looking at this more closely.

The left panels in Figure 3 show the FT of our March 2008 observations (omitting run S7813, where the spiky superhumps cause some problems) in the vicinity of the 57 s modulation. The dominant signal is an FT window pattern centred on 56.825 ± 0.001 s, with amplitude 4.2 mmag. Prewhitening at that period leaves a signal (visible as a sideband $70.5 \pm 0.6\,\mu$Hz away on the low frequency side in Fig. 3) at 57.054 ± 0.002 s and amplitude 1.6 mmag. Prewhitening the light curve with these signals leaves no significant signals in their vicinity in the FT (bottom left panel of Fig. 3), which shows that both are stable over the five day baseline.

The February FT is shown in Figure 3 (right panels), and shows both the principal signal and the longer period sideband, but in addition a short period sideband is present. The mean period of the principal signal, over the three day baseline, is 56.828 ± 0.002 s with amplitude 3.9 mmag, which further establishes its stability. By simultaneously fitting three sinusoids to the light curves, by least squares, the two side bands are found to be at 57.055 ± 0.005 s, amplitude 1.7 mmag and 56.598 ± 0.005 s, amplitude 1.7 mmag. After prewhitening with these three periodicities there is no significant remaining signal (bottom section of the right panel of Fig. 3). The frequency difference between the additional and the principal signals is $71.6 \pm 2.3\,\mu$Hz, which is equal within errors to the splitting on the low frequency side of the principal. Thus in February 2008 we have equally split sidebands but in March 2008 only the lower

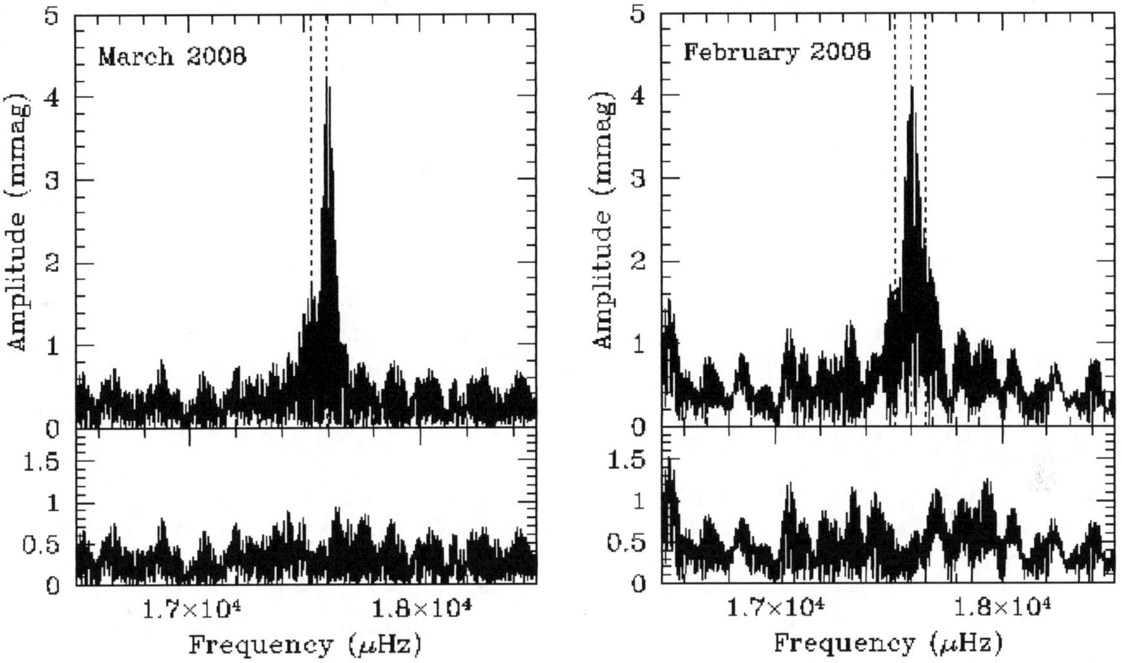

Figure 3. Fourier transforms in the vicinity of the 57 s signals.

frequency side band is present.

There is no sign in the FTs of a harmonic to the 56.83 s modulation.

Seeking the 70.5 μHz frequency at the low frequency end of the FT reveals an interesting set of long period modulations – full details will be given elsewhere, but, in summary, the largest signal is at 2.886 h (96.26 μHz) and the next largest is at the 3.780 h (73.5 μHz) period mentioned above. This latter modulation had strong higher harmonics, as is evident from the profiles seen in Fig. 1; this is characteristic of a superhump, not an orbital modulation. If the 70.5 μHz sideband separation of components in the 57 s modulation is an orbital sideband it implies an orbital period of 3.94 h, which is not itself directly seen in the FT.

If $P_{\rm orb}$ is indeed the value we have deduced, then the superhump period is shorter than the orbital period, implying that V842 Cen possesses a negative superhump, which is not unusual for high $\dot{M}$ systems. In particular, TT Ari, which we previously pointed out (WW03) has a very active light curve similar to V842 Cen, usually has a negative superhump which is 3.4% shorter than its $P_{\rm orb}$ of 3.30 h (Skillman et al. 1998); in V842 Cen we find the difference is 4.1%.

In the FTs there are also quasi-periodic oscillations with time scales ~ 2000 s. These are related to conditions in the disc and are not directly relevant to the white dwarf, but we note that TT Ari, as well as having a negative superhump, has QPOs on time scales ranging over 900 to 1500 s (Kim et al. 2008).

3. Discussion

The derivation of $P_{\rm orb}$ that we have given above implies that the 57.04 s sideband is caused by reprocessing of the 56.825 s signal off a surface sharing in the orbital motion; this is the classic situation for an intermediate polar where gas accretes onto a rotating white dwarf primary and radiation from the accretion zones is reprocessed from the bright spot region of the accretion disc or from the secondary star (e.g. Warner 1995). The 56.60 s sideband adds weight to this interpretation – denoting the orbital frequency by Ω and the white dwarf rotational frequency by ω, we have detected three signals at ω, $\omega - \Omega$ and $\omega + \Omega$, which are characteristic of an intermediate polar (IP) - see Warner (1986).

Thus V482 Cen joins the group IPs that are known novae (four definite, 13 possible – Mukai 2008), but has the shortest rotation period of them all, the previously known shortest being V533 Her (Nova Her 1963) with $P_{rot} = 63.33$ s (Patterson 1979), and as such may be clasified as a member of the DQ Her subclass of IPs (Chapter 8 of Warner 1995), and is the first in the southern sky.

Recognising that the long-known 71 s signal in DQ Her (Nova Her 1934) is now sometimes attributed to two-pole accretion onto a 142 s rotating white dwarf (Zhang et al. 1995), we looked for a 113.6 s signal in our FTs, but without success. The next shortest solid body rotations known for white dwarfs in cataclysmic variables are 33.06 s in the nova-like AE Aqr and 27.87 s in the dwarf nova WZ Sge (Warner & Pretorius 2008), so V842 Cen contains the third fastest known rotating white dwarf.

An obvious question is why the rotation modulation has appeared between our observation in 2000 and those in 2008. It is probable that in the earlier observations, made 14 years after the nova eruption, the irradiation enhanced $\dot{M}$ from the secondary was high enough to compress the primary's magnetosphere down to its surface, whereas eight years later the white dwarf had cooled and $\dot{M}$ had fallen sufficiently to allow the magnetospheric radius to exceed the primary's radius, establishing an IP structure. The lowered $\dot{M}$ suspected from recent spectra (Schmidtobreick et al. 2005) is compatible with this model. There is, however, some conflict in interpretation of the observations – the existence of superhumps normally implies a high $\dot{M}$ disc, especially for a relatively large value of $P_{\rm orb}$.

Finally, the mysterious 2.886 h modulation is probably an example of the 'GW Lib' phenomenon, seen in some cataclysmic variables (see Table 2 of Woudt, Warner & Pretorius 2004).

Acknowledgments
BW's research is supported by the University of Cape Town and the University of Southampton. PAW's research is supported by the National Research Foundation and the University of Cape Town.

References

Kim Y et al. 2008 *A&A* in press (arXiv:0810.1489)
McNaught R H 1986 *IAUC* **4274**
Mukai K 2008 (`http://asd.gsfc.nasa.gov/Koji.Mukai/iphome/catalog/alpha.html`)
Patterson J 1979 *ApJL* **233** 13
Schmidtobreick L Tappert C Bianchini A Mennickent R E 2005 *A&A* **432** 199
Skillman D R et al 1998 *ApJL* **503** 67
Warner B 1986 *MNRAS* **219** 347
Warner B 1995 *Cataclysmic Variable Stars* Cambridge University Press
Warner B and Pretorius M L 2008 *MNRAS* **383** 1469
Woudt P A and Warner B 2003 *MNRAS* **340** 1011
Woudt P A, Warner B and Pretorius M L 2004 *MNRAS* **351** 1015
Zhang E Robinson E L Stiening R F and Horne K 1995 *ApJ* **454** 447

Journal of Physics: Conference Series **172** (2009) 012039 doi:10.1088/1742-6596/172/1/012039

The accreting white dwarfs in VY Scl nova-like variables

Edward M Sion, Trisha Mizusawa and Ronald-Louis Ballouz

Department of Astronomy and Astrophysics, Villanova University, Villanova, PA 19085, USA

E-mail: edward.sion@villanova.edu, trisha.mizusawa@villanova.edu, ronald-louis.ballouz@villanova.edu

Abstract. Accurate distances for nova-like variables offer the possibility of extracting information on nova-like accretion rates during high states of optical brightness and on their underlying accretion-heated white dwarfs during intermediate and low brightness states. The modeling technique which is employed is discussed and a representative example, the nova-like variable KR Aur, is presented. Although KR Aur was in a fainter high state when its far ultraviolet spectrum was obtained, roughly one-half of its FUV radiation is due to the light of an accretion disk and the other half is contributed by a hot white dwarf with $T_{\rm eff} = 29,000{\pm}2,000$K. However, this best-fit solution corresponds to a distance of 180 pc which was an early distance estimate due to Patterson (1984).

1. Introduction

The nova-like variables are a non-magnetic subclass of CVs in which the mass-transfer rate tends to be high and the light of the system is typically dominated by a very bright accretion disk (Warner 1995). The spectra of nova-like variables resemble those of classical novae (CNe) that have settled back to quiescence. However, nova-like variables have never had a recorded CN outburst or any outburst. Hence their evolutionary status remains unknown. They could be close to having their next CN explosion, or they may have had an unrecorded explosion, in the recent past. Their distribution of orbital periods reveals a large concentration of systems in the range between three and four hours, the former period being the upper boundary of the CV period gap where very few CVs are found. Some nova-likes (classified as the VY Sculptoris systems) show the behavior of being in a high optical brightness state for most of the time, but then, for no apparent reason, plummeting into a deep low optical brightness state with little or no ongoing accretion. Then, just as unpredictably, their optical brightness returns to the high state (cf. Honeycutt & Kafka 2004 and references therein). These precipitous drops in brightness are possibly related to the cessation of mass transfer from the K-M dwarf secondary star either by starspots that drift into position under the inner Lagrangian point, L1 (Livio & Pringle 1998) or irradiation feedback in which an inflated outer disk can modulate the mass transfer from the secondary by blocking its irradiation by the hot inner accretion disk region (Wu et al. 1995). When VY Scl nova-likes drop into a deep low state, the underlying accretion heated white dwarf is exposed and its surface temperature can be determined with model photospheres.

Until recently, the accretion rates of nova-likes have been reported for only a few individual systems from a variety of model analyses of their optical, FUV spectra or X-ray spectra.

Journal of Physics: Conference Series **172** (2009) 012039 doi:10.1088/1742-6596/172/1/012039

Optical determinations of the accretion rates in nova-likes are based upon estimates of their disk luminosity using distance estimates or clues (Patterson 1984). The absolute magnitudes of the accretion disks in nova-likes reveal that their accretion rates are similar to those derived for dwarf novae during their outbursts (Warner 1995). Unfortunately, the distances of nova-like variables remain uncertain due to the scarcity of trigonometric parallaxes and the absence of a reliable usable relation for nova-like variables between their absolute magnitude at maximum light versus orbital period similar to what exists for dwarf novae. A more systematic study of a larger number of systems is clearly needed in order to compare accretion rates among different subgroups of CVs. One recent statistical study explored how well current optically thick accretion disk models fit the FUV spectra of nova-likes and old novae in a sample of 33 nova-like and old novae by Puebla, Diaz & Hubeny (2007). The average value of $\dot{M}$ for nova-like systems was $\sim 9.3 \times 10^{-9} M_{\odot}$ yr^{-1}. Interestingly, this study included an assessment of the contribution of an assumed 40,000 K white dwarf to the FUV spectra. The authors did not include the white dwarf flux contribution in each system explicitly as a free parameter in the fitting. Rather, they used a parameter defined as the ratio of the integrated spectral flux between 1500 Å and 3250 Å of the white dwarf model and disk model to estimate the flux contribution of the white dwarf. If this ratio was < 0.1, then the system was defined as disk-dominated. Indeed, the white dwarf contribution is expected to be minimal in these systems because their disks are thick and luminous and because at high inclination the inner disk, boundary layer and white dwarf should be significantly obscured by vertical structure in the disk. However, Puebla et al. (2007) found nine systems which according to the above criterion should have a considerable contribution from the accreting white dwarf. They are BK Lyn, V442 Oph, SW Sex, V1315 Aql, PX And, V794 Aql, LX Ser, KR Aur, and AC Cnc. Five of these nine nova-like variables have high orbital inclinations and undergo eclipses. Here we use the IUE archival spectra of KR Aur as an example of our approach. to modeling nova-like variables.

Our long term objectives are: (1) to see if we can characterize or constrain the temperature of the white dwarf in nova-like varibles using a different approach than Puebla et al. (2007) and (2) to carry out a systematic synthetic spectral analysis of IUE and HST archival spectra of nova-like variables based upon a uniformly applied new method of distance determination to obtain accretion rates from fitting the observed spectral slopes and Lyman alpha profiles.

2. Distances to nova-likes

There is only one nova-like, RW Tri, with a reliable trigonometric parallax measurement. Unlike dwarf novae where there exists a correlation between their absolute magnitude at maximum and their orbital period (Warner 1995), there is no such relation for the nova-likes. A new method (Knigge 2006) utilizing 2MASS JHK photometry and the observed properties of CV donor stars has proven useful for constraining nova-like distances. At present, this may be the only reliable handle one has on nova-like distances. For each system, we obtained the J, H, K apparent magnitudes from 2MASS. For a given orbital period, Knigge (2006) provides absolute J, H and K magnitudes based upon his semi-empirical donor sequence for CVs. If it is assumed that the donor provides 100% of the light in J, H and K, then the distance is a strict lower limit. If the donor emits 33% of the light (the remainder being accretion light), then an approximate upper limit is obtained. The latter limit is a factor of 1.75 times the lower limit distance. These distances using IR photometry provide useful comparisons with distance estimates published by Patterson (1984) from the mean light levels in AAVSO light curves of nova-like variables on the assumption that the accretion disk provides 100% of the optical light.

3. Synthetic spectral fitting models

We adopted model accretion disks from the optically thick disk model grid of Wade & Hubeny (1998). In these accretion disk models, the innermost disk radius, $R_{\rm in}$, is fixed at a fractional

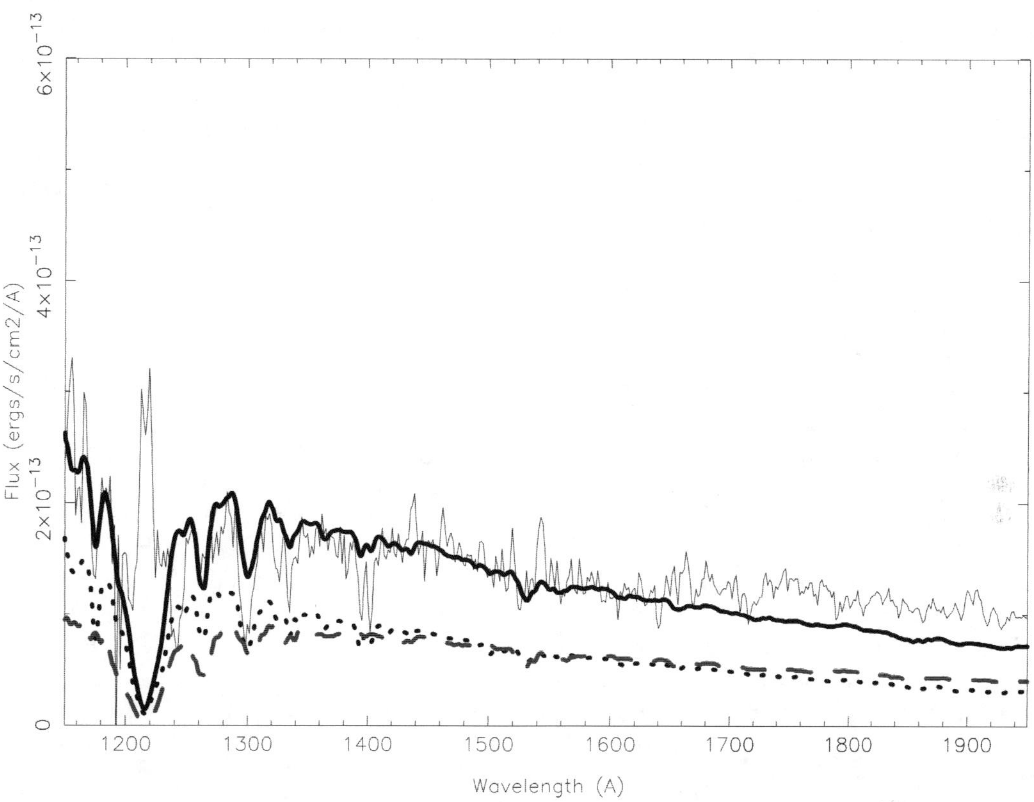

Figure 1. The best-fitting combination accretion disk + white dwarf photosphere synthetic fluxes to the spectrum SWP27096 of the VY Scl-type nova-like variable KR Aur during an fainter high state. The accretion disk corresponds to $\dot{M} = 3 \times 10^{-10} M_{\odot}$ yr^{-1}, $i = 41°$, and $M_{\rm WD} = 0.55\,M_{\odot}$. The top solid curve is the best-fitting combination, the dotted curve is the contribution of the white dwarf alone and the dashed curve is the accretion disk synthetic spectrum alone. In this fit, the white dwarf with $T_{\rm eff} = 29,000 \pm 2,000$K, log $g = 8$ contributes 50% of the FUV light and the accretion disk contributes 50% of the far UV flux.

white dwarf radius of $x = R_{\rm in}/R_{\rm WD} = 1.05$. The outermost disk radius, $R_{\rm out}$, was chosen so that $T_{\rm eff}(R_{\rm out})$ is near 10,000 K since disk annuli beyond this point, which are cooler zones with larger radii, would provide only a very small contribution to the mid and far UV disk flux, particularly the SWP FUV bandpass. The mass transfer rate is assumed to be the same for all radii.

Theoretical, high gravity, photospheric spectra were computed by first using the code TLUSTY (Hubeny 1988) to calculate the atmospheric structure and SYNSPEC (Hubeny and Lanz 1995) to construct synthetic spectra. We compiled a library of photospheric spectra covering the temperature range from 15,000 K to 70,000 K in increments of 1,000 K, and a surface gravity range, log $g = 7.0 - 9.0$, in increments of 0.2 in log g.

4. KR Aur

This nova-like system is classified as a VY Scl subtype with high states at magnitude 12.7 and the deepest low state being 17.9. The system underwent a deep minimum during 1994–95 when optical spectra revealed emission lines possibly indicative of a disk still being present (Antov et al. 1996). The minimum lasted eight months. However, usually the system brightness is magnitude 12 to 14 with 13.5 being the most typical brightness. KR Aur is one of six out of 23 VY Scl nova-likes which have negative superhumps. Negative superhumps may arise when a disk becomes tilted due to intense radiation from every hot white dwarf and thus could be an indirect indicator of nuclear burning (Kozhevnikov 2007). The other systems with negative superhumps are V442 Oph, DW UMa, TT Ari and V751 Cyg. The orbital inclination is low with an upper limit of 38 degrees and a lower limit of 10 degrees (Ritter & Kolb 2003). The orbital period is 0.1628 days. The white dwarf mass has been estimated to be in the range of 0.59 to 0.17 $M_\odot$. Its two far UV spectra were obtained during a fainter high state and are dominated by absorption features due to metals with the strongest features due to N V (1240), Si III (1300), C II (1335), Si IV (1400), Of the two SWP spectra, SWP 13584 has a slightly higher flux level and a better signal to noise. There are no P Cygni profiles evident indicating wind outflow. KR Aur has a reddening value, $E(B-V) = 0.05$ (Bruch & Engel 1994).

The white dwarf contribution is expected to be minimal in these systems because their disks are thick and luminous and because at high inclination the inner disk, boundary layer and white dwarf should be significantly obscured by vertical structure in the disk. For KR Aur, Puebla et al. (2007) explored a range of white dwarf masses $0.4 - 0.8\,M_\odot$, inclinations $i = 30^\circ - 50^\circ$ and an adopted Patterson (1984) distance of $d = 180$ pc. They found an accretion rate $\dot{M} = 4.5 \times 10^{-10}\,M_\odot$/yr which is atypically low for a VY Scl nova-like variable in its high state. Moreover, Puebla et al. (2007) found that KR Aur should have a significant flux contribution of a hot white dwarf in addition to an accretion disk.

Using our different modeling approach to KR Aur, we found that neither an accretion disk alone nor a white dwarf photosphere alone provided a satisfactory fit to the observed continuum slope and absorption spectrum of KR Aur. However, we found that a statistically significant improvement (lower χ^2) in the fitting was evident when we combined an accretion disk model with a white dwarf model. We find that the optimal combination for KR Aur at the Patterson (1984) distance of 180 pc consists of an accretion disk model with a white dwarf mass of $0.55\,M_\odot$, an inclination angle $i = 41^\circ$ and an accretion rate of $3 \times 10^{-10}\,M_\odot$/yr together with a white dwarf photosphere model with $T_{\rm eff} = 29,000 \pm 2,000$ K, log $g = 8$. This best combination fit to the IUE spectrum of KR Aur is is displayed in Fig. 1. In this fit, the accretion disk provides 50% of the FUV flux and the white dwarf contributes 50% of the flux.

5. Conclusion

For the VY Scl nova-like variable KR Aur, we find that a combination of models consisting of a hot white dwarf and an optically thick steady state accretion disk provides a significantly improved model fit to its FUV spectrum. The best-fitting model is a combination of a white dwarf with $T_{\rm eff} = 29,000 \pm 2,000$ K, log $g = 8$, contributing 50% of the FUV flux and an accretion disk with accretion rate $\dot{M} = 3 \times 10^{-10}\,M_\odot$/yr at an inclination of 41 degrees, contributing 50%. This fitting solution was achieved for a fixed distance of 180 pc which was the early distance estimate by Patterson (1984). However, the distance implied by this combination fit is at odds with the Knigge (2006) distance. This difference is explored elsewhere (Mizusawa et al. 2009).

Acknowledgments

This work is supported by NSF grant AST-0807892 to Villanova University and by the Delaware Space Grant Consortium.

References

Antov A P et al. 1996 IBVS # 4315
Bruch A & Engel A 1994 *A&A Suppl. Ser.* **104** 79
Honeycutt R K & Kafka S 2004 *AJ* **128** 1279
Hubeny I 1988 *Comput. Phys. Comun.* **52** 103

Hubeny I & Lanz T 1995 *ApJ* **439** 875
Knigge C 2006 *MNRAS* **373** 484
Kozhevnikov V P 2007 *MNRAS* **378** 955
Livio M & Pringle J E 1998 *ApJ* **505** 339
Mizusawa T et al. 2009 in preparation
Patterson J 1984 *ApJS* **54** 443
Puebla R E Diaz M P & Hubeny I 2007 *AJ* **134** 1923
Ritter H & Kolb U 1998 *A&A* **129** 83
Wade R A & Hubeny I 1998 *ApJ* **509** 350
Warner B 1995 *"Cataclysmic Variable Stars"* (Cambridge: Cambridge Univ. Press)
Wood M 1995 in *"White Dwarfs"*, Proceedings of the 9th European Workshop on White Dwarfs, Lecture Notes in Physics **443** Eds.: Detlev Koester & Klaus Werner (Springer-Verlag: Berlin) 41
Wu K Warner B Wickramasinghe D T 1995 *Pub. Ast. Soc. Aus.* **12** 60

Journal of Physics: Conference Series **172** (2009) 012040 doi:10.1088/1742-6596/172/1/012040

Where are the Progenitors of Magnetic Cataclysmic Variables (Polars)?

James Liebert

Steward Observatory, University of Arizona, Tucson AZ 85721

E-mail: liebert@as.arizona.edu

Abstract.
As over a thousand white dwarf plus M dwarf composite spectrum binaries with high quality spectra are now tabulated from the Sloan Digital Sky Survey alone, a remarkable discovery has been made. No pairing of a magnetic white dwarf of two megagauss or greater field strength with an M dwarf has been found. At this field strength, Zeeman splitting of hydrogen or helium lines would be detectable. Thus there are no known progenitors of magnetic cataclysmic variables – AM Herculis systems or polars. The formation of the strong magnetic field by a dynamo mechanism within the common envelope of a binary system has been proposed as the possible explanation.

1. Introduction

Dozens of close white dwarf plus nondegenerate (usually M) dwarf pairs have been found in the Sloan Digital Sky Survey (SDSS) which are close enough to be pre-cataclysmic variables. Dozens more have been catalogued by Ritter and Kolb (2003, 2006). In none of these CV progenitors, however, is the white dwarf strongly magnetic. Any field of about 2 megagauss (MG) or larger should have been detectable via Zeeman splitting of the Balmer or He I lines – we arbitrarily define these as "strongly magnetic". Yet CV systems involving a strongly magnetic white dwarf account for a lower limit of 10% of known systems (Araujo-Belancor et al. 2005, Thomas and Beuermann 1998). Double this percentage if you include the Low Accretion Rate Polars (LARPs) to be discussed later. To be sure, a few pre-CV systems involve a white dwarf with an apparent, weaker magnetic field, perhaps the best-studied case being V471 Tauri. The presence of a field is inferred because the X-ray emission varies with the apparent rotation period of the white dwarf in this binary system, as it presumably accretes some of the wind from the K secondary star to the magnetic poles. However, the field strength has not been measured, and is evidently too weak to show Zeeman-split lines (probably $<<$ 1 MG).

In SDSS Data Release Five, some 1,253 spectroscopically-observed close binary systems involving a white dwarf and a nondegenerate (nearly always M) dwarf (hereafter WD+M) have been identified and catalogued by Silvestri et al. (2007, hereafter S07). Only those with signal-to-noise ratio of at least 5 per pixel were included, and having DA or DB white dwarf spectra. Of all these pairs, none has been found where the white dwarf is observed to be magnetic. A working catalog of WD+M binaries is also maintained by Boris Gaensicke and collaborators (cf. Dillon et al. 2008) and the total may now exceed 1,600 pairs (B. Gaensicke, private communication, 2008).

These investigations extend the size of the sample and strengthen greatly the statistics on the apparent absence of magnetic white dwarf pairings with a nondegenerate dwarf star, as first discussed at length in Liebert et al. (2005, hereafter L05). In addition to the polars which allow no accretion disk to form, also relevant are some of the intermediate polars or DQ Herculis systems in which a partial accretion disk forms, and the white dwarf gets spun up to a rotation period shorter than the orbital period. Some intermediate polars have primaries with smaller magnetic field strengths than polars. Others in longer period systems may have field strengths comparable to the polars.

The best explanation for this apparent absence of magnetic white dwarfs associated with M dwarfs put forth in L05 is that there has been evidence that at least some strongly magnetic white dwarfs have higher than average masses (see below). As demonstrated by representative simulations in S07, the smaller radius and luminosity of a massive magnetic primary could cause its spectrum to be masked more easily by an M companion.

In Figure 1 – which is Figure 3 from S07 – simulated magnetic and nonmagnetic WD+M composite spectra are shown by coadding single WD and M dwarf spectra. The listed ratio is that of the fluxes of the two components at the halfway point in wavelength of the spectra, 6500Å. Clearly the more massive and cooler the white dwarf, and the more luminous the M dwarf, the more easily the Zeeman spectral features of the former may be masked.

Liebert (1988) summarized the evidence that several, nearby magnetic white dwarfs with trigonometric parallaxes had relatively large masses, small radii and lay below the sequence of most white dwarfs in an HR Diagram. These objects included several known since the 1970s – Grw+70 8247, G 227-35, G 240-72 and GD 229. Since then, many more magnetic white dwarfs have been shown to be massive (Wickramasinghe and Ferrario 2005). However, there is also evidence that some of them have more ordinary masses near $0.6M_\odot$ or less. The presence of a very strong field generally prevents any direct measurement of the mass through *log g*. Therefore, the mass estimates for magnetic white dwarfs have been possible for only a modest fraction of the known objects (Wickramasinghe and Ferrario 2000).

It is of interest to note that decided differences are apparent in the fraction of white dwarfs that are magnetic between apparent magnitude-limited samples vs a volume-limited sample. In the Palomar Green Survey (Green, Schmidt, and Liebert 1986, hereafter PG) – an example of the former, only 2% are strongly magnetic. In the Holberg et al. (2008) "local" sample, of the 126 white dwarfs, 16 objects or 12.7(+/-4)% are strongly magnetic. The local sample is arguably close to volume-limited, since all lie within 20 pc according to best distance estimates. This distinction has been discussed in Liebert, Bergeron, and Holberg (2005), in the PhD dissertation of A. Kawka (2004) and in Kawka et al. (2007). Both groups argue that that the likely true frequency of strong magnetism in white dwarfs approaches or exceeds 10%.

2. Assessing the Statistics

At the same time the known number of WD+M pairs now exceeds 1,200, the number of magnetic white dwarfs has grown to over 160, with some 100 of these found in the SDSS (Vanlandingham et al. 2005), with a few additions since then. In the most recent compilation by Kawka et al. (2007), 149 magnetic white dwarfs are listed with B = 3 MG or larger.

How many of these magnetic white dwarfs would be expected to have a companion, if the assumption were made that WD+M pairings occur with the same frequency as for nonmagnetic white dwarfs? The "local"sample of white dwarfs within 20 pc (Holberg et al. 2008) has been searched for companions thoroughly. In fact 19.6 (+/-4.5)% have main sequence companions. Assuming that the binary frequency were similar, we might therefore expect 14-24% of the 149 magnetic white dwarfs – a total of 21-35 WDs – to have such companions. There are *none*. By the square-root-of-N as one sigma, the absence of any pairings yields already a four sigma level of significance.

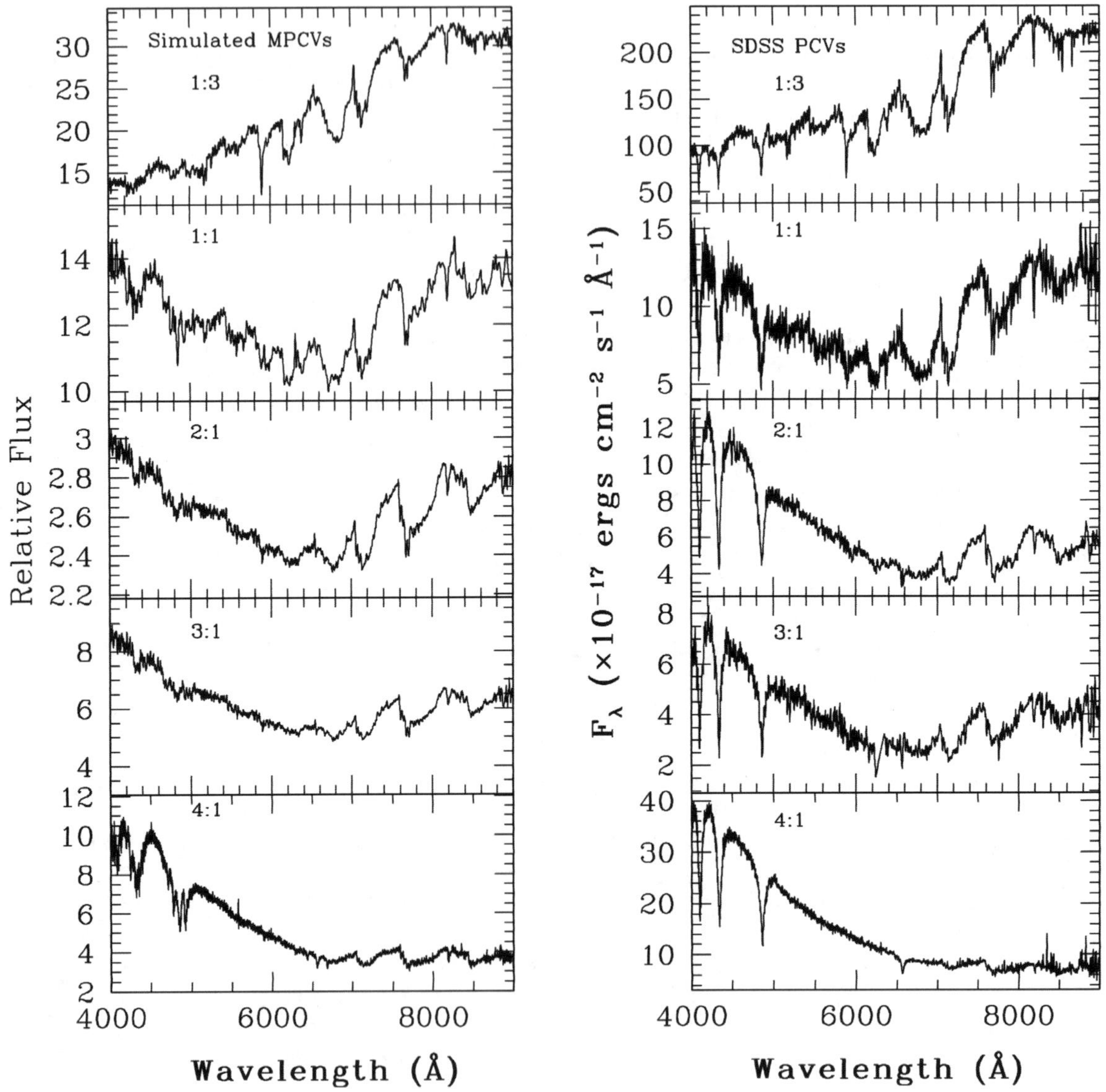

Figure 1. Combination of a magnetic (left panel) and nonmagnetic (right panel) white dwarf spectrum with an M dwarf where the ratio listed is that of the fluxes of the two components at the halfway point in wavelength (6500 Angstroms). The purpose is to simulate what a spectrum would look like if a strongly magnetic white dwarf were to be paired with an M dwarf. The Zeeman-split features are less deep than the nonmagnetic hydrogen lines, and are more easily masked by the M dwarf. If the magnetic white dwarf has a small radius (large mass), it is more easily masked. This may be an important selection effect, but one doubts that this can be the entire explanation.

Journal of Physics: Conference Series **172** (2009) 012040 doi:10.1088/1742-6596/172/1/012040

One can also use larger, more distant samples to improve on these statistics. Many of the several hundred hot white dwarfs found in the PG Survey show the existence of a companion in the optical spectrum. Further improvement was provided by Holberg and Magargal (2005). Using the Two Micron All Sky Survey detections at 1.2-2.0 micron JHK bands to look for excess over the white dwarf Rayleigh-Jeans tail, the fraction having apparent companions is increased. These authors found that 23% of the PG sample had "definite" and 29% had "definite or probable" cool companions. Assuming now that the 149 strongly magnetic WDs had binary frequency similar to the PG, 34 to 43 should have had companions. We now encroach on a six sigma level of significance.

3. Low Accretion Rate Polars (LARPs)

Technically, a recently-discovered class of close, accreting binaries with strongly-magnetic primary stars may be detached systems (Schmidt et al. 2007). The M dwarf is arguably not filling its Roche lobe, though it is losing mass through a wind, and it is believed that the field of the white dwarf captures most or all of this wind. The accretion rate is therefore fixed by the amount of the secondary's wind. Typically the loss/transfer rates are only of the order of 10^{-13} or 10^{-14} $M_{\odot}$ yr^{-1}. The magnetic fields of the primaries are generally rather large (40 to perhaps 80 MG), and the secondaries are mid-to-late M dwarfs. Periods range from 1.37 hours, where the secondary is later than M6, to over four hours.

These may be regarded as close binaries which have eluded full Roche-lobe overflow, but they are in some sense near misses. Schmidt et al. argue that their optical and X-ray luminosities are so low that they weren't discovered in X-ray surveys and are scarcely more luminous than fully-detached WD+M binaries. They are known to be accreting because the appearance of cyclotron harmonics. Given the low luminosities, the true space density of the LARPs could be comparable to that of the polars.

4. The Origin of Strong Magnetic Fields in White Dwarfs

For decades the presumption had been made that the magnetic fields in white dwarfs are fossil in nature, and that probably the white dwarfs are descended from magnetic Ap/Bp main sequence stars (Angel, Borra and Landstreet 1981). This interpretation has recently been called into question, because the fraction of of strongly magnetic white dwarfs seems to be too high to be covered by this limited range of progenitors (Wickramasinghe and Ferrario 2005, Kawka et al. 2007).

In Tout et al. (2008), a novel, new hypothesis has been put forth – due mainly to the first two authors. Strong magnetic fields in white dwarfs are formed in the common envelope phase of a close binary where at least the primary core is degenerate. Two stellar cores circle at close range in a differentially-rotating, convective envelope, which generates a strong dynamo. The strong magnetic field forms around the primary core. Frictional braking brings the cores together before the envelope is lost. If both cores are degenerate, they merge with the formation of a single magnetic white dwarf which may therefore generally be massive. If the secondary core is nondegenerate, they come into contact as a magnetic CV – either a polar, intermediate polar, or LARP.

I have to raise some issues about this hypothesis which require further attention:

–The generation of the magnetic field in the common envelope always must bring the two cores into a merger or an accretion phase. Yet most post-common envelope binaries are *not* precataclysmic systems. The orbital period is shortened but not by enough.

–Polars are always observed to be old systems with rather cool white dwarf primaries, with cooling times approaching or exceeding 1 Gyr (Sion et al. 1999). The LARPs appear to have cool white dwarf primaries as well. Yet if they come out of the common envelope already in contact, at least a few primaries of AM Herculis / LARP systems should be hot.

–Is it reasonable to assume that at least 10% of pre-white dwarfs are close, common envelope binaries?

Nonetheless, this is the first new idea for the origin of strong magnetism in white dwarfs since their original discovery. It deserves further study. There is evidence that the magnetic field distribution in white dwarfs is bimodal, with many having kilogauss fields (Aznar Cudrado et al. 2004, Wickramasinghe and Ferrario 2005). It may well be that the fossil field idea is applicable to white dwarfs with weaker fields, as the primary of V 471 Tauri could be a descendent of an Ap/Bp progenitor in the Hyades cluster.

4.1. Acknowledgments

Parts of this work have been done in collaboration with Nicole Silvestri, Dayal Wickramasinghe, Gary Schmidt, Suzanne Hawley, Paula Szkody, Lilia Ferrario, Ronald Webbink, Mara Lemagie, Terry Oswalt, J. Allyn Smith, and now Chris Tout. This work was supported by the National Science Foundation through the renewal of grant AST-0307321.

References

Angel, J.R.P., Borra, E., and Landstreet, J.D. 1981, *Astrophys. J. Suppl.*, **45**, 457

Araujo-Belancor, S., Gaensicke, B., Long, K.S., Beuermann, K., De Martino, D., Sion, E.M., and Szkody, P. 2005, *Astrophys. J.*, **622**, 589

Aznar Cuadrado, R.A., Jordan, S., Napiwotzki, R., Schmid, H.M. Solanki, S.K., & Mattys, G.2004, *Astron. Astrophys.*, **423**, 1081

Dillon, M., Gaensicke, B., Aungwerojwit, A., Rodrigues-Gil, P., Marsh, T.R., Barros, S., Szkody, P., Brady, S., Krajci, T., and Oksanen, A. 2008, *Mon. Not. R.A.S.*, **386**, 1568

Green, R.F., Schmidt, M., and Liebert, J. 1986, *Astrophys. J. Suppl.*, **61**, 305 (PG)

Holberg, J.B. and Magargal, K. 2005, *ASP Conf. Series*, **v334**, eds. D. Koester and S. Moehler, San Francisco: ASP, p. 419

Holberg, J.B., Sion, E.M., Sion, E.M., Oswalt, T.D., McCook, G.P., Foran, S., and Subasavage, J.P. 2008, *Astron. J.*, **135**, 1225

Kawka, A., 2004, Ph.D. thesis, Murdoch University

Kawka, A., Vennes, S., Schmidt, G.D., Wickramasinghe, D.T., and Koch, R. 2007, *Astrophys. J.*, **654**, 499

Liebert, J. 1988, *Publ. Astron. Soc. Pacific*, **100**, 1032

Liebert, J., Bergeron, P., and Holberg, J.B. 2005, **Astrophys. J. Suppl.**, **129**, 2376

Liebert, J., 2005, Wickramasinghe, D.T., Schmidt, G.D., Silvestri, N.M., Hawley, S.L., Szkody, P., Ferrario, L., Webbink, R.F., Oswalt, T.D., Smith, J.A., Lemagie, M.P. *Astron. J.*, **129**, 2376 (L05)

Ritter, H. and Kolb, U. 2006, *VizieR On-Line Data Catalog*

Ritter, H. and Kolb, U. 2003, *Astron. Astrophys.*, **404**, 301

Schmidt, G.D., Szkody, P., Henden, A., Anderson, S.F., Lamb, D.Q., Margon, B., and Schneider, D.P. 2007, *Astrophys. J.*, **654**, 521

Silvestri, N., Lemagie, M.P., Hawley, S.L., West, A.A., Schmidt, G.D., Liebert, J., Szkody, P., Mannikko, L., Wolfe, M.A., Barentine, J.C., Brewington, H.J., Harvanek, M., Krzesinski, J., Long, D., Schneider, D.P., and Snedden, S.A. 2007, *Astron J.*, **134**, 741 (S07)

Sion, E.M. 1999, *Publ. Astron. Soc. Pacific*, **111**, 532

Thomas, and Beuermann, K. 1998, *Lecture Notes in Physics*, **506**, 247

Tout, C.A., Wickramasinghe, D.T., Liebert, J., Ferrario, L., and Pringle, J.E. 2008, *Mon. Not. R.A.S.*, **387**, 897

Vanlandingham, K.M., Schmidt, G.D., Eisenstein, D.J., Harris, H.C., Anderson, S.F., Hall, P.B., Liebert, J., Schneider, D.P., Silvestri, N., Stinson, G.S., and Wolfe, M.A. 2005, *Astron. J.*, **130**, 734

Wickramasinghe, D.T., and Ferrario, L. 2000, *Publ. Astron. Soc. Pacific*, **112**, 873

Wickramasinghe, D.T., and Ferrario, L. 2005, *Mon. Not. R.A.S.*, **356**, 1576

Journal of Physics: Conference Series **172** (2009) 012041 doi:10.1088/1742-6596/172/1/012041

The not-so-extreme white dwarf of the CV GD 552

E Unda-Sanzana[1], T R Marsh[2], B T Gänsicke[2], P F L Maxted[3], L Morales-Rueda[4], V S Dhillon[5], T D Thoroughgood[5], E Tremou[6], C A Watson[5] and R Hinojosa-Goñi[1]

[1] Instituto de Astronomía, Universidad Católica del Norte, Antofagasta, Chile
[2] Dept. of Physics, University of Warwick, Coventry, United Kingdom CV4 7AL
[3] Dept. of Physics and Astronomy, Keele University, Keele, Staffordshire ST5 5BG, United Kingdom
[4] Dept. of Astrophysics, University of Nijmegen, 6500 GL Nijmegen, The Netherlands
[5] Dept. of Physics and Astronomy, University of Sheffield, Sheffield S3 7RH, United Kingdom
[6] Dept. of Physics, Sect. of Astrophysics, Astronomy & Mechanics, Univ. of Thessaloniki, 541 24 Thessaloniki, Greece

E-mail: eundas@almagesto.org; rhg002@ucn.cl, t.r.marsh@warwick.ac.uk, boris.gaensicke@warwick.ac.uk, pflm@astro.keele.ac.uk, lmr@astro.ru.nl, vik.dhillon@sheffield.ac.uk, C.Watson@sheffield.ac.uk, etremo@physics.auth.gr

Abstract. GD 552 is a cataclysmic binary which was previously believed to be composed of an M-star and a white dwarf, the latter having an extreme mass of 1.4 solar masses. In a recent paper we showed that this is not compatible with new observational evidence and presented an alternative model in which the white dwarf has a typical mass and the companion is a brown dwarf, making the system a likely member of the elusive group of CVs which have already evolved through minimum orbital period. Here we present additional spectroscopical evidence supporting this conclusion by means of skew mapping.

1. Introduction

GD 552 is a blue, high proper motion star (0.18″/yr) discovered by Giclas et al. (1970). It was first observed spectroscopically by Greenstein & Giclas (1978), who found that it is a cataclysmic variable star (CV). Up to very recently the main spectroscopic study of GD 552 had been carried out by Hessman & Hopp (1990). They measured the orbital period of GD 552 to be 102.7 min. They also measured the white dwarf's projected orbital velocity to be $K_1 = 17 \pm 4$ km s^{-1}. This suggested a low inclination system, since for edge-on systems with orbital periods similar to GD 552 K_1 is typically $\sim 60 - 80$ km s^{-1}. However, the emission lines from the accretion disc suggested instead a moderately inclined system as they display clearly separated double-peaks (which come from the outer disc, Horne & Marsh 1986) with velocities of ± 450 km s^{-1} which can be compared to typical peak velocities of ~ 600 km s^{-1} for edge-on systems. Hessman & Hopp (1990) explained this by saying that the white dwarf in GD 552 is close to the Chandrasekhar limit. This alllowed a low orbital inclination ($\sim 20°$), a small K_1 value, and large disc velocities. This explanation suited well their assumption that the companion to the white dwarf had to be a main-sequence star. However, Unda-Sanzana et al. (2008) provided evidence supporting an alternative model in which the system is composed of a brown dwarf and an ordinary white dwarf. They used spectroscopic templates to estimate the maximum possible flux contribution of an hypothetical main-sequence companion to the spectra of the binary. They estimated this to be 10% and thus were able to estimate the magnitude of the companion under that assumption. They showed that this was incompatible with a main-sequence star. In this work we carry out skew mapping of the spectroscopic data acquired for GD 552, offering an alternative means of estimating the contribution of a main-sequence mass donor to the flux of the system, which in turn allows to constrain the mass of the white dwarf.

2. Observations

The observations and its reduction are fully described in Unda-Sanzana et al. (2008). For the analysis in section 3 we use the dataset acquired in August 2001 using the 4.2-m William Herschel Telescope (Roque de los Muchachos) and the ISIS spectrograph.

3. Skew mapping

The aim of this section is to use an alternative means to put a limit on the contribution of the mass donor to the flux of the system. To try to detect the mass donor in our spectra we used skew mapping (e.g. Smith, Dhillon and Marsh 1998; Putte et al. 2003) with Gl 65A, an M5.5 star. A skew map is produced by applying Doppler map techniques (Marsh & Horne, 1988) to the cross-correlation functions of the CV data with a suitable template. If there is a consistent cross-correlation between the template and a pattern following the orbital motion of the binary system, then the skew map will show a spot at the velocity coordinates of the source of the pattern. For skew mapping we used the range 7850-8400 Å, which is devoid of emission lines. In order to preserve the relative line strengths along the spectrum we did normalise it carefully. First of all we fitted a constant to the continuum of both GD552 and Gl 65A data (for scaling). Then we fitted a 5-knot spline to both continua (the number of knots set due to the complex shape of Gl 65A's continuum). Dividing by these fits and subtracting unity, we obtained flattened continua. Then Gl 65A had to be prepared as a suitable template for a hypothetical mass donor present in the spectra of GD552, accounting for both instrumental and physical effects. First we smeared the average spectrum of Gl 65A to account for its orbital motion during the exposure using

$$V_{\rm smear} = \frac{2\pi K T_{\rm exp}}{P}, \tag{1}$$

where K is the velocity of the moving object (160 $\rm km\,s^{-1}$ assuming Hessman & Hopp (1990)'s parameters for GD 552), $T_{\rm exp}$ is the exposure time (300 s in our case) and P is the period (0.07134 d for GD 552). From this we obtained $\sim$ 49 $\rm km\,s^{-1}$ for the smearing to be applied. Also, we rotationally broadened this spectrum by 160 $\rm km\,s^{-1}$, which is the velocity of the edge-on system (460 $\rm km\,s^{-1}$) projected to the line of sight with Hessman & Hopp (1990)'s inclination of 20°. Finally, we shifted the Gl 65A spectrum by 29 $\rm km\,s^{-1}$ to account for its systemic radial velocity (Wilson 1953).

In Figure 1 the grey scale represents the degree of cross-correlation between data and template, the blacker the shade of gray the better the cross-correlation. We normalized the scale of each panel so that white/black corresponds to minimum/maximum value in each skew map. In the figure, the top left panel is the skew map produced by cross correlation of GD552 with Gl 65A prepared as described above. There are no discernible peaks in this panel. We added percentages of Gl 65A to the original GD552 data, after applying a sinusoidal velocity factor to Gl 65A ($\gamma = -55$ $\rm km\,s^{-1}$, $K_{\rm x} = 0$ $\rm km\,s^{-1}$, $K_{\rm y} = 400$ $\rm km\,s^{-1}$), using the orbital phases of the GD552 data. The top right panel of Figure 1 is the result of skew mapping the GD552 data + 10% Gl 65A; bottom-left panel with 20% Gl 65A and bottom-right panel with 30% Gl 65A.

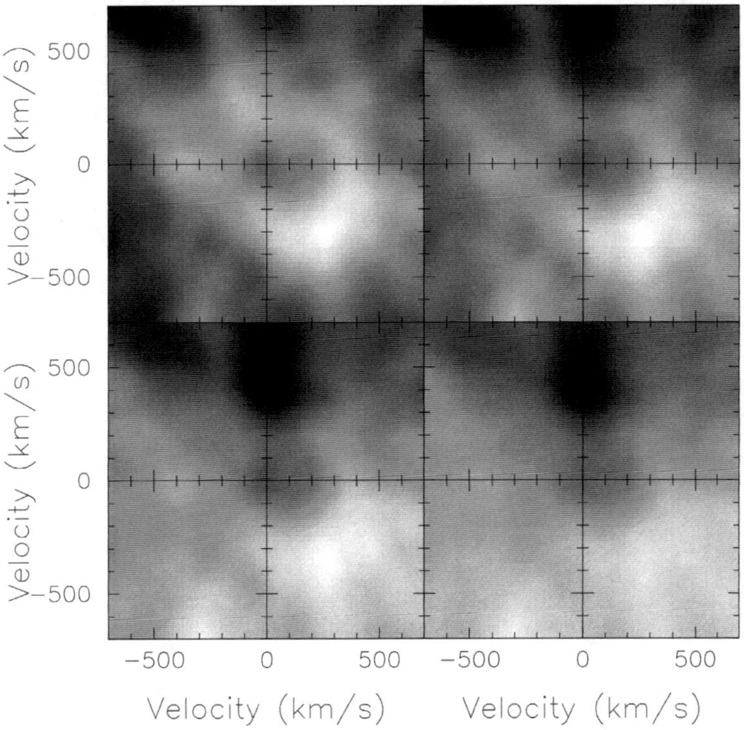

Figure 1. Skew mapping of GD 552 and a template M star. The top left panel uses only GD 552 data for cross correlation. The other panels use a mixture of GD 552 and increasing percentages of the template star. See text for details.

We found that we could recover the position of the artificial mass donor by adding between 10% to 20% of Gl 65A to the GD552 data prior to cross-correlation and skew mapping. This is compatible with the conclusion reached by Unda-Sanzana et al. (2008) obtained by subtracting increasing amounts of Gl 65A to GD552 data until the features of the M star became distinguishable.

4. Conclusions

We estimated the maximum contribution of an hypothetical main sequence donor to the flux of GD 552. This was done by an alternative means to that used by Unda-Sanzana et al. (2008) but the result was nonetheless compatible with theirs. By considering either the distance $d \sim 70\,\mathrm{pc}$ from Greenstein & Giclas (1978) or the value $d \sim 125\,\mathrm{pc}$ from Unda-Sanzana et al. (2008) it is possible to set a limit on the absolute magnitude of the mass donor and on the mass of the white dwarf. Unda-Sanzana et al. (2008) show that for a 10% contribution the mass of the mass donor is $M_2 < 0.08\,M_\odot$ (a brown dwarf) and the mass of the white dwarf is $M_1 = 0.6 M_\odot$, a rather average specimen.

Acknowledgements

EU was supported by Proyecto FONDECYT de Iniciación en Investigación 11060015 through CONICYT-Chile. EU and RH-G were also supported by Universidad Católica del Norte's Proyecto de Investigación en Docencia 220401-10602015. TRM was supported under a PPARC SRF during some of the period over which this work was undertaken. LM-R is supported by

NWO VIDI grant 639.042.201 to P. J. Groot. BTG was supported by a PPARC Advanced Fellowship.

The authors acknowledge facilities and data provided by the Starlink Project, NASA's Astrophysics Data System Bibliographic Services, AAVSO International Database and the Isaac Newton Group of telescopes located at Observatorio del Roque de los Muchachos of the Instituto de Astrofisica de Canarias. This research was partly based on observations made with the NASA/ESA Hubble Space Telescope, obtained from its data archive. STScI is operated by the Association of Universities for Research in Astronomy, Inc. under NASA contract NAS 5-26555; these observations are associated with program 9406.

References

Giclas H Burnham R and Thomas N 1970 *Lowell Obs. Bull.* 153
Greenstein J and Giclas H 1978 *PASP* **90** 460
Hessman F Hopp U 1990 *A&A* **228** 387
Horne K and Marsh T 1986 *MNRAS* **218** 761
Marsh T R and Horne K 1988 *MNRAS* **235** 269
Putte D Smith R Hawkins N and Martin J 2003 *MNRAS* **342** 151
Smith M Dhillon V and Marsh T 1998 *MNRAS* **296** 465
Unda-Sanzana E Marsh T R Gänsicke B T Maxted P F L Morales-Rueda L Dhillon V S Thoroughgood T D Tremou E Watson C A and Hinojosa-Goni R 2008 *MNRAS* **388** 889
Wilson R 1953 *Carnegie Inst. Washington D.C. Publ.* 601

Journal of Physics: Conference Series **172** (2009) 012042 doi:10.1088/1742-6596/172/1/012042

The fainter the better: cataclysmic variable stars from the Sloan Digital Sky Survey

J Southworth, B T Gänsicke and T R Marsh

University of Warwick, Coventry CV4 7AL, UK

E-mail: `J.K.Taylor@warwick.ac.uk`

Abstract. The Sloan Digital Sky Survey has identified a total of 212 cataclysmic variables, most of which are fainter than 18th magnitude. This is the deepest and most populous homogeneous sample of cataclysmic variables to date, and we are undertaking a project to characterise this population. We have found that the SDSS sample is dominated by a great "silent majority" of old and faint CVs. We detect, for the first time, a population spike at the minimum period of 80 min which has been predicted by theoretical studies for over a decade.

1. Introduction

Cataclysmic variables (CVs) are interacting binary stars containing a white dwarf accreting material from a low-mass main sequence star via an accretion disc or stream. Theoretical studies have consistently predicted that the population of CVs should be dominated by short-period systems, with many piling up up at a minimum period of 60–70 minutes (Kolb 1993; Politano 1996). Unfortunately, the known population of CVs, visciously biased by observational selection effects, bears no resemblance to theoretical predictions (Ritter & Kolb 2003). We are therefore undertaking a project to characterise the population of CVs discovered by the Sloan Digital Sky Survey (SDSS) (Szkody et al., 2002-6). This sample was identified spectroscopically and extends to faint magnitudes, so is much less biased than all previous samples.

We have identified the first CVs known to have a secondary star of brown dwarf mass (Southworth et al., 2006; Littlefair et al., 2006a, 2008), and for the first time are finding evidence for the long-predicted pile-up at the minimum period (Gänsicke et al., 2008). The orbital period distribution of the SDSS CVs is compared to the distribution for the previously known CVs in Fig. 1. Results for individual systems, based on observations with the VLT, NTT, WHT, INT and NOT, can be found in Gänsicke et al. (2006, 2008), Southworth et al. (2006, 2007ab, 2008ab), Dillon et al. (2008) and Littlefair et al. (2006ab, 2007, 2008).

Our work on the SDSS population of CVs is intended to shed light on how these weird objects evolve. For contrast, we are also running a program to measure the orbital periods of a sample of pre-CVs which has been spectroscopically identified by the SDSS (Rebassa-Mansergas et al. 2007, 2008; Schreiber et al. 2008; see also Gänsicke & Schreiber 2003). Below we pick out two recent highlights of our characterisation of the SDSS CVs.

2. SDSS J220553.98+115553.7: the pulsator which stopped

SDSS J2205 has twice been observed to vary in brightness (Woudt & Warner 2004; Szkody et al., 2007), with periods of 575 s, 475 s and 330 s and amplitudes of ∼10 mmag, which is typical

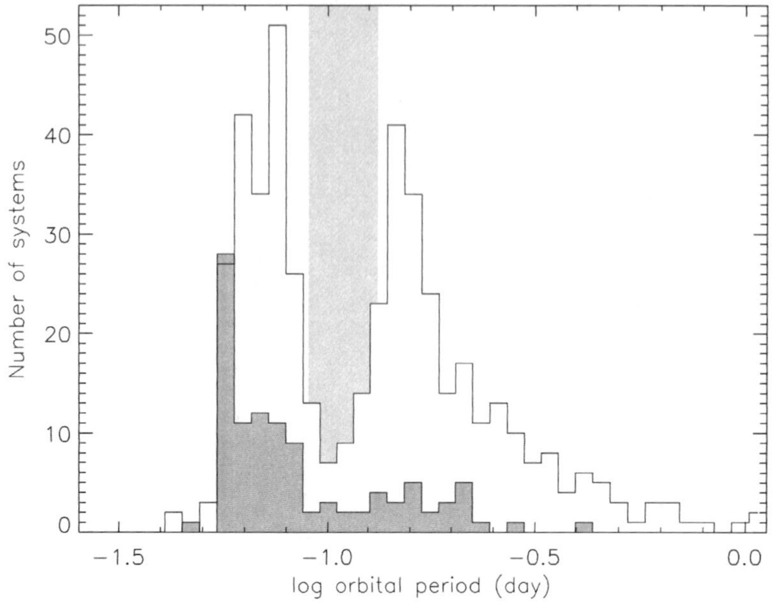

Figure 1. Orbital period distribution of the SDSS CVs (grey histogram) compared to all other known CVs (white histogram). The latter population has been taken from the catalogue of Ritter & Kolb (2003). The 2–3 hour period gap in the known CV population is shown with light shading.

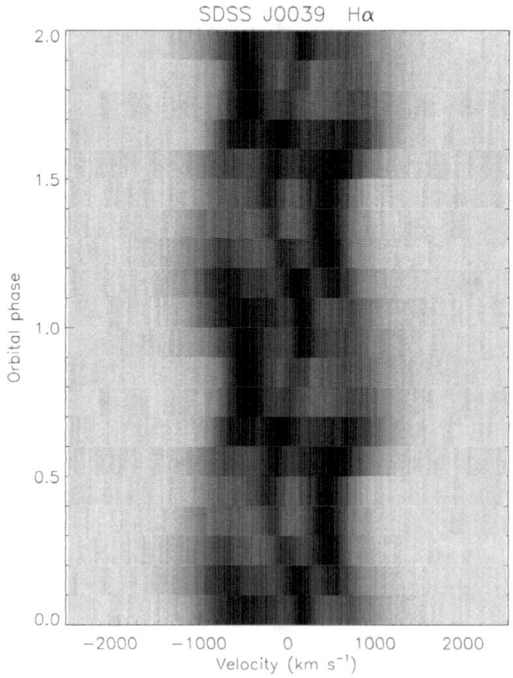

Figure 2. Trailed spectra of the Hα emission line from SDSS J0039. Two orbital phases are plotted for clarity.

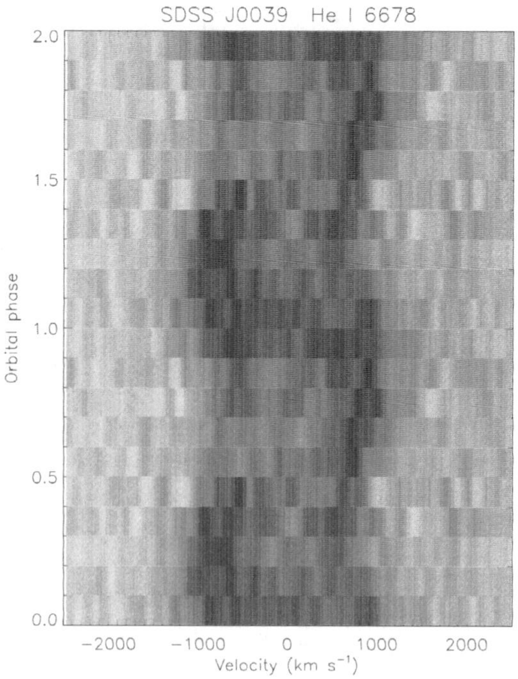

Figure 3. Trailed spectra of the He I 6678 Å emission line from SDSS J0039. The spectra have been smoothed for display.

of ZZ Ceti-type nonradial pulsations. Over two nights in 2008 August we obtained a light curve with the NTT. We were unable to detect these pulsations to a limit of 5 mmag, but instead found a previously unseen photometric period of 44.8 min. Our VLT spectroscopy yielded an orbital period of 82.83 ± 0.09 min (Southworth et al., 2008a). The vanishing pulsation periods cannot

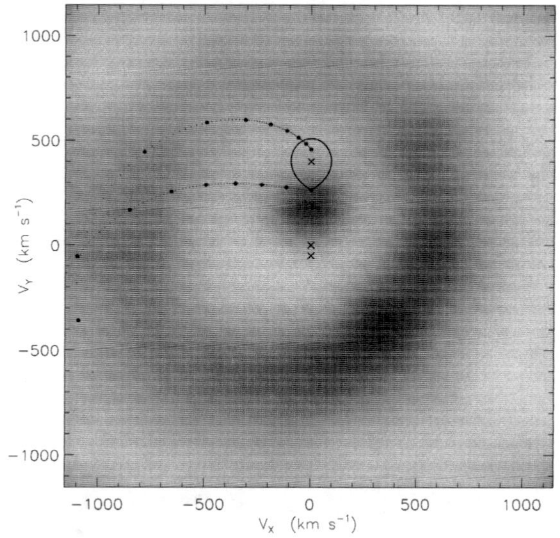

Figure 4. Doppler map of the Hα emission line from SDSS J0039.

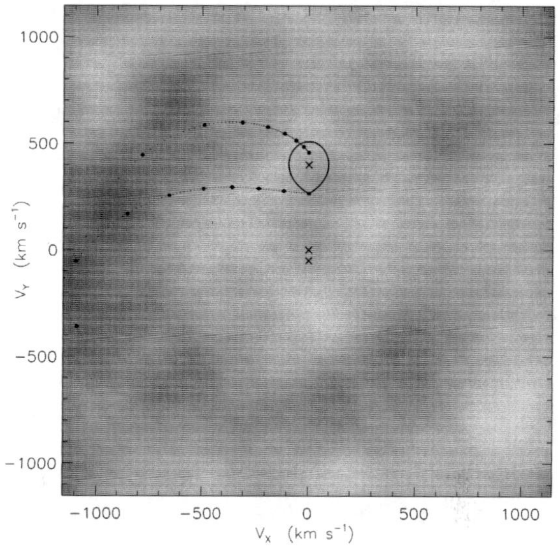

Figure 5. Doppler map of the He I 6678 Å emission line from SDSS J0039.

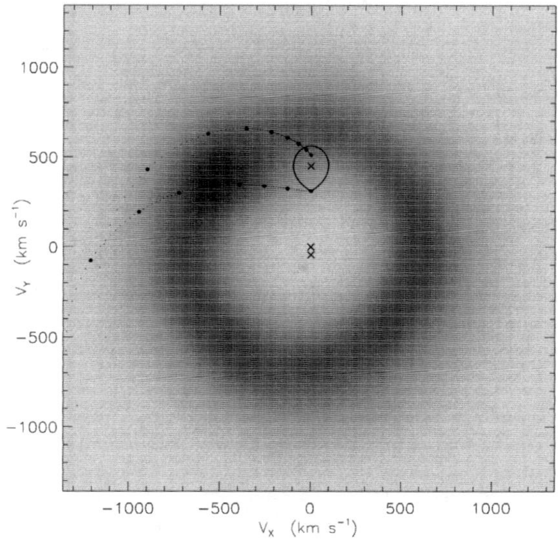

Figure 6. Doppler map of the Hα emission line from SDSS J2205.

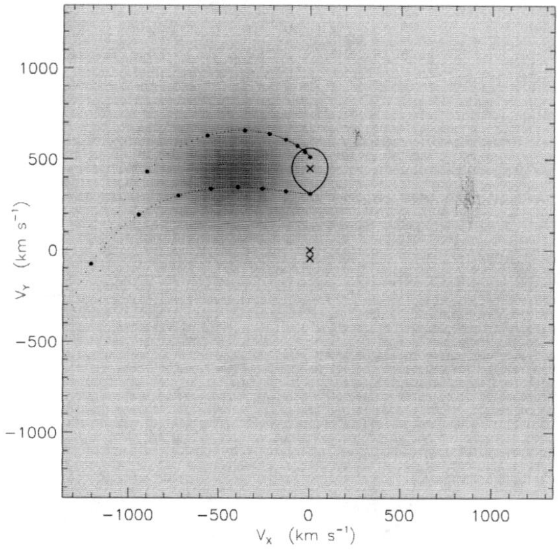

Figure 7. Doppler map of the He I 6678 Å emission line from SDSS J2205.

be attributed to destructive interference or changes in accretion rate, but may be explicable by changes in the white dwarf temperature or the surface visibility of the pulsation modes.

3. Triple-peaked Hα emission from SDSS J003941.06+005427.5

We obtained 29 spectra of SDSS J0039 over two nights in 2007 August using VLT/FORS2, finding an orbital period of 91.395 $\pm$ 0.093 min. The spectra of this system show a remarkable and unique triple emission peak at Hα (Southworth et al., 2008c). We have used the technique of Doppler tomography (Marsh & Horne 1988) to generate velocity maps of the emission. These show that the inner peak moves in velocity with an amplitude of 180 km s^{-1}, so cannot easily be attributed to either the white dwarf, secondary star, or accretion disc. Its existence remains a mystery. The Hα Doppler maps also show that the accretion disc is very elliptical, which is not expected for a system in a state of very low mass transfer. Finally, there is also strong emission from the accretion disc in the [$+V_X$,$-V_Y$] quadrant, which has been seen in some ultra-compact binaries but is very odd for standard hydrogen-rich CVs. These aspects of SDSS J0039 defy explanation in the current picture of the properties of CVs, and demand further follow-up observations.

Doppler maps of the Hα and He I 6678 Å emission lines from SDSS J0039 are shown in Figs. 4 and 5. For comparison, we show Doppler maps from SDSS J2205 in Figs. 6 and 7, which have a behaviour which is much more like that expected from short-period CVs: a circular accretion disc in Hα emission and a bright spot visible in He I 6678 Å light where the mass transfer stream from the secondary star collides with the accretion disc. Overlaid on each Doppler map is a solid line indicating the Roche lobe of the secondary star in velocity space, crosses indicating the velocities of the centres of mass of the two stars and of the system, and dotted lines showing the velocity of the mass transfer stream as it follows a ballistic trajectory from the inner Lagrangian point.

References

Dillon M Gänsicke B T Aungwerojwit A Rodríguez-Gil P Marsh T R Barros S C C Szkody P Brady S Krajci T & Oksanen A 2008 *MNRAS* **386** 1568
Gänsicke B T et al. 2006 *MNRAS* **365** 969
Gänsicke B T et al. 2008 *MNRAS* in preparation
Kolb U 1993 *A&A* 271 149
Littlefair S Dhillon V S Marsh T R Gänsicke B T Southworth J & Watson C A 2006a *Science* **314** 1578
Littlefair S Dhillon V S Marsh T R & Gänsicke B T 2006b *MNRAS* **371** 1435
Littlefair S Dhillon V S Marsh T R Gänsicke B T Baraffe I & Watson C A 2007 *MNRAS* **381** 827
Littlefair S Dhillon V S Marsh T R Gänsicke B T Southworth J Baraffe I Watson C A & Copperwheat C 2008 *MNRAS* **388** 1582
Marsh T R & Horne K D 1988 *MNRAS* **235** 269
Politano M 1996 *ApJ* **465** 338
Rebassa-Mansergas A Gänsicke B T Rodríguez-Gil P Schreiber M R & Koester D 2007 *MNRAS* **382** 1377
Rebassa-Mansergas A et al. 2008 `arXiv:0808.2148`
Ritter H & Kolb U 2003 *A&A* **404** 301
Schreiber M R & Gänsicke B T 2003 *A&A* **406** 305
Schreiber M R Gänsicke B T Southworth J Schwope A D & Koester D 2008 *A&A* **484** 441
Southworth J Gänsicke B T Marsh T R de Martino D Hakala P Littlefair S & Rodríguez-Gil P 2006 *MNRAS* **373** 687
Southworth J Gänsicke B T Marsh T R de Martino D & Aungwerojwit A 2007a *MNRAS* **378** 635
Southworth J Marsh T R Gänsicke B T Aungwerojwit A Hakala P de Martino D & Lehto H 2007b *MNRAS* **382** 1145
Southworth J Townsley D M & Gänsicke B T 2008a *MNRAS* **388** 709
Southworth J Gänsicke B T Marsh T R Torres M A P Steeghs D Hakala P Copperwheat C Aungwerojwit A & Mukadam A 2008b `arXiv:0809.1753`
Southworth J Marsh T R & Gänsicke B T 2008c in preparation
Szkody P et al. 2002 *AJ* **123** 430
Szkody P et al. 2003 *AJ* **126** 1499
Szkody P et al. 2004 *AJ* **128** 1882
Szkody P et al. 2005 *AJ* **129** 2386
Szkody P et al. 2006 *AJ* **131** 973
Szkody P et al. 2007 *ApJ* **658** 1188
Woudt P Warner B 2002 *Ap&SS* **282** 433

Short-term variability of dwarf nova SS Cyg during outbursts

I Voloshina[1], H Rovithis-Livaniou[2] and V Metlov[1]

[1] Sternberg Astronomical Institute, Moscow State University,
Universitetskij prospect 13, Moscow 119991, Russia
[2] Section of Astrophysics, Astronomy and Mechanics, Department of Physics,
Athens University, Zagrafos 15784, Athens, Greece

E-mail: vib@sai.msu.ru

Abstract. Here we report the results of CCD observations of classical dwarf nova SS Cyg carried out with the two 60-cm telescopes in Crimea during the last years. These observations cover a few outbursts in 2006, 2007 and 2008. Power spectrum analysis of our CCD data clearly shows the existence of rapid periodic oscillations in the light curve of SS Cyg at the stage of decline after maximum. CCD observations of SS Cyg in autumn 2006 outburst revealed oscillations with the two periods 10 s and 76 s, in November 2007 – with 41 s period and in January 2008 with 98 s. We interpret detected variations as quasi-periodic oscillations.

1. Introduction

Dwarf nova SS Cyg belongs to U Gem sub-type (classical dwarf novae with quick brightening of large amplitude). SS Cyg undergoes outbursts once every 50 days. During an outburst the magnitude of SS Cyg increases up to 8^m. After outburst it returns to the previous brightness12^m and remains in quiescent state a few weeks. Three types of outbursts are observed in this system depending of outburst amplitude and duration: ***normal, long and anomalous ones.*** SS Cyg system consists from white dwarf and a cool late type optical star that fills its Roche lobe and transfers mass to the white dwarf, forming an accretion disk around it. The orbital period is $0^d.275130$ and masses of the components $M_{rd} \sim 0.46\ M_\odot$ and $M_{wd} \sim 0.66\ M_\odot$, orbital inclination of the system is about 50° Voloshina and Khruzina (2000) .

2. Observations

Observations of dwarf nova SS Cyg were performed during different outbursts in 2006, 2007 and 2008. Photoelectric observations of SS Cyg in August and October 2007 were made with ***UBV*** photometer on the 60-cm and 50 cm telescopes of Sternberg Astronomical Institute in Crimea. CCD observations of SS Cyg during outburst in October 2006 and November-December 2007 were taken with Apogee 47 device on the 60 cm telescope also with 1 s and 2 s time resolution (***V*** band). Observations of SS Cyg in January 2008 were made on the 60-cm telescope with UBV photometer.
In figure 1 we illustrate the overall light curves of SS Cyg for 2 outbursts when our observations were carried out. The visual data for them were kindly given to us by American Association of Variable

[1] To whom any correspondence should be addressed.

Star Observers (AAVSO). Figure 2 shows some of the daily light curves of SS Cyg obtained in maximum, decline and quiescence.

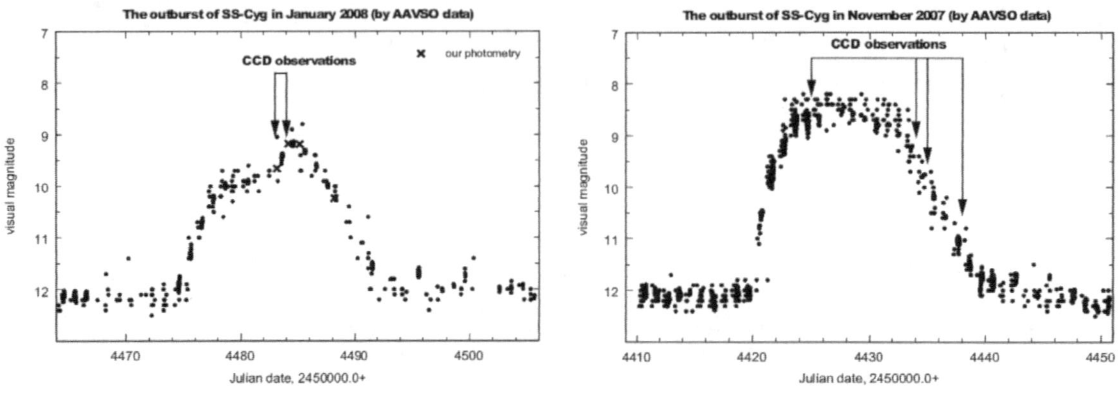

Figure 1. The latest outbursts of SS Cyg in 2007 and 2008.

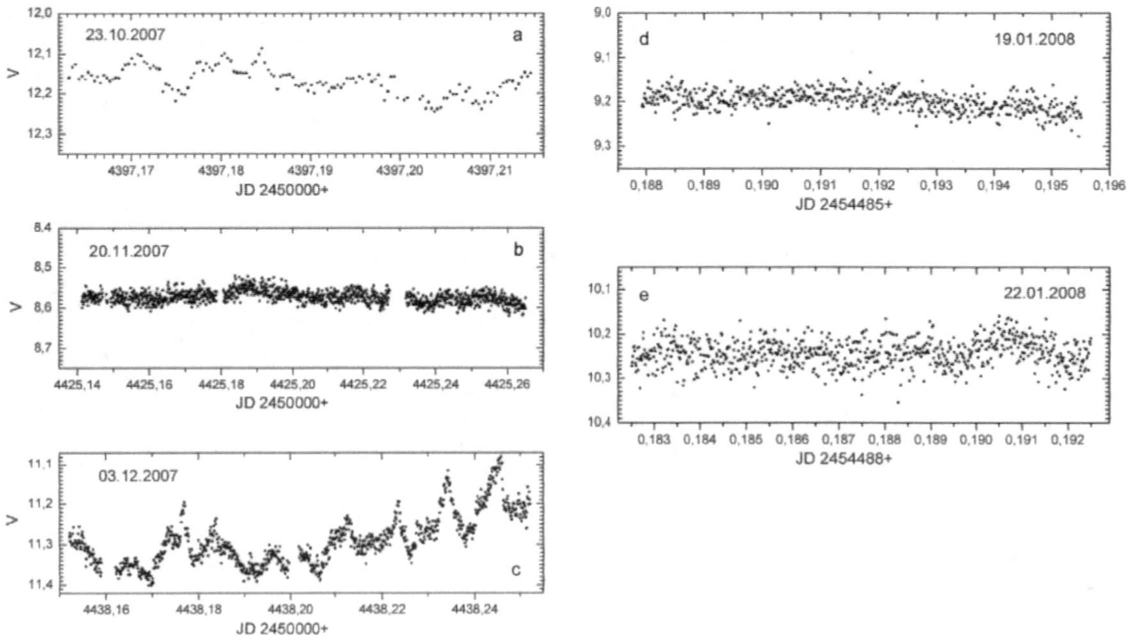

Figure 2. Some of SS Cyg daily light curves obtained during two recent outbursts: in autumn 2007 (quiescence, maximum and decline, left) and in anomalous outburst in January 2008 (maximum and decline, right).

3. Short-term variability

Different types of short-term variations occur in dwarf novae (Hack and La Dous 1993). They are:

- ***the flickering*** – random light variations on the time-scales of minutes with amplitudes of tenths of a magnitude;
- ***the coherent oscillations*** (DNO) with periods of a few tens of seconds and amplitudes on the order of $0^m.002$;
- ***quasi-periodic oscillations*** (QPO) with periods from some 20 s to several hundred s and amplitudes larger than the coherent oscillations. Quasi-periodic oscillations are not strictly periodic. Very different values of periods are possible in one object during different outbursts. They may or may not appear simultaneously with coherent oscillations.

Three groups of short-term oscillations were found during the outburst of SS Cyg: rapid variations first were detected in October 1976 with P=9.735±0.002 s and amplitude $0^m.0002$ (Patterson 1978) and later oscillations with periods **8.5-10.9 s** and **32-36 s** with amplitude up to $0^m.001$ were found in the light curve of this dwarf nova (Horne and Gomer 1980). In 1982 the third group of quasi-periodic oscillations with **P=86 s** and full amplitude of 1.5 % have been detected for SS Cyg by Voloshina and Lyuty (1988). Quasi-periodic oscillations of this group early were detected in other cataclysmic variables.

4. Data analysis

We use Fourier method for our data analysis. Periodograms obtained from this analysis which demonstrate our findings are shown in Figure 3.We find oscillations with periods 10 s and 76 s in 2006 outburst. Close examination of the power density spectra (PDS) related to the 2007 outburst light curves reveals a few peaks. The most prominent PDS peak corresponds to persistent coherent oscillation of period about 41 s for all nights of our observations. For the first time we detect this oscillation in November 18 light curve. The amplitudes of these oscillations increased from A=0.004 (on November 20, the maximum of the outburst) to A=0.007 (on December 3, decline stage). Oscillations of P=35 s, P=72 s take place in the light curves on November 18 and December 3 respectively. In the latter case the period is very close to that in 2006. One can also see a prominent peak of period 98 s in PDS of the 2008 outburst (lowest panel in figure 3). Note the other period (21 s) is not real. Observations of SS Cyg folded with periods of 98 s and 41 s are shown in figure 4. Each point on these curves represent the mean value of all data points in the 0.1 phase range.

5. Summary

The following rapid periodic oscillations were detected in SS Cyg light curves:

- ***outburst in October 2006*** - 10 s and 76 s
- ***outburst in November 2007*** - 41.5 s, 35 s and 72 s .
- ***outburst in January 2008 -*** ~98 s.
- according to their properties the oscillations could be interpreted as quasi-periodic oscillations.
- no oscillations were detected by UBV observations in 1996 and 2000 outbursts.
- flickering is also present in SS Cyg light curve in quiescence.

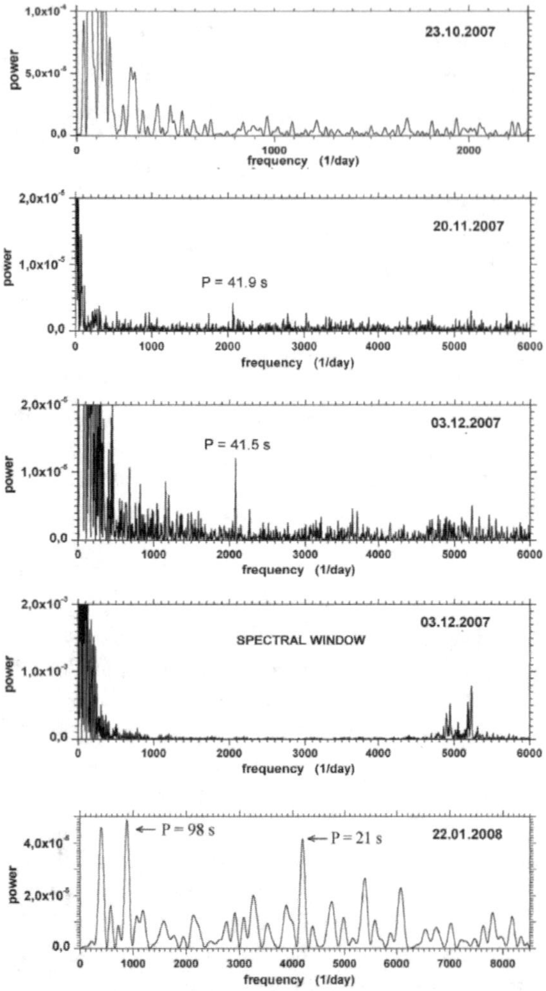

Figure 3. Power spectra of light curves of SS Cyg for different observational nights in 2007-2008 (left) obtained from analysis of CCD observations

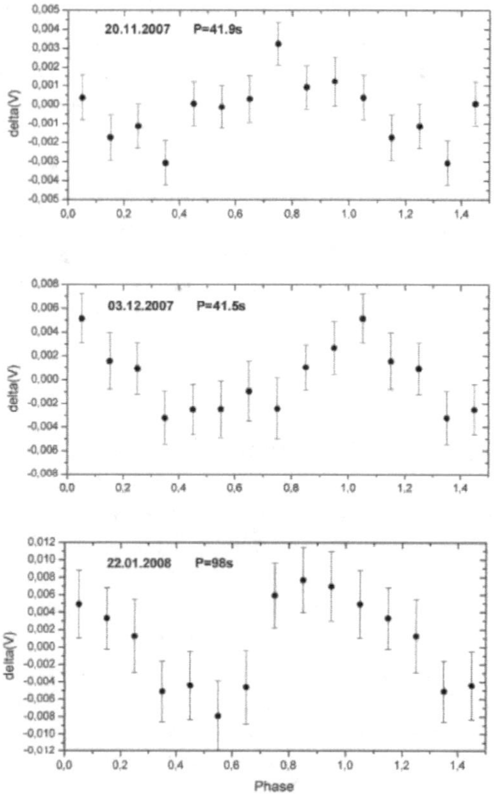

Figure 4. Observations of SS Cyg folded with values of obtained periods.

Acknowledgments

We express our gratitude to American Association of Variable Star Observers for the visual data of SS Cyg. This research was supported by RFBR under grant № 06-02-16411 and grant NSh - 5218.2006.2

References

http:// American Association of Variable Star Observers (AAVSO)
Hack M and La Dous C 1993 *Cataclysmic Variables and Related Objects* (Washington: NASA SP-507) p 59
Horne K and Gomer R 1980 *ApJ* **237** 845
Patterson J, Robinson E L and Kiplinger A L 1978 *ApJ Let.* **226** L137
Voloshina I B and Lyuty V M 1988 *Astrofisika* **29** 433
Voloshina I B and Khruzina T S 2000 *Astronomy Reports* **44** 89

V455 And — a life before the outburst

N Katysheva[1] **and S Shugarov**[1,2]

[1]Sternberg Astronomical Institute, Moscow, 119991, Russia
[2]Astronomical institute of the SAV, 05960, Tatranska Lomnica, Slovakia

E-mail: nk@sai.msu.ru

Abstract. V455 And (= HS 2331+3905) is a WZ Sge-type cataclysmic variable which shows a double-humped profile in its orbital light curve. It is also one of the few ZZ Cet pulsators among cataclysmic variables (CVs). In quiescence V455 And shows several types of variability in accordance with Araujo-Betancor et al. (2005). We present the results of observations of V455 And during the autumn seasons of the years 2004-2006, just before the superoutburst in 2007.

1. Introduction

The WZ Sge-type binary systems are a subclass of dwarf nova cataclysmic variables — consisting of a red dwarf and a white dwarf — that have the orbital periods close to the period minimum (80–90 min). The outbursts of WZ Sge-stars are very rare and more powerful than the usual dwarf nova outbursts. One of the WZ Sge-features is a double-humped orbital light curve in quiescence. At present about 20 of such systems are known. Usually they are discovered during their outbursts because their brightness in quiescence is very weak and their photometric characteristics are very similar to normal dwarf novae.

HS2331+3905 was discovered (Gänsicke et al., 2007) in the Hamburg Quasar Survey. The results of the observations during three years and the main characteristics of the star were described in the paper by Araujo-Betancor et al. (2005) "HS 2331+3905: The cataclysmic variable that has it all". In this paper they defined HS2331+3905 as a cataclysmic variable with a very wide spectra of variability: orbital (81.08 min) and superhump periods (83.38 min), non-radial pulsations of the white dwarf (5–6 min) and a coherent signal at 1.12 min. The spectroscopic period was about 3.5 hour. Besides, HS 2331+3905 is "the brightest CV white dwarf pulsator" (Gänsicke et al., 2007). Araujo-Betancor et al. (2005) proposed a possible superoutburst of HS2331+3905 as a WZ Sge-type binary. And such outburst (precisely, the superoutburst) happened in September, 2007. After the beginning of superoutburst it was named V455 And (Samus et al., 2007). Thus, V455 And was classified as WZ Sge-type system long before its outburst. Here we present the result of our observations obtained before the 2007 superoutburst.

2. Observations

At first we studied the photo-plates of the Moscow archive. The limit of these Moscow plates is about of 12–13 magnitude. There were 43 plates from 1907 till 1957. We did not see any outbursts on them. Due to the fact that the 2007 superoutburst was about 8.5 mag, any superoutburst had to be visible on these plates. The Julian Dates of plates are given in Table 1

(JD 2400000+). We carried out our observations of V455 And at the Crimean Laboratory of Sternberg Institute and at the Crimean Astrophysical Observatory in the autumn seasons of 2004 to 2007 with the 38-, 50-, 60-cm and 1.25-m telescopes. Below we present the results of our observations during the quiescence state of this system, before the 2007 superoutburst.

Table 1. Julian dates of archive plates (JD 2400000+).

JD	JD	JD	JD	JD
16708,44	17861.23	18925.30	29284.16	35718.45
16730,41	17939.17	18929.44	29285.45	35721.46
16880,21	18219.34	28043.44	29286.19	35723.49
17469,45	18241.21	28082.38	29287.15	35724.42
17587,21	18267.26	28753.42	29310.18	35724.44
17794,40	18564.41	28805.34	29318.21	36084.47
17826,39	18568.36	29147.49	35718.45	36185.17
17849,31	18598.23	29167.40	35718.45	
17860.23	18600,21	29172.42	35718.48	

3. The orbital light curves

The aggregate light curves of V455 And in V- and R-bands for 2004–2006 are given in Figs. 1 and 2. The average brightness of the system is practically constant during the period 2004–2006. This average brightness is about 16.0 mag in R-band and 16.4 mag in V-band. In Figs. 3 and 4 the mean phase-folded light curves in the R- and V-bands are presented. The double-hump structure of the light curves is clearly seen. Similar light curves are characteristic of WZ Sge-type stars and reflect the spiral structure of the accretion disc around the white dwarf. Gänsicke et al. (2007) proposed that the very complicated form of the warped accretion disc of V455 And is responsible for the different variabilities of this system. The period analysis of the data of 2006 showed the strongest peak at 17.76 d^{-1}, corresponding to the orbital period $0.^{d}0563$, while Araujo-Betancor et al. (2005) found the stronger peak at 35.52 d^{-1} corresponding to half of the period.

It can be noticed that sometimes the broad minimum (which is the deepest) occurs at phase 0.5, and the sharp minimum happens at phase near 0.0. However, sometimes the minimum at phase 0.0 is deeper. The double-humped light curve fitting to the ephemeris of Araujo-Betancor et al. (2005) allow to consider this modulation as the orbital one. The mean 2005–2006 R-light curve is more noisy than the mean 2004–2006 V-curve. This is due to the circumstances under which the observations in 2005 were obtained — the bulk of R-data was obtained with the smallest telescope. Concerning the individual light curves, an example in the B-band is shown in Fig. 5. Practically, the average 2006 B-light curve (Fig. 6) was obtained during three November nights of 2006. Note as well that the shape of both light curves (corresponding to years 2005 and 2006) is different. In particular, the humps switched their places and the minimum at phase 0.5 became smoother. In contradiction with the dwarf nova light curves where the usual orbital hump before the eclipse is generated by "a hot spot" on the accretion disk, the light curves of WZ Sge-stars in quiescence have a more complicated form. The disc in these systems has a complex structure and the pulsation (for example, V455 And, SDSS 0804) and the rotation of the white dwarf (WZ Sge, V455 And) can change the light curves.

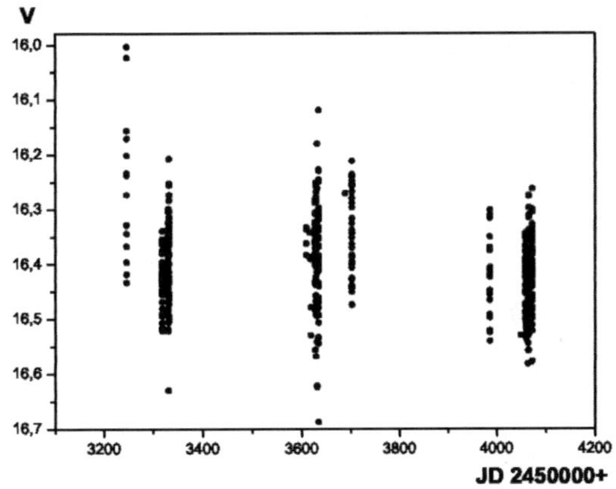

Figure 1. The aggregate light curve of V455 And during the years 2004–2006 in V-band.

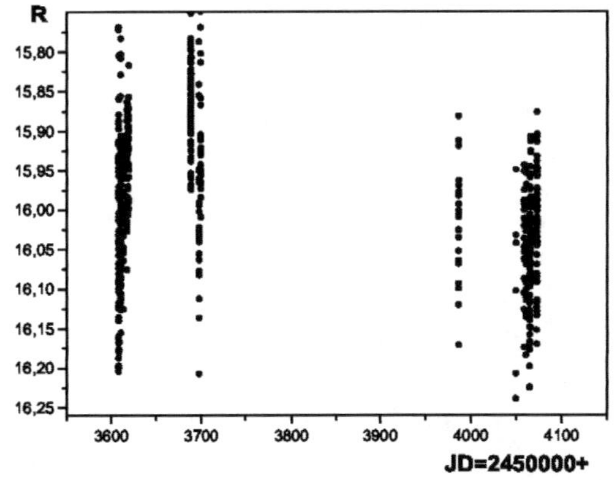

Figure 2. The aggregate light curve of V455 And during the years 2005–2006 in R-band.

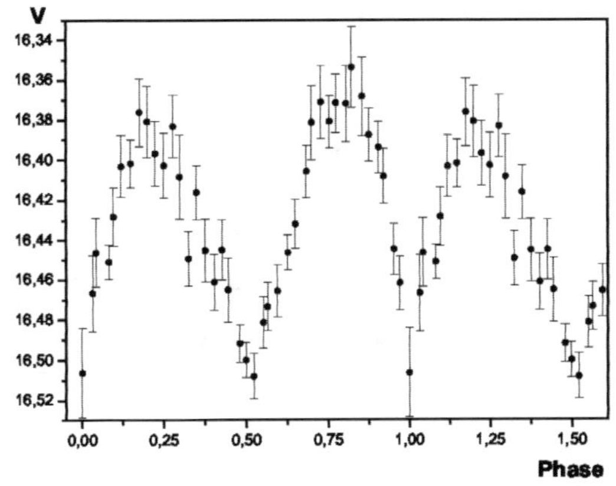

Figure 3. The mean V-orbital light curve of 2004–2006 years.

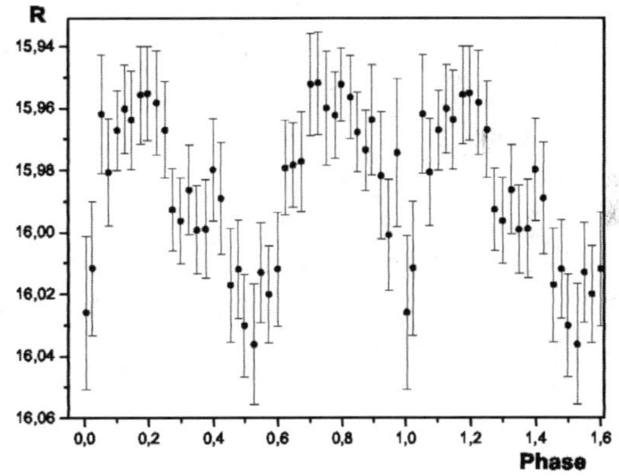

Figure 4. The mean R-orbital light curves of 2005–2006 years.

The search of white dwarf pulsations in binary systems is difficult because of hasting some quasi-periodical signals due to the orbital and rotational motions, the instabilities in the accretion stream and in the disc. As Gänsicke (2007) emphasized there were many additional signals. We found one of the signals at a frequency near 260 d^{-1} corresponding to the period about 5 min (Araujo-Betancor et al., 2005) which was proposed to be caused by non-radial pulsations of the white dwarf.

4. Discussion

In this paper we have exemplified only a small part of the light curves we have obtained. The orbital light curves of V455 And have a typically double-humped profile which is one of the WZ

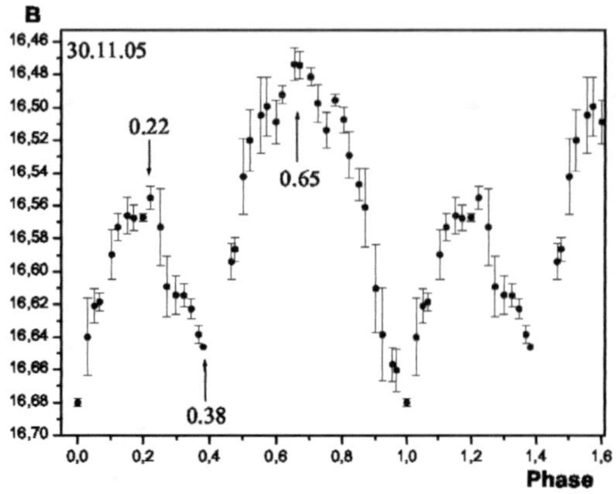

Figure 5. The orbital light curve for the 13th of November, B-band.

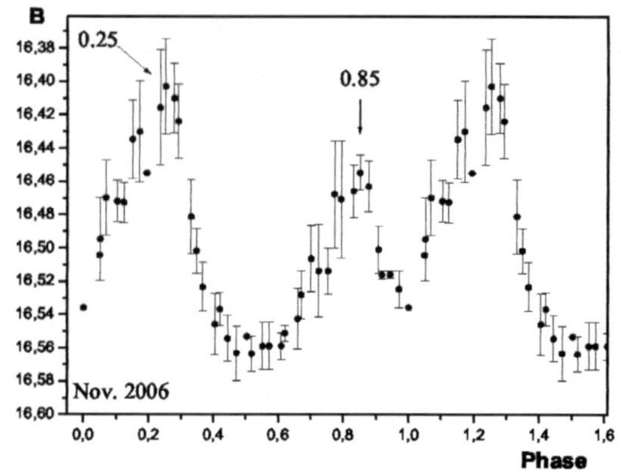

Figure 6. The mean orbital light curve in B-band, 2006, November.

Sge-type features, but sometimes each hump can split in two humps. The light curves of another WZ Sge-type star SDSS J080434.20+510349.2 show the same behaviour (Pavlenko et al., 2007; Pavlenko, 2008). The variable and sometimes multi-humped orbital modulation point at the complex and variable structure of accretion disc. However, we noticed no relevant changes in the light curves of V455 And nine months before the 2007 superoutburst.

Due to their brightness, V455 And and WZ Sge (as a prototype) are the only systems which were observed before their superoutbursts. The observations of the WZ Sge-type CVs during the superoutbursts and in quiescence are very important for elucidating the physics of accretion discs and the nature of both stellar components — a red and a white dwarf. One of the aspects of interest is the existence of non-radial pulsations of the white dwarf before and after the superoutburst because it is connected with the temperature of the white dwarf and the influence of the outburst on it. The second is rotation of the white dwarf. At least, there are two known spinning white dwarfs — WZ Sge and V455 And. And one more question still remains to be answered: are these systems intermediate polars or not? And, is it possible that the most of WZ Sge-type stars are intermediate polars?

Acknowledgments

This work was partially supported by the grants of Russian president NSh-1685.2008.2 and RFBR 06–02–16411. We are also grateful to the SOC and LOC for the hospitality during the conference.

References

Araujo-Betancor S Gänsicke B T Hagen J-H et al 2005 *A&A* **430** 629
Gänsicke B T 2007 *ASP Conf. Ser.* **372** Eds.: Ralf Napiwotzki and Matthew R Burleigh (San Francisco) 597
Pavlenko E P 2008 (this volume)
Pavlenko E P et al. 2007 *ASP Conf. Ser.* **372** Eds.: Ralf Napiwotzki and Matthew R Burleigh (San Francisco) 511
Samus N N Broens E Diepvens A et al. 2007 *IAUC* **8868**

Journal of Physics: Conference Series **172** (2009) 012045 doi:10.1088/1742-6596/172/1/012045

Diagnostics of very early stages of the classical nova explosion by the modeling of its X-ray emission

E Filippova, M Revnivtsev and A Lutovinov

Space Research Institute, Profsoyznaya str. 84/32, Moscow, Russia

E-mail: kate@iki.rssi.ru

Abstract. We present a spherically symmetric model of the interaction of classical nova ejecta with the matter of the companion star stellar wind. We compare its predictions with the data of observations of the unusual bright X-ray transient CI Cam and demonstrate that the outburst in 1998 can be explained as the X-ray emission at early stages of the classical nova explosion in the binary system. The X-ray outburst was observed almost immediately after the start of the motion of the ejecta (within 0.1-0.5 days) which give us the possibility to constrain two parameters of the explosions - the velocity and the mass of the ejecta. We show that the ejecta velocity was ~2700 km/s and remained at this value over ~2 day after the beginning of the explosion, that indicates the presence of the external forces acting on the ejecta. The mass of the ejecta was $\sim 10^{-7} - 10^{-6} M_{\odot}$.

1. Introduction

Classical nova can be a source of the standard (~2–10 keV) and hard (> 20 keV) X-rays. The most popular model for the generation of the X-ray emission is shock wave model: an envelope (up to $\sim 10^{-4} M_{\odot}$ Prialnik and Kovetz 2005) expelled from the white dwarf (WD) surface moves in the dense stellar wind of the companion star with high velocities (~1000-4000 km/s, Prialnik and Kovetz 2005) and forms the double shock wave structure (Figure 1). The forward shock can heat the matter up to ~5-80 keV ($h\nu \sim mU^2/2$). The reverse shock can also heat up the matter, but it is weaker and the radiation cooling in the ejecta is very effective; therefore in the following analysis we neglect it and consider the ejecta like a piston of the certain mass. To check the law of envelope expansion one needs observations of the first few hours or even minutes after the explosion beginning, i.e. long before of the maximum of the light curve in the optical band. Observations of the several Nova in hard X-rays in several days after the maximum in the optical band are not enough.

We argue that observations of the unusual X-ray transient XTE J0421+560/CI Cam can give us an opportunity to check models of the envelope expansion during the Nova outburst. More detailed studing can be found in the work of Filippova et al. (2008).

2. Outburst of the system CI Cam

The system XTE J0421+560/CI Cam was discovered as an X-ray transient on 31 March 1998 (Smith et al. 1998). Its properties were dramatically differed from the properties of the ordinary X-ray transients. The system spectrum contained emission lines, which are typical for the emission of the optically thin emission plasma with the temperature of $1 - 10$ keV (Orr et al.

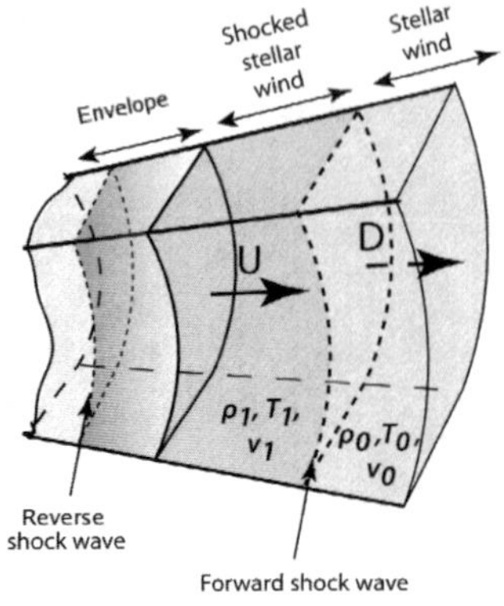

Figure 1. Scheme of the shock waves. D - velocity of the forward shock. U - velocity of the ejecta (read the text).

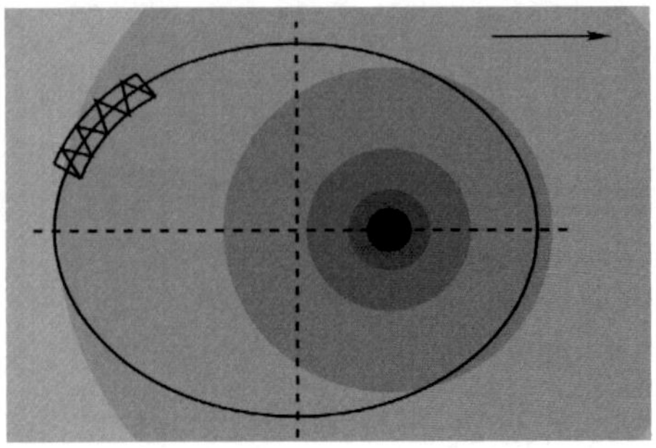

Figure 2. The companions position in the beginning of the outburst. The hatched region corresponds to the possible position of the white dwarf; grey rings are stellar wind.

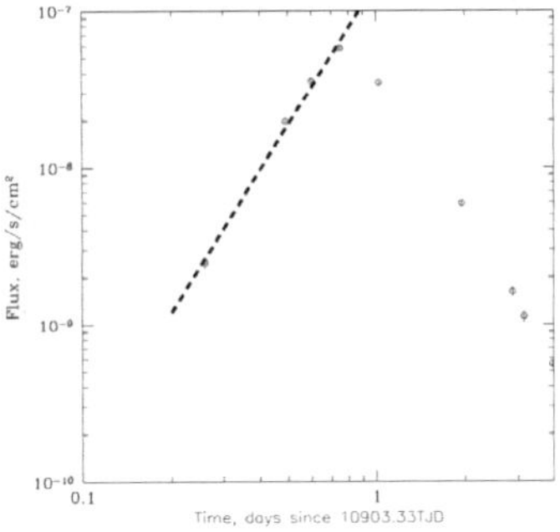

Figure 3. Light curve of the system CI Cam in the 3 - 20 keV energy band. Dashed line corresponds to the law of $L_{\rm x} \sim t^3$.

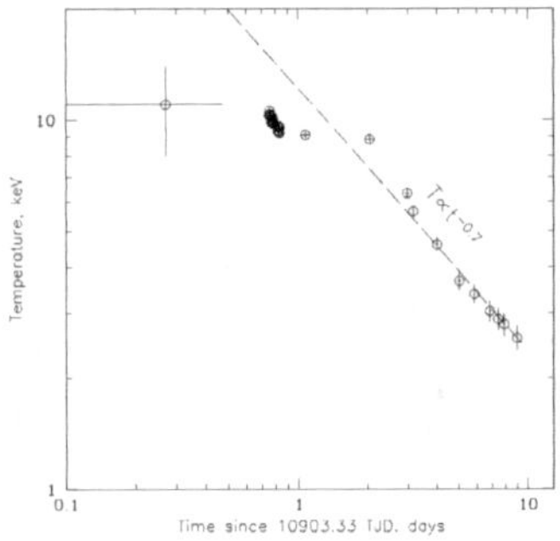

Figure 4. Dependence of the emission mean temperature on the time.

1998, Revnivtsev et al. 1999). A study of the system emission on the later stages of the evolution (Ishida et al. 2004, Boirini et al. 2002, Mioduzewski and Rupen, 2004) allowed to suggest that this bright outburst ($F_{\rm max\ 3-20\,keV} \sim 2$ Crab) is a classical Nova explosion.

An optical companion of the WD is B[e] star of the B4 III-V type (Barsukova et al. 2006). Using parameters from this work we calculate the position of the WD in the binary system at the moment of the explosion: the orbital phase was $\varphi_{\rm orb}$=0.2-0.34 and the distance between stars was $r = (5.47 - 7.54) \times 10^{12}$ cm (Figure 2, the hatched region).

3. Numerical method

We use a spherically symmetric Lagrangian code with the staggered mesh and artificial viscosity (more detail description can be found in Janka et al. 1993). We take into account the radiation cooling of the matter heated by the shock wave in the optically thin regime.

To obtain the simulated mean temperature of the emission we calculate the ratio of fluxes in the $3-5$ and $5-20$ keV energy bands, which corresponds to the certain temperature in the model of the one temperature optically thin plasma emission.

For the simplest estimations the volume averaged distribution of the stellar wind density relative to the WD position can be approximated by the constant near WD, which transforms to the $n \sim r^{-2}$ dependence of the density on the distance at $r > r_{\rm c}$.

4. Ejecta velocity

The constant temperature during the first ~2 days (Figure 4) indicates that the velocity of the ejecta at this time was constant. For the strong shock wave its temperature equals to $kT_1 = 2\mu m_p D^2 \frac{\gamma-1}{(\gamma+1)^2}$, where γ is an adiabatic index of the stellar wind. For the observed kT~10 keV we obtain D~2900 km/s and, for the constant density, U~2640 km/s. For the following calculations we adopte U=2700 km/s. The difference between the mean temperature behind the shock wave and the temperature on the shock wave is illustrated in Figure 5 for different initial densities of the stellar wind (dash-dotted line - $n_0 = 10^9\,\mathrm{cm}^{-3}$, long dashed line - $n_0 = 5 \times 10^9\,\mathrm{cm}^{-3}$, dashed line - $n_0 = 10^{10}\,\mathrm{cm}^{-3}$). It is clear that if the radiation cooling is low, than our estimations are valid.

5. Density of the stellar wind near WD

Observations indicate that during the rising part of the outburst flux rose as $L_{3-20\,\mathrm{keV}} \sim 1.1 \times 10^{23} t_c^3 d_{2\,\mathrm{kpc}}^2$ erg (Figure 3). If the piston moves with the constant velocity in the medium with the constant density than the luminosity increases as an emissive volume: $L = \Lambda n^2 \frac{4\pi}{3}(D^3 - U^3)t^3$, so we can estimate the stellar density near WD $n_0(r < r_c) \sim 8.6 \times 10^9 d_{2\,\mathrm{kpc}} U_{2700\,\mathrm{km/s}}^{-3/2}\,\mathrm{cm}^{-3}$ and $r_c = D\,t_{\mathrm{peak}} \sim 1.9 \times 10^{13}(t_{\mathrm{peak}}/0.75\,\mathrm{day}) D_{3000\,\mathrm{km/s}}\,\mathrm{cm}$.

6. Temperature on the later stages of the expansion

When the piston transfers its kinetic energy to the surrounding matter it starts to decelerate and the shock wave begins to move in the Sedov regime where the temperature on the shock wave depends on time as $T \sim t^{-0.66}$ (in the medium with $n \sim r^{-2}$), which is close to the observed law $T \sim t^{-0.7-0.6}$. We find that the mean temperature behind the shock wave in the Sedov regime is also can be approximated by the law of $T \sim t^{-0.6}$, i.e. the shock wave was in the Sedov regime on 4-10 days after the explosion.

7. Mass of the shell

We calculate models where the shell expands freely (Figure 6). These calculations demonstrate, that to have a constant temperature during first 1-2 days after the outburst the shell should be massive, but in order to be decelerated on the later stages of the expansion, the shell should be light enough and first 1-1.5 days should move under the external pressure. From our calculations it follows that several models can describe observations equally well (Figure 7). In these models the time of pushing varies between 1 and 1.5 days, the mass of the shell should be about $10^{-7} - 10^{-6} M_\odot$.

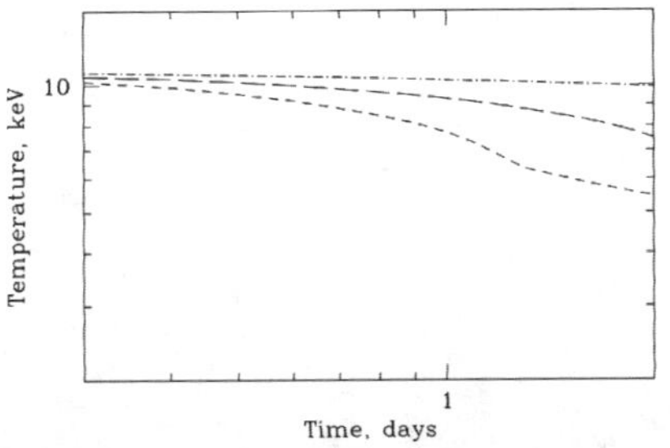

Figure 5. Dependence of the mean temperature on time for different initial densities of the stellar wind. Thin solid line - temperature on the shock wave.

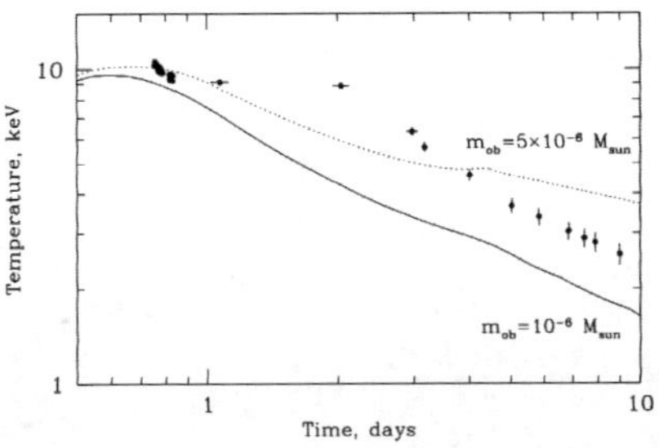

Figure 6. Dependence of the mean temperature on time for different mass of the shell.

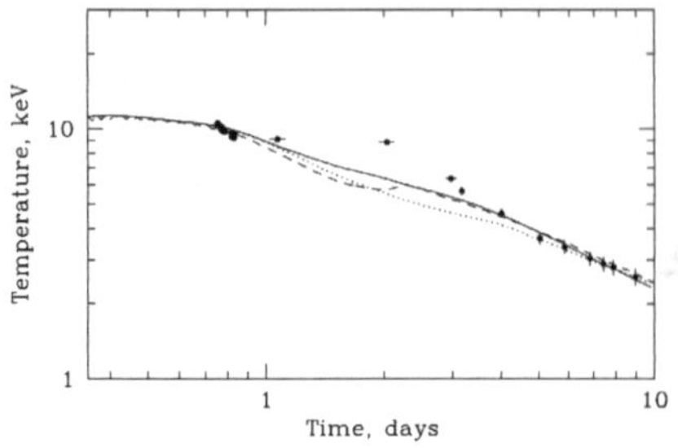

Figure 7. Models of the best approximation the observed temperature. Points - observations.

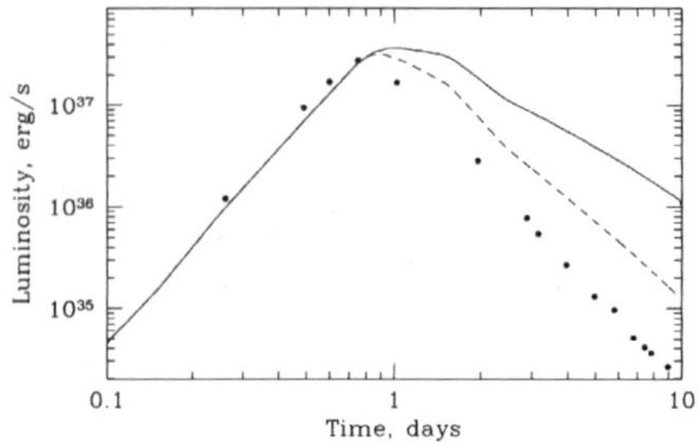

Figure 8. Light curves: solid line - spherically symmetric case, dashed line - disk like structure, points - observations.

8. Light curve approximation

The light curve predicted by one of our best models is presented in the Figure 8. The discrepancy in the decay phase can be result of the non-spherical symmetry of the stellar wind; particularly, it can be followed from the disk-like dense stellar wind of the B[e] star in the equatorial plane. We calculate the qualitative light curve on the phase of the luminosity decay assuming that emission comes from the dense disk with the constant height but dynamics of the shock wave is the same as in the spherically symmetric case. The slope of the luminosity decline in this model is very close to the observed one (Figure 8).

9. Conclusions

(i) the X-ray outburst of the CI Cam system in 1998 can be explained by the model of the classical Nova explosion in the dense stellar wind;

(ii) we have measured the velocity of the shell on 0.1-0.5 day (during first several hours) after outburst $\sim$2700 km/s;

(iii) the velocity remained approximately constant during first 1-1.5 days, which indicates an importance of the external force acting on the ejecta, e.g., the radiation pressure from the surface of the WD;

(iv) the estimated mass of the ejecta is about $10^{-7} - 10^{-6} M_{\odot}$.

Acknowledgments

E. Filippova wish to thank D. Docenko and D. Giannous for helpful discussions. This work was supported by grants RFFI (07-02-01051), "Russian Science Support Foundation" and the PRAS ("Origin and evolution of stars and galaxies" program).

References

Barsukova E A Borisov N V Burenkov A N et al. 2006 *Astron. Rep* **50** 664
Boirin L Parmar A Oosterbroek T Lumb D Orlandini M and Schartel N 2002 *Astron.Astrophys.* **394** 205
Ishida M Morio K and Ueda Y 2004 *Astrophys.J.* **601** 1088
Filippova E Revnivtsev M and Lutovinov A 2008 *Astron.Letters* **34** 797
Mioduszewski A J and Rupen M P 2004 *Astrophys.J.* **615** 432
Orr A Parmar A Orlandini M et al. 1998 *Astron.Astrophys* **340** L19
Prialnik D and Kovetz A 2005 *AIP Conf. Proc.* **797** 319
Revnivtsev M G Emel'Yanov A N and Borozdin K N 1999 *Astron. Lett.* **25** 294
Robinson E L Ivans I I and Welsh W F 2002 *Astrophys.J.* **565** 1169
Smith D Remillard R Swank J Takeshima T and Smith E 1998 *IAUCirc.* 6855
Janka H-T Zwerger T and Moenchmeyer R 1993 *Astron.Astrophys* **268** 360

Journal of Physics: Conference Series **172** (2009) 012046 doi:10.1088/1742-6596/172/1/012046

Parameter-free Stark Broadening of Hydrogen Lines in DA White Dwarfs

P-E Tremblay[1], P Bergeron[1] and J Dupuis[2]

[1] Département de Physique, Université de Montréal, C.P. 6128, Succ. Centre-Ville, Montréal, Québec H3C 3J7, Canada

[2] Canadian Space Agency, 6767 Route de l'Aéroport, Longueuil, Québec J3Y 8Y9, Canada

E-mail: tremblay@astro.umontreal.ca, bergeron@astro.umontreal.ca, jean.dupuis@space.gc.ca

Abstract. We present new calculations for the Stark broadening of the hydrogen line profiles in the dense atmospheres of white dwarf stars. Our improved model is based on the unified theory of Stark broadening from Vidal, Cooper & Smith (1970), but it also includes non-ideal gas effects from the Hummer & Mihalas (1988) occupation probability formalism directly inside the line profile calculations. This approach improves upon previous calculations that relied on the use of an ad-hoc free parameter to describe the dissolution of the line wing opacity in the presence of high electric microfields in the plasma. We present here the first grid of model spectra for hot ($T_{\rm eff} \geq 12,000\,$K) DA white dwarfs that has no free parameters. The atmospheric parameters obtained from optical and UV spectroscopic observations using these improved models are shown to differ substantially from those published in previous studies.

1. Introduction

The spectroscopic technique has been the most successful method for determining the atmospheric parameters — $T_{\rm eff}$ and $\log g$ — of hydrogen-line DA white dwarfs (Bergeron, Saffer & Liebert 1992). The technique consists in comparing the observed and predicted Balmer line profiles, and it has recently been expanded to the Lyman line series in the ultraviolet (Hébrard et al. 2003). The advantage of this method is that the theoretical profiles are very sensitive to variations of the atmospheric parameters. Since about 80% of the white dwarf population is of the DA type, the spectroscopic technique coupled with high signal-to-noise spectroscopic observations of Lyman or Balmer lines for large samples of DA stars can reveal important details about the mass distribution, the luminosity function, and the formation rate of white dwarfs. Even though this technique yields very accurate results with uncertainties less than 2% in terms of the *relative* parameters between stars, the *absolute* values of these parameters may still depend on the physics included in the models. One important aspect of the model atmosphere calculations is the opacity of the hydrogen lines, which are the dominant features observed in optical and UV spectra.

2. Models with a free parameter

In their preliminary analysis of DA white dwarfs, Bergeron, Saffer & Liebert (1992) found a lack of internal consistency between the spectroscopic solutions obtained when an increasing number of Balmer lines was included in the fitting procedure. This is illustrated in the top panel of Figure

1 for a typical DA star where we can see the solution gradually drifting in the $T_{\rm eff} - \log g$ diagram as more lines are included in the fit. Bergeron (1993) traced back the problem to the neglect of non-ideal gas effects inside the line profile calculations. Indeed, all white dwarf models rely on the theory of Stark broadening from Vidal, Cooper & Smith (1970) (hereafter VCS), which assumes that the absorber is in an ideal gas. However, Seaton (1990) has argued that non-ideal effects resulting from the perturbations of neighboring particles should be included *directly* in the line profile calculations. As a simpler alternative, Bergeron (1993) proposed instead to parameterize the value of the critical field ($\beta_{\rm crit}$) used in the non-ideal equation of state of Hummer & Mihalas (1988) (hereafter HM88) to *mimic* the non-ideal effects in the line opacity, and in particular in the regions where the line wings overlap. Model spectra calculated with twice the value of the critical field ($\beta_{\rm crit} \times 2$) were shown to improve the internal consistency significantly, as can be observed in the middle panel of Figure 1. It should be stressed, however, that the use of this free parameter does not imply that HM88 have underestimated the value of the critical field. As explained in detail by Bergeron (1993), this is just a way to simulate the non-ideal effects discussed in Seaton (1990) by *reducing* the line wing opacity near the Balmer limit. The effect of this additional free parameter on the line wing opacity is illustrated in Figure 2 where two model spectra are compared. The problem with this approach, of course, is that it has no physical basis, and there is also no reason to believe that this free parameter is the same for various hydrogen states or for different atmospheric parameters. It also has the disadvantage of modifying artificially the various populations in the equation of state. We note that all currently available model spectra of DA white dwarfs (Bergeron, Wesemael & Fontaine 1991, Hubeny & Lanz 1995, Finley, Koester & Basri 1997, Vennes et al. 2005) make use of this free parameter. Finally, even though the same parameterization is also used for transitions in the UV, it has never been tested empirically in an analysis similar to that displayed in Figure 1.

3. Improved Stark profiles

In this work, we provide a framework with no free parameter by improving the physics in the line profile calculations. We briefly describe here our procedure; further details will be provided elsewhere. Our work is based on the unified theory of Stark broadening from VCS. An extension of the VCS calculations in the white dwarf regime has been published by Lemke (1997), and these tables have been the most commonly used to model the Balmer and Lyman line profiles in DA stars. The unified VCS theory relies on the quasi-static approximation to describe the broadening by protons and a unified classical path theory for the broadening by electrons, which reduces to the impact approximation in the line cores and to the quasi-static approximation in the wings. One major problem for the modeling of DA atmospheres, however, is that the theory considers the absorber to be in an ideal gas, which means that line dissolution due to plasma perturbations is not allowed. Such perturbations are described at length in the HM88 non-ideal equation of state for astrophysical applications. They include, in particular, the high amplitude electric microfields produced by protons and the frequent electronic collisions.

The occupation probability formalism of HM88 has been introduced in the context of white dwarf spectra since the work of Bergeron, Wesemael & Fontaine (1991). According to this formalism, the probability that one electronic state is bound to the atom is obtained from the integral of the electric microfield probability distribution up to a given value for the critical field $\beta_{\rm crit}$. The critical field is roughly defined as the point where one Stark state for one particular level with a principal quantum number n "crosses" another Stark state associated with a higher level $n+1$. At the critical field, we expect that a bound electron on a level n, and a superposition of the accessible Stark states due to the electronic collisions and fluctuations in the direction of the microfield, will undergo a cascade of transitions to higher levels all the way up to the continuum. The occupation probability obtained with this formalism can then be used to compute the hydrogen state populations. The treatment of the amplitude of the line opacity

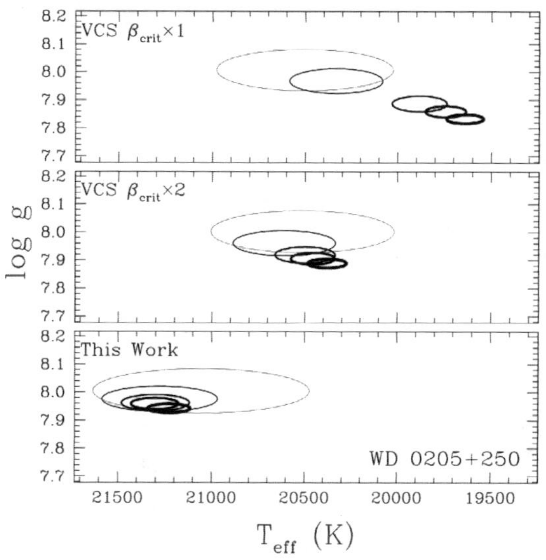

Figure 1. Solutions in a $T_{\rm eff} - \log g$ diagram for a typical DA star using 1 line (Hβ), 2, 3, 4 and 5 lines (up to H8) in the fitting procedure (represented by thicker 1σ uncertainty ellipses). The top panel shows the results with the original VCS profiles, while the middle panel also includes the free parameter proposed by Bergeron (1993). The bottom panel is with our new line profiles (see § 3).

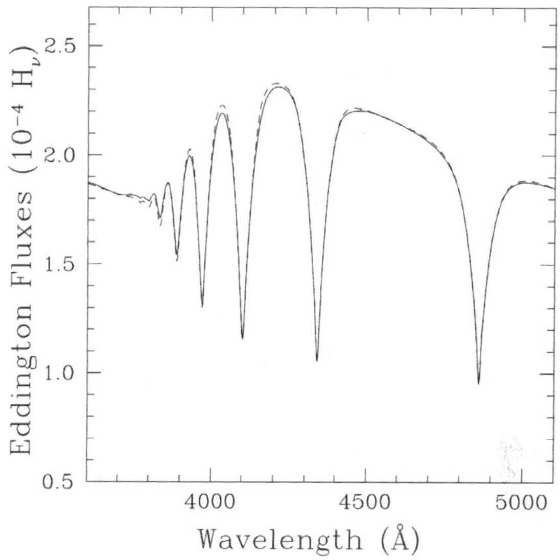

Figure 2. Comparison of synthetic spectra at $T_{\rm eff}$ = 20,000 K and $\log g$ = 8 without (solid line) and with (dashed line) the parameterization of the critical field ($\beta_{\rm crit} \times 2$) proposed by Bergeron (1993).

must also be modified to take into account the fact that a transition to an upper "dissolved" atomic level must be treated as a bound-free opacity rather than a bound-bound transition — the so-called pseudo-continuum opacity (Dappen, Anderson & Mihalas 1987).

The late Mike Seaton was the first to introduce the HM88 non-ideal effects directly inside the line profile calculations so that both the spectral shape and the amplitude are affected in a consistent way (Seaton 1990). However, his calculations made for the Opacity Project rely on an approximate electronic broadening theory that is inappropriate in the context of white dwarf atmospheres. We have therefore performed our own implementation of these non-ideal effects in a version of the VCS code provided to us by M. Lemke. In the case of proton perturbations, the line wing opacity will be reduced with respect to the ideal gas case since the high electric microfields contribute more importantly to the bound-free opacity by dissociating the atoms. We therefore use here a truncation of the proton microfields in the calculations of the quasi-static proton line broadening and include corrections due to electron perturbations as well. This is the first time these non-ideal effects are included directly into the VCS theory, and that DA models are calculated with line profiles that are coherent with the equation of state, *without any free parameter*. In Figure 3, we compare our improved line profile calculations with the original VCS profiles. Also shown for comparison are the approximate calculations of Seaton (1990).

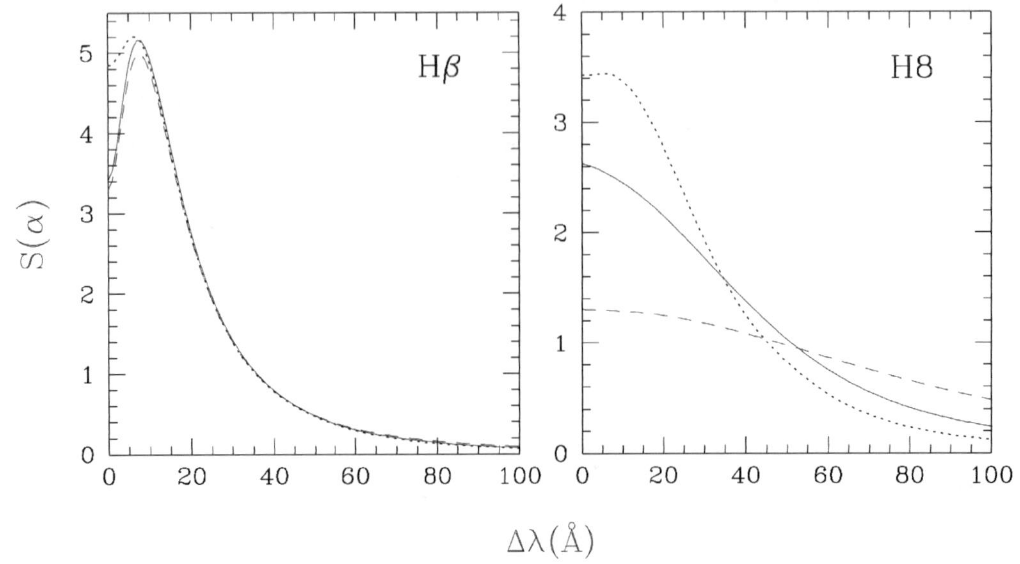

Figure 3. Stark broadening profiles as a function of the distance from the line center (in Å) at $T = 10,000$ K and $\log N_e = 17$. Results are shown for $H\beta$ (left) and $H8$ (right). In each panel, we compare the results from this work (solid lines), from the original VCS profiles (dashed lines), and from the approximate calculations of Seaton (1990) (dotted lines).

4. Results

4.1. Balmer lines

We now discuss the astrophysical implications of our improved line profiles. We calculated two grids of model atmospheres using (1) our new line profile calculations and (2) the original VCS theory using the tables computed by Lemke (1997), but with twice the value of the critical field ($\beta_{\rm crit} \times 2$) as proposed by Bergeron (1993). Both grids range from 13,000 to 40,000 K in $T_{\rm eff}$ and from 6.5 to 9.5 in $\log g$. In this analysis, we chose to restrain our calculations to this range of effective temperature to avoid the uncertainties related to convective energy transport, neutral broadening, and non-LTE effects.

First of all, we can already see in the bottom panel of Figure 1 that our line profiles have significantly improved the internal consistency between the solutions obtained from various Balmer lines compared to the previous calculations displayed in the two upper panels. We find, however, that the quality of the fits (not shown here) are nearly identical for all grids. In the following, we analyze the PG spectroscopic sample of DA white dwarfs from Liebert, Bergeron & Holberg (2005), but we restrict the range of temperature to that of our model grid. This yields a sample of 250 objects out of the 348 DA stars for the complete PG sample. Our spectroscopic data and fitting procedure are identical to those described in Liebert, Bergeron & Holberg (2005). The atmospheric parameters for each star are obtained using both model atmosphere grids, and the $\log g$ values are then converted into masses using evolutionary models appropriate for white dwarfs with thick hydrogen layers (see Liebert, Bergeron & Holberg 2005 for details). Our results are presented in Figure 4 in a mass versus $T_{\rm eff}$ diagram. We can see that the distributions with both grids are significantly different. In particular, the masses derived with our improved models are globally larger and the $T_{\rm eff}$ values are also higher (see also Fig. 1). As seen from Figure 5, the mean mass of this PG sample is shifted by $\Delta M/M_\odot = +0.034$ when our new models are used. The shape of the mass distribution remains statistically equivalent

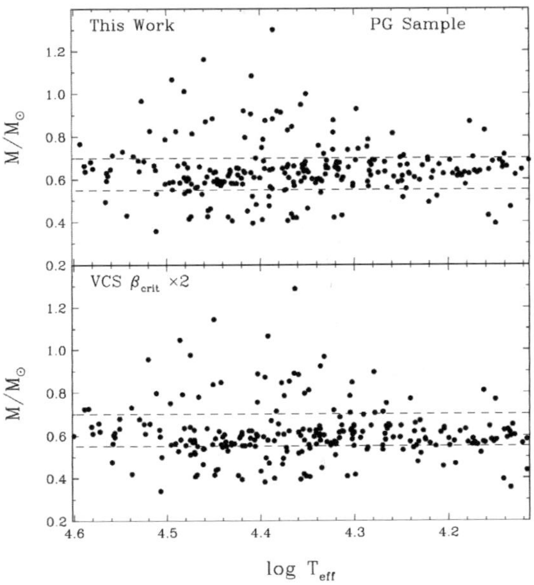

Figure 4. Mass versus $T_{\rm eff}$ distributions for the PG sample in the range 40,000 K > $T_{\rm eff}$ > 13,000 K. Results are shown for both our improved line profiles (top panel) and our old models (bottom panel). Lines of constant masses at 0.55 and 0.70 $M/M_\odot$ are shown as a guide.

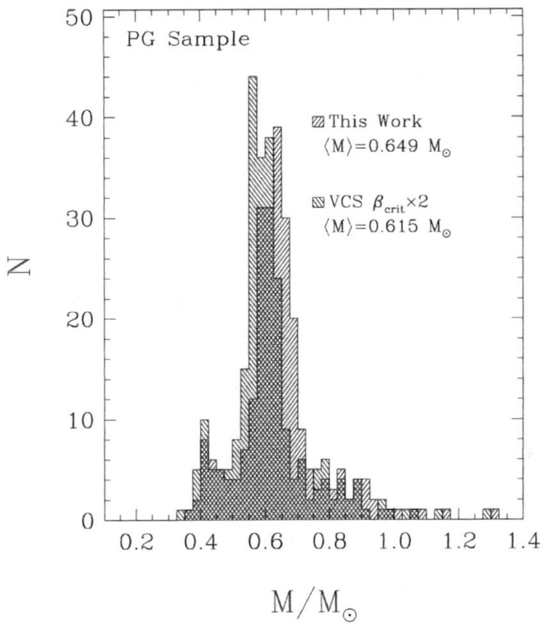

Figure 5. Mass distributions for the PG sample used in Figure 4. The mean masses are reported in the figure.

for both grids, however, and the dispersions are comparable.

As discussed by Bergeron, Saffer & Liebert (1992), even though the spectroscopic technique provides very accurate individual determinations of atmospheric parameters, the *absolute* values of log g may suffer from a zero-point offset and it is therefore important to compare the individual measurements with results obtained from independent techniques. The only reliable independent observational constraints for large samples of white dwarfs are trigonometric parallax measurements. We thus compare the absolute visual magnitudes M_V derived from the measured parallaxes (combined with V) with the absolute magnitudes obtained from the spectroscopic values of $T_{\rm eff}$ and $\log g$ following the procedure described at length in Holberg & Bergeron (2006). The results presented here are based on the high signal-to-noise spectroscopic sample of Bergeron, Gianninas & Boudreault (2007) for 92 DA white dwarfs with known parallaxes. The comparisons are shown in Figure 6 for both model grids. Again, we have restrained the analysis to stars above $T_{\rm eff} = 13,000$ K. The agreement is satisfactory within the parallax uncertainties for both model grids. Hence, despite the fact that our new models yield higher values of $T_{\rm eff}$ and $\log g$ (i.e., smaller radii), these two effects almost cancel each other and the predicted luminosities (or M_V) remain unchanged. Another important constraint is provided by the bright white dwarf 40 Eri B for which a very precise trigonometric parallax and visual magnitude have been measured by Hipparcos. These measurements yield $M_V = 11.01 \pm 0.01$, while we predict $M_V = 11.02 \pm 0.07$ and 10.97 ± 0.07 based on our new and old models, respectively. Although both determinations agree within the uncertainties with the observed value, our new grid provides an exact match to the measured M_V value.

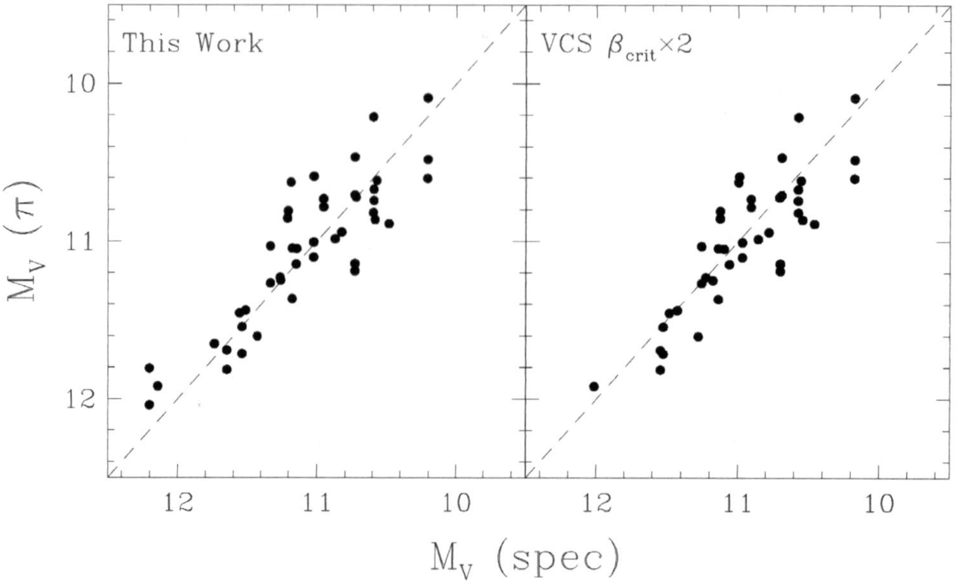

Figure 6. Comparison of the absolute visual magnitudes obtained from trigonometric parallax measurements and from the spectroscopic analysis of DA stars above $T_{\rm eff} \sim 13,000$ K using both model grids.

4.2. Lyman lines

This section presents a preliminary investigation of our model predictions for the Lyman lines. We analyze the FUSE spectra of two bright white dwarfs, GD 71 and Sirius B, both of which are in the temperature range considered in this work. In Figure 7, we show χ^2 fits of the entire FUSE spectra obtained with our new models to determine the atmospheric parameters given in the figure. The quality of the fits is very good. We performed similar fits with our old grid and we find that the quality of the fit is comparable, although our new atmospheric parameters are shifted by +60 K in $T_{\rm eff}$ and by +0.06 in $\log g$ for both stars. Thus, the shifts in $\log g$ are similar to those observed in the optical. It should be noted, however, that an increase of both $T_{\rm eff}$ and $\log g$ by $\sim$ 1% results in fits that are equally good from visual inspection, so the values of the best fit parameters reported here should be taken with caution. Recent analyses by Vennes et al. (2005) and Barstow et al. (2003) suggest that the $\log g$ values derived from the optical and the UV agree in this range of $T_{\rm eff}$, and this conclusion should not be affected by our new models.

The temperature shifts in the UV, however, are much lower than those in the optical (see, e.g., Fig. 1). Since the UV spectra were calibrated using DA models with the older line profiles, the data reduction might have to be revisited before we confirm these results. For instance, we show in Figure 8 the predicted spectra over the visible and UV regions for two flux standards (GD 71 and G191-B2B) used for the FUSE and HST calibrations among others. The atmospheric parameters for these LTE models are obtained from the best fits to the observed Balmer lines of each star using both grids (hotter models have been calculated for this exercise). The results indicate that the absolute predicted fluxes, as well as the slopes of the energy distributions, are very different even up to the high temperature of G191-B2B ($T_{\rm eff} \sim 60,000$ K).

We finish this section by noting that the non-ideal effects for the lower Lyman lines are still not well understood. The critical field in this case becomes so large that it must have been created by a single proton close to the absorber in such a way that the approximation of the

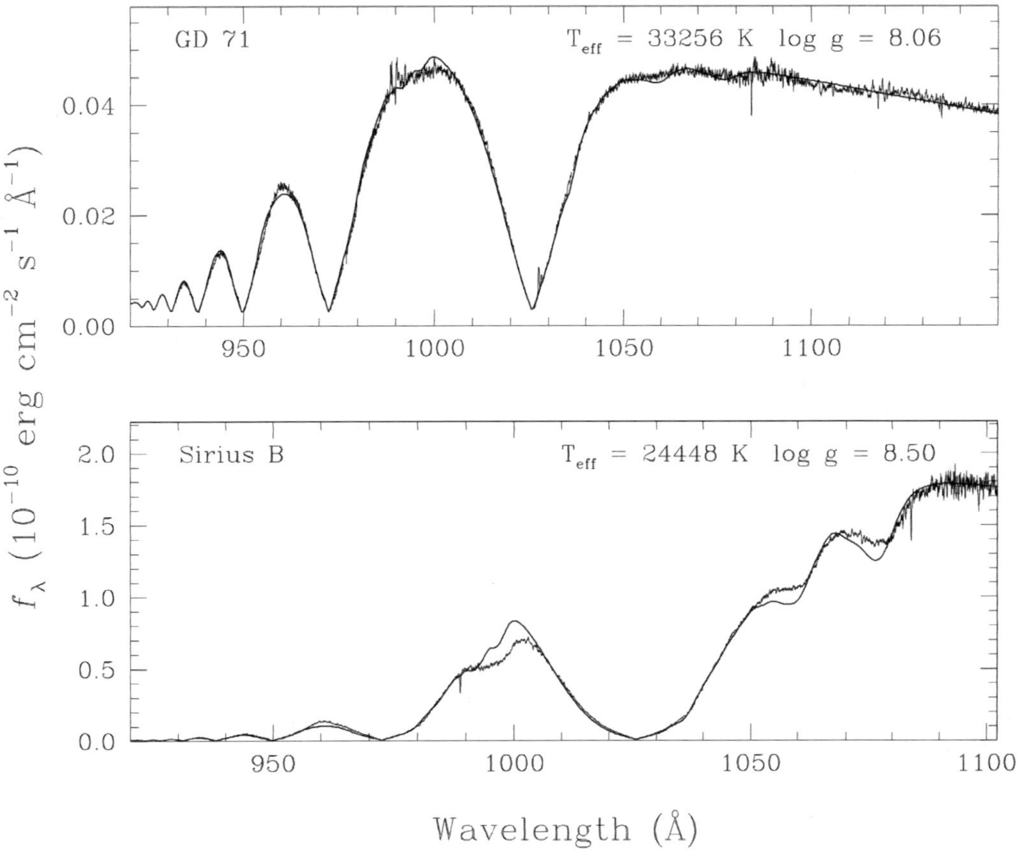

Figure 7. Fits to two FUSE spectra using our improved model spectra; $T_{\rm eff}$, $\log g$, and the solid angle are considered free parameters. Interstellar features, including all cores of the lines, have been removed from the observed spectra. The atmospheric parameters obtained from the minimization procedure are given in each panel.

dipolar proton-absorber interaction fails completely. For instance, exact calculations show that $n = 1, 2$ and 3 have no crossings. Current white dwarf model atmospheres (this code, TLUSTY) neglect non-ideal effects for Lyman α, Lyman β and in some cases Lyman γ. In other words, there is no pseudo-continuum opacity, and ideal gas profiles are used for these lines. This significantly changes the predicted UV spectra for $T_{\rm eff} < 30,000$ K. For Hα, non-ideal effects are already negligible with the HM88 theory and no modification is necessary. We plan to study further this problem with a larger sample of FUSE spectra.

5. Conclusion

We have combined for the first time the unified theory of Stark broadening from VCS with the non-ideal equation of state of HM88 to build model atmospheres for DA white dwarfs that are consistent to first order in terms of the physics of the equation of state and the line profiles, without a free parameter. We have shown that in the range 40,000 K $> T_{\rm eff} >$ 13,000 K, the revised mean mass for the PG sample is significantly larger than previous determinations. The atmospheric parameters we obtain with our new model spectra are in excellent agreement with trigonometric parallax measurements. Future work will extend our analysis to the more complex

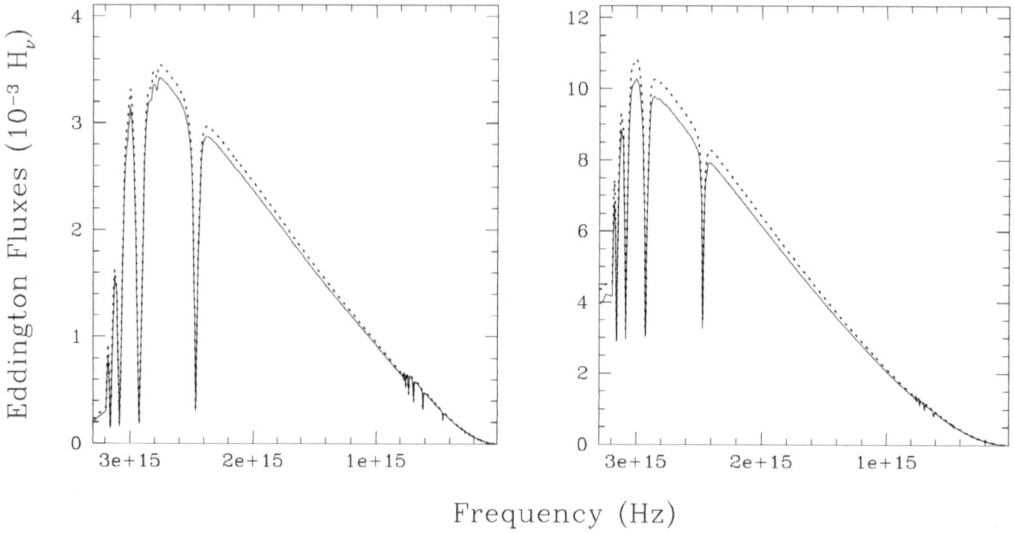

Figure 8. Theoretical spectra for the flux standards GD 71 (left) and G191-B2B (right) with our improved models (dotted lines) and our old grid (solid lines) using the atmospheric parameters obtained from fits to the Balmer lines.

regime of white dwarf stars at higher and lower effective temperatures.

Acknowledgments

We thank M. Lemke for providing us with his computer version of the VCS code. This work was supported in part by the NSERC Canada and by the Fund FQRNT (Québec). P. Bergeron is a Cottrell Scholar of Research Corporation.

References

Barstow M A Good S A Burleigh M R Hubeny I Holberg J B and Levan A J 2003 *MNRAS* **344** 562
Bergeron P 1993 *NATO ASIC Proc. 403: White Dwarfs: Advances in Observation and Theory* (Dordrecht: Kluwer) p 267
Bergeron P Gianninas A & Boudreault S 2007 *15th European Workshop on White Dwarfs ASP Conference Series* vol 372 (San Francisco: ASP) p 29
Bergeron P Saffer R A and Liebert J 1992 *ApJ* **394** 228
Hébrard G Allard N F Kielkopf J F Chayer P Dupuis J Kruk J W & Hubeny I 2003 *A&A* **405** 1153
Bergeron P, Wesemael F & Fontaine G 1991 *ApJ* **367** 253
Dappen W Anderson L & Mihalas D 1987 *ApJ* **319** 195
Finley D S Koester D & Basri G 1997 *ApJ* **488** 375
Holberg J B & Bergeron P 2006 *ApJ* **132** 1221
Hubeny I and Lanz T 1995 *ApJ* **439** 875
Hummer D G & Mihalas D 1988 *ApJ* **331** 794
Liebert J Bergeron P and Holberg J B 2005 *ApJ Supp. Ser.* **156** 47
Lemke M 1997 *A&A Supp. Ser.* **122** 285
Seaton M J 1990 *J. of Phys. B* **23** 3255
Vennes S Chayer P Dupuis J & Lanz T 2005 *14th European Workshop on White Dwarfs ASP Conference Series* vol 334 (San Francisco: ASP) p 185
Vidal C R Cooper J & Smith E W 1970 *J. of Quant. Spect. and Rad. Trans.* **10** 1011

Analysis of the Hydrogen-rich Magnetic White Dwarfs in the SDSS

Baybars Külebi[1], Stefan Jordan[1], Fabian Euchner[2], Heiko Hirsch[3] and Wolfgang Löffler[1]

[1]Astronomisches Rechen-Institut, Mönchhofstrasse 12-14, 69120 Heidelberg, Germany
[2]Swiss Seismological Service, ETH Zurich, Sonneggstrasse 5, 8092 Zurich, Switzerland
[3]Dr.-Remeis-Sternwarte, Bamberg, Sternwartstrasse 7, D-96049 Bamberg

E-mail: bkulebi@ari.uni-heidelberg.de

Abstract. We have calculated optical spectra of hydrogen-rich (DA) white dwarfs with magnetic field strengths between 1 MG and 1000 MG for temperatures between 7000 K and 50000 K. Through a least-squares minimization scheme with an evolutionary algorithm, we have analyzed the spectra of 114 magnetic DAs from the SDSS (95 previously published plus 14 newly discovered within SDSS, and five discovered by SEGUE). Since we were limited to a single spectrum for each object we used only centered magnetic dipoles or dipoles which were shifted along the magnetic dipole axis. We also statistically investigated the distribution of magnetic-field strengths and geometries of our sample.

1. Introduction

The Sloan Digital Sky Survey (SDSS) project covering 10^4 deg^2 of the sky has discovered thousands of new white dwarfs, among them 105 with magnetic fields (Gänsicke et al. 2002; Schmidt et al. 2003; Vanlandingham et al. 2005). White dwarfs with magnetic fields between 10^4 and 10^9 G are thought to represent more than 10% of the total population of white dwarfs (Liebert et al. 2003). The SDSS Data Release 3 (DR3, $\approx 4200\,\mathrm{deg}^2$) increased the number of known magnetic white dwarfs from 65 (Wickramasinghe & Ferrario 2000; Jordan 2001) to 170 (Kawka et al. 2007).

In this work we present the re-analysis of the 95 DA magnetic white dwarfs (MWDs) discovered by Schmidt et al. (2003) and Vanlandingham et al. (2005), plus the analysis of 19 additional objects from SDSS up to DR6[1] ($9583\,\mathrm{deg}^2$).

Schmidt et al. (2003) and Vanlandingham et al. (2005) determined the field strengths and the inclinations of magnetic dipoles by comparing visually the observed spectra with model spectra calculated using a simplified radiation transfer code (Latter et al. 1987). Their analyses resulted in dipolar field strengths for the SDSS MWDs between 1.5 MG and ~ 1000 MG. From their sample of 105 MWDs, 97 were classified as DAs.

2. SDSS data

The spectroscopic targets of the SDSS are selected based on color criteria optimized for galaxies and quasars. However, follow-up spectroscopy of many stars is also performed with the twin dual

[1] http://www.sdss.org/dr6/

beam spectrographs (3900 - 6200 and 5800 - 9200 Å, $\lambda/d\lambda \sim 1800$), in particular of blue objects like white dwarfs and hot subdwarfs (Harris et al. 2003; Kleinman et al. 2004). Since the energy distribution of strongly magnetic white dwarfs can differ from nonmagnetic ones, MWDs are not only found in the SDSS color categories for white dwarfs or blue horizontal-branch stars, but may also fall into the color categories for quasars (QSOs), "serendipitous blue objects", and hot subdwarfs. Based on their colors, objects are assigned to fibers for spectroscopic investigation (for spectroscopic target selection, see Stoughton et al. (2002).

In order to identify magnetic white dwarfs among the white dwarfs or other categories, different techniques were used: From the sample of white dwarfs, selected by color cuts, Gäansicke et al. (2002) and Schmidt et al. (2003) used visual inspection. Vanlandingham et al. (2005) inspected only those objects visually for which bad χ^2 were obtained by their *autofit* process. This procedure selects in particular white dwarfs with magnetic fields above 3 MG and misses objects with weaker magnetic fields.

In addition to these already known objects, we have analyzed data of nineteen additional SDSS objects discovered to be magnetic due to suspicious radial velocity measurements by the HYPERMUCHFUSS project (see, Geier et al., these proceedings and Tillich et al., these proceedings). The one-dimensional spectra which we used in this work were generated by SDSS's spectroscopic pipeline `spectro2d` and downloaded from Data Archive Server.

3. Analysis

In order to increase efficiency, we pre-computed a three-dimensional grid of Stokes I and V model spectra with effective temperature $7000\,\mathrm{K} \leq T_{\rm eff} \leq 50000\,\mathrm{K}$, magnetic field strength $1\,\mathrm{MG} \leq B \leq 1\,\mathrm{GG}$, and 18 different directions ψ relative to the line of sight as the independent variables (9 entries, equally spaced in $\cos\psi$) using the radiative transfer code for magnetized white dwarf atmospheres, see Jordan (1992) and Jordan & Schmidt(2003). All spectra are calculated for a surface gravity of $\log g = 8$. Since no polarization information is available for the SDSS, our analysis is limited to the flux spectra (Stokes parameter I).

The magnetic field geometry of the magnetic white dwarfs was determined with a modified version of the code developed by Euchner et al. (2002). This code calculates the total flux (and circular polarization) spectra for an arbitrary magnetic field topology by adding up appropriately weighted model spectra for a large number of surface elements. Complex magnetic field geometries are accounted for by a multipole expansion of the scalar magnetic potential, but in our paper only centered and offset dipoles were considered. The observed spectra are fitted using an evolutionary algorithm (Rechenberg 1994) with a least-squares quality function.

Due to the lack of a consistent theory for Zeeman and Stark broadening, the latter is only taken into account by a rather simple approximation (Jordan 1992) so that systematic uncertainties, particularly in the low-field regime ($\leq 5\,\mathrm{MG}$), are unavoidable. Consequently, effective temperatures and surface gravities derived from fitting the Balmer lines alone are less reliable than in the case of non-magnetic white dwarfs. This may also result in disagreements with temperatures estimates derived from the continuum slope.

Our fitting procedure had to adjust three or four free parameters: the magnetic dipole field strength B_d, the effective temperature $T_{\rm eff}$, the inclination of the dipole axis i, and an offset along the magnetic axis $z_{\rm off}$, if offset dipoles are used. $T_{\rm eff}$ needed to be independently calculated for our purposes. For the 95 MWDs analyzed, we used the literature values for $T_{\rm eff}$ which were determined by comparison to the theoretical non-magnetic DA colors in the $u - g$ vs $g - r$ plane (Schmidt et al. 2003, Vanlandingham et al 2005). The temperature of the new MWDs presented in Table 1 were estimated by the synthetic SDSS color-color diagrams by Holberg & Bergeron (2006).

All fits resulted in reduced χ^2 values between 0.8 and 3.0 except for some high-field objects. In Fig. 1 we show fits of 6 MWD spectra as an example.

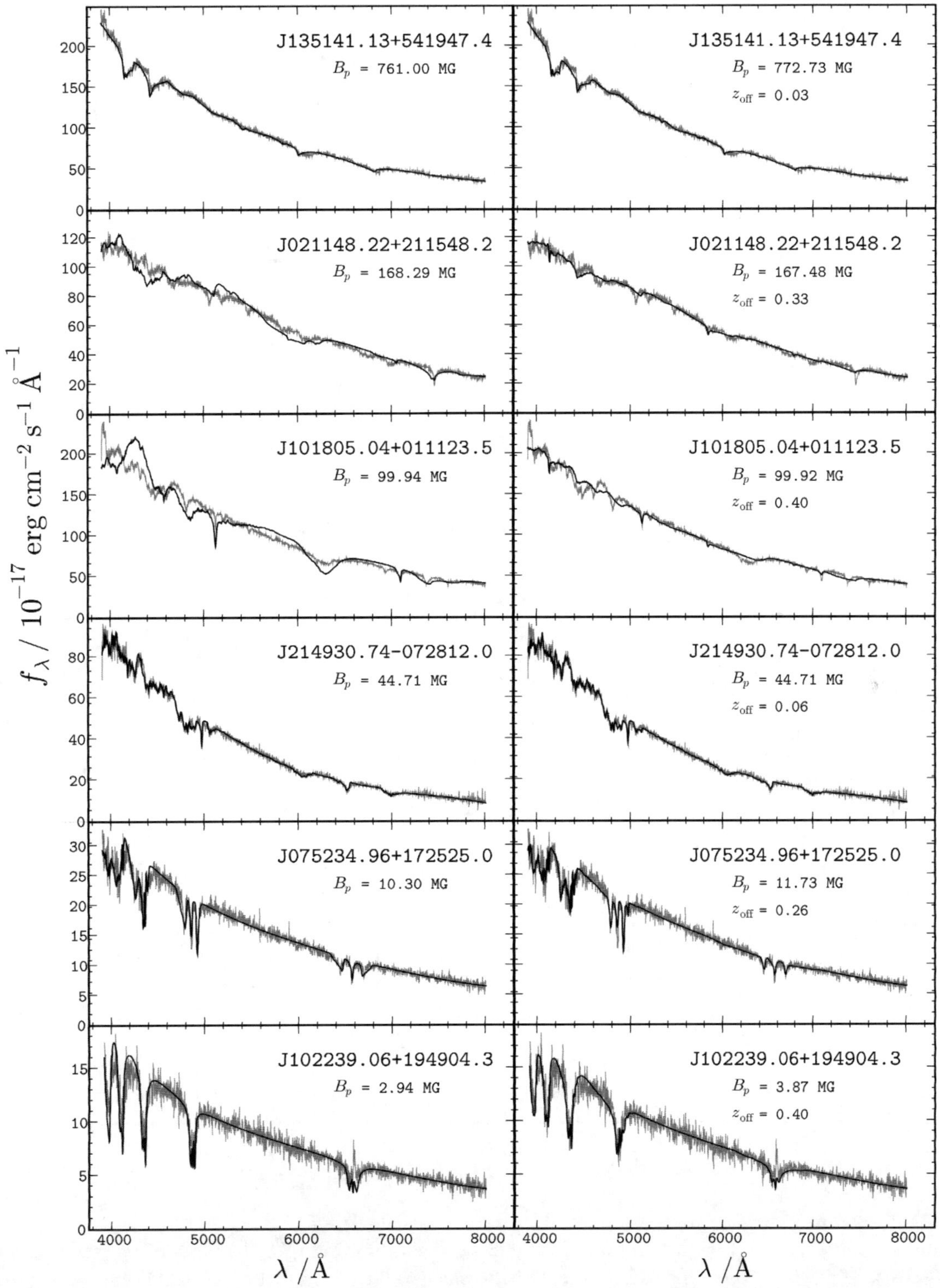

Figure 1. Representative sample of fits of observed spectra of MWDs from the SDSS to centered magnetic dipoles with a polar field strength B_p (left) and dipoles shifted by $z_{\rm off}$ stellar radii along the dipole axis (right).

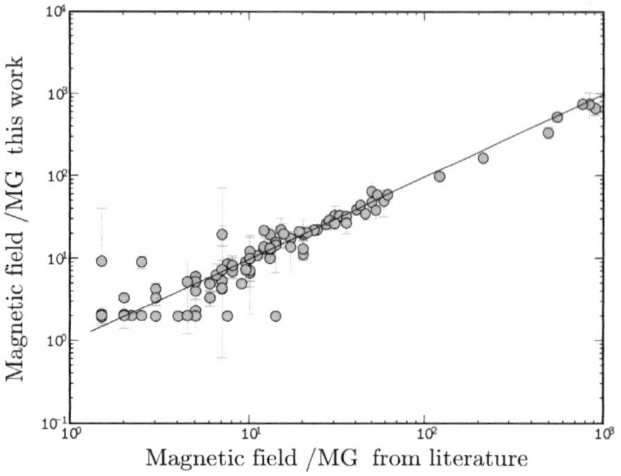

Figure 2. Comparison of dipole magnetic field fit values in this work versus those of Schmidt et al. (2003) and Vanlandingham et al. (2005).

Table 1. Best fit parameter values for centered dipole and offset dipole models for the objects discovered by HYPERMUCHFUSS.

MWD (SDSS+)	B_p / MG	i / deg	$B_{\rm off}$ / MG	$z_{\rm off}$ / $r_{\rm WD}$	i / deg	$T_{\rm eff}$
J031824.19+422651.0	10.12±0.10	54.6 ± 4.7	10.77±0.10	0.29±0.05	61.1 ± 10.0	10500
J075234.96+172525.0	10.30±1.23	72.4 ± 22.4	11.73±1.05	0.26±0.05	58.7 ± 38.1	9000
J083945.56+200015.7	3.38±0.49	48.6 ± 7.7	2.15±0.10	0.29±0.08	49.9 ± 90[1]	15000
J085106.12+120157.8	2.03±0.10	81.9 ± 90[1]	2.47±0.10	0.35±0.06	72.8 ± 18.8	11000
J085523.87+164059.0	12.23±2.92	48.6 ± 8.6	7.86±1.63	0.36±0.06	10.8 ± 6.1	15500
J091833.32+205536.9	2.04±0.10	87.2 ± 41.9	2.66±1.71	0.39±0.17	70.3 ± 61.9	14000
J100657.51+303338.1	1.00±0.10	82.5 ± 30.8	1.30±1.23	0.37±0.39	14.3 ± 13.1	10000
J102239.06+194904.3	2.94±0.71	49.0 ± 13.0	3.87±1.11	0.40±0.07	51.3 ± 40.3	9000
J112257.10+322327.8	11.38±3.42	49.0 ± 12.3	7.46±1.68	0.37±0.11	4.2 ± 6.3	12500
J125434.65+371000.1	4.10±0.35	41.9 ± 19.2	4.89±0.42	0.40±0.03	50.1 ± 21.0	10000
J125715.54+341439.3	11.45±0.71	0.5 ± 0.6	13.70±1.69	0.07±0.02	7.7 ± 12.0	8500
J134820.79+381017.2	13.65±2.66	89.4 ± 90[1]	14.45±4.65	0.22±0.04	54.8 ± 25.3	35000
J140716.66+495613.7	12.49±6.20	88.1 ± 90[1]	13.20±4.21	0.24±0.10	63.3 ± 81.1	20000
J141906.19+254356.5	2.03±0.10	81.2 ± 8.7	2.56±0.10	0.38±0.03	54.8 ± 10.4	9000
J143019.05+281100.8	9.34±1.44	5.6 ± 4.5	6.25±0.75	0.16±0.03	5.6 ± 4.5	9000
J151130.17+422023.0	22.40±9.41	48.6 ± 19.5	8.37±1.07	0.31±0.06	5.8 ± 21.1	9500
J202501.10+131025.6	10.10±1.76	68.5 ± 9.1	10.72±1.71	0.29±0.04	53.7 ± 9.0	17000
J220435.05+001242.9	1.02±0.10	71.2 ± 90[1]	2.50±5.47	0.36±0.69	3.1 ± 13.6	22000
J225726.05+075541.7	16.17±2.81	74.9 ± 16.1	17.39±3.21	0.15±0.05	78.5 ± 34.8	40000

[1]Inclination errors are estimated as large, for explanation see Sec. 4

4. Discussion

Three objects analyzed by Schmidt et al. (2003) and Vanlandingham et al. (2005) are omitted in this work. *SDSSJ05959.56+433521.3, G111-49* was listed by Schmidt et al. (2003) as a magnetic DA, but is a Carbon-rich (DC) MWD (Putney 1995); *SDSSJ084716.21+148420.4* is a DAH+DB binary, hence we were unable to analyze these two objects with our code. Finally, neither the coordinates nor fiber identifier of *SDSSJ234605.44+385337.7* mentioned by Vanlandingham et al. (2005) was accessible through the SDSS database.

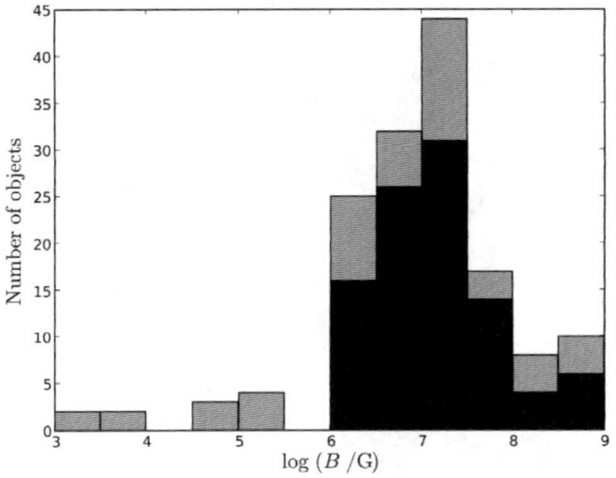

Figure 3. Histogram of magnetic white dwarfs in equal intervals of $\log B$. Gray columns represent the number of all DA MWDs and black shades represent the the contribution of SDSS to DA MWDs.

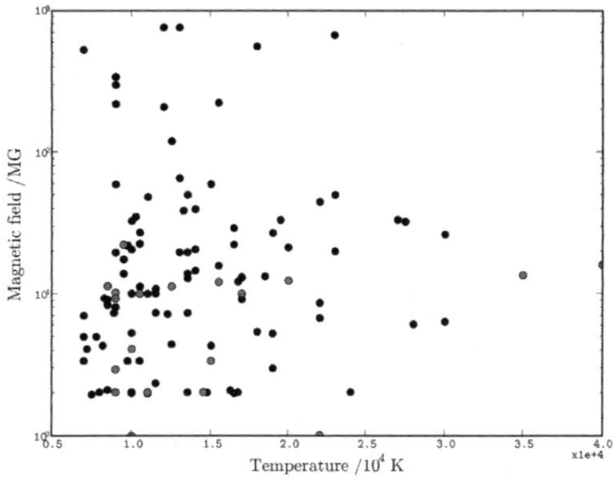

Figure 4. Scatter plot of dipole magnetic field value vs. temperature in this work. The random distribution of field strengths with respect to the age indicator temperature is consistent with a long decay timescale of MWDs. Gray dots indicate the objects from HYPERMUCHFUSS.

Emission lines were found in *SDSSJ102220.69+272539.8* and *SDSSJ102239.06+194904.3* (the latter is shown in Fig. 1), very similar to *SDSSJ121209.31+013627.7* which could indicate that these objects may be EF Eri like, magnetic cataclysmic variables with a brown dwarf companion (Burleigh et al. 2007).

The spectra of the high-field objects *SDSSJ224741.41+145638.8* and *SDSSJ101805.04-+01123.5* (PG 1015+014, shown in Fig. 1) do not fit particularly well. At higher field strengths ($>$ 50 MG) the spectra become very sensitive to the details of the magnetic field geometry, as was demonstrated by Euchner et al. (2002, 2005, 2006). The deviations of the observed spectra from our theoretical spectra assuming (offset) dipole models hint therefore to a magnetic field geometry that is more complex than a shifted dipole. A more comprehensive analysis of PG 1015+014 (Euchner et al. 2006) showed that individually tilted and off-centered zonal multipole components with field strengths between 50-90 MG are needed to represent the global magnetic field, consistent with our analysis. On the other side, some high-field objects in our sample, like *SDSSJ135141.13+541947* (Fig. 1) are well fitted.

SDSS has proven to be extremely important in increasing the total number of MWDs. Out of 8000 WDs of the DR4 (Eisenstein et al. 2006), a total 139 MWDs were discovered (DR6; Kulebi et al. 2009), still a very low percentage compared to the sample of MWDs in the solar neighborhood (13% $\pm$ 4%; Kawka et al. 2007). Note, however, that no systematic search for MWDs in DR4-DR6 was performed yet. SDSS is designed to investigate galaxies and quasars, thus stars are not its primary targets. Formerly the WD acquisition efficiency of the SDSS was partly defined by the probability of WD spectra to be target labelled as QSO. With each new data release, target selection criteria has improved in separating the QSO from non-QSO targets. Unfortunately the improvement in extragalatic target selection impairs SDSSs capability to detect serendipitous MWDs. This was exemplified by the inability to recover most of the formerly known MWDs in regions of the sky covered in DR2-3 by Vanlandingham et al. (2005).

Overall, SDSS has nearly tripled the number of MWDs, conversely the completeness of the total MWD population is affected significantly by the SDSS biases because of this high impact of SDSS. The most important selection effect arises from the priority selection of the spectroscopic targeting, which is determined from color categories of each object (Stoughton et al. 2002). Due to unconventional spectrum and absorption features, MWDs with high field strengths have a tendency to be labeled as interesting objects (e.g., BLUE_SERENDIPITY, which is an object in undocumented coordinates with blue colors). This contributes as a positive bias for MWDs with high magnetic field strengths ($B > 100\,\mathrm{MG}$). On the other side, seven objects from the HYPERMUCHFUSS contributed to the $B \leq 3.5\,\mathrm{MG}$ range . This limit was noted to be a negative bias by Vanlandingham et al. (2005), due to the insensitivity of the *autofit* process. Since spectra from objects with such low fields only slightly differ from the non-magnetic cases, they were not found by the automatic fitting process and therefore no visual inspection was triggered. Since some HYPERMUCHFUSS objects were found in DR1 to DR3, the radial velocity method has proven to be complementary in finding some of the missing low-field MWDs.

These selection effects do not apply to the second phase of the SDSS, Sloan Extension for Galactic Understanding and Exploration (SEGUE), which is targeting the stellar population specifically. This includes the white dwarfs and blue horizontal-branch stars[2] (that are also in MWD color criteria). In our sample, five out of the nineteen new MWDs are from SEGUE.

5. Conclusion

Overall, our results are consistent with the former analyses of MWDs (see Fig. 3), which shows that simple atmosphere models with preassumed dipole magnetic values are good approximations for these objects, if only single-phase spectroscopy without polarization information is considered. Our method is able to account for a diversity in low magnetic field strengths ($B < 20\,\mathrm{MG}$). This is apparent in the deviation of our best-fit magnetic field strength values from literature values in the horizontal ~2 MG axis of Figure 3. In many cases offset dipole models resulted in significantly better fits than the models with centered dipoles, however, at smaller field strength this can be in part due to the simplifications in the treatment of Stark broadening.

The distribution of the magnetic field strengths of the MWDs from the Schmidt et al. (2003) sample had a maximum around ~5 - 30 MG. We have updated the values of DA white dwarfs and created a histogram of all known magnetic DA WDs (Fig. 3) and added results for the new objects. The magnetic field strengths of non-SDSS MWDs are from the compilation by Kawka et al. (2007). In spite of the SDSS biases, the same ~5 - 30 MG peak as in Schmidt et al (2003) is apparent in Fig. 3.

The dipole magnetic field ohmic decay timescale is 10^{10} yr. Even the higher multipoles can live for such a long period of time (see, e.g. Muslimov et al. 1995). Therefore, no significant correlation between temperature and magnetic field strength is expected if temperature is assumed as an indicator of age (Fig. 4). This lack of correlation supports the fossil ancestry of these fields inherited from earlier stages of stellar evolution.

High field MWDs are thought to be remnants of magnetic Ap and Bp stars. If flux conservation is assumed, the distribution of the polar field strengths of high field MWDs should be largest in the interval 50–500 MG. In our sample, objects with magnetic field strengths lower than 50 MG are more numerous than the objects with higher magnetic field strengths (see Fig. 3), part of this effect is due to our biases (see Section 2). Nevertheless it is consistent with previous results and supports the hypothesis that magnetic fossil fields from Ap/Bp stars only are not sufficient to produce high field MWDs (Wickramasinghe & Ferrario 2005). Aurière et al. (2007) argued that dipole magnetic field strengths of magnetic Ap/Bp stars have a "magnetic threshold" due to large scale stability conditions, and this results in a steep decrease in the number of magnetic Ap/Bp stars below polar magnetic fields of 300 G.

A possible progenitor population for MWDs with dipolar field strengths below 50 MG is the currently unobserved population of A and B stars with magnetic field strengths of 10 - 100 G. Wickramasinghe & Ferrario (2005) suggested that if ~ 40% of A/B stars have magnetism, this would be sufficient to explain the observed distribution of MWDs. However, the existence of this population seems to be highly unlikely since investigations of Shorlin et al. (2002) and Bagnulo

[2] http://segue.uchicago.edu/targetsel.html#v4.2

et al. (2006) for magnetism in this population yielded null results, for median errors of 15 - 50 G and 80 G respectively. Another candidate group is the yet undetected magnetic F stars for these MWDs with lower field strengths (Schmidt et al. 2003). But this conclusion is strongly affected by SDSS MWD discovery biases.

Acknowledgments

Part of this work is supported by the DLR under grant 50 OR 0802. Funding for the creation and distribution of the SDSS archive has been provided by the Alfred P. Sloan Foundation, the SDSS member institutions, the National Aeronautics and Space Administration, the National Science Foundation, the U.S. Department of Energy, Monbusho, and the Max Planck Society. The SDSS World Wide Web site is http://www.sdss.org/.

References

Aurière M et al. 2007 *A&A* **475** 1053
Bagnulo S Landstreet J D Mason E Andretta V Silaj J & Wade G A 2006 *A&A* **450** 777
Burleigh M R et al. 2007 in "Proc. of the 15th European Workshop on White Dwarfs" *Astronomical Society of the Pacific Conference Series* **372** Eds.: Napiwotzki R & Burleigh M R 477
Eisenstein D J et al. 2006 *ApJS* **167** 40
Euchner F Jordan S Beuermann K Gänsicke B T & Hessman F V 2002 *A&A* **390** 633
Euchner F Reinsch K Jordan S Beuermann K & Gänsicke B T 2005 *A&A* **442** 651
Euchner F Jordan S Beuermann K Reinsch K & Gänsicke B T 2006 *A&A* **451** 671
Gänsicke B T Euchner F & Jordan S 2002 *A&A* **394** 957
Harris H C et al. 2003 *AJ* **126** 1023
Holberg J B & Bergeron P 2006 *AJ* **132** 1221
Jordan S 1992 *A&A* **265** 570
Jordan S 2001 in "Proc. of the 12th European Workshop on White Dwarfs" *Astronomical Society of the Pacific Conference Series* **226** Eds.: Provencal J L Shipman H L, MacDonald J & Goodchild S 269
Jordan S & Schmidt H 2003 in "Stellar Atmosphere Modeling" (*Astronomical Society of the Pacific Conference Series* **288** Eds.: Hubeny I Mihalas D & Werner K 625
Kawka A Vennes S Schmidt G D Wickramasinghe D T & Koch R 2007 *ApJ* **654** 499
Kleinman S J et al. 2004 *ApJ* **607** 426
Külebi B Jordan S Euchner F Hirsch H & Gänsicke B 2009 in preparation
Latter W B Schmidt G D & Green R F 1987 *ApJ* **320** 308
Liebert J Bergeron P & Holberg J B 2003 *AJ* **125** 348
Muslimov A G van Horn H M & Wood M A 1995 *ApJ* **442** 758
Putney A 1995 *ApJ* **451** L67
Rechenberg I 1994 *Werkstatt Bionik und Evolutionstechnik No. 1 (Stuttgart: frommann-holzboog)*
Schmidt G D et al. 2003 *ApJ* **595** 1101
Shorlin S L S Wade G A Donati J F Landstreet J D Petit P Sigut T A A & Strasser S 2002 *A&A* **392** 637
Stoughton C et al. 2002 *AJ* **123** 485
Vanlandingham K et al. 2005 *AJ* **130** 734
Wickramasinghe D T & Ferrario L 2000 *PASP* **112** 873
Wickramasinghe D T & Ferrario L 2005 *MNRAS* **356** 1576

Journal of Physics: Conference Series **172** (2009) 012048 doi:10.1088/1742-6596/172/1/012048

Non-LTE spectral analyses of the lately discovered DB-gap white dwarfs from the SDSS

S D Hügelmeyer and S Dreizler

Institut für Astrophysik, Georg-August-Universität Göttingen, Friedrich-Hund-Platz 1, 37077 Göttingen, Germany

E-mail: shuegelm@astro.physik.uni-goettingen.de

Abstract. For a long time, no hydrogen-deficient white dwarfs have been known that have effective temperature between 30 kK and < 45 kK, i. e. exceeding those of DB white dwarfs and having lower ones than DO white dwarfs. Therefore, this temperature range was long known as the DB-gap. Only recently, the SDSS provided spectra of several candidate DB-gap stars. First analyses based on model spectra calculated under the assumption of local thermodynamic equilibrium (LTE) confirmed that these stars had 30 kK < $T_{\rm eff}$ < 45 kK (Eisenstein et al. 2006). It has been shown for DO white dwarfs that the relaxation of LTE is necessary to account for non local effects in the atmosphere caused by the intense radiation field. Therefore, we calculated a non-LTE model grid and re-analysed the aforementioned set of SDSS spectra. Our results confirm the existence of DB-gap white dwarfs.

1. Introduction

The cooling sequence of hydrogen-deficient white dwarfs (WDs) is populated by different subtypes of this species. At $T_{\rm eff} > 45$ kK, stars are called DO WDs. These objects display mainly He II at high and a combination of He II and He I lines at lower temperatures. Stars with spectra showing only He I lines are called DB WDs. Prior to the data releases of the Sloan Digital Sky Survey (SDSS), the hottest DB white dwarf (WD) known was PG 0112+104 with $\sim 30\,000$ K. The coolest DO WD prior to the SDSS on the other hand was PG 1133+489 with 47 500 K (Wesemael et al. 1985). These two stars constituted the cool and hot end of the so-called DB-gap (Liebert et al. 1986), a temperature region with a deficiency in helium-rich white dwarfs. The SDSS survey was rich in new white dwarfs and brought forth a large number of DB and DO white dwarfs. In the paper of Eisenstein et al. (2006), the authors report a considerable number of H-deficient WDs with temperatures that put them in the DB-gap. To derive WD temperatures, they used model atmospheres assuming local thermodynamic equilibrium (LTE). Dreizler & Werner (1996) showed that even for stars with $T_{\rm eff} = 50$ kK non-local effects are still important. Therefore we re-analysed the sample of Eisenstein et al. (2006) with our non-LTE model spectra.

2. Models and fitting

For our non-LTE spectral analysis, we have calculated a grid of model atmospheres with the following parameters:

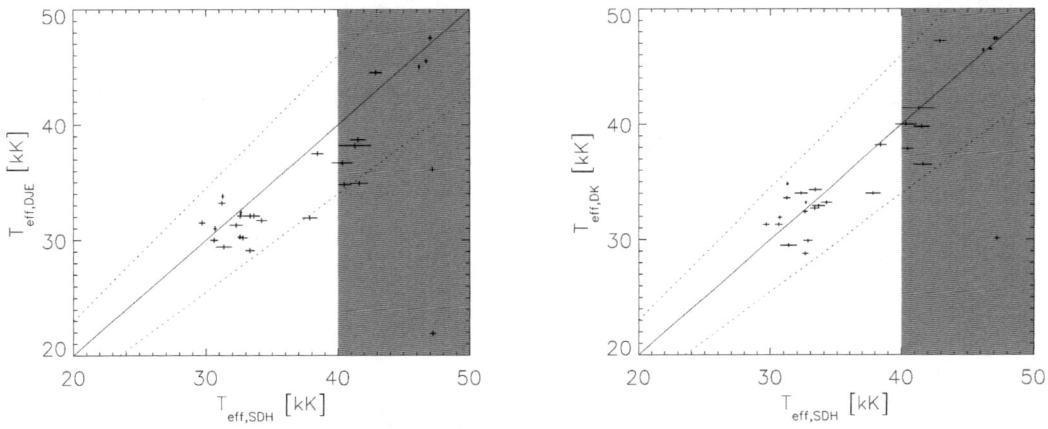

Figure 1. Comparison of LTE effective temperatures from the Eisenstein (left panel) and Koester (right panel) fits to our NLTE fits for all analysed WDs. The Eisenstein fits show systematically lower temperatures than the Koester fits. The latter ones agree better with our fitting results. The shaded area is the temperature range where we expect deviations from LTE, i. e. differences in fitted temperatures. The dotted lines are 15% deviations from equality (solid lines).

$T_{\rm eff}$:	27 500 – 50 000 K	steps of 2 500 K
$\log g$:	7.60 – 8.80 (cgs)	steps of 0.20 dex
X(He) :	99.0 & 99.9	

Our fitting method applied is the same as described in Hügelmeyer et al. (2006). It is similar to the one used by Koester in Eisenstein et al. (2006). We first normalized the SDSS spectrum by determining the continuum points of the observations by the continuum of the normalized model spectrum and then fitted a third order polynomial to the double logarithmic data. We then applied a χ^2-fit to the normalized models in certain wavelength regions, i. e. 3900 Å – 5100 Å, 5800 Å – 6000 Å, and 6500 Å – 6800 Å.

3. Results & discussion

Our best fit models are shown in Figure 2 and the photospheric values are presented in Table 1. The comparison of our NLTE model fits to the temperature values obtained with LTE Koester models is shown in Figure 1. Our derived temperatures are in better agreement with the fitting method of Koester. This is not very surprising since the procedures are very similar. The Eisenstein `autofit` method fits photometric data additionally to the spectroscopy.

The left panel plot in Figure 1 shows that our fitting method, especially in the high temperature regime, produces higher $T_{\rm eff}$ values than the Eisenstein method. However, all but two fits – one of which is for an object with a significant amount of hydrogen and therefore cannot be reliably fit with our He-rich models– are within the 15% deviation range. The Koester fits are in even better agreement with our model fitting. Direct comparison of our non-LTE to the LTE Koester models show some obvious deviations in the continuum flux which increase towards lower temperatures. Currently, we do not have an explanation for these discrepancies.

Overall, we can confirm the results of Eisenstein et al. (2006). Especially the nice agreement of our fits to those of Koester at effective temperatures > 45 K suggests that LTE is a valid assumption for the analysis of DB-gap white dwarfs.

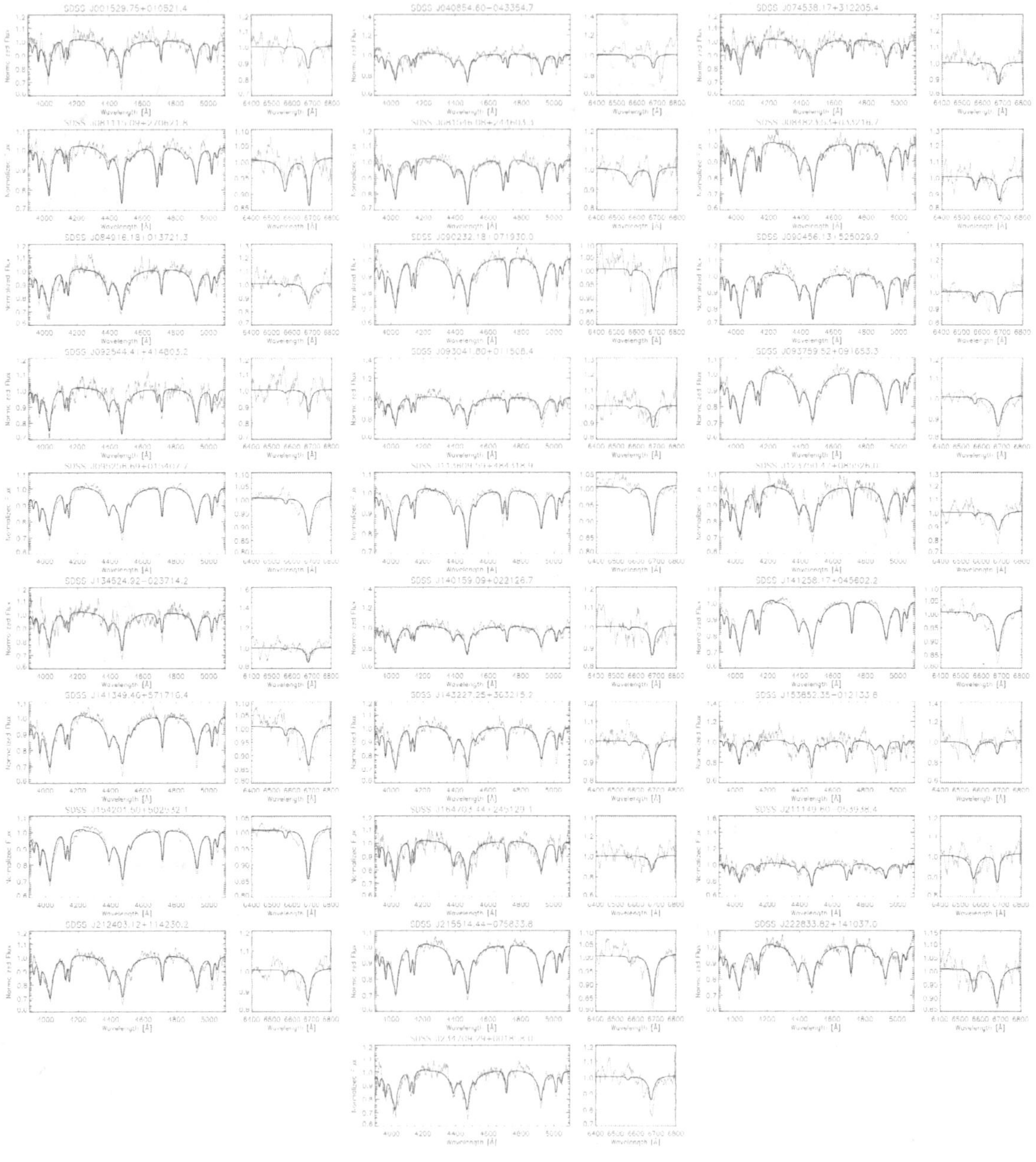

Figure 2. SDSS spectra (grey lines) and best-fit models (black lines). SDSS J153852.35-012133.8 clearly shows hydrogen Balmer lines and cannot be fit with our model grid. Instead the fitting method tries to compensate the missing hydrogen in the model by increasing effective temperature to fill Hα with HeII. For clarity the SDSS spectra are smooth for these plots.

Acknowledgments
SDH is supported by a scholarship of the DFG Graduiertenkolleg 1351 "Extrasolar Planets and their Host Stars"

Table 1. Best-fit model values for the analyzed SDSS spectra. SDSS J153852.35-012133.8 clearly shows hydrogen Balmer lines and cannot be fit with our model grid. The real effective temperature is lower than the value given in this table. All errors are 1-σ statistical errors.

Name	$T_{\rm eff}$ [K]	$\log g$ (cgs)	X(He)	χ^2
SDSS J001529.75+010521.4	40325 ± 821	7.88 ± 0.07	99.9	1.1601
SDSS J040854.60−043354.7	41607 ± 697	7.76 ± 0.04	99.0	1.2157
SDSS J074538.17+312205.4	42890 ± 483	7.72 ± 0.06	99.9	1.2050
SDSS J081115.09+270621.8	47007 ± 168	7.91 ± 0.03	99.0	1.1289
SDSS J081546.08+244603.3	46692 ± 157	7.92 ± 0.04	99.0	1.1678
SDSS J084823.53+033216.7	33327 ± 495	7.75 ± 0.07	99.0	1.1150
SDSS J084916.18+013721.3	31347 ± 618	8.03 ± 0.04	99.9	1.1431
SDSS J090232.18+071930.0	32787 ± 360	7.96 ± 0.03	99.9	1.1815
SDSS J090456.13+525029.9	38435 ± 461	7.60 ± 0.01	99.0	1.2014
SDSS J092544.41+414803.2	41517 ± 596	7.83 ± 0.10	99.9	1.1468
SDSS J093041.80+011508.4	37850 ± 585	7.60 ± 0.01	99.9	1.1464
SDSS J093759.52+091653.3	29705 ± 247	8.23 ± 0.01	99.9	1.2879
SDSS J095256.69+015407.7	31257 ± 112	8.23 ± 0.01	99.9	1.2535
SDSS J113609.59+484318.9	46152 ± 90	8.17 ± 0.01	99.9	1.4391
SDSS J123750.47+085526.0	32270 ± 495	8.24 ± 0.03	99.9	1.1364
SDSS J134524.92−023714.2	41292 ± 1260	7.96 ± 0.03	99.9	1.2249
SDSS J140159.09+022126.7	40460 ± 461	7.99 ± 0.03	99.9	1.1346
SDSS J141258.17+045602.2	30695 ± 135	8.19 ± 0.01	99.9	1.3531
SDSS J141349.46+571716.4	30627 ± 292	8.17 ± 0.02	99.9	1.1883
SDSS J143227.25+363215.2	33305 ± 360	7.91 ± 0.04	99.9	1.1576
SDSS J153852.35−012133.8	47210 ± 225	7.60 ± 0.01	99.0	1.1996
SDSS J154201.50+502532.1	32630 ± 101	7.96 ± 0.01	99.9	1.4335
SDSS J164703.44+245129.1	34182 ± 427	7.60 ± 0.01	99.9	1.1683
SDSS J211149.60−053938.4	47165 ± 202	7.74 ± 0.05	99.0	1.1400
SDSS J212403.12+114230.2	32562 ± 202	7.99 ± 0.02	99.9	1.2352
SDSS J215514.44−075833.8	32585 ± 225	8.05 ± 0.02	99.9	1.1418
SDSS J222833.82+141037.0	31212 ± 270	8.32 ± 0.03	99.9	1.1198
SDSS J234709.29+001858.0	33597 ± 472	7.96 ± 0.05	99.9	1.2369

References

Dreizler S & Werner K 1996 *A&A* **314** 217
Eisenstein D J Liebert J Koester D et al. 2006 *AJ* **132** 676
Hügelmeyer S D Dreizler S Homeier D et al. 2006 *A&A* **454** 617
Liebert J Wesemael F Hansen C J et al. 1986 *ApJ* **309** 241
Wesemael F Green R F& Liebert J 1985 *ApJS* **58** 379

Journal of Physics: Conference Series **172** (2009) 012049 doi:10.1088/1742-6596/172/1/012049

Profiles for density and temperature dependence of the far red wing of the Lyman α line compared

N F Allard[1] and J F Kielkopf[2]

[1] Observatoire de Paris, GEPI, UMR 8111, CNRS, 61, Avenue de l'Observatoire, F-75014 Paris, France

[2] Department of Physics and Astronomy, University of Louisville, Louisville, KY, 40292, USA

E-mail: nicole.allard@obspm.fr

Abstract. A reliable determination of the line profiles of atomic hydrogen for physical conditions of cool white dwarfs requires a unified theory which takes into account both the singlet and triplet transitions contributing to Lyman alpha using accurate interaction potentials and radiative dipole transition moments. We show that multiple perturber effects have to be considered using the autocorrelation formalism. A comparison with experimental laboratory spectra of cool dense plamas also shows that indeed multiple H-perturbers are an important influence in the far wing extending to 3000 Å in the ultraviolet. Under cool dense white dwarf atmosphere conditions, multiple perturber collisions for the singlet states contribute more to the ultraviolet opacity than binary collisions of the triplet states.

1. Introduction

An accurate determination of the line broadening of the Lyman series of atomic hydrogen has been shown to be fundamental in the interpretation of UV and FUV spectra of DA white dwarfs. The pseudocontinuum due to the $b\ ^3\Sigma_u^+ \rightarrow a\ ^3\Sigma_g^+$ triplet transition which contributes to the red wing of the Lyman α line perturbed by neutral hydrogen was recently investigated by Kowalski (2006) and Kowalski & Saumon (2006) to improve previous calculations of atmosphere models of white dwarfs.

We will show that at the low temperature of cool white dwarfs the contribution of the singlet $X^1\Sigma_g^+ \rightarrow B^1\Sigma_u^+$ transition cannot be neglected in the calculation of Lyman α profile perturbed by neutral hydrogen. A comparison with experimental spectra also shows that indeed multiple H-perturbers contribute to the broadening that makes the far wing. The collisional effects must be treated by using the autocorrelation formalism described in Allard et al (1999).

2. Line profiles for T=10000K, variation with density of neutral hydrogen

Many of the problems in collision induced radiative transitions have been solved within the one-perturber approximation. At very low densities, the binary model for an optically active atom in collision with one perturber is valid for the whole profile except the central part of the line.

Figure 1 shows the prominent satellite due to interacting neutral hydrogen atoms in the $X\ ^1\Sigma_g^+$ ground electronic state making free-free transitions to the $B\ ^1\Sigma_u^+$ excited state.

In this work we use the most recent molecular data of Spielfiedel 2003, Spielfiedel et al 2004.

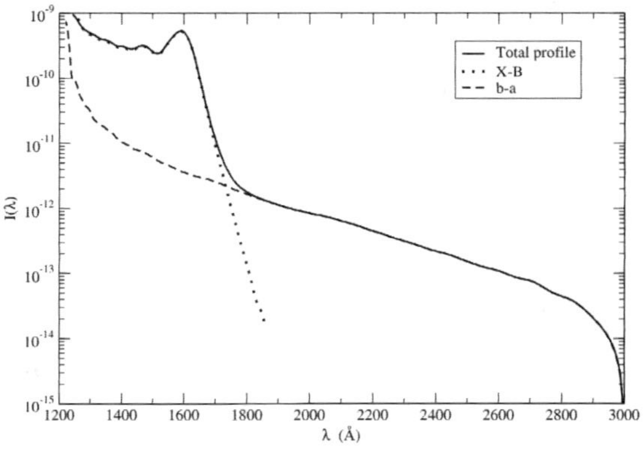

Figure 1. Contribution of the $X \rightarrow B$ and $b \rightarrow a$ transitions to the red wing for T=10000 K and n_H= 10^{17} atoms cm^{-3}.

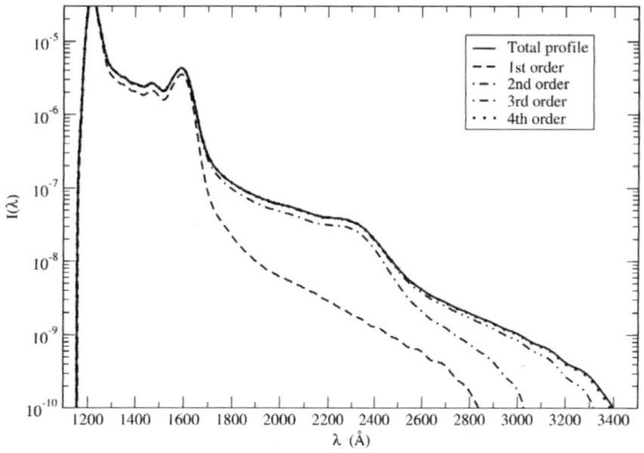

Figure 2. Contribution of the different orders of the expansion to the whole profile for T=10000 K and n_H= 10^{21} atoms cm^{-3}.

The contribution of the $X \rightarrow B$ transition is identical to the first order of the expansion shown in Fig. 1.

For this low density (n_H= 10^{17} atoms cm^{-3}) the one-perturber approximation is valid, and the total profile above 1800 Å is due to the triplet $b \rightarrow a$ transition.

The semi-classical methods also can be extended by unified theories of line broadening to permit an inclusion of the effects of multiple collisions (Allard 1082). Figure 2 shows the total profile which takes into account the 4 allowed transitions, the $X \rightarrow B$ and $X \rightarrow C$ singlet transitions, and the $b \rightarrow a$ and $b \rightarrow i$ triplet transitions.

The red wing is dominated by the contribution of multiple-perturber collisions and the expansion of the autocorrelation function in density (Royer 1971, 1978) has to be made to the fourth order to be comparable with the result obtained from a unified calculation.

When the density is as high as n_H= 10^{21} atoms cm^{-3}, the one-perturber binary approximation is no longer valid.

The total profiles shown in the figures are the Fourier transforms of the correlation function given by Eq. 121 of Allard (1999), in which the contributions from different components of the transition enter with their statistical weights. The figures also show the individual components for comparison, weighted as if they were the only contribution to the profile.

3. Variation with temperature

The appearance of an extended wing shown in Fig. 3 is sensitive to temperature because of the variation of the modulated transition dipole moment with temperature. Figure 4 illustrates that there is no contribution from triplets above 1900 Å (-30000 cm^{-1}) for T=5000 K because the modulated transition dipole moment is too weak.

Figure 5 shows that for 5000 K the $X \rightarrow B$ and $b \rightarrow a$ quasi-molecular transitions both contribute and have to be included.

Because of the weighting, the singlet $X \rightarrow B$ makes only 1/3 of the total line far wing strength. Nevertheless, its contribution through multiple perturber collisions dominates the 1800 to 3000 Å region.

Consequently, the ultraviolet triplet $b \rightarrow a$ transition does not, by itself, account for the far red wing of Lyman α profile broadened by collisions with atomic hydrogen. Kowalski & Saumon (2006) in their analysis of the missing opacity in cool dense white dwarf stars considered the triplet contribution but did not include multiple perturber effects.

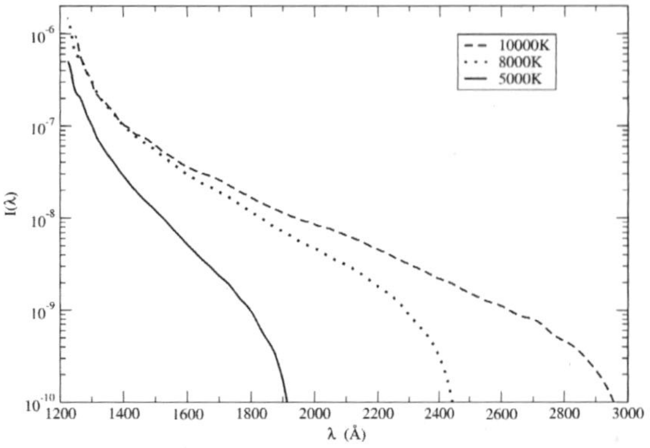

Figure 3. Variation of the extension of the individual profile due to the $b \rightarrow a$ transition in the red wing for T=10000 K, T=8000 K, T=5000 K for $n_H = 10^{21}$ atoms cm^{-3}.

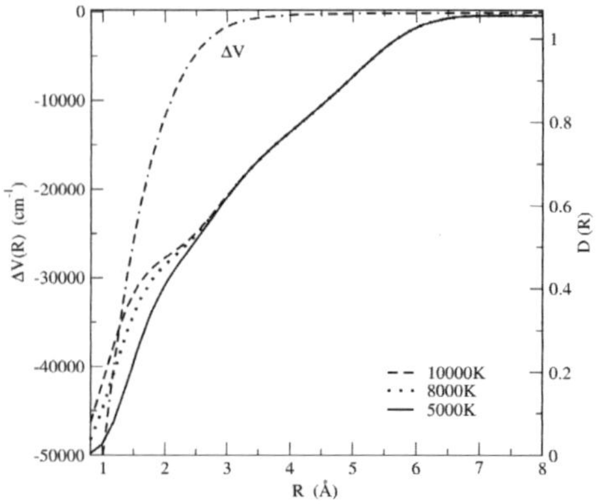

Figure 4. Variation with temperatures of the modulated dipole and the difference potential of the $b \rightarrow a$ transition.

4. Laboratory spectra

Satellites in the far wings of Lyman α are seen in the emission spectra of plasmas produced when a pulsed laser excites an H_2 gas target (Kielkopf & Allard, 1998). The laser-produced plasma sources in atmospheric density H_2 were shown to generate profiles of Lyman α that extend in the wing up to above 1600 Å. Those experiments cover conditions over which only binary collisions were important in the spectral regions shown.

Kielkopf & Allard (2000) described experiments at higher density extending farther into the Lyman α wing where the effects of more than one perturber acting simultaneously on the radiating atom appear.

The spectrum of Lyman α from a dense atomic hydrogen plasma exhibits a far wing that can be followed to 3000 Å as shown in Fig. 6. A shoulder in the wing above 2500 Å is evidence of multiple neutral H perturbers acting simultaneously on an excited $2p$ H atom.

5. Synthetic spectra of cool white dwarfs

Previous Lyman α line profile calculations were done at low temperatures and were included in atmosphere models by Wolff et al. (2002), and by Kowalski & Saumon (2006). Both papers presented a comparison of synthetic spectra to observed spectra of BPM4729. Unfortunately, neither paper shows the actual line profiles that were used. In Fig.3 of Wolff et al. (2002), the change of slope of the synthetic spectrum may be due to the second satellite. Since the transitions and the corresponding potentials which were used were not specified in Wolff et al.(2002), we cannot know if the $b \rightarrow a$ transition was included. Their predicted profile in a pure hydrogen atmosphere does not agree with the observation. Kowalski & Saumon (2006) were successful in reproducing the spectrum of BPM4729 from the ultraviolet to near IR. They used a line profile calculated in the quasi-static approximation for only the $b \rightarrow a$ transition in H-H collisions, and they also included collisions with H_2, but the transitions and symmetries for the corresponding H_3 potentials which have been taken into account in this work are not specified in their paper. We have seen in our work reported here and in Allard and Kielkopf (2009) that the $b \rightarrow a$ transition alone is inadequate to account for the spectrum above 1600 Å at elevated

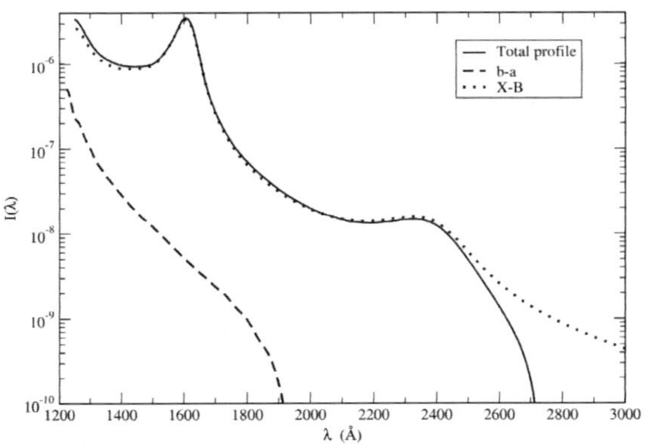

Figure 5. Contribution of the $X \rightarrow B$ and $b \rightarrow a$ transitions to the red wing for T=5000 K and $n_H = 10^{21}$ atoms cm^{-3}.

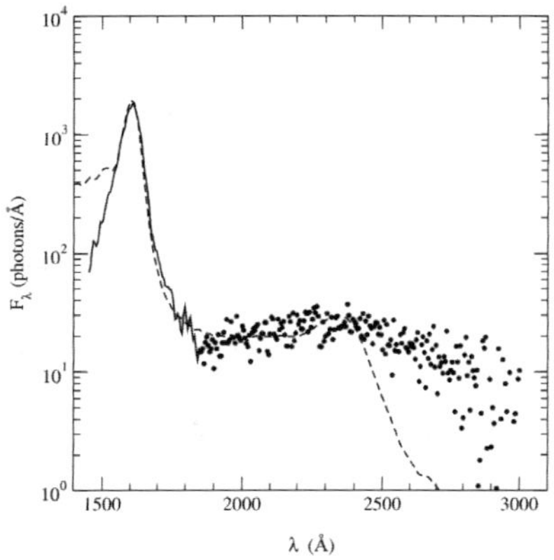

Figure 6. Comparison of the experimental profile to the theoretical one obtained for T=4100 K and $n_H = 2.5 \times 10^{21}$ at cm^{-3}.

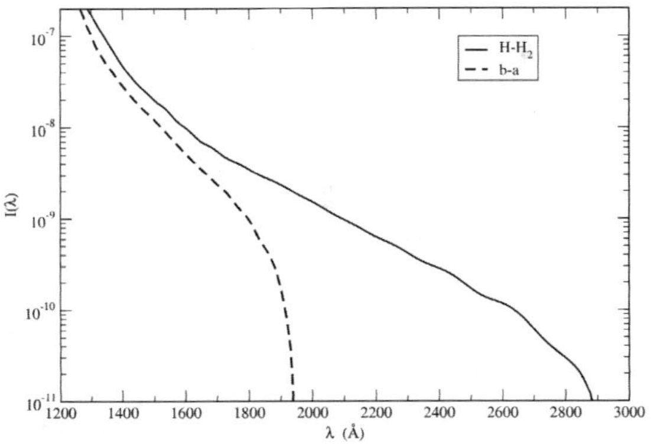

Figure 7. Comparison of the individual profile due to the $b \rightarrow a$ transition to the obtained H-H_2 profile using the potentials shown in Fig.1 of Kowalski & Saumon 2006 (T=5000 K and $n_H = 10^{21}$ atoms cm^{-3}).

atomic hydrogen densities. Moreover, we used the H_3 potentials shown in Figure 1 of their paper to compute far wing spectra using the autocorrelation formalism. Fig. 7 shows that the red wing that would be produced by these potentials does not extend beyond 3000 Å.

References

Allard N F & Kielkopf J F 1982 *Rev. Mod. Phys.* **54** 1103
Allard N F & Kielkopf J F 2009 *A&A* **493** 1155
Allard N F Royer A Kielkopf J F & Feautrier N 1999 *Phys. Rev. A* **60** 1021
Kielkopf J F & Allard N F 1998 *Phys. Rev.* A **58** 4416
Kielkopf J F & Allard N F 2000 *Connecticut meeting of the APS, Division of Atomic, Molecular, and Optical Physics* June 14-17
Kowalski P M 2006 *ApJ* **651** 1120
Kowalski P M & Saumon D 2006 *ApJ* **651** L137
Royer A 1971 *Phys. Rev.* A **3** 2044
Royer A 1978 *Acta Phys. Pol.* A **54** 805
Spielfiedel A 2003 *J. Mol. Spectrosc.* **217** 162
Spielfiedel A Palmieri P & Mitrushenkov A O 2004 *Molec. Phys.* **102** No. 21 2249
Wolff B Koester D & Liebert J 2002 *A&A* **385** 995

FUSE observations of intermediate temperature DA: atmospheric parameters and metal abundances

J Dupuis[1], V Hénault-Brunet[2], P Chayer[3], S Vennes[4] and J W Kruk[5]

[1] Agence Spatiale Canadienne,6767 Route de l'Aéroport, Longueuil, Québec J3Y 8Y9, Canada
[2] Département de Physique, Université de Montréal, C.P. 6128, Succ. Centre-Ville, Montréal, Québec H3C 3J7, Canada
[3] Space Telescope Science Institute, 3700 San Martin Drive, Baltimore, MD 21218, USA
[4] Department of Physics and Space Sciences, Florida Institute of Technology, Melbourne, FL 32901, USA
[5] Department of Physics and Astronomy, The Johns Hopkins University, 3400 N. Charles St., Baltimore, MD 21218, USA

E-mail: jean.dupuis@asc-csa.gc.ca, henault@astro.umontreal.ca, chayer@ststci.edu, svennes@fit.edu, kruk@pha.jhu.edu

Abstract. We present results from our analysis of a sample of DA white dwarfs having effective temperature below 25,000 K observed with the *FUSE* satellite with the goals of better understanding the origin of metals detected in the atmosphere of these stars. When possible, we combine the *FUSE* spectra with the *IUE* spectra and determine atmospheric parameters by fitting the Lyman line profiles. In general we find a good agreement with published values based on fits of the Balmer series. We observe that the continuum in the blue wing of the Lyman α line profile is generally lower in comparison with model spectra and that the discrepancy appears to become less important at higher effective temperature. The agreement between models and observations is excellent at wavelengths shorter than 1100 Å, which gives us confidence in the determination of atmospheric parameters. Finally, using adopted atmospheric parameters; we have performed a detailed analysis of the composition of these stars. In several instances, we have observed the presence of silicon and in one case that of carbon. For each star in the sample we have either measured or set an upper limit on the presence of key species such as C II, C III, Si III, and Si IV. We then compare the measured abundances with equilibrium abundance predicted by radiative levitation theory for each star. In this limited sample, we find that when detected, the abundance of silicon is in good agreement with theory. However there are several cases where the upper limits are smaller than the predictions and one case where it is considerably larger.

1. Introduction

White dwarfs are generally believed to have metal-poor atmospheres because of short timescales for gravitational settling at their surface. Radiative levitation can effectively offset gravitational settling down to effective temperatures as low as 20,000 K according to Chayer et al. (1995). In view of the recent discoveries of metal lines in white dwarfs cooler than 20,000 K by Zuckerman et al. (2003), it is of interest to assess the role played by radiative levitation in intermediate temperature white dwarfs

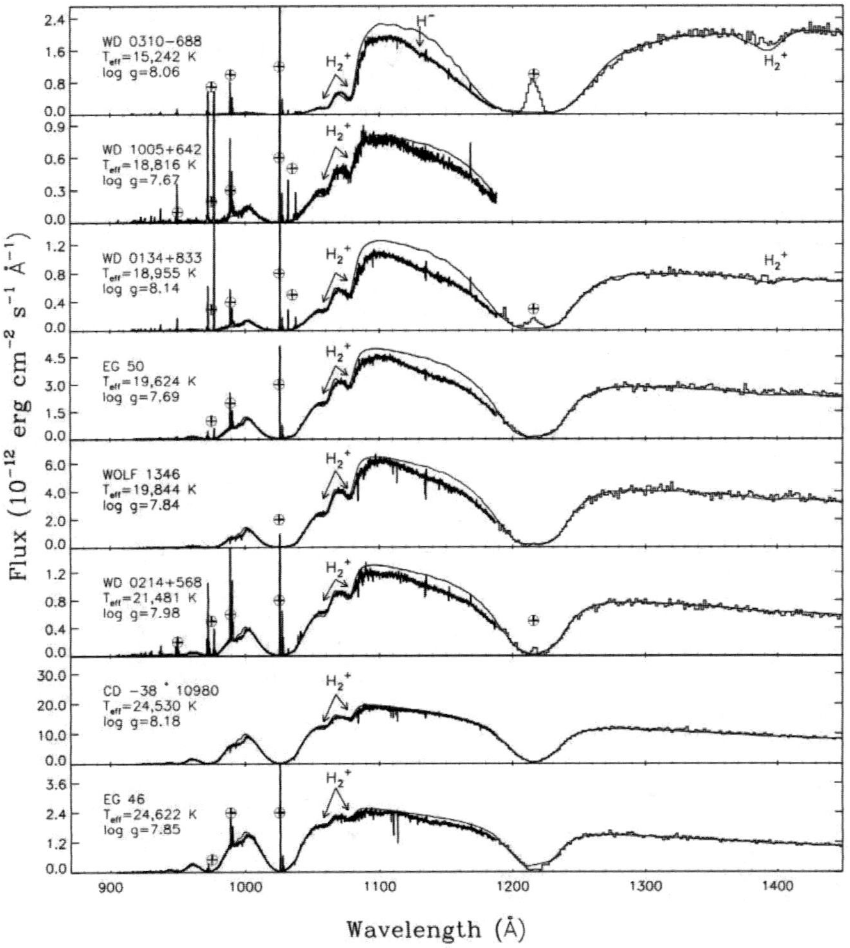

Figure 1. Combined *FUSE* and *IUE* DA spectra (in histogram) compared to the best-fit models (full black line). Airglow emission lines are identified by the Earth symbol. When visible, H^- absorption features and quasi-molecular H_2^+ absorption features are identified. Stellar parameters determined from Lyman series fit are shown on the left side of each graph.

2. Analysis

We observed several DA white dwarfs with temperature lower than 25,000 K down to 12,000 K with the *FUSE* satellite under guest observer time. We also augmented the sample with observations of other objects available in the *FUSE* archive. When available, low dispersion *IUE* spectra of these stars were combined with the *FUSE* spectra.

The Lyman series profiles are a powerful diagnostic to measure the stellar parameters of DA white dwarfs. We therefore fitted a grid of model spectra to the observed far-UV spectra (see Fig.1 for a representative subset) of our sample of stars in order to determine these atmospheric parameters and compare the results with those obtained from fits of the Balmer series profiles. In general, we find a satisfactory agreement between the two techniques, although some inconsistencies still persist in the blue wing of Lyman-α.

With these validated stellar parameters, we computed grids of synthetic spectra for key species likely to be present in the far-UV spectrum of white dwarfs, and used them to measure or set upper limits on the photospheric abundance of C, Si, P, and S. Silicon is detected in four of the stars of the sample (see best-fit spectra in Fig. 2 and Fig. 3) and also carbon in one of them.

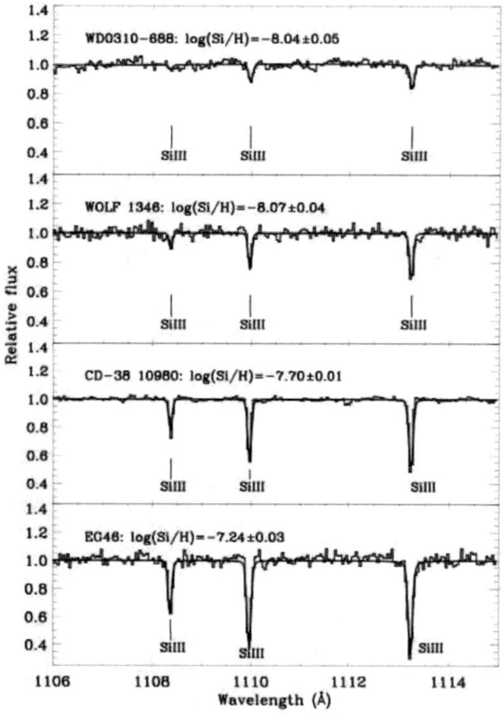

Figure 2. Sections of far-UV spectra showing detected Si III lines (in histogram) and the best-fit synthetic spectra (full curve).

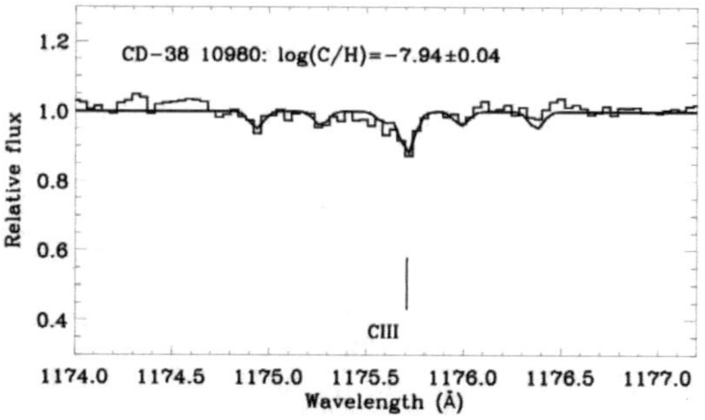

Figure 3. Best-fit of the C III λ1175 multiplet in the *FUSE* spectrum of CD-38° 10980.

3. Discussion

o Figure 4 shows the upper limits and abundances determined with the method discussed above for the larger sample of DA considered in our study. These results are compared with the equilibrium abundance predicted by radiative levitation theory (Chayer et al. 1995).

In the few stars where silicon is detected, there is relatively good agreement with the theory. However, when silicon is not detected, our upper limits often fall well below the predicted level. Although phosphorus is expected to be supported at a similar level, it is not detected in any of the stars in this sample, despite stringent upper limits in some cases. These very low upper limits call for the interplay of a mechanism capable of depleting the atmospheric reservoir of silicon and phosphorus. An hypothetical wind may be an interesting process to consider over evolutionary timescales, even though there is no strong evidence for significant mass loss in white dwarfs in this intermediate temperature range.

Carbon is detected in the spectrum of CD-38° 10980, well in excess of the abundance predicted by radiative levitation theory if we assume a photospheric origin for the detected carbon. We note

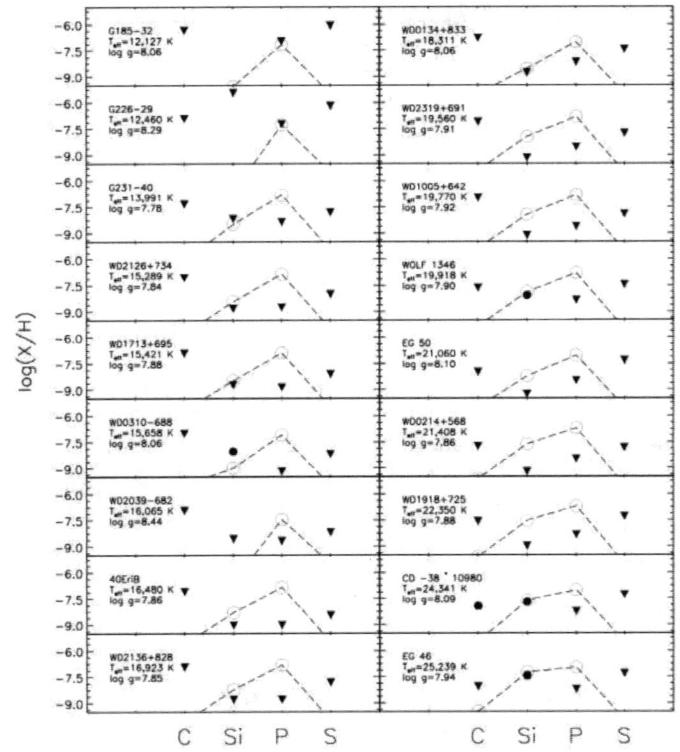

Figure 4. Logarithm of the number density of C, Si, P, and S relative to hydrogen for a sample of DA stars, assuming a photospheric origin. Upper limits are represented by upside down triangles and abundance measurements are represented by filled circles. Empty circles joined by dashed lines represent the abundances predicted by the equilibrium radiative levitation theory.

that the presence of carbon in the *FUSE* spectrum of CD-38° 10980 went unnoticed by Wolff et al. (2001). In case of an excess abundance with respect to radiative levitation predictions, we may explain the carbon abundance by accretion but this may ask for a peculiar abundance pattern of elements for the accreted material. It has also been proposed in Holberg et al. (1995) that the origin of carbon and silicon in this star may be circumstellar . For the other stars, the upper limits on the carbon abundance are not stringent enough to confront the theory.

Acknowledgments

JD wish to thank the Canadian Space Agency for its financial support to participate to this conference. We also thank NASA for supporting this project through *FUSE* guest investigators grants.

References

Chayer P Fontaine G and Wesemael F 1995 *ApJS* **99** 189
Holberg J B Bruhweiler F C and Andersen J 1995 *ApJ* **443** 753
Wolff B Kruk J W Koester D Allard N F Ferlet R and Vidal-Madjar A 2001 *A&A* **373** 674
Zuckerman B Koester D Reid I N and Hünsch M 2003 *ApJ* **596** 477

Journal of Physics: Conference Series **172** (2009) 012051 doi:10.1088/1742-6596/172/1/012051

Non-thermal electromagnetic radiation from pulsating and collapsing magnetized white dwarfs

Volodymyr Kryvdyk

Dept. Astronomy, Kyiv National University,
av. Glushkova 2/1, Kyiv 03022, Ukraine

E-mail: kryvdyk@univ.kiev.ua

Abstract. The non-thermal electromagnetic radiation from pulsating and collapsing magnetized white dwarfs is considered. This radiation generates when the star magnetosphere compresses during collapse and its magnetic field greatly increases. The electric field produced involves acceleration of charged particles, which generates radiation when moving in the magnetic field. Thus the pulsating and collapsing magnetized white dwarfs can be powerful sources of electromagnetic pulses. These pulses can be observed by means of modern instruments (radio, X- and gamma- telescopes).

1. Introduction

During their collapse stars emit gravitational and electromagnetic waves. In Dar et al. (1992), the acceleration non-relativistic nuclei and the generation gamma-ray bursts from accretion-induced collapse white dwarfs to neutron star was considered. As follows from this paper results, accretion induced collapse may be an important source of cosmic ray and cosmological gamma-ray bursts. De Couveia Dal Pino and Lazarian (2001a,b) have proposed that ultra high-energy cosmic rays observed above the Greisen-Zatsepin-Kuzmin (GZK) limit could be mostly protons accelerated in reconnection sites just above the magnetosphere of newborn millisecond pulsars, originated by accretion-induced collapse. Dermer and Atoyan (2006) considered the collapse of neutron stars to black holes in binary systems as a model for short gamma ray bursts. They found that the accretion of $\approx 0.1 - 1\ M_{\odot}$ of material by a neutron star through Roche lobe overflow of its companion, or through white-dwarf/neutron-star coalescence in a low mass binary system, could be enough to exceed the critical mass of a neutron star and trigger its collapse to a black hole, leading to the production of a short gamma-ray burst. Henriksen et al. (1979) calculated the dipole radiation from an exploding (or collapsing), homogeneous, uniformly rotating spheroid. They found that $\sim 2.4 \times 10^{40}$ ergs are radiated by an object with a mass of $1.4\ M_{\odot}$, an initial magnetic field of 10^8 gauss, and an initial density of 10^9 g cm^{-3} collapsing to a black hole. The radiation frequency is ~ 1 kHz, and such low frequency radiation cannot be observed directly near Earth. MacFadyen and Woosley (1999) explored the evolution of rotating helium stars with mass $M_s > M_{\odot}$, in which iron-core collapse does not produce a successful outgoing shock but instead forms a black hole. These are the best candidates for producing gamma-ray bursts (GRBs). Authors studied the formation of an accretion disk and the strong relativistic outflow jets in the polar regions. These outflows, powered by viscous dissipation in the disk, have energy of up to a few times 10^{51} erg/s.

The electromagnetic pulse from final gravitational stellar collapse is computed by Morley and Schmidt (2002). In this paper they calculated the electromagnetic energy radiated by stellar objects that bounce and become stable neutron stars, and stellar objects so massive that they become black holes. Due to the peak of the spectral curve around the wavelength 2 km, the receiver will have to be a satellite able to detect a broadband spectrum within the electromagnetic pulse. The theoretical predictions for the signatures of the electromagnetic radiation emitted during the process of the gravitational collapse of a stellar core to a black hole are studied by Ruffini at al. (2005). The last phases of this gravitational collapse are studied, leading to the formation of a black hole with a subcritical electromagnetic field, and an outgoing pulse of an initially optically thick e+e−-photon plasma. Such a pulse reaches transparency at Lorentz gamma factors of 102–104. Authors find a clear signature in the outgoing electromagnetic signal.

In this paper we consider the generation of non-thermal radiation from collapsing magnetized white dwarfs with heterogeneous magnetospheres. This radiation will be generated when the stellar magnetosphere is compressed during the collapse and its magnetic field increases considerably. A cyclic electric field is produced and the charged particles will accelerate. Moving in the magnetic field, these particles will generate radiation. The frequencies of this radiation span a wide range (from gamma-rays to radio waves) and therefore they can be detected near Earth.

2. Magnetosphere of a collapsing star

The external electromagnetic fields of a collapsing white dwarf will change as (Ginzburg 1964, Kryvdyk 1999)

$$B(r,\theta,R) = (1/2)F_0 R r^{-3}(1+3\cos^2\theta)^{1/2},$$
$$E_\varphi = -\frac{1}{cr^2}\frac{\partial\mu}{\partial t}\sin\theta. \tag{1}$$

where θ is the polar angle, $R(t)$ the radius of the collapsing star, $\mu(t) = (1/2)R_0 B_0^2 R(t)$ the magnetic momentum of the collapsing star.

The particle energy will change as a result of two mechanisms: 1) a betatron acceleration in the variable magnetic field and 2) bremsstrahlung energy losses in this field.

The equation of particle transitions in the regular magnetic field of the heterogeneous magnetosphere was solved by Kryvdyk (2005) for two special cases: (i) when energy losses do not influence the particle spectrum in the magnetosphere and (ii) when the energy losses determine the particle spectrum.

The solution of this equation, for the first case and for three initial particle heterogeneous distributions in the magnetosphere (power-series (P), relativistic Maxwell (M) and Boltzmann (B) distributions) is given by Kryvdyk (2005)

$$N^i{}_P(E,R,r) = K_P r_*^{-3} E_*^{-\gamma} R_*^{-\beta_P}, \tag{2}$$
$$N^i{}_M(E,R,r) = K_M r_*^{-3} E_*^{2} R_*^{-\beta_M} e^{-E/kT} \tag{3}$$
$$N^i{}_B(E,R,r) = K_B r_*^{-3} R_*^{-\beta_B} e^{-E/kT} \tag{4}$$

Here $E_* = E/E_o$; $R_* = R_0/R$; $r_* = r_o/r$; $\beta_P = A_1(\theta)(\gamma-1)$; $\beta_M = A_1(\theta)(E/kT\ln E_* - 3)$; $\beta_B = A_1(\theta)(E/kT\ln E_* - 1)$; $A_1(\theta) = (5/3)k_1(3\cos^4\theta + 1.2\cos^2\theta - 1)(1+3\cos^2\theta)^{-2}$;

$k_1 = 2$ and $k_1 = 1$ for relativistic and non-relativistic particles respectively.

Eqs. (2) - (4) determine the particle spectrum in the magnetosphere and its evolution during collapse for the initial stage of the collapse, when the energy losses can be neglected. We will consider this case in this paper. During the final collapse, the magnetic field grows to extreme values, and the energy losses will influence the particle spectrum considerably. This case will be consider in a future paper.

3. Non-thermal emission from collapsing magnetized white dwarfs

The ratio $I_{*\nu} = I_\nu / I_{\nu 0}$ between the radiation flux I_ν from collapsing stars with radius R and its initial radiation flux $I_{\nu 0}$ (by $R = R_0$) for the power-series (P), relativistic Maxwell (M) and Boltzmann (B) distributions are (Kryvdyk 2005)

$$I_{*P} = r_*^{-3}(\nu_*)^{(1-\gamma)/2} R_*^{\gamma-2} \int_0^{\pi/2}\int_0^{\infty} R_*^{-\beta_P} \sin\theta d\theta dE, \tag{5}$$

$$I_{*M} = r_*^{-3}\nu_* R_*^{-3}(kT)^{-1} \int_0^{\pi/2}\int_0^{\infty} R_*^{-\beta_M} e^{-E/kT} \sin\theta d\theta dE \tag{6}$$

$$I_{*B} = r_*^{-3}\nu_* R_*^{-3}(kT) \int_0^{\pi/2}\int_0^{\infty} R_*^{-\beta_B} E^{-2} e^{-E/kT} \sin\theta d\theta dE. \tag{7}$$

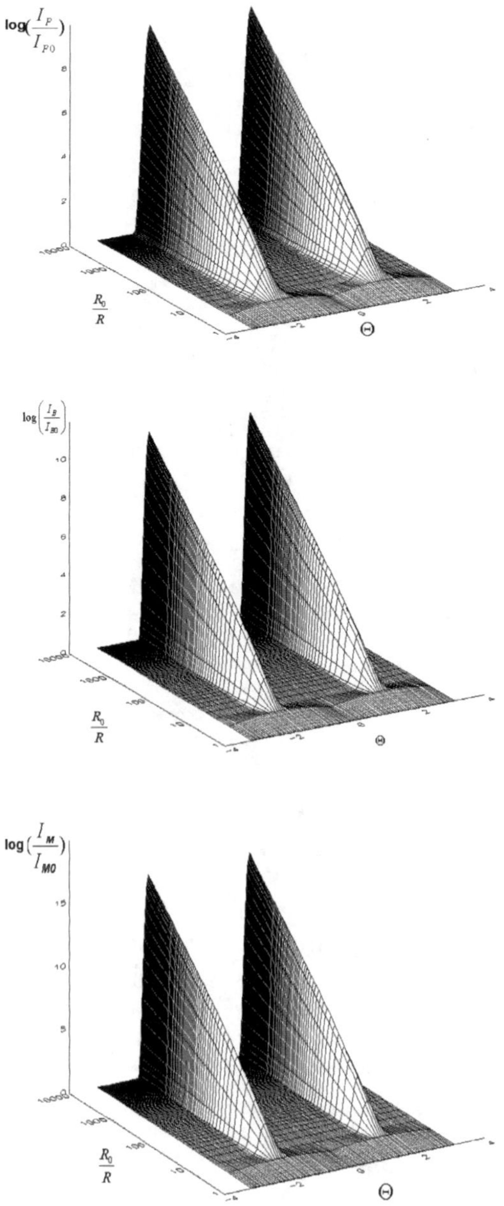

Figure 1. Pulses of the electromagnetic radiation from WD collapse for different R_0/R and θ

Using Eqs. (5) - (7), the radiation flux from the magnetospheres of collapsing stars with variable dipole magnetic fields can be calculated. The temporal changes of the flux radiation during collapse are shown in Fig.1. As follows from these results, the flux radiation increases during collapse, and during the final collapse stage the stars are very powerful sources of non-thermal radiation.

4. Conclusions

We can extract the following conclusions from the results obtained. The magnetic field will increase during the collapse. The charged particles will accelerate to relativistic energy in the magnetospheres of collapsing stars. These particles will emit electromagnetic waves in the wide frequency range from radio waves to gamma rays. The radiation from collapsing magnetized white dwarfs can be observed as electromagnetic pulses in a wide frequency range, from radio to gamma rays. The pulse duration is equal to the time of stellar collapse defined from the mass and the radius of the collapsing star. The intensity of this pulse is very strong. The radiation flux from collapsing stars exceeds the initial flux by factors of millions at the final stage of the collapse.

References

Dar A, Kozlovsky B Z, Nussinov S and Ramaty R 1992, *Astrophys.J.* **388** 164
De Gouveia Dal Pino E M and Lazarian A 2001a, *Astrophys. J.* **560** 358
De Gouveia Dal Pino E M and Lazarian A 2001b, *Proceedings of ICRC* **2105**
Dermer C D and Atoyan A 2006, *Astrophys. J.* **643** L13
Henriksen R N , Chau W Y and Chau K L 1979, *Astrophys. J.* **227** 1013
MacFadyen A I and Woosley S E 1999, *Astrophys. J.* **524** 262-289
Morley P D and Schmidt I 2002, *A&A* **384** 899
Ruffini R, Fraschetti F, Vitagliano L and She-Sheng Xue 2005, *Intern. Journ.Mod. Phys.D.* **14** 131
Ginzburg V L and Ozernoy L M 1964, *Zh.. Exper. i Theor. Fiz.(ZhEThF)* **47**, 1030 (In Russian).
Kryvdyk, V 1999, *Mon. Not. Roy. Astron. Soc.* **309** 593
Kryvdyk V and Agapitov A 2005, in *14th EW on White Dwarfs. ASP Conf. Series* **334** 277

Journal of Physics: Conference Series **172** (2009) 012052 doi:10.1088/1742-6596/172/1/012052

Strong linear polarization of bremsstrahlung emissivity in photospheres of magnetic white dwarfs

I I Bubukina and S A Koryagin

Institute of Applied Physics of the Russian Academy of Sciences,
46 Uljanova str., Nizhny Novgorod 603950, Russia

E-mail: `koryagin@appl.sci-nnov.ru`

Abstract. We discuss the strong linear polarization and the appreciable decrease of the bremsstrahlung emissivity at frequencies below the electron cyclotron frequency (infrared wave band) in the photospheres of the isolated magnetic white dwarfs. In the photospheres of strongly magnetized white dwarfs ($B > 10^7$ G, $T \sim 10^4$ K), the electron's Larmor radius becomes smaller than the characteristic impact parameter of close Coulomb collisions in a non-magnetized plasma. Thus, the cyclotron period of the electron becomes smaller than the duration of all distant collisions and of most close collisions. The magnetic field effectively "freezes" the electron motion in the plane transverse to the magnetic field lines. The resulting motion is nearly one-dimensional and parallel to the magnetic field, inducing a strong linear polarization of the bremsstrahlung emission. Being attached to a magnetic field line, an electron cannot approach an ion as closely as it does in the case in which the magnetic field is absent. Thus, the bremsstrahlung emissivity appreciably decreases. We analytically compute an approximation to the spectrum of the strongly linear polarized bremsstrahlung emissivity at the frequencies below the electron cyclotron frequency.

1. Introduction

In the photospheres of the strongly magnetized white dwarfs, the electron's Larmor radius

$$r_B = v_T/\omega_B \tag{1}$$

becomes smaller than the characteristic impact parameter of close Coulomb collisions in a non-magnetized plasma

$$r_\mathrm{s} = Ze^2/(m_\mathrm{e}v_T^2). \tag{2}$$

Here, $v_T = (k_\mathrm{B}T/m_\mathrm{e})^{1/2}$ is the thermal velocity of electrons, ω_B is the electron cyclotron frequency, m_e is the electron mass, e is the elementary charge, Z is the ion charge number, and k_B is the Boltzmann constant. The ratio

$$r_B \ll r_\mathrm{s} \tag{3}$$

is realized in magnetic fields $B > 10^7$ G at temperatures $T \sim 10^4$ K:

$$r_B/r_\mathrm{s} = 1{,}3Z^{-1}\,(T/10^4\ \mathrm{K})^{3/2}\,(B/10^7\ \mathrm{G})^{-1} < 1.$$

Under the condition of Eq. (3), the Coulomb collisions dramatically change in comparison with the collisions in a non-magnetized plasma (Zheleznyakov 1997; Zheleznyakov et al. 1999;

Koryagin 2000). The collision transport frequencies (which characterize the non-radiative relaxation processes) are determined by close collisions with impact parameters of the order or smaller than the distance

$$L_{\rm u} = (Zm_{\rm e}c^2/B^2)^{1/3}. \tag{4}$$

In these collisions, an electron transits the region near an ion for a time of the order or smaller than the cyclotron period. The distance of Eq. (4) coincides with the radius of a highly excited non-magnetized atom in which an electron orbits a nucleus for a time equal to the given electron cyclotron period. The distance given by Eq. (4) is smaller than the parameter $r_{\rm s}$ and is larger than the Larmor radius given by Eq. (1). In the case of close collisions, the electron trajectory becomes quasi-trapped. The particle returns to the ion many times before it overcomes the attraction of the scattering center (Zheleznyakov et al. 1999; Schmidt 2000; Hu et al. 2002a; Hu et al. 2002b; Correa et al. 2005).

In the case of distant collisions with impact parameters $p_{\rm h} \gg L_{\rm u}$, the ion Coulomb field accelerates an electron along a magnetic field line at the initial path to a scattering center and, then, deccelerates it down to the initial velocity at the final part of the trajectory. The component transverse to the magnetic field of the ion Coulomb field only causes a slow drift of the electron in the crossed electric and magnetic fields. The resulting change of the electron pitch-angle is exponentially small.

The qualitative change of the electron-ion collisions must influence the bremsstrahlung. We consider the bremsstrahlung under the criterion given by Eq. (3) at the frequencies below the electron cyclotron frequency:

$$\omega \ll \omega_B, \tag{5}$$

which fall into the infrared or optical wavebands for the strongly magnetized white dwarfs:

$$\lambda \gg 2\pi c/\omega_B = 11\,(B/10^7~{\rm G})^{-1}~\mu{\rm m}.$$

2. Polarization

It is known that at the frequencies which satisfy Eq. (5), a magnetic field significantly suppresses the high frequency plasma conductivity in a plane transverse to the magnetic field. This suppression does not require the qualitative change of collisions. It stems from the limited displacement of an electron in a direction transverse to the magnetic field plane. This displacement is smaller than twice the Larmor radius. Then, in accordance with the Kirchoff law (Bekefi, 1966), the bremsstrahlung emissivity for the electromagnetic waves polarized in a plane transverse to the magnetic field is also suppressed. As a result, the bremsstrahlung emissivity becomes strongly linear polarized (Gnedin & Pavlov 1974). The plane of polarization is parallel to the magnetic field vector and the wave vector. The beam pattern of the linear polarized bremsstrahlung emissivity corresponds to the radiation of a dipole directed along the magnetic field:

$$\eta_\omega(\omega,\Theta) = \eta_{\omega 0}(\omega)\sin^2\Theta, \tag{6}$$

where Θ is the angle between the wave vector and the magnetic field, $\eta_{\omega 0}$ is the spectral emissivity per unit solid angle in a direction transverse to the magnetic field.

3. Spectrum

We divide the considered frequencies into several sub-bands. At frequencies lower than the inverse time of the electron's transit of a region near the ion in a collision with the impact parameter given by Eq. (2):

$$\omega \ll \omega_{\rm s} = mv_T^3/(Ze^2), \qquad \lambda \gg 2\pi c/\omega_{\rm s} = 8\,(T/10^4~{\rm K})^{-3/2}~\mu{\rm m}, \tag{7}$$

the bremsstrahlung is determined by distant collisions with impact parameters $p_\mathrm{h} \gg r_\mathrm{s}$. In these collisions, the ion weakly disturbs the electron motion. However, the spectral emissivity of an electron diverges with the decrease of its velocity $v_\parallel$ along the magnetic field proportionally to $v_\parallel^{-1}$. As a result, the bremsstrahlung emissivity integrated over the Maxwell velocity distribution contains the logarithmic factor which coincides with the Coulomb logarithm without the magnetic field:

$$\eta_{\omega 0}^{(\mathrm{d})} = \frac{(\sqrt{2}/3)\, n_\mathrm{i} n_\mathrm{e} Z^2 e^6}{\pi^{3/2} m_\mathrm{e}^2 c^3 v_T} \ln\left(\frac{m v_T^3}{\omega Z e^2}\right). \tag{8}$$

This spectral emissivity for only one polarization in a direction transverse to the magnetic field is 2 times smaller than the analogous isotropic value in a non-magnetized plasma. The total emissivity integrated over all directions and summed over the two polarizations spectral power $8\pi\eta_{\omega 0}/3$ is 6 times smaller than the corresponding value in the absence of a magnetic field. The decrease of the bremsstrahlung emissivity in comparison with the case of a non-magnetized plasma is conditioned by the exponential suppression of the change of the electron pitch-angle in the distant collisions.

At the higher frequencies

$$\omega \gg \omega_\mathrm{s} = m v_T^3/(Ze^2), \qquad \lambda \ll 2\pi c/\omega_\mathrm{s} = 8\,(T/10^4\ \mathrm{K})^{-3/2}\ \mu\mathrm{m}, \tag{9}$$

the contribution from distant collisions decreases for increasing frequencies as $\omega^{-2/3}$:

$$\eta_{\omega 0}^{(\mathrm{d})} = \frac{0{,}57 n_\mathrm{i} n_\mathrm{e} Z^2 e^6}{\pi^{3/2} m_\mathrm{e}^2 c^3 v_T} \left(\frac{\omega}{\omega_\mathrm{s}}\right)^{-2/3}. \tag{10}$$

In absence of magnetic field, the bremsstrahlung emissivity spectrum is flat at the frequencies given by Eq. (9). The flat spectrum is determined by hyperbolic trajectories with impact parameters $p \ll r_\mathrm{s}$ in which an electron approaches an ion much closer than the impact parameter. The strong magnetic field prevents the electron displacement in a direction transverse to the magnetic field plane, and the distance of the closest electron-ion approach equals the impact parameter. This effect explains the power-law decrease of the bremsstrahlung emissivity from the distant collisions.

The bremsstrahlung from close collisions with impact parameters $p_\mathrm{h} \sim L_\mathrm{u}$ is determined by the parts of the quasi-bound electron trajectory for which the electron moves from the ion to a turning point and back to the ion for a time of the order or longer than the period of the emitted electromagnetic wave $2\pi/\omega$. The number of such parts increases with the increase of the frequency ω. This determines the increase of the bremsstrahlung emissivity from the close collisions as $\omega^{4/3}$:

$$\eta_{\omega 0}^{(\mathrm{c})} = \frac{1.6 n_\mathrm{i} n_\mathrm{e} Z^2 e^6}{\pi^{3/2} m_\mathrm{e}^2 c^3 v_T} \left(\frac{\omega}{\omega_B}\right)^{4/3} \ln^{4/3}(r_\mathrm{s}/r_B). \tag{11}$$

The minimum emissivity from the distant and close collisions $\eta_{\omega 0} = \eta_{\omega 0}^{(\mathrm{d})} + \eta_{\omega 0}^{(\mathrm{c})}$ falls onto the frequency

$$\omega_\mathrm{i} = 0.42 \omega_\mathrm{s}^{1/3} \omega_B^{2/3} / \ln^{2/3}(r_\mathrm{s}/r_B), \tag{12}$$
$$\lambda_\mathrm{i} = 2\pi c/\omega_\mathrm{i} = 23\,(B/10^7\ \mathrm{G})^{-2/3}\,(T/10^4\ \mathrm{K})^{-1/2} \ln^{2/3}(r_\mathrm{s}/r_B)\ \mu\mathrm{m}.$$

The extrapolation of the emissivity of Eq. (11) up to the frequencies of the order of the cyclotron frequency — the upper limit of the considered band, see Eq. (5) — yields the value of the order of the bremsstrahlung emissivity without the magnetic field.

Thus, the bremsstrahlung spectrum is nearly flat at low frequencies — those which fullfil Eq. (7) — and under these conditions it is described by Eq. (8). At higher frequencies, $\omega_\mathrm{s} \ll \omega \ll \omega_\mathrm{i}$, the spectrum decreases as the frequency increases following the power law of Eq. (10). It achieves the minimum value at a frequency given by Eq. (12), whereas within the interval $\omega_\mathrm{i} \ll \omega \ll \omega_B$, it increases following the power law of Eq. (11).

4. Discussion

In this contribution we have shown that the linear polarization of the bremsstrahlung emissivity coincides with the polarization of one of the normal electromagnetic waves in a magnetized plasma (ordinary wave). Thus, the propagation effects (the transverse Faraday effect) do not destroy the linear polarization. Applying the Kirchoff law (Bekefi 1966), we conclude that the extraordinary electromagnetic wave is less reabsorbed than the ordinary one. Thus, the outgoing waves of the extraordinary polarization come from the deeper and hotter layers of the photosphere. So, the mean polarization vector of the observed continuum radiation at the frequencies below the minimum electron cyclotron frequency at the star equator is expected to be transverse to the average projection of the white dwarf magnetic field onto the plane of the sky.

5. Conclusions

In the photospheres of the strongly magnetized white dwarfs, the bremsstrahlung emissivity at the frequencies below the electron cyclotron frequency becomes 1) strongly linear polarized, 2) anisotropic, and 3) decreases in magnitude compared with its value in a non-magnetized plasma.

Acknowledgments

This work was supported by the Russian Foundation for Basic Research (project No. 08-02-00163), the Council of the President of the Russian Federation for Support of Young Russian Scientists and the Leading Scientific Schools of the Russian Federation (projects NSh-4485.2008.2 and MK-4925.2007.2), and the program "Origin and Evolution of Stars and Galaxies" of the Presidium of the Russian Academy of Sciences.

References

Bekefi G 1966 *Radiation Processes in Plasmas* (New York, Wiley) Ch. 2
Correa C E Correa J R Ordonez C A 2005 *Phys. Rev.* E. **72** 046406
Gnedin Yu N Pavlov G G 1974 *Sov. Phys. JETP* **38** 903
Hu B Horton W Chiu C & Petrosky T 2002a *Phys. Plasmas* **9** 1116
Hu B Horton W & Petrosky T 2002b *Phys. Rev.* E **65** 056212
Koryagin S A 2000 *J. Exp. Theor. Phys.* **90** 741.
Schmidt G Kunhardt E E & Godino J L 2000 *Phys. Rev.* E **62** 7512
Zheleznyakov V V 1997 *Radiation in Astrophysical Plasmas* [in Russian] (Moscow: Yanus-K) §13.1
Zheleznyakov V V Koryagin S A & Serber A V 1999 *Astron. Lett.* **25** 437

Peculiarities of electron-ion collisions in strong magnetic field white dwarfs

V V Zheleznyakov and S A Koryagin

Institute of Applied Physics of the Russian Academy of Sciences,
46 Uljanova str., Nizhny Novgorod 603950, Russia

E-mail: koryagin@appl.sci-nnov.ru

Abstract. We consider electron-ion collisions in strong magnetic field ($B > 10^7$ G) white dwarfs at a photospheric temperature $T \sim 10^4$ K. Under these unique conditions, the electron's Larmor radius becomes less than the characteristic impact parameter of the close Coulomb collisions in a non-magnetized plasma. We analyze electron-ion collisions in the classic and in the strong quantum limits which both can be observed in the photospheres of the isolated magnetic white dwarfs. In the both limits, the collision transport frequencies are determined by the close collisions in which an electron transits the region near an ion for a time of the order or less than the cyclotron period. In these effective collisions, an electron returns to an ion many times. The final state of an electron non-regularly (chaotically) depends on its initial state. The strong magnetic field significantly reduces the values of the collision transport frequencies in comparison with the analogous values in a non-magnetized plasma.

1. Physical conditions

In a completely ionized non-magnetized plasma, the collision transport frequencies are determined by the distant Coulomb collisions with the impact parameters varying from the distance

$$r_\mathrm{s} = Ze^2/(2m_\mathrm{e}E) = r_\mathrm{a}\,(E/E_\mathrm{a})^{-1} \tag{1}$$

up to the Debye screening length $r_\mathrm{D} \gg r_\mathrm{s}$. A magnetic field qualitatively changes the Coulomb collisions when the electron's Larmor radius

$$r_B = (2E/m_\mathrm{e})^{1/2}/\omega_B = r_\mathrm{a}\,(E/E_\mathrm{a})^{1/2}\,(B/B_\mathrm{cr})^{-1} \tag{2}$$

becomes smaller than the typical impact parameter of close collisions in a non-magnetized plasma given by Eq. (1) — see Zheleznyakov (1997), Zheleznyakov et al. (1999a), Koryagin (2000), Schmidt et al. (2000), and Hu et al (2002a, b). Here, E is the electron kinetic energy far from an ion, m_e is the electron mass, $\omega_B = eB/(m_\mathrm{e}c)$ is the cyclotron frequency, e is the electron elementary charge, $Z > 0$ is the ion charge number, B is the magnetic field strength, and c is the speed of light. For the sake of comparison, the length scales given by Eqs. (1) and (2) are expressed using the Bohr radius $r_\mathrm{a} = Ze^2/(2E_\mathrm{a}) = 0.53Z^{-1}$ Å. In this expression the ionization potential is given by $E_\mathrm{a} = Z^2e^4m_\mathrm{e}/(2\hbar^2) = 13.6Z^2$ eV. Furthermore, the magnetic field is expressed in terms of $B_\mathrm{cr} = Z^2e^3m_\mathrm{e}^2c/\hbar^3 = 2.35\cdot 10^9 Z^2$ G, for which the energy of the fundamental Landau level, $\hbar\omega_B/2$, equals E_a. The ratio of the length scales

$$r_B/r_\mathrm{s} = (E/E_\mathrm{a})^{3/2}\,(B/B_\mathrm{cr})^{-1} \ll 1 \tag{3}$$

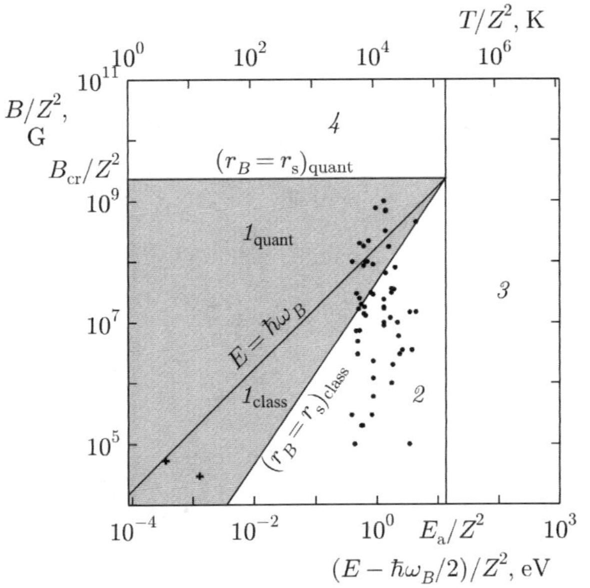

Figure 1. The plane of parameters $E - \hbar\omega_B/2$, B. The condition given by Eq. (3) is fulfilled in the grey-shaded region. The dots designate the magnetic fields B and the thermal energies $k_B T = E - \hbar\omega_B/2$ in the photospheres of the isolated magnetic white dwarfs (Wickramasinghe & Ferrario 2000), the crosses mark the analogous parameters in the experiments of antihydrogen creation (Amoretti et al. 2002, Gabrielse et al. 2002). In the "classic" region *2* of the weaker magnetic field and in the "quantum" region *3* of larger energies, the Coulomb collisions were analyzed by Silin (1960) and Pavlov & Yakovlev (1976). The region *4* requires further investigation.

is realized in the photospheres of magnetic white dwarfs whose temperatures and magnetic fields fall into the regions 1_{class} and 1_{quant} in Fig. 1.

In the region 1_{class}, the electron energy E exceeds the quantum one $\hbar\omega_B$, therefore the electron motion can be described classically. In the quantum region 1_{quant}, the electron energy is smaller than $\hbar\omega_B$. Nevertheless, the definitions of the parameter in Eq. (1) and of the Larmor radius of Eq. (2) keep valid in the region 1_{quant}. For this, one should use the energy of the fundamental Landau level $\hbar\omega_B/2 = E_a\,(B/B_{cr})$ as the kinetic energy E (Koryagin 2008a). Then, the electron energy E is not a free parameter in the quantum limit: $E = \hbar\omega_B/2$, the distance r_s of Eq. (1) starts decreasing for increasing magnetic field strengths: $r_s \propto B^{-1}$, while the Larmor radius decreases slower: $r_B \propto B^{-1/2}$. As a result, the ratio of Eq. (3) can be met only for magnetic fields $(r_B/r_s)_{E=\hbar\omega_B/2} = (B/B_{cr})^{1/2} \ll 1$.

2. Close collisions

In the region for which Eq. (3) is valid, the collision transport frequencies are determined by the effective close collisions, in which an electron transits the region near an ion for a time of the order or less than the cyclotron period (Zheleznyakov 1997, Zheleznyakov et al. 1999a). The impact parameters of these collisions are of the order of $L_u = (Zm_e c^2/B^2)^{1/3} = r_a\,(B/B_{cr})^{-2/3}$. The length scale L_u coincides with the radius of a non-magnetized hydrogen-like ion in a highly excited (Rydberg) state in which an electron orbits a nucleus for a time equal to the given cyclotron period. If the condition of Eq. (3) is fulfilled, the distance L_u falls into the interval $r_B \ll L_u \ll r_s$. In accordance with this equation, the ion's Coulomb field strongly accelerates an electron at the path to an ion, so that the electron kinetic energy near the ion achieves the typical value $E_u = Ze^2/L_u = 2E_a\,(B/B_{cr})^{2/3}$. This energy significantly exceeds the electron kinetic energy E far from an ion and, therefore, is greater than the quantum $\hbar\omega_B \geq 2E$. Thus, an electron moves classically near an ion even if its initial and final motions are quantized: $E \sim \hbar\omega_B$. The classical motion near an ion unifies the regions 1_{class} and 1_{quant} in Fig. 1 and explains the identical values of the characteristic impact parameters of the effective close collisions in these regions (Koryagin 2008a).

In effective close collisions, the motion of an electron becomes quasi-bound — a particle returns to an ion many times before it goes away from a scattering center. Figures 2*a* and

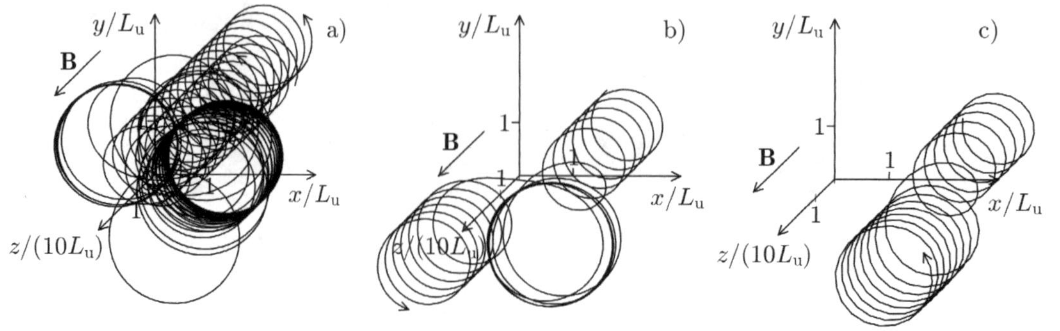

Figure 2. The electron trajectories in the head-on (a), close (b), and distant (c) collisions.

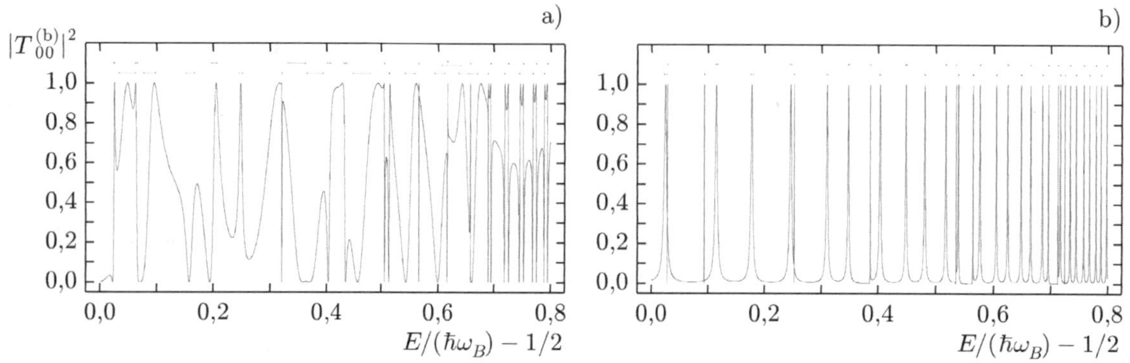

Figure 3. The dependence of the backscattering probability $|T_{00}^{(b)}|^2$ on the electron energy E in the close (a) and distant (b) collisions. The bars under the upper horizontal axis mark the energies and the energy widths of the resonant autoionizing states.

2b show examples of the quasi-bound classic trajectories in a head-on collision and in a close collision with the impact parameter $p_h = L_u$; the initial angle between the electron velocity and the magnetic field (pitch-angle) $\theta = \pi/4$ and the energy $E = E_u/2$ (this corresponds to the equality $r_B = r_s$). The final pitch-angle non-regularly (chaotically) depends on the initial electron energy, the magnetic field strength, and the impact parameter (Hu et al. 2002a, b).

In the quantum-mechanical approach, the quasi-bound electron motion corresponds to the particle transition to the highly excited Landau levels at which the kinetic energy of the transverse to the magnetic field motion $(n + 1/2)\,\hbar\omega_B$ exceeds the electron kinetic energy far from an ion E. At these levels, an electron can not go away from an ion and has to return back to a scattering center. In the quantum limit ($E \sim \hbar\omega_B$), the electron backscattering probability keeps non-regular (chaotic) dependence on the initial electron energy and the magnetic field strength (Koryagin 2008b). Figure 3a shows such a non-regular dependence on the energy of the electron motion along the magnetic field for a low-energy particle which can occupy only the fundamental Landau level before and after a collision; the impact parameter equals the double Larmor radius at the fundamental Landau level (this corresponds to the electron state with the magnetic azimuthal number $m = -1$), $B = 0.005B_{cr}$.

3. Distant collisions

In distant collisions with impact parameters $p_h \gg L_u$, an electron moves along the whole trajectory under the conditions of the drift approximation. Thus, in the classical description, the final change of the electron pitch-angle exponentially decreases with increasing impact parameter as $\Delta\theta \propto \exp(-0{,}62 p_{\mathrm{h}}^{3/2}/L_{\mathrm{u}}^{3/2})$ (Koryagin 2000). Figure 2*c* shows an example of a distant collision.

In the quantum limit ($E \sim \hbar\omega_B$), an electron can completely scatter backwards even in a distant collision (Koryagin 2008b). For this, the electron energy must coincide with the energy of a so-called autoionizing state (Friedrich & Chu 1983, Chu & Friedrich 1984). The dependence of the backscattering probability on the electron energy acquires the form of a sequence of the resonant peaks (Fig. 3*b*). However, the energy width of the resonances exponentially decreases with increasing impact parameters. Thus, the contribution of distant collisions to the backscattering cross-section is small compared to that of close collisions.

4. Transport frequencies

The typical transport cross-sections are determined by close collisions. Therefore, their values are of the order of πL_{u}^2 which is much less than their typical values πr_{s}^2 without a magnetic field. For example, the deceleration of the non-zero hydrodynamic (mean) velocity of the thermal electrons along a magnetic field is described by the conventional transport frequency $\nu_{\|} = 2.2 n_{\mathrm{i}} v_T L_{\mathrm{u}}^2 \ln^{4/3}(r_{\mathrm{s}}/r_B)$ both in the classic (Koryagin 2000) and in the quantum (Koryagin 2008b) limits, where n_{i} is the ion space density and $v_T = (k_{\mathrm{B}}T/m_{\mathrm{e}})^{1/2}$ is the thermal velocity of electrons.

The decrease of the transport frequencies in a strong magnetic field conditions the increase of the required photospheric temperature (up to $T = 4 \cdot 10^4$ K) at which the cyclotron radiation pressure can exceed gravity, expel plasma from a photosphere, and, thus, inflate the extensive plasma envelope in a magnetosphere — the so-called radiation-driven diskon (Bespalov & Zheleznyakov 1990, Zheleznyakov et al. 1999b).

Acknowledgments

This work was supported by the Russian Foundation for Basic Research (project No. 08-02-00163), the Council of the President of the Russian Federation for Support of Young Russian Scientists and the Leading Scientific Schools of the Russian Federation (projects NSh-4485.2008.2 and MK-4925.2007.2), and the program "Origin and Evolution of Stars and Galaxies" of the Presidium of the Russian Academy of Sciences.

References

Amoretti M Amsler C Bonomi G et al. 2002 *Nature* **419** 456
Bespalov P A & Zheleznyakov V V 1990 *Sov. Astron. Lett.* **16** 442
Chu M-C & Friedrich H 1984 *Phys. Rev.* A **29** 675
Friedrich H & Chu M 1983 *Phys. Rev.* A **28** 1423
Gabrielse G Bowden N S Oxley P et al. 2002 *Phys. Rev. Lett.* **89** 213401
Hu B Horton W Chiu C & Petrosky T 2002a *Phys. Plasmas* **9** 1116
Hu B, Horton W & Petrosky T 2002b *Phys. Rev.* E **65** 056212
Koryagin S A 2000 *J. Exp. Theor. Phys.* **90** 741.
Koryagin S A 2008a *Radiophys. Quantum Electron.* **51** 462
Koryagin S A 2008b *Radiophys. Quantum Electron.* **51** 616
Pavlov G G & Yakovlev D G 1976 *Sov. Phys. JETP* **43** 389
Schmidt G Kunhardt E E & Godino J L 2000 *Phys. Rev.* E **62** 7512
Silin V P 1960 *Sov. Phys. JETP* **11** 1277
Wickramasinghe D T & Ferrario L 2000 *Publ. Astron. Soc. Pacific* **112** 873
Zheleznyakov V V 1997 *Radiation in Astrophysical Plasmas* (Moscow: Yanus-K) §13.1
Zheleznyakov V V Koryagin S A & Serber A V 1999a *Astron. Lett.* **25** 437
Zheleznyakov V V Koryagin S A & Serber A V 1999b *Astron. Lett.* **25** 445

Journal of Physics: Conference Series **172** (2009) 012054 doi:10.1088/1742-6596/172/1/012054

Spectral modeling of gaseous metal disks around DAZ white dwarfs

Klaus Werner, Thorsten Nagel and Thomas Rauch

Institute for Astronomy and Astrophysics, Kepler Center for Astro and Particle Physics, University of Tübingen, Germany

E-mail: werner@astro.uni-tuebingen.de

Abstract. We report on our attempt for the first non-LTE modeling of gaseous metal disks around single DAZ white dwarfs recently discovered by Gänsicke et al. and thought to originate from a disrupted asteroid. We assume a Keplerian rotating viscous disk ring composed of calcium and hydrogen and compute the detailed vertical structure and emergent spectrum. We find that the observed infrared Ca II emission triplet can be modeled with a hydrogen-deficient gas ring located at $R = 1.2\,R_{\odot}$, inside of the tidal disruption radius, with $T_{\rm eff} \approx 6000$ K and a low surface mass density of $\approx 0.3\,{\rm g/cm^2}$. A disk having this density and reaching from the central white dwarf out to $R = 1.2\,{\rm R_{\odot}}$ would have a total mass of $7 \cdot 10^{21}$ g, corresponding to an asteroid with ≈160 km diameter.

1. Introduction: Dust around DAZ white dwarfs

More than two decades ago Zuckerman & Becklin (1987) announced the discovery of an IR excess around the DAZ white dwarf G29−38. The white dwarf itself is enriched in metals. Considering the short sedimentation timescales in the photosphere this implies that the star is accreting matter at a relatively high rate (Koester et al. 1997). Since no cool companion has been found at G29−38, the hypothesis was put forward that a dust cloud around the white dwarf causes the IR excess. In fact, the presence of dust has been confirmed by *Spitzer* observations (Reach et al. 2005). Graham et al. (1990) concluded that the dust is located in the equatorial plane. Subsequently, further DAZ white dwarfs with potential dust disks were found (Becklin et al. 2005, Kilic et al. 2005, 2006). As a possible origin of these disks tidally disrupted comets were discussed (Debes & Sigurdsson 2002) and, more likely because of the absence of H and He, disrupted asteroids (Jura 2003).

2. Gas disks around DAZ white dwarfs

Recently, signatures of a gas disk were discovered in Sloan Digital Sky Survey (SDSS) spectra of two DAZ white dwarfs (Gänsicke et al. 2006, 2007). The spectra display double-peaked emission lines of the infrared Ca II triplet $\lambda\lambda$ 8498, 8542, 8662 Å.

In the present paper, we concentrate on one of these two white dwarfs, namely SDSS 1228+1040, because its emission line profiles are more prominent. The white dwarf's atmospheric parameters are $T_{\rm eff} = 22\,020$ K and $\log g = 8.24$. The derived stellar mass $M_{\rm WD} = 0.77\,{\rm M_{\odot}}$ and radius $R_{\rm WD} = 0.011\,{\rm R_{\odot}}$ are quantities that enter our disk model. The photospheric magnesium abundance is 0.8 times solar. The spectrum neither exhibits radial

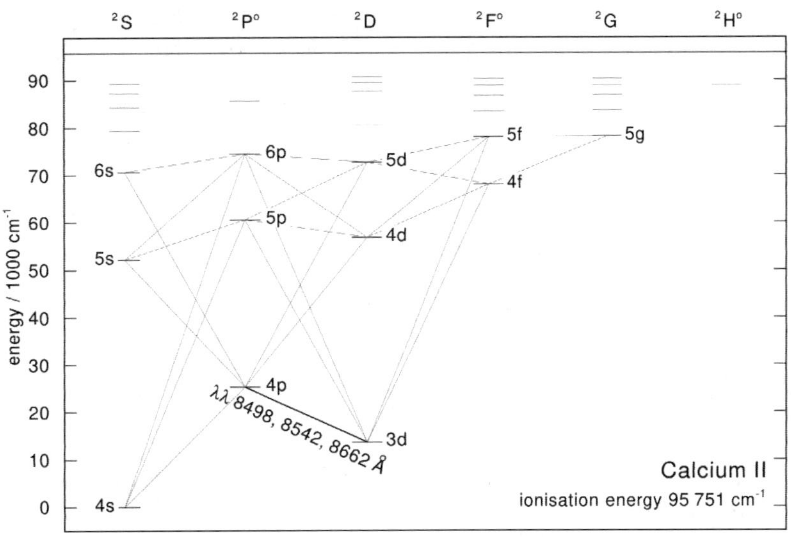

Figure 1. Grotrian diagram of our Ca II model ion. All highly excited levels that are not linked by radiative line transitions are treated in LTE. The 3d–4p transition causes the observed IR triplet.

velocity variations nor photometric variability. Besides the calcium emission lines only two other weaker emission features are seen (Fe II λλ 5018, 5169 Å). In particular, hydrogen and helium emissions are not discovered. It is concluded that the Ca and Fe emission lines stem from a metal-rich Keplerian disk around a single white dwarf. The Ca II line profiles are double peaked emission lines with a peak-to-peak separation of 630 km/s, i.e., the Keplerian rotation velocity is $v \sin i = 315$ km/s. There is a clear violet/red asymmetry in the double-peaked profiles, well known from a similar phenomenon in Be star disks that is ascribed to one-armed spiral waves.

From a spectral analysis with a kinematical LTE emission model Gänsicke et al. (2006) conclude that we see a geometrically thin, optically thick disk at high inclination ($i = 70°$). The inner and outer disk radii are $R_{\rm in} = 0.64$ R$_\odot$ and $R_{\rm out} = 1.2$ R$_\odot$, respectively. While the outer disk radius is quite sharply confined because of the steep line wings, the value derived for the inner disk radius possibly just marks the inner edge of the Ca II line emission region and not the physical inner disk edge, i.e., $R_{\rm in}$ could reach down to the white dwarf's surface. They also conclude that the gas temperature in the disk is around 4500–5500 K. Since the tidal disruption radius for a rocky asteroid at the white dwarf is $R \approx 1.5$ R$_\odot$, it is possible that the material in the gas disk may be a disrupted asteroid whose dust was sublimated by the white dwarf's radiation field. It seems that SDSS 1228+1040 could be the hot counterpart to G29−38 and other cool DAZ stars harboring dust disks, but in the meantime DAZ white dwarfs with a dust disk and as hot as SDSS 1228+1040 were discovered (e.g., Jura et al. 2007).

It is our aim to compute a more realistic model for the gas disk around SDSS 1228+1040 and we are reporting here very first provisional results.

3. A viscous disk ring model

Basically we assume a Shakura & Sunyaev (1973) α-disk model, i.e., a geometrically thin Keplerian disk heated by viscosity. For numerical calculations we employ AcDc, our Ac*cretion* D*isk* c*ode* that we developed for modeling disks in cataclysmic variables and low-mass X-ray binaries (Nagel et al. 2004). The code computes a detailed vertical structure and the spectrum of

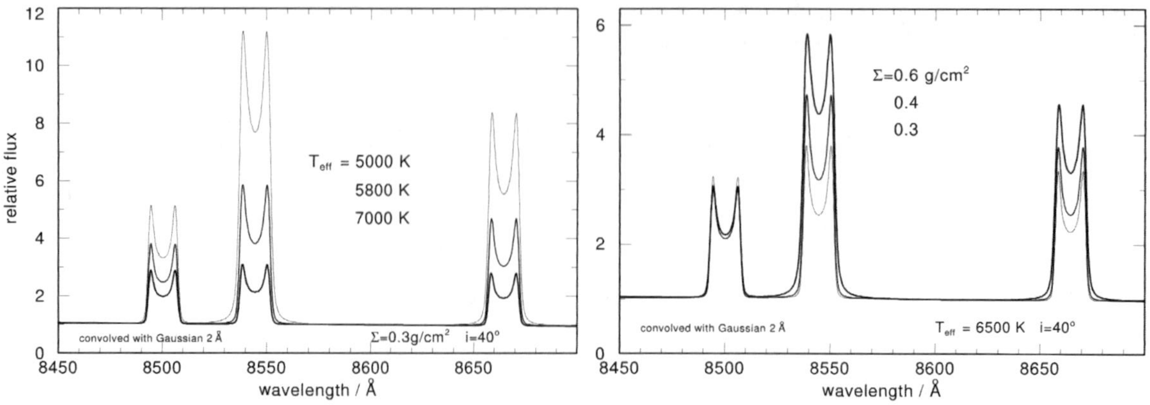

Figure 2. *Left:* T_{eff} dependence of the Ca II triplet emitted by a Keplerian rotating gas ring seen at an inclination angle $i = 40°$. The emission becomes weaker with increasing T_{eff}. At the same time, the relative strengths of triplet components become equal. *Right:* A similar effect is seen when the surface mass density Σ decreases. Values for the ring parameters T_{eff} and Σ are given in the panels. All profiles are convolved with a 2 Å FWHM Gaussian.

a disk which is being built from radial-symmetric annuli. For any disk annulus we assume that it radiates like a plane-parallel slab in non-LTE, and in radiative and hydrostatic equilibrium. We are presenting here the results of a single disk ring using different values for the input parameters T_{eff} (as a measure for the viscously dissipated energy) and surface mass density Σ (the vertical mass column from the disk midplane to the surface). Further input parameters, which are kept fixed, are white dwarf's mass and radius as given above, and $R_{\mathrm{in}} = 91\,\mathrm{R_{WD}}$ and $R_{\mathrm{out}} = 109\,\mathrm{R_{WD}}$ for the inner and outer disk radii, respectively. The chemical composition is calcium dominated with an admixture of hydrogen with varying H/Ca abundance ratios in order to determine the extent of H-deficiency.

The principal problem for any modeling attempt is posed by the question: what heats the Ca II emission line region? Although being hot, it cannot be the white dwarf because it is too distant. Also, it cannot be gravitational energy released through viscosity because the required mass-accretion rate would be $\approx 10^{-8}\ \mathrm{M_{\odot}/yr}$ which is by many orders of magnitude larger than the accretion rate invoked for the presence of settling metals in DAZ photospheres ($\approx 10^{-15}\ \mathrm{M_{\odot}/yr}$, Koester & Wilken 2006). A speculation by Jura (2008) is additional heating by energy dissipation through disk asymmetries, being driven by some external unseen planet. As mentioned above, such an asymmetry is obvious from the line profiles. Whatever, at the moment we do not know the heating mechanism. Therefore we need to use T_{eff} as a free parameter for the disk ring. *In praxi* this means that we set the accretion rate $\dot{M}$, which is related to T_{eff} through

$$T_{\mathrm{eff}}^4(R) = [1 - (R_{\mathrm{WD}}/R)^{1/2}]\, 3GM_{\mathrm{WD}}\dot{M}/8\sigma\pi R^3,$$

to an artificially high value.

The calcium model atom used for the non-LTE calculations comprises three ionisation stages, Ca I – III with 7, 12, 1 non-LTE levels and 3, 26, 0 radiative line transitions, respectively. For the disk model calculations we do not account for level fine structure splitting. For the Ca II IR triplet this is done in the final formal solution of the transfer equation by distributing the level populations over the fine structure levels according to their statistical weight. Fig. 1 is a Grotrian diagram of our Ca II model ion with all implemented levels and lines, indicating the transition responsible for the observed IR triplet.

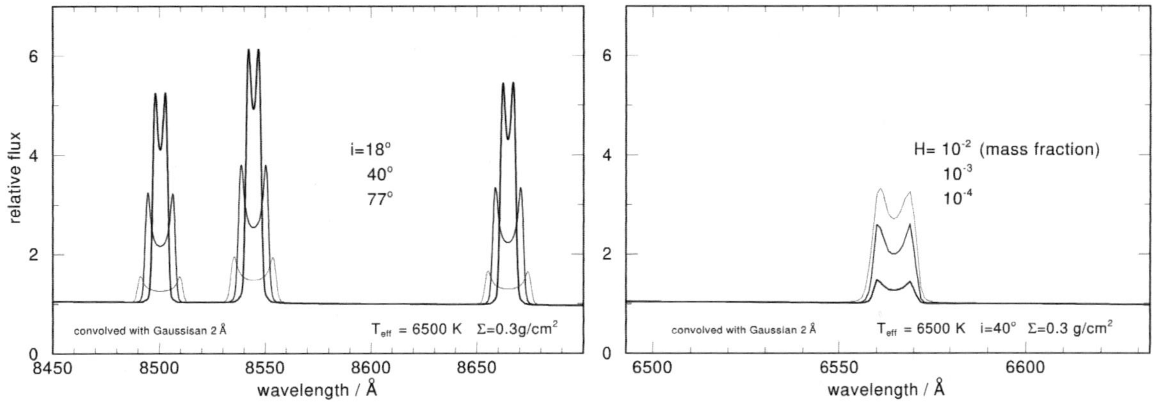

Figure 3. *Left:* The Ca II line shape of a particular model ($T_{\rm eff} = 6500$ K, $\Sigma = 0.3$ g/cm^2) seen under different inclination angles i. The relative strength of the profile depression between double peaks increases with i. *Right:* Hα line shape of the same model but with different hydrogen content. H $> 1\%$ would be detectable in SDSS 1228+1040.

Energy levels were taken from NIST, and oscillator strengths and photoionisation cross-sections from the Opacity and Iron Projects. Electron collisional rates are computed from usual approximation formulae. Rates for collisions with H are currently neglected, probably without consequences. That is because, first, Mashonkina et al. (2007) have shown that these collisions are unimportant in F/G type stars (having similar physical conditions in their photospheres) and, second, we are faced here with a H-deficient environment. Collisions of Ca with heavy ions are neglected, too. Their rates are unknown but probably unimportant because of the low thermal velocities of the perturbers. For the spectral lines we assume Voigt profiles with radiation damping parameters. Future improvement should include Van der Waals damping, although we don't expect it to be important in the disk environment. The hydrogen model atom is a standard configuration with ten non-LTE levels that we use routinely for stellar spectral modeling.

4. Results

We present first results from disk ring models for SDSS 1228+1040, discussing:
- the influence of $T_{\rm eff}$, Σ, and inclination i on emergent line profiles of the Ca II IR triplet;
- an upper limit for hydrogen content from Hα;
- the characteristics of one representative model, in particular its vertical structure;
- a comparison of the computed spectrum with observation.

4.1. Influence of $T_{\rm eff}$ Σ, i on the infrared Ca II triplet

In Fig. 2 (left panel) we see that the emission strength of the Ca II triplet decreases with increasing $T_{\rm eff}$ (with $\Sigma = 0.3$ g/cm^2 kept fixed). This is because of the shifting Ca II/Ca III ionisation balance. A closer comparison of the three triplet components shows that the relative strengths of the components become equal with increasing $T_{\rm eff}$, a behaviour that constrains $T_{\rm eff}$ from the observed line strengths. A similar trend is seen (Fig. 2, right panel) when Σ is reduced (with $T_{\rm eff} = 6500$ K kept fixed). We stress that the models have a considerable continuum flux compared to line emission peak heights.

In Fig. 3 (left panel) we demonstrate that in principle the inclination angle can be constrained from the Ca II triplet line shape. Obviously, the relative strength of the profile depression between

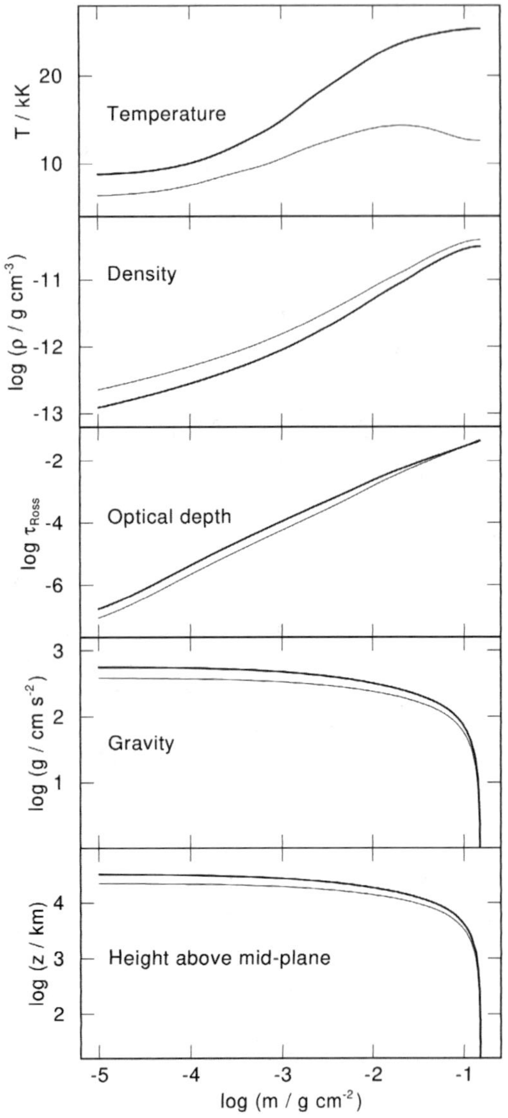

Figure 4. *Left:* Vertical structure of two models ($T_{\rm eff} = 5000\,$K, thick lines, and 7000 K) with $\Sigma = 0.3\,{\rm g/cm^2}$. The run of various physical quantities is shown on a column-mass scale measured from the disk surface (left) to the disk midplane (right); see discussion in the text (Sect. 4.3). The disks are optically thin ($\tau_{\rm Ross} < 0.1$).

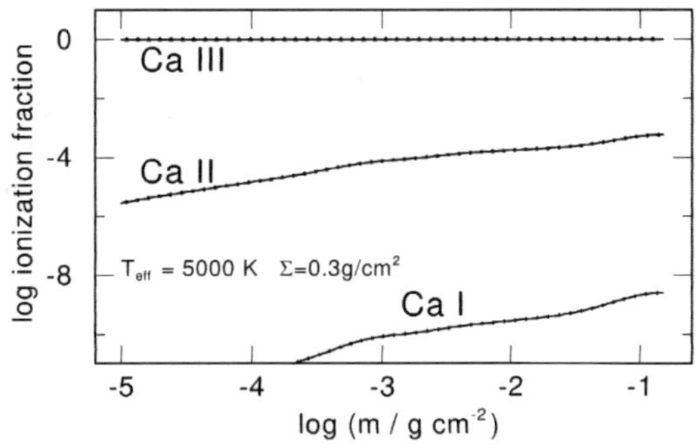

Figure 5. *Above:* Vertical run of the calcium ionisation fractions in the model with $T_{\rm eff} = 5000\,$K and $\Sigma = 0.3\,{\rm g/cm^2}$. Ca III is the dominant ionisation stage.

the double peaks increases with inclination.

4.2. Upper limit for hydrogen

From the lack of Hα emission in the spectrum of SDSS 1228+1040 we can make a quantitative estimate of the hydrogen-deficiency in the gas disk. Fig. 3 (right panel) depicts the Hα line shape of a particular model in which we varied the H content (H = 1%, 0.1%, 0.01%; mass fraction). With an abundance of 1% the Hα peak height is comparable to that of the Ca II triplet and, hence, would be detectable in the spectrum of SDSS 1228+1040.

4.3. Vertical structure of disk ring

Let us inspect the vertical structure of two representative ring models with different effective temperature displayed in Fig. 4. The range of temperature T and mass density ρ is comparable to the circumstances encountered in the atmospheres of F/G-type giants. Accordingly, the gravity increases from the disk midplane toward the disk surface, up to $\log g \approx 2.5$. The run

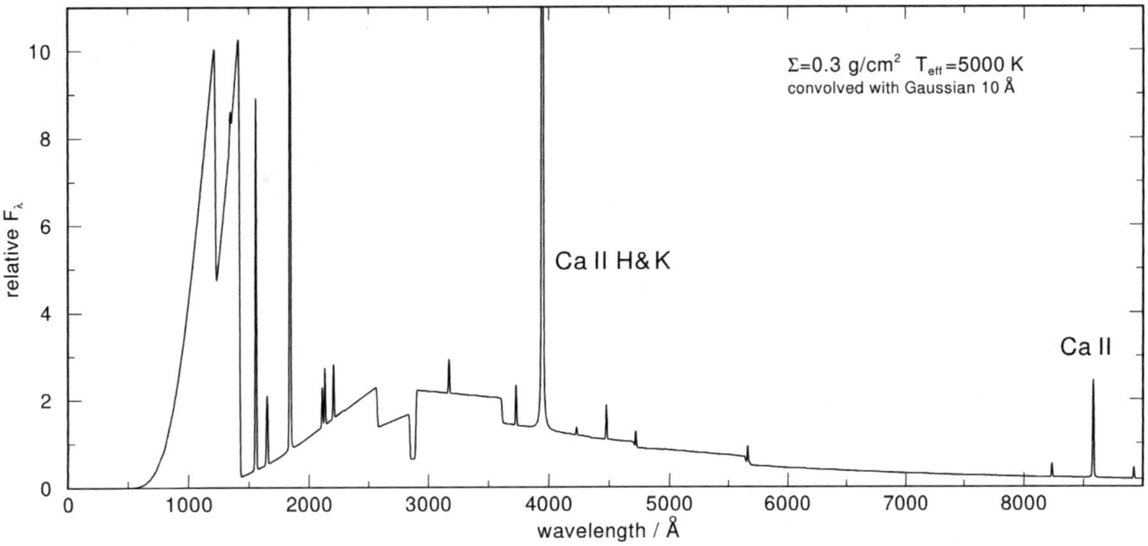

Figure 6. Flux distribution in the UV/optical range of the $T_{\rm eff} = 5000$ K, $\Sigma = 0.3\,{\rm g/cm^2}$ model (no fine structure splitting applied to lines). The spectrum is dominated by Ca II emission lines and bound-free emission edges. The strongest feature is the H&K line. For the sake of clarity, the spectrum is convolved with a 10 Å FWHM Gaussian.

of the Rosseland optical depth $\tau_{\rm Ross}$ shows that the disks are optically thin. The geometrical height z above the midplane shows that the thickness of the disk ring is about $\Delta R = 50\,000$ km, hence, $\Delta R/R \approx 15$. The non-LTE departure coefficients (not plotted) of the atomic population numbers in these two models deviate significantly from the LTE value $b = 1$. For the lower levels of the Ca II IR triplet in the $T_{\rm eff} = 5000$ K model, for example, they range between $b \approx 0.001$ in the upper layers of the disk and $b \approx 50$ in deeper regions.

Fig. 5 shows the vertical calcium ionisation stratification for the cooler ($T_{\rm eff} = 5000$ K) of these two models. Ca III is by far the dominant ionisation stage everywhere in the model. This advises us to extend the model atom to include the next higher ionisation stage in future calculations, although we do not expect a dramatic depopulation of Ca III (and Ca II) because its ionisation potential is rather high (50.9 eV).

It is interesting to look at the overall flux spectrum of the models, e.g., Fig. 6. The spectrum is dominated by Ca II emission lines. The Ca II H&K resonance doublet is by far the strongest emission feature that should be detectable in the spectrum of SDSS 1228+1040, although the WD flux increases by a about an order of magnitude from the location of the IR triplet to the H&K line. A close inspection of the residual disk spectrum of SDSS 1228+1040 (i.e., observation minus WD model spectrum) presented by Gänsicke et al. (2008) indeed displays an emission feature at the location of the H&K line, although not explicitly described by the authors. In a new high-quality spectrum of SDSS 1228+1040 (Gänsicke, this meeting) the presence of this emission line is clearly confirmed. It seems, however, that the observed strength is strongly overestimated by our models, a result that also holds for the LTE emission models presented by Gänsicke.

Another remarkable feature of the overall model flux spectrum in Fig. 6 is the occurrence of two prominent Ca II *emission edges* at 1219 Å and 1420 Å, created by recombination processes into the first and second excited states, respectively. These features should be easily detectable in UV spectra as well as many other emission lines from other metals as predicted and presented

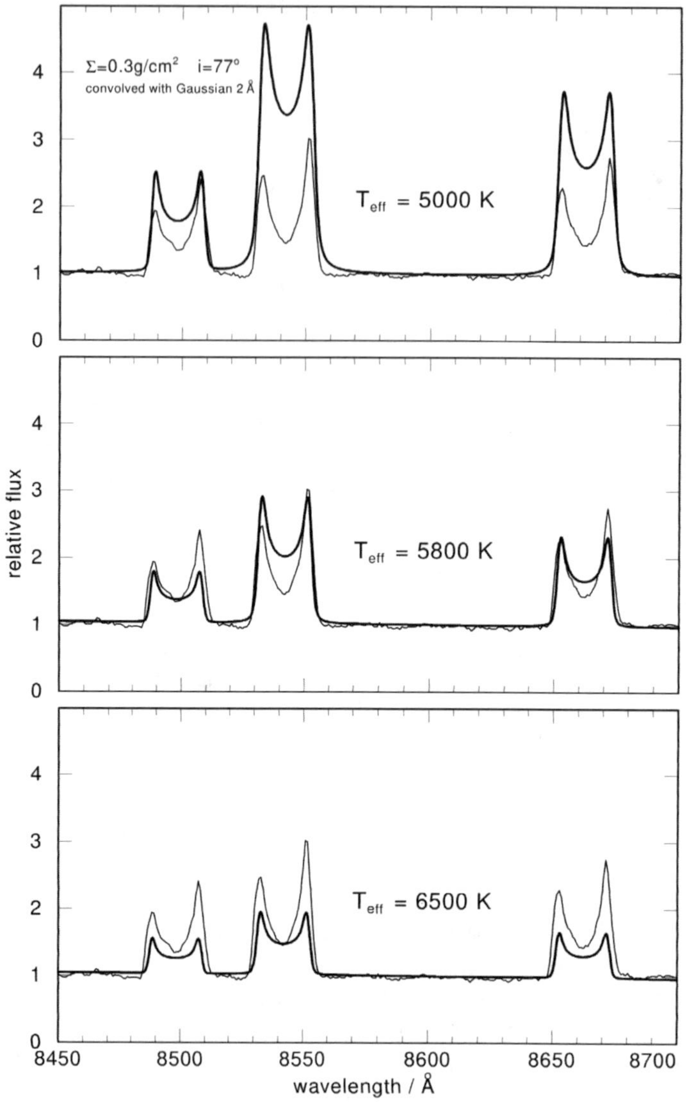

Figure 7. Normalized spectra of three models (thick lines) with different $T_{\rm eff}$ compared to the observed spectrum of SDSS 1228+1040. The fitting procedure is ambiguous because the observation is normalized to the continuum that is probably dominated by the white dwarf. All models have $\Sigma = 0.3\,{\rm g/cm^2}$ and are viewed at an inclination of $i = 77°$.

by Gänsicke at this meeting. Our models promise that the proposed (and approved) HST/COS spectroscopic observation of SDSS 1228+1040 by Gänsicke and collaborators will deliver an exciting dataset that will allow important conclusions on the chemical composition of the gas disk.

4.4. Models vs. observation

In Fig. 7 we show a direct comparison of the normalised spectrum of SDSS 1228+1040 to three models with different effective temperatures. This comparison is, however, complicated by the fact that Gänsicke et al. (2006) suggest that the continuous radiation in the vicinity of the Ca II triplet stems from the white dwarf alone, while in contrast our models predict a non-negligible contribution by the disk. What we compare in Fig. 7 is the observed spectrum normalized to the WD (+disk) continuum and the computed disk spectrum alone, normalized to its emission continuum. In any case, it can be seen that the effective temperature of the disk at SDSS 1228+1040 is well constrained by the three models, being $T_{\rm eff} \approx 5800$ K. The cooler model ($T_{\rm eff} = 5000$ K) is perhaps more favorable because of the larger line-to-continuum emission ratio, while the hotter model ($T_{\rm eff} = 6500$ K) has the advantage that the relative strengths of the three line components are matched better.

5. Summary and conclusions

The infrared Ca II emission triplet in the spectrum of the DAZ white dwarf SDSS 1228+1040 can be modeled with a geometrically and optically thin, Keplerian viscous gas disk ring at a distance of 1.2 $R_\odot$ from the WD, with $T_{\rm eff} \approx 6000$ K and a low surface mass density $\Sigma \approx 0.3\,{\rm g/cm^2}$. One serious open problem is the unknown disk-heating mechanism. The disk is hydrogen-deficient (H $\leq 1\%$ by mass) and it is located within the tidal disruption radius ($R_{\rm tidal} = 1.5\ R_\odot$). If one assumes that the disk reaches down to the WD and that it has uniformly this surface density, then its total mass would be $7 \cdot 10^{21}$ g. A rocky asteroid ($\bar{\rho} = 3\,{\rm g/cm^3}$) with this mass would have a diameter of about 160 km. An asteroid of this size in our solar system is, e.g., 22 Kalliope.

Future work will include other abundant metal species, with a composition appropriate for asteroids. However, at the moment there are only two Fe II lines detected. Real progress for an abundance analysis of the gas disk around SDSS 1228+1040 is only possible with future UV spectroscopy.

While the determination of the composition of accreted material from abundance analyses in DAZ atmospheres is an indirect method that depends on our theoretical knowledge about metal settling timescales, the analysis of the gas disks has the obvious advantage that is a direct composition measurement.

Acknowledgments

We thank Boris Gänsicke for sending us his SDSS 1228+1040 spectrum in electronic form and for useful discussions. T.R. is supported by the *German Astrophysical Virtual Observatory* project of the Federal Ministry of Education and Research (grant 05 AC6VTB).

References

Becklin E E Farihi J Jura M Song I Weinberger A J & Zuckerman B 2005 *ApJ* **632** L119
Debes J H & Sigurdsson S. 2002 *ApJ* **572** 556
Gänsicke B T Marsh T M Southworth J & Rebassa-Mansergas A 2006 *Science* **314** 1908
Gänsicke B T Marsh T M & Southworth J 2007 *MNRAS* **380** L35
Gänsicke B T Marsh T M Southworth J & Rebassa-Mansergas A 2008 *Preprint* ArXiv:0710.2807v1
Graham J R Matthews K Neugebauer G & Soifer B T 1990 *AJ* **357** 216
Jura M 2003 *ApJ* **584** L91
Jura M Farihi J & Zuckerman B 2007 *ApJ* **663** 1285
Jura M 2008 *AJ* **135** 1785
Kilic M von Hippel T Leggett S K & Winget D E 2005 *ApJ* **632** L115
Kilic M von Hippel T Leggett S K & Winget D E 2006 *ApJ* **646** 474
Koester D & Wilken D 2006 *A&A* **453** 1051
Koester D Provencal J & Shipman H L 1997 *A&A* **320** L57
Mashonkina L Korn A J & Przybilla N 2007 *A&A* **461** 261
Nagel T Dreizler S Rauch T & Werner K 2004 *A&A* **428** 109
Reach W T Kuchner M J von Hippel T et al. 2005 *ApJ* **635** L161
Shakura N I & Sunyaev R A 1973 *A&A* **24** 337
Zuckerman B & Becklin E E 1987 *Nature* **330** 138

16th European White Dwarfs Workshop IOP Publishing
Journal of Physics: Conference Series **172** (2009) 012055 doi:10.1088/1742-6596/172/1/012055

24μm Excesses of Hot WDs - Evidence of Dust Disks?

Jana Bilikova[1], You-Hua Chu[1], Kate Su[2], Robert Gruendl[1], Thomas Rauch[3], Orsola De Marco[4] and Kevin Volk[5]

[1]Astronomy Department, University of Illinois, 1002 W. Green St., Urbana, IL 61801, USA
[2]Steward Observatory, University of Arizona, 933 N. Cherry Ave., Tuscon, AZ 85721, USA
[3]Institute for Astronomy and Astrophysics, Kepler Center for Astro and Particle Physics, Eberhard Karls University, Tübingen, Germany
[4]American Museum of Natural History, Dept. of Astrophysics, Central Park West at 79th St., New York, NY 10024, USA
[5]Gemini Observatory, Northers Operations Center, 670 N. A ohoku Place, Hilo, HI 96720, USA

E-mail: jbiliko2@astro.uiuc.edu

Abstract. *Spitzer Space Telescope* observations of the Helix Nebula's hot ($T_{\rm eff} \approx 110\,000$ K) central star revealed mid-IR excess emission consistent with a continuum emission from a dust disk located at 35−150 AU from the central white dwarf (WD), and the dust is most likely produced by collisions among Kuiper Belt-like objects (Su et al. 2007). To determine how common such dust disks are, we have carried out a *Spitzer* 24 μm survey of 72 hot WDs, and detected at least 7 WDs that exhibit clear IR excess, all of them still surrounded by planetary nebulae (PNe). Inspired by the prevalence of PN environment for hot WDs showing IR excesses, we have surveyed the *Spitzer* archive for more central stars of PN (CSPNs) with IR excesses; the search yields four cases in which CSPNs show excesses in 3.6−8.0 μm, and one additional case of 24 μm excess. We present the results of these two searches for dust-disk candidates, and discuss scenarios other than KBO collisions that need to be considered in explaining the observed near and/or mid-IR excess emission. These scenarios include unresolved companions, binary post-AGB evolution, and unresolved compact nebulosity. We describe planned follow-up observations aiming to help us distinguish between different origins of observed IR excesses.

1. Dust Disk around the Central Star of the Helix Nebula

The central star of the Helix Nebula is a hot white dwarf (WD) with an effective temperature of $\sim 100\,000$ K. Its *Spitzer Space Telescope* observations have revealed a new phenomenon. While the IRAC 3.6, 4.5 and 5.8 μm observations show a point source well accounted for by the photospheric emission from the hot WD, the 8 μm flux is twice the value expected from a single WD, and the MIPS 24 and 70 μm observations also show a compact source coincident with the WD. A follow-up Infra-Red Spectrograph (IRS) observation of the central point source has confirmed that the mid-IR emission originates from a dust continuum. The spectral energy distribution (SED) of the IR emitter indicates a temperature of 90−130 K, too cold for a stellar companion. The luminosity requires an emitting area of 4−40 AU^2, too large for planets. Only an extended object, such as a dust disk, can explain these properties. The location of the dust, 40−100 AU, corresponds to the location of the Kuiper Belt in the solar system, and the dust disk was suggested to originate from collisionally disrupted Kuiper Belt-like objects (KBOs) that were dynamically rejuvenated in the AGB and post-AGB evolutionary stages of the central star (Su et al. 2007).

To search for more dust disks similar to that around the WD in the Helix Nebula, we have carried out a *Spitzer* MIPS 24 μm survey of 72 hot WDs. The results of this survey have inspired us to search the *Spitzer* archive for central stars of planetary nebulae (CSPNs) with IR excesses. Here we present the results of the 24 μm survery and the *Spitzer* archival search. In addition to dust disks produced by KBO collisions, we also consider other scenarios that could produce IR excesses, such as remnants of post-AGB binary disks or H-poor compact nebulosities produced in born-again planetary nebulae (PNe).

2. Spitzer MIPS 24 μm Survey of Hot WDs

Could the dust disk around the Helix central star be a common phenomenon? Dust disks around low- and intermediate-mass main sequence stars have been identified via IR excesses in many surveys (e.g. Rieke et al. 2005; Bryden et al. 2006; Beichman et al. 2006; Trilling et al. 2008), and the dust in these disks is produced by collisions in planetisimal belts (Rieke et al. 2005), the same mechanism as suggested for the Helix dust disk. KBOs at 40–100 AU are sufficiently far to survive through stellar evolution (Debes & Sigurdsson 2002), and therefore, dust disks around WDs could be common.

These dust disks can, however, be detected through their IR emission only for a limited time. Figure 1 shows curves of constant dust temperatures as a function of WD's effective temperature and the dust's orbital radius. This figure demonstrates that only the hottest WDs ($T_{\rm eff} \approx 100\,000$ K) can sufficiently heat the dust at 40−100 AU to emit in the mid- to far-IR wavelengths, where they would be detectable with *Spitzer*. As the WDs cool, the dust temperatures drop, shifting the peak of excess emission out of the *Spitzer* wavelength range. Furthermore, the collisional rate decreases with time, and the dust disk will eventually dissipate below the detection limit.

We have therefore selected hot, young WDs for the *Spitzer* 24 μm survey. Out of 72 targets, 34 are still CSPNs, and are the closest analogs of the Helix Nebula and its central hot WD. The remainder of our sample included hot WDs with the broadest possible range of spectral classes.

We observe unresolved emission coincident with the WD is convincingly detected in 7 cases, all of them in PNe. In 5 additional cases, the possible point source is superposed on a bright nebular background. We have constructed SEDs of these WDs using optical and near-IR values from the literature. All 7 clear detections show excess emission at 24 μm.

Figure 2 shows examples of images and SEDs of WDs with 24 μm excesses. WD 0726+133 is the central star of the PN Abell 21. Its SED from optical to IRAC 4.5 μm can be well approximated a blackbody with WD's effective temperature (140 000 K, Werner et al. 1997). The 24 μm flux is, however, ~1500 times higher than the expected value. The upper limit on the dust temperature imposed by nondetection at 5.8 and 8.0 μm is ~120 K. Given the effective temperature of the WD, the dust would be heated to 120 K at a distance of ~50 AU. The emitting area required to account for the 24 μm flux is ~4 AU^2. These values are very similar to the properties of the dust disk around the Helix CSPN. While further observations are still needed to confirm the nature of the mid-IR emitter and constrain the SED at longer wavelengths, Abell 21 likely has a dust disk similar to that of the Helix CSPN.

Another dust disk candidate with SED similar to that of Helix CSPN is the central star of EGB 1, WD 0103+732. Optical and near-IR data points follow the blackbody tail of the hot central WD ($T_{\rm eff} = 147\,000$ K, Napiwotzki 2001), and the MIPS 24 μm flux is ~3700 times higher than the expected value. Follow-up IRAC observations from *Spitzer's* Cycle 5 show that IRAC 3.6, 4.5 and 5.8 fluxes lie on the blackbody tail, but the 8 μm data point shows flux ~60 times higher than expected. The IR excess can be fitted by a blackbody with a temperature of 170 K and an emitting area of ~5 AU^2. The dust temperature corresponds to the distance of ~40 AU from the central star. This CSPN also likely posesses a dust disk similar to Helix CSPN.

WD 0950+139, the CSPN of EGB 6, shows an SED that is somewhat different from that

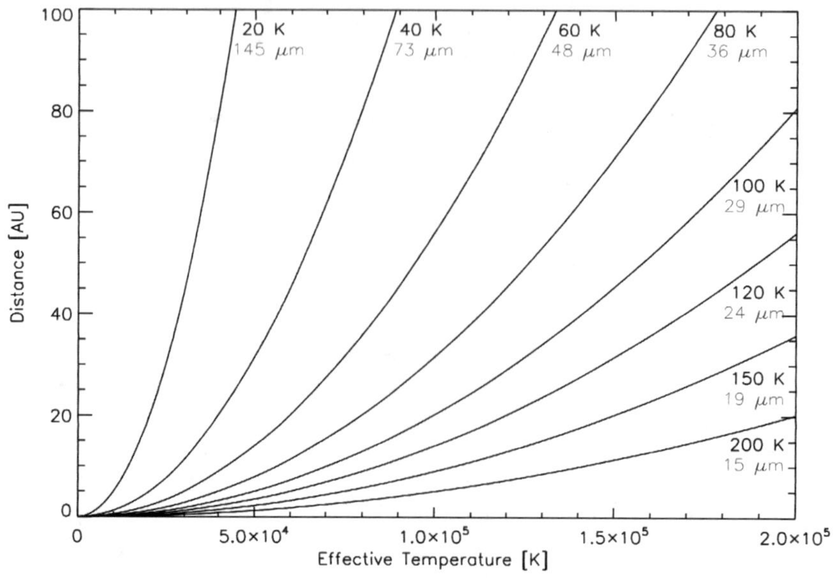

Figure 1. Curves of constant dust temperature in the $T_{\rm eff} - D$ (effective temperature - distance) parameter space. The dust temperatures and corresponding blackbody peak wavelengths are labelled. Dust temperatures ($T_{\rm dust}$) are related to D and $T_{\rm eff}$ via $T_{\rm dust} = (1-a)^{\frac{1}{4}} \sqrt{\frac{R_\star}{2D}}\, T_{\rm eff}$, where a is dust's albedo and R is the stellar radius. We have used the value of 0.1 for a, typical of asteroids and cometary dust, and 1 $R_\oplus$ for stellar radius, representative of WD radii. The dust at distances of 30−100 AU from the star will peak in mid-IR (24−70 μm) only if its central star is very hot (>80,000 K).

of the Helix CSPN. Emission in excess of the central star can be seen in J, H and K bands as well as at 24 μm, unlike the Helix CSPN, which showed excess starting from 8 μm. In addition, optical observations of this CSPN revealed a strong emission line component superposed on the continuum emission from the central WD. A spectrum of the sky 10″ N from the PN nucleus did not show any emission lines, thus suggesting that they arise in a compact unresolved nebulosity around the central WD (Liebert et al. 1989). High-resolution *HST* emission-line images revealed that the emission actually originates from a point-like source that is offset from the PN nucleus by 0.18″ (120 AU), and later continuum images confirmed the presence of a dM companion (Bond 1994). A dM ($T_{\rm eff} \approx 2000$ K) star can account for the near-IR excess, but not the MIPS 24 μm excess. It has been suggested that the companion has accreted material from the progenitor of WD 0950+139 and that this accretion disk is responsible for the nebular line emission (Bond 1994). The emission at 24 μm suggests that dust is likely present either around the WD or the dM companion, but IR spectra are needed to assess the nebular line contribution to the 24 μm emission.

3. Spitzer Archival survey of CSPNs

Interestingly, all hot WDs that exhibit 24 μm excesses are very young, still surrounded by PNe. Inspired by the prevalence of PN environment for hot WDs showing mid-IR excesses, we have used archival *Spitzer* IRAC (3.6, 4.5, 5.8, and 8.0 μm) and MIPS (24, 70, and 160 μm) observations of PNe to search for CSPNs with IR excesses.

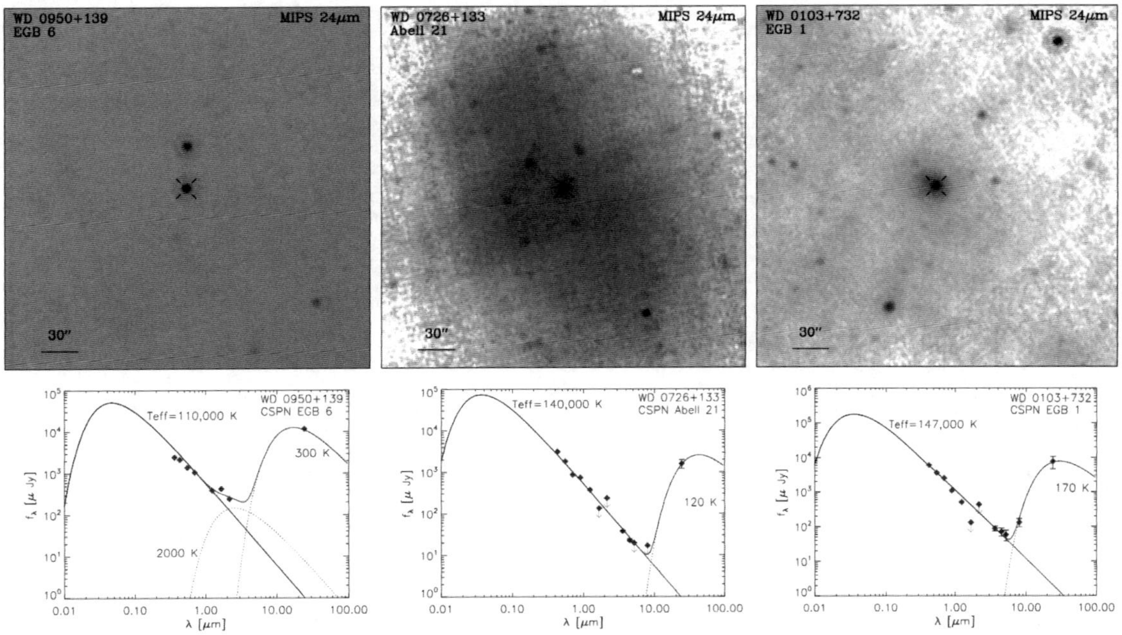

Figure 2. *Spitzer* 24 μm images (top) and SEDs (bottom) of three hot WDs from the 24 μm survey that exhibit 24 μm excesses. Each panel is labeled with WD and PN names, passband, and image scale. The SEDs are constructed with optical photometry from McCook & Sion (1999), 2MASS *JHK*, IRAC photometry (where available) and *Spitzer* MIPS 24 μm data.

We have examined images of 55 resolved PNe, and selected 13 in which the nebular emission was not too confusing or dominant in the central region, and the CSPN was detected in most IRAC bands. For these 13 cases, we have carried out photometric measurements. Supplemented with optical fluxes and near-IR photometry from 2MASS, we have constructed SEDs for these targets, and fitted them with blackbody curves. Four CSPNs show clear excess emission in the IRAC bands, and one shows excess in the MIPS 24 μm band.

The images and SEDs of three of these CSPNs are shown in Figure 3. The SED of NGC 2438 shows excess emission starting from the J-band across all IRAC bands. Furthermore, NGC 2438 was also imaged in the MIPS 24 μm survey of hot WDs, and the IR flux from the central source is ~6 orders of magnitude higher than what is expected from the CSPN alone. The excess emission in JHK and IRAC 3.6, 4.5 and 5.8 μm bands can be fitted by an M3 dwarf (0.5 $R_\odot$, 3470 K). The IR excess at 8 and 24 μm can be fitted with a 120 K dust disk with an emitting surface area of ~1000 AU^2.

In the case of NGC 6804, prominent IR excess is seen over the entire 1–8 μm range plotted. Fitting of the IR excess requires more temperature components (~1500 K, ~600 K), likely indicating a range of dust temperartures, if dust continuum is indeed the source of the observed IR emission. The SED does not exclude the possible presence of a cool ($T_{\rm eff} \approx 1500$ K) companion; however, a companion alone cannot account for all of the observed excess, and furthermore, no companion has been detected around this CSPN (Ciardullo et al. 1999).

The CSPN of NGC 7139 is too faint to be detected in 2MASS, but is clearly seen across all IRAC bands. We have used the faintest stars detected near the CSPN to place upper limits on the CSPN's JHK fluxes. The stellar emission was modelled based on the CSPN's effective temperature (125 000 K) and B magnitude from Walton et al. (1990). The SED shows clear

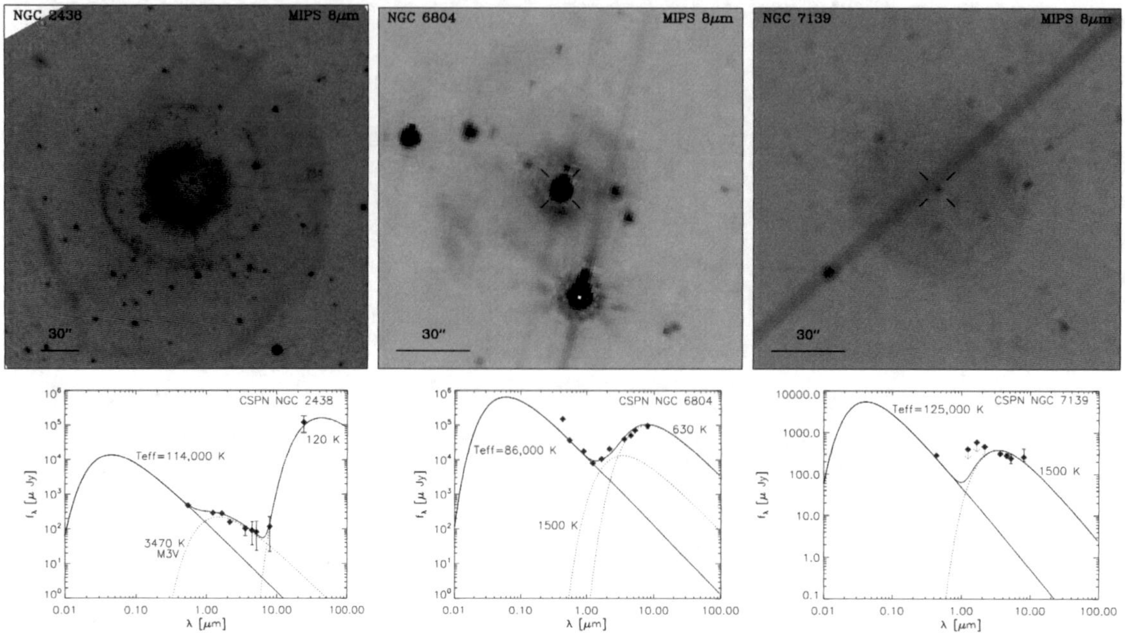

Figure 3. IRAC 8 μm images (top), and SED plots (bottom) of NGC 2438 (left), NGC 6804 (center), and NGC 7139 (right). The hot CSPNs are marked with crosslines. Image scales are marked in each panel. The data points in SEDs represent optical fluxes gathered from Ciardullo et al. (1999), Acker et al. (1990), Gathier & Pottasch (1988) and Walton et al. (1990) near-IR fluxes from 2MASS, and IRAC and MIPS photometry. The CSPN of NGC 7139 was below 2MASS detection limit, and the plotted upper limits represent fluxes of the faintest stars in the surrounding field of view.

excess in all IRAC bands, and can be fitted with a 1500 K blackbody for an emitting area of 0.002 AU^2. A 1500 K brown dwarf alone cannot account for all of the observed excess, as the emitting area required is higher then what a brown dwarf can provide.

4. Possible Origins of IR excesses

The IR excess of a given WD/CSPN is a necessary, but not sufficient, condition to conclude that the star is surrounded by a dust disk. Indeed, the differences between the SEDs of our targets imply different properties of the disks, if not even different origins of excess emission. We therefore need to consider other potential mechanisms that can produce IR excess.

4.1. Post-AGB Binary Evolution

It has been suggested from an analysis of broad-band SEDs of 51 post-AGB stars that Keplerian rotating dust disks are commonly present around binary post-AGB stars (De Ruyter et al. 2006). van Winckel (2003) showed that post-AGB stars displaying SEDs of warm dusty disks are all single-lined spectroscopic binaries. A tight correlation between the presence of hot dust and a binary companion (Waters et al. 1997, van Winckel et al. 1999) suggests that the hot dust persists in the system because it is trapped in a stable circumstellar or circumbinary orbit (van Winckel 2003).

The 7 hot WDs with 24 μm excesses from our survey and the 5 CSPNs with IRAC and/or MIPS excesses found in the *Spitzer* archive are all in PNe and thus represent the youngest

WDs that have just evolved past the post-AGB phase. One of them (NGC 2346) has a confirmed binary companion (Mendez & Niemela 1981) and SED that resembles those of post-AGB binaries; thus, one cannot help asking whether some of these 24 μm and/or IRAC excesses are also related to the IR excesses of binary post-AGB stars reported by De Ruyter et al. (2006). The fact that the near-IR SEDs of EGB 6 and NGC 2438 can be fitted by a cool companion, and the SEDs of NGC 6804 and 7139 do not exclude a companion lends support to this possibility. On the other hand, the SEDs of, e.g., the CSPNs of the Helix Nebula, Abell 21, and EGB 1 do not show any signature of a low-mass companion.

The binarity of AGB stars may play an important role in shaping the PNe they later produce, especially the bipolar PNe (Soker 1998; De Marco 2006). However, it is difficult to indentify and confirm the presence of a close companion of a CSPN via direct imaging (e.g., Ciardullo et al. 1999); furthermore, irregular spectral variations due to winds hamper the detection of periodic radial velocity variations (De Marco et al. 2007). If a dust disk trapped in a stable orbit around a binary system persists throughout the PN phase, its presence can serve as a powerful diagnostic for the binarity of a CSPN, and further our understanding of the role of binaries in the shaping of PNe.

4.2. Compact Unresolved Nebulosity

Spitzer's MIPS and IRAC passbands contain a number of spectral lines, for example, [Ar II] λ6.98 μm and [Ar III] λ8.99 μm in the IRAC 8 μm band, and [Ne V] λ24.32 μm and [O IV] λ25.89 μm in the MIPS 24 μm band. It is possible that the IR excesses of the hot WDs and CSPNs originate from these nebular lines, as for the rest of the PNe. In such a case, the central diffuse emission must be associated with density enhancement, but density enhancement is not expected near the centers of a large evolved PNe, unless they are born-again PNe like Abell 30 and Abell 78.

Abell 30 and Abell 78 are two large and faint PNe with hot nuclei that were both found to exhibit IR emission at 10 and 18 μm (Cohen & Barlow 1974). This central emission enhancement originates from knots seen in [OIII] λ5007 Å and He II λ4686 Å, but undetected in Hα (Jacoby 1979; Hazard et al. 1980). This observed morphological difference was suggested to be caused by H depletion in the central region (Jacoby & Ford 1983), which occurs in a born-again PN whose WD experiences a final thermal pulse during which H is completely burnt (Iben et al. 1983). It is possible that some of the observed IR excesses originate from an unresolved H-depleted nebulosity, thus, spectroscopic observations are necessary to unambiguously assess the nature of a mid-IR excess.

5. Future Work - Follow-up Observations

For a number of targets, we have very few, sometimes only one, data point in excess of the expected WD's photospheric emission. With multiple possible origins of the excess, it is imperative that we characterize the excess emission over a wider wavelength range, and carry out observations that would serve as diagnostics to distinguish among the possible origins. Some of these observations are outlined below.

5.1. Infra-Red Spectra and Photometry

CSPNs are, by definition, surrounded by ionized nebulae, and thus nebular emission lines will be present in the IRAC and MIPS bands. Spectral observations are therefore crucial to separate the dust continuum from the line emission and to evaluate their relative contributions to the observed excesses. We were awarded *Spitzer* priority 1 time to get IRS spectra and additional IRAC and MIPS imaging of CSPNs K 1-22, EGB 1, Sh 2-188 and Sh 2-216 and priority 3 time for CSPNs NGC 2438, NGC 6804 and NGC 7139. The extended SED coverage, provided by additional IRAC and MIPS observations, will help us better constrain the the SED and find the range of temperatures in the disk. With reasonable assumptions on grain properties (size distribution and composition), these observations will allow us to determine the radial distribution of the disk, which can help us differentiate between dust originating from binary post-AGB evolution or KBO collisions. The most interesting aspect of the spectra is, however, the possible detection of mineralogical features of dust grains. For example, if the silicate emission feature at 10 μm is detected, connections with crushed asteroids can be made.

5.2. High-resolution Optical Imaging

The *Spitzer* spectra will allow us to distinguish between the dust continuum and nebular line contributions to the excess IR emission, but *Spitzer*'s angular resolution limits our ability to conclude definitively whether the excess emission is associated with the CSPNs, their companions, or chance superposition of background sources. High-resolution view of the region within *Spitzer*'s point spread function is therefore needed to properly interpret the IR observations.

Broad-band imaging would provide the clearest view of continuum sources in the vicinity of the WDs with IR excesses and should thus help us find previously unresolved companions or background galaxies. Narrow-band nebular line (Hα, [O III] and He II) images would allow us to diagnose the compositions of the ionized gas in the central region. H-deficient gas would be present in the [O III] and He II images, but not in Hα images. Comparisons between Hα and [O III] images will therefore reveal the presence of H-deficient ejecta such as those observed at centers of born-again PNe.

5.3. High-dispersion Optical Spectroscopy

High-dispersion (R$\sim$30,000) echelle spectroscopy of hot WDs and CSPNs with IR excesses can help us find spatially unresolved emission line sources and measure their velocities. The receding and approaching sides of the PN should average to the WD velocity. An unresolved nebula centered on the WD would have the same central velocity but a different FWHM, and a nebula centered on a companion would have a different central velocity from the WD. If the Hα component is absent at velocities corresponding to these emission lines, H-poor ejecta is likely present.

In addition, the high-dispersion echelle observations would also help us search for unresolved companions of the hot WDs and CSPNs. Hot WDs may show a non-LTE inversion in the Hα absorption line, such as the one seen in the Helix central star. Monitoring temporal variations of stellar Hα profiles could help us search for companions, similar to the work done for the Helix central star (Gruendl et al. 2001).

6. Summary

The discovery of a dust disk around the CSPN of the Helix nebula through its 24 μm excess has inspired us to conduct a *Spitzer* 24 μm survey of hot ($T_{\rm eff} \approx 100\,000$ K) WDs. Out of 72 targets observed, 7 show 24 μm excesses, all of them in PNe. To find more cases of CSPNs with IR excess, we have searched the *Spitzer* archive, and found 4 targets with IRAC excesses, and 1 additional target with MIPS 24 μm excess. While some SEDs are similar to that of the Helix CSPN and show excess emission only at wavelengths longwards of 8 μm, others show excess starting at shorter wavelength.

The dust around the Helix central star was suggested to be produced by collisions among KBOs (Su et al. 2007). However, other mechanisms that could produce IR excess need to be considered as well. Since 24 μm excesses were found only in CSPNs, it is possible that the observed dust is a remnant from disks formed during post-AGB binary evolution. If the IR excesses originate from nebular emission lines, the central emission enhancement suggests a

Journal of Physics: Conference Series **172** (2009) 012055 doi:10.1088/1742-6596/172/1/012055

presence of compact nebulosity, which is not expected in large, evolved PNe, unless they are born-again PNe. More observations, such as IR spectra, high-resolution imaging, or high-dispersion optical spectroscopy, are needed to help us identify the origins of these IR excesses.

References

Beichman C A et al. 2006 *ApJ* **652** 1674
Bond H E 1994 *Interacting Binary Stars* **56** 179
Bryden G et al. 2006 *ApJ* **636** 1098
Ciardullo R Bond H E Sipior M S et al. 1999 *AJ* **118** 488
Cohen M & Barlow M J 1974 *ApJ* **213** 737
Debes J H & Sigurdsson S 2002 *ApJ* **572** 556
De Marco O 2006 *Planetary Nebulae in our Galaxy and Beyond* **234** 111
De Marco O Wortel S Bond H E & Harmer D 2007 *arXiv:0709.1508*
De Ruyter S van Winckel H Maas T et al. 2006 *A&A* **448** 641
Gruendl R A Chu Y-H O'Dwyer I J & Guerrero M A 2001 *AJ* **122** 308
Iben I Jr Kaler J B Truran J W & Renzini A 1983 *ApJ* **264** 605
Jacoby G H 1979 *PASP* **91** 754
Jacoby G H & Ford, H.C. 1983, ApJ, 266, 298
Liebert J Green R Bond H E et al. 1989 *ApJ* **346** 251
Mendez R H & Niemela V S 1981 *ApJ* **250** 240
Rieke G H et al. 2005 *ApJ***620** 1010
Soker N 1998 *V* **496** 833
Su K Y L et al. 2007 *ApJL* **657** L41
Trilling D E et al. 2008 *ApJ* **674** 1086
Walton N A Walsh J R & Sahu K C 1990 *A&A* **230** 445
Waters L B F M Waelkens C & Van Winckel H 1997 *Planetary Nebulae* **80** 313
Van Winckel H 2003 *Astronomical Society of the Pacific Conference Series* **303** 294
van Winckel H 1999 *Asymptotic Giant Branch Stars* **191** 465
Werner K Dreizler S Heber U et al. 1997 *Reviews in Modern Astronomy* **10** 219

16th European White Dwarfs Workshop IOP Publishing
Journal of Physics: Conference Series **172** (2009) 012056 doi:10.1088/1742-6596/172/1/012056

Spitzer Planet Limits around a Pulsating White Dwarf

Fergal Mullally

Dept. Astrophysical Sciences, Princeton University, Peyton Hall, Princeton, NJ 08544-1001 USA

E-mail: `fergal@astro.princeton.edu`

Abstract. We present infrared observations in search of a planet around the white dwarf GD66. Time-series photometry of GD66 shows a variation in the arrival time of stellar pulsations consistent with the presence of a planet with mass $\geq 2.4\,M_J$. Any such planet is too close to the star to be resolved, but the planet's light can be directly detected as an excess flux at $4.5\,\mu m$. We observed GD66 with the two shorter wavelength channels of IRAC on Spitzer but did not find strong evidence of a companion, placing an upper limit of 5–7 M_J on the mass of the companion, assuming an age of 1.2–1.7 Gyr.

1. Introduction

The advantage of searching for sub-stellar companions around white dwarf stars (WDs) has long been recognized. The improved contrast between a WD and a companion was first exploited by Probst (1983) and the first L dwarf around a WD (or indeed any star) was discovered around GD165 by Becklin & Zuckerman (1988). In recent years, improved instrumentation has enabled searches for lower mass objects, to the point where direct detection searches for planetary mass objects are possible from the ground (e.g. Debes et al. 2005; Hogan et al. 2008, in press). Searches utilizing the *Spitzer* Space Telescope (Werner et al. 2004) can obtain mass limits as low as $\sim 10\,M_J$ (Farihi et al. 2008a; Mullally 2007).

WDs are interesting targets for more reasons than just their luminosity. By looking at WDs we can sample a range of progenitor masses from about 1-8 $M_\odot$, including higher mass stars not amenable to radial velocity surveys. Higher mass main-sequence stars form more, and more massive planets (Johnson et al. 2007), but suffer from higher constrast ratios with those planets. However, those stars die faster, and produce more massive (and hence lower radius) WDs. WDs formed from high mass progenitors therefore make the best targets for direct detection searches.

Currently, the best main-sequence targets for direct detection of planetary companions are either transiting systems or very young stars. Detection of a planet around a WD would open up a older, more sedate region of parameter space where the predictions of planet models could be tested against observation for ages ranging from hundreds of millions to billions of years. With lower constrasts and older ages, WD planets are key to testing our models of planetary evolution and the prime target for direct detection searches. Moreover, the lower contrast means future instruments will always be able to extract more detail from spectra of WD companions than main-sequence.

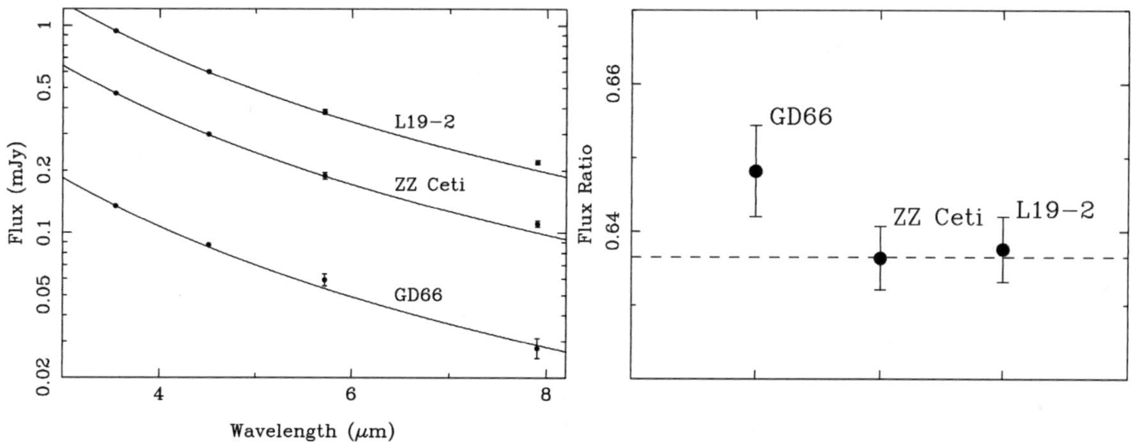

Figure 1. Left: Fluxes of three WDs in all 4 IRAC bands. The solid line connecting the points is a 12,000 K DA white dwarf atmosphere model. The model is normalised to agree with the flux at 3.6 μm. **Right:** Flux ratio for three WDs. The dash line is the weighted average flux of the two comparison WDs.

WDs also lend insight into the ultimate fate of planetary systems. Livio & Soker (1984) first estimated the conditions necessary for a gas giant planet to survive the red giant phase of its parent star. Sackmann et al. (1993) concluded that the Earth would survive the red giant phase of the Sun, but more recent calculations by Schröder & Connon Smith (2008) concluded the opposite. These calculations depend on stellar mass loss rates that are weakly constrained by observation due to the rapid evolution of evolved stars. Determining the distribution of planets around WDs will provide a key test for these models. The winds from evolved stars pollute the interstellar medium with the material that creates the next generation of terrestrial planets, so surveys for planets around dead stars will inform our understanding of planets around stars not yet born.

Early empirical success comes from the detection of a planet around a sub-dwarf star using pulsation timings (Silvotti et al. 2007), and the detection of silicate rich debris disks around WDs (e.g. Reach et al. 2005b; Farihi et al. 2007). Analysis of the composition of material raining down onto the surface of a WD from such a disk by Zuckerman et al. (2007) concluded that the pattern of elemental abundances is consistent with a species of asteroid. Asteroids are markers of planetary systems, and so the frequency of debris disk WDs may eventually provide an independent measure of the fraction of stars that bear (or bore) planetary systems.

Mullally et al. (2008, hereafter M08) published the results of a survey that sought to detect evidence of planets as variations in the arrival time of pulsations from variable WDs (in a manner analogous to sub-dwarf and pulsar planets). For one object, GD66, they discovered a variation consistent with a 2.1 M_J planet in a 4.5 yr orbit. Unfortunately, the span of their data was not long enough to cover an entire orbit and their detection remains provisional. We observed GD66 with the two short wavelength bands of IRAC on *Spitzer* in an attempt to confirm the existence of this object by direct detection in the infrared. Although the candidate system is too distant to resolve into individual components, the low luminosity of the star means that the flux from the planet at 4.5 μm rivals that of the WD. The goal of these observations is to measure an excess flux at 4.5 μm over the bare photosphere of GD66 due to the light from the planet. We observed the flux ratio (i.e. the color) between channels 2 and 1 on IRAC (4.5 and 3.6 μm respectively, Fazio et al. 2004) and comparing this ratio to a sample of other similar stars.

Table 1. Observed fluxes. Note. – Integration times are for channels 2 and 4, and are four times shorter for channels 1 and 3. The flux from G238−53 is probably overestimated due to the presence of a diffraction spike from a nearby star. The quoted uncertainties in these measurements include a 5% absolute calibration uncertainty added in quadrature.

Star	J (mag)	Integration (s)	Ch 1 (mJy)	Ch 2 (mJy)	Ch 3 (mJy)	Ch 4 (mJy)
GD66	15.7	6400	0.1349(68)	0.0875(44)	0.0593(50)	0.0277(35)
L19-2	13.6	768	0.941(47)	0.600(30)	0.383(22)	0.219(12)
ZZCeti	14.4	1920	0.470(24)	0.299(15)	0.189(12)	0.1105(67)
G238-53	15.7	6400	0.1385(69)	0.0911(46)	0.0573(39)	0.0405(26)

2. Observations

We observed GD66 (WD0517+307) with all four IRAC channels on 17th October 2007 (AORs 22855424 and 22856448) as part of program 40255. We recorded sixteen 100 second exposures simultaneously in channels 1 and 3, using a spiral dither pattern. We repeated the procedure for channels 2 and 4, recording 4 sets of sixteen images for a total exposure time of 6,400 seconds. The raw frames were calibrated using version S16.10.0 of the IRAC pipeline to create bcd frames.

We also observed 3 WDs with similar temperatures to calibrate our observations. L19-2 (AORs 22856192 and 22857216), ZZ Ceti (AORs 22855936 and 22856960) and G238-53 (AORs 22855680 and 22856704) are all DAVs, and Bergeron et al. (2004) reports temperatures within 120 K of GD66 based on optical spectroscopy. This agreement is better than the quoted uncertainty in the measurements, and so we can consider our sample to have a uniform temperature. As an independent, albeit cruder approach, we also insisted on broadly similar pulsation properties for each star, as Mukadam et al. (2006) showed that pulsation properties are correlated with temperature. We observed each of these stars in a similar pattern as GD66, scaling our total exposure time with magnitude.

We extracted photometry from these frames using the astrolib package in IDL. We first convert the recorded flux into units of electrons. Pixels with bits set in any of the pmask, imask or rmask images were marked as bad and their values set to the median value of pixels in a box 21 pixel on a side centered on the bad pixel in question. We ignored bit 13 of the imask, indicating crosstalk, which was set in a large number of pixels that did not seem to be affected by any noticeable problem. We see no evidence of a systematic drift in the measured flux in any channel.

We extract the flux using a sky annulus of 10-20 pixels and series of apertures ranging from 2 to 7 pixels radius. We choose the aperture that gives the greatest signal to noise (2 pixels), adding in quadrature a contribution to the error term due to the read noise, as listed in the header of the bcd frame. We then apply corrections to the observed flux as suggested by the Data Handbook: array location dependence, aperture correction, color correction and a pixel phase correction for channel 1.

3. Results

In Figure 1 we show spectral energy distributions (SEDs) for GD66, L19-2 and ZZ Ceti. The close correspondence between the observed and expected SED rules out the presence of debris disks similar to those found around other WDs. We show a table of observed fluxes for all observed stars in Table 1.

Table 2. Observed fluxes ratios and uncertainties

Star	Ratio (Ch 2/Ch 1)	Photometric Uncertainty	Systematic Uncertainty	Total Uncertainty
GD66	0.6482	0.0051	0.0035	0.0062
ZZCeti	0.6364	0.0026	0.0034	0.0043
L19-2	0.6376	0.0028	0.0034	0.0044
G238-53	0.6574	0.0030	0.0036	0.0046

We show the flux ratio between channels 2 and 1, r, for each star in Figure 1 and Table 2. GD66 and G238-53 are noticeably redder (higher value of r) than the other two stars. However, careful analysis of the images shows that this flux excess is most likely an artefact for G238-53. The diffraction spike from a nearby bright star passes very close to the position of G238-53 on the chip. Unlike the other three stars, the observations for the two channels for this star were not take sequentially, but separated by $\sim$ 1 week. The slightly different viewing angle between the two visits means that the diffraction spike is closer to the star in the epoch of channel 2. Although some of these pixels are masked, we consider the measured flux ratio to be untrustworthy. The flux ratio is significantly higher (0.675 versus 0.657) for a 3 pixel aperture, a further indication that flux is contaminated. We therefore ignore G238-53 in our analysis.

3.1. Error Analysis

We now determine whether this excess in r for GD66 is statistically significant. The error budget of an IRAC observation is divided to photometric, systematic and calibration components. The IRAC data handbook gives a value of 3% for the calibration uncertainty while Farihi et al. (2008) concludes, based in part on a literature survey that 5% is more appropriate. However, because we are measuring a flux ratio, we need not concern ourselves with the absolute calibration error in converting from electrons to mJy necessary to compare IRAC fluxes to other photometric systems. We do, however, include a conservative uncertainty term of 5% in Table 1 to facilitate comparison with other stars.

The photometric uncertainty in channel 1 can be computed from the rms scatter of individual bcd frames. Adding the fractional uncertainties in quadrature gives the fractional uncertainty in the flux ratio. Our observations were designed to achieve <1% photometric accuracy, and this goal was comfortably exceeded, as shown in Table 2.

The systematic uncertainty is measure of the variability in the performance of the instrument, and the spread in values we would obtain by performing the same experiment multiple times. The Observers' Manual quotes a stability for IRAC of <1%. To more accurately measure this value, we examined data on BD +60 1753, a calibration star that has been repeatedly observed by IRAC over the lifetime of the instrument (Reach et al. 2005a) and determined the systematic repeatability limit to be 0.54%.

We add our independent estimate of the systematic uncertainty in quadrature to the photometric uncertainty to get a total ($1\,\sigma$) uncertainty in the ratio, δr, which is given in the third column of Table 2. The *excess flux*, E is the difference between r for GD66, 0.6482(62) and the mean value of r for the other two stars, 0.6365(31), or E=0.0117(69). The excess flux is only 1.7 times greater than the uncertainty and therefore not statistically significant.

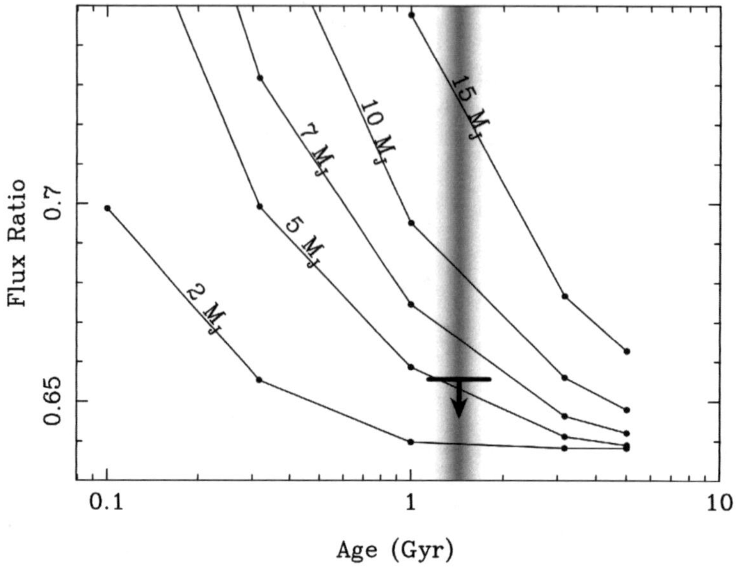

Figure 2. Mass limits on a planetary companion as a function of the assumed age. The solid lines show the expected flux ratio for WD+planet systems of different ages. The shaded region indicates the best estimate of the age of GD66. The downward arrow indicates the 3σ upper bound on the flux ratio.

4. Limits on Planet Mass

While we did not detect an excess, this does not rule out the evidence from timing measurements, and we can use our observations to constrain the mass of any companion. We can state with 3σ confidence that the flux ratio is less than 0.6556. To compare this number to planet models and place an upper bound on the mass of the companion we first need the age of the system.

Based on the measured surface temperature and gravity (11,989 K and $\log g$=8.05 Bergeron et al. 2004), and comparing to models from Wood (1992), we determine a mass of 0.64(03) $M_\odot$ and a cooling age of 500 Myr. The age of the progenitor main-sequence star is more difficult to estimate. Estimates of the initial-final mass relationship (IFMR) for WDs are based on comparing the masses of WDs in open clusters to the main-sequence turn-off age of the cluster itself. Until recently, this relationship was only empirically constrained for young, and hence high mass progenitors. However, newer observations have begun to extend the IFMR down to the relatively lower mass of GD66. Using the linear fits to the IFMR from Dobbie et al. (2006), Kalirai et al. (2007) and Catalán et al. (2008) we estimate an initial mass of 2.20(46), 2.26(46) and 2.64(57) $M_\odot$ respectively, which agree within the uncertainties. A theoretical relationship from Meng et al. (2007) gives an intermediate value of 2.49(41) $M_\odot$, while the semi-empirical method of Salaris et al. (2008) suggests 2.31(22) $M_\odot$. Calculations from Pols et al. (1998) indicate that the lifetime of a 2.5 $M_\odot$ star is 830 Myr. The lifetime of a main sequence star scales roughly as $M^{-2.5}$ in this mass regime, so the calculated main-sequence lifetime of GD66 is 700 Myr to 1.2 Gyr. Including the WD cooling age, the total age of the system is 1.2–1.7 Gyr.

We take atmosphere models of a 12,000 K DA courtesy of D. Koester (Finley et al. 1997), and combine them with models of planetary mass objects from Burrows et al. (2003). We perform synthetic photometry on the combined model to determine the flux ratio between channels 1 and 2 for planets of different masses. We repeat the procedure for models with ages from 0.3 to 5 Gyr and show the results in Figure 2. Given our limit on the flux excess, and the uncertainty in the age, the 3σ upper bound on the mass of the planet is approximately 5–6 M_J.

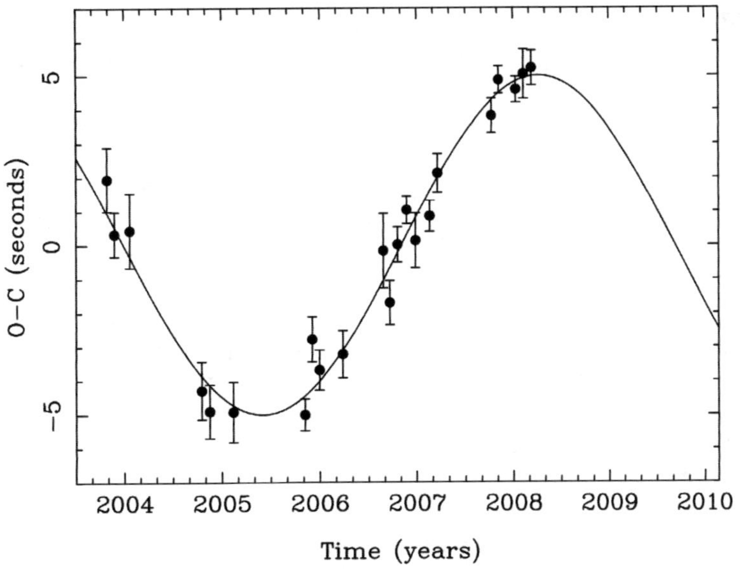

Figure 3. O-C diagram for GD66. Filled circles are the observed arrival times of the pulsations. Values greater than zero indicate arrival times that were later than expected based on the assumption of a constant pulsation period and no companions. The solid line is the best fit sinusoid to the data.

5. Additional Time Series Observations

We also present additional ground-based optical time-series observations of GD66 taken with Argos (Nather & Mukadam 2004) on the 2.1 m telescope at McDonald Observatory. A full description of our observational technique is given in M08, but we repeat a brief summary here. We performed time series differential photometry on GD66 with a continual sequence of exposures over a period of 4–8 hours per night. We used Wqed (Thompson & Mullally, this volume) to divide the lightcurve of the target star by the sum of one of more reference stars to remove the effects of seeing and transparency variations and we corrected our times for the motion of the earth around the sun using the method of Stumpff (1980). We combined all observations over the span of a week into a single data set, and measured the difference between the observed time of arrival of the 302s pulsation mode and that calculated based on the assumption of a constant period (O-C). We show our updated O-C diagram in Figure 3.

M08 predicted that the orbit would turn over in late 2007 and the slope of the O-C diagram would be negative in 2008. This has clearly not happened and the orbital period is longer than previously expected. Our current best fit parameters are a period of 5.69(30) yr, an orbital separation of 2.75(11) AU and $m_p . \sin i = 2.36(17)\,M_J$, where m_p is the mass of the planet and i the inclination of the orbit to the line of sight. With previous experience in mind we are cautious in predicting the future behavior of the orbit, but we note that our current circular best fit orbit appears to be on the verge of turning over. Inspection of the fit, especially in the years 2004/2005 also suggest that an eccentric fit would better match the data. However, without a full orbit, the eccentric fit is unconstrained. Consequently, the true orbital period may be very much longer than 6 years.

6. Discussion

Although confirmation of the existence of this planet remains frustratingly elusive, it is interesting to assume that it does exist and speculate on its history. Livio et al. (2005) suggest that planets could form during a binary WD merger, but the low mass of GD66 makes this scenario unlikely. It is possible that there is a second epoch of planet formation around stars after the red giant stage. Only a tiny fraction of the carbon and silicon rich material ejected by a star needs to fall back onto a circumstellar disk to create a minimum mass solar nebula, although such a disk would have to survive the high UV flux of the newly born WD (Villaver & Livio 2007). In this scenario, the planet's age would be less than 0.5 Gyr, and our observations would constrain the mass to $<3\,M_J$, uncomfortably close to the lower mass limit set by our timing measurements.

If the planet was born co-eval with the star, conservation of angular momentum would have caused it to drift outward linearly with the mass loss that pressaged the birth of the WD (Jeans 1924; Burleigh et al. 2002). If this were the *only* force acting on the orbit, the initial orbit would have been only 0.6–0.8 AU. At maximum extent, the envelope of the AGB phase of a 2.5 AU star extends out as far as 4–4.5 AU (Pols et al. 1998), engulfing the planet and presumably destroying it. Farihi et al. (2008) propose the idea of a sacrifical interior planet which donates angular momentum to the AGB envelope. The envelope is ejected before it can engulf the outer planet(s), while the interior planet spirals onto the red giant core and it is destroyed.

A star goes through a series of pulses during the final stages of evolution, where the radius increases dramatically for a short period of time. If the planet is engulfed (or nearly so) during one of these pulses, viscous drag (or tidal forces) will cause the orbit to spiral inward. For a planet that was only engulfed for short periods of time, its orbits would spiral in a small amount, which might explain the GD66 system.

7. Conclusion

We report on *Spitzer* IRAC observations of the white dwarf GD66 in an attempt to directly detect a companion planet. We fail to detect an excess with any statistical significance. However, combined with our ground based timing observations we can now constrain the planet candidate's mass to between 2.4 and $\approx$5–6 M_J.

Acknowledgments

We thank Mike Montgomery for donating some of the telescope time used to make ground based observations in this work. We thank Detlev Koester for the use of his atmosphere models. We also thank D. E. Winget, Ted von Hippel, Mukremin Kilic and Marc Kuchner for their assistance at various stages of this project. This work is based in part on observations made with the Spitzer Space Telescope, which is operated by the Jet Propulsion Laboratory, California Institute of Technology under a contract with NASA. Support for this work was provided by NASA through an award issued by JPL/Caltech.

References

Becklin E E & Zuckerman B 1988 sl Nature **336** 656
Bergeron P Fontaine G Billères M Boudreault S & Green E M 2004 *ApJ* 600 404
Burleigh M R Clarke F J & Hodgkin S T 2002 *MNRAS* **331** L41
Burrows A Sudarsky D & Lunine J I 2003 *ApJ* **596** 587
Catalán S Isern J García-Berro E Ribas I Allende Prieto C & Bonanos A Z 2008 *A&A* **477** 213
Debes J H Sigurdsson S & Woodgate B E 2005 *AJ* **130** 1221
Dobbie P D et al. 2006 *MNRAS* **369** 383
Farihi J Becklin E E & Zuckerman B 2008a *ArXiv 0804 0237*
Farihi J Zuckerman B & Becklin E E 2008b *ApJ* **674** 431
Farihi J Zuckerman B Becklin E E & Jura M 2007 *15th European Workshop on White Dwarfs* ed Napiwotzki R & Burleigh M R 372 315
Fazio G G et al. 2004 *ApJS* **154** 10
Finley D S Koester D & Basri G 1997 *ApJ* **488** 375
Jeans J H 1924 *MNRAS* **85** 2

Johnson J A Butler R P Marcy G W Fischer D A Vogt S S Wright J T & Peek K M G 2007 *ApJ* **670** 833
Kalirai J S Hansen B M S Kelson D D Reitzel D B Rich R M & Richer H B 2008 *ApJ* **676** 594
Livio M & Soker N 1984 *MNRAS* **208** 763
Livio M Pringle J E & Wood K 2005 *ApJL* **632** L37
Meng X Chen X & Han Z 2007 *ArXiv 0710 2397*
Mukadam A S Montgomery M H Winget D E Kepler S O & Clemens J C 2006 *ApJ* **640** 956
Mullally F Winget D E Degennaro S Jeffery E Thompson S E Chandler D & Kepler S O 2008 *ApJ* **676** 573
Mullally F R 2007 *PhD Thesis* The University of Texas at Austin
Nather R E & Mukadam A S 2004 *ApJ* **605** 846
Pols O R Schroder K P Hurley J R Tout C A & Eggleton P P 1998 *MNRAS* **298** 525
Probst R G 1983 *ApJS* **53** 335
Reach W T et al. 2005a *PASP* **117** 978
Reach W T Kuchner M J von Hippel T Burrows A Mullally F Kilic M & Winget D E 2005b *ApJL* **635** L161
Sackmann I J Boothroyd A I & Kraemer K E 1993 *ApJ* **418** 457
Salaris M Serenelli A Weiss A & Miller Bertolami M 2008 *ArXiv 0807 3567*
Schröder K P & Connon Smith R 2008 *MNRAS* **386** 155
Silvotti R et al. 2007 *Nature* **449** 189
Stumpff P 1980 *A&AS* **41** 1
Villaver E & Livio M 2007 *ApJ* **661** 1192
Werner M W et al. 2004 *ApJS* **154** 1
Wood M A 1992 *ApJ* **386** 539
Zuckerman B Koester D Melis C Hansen B M & Jura M 2007 *ApJ* **671** 872

Journal of Physics: Conference Series **172** (2009) 012057 doi:10.1088/1742-6596/172/1/012057

Transit detection limits for sub-stellar and terrestrial companions to white dwarfs

***F Faedi[1], R West[1], M R Burleigh[1], M R Goad[1] and L Hebb[2]**

[1] Department of Physics and Astronomy, University of Leicester, Leicester, LE1 RH7 UK
[2] School of Physics and Astronomy, University of St Andrews, North Haugh, St Andrews, Fife KY16 9SS UK

E-mail: *ff20@star.le.ac.uk

Abstract. The SuperWASP project is a ground-based ultra wide angle search for extra-solar planetary transits that has successfully detected 15 previously unknown planets in the last two years. We have used SuperWASP photometric data to investigate the transit characteristics of and detection limits for brown dwarfs, gas giants and terrestrial companions in orbit around white dwarfs. The relatively small size of a white dwarf host star (approximately 1 Earth radius), implies that any sub-stellar or gas giant companion will completely eclipse it, while terrestrial bodies smaller than the Moon will produce relatively large ($> 1\%$) transits, detectable in good S/N light-curves. We performed extensive simulations using SuperWASP photometric data and we found that for Gaussian random noise we are sensitive to companions as small as the Moon. Our sensitivity drops in the presence of co-variant noise structure, nevertheless Earth-size bodies remain readily detectable in relatively low S/N data. We searched for eclipses and transit signals in a sample of 174 WASP targets, resulting from a cross-correlation of the McCook & Sion catalogue and the SuperWASP data archive. This study found no evidence for sub-stellar or planetary companions in close orbits around our sample of white dwarfs.

1. Introduction

The SuperWASP project is a Wide Angle Search for transiting Extra-solar Planets, which consists of two completely robotic telescopes, SuperWASP-North (La Palma, Canary Islands, Spain) and SuperWASP-South (Sutherland, South Africa). Each telescope comprises eight f/1.8 Canon lenses each with a 2048^2 13.5μm pixels CCD, giving a field of view of 7.8 degrees square for each camera. The instruments cyclically raster the sky in a series of fields centred on the current local sidereal time and spaced by 1 hour in right ascension to obtain time series of images with typical sampling rate of 8-10 minutes. SuperWASP is the most successful ground-based transit survey with 15 previously unknown extra-solar planets detected in the last two years. More information on the project can be found in Pollacco et al. (2006) and on-line at http://www.superwasp.org/.

The transit technique involves searching for periodic dips in stellar light-curves due to the orbital revolution of a transiting planet, blocking a fraction of the stellar light. The planet transit probability (P_{tr}) is directly related to the stellar and planetary radii $R_\star$, R_{pl} and the semi−major axis of the orbit a by $P_{tr}{=}(R_\star + R_{pl})/a$. Therefore, only planets with their orbital plane aligned within a few degrees to the line of sight have their transit visible. For a given planetary radius, the transit depth (δ) is proportional to $(R_{pl}/R_\star)^2$. Thus, planets orbiting main-

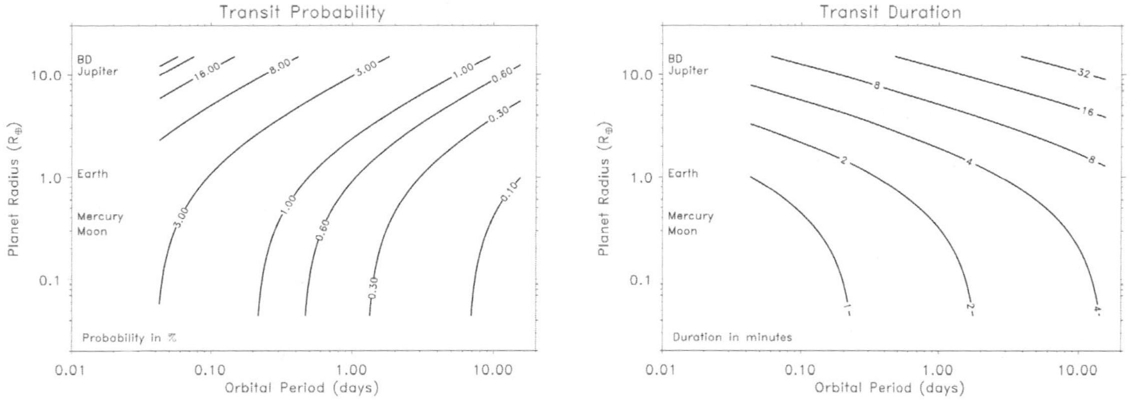

Figure 1. Left panel: geometric transit probability in percent, in the parameter space defined by companions radius in $R_\oplus$ and orbital periods in days. Right panel: transit duration in minutes, same parameter space as left panel

sequence solar-type stars have extremely shallow transits, about 10^{-2} magnitudes for a giant planet and 10^{-4} magnitudes for an Earth-size planet. A strong advantage over main sequence star primaries is offered by white dwarf stars. White dwarfs (WD) are compact degenerate objects, with approximately the same radius as the Earth, and represent the final stage of evolution of main−sequence stars with masses $\leq 8\,M_\odot$ (i.e. $\sim 97\%$ of all stars in our galaxy). Any sub-stellar or gas giant companion in orbit around a white dwarf will completely eclipse it, while bodies as small as the Moon will have relatively large transit depths ($\sim 3\%$), assuming that such systems survive the later stages of stellar evolution. Sub-stellar companions to white dwarfs are rare ($< 0.5\%$, e.g. Farihi et al. 2005). Only one wide binary system GD165 (Becklin and Zuckerman, 1988), and two detached non-eclipsing, short-period white dwarf+brown dwarf (BD) systems are currently known, GD1400 (Burleigh et al. 2008 in preparation, orbital period ~10 hours) and WD0137-349 (Maxted et al. 2006, orbital period ~116 min). The latter is the lowest mass object ($\sim 50\,M_{JUP}$) known to have survived the common envelope (CE) phase of stellar evolution. These systems remain difficult to identify either as infra-red excesses or through radial velocities measurements. The detection of more such systems will allow us to place observational upper limits on the mass of sub-stellar companions that can survive CE evolution. For example, can Hot Jupiters survive their parent star's evolution to a white dwarf? The detection of a significant number of eclipsing WD+BD binary systems might help uncover the hypothesised population of 'old' cataclysmic variables (CV) in which the companion has been reduced to a sub−stellar mass (e.g. Patterson 1998; Littlefair et al. 2003). These systems are undetectable as X-ray sources and difficult to identify in optical and infra-red surveys. The first such system was finally confirmed by Littlefair et al. (2006), although not through a direct infra-red spectroscopic detection of the brown dwarf companion.

Several theoretical studies discuss post-main sequence evolution of planetary systems and show that planetary survival is not beyond possibility (Duncan and Lissauer 1998; Debes and Sigurdsson 2002; Burleigh et al. 2002; Villaver and Livio 2007). Observations indicate that planets in wide orbits can survive stellar evolution (see Sato et al. 2003 and Silvotti et al. 2007). Moreover, Mullally et al. (2008) found convincing evidence of a $2\,M_{JUP}$ planet in a 4.5 year orbit around a pulsating white dwarf. If confirmed, this will show that planets can indeed survive the death of their parent star. Short-period planetary companions to white dwarfs may seem less likely. Villaver and Livio (2007) suggest that planets in orbit within the reach of the red giant's envelope will either evaporate or in rare cases, will accrete mass and become a close, low-mass companion to the star. Nevertheless, the recent detection of silicate-rich dust discs around a growing number of white dwarfs at orbital radii up to $\sim$1 $R_\odot$ (e.g. Reach et al. 2005, Farihi et al. 2007, 2008; Jura 2003) suggests that asteroids and larger objects can migrate into such orbits during the final stages of solar system evolution, perhaps because of dynamical instabilities (Debes and Sigurdsson 2002).

In §2 we discuss the characteristics of the transit signals, the parameter space investigated and our detection method. In §3 we analysed a sample of 174 WDs in the superWASP archive. Finally in §4 we discuss our conclusions.

2. Characteristics of the transit signal

A transit signal is described by its *duration*, its *depth* and its *shape*. The known transiting extra-solar planets show signals characterised by an ingress, a flat bottom and an egress, with durations between 2-3 hours and depths of about 1% (see for example Cameron et al. 2007). Our simulated transit events have been modelled assuming circular orbits and fixed stellar parameters. We considered a typical 1 Gyr old Carbon core white dwarf of mass 0.6 $M_\odot$ and radius of $\sim$ 0.013 $R_\odot$. We investigated transits for planets ranging from the Moon size to Jupiter and/or a brown dwarf (which have approximately the same radius), with orbital periods in the range $\sim$2 h - 15 days (i.e. $\sim$0.003 - 0.1 A.U.). For this choice of parameters our simulated transits show very short durations ($\sim$1 to 32 min) and very deep transit signals ($\sim$3-100% depth). Figure 1 shows the transit probability and duration for our choice of parameters. Figure 2-left panel shows the transit depth in the same parameter space as Figure 1.

2.1. Synthetic SuperWASP light-curves

SuperWASP light-curves are well sampled light-curves of $\sim 120 - 150$ days/season, taken with a sampling rate of $\sim$8-10 minutes. To create our set of synthetic data we used the time sampling of SuperWASP light-curves from a frequently observed field of 2004 observing season. Due to our choice of parameters and the SuperWASP observing strategy the injected transit signals appear very deep, of very short durations and with no visible points during ingress or egress. Figure 2 right panel, shows an example of two simulated transit light-curves. Top panel: shows the synthetic light-curve of a hypothetical eclipsing WD+BD binary system with an orbital period of $P \sim 116$ min similar to WD0137-349 (Maxted et al. 2006) (which itself does not show eclipses). The lower panel shows the simulated transit light-curve for a rocky body of the size of the Earth in a $\sim$5 h orbit.

2.2. Detecting transit signals

To recover the transit signals injected in our set of synthetic light-curves we used a Box Least Square (BLS) algorithm, see Kovács et al. 2002. The BLS utilises the anticipated squared shape of the transit signal and it performs least square fits of step functions to the light-curve, after folding at several trial periods. To better account for the shape of our set of synthetic data we used the BLS with a grid of fixed transit durations $D_{tr} = 1, 2, 4, 8, 16, 32$ min. This covers our transit durations for orbital periods between $\sim$2 h and 15 days (see Figure 1). For each trial period a $\Delta\chi^2$ is evaluated where $\Delta\chi^2$ is the improvement in the fit χ^2 when compared to that of a constant light-curve χ^2_0 (see Collier Cameron 2006 for more details). Therefore the $\Delta\chi^2$ is proportional to the S/N value of the transit event $\Delta\chi^2 = (S/N)^2$ and S/N is given by:

$$S/N = \frac{\delta}{\sigma_{tr}}\sqrt{N_{tr}} \qquad (1)$$

Where δ is the transit depth, σ_{tr} is the *rms* of the in-transit points and N_{tr} is the number of points in transit. Because planetary detections are strongly affected by the signal-to-noise value of the transit event (see Pont et al. 2006 and Collier Cameron 2006) we defined the BLS power spectra in terms of the S/N value of the transit. For each trial period we evaluate the *signal detection efficiency* (SDE) (Kovács et al. 2002) as:

$$SDE = \frac{S/N_{peak} - < S/N >}{\sigma_{S/N}} \tag{2}$$

Where S/N_{peak} is the S/N at the highest peak, $< S/N >$ is the average, and $\sigma_{S/N}$ is the standard deviation over the frequency band tested. A detection is represented by the highest peak in the BLS power spectrum. We assigned a statistical significance to the highest peak in the periodogram by testing the BLS response to pure noise. This allowed us to set a detection threshold which we used to define our set of empirical criteria for successful detection. A transit was considered successfully recovered if a peak in the BLS power spectrum was within 0.005 d of the correct (inserted) period and at least 7.3σ above the average level of the BLS power spectrum.

The results of our simulations show that for bright white dwarfs (V ~12) in the presence of Gaussian random noise we are sensitive to objects as small as the Moon. For fainter white dwarfs, we are sensitive to increasingly larger radius rocky bodies. In the presence of covariant noise structure in the data our detection sensitivity is reduced. Nevertheless, Earth size objects remain detectable in low signal-to-noise data. Supported by these results we have investigated a sample of 174 white dwarfs brighter than V ~15 for which we have SuperWASP data.

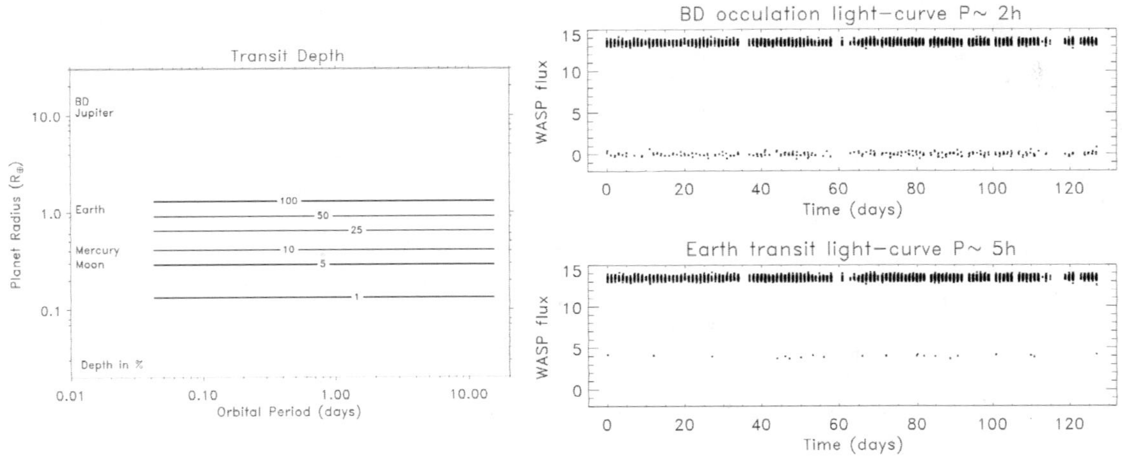

Figure 2. Left panel: transit depth in percent in the same parameter space as Figure 1. Right panel: Two examples of simulated transit light curves. Top, a brown dwarf in ~2 h orbit. Bottom, an Earth-sized companion with a period of ~5 h

3. Searching for transit signals in real SuperWASP data

Our white dwarf sample consists of all objects resulting from a cross-correlation of the SuperWASP data archive and the McCook & Sion catalogue (2006). We only consider objects with more than ~ 600 photometric points (equivalent to ~11 observing nights, for a night of ~8 hour and sampling cadence of ~8-10 min). We considered data with multi-season observations (from 2004 to 2008) in order to increase our S/N (see equation 1) and thus our detection probability. The sample of 174 WASP targets has been searched for transits and eclipses using the BLS algorithm. In addition we have inspected the data by eye, individually, for each light-curve. Transits studied in this work can have durations as short as 1 minute, this might correspond to an individual point, dropped periodically, in a SuperWASP light-curve. Our key result is that we find no evidence of sub-stellar or smaller companions. A detailed discussion of our results can be found in Faedi et al. 2009 (in prep.).

4. Conclusions

We have investigated the transit characteristics and detection limits for sub-stellar and terrestrial companions around white dwarfs. For our chosen range of orbital periods ($\sim$2 h - 15 d) and companion sizes (Moon - Jupiter/BD) the simulated transit events appear extremely different from the transit signals due to known extra-solar planets around solar-type stars. Due to the small radii of white dwarfs ($\sim 1\,R_\oplus$), transits have very short durations and appear extremely deep (100% for gas giants and any body $\geq 1\,R_\oplus$). We have investigated our ability to detect transits in SuperWASP light-curves of white dwarfs. We found that for Gaussian random noise we are sensitive to Moon size companions in orbit around a $V \sim$12 white dwarfs. For fainter white dwarfs we are sensitive to increasingly larger radius rocky bodies. Our sensitivity is reduced in light-curves dominated by covariant noise structure. Nevertheless, Earth-size companions remain detectable in low signal-to-noise light-curves. We then analysed a sample of 174 white dwarfs in the SuperWASP data archive for transits and eclipses using a slightly modified BLS routine. We found no evidence for sub-stellar or smaller companions in the sample of 174 WDs. Extensive details on the simulations, the detection algorithm and an interpretation of our null result can be found in Faedi et al. 2009 (in prep.).

References

Becklin E E & Zuckerman B 1988 *Nature* **336** 656
Burleigh M R Clarke F J & Hodgkin S T 2002 *MNRAS* **331** L41
Cameron A C et al. 2007 *MNRAS* **375** 951
Collier Cameron A et al. 2006 *MNRAS* **373** 799
Debes J H & Sigurdsson S 2002 *ApJ* **572** 556
Duncan M J & Lissauer J J 1998 *Icarus* **134** 303
Farihi J Becklin E E & Zuckerman 2005 *ApJ* **161** 394
Farihi J Zuckerman B Becklin E E & Jura M 2007 *15th European Workshop on White Dwarfs* **372** 315
Farihi J Becklin E E & Zuckerman B 2008 *ApJ* **681** 1470
Jura M 2003 *ApJ* **584** L91
Kovács G Zucker S & Mazeh T 2002 *A&A* **391** 369
Littlefair S P Dhillon V S & Martín E L 2003 *MNRAS* **340** 264
Littlefair S P Dhillon V S Marsh T R Gänsicke B T Southworth J & Watson C A 2006 *Science* **314** 1578
McCook G P & Sion E M 2006 *VizieR Online Data Catalog* **3235** 0
Maxted P F L Napiwotzki R Dobbie P D & Burleigh M R 2006 *Nature* **442** 543
Mullally F Winget D E Degennaro S Jeffery E Thompson S E Chandler D & Kepler S O 2008 *ApJ* **676** 573
Patterson J 1998 *PASP* **110** 1132
Pollacco D L et al. 2006 *PASP* **118** 1407
Pont F Zucker S & Queloz D 2006 *MNRAS* **373** 231
Reach W T Kuchner M J von Hippel T Burrows A Mullally F Kilic M & Winget D E 2005 *ApJ* **635** L161
Sato B et al. 2003 *ApJ* **597** L157
Silvotti R et al. 2007 *Nature* **449** 189
Villaver E & Livio M 2007 *ApJ* **661** 1192

Journal of Physics: Conference Series **172** (2009) 012058 doi:10.1088/1742-6596/172/1/012058

White dwarfs with unresolved substellar companions and debris disks in the UKIDSS Survey

Paul Steele[1], Matt Burleigh[1], Martin Barstow[1], Richard Jameson[1] and Paul Dobbie[2]

[1]Department of Physics & Astronomy, University of Leicester, University Road, Leicester, LE1 7RH, UK

[2]Anglo-Australian Observatory, PO Box 296, Epping, Sydney, NSW 1710, Australia

E-mail: prs15@star.le.ac.uk

Abstract. We present a near-infrared (NIR) photometric search for substellar companions and debris disks around white dwarfs in the UKIRT Infrared Deep Sky Survey (UKIDSS). We cross-correlate the SDSS DR4 and McCook & Sion catalogues of white dwarfs with the UKIDSS DR3 database producing 408 and 133 matches respectively. Models are then fitted to Sloan photometry to identify those with NIR photometric excesses consistent with an unresolved sub-stellar companion or a debris disk. We present follow up photometry from targets previously identified in UKIDSS DR2 and the first results from DR3. In total we identify 8 potential white dwarf + very low mass binary systems, 2 potential white dwarf + disk systems, and 2 systems which require further investigation to clarify the source of the excess.

1. Introduction

The detection and observation of very low mass companions to white dwarfs allows us to investigate an extreme of binary formation and evolution. White dwarf + brown dwarf binaries also allows us also to investigate the fraction of main-sequence stars with substellar companions. This will give us further insight into the possible discrepancy between the brown dwarf companion fraction at small separations ($1 \pm 1\%$ at < 10 AU, Marcy & Butler 2000; McCarthy & Zuckerman 2004) and large radii ($a > 1000$ AU; 10–30%, Gizis et al. 2001). In addition there are very few observational constraints on brown dwarf evolutionary models at large ages, such as those expected for most white dwarfs (> 1 Gyr). Thus wide, detached white dwarf + brown dwarf binaries can provide "benchmarks" for testing these models (Pinfield et al. 2006).

Near-infrared (NIR) photometry and spectroscopy allows for the detection and study of such systems. The spectral energy distribution of the white dwarf in this region is markedly different to that of a brown dwarf with the latter peaking at these wavelengths. As the white dwarf falls off in flux the two components of a binary system are easily separated in broadband photometry, facilitating detection and enabling straightforward spectroscopic followup. Close WD+BD are of interest as an outcome of common envelope (CE) evolution, another channel for cataclysmic variable (CV) evolution, or the end state of CV evolution. In the latter case the secondary will have become highly evolved through mass transfer, perhaps beginning its life at a much higher mass and earlier stellar class.

Unfortunately, brown dwarf companions to white dwarfs are rare ($< 0.5\%$: Farihi et al. 2005) with only 3 systems confirmed; GD165 (DA+L4, Becklin & Zuckerman 1988), GD 1400 (DA+L6/7, Farihi & Christopher 2004) and WD0137−349 (DA+L8, Burleigh et al. 2006). GD165 is the only resolved widely separated system (120 AU) whereas WD 0137 and GD 1400 (Burleigh et al., these proceedings) are close orbit binaries. Hence, large, deep near-IR surveys such as the UKIRT Infrared Deep Sky Survey (UKIDSS) are excellent databases to search for these objects.

Recently a number of white dwarfs have been found to harbour metal-rich debris disks (e.g. GD362; Kilic et al. 2005). These can be revealed optically by looking for distinct, double-peaked Ca II emission triplet at 850-866nm (e.g. Gaensicke et al. 2007), or through K-band excesses in the near-IR. Therefore, near-IR photometry and spectroscopy can also be used to search for these debris disks and, for those white dwarfs showing heavy element contamination in their photospheres, may provide an important diagnostic tool for distinguishing between disk accretion and accretion from very-low mass companions.

1.1. The UKIDSS survey

UKIDSS is the near-infrared counterpart to the Sloan Digital Sky Survey (SDSS: York et al. 2000) and the successor to the previous 2 Micron All Sky Survey (2MASS: Skrutskie et al. 2006). The UKIDSS began in May 2005 and aims to cover 7500 deg^2 of the Northern Sky in $YJHK$ down to $K = 18.4$, 3 magnitudes deeper than the 2MASS. The UKIDSS actually consists of a set of 5 complimentary surveys; the Large Area Survey (LAS), the Galactic Plane Survey (GPS), the Galactic Clusters Survey (GCS), the Deep Extragalactic Survey (DXS) and the Ultra-deep Extragalactic survey (UDS). Following a release of science verification data and an Early data release (Dye et al. 2006), UKIDSS has had 3 data releases (DR1: Warren et al. 2007b, DR2: Warren et al. 2007a, DR3: Warren et al. 2008 in prep.) in July 2006, March 2007, and December 2007. Each subsequent data release contains all data from the previous releases and at present represents 15% of the complete data set; that is 1100 deg^2 of sky coverage. The survey of most importance to this work is the Large Area Survey.

The LAS (PI: Richard Jameson) aims to cover 4000 deg^2 upon it's 7 year completion down to 5σ limiting magnitudes of $Y = 20.5$, $J = 20.0$, $H = 18.8$ and $K = 18.4$ (Vega). The target 4000 deg^2 is a subsection of the SDSS. As of DR3 approximately 900 deg^2 has been completed.

2. Identifying white dwarfs

The UKIDSS LAS DR3 catalogue was cross correlated with two catalogues of white dwarfs: the on-line August 2006 version of the McCook & Sion catalogue of spectroscopically identified white dwarfs (McCook & Sion 1999) which contains 5557 stars, and the Eisenstein et al. (2006) catalogue of 9316 spectroscopically confirmed white dwarfs from the SDSS DR4. The McCook & Sion catalogue contains the brightest white dwarfs known, whilst the white dwarfs identified in the SDSS are constrained by an upper brightness limit ($g' \approx 15$) and the lower sensitivity limit ($g' \approx 21$) of that survey. The majority of the SDSS white dwarfs are too faint to have been detected by the 2MASS survey, therefore UKIDSS allows the investigation of the NIR properties of a large number of new white dwarfs discovered in the SDSS, as well as significantly reducing the measurement errors on the NIR photometry of the brighter white dwarfs detected in 2MASS.

Many of the relatively bright white dwarfs have large and unmeasured proper motions, meaning their astrometry is often unreliable. For the McCook & Sion white dwarfs, a set of improved coordinates supplied by Jay Holberg (private communication) were utilised in the cross correlation with the UKIDSS DR3. The SDSS stars, being fainter and therefore more distant have more reliable astrometry. The cross correlation was performed using the on-line "CrossID" tool at the WFCAM science archive. The results were constrained to stars brighter than the 5σ limiting magnitudes (Section 1.1) using a pairing radius of 5″ taking only the nearest IR source to each white dwarf's coordinates as the primary result. In general, the UKIDSS and white dwarf coordinates agreed within 5″ and the SDSS images revealed the same object as in the corresponding UKIDSS images. The images of each cross matched object in both databases were inspected by eye to ensure the correct object had been identified. Six of the sample were rejected as being the wrong source in the UKIDSS when compared to the SDSS image.

The cross correlation produced 493 matches with the SDSS DR3, and 201 with the McCook & Sion white dwarf catalogues. Only 77 of the McCook & Sion white dwarfs were not found in the SDSS.

3. Identifying candidate systems

In order to efficiently process the cross-matched catalogue of white dwarfs it was decided to model each white dwarf independently and then inspect each for evidence of a NIR excess. To that end, a program was written to fit a blackbody continuum to the SDSS photometry of each white dwarf using a grid of models and performing a Chi squared minimisation to achieve the best fit. Candidate stars were identified as those that showed a significant ($> 3\sigma$) excess in one or more of the UKIDSS bands. 12 such stars were identified with the blackbody fitting and the corresponding photometry is given in table 1. Out of these 12, 10 show evidence of multiple band excesses potentially indicating white dwarf brown + dwarf binaries, and 2 show only K-band excesses indicative of a debris disk.

For candidate white dwarfs with $T_{\rm eff} > 16,000$ K we improved the determination of the size of the excess by calculating a synthetic spectrum at the temperature and gravity determined by Eisenstein et al. (2006) (Table 2). For the candidates with effective temperatures $\lesssim 16\,000$ K we have generated blackbody models at the temperatures given in table 2. The synthetic spectra were then normalised to the Sloan g' magnitude of the corresponding white dwarf (Table 2).

3.1. Identifying sub-stellar companions

For each candidate white dwarf + brown dwarf binary the SDSS optical spectrum was first examined for evidence of a low mass main-sequence companion. An increase in flux of the spectral energy distribution towards longer wavelengths or spectral features indicative of early-mid M dwarfs (e.g. Hα emission and/or TiO absorption bands) were searched for by eye. None of our candidates exhibited these features definitively and so none were removed from our sample at this stage. Empirical models for late M to T spectral types were then added to the white dwarf synthetic spectrum and these composites were compared to UKIDSS photometry to obtain an approximate spectral type for the potential companion.

The empirical models were constructed using the NIR spectra of late M, L and T dwarfs from the IRTF spectral library (Cushing et al. 2005, Rayner et al. in prep.). Subsequently, these fluxes were scaled to the distance of the relevant white dwarf (Table 2) and, starting with T5, earlier types were progressively added until a match could be made with the UKIDSS photometry.

3.2. Identifying debris disks

We could not match any substellar model to the white dwarfs displaying only K-band excess emission. For these objects, it is plausible that the excess is due to the presence of a warm debris disk (e.g. GD362). We estimate the temperature of the putative disk by fitting combined white dwarf + blackbody models.

4. Results

After matching these models to the stars showing near-IR excesses, we confirmed that 8 potentially harbour very low mass or substellar companions, 2 show K-band only excesses

Journal of Physics: Conference Series **172** (2009) 012058 doi:10.1088/1742-6596/172/1/012058

Table 1. UKIDSS Photometry of candidate binary/disk systems

SDSS J	WD	Y	J	H	K
003902.47−003000.3	-		18.39 ± 0.08	17.87 ± 0.11	17.42 ± 0.10
003923.04+003534.7	-	-	-	17.49 ± 0.05	17.05 ± 0.06
090759.59+053649.7	-	18.79 ± 0.05	18.64 ± 0.08	18.61 ± 0.16	18.39 ± 0.19
092452.73+020712.2	0922+006	-	-	18.21 ± 0.12	17.63 ± 0.11
101532.10+042507.5	-	-	-	18.42 ± 0.17	18.17 ± 0.19
120212.00+082244.6	1159+086	18.02 ± 0.04	18.14 ± 0.07	18.23 ± 0.10	17.88 ± 0.13
120503.99+070031.0	-	-	-	17.76 ± 0.08	17.72 ± 0.13
131724.75+000237.3	1314+003	16.77 ± 0.01	16.91 ± 0.02	-	16.84 ± 0.08
132044.68+001855.0	1318+005	17.64 ± 0.02	17.64 ± 0.04	17.76 ± 0.11	17.40 ± 0.11
155720.77+091624.7	-	18.82 ± 0.04	18.82 ± 0.06	19.03 ± 0.14	18.34 ± 0.15
222030.68−004107.9	-	16.97 ± 0.01	16.76 ± 0.02	16.45 ± 0.03	16.38 ± 0.04
222551.65+001637.7	-	18.60 ± 0.05	18.52 ± 0.07	18.16 ± 0.13	17.63 ± 0.13

Table 2. Physical parameters of the candidate white dwarfs showing near-infrared excesses. Temperatures and surface gravities are from Eisenstein et al. (2006) unless otherwise stated. Distances are calculated through evolutionary models.

SDSS J	WD	g'	$T_{\rm eff}$ (K)	$\log g$	D (pc)
003902.47−003000.3	-	18.74	12314 ± 268	7.39 ± 0.08	
003923.04+003534.7	-	19.09	10080 ± 81	7.99 ± 0.11	269
090759.59+053649.7	-	18.17	19474 ± 293	7.82 ± 0.05	323
092452.73+020712.2	0922+006	18.55	15231 ± 147^1	8.49 ± 0.10^1	209
101532.10+042507.5	-	17.63	35073 ± 195	7.38 ± 0.04	604
120212.00+082244.6	1159+086	17.63	16400^2	-	-
120503.99+070031.0	-	18.89	8688 ± 61	8.21 ± 0.12	185
131724.75+000237.3	1314+003	15.75	60000^3	7.8^3	262
132044.68+001855.0	1318+005	17.10	19649 ± 130	8.36 ± 0.02	133
155720.77+091624.7	-	18.43	21990 ± 403	7.67 ± 0.06	437
222030.68−004107.9	-	17.44	8037 ± 31	8.28 ± 0.07	84
222551.65+001637.7	-	18.80	10640 ± 94	8.16 ± 0.09	216

[1] Helium fit.
[2] Blackbody fit of SDSS photometry.
[3] Werner et al. (2004).

indicative of debris disks, and 2 objects show H and K-band excesses that can be explained by either a companion or a disk. Figures 1-12 show each white dwarf's modelled spectrum plotted with both SDSS and UKIDSS photometry. The model used to match the excess(es) are identified in each plot (the vertical order of the labels matches the vertical order of the emprical models). The results are summarised in table 3.

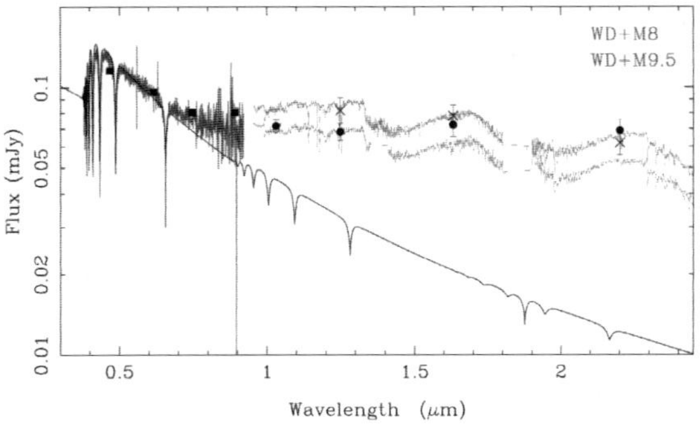

Figure 1. Model spectrum (——) photometry (SDSS: ■ UKIDSS: ● SOFI: **X**) and spectroscopy (SDSS:——) of SDSS J 003902.47−003000.3

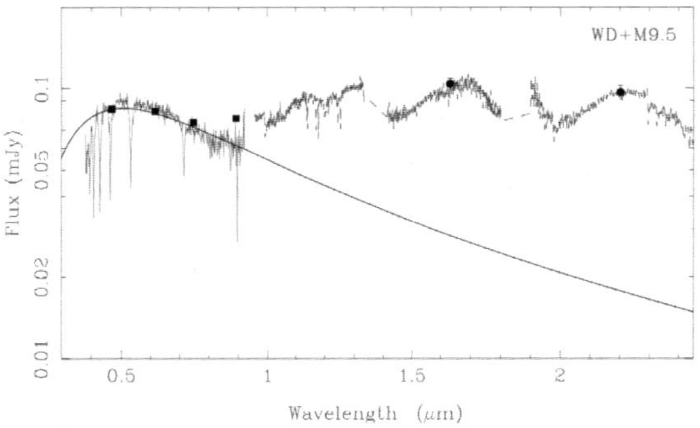

Figure 2. Model spectrum (——) photometry (SDSS: ■ UKIDSS: ●) and spectroscopy (SDSS:——) of SDSS J 003923.04+003534.7

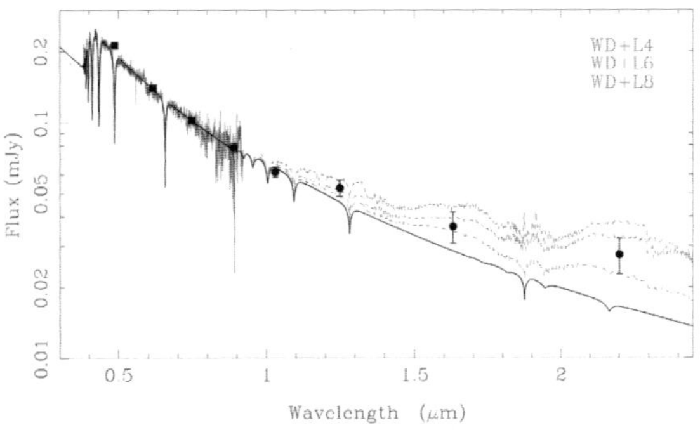

Figure 3. Model spectrum (——) photometry (SDSS: ■ UKIDSS: ●) and spectroscopy (SDSS:) of SDSS J 090759.59+053649.7

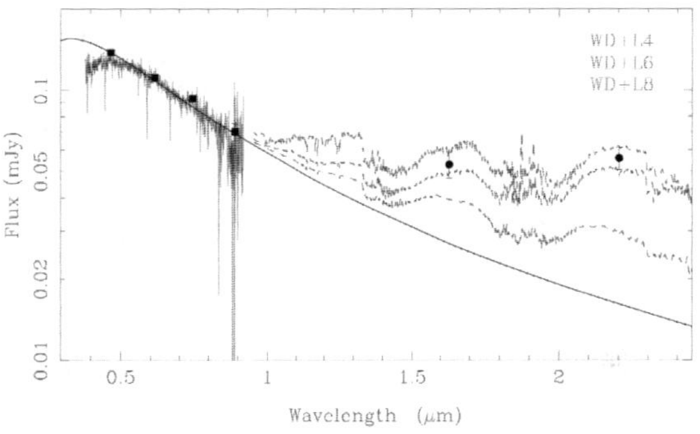

Figure 4. Model spectrum (——) photometry (SDSS: ■ UKIDSS: ●) and spectroscopy (SDSS:) of SDSS J 092452.73+020712.2

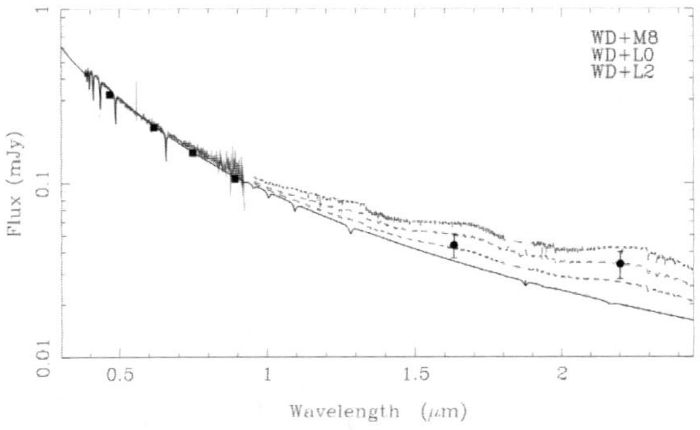

Figure 5. Model spectrum (——) photometry (SDSS: ■ UKIDSS: ●) and spectroscopy (SDSS:——) of SDSS J 101532.10+042507.5

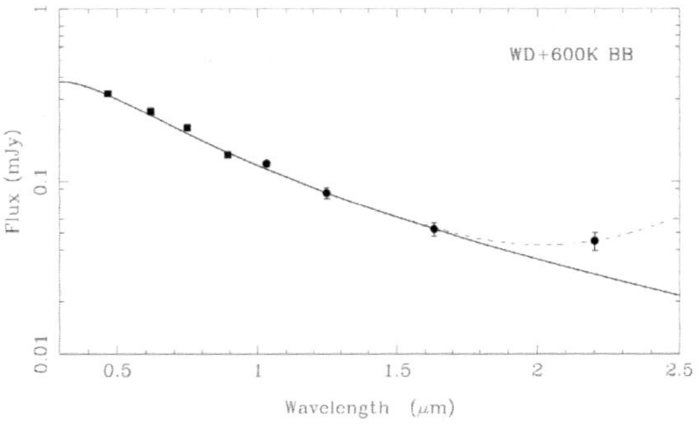

Figure 6. Model spectrum (——) and photometry (SDSS: ■ UKIDSS: ●) of SDSS J 120212.00+082244.6

5. Notes on Individual Objects

5.1. SDSS J 003902.47−003000.3

This white dwarf's optical spectrum (Figure 1) shows no evidence for an M dwarf companion. When compared to a model spectrum (Figure 1) a small excess emission can be identified in

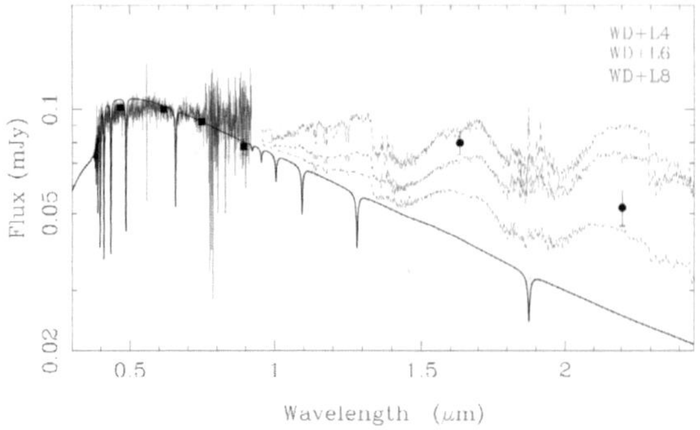

Figure 7. Model spectrum (——) photometry (SDSS: ■ UKIDSS: ●) and spectroscopy (SDSS:) of SDSS J 120503.99+070031.0

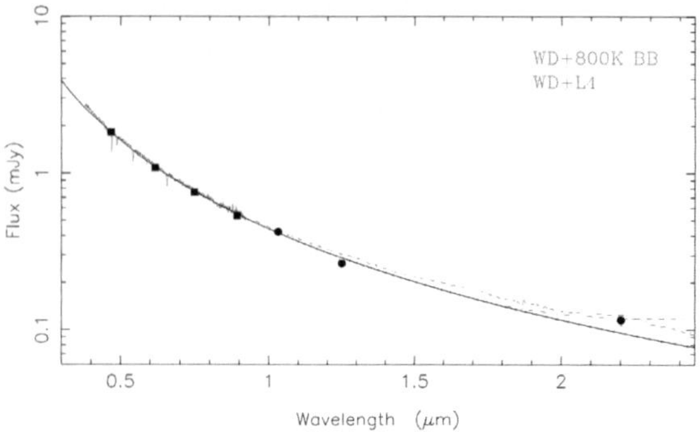

Figure 8. Model spectrum (——) photometry (SDSS: ■ UKIDSS: ●) and spectroscopy (SDSS:——) of SDSS J 131724.75+000237.3

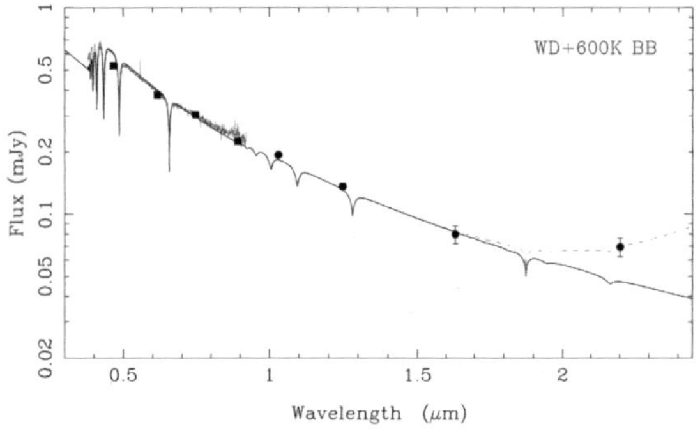

Figure 9. Model spectrum (——) photometry (SDSS: ■ UKIDSS: ●) and spectroscopy (SDSS:——) of SDSS J 132044.68+001855.0

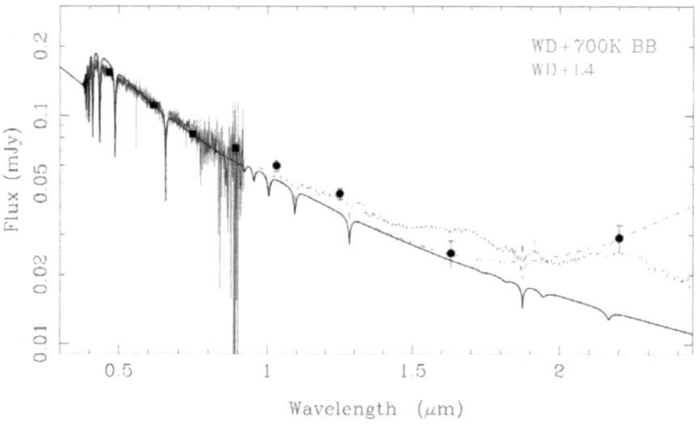

Figure 10. Model spectrum (——) photometry (SDSS: ■ UKIDSS: ●) and spectroscopy (SDSS:) of SDSS J 155720.77+091624.7

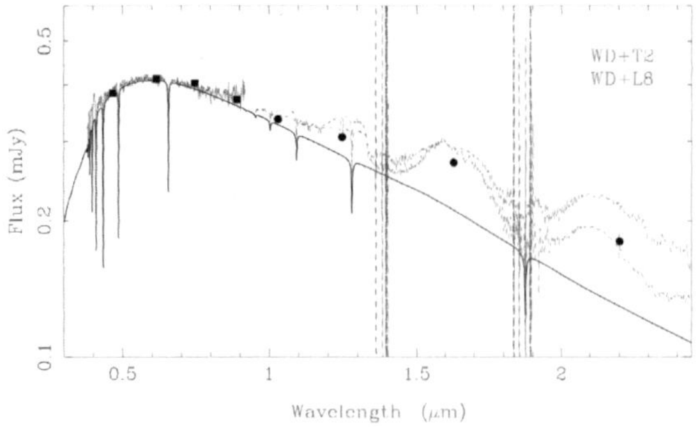

Figure 11. Model spectrum (——) photometry (SDSS: ■ UKIDSS: ●) and spectroscopy (SDSS:——) of SDSS J 222030.68−004107.9

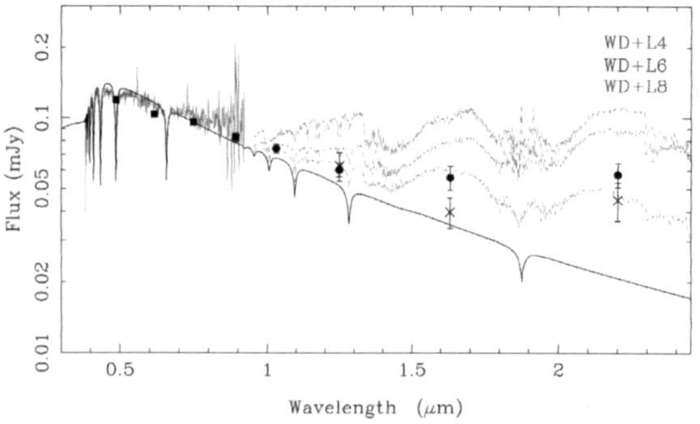

Figure 12. Model spectrum (——) photometry (SDSS: ■ UKIDSS: ● SOFI: **X**) and spectroscopy (SDSS:) of SDSS J 222551+001637.7

the optical z-band, with strong excesses clearly visible in the UKIDSS $YJHK$ photometry. The photometry is best matched by the addition of either an M8 or M9.5 dwarf, placed at the distance predicted by the star's temperature and surface gravity (Table 2).

This result was first identified in UKIDSS DR2 (Burliegh et al. 2008, in prep.) and follow-up

Table 3. Table of Results identifying binary and disk candidates.

SDSS J	WD	Binary	Spectral Type	Disk	Temperature (K)
003902.47−003000.3	-	Yes	M8±1	-	-
003923.04+003534.7	-	Yes	M9±1	-	-
090759.59+053649.7	-	Yes	L6±2	-	-
092452.73+020712.2	0922+006	Yes	L6±2	-	-
101532.10+042507.5	-	Yes	L0±2	-	-
120212.00+082244.6	1159+086	-	-	Yes	< 600 K
120503.99+070031.0	-	Yes	L6±2	-	-
131724.75+000237.3	1314+003	Yes	L4±1	Yes	< 800 K
132044.68+001855.0	1318+005	-	-	Yes	< 600 K
155720.77+091624.7	-	Yes	L4±1	Yes	< 700 K
222030.68−004107.9	-	Yes	T0±1	-	-
222551.65+001637.7	-	Yes	L8±1	-	-

JHK photometry was obtained using SOFI on the NTT on October 25th 2007 (Figure 1). No exact match is obtained with either model but the SOFI photometry confirms the presence of a NIR excess and indicates the companion is of spectral type M8±1. As this is a late-M it is doubtful that this companion can be identified as a brown dwarf.

5.2. SDSS J 092452.73+020712.2

This is only one of two helium atmosphere (DB) white dwarfs in the list of candidates for low mass sub-stellar companions. The optical photometry shows no excess but there is a clear rise in the H and K band photometry. This is best matched by the addition of an L4-L6 dwarf scaled to the distance of the primary (Table 2), and would thus be the first DB white dwarf to potentially harbour a brown dwarf companion.

5.3. SDSS J 222030.68−004107.9

This white dwarf has the brightest NIR photometry out of the sample of candidates for sub-stellar binarity. Its optical spectrum reveals no evidence of a main-sequence companion. Eisenstein et al. (2006) give the physical parameters of this star as $T_{\rm eff} = 8037 \pm 31\,{\rm K}$ and $\log g = 8.28 \pm 0.07$ from an automated fit, also making this the coolest star in the sample. A comparison to a model for these parameters reveals a NIR excess which is best matched by the addition of an Late-L/early-T dwarf scaled to the distance of the primary star (Table 2). This would make this system the first white dwarf T dwarf binary detected and certainly the lowest mass object detected around a white dwarf to date.

However, UKIDSS classified this object as a merged source possibly indicating background contamination. A closer inspection of the UKIDSS images reveals the possibility that the system is becoming resolved in the longer wavelength HK-bands. This was indeed the case for a number of white dwarf M-dwarf binaries. Therefore this system could be a wider orbit Late-L/Early-T dwarf binary, however high resolution spectroscopy is needed to rule out the possibility of background sources.

5.4. SDSS J 222551.65+001637.7

This white dwarf shows no sign of an M dwarf in its optical spectrum and an atmospheric model (a cooler model was obtained in this case with parameters of $T_{\rm eff} = 10,000$ K and $\log g = 8.00$, a close matched to the Eisenstein et al. (2006) autofit parameters of $T_{\rm eff} = 10640 \pm 94$ K and $\log g = 8.16 \pm 0.09$) matches the SDSS photometry (Figure 12). A clear excess is then seen in the H and K-band photometry that is best matched by the addition of companion of spectral type L6 scaled to the distance of the star.

This result was first identified in UKIDSS DR2 (Burliegh et al. 2008, in prep.) and follow-up JHK photometry was obtained using SOFI on the NTT on October 2nd 2007 (K) and October 3rd 2007 (JH) — Figure 12. The K-band photometry agrees with the UKIDSS data within errors and indeed indicates an excess. However, the JH-band data matches the white dwarf model more closely when compared with the UKIDSS data. This could either indicate that there is only a K-band excess and therefore this system is most likely a disk candidate, or as the data was taken on separate nights the companion is in a close orbit and is variable due to heating on the hemisphere facing the white dwarf. Follow-up spectroscopy or time series NIR photometry is needed to confirm this scenario.

6. Conclusions

By cross-correlating the McCook & Sion and the SDSS DR4 catalogues of spectroscopically identified white dwarfs with the UKIDSS DR3 database, we have identified 12 systems with significant near-infrared excesses. 8 of these have multiple band excesses consistent with the presence of a low mass companion, these range in spectral type from M8-T0. Of notable interest are SDSS J 092452.73+020712.2 which is a potential DB+L binary and SDSS J 222030.68−004107.9 which is our latest typed companion (L8-T0). Two have a single K-band excess indicating they may harbour a debris disk, and 2 require further investigation to identify the exact source of the excess infrared emission.

We are now obtaining follow-up spectra for a number of these objects with Gemini North to confirm the nature of the excesses. This will allow us to place tighter constraints on the fraction of white dwarfs with substellar (L and T) companions.

References

Cushing M C Rayner J T & Vacca W D 2005 *ApJ* **623** 1115
Becklin E E & Zuckerman B 1988 *Nature* **336** 656
Burleigh M R Hogan E Dobbie P D Napiwotzki R Maxted P F L 2006 *MNRAS* **373** 55
Dye S et al. 2006 *MNRAS* **372** 1227
Eisenstein D J et al. 2006 *ApJS* **167** 40
Farihi J & Christopher M. 2004 *AJ* **128** 1868
Farihi J Becklin E E & Zuckerman B 2005 *ApJS* **161** 394
Gaensicke B T Marsh T R Southworth J Rebassa-Mansergas A 2007 `arXiv:astro-ph/07102807`
Gizis J E Kirkpatrick J D Burgasser A Reid I N Monet D G Liebert J & Wilson J C 2001 *ApJ* **551** 163
Hubeny I 1988 *Computer Physics Communications* **52** 103
Hubeny I & Lanz T 1995 *ApJ* **439** 875
Kilic M von Hippel T Leggett S K & Winget D E 2005 *ApJ* **632** 115
Marcy G W & Butler P R 2000 *PASP* **112** 768
McCarthy C & Zuckerman B 2004 *AJ* **127** 2781
McCoo, G P & Sion E M 1999 *ApJS* **121** 1
Pinfield D J Jones H R A Lucas P W Kendall T R Folkes S L Day-Jones A C Chappelle R J & Steele I A 2006 *MNRAS* **368** 1281
Skrutskie et al. 2006 *AJ* **131** 1163
Warren S J et al. 2007a `arXiv:astro-ph/0703037`
Warren S J et al. 2007b *MNRAS* **375** 213
Werner K Rauch T Napiwotzki R Christlieb N Reimers D Karl C A 2004 *A&A* **424** 657
York D G et al. 2000 *AJ* **120** 1579

Photospheric Ca and Mg line-strength variations in G29-38

Ted von Hippel[1,2], Susan E Thompson[3,4], W T Reach[5], F Mullally[6], Mukremin Kilic[7] and Atsuko Nitta[8]

[1]Physics Department, Siena College, Loudonville, NY, USA
[2]McDonald Observatory, University of Texas, Austin, TX, USA
[3]Department of Physics and Astronomy, University of Delaware, Newark, DE, USA
[4]Delaware Astroseismic Research Center, Mt. Cuba Observatory, Greenville, DE, USA
[5]IPAC, California Institute of Technology, Pasadena, CA, USA
[6]Department of Astronomy, Princeton University, Princeton, NJ, USA
[7]Center for Astrophysics, Harvard University, Cambridge, MA, USA
[8]Gemini Observatory, Hilo, HI, USA

E-mail: tvonhippel@siena.edu, sthomp@physics.udel.edu, reach@ipac.caltech.edu, fergal@astro.princeton.edu, kilic@astronomy.ohio-state.edu, anitta@gemini.edu

Abstract. Temporal variations in metal-line strengths in H-atmosphere white dwarfs hold the potential to test the timescales of graviational settling theory. These short timescales, in turn, require that DAZs are currently accreting. Such temporal variations would also indicate that accretion from a circumstellar dust disk can be episodic. We are compiling increasing evidence for time-variable Ca and Mg line-strength variations in the best studied DAZ, G29-38. Our evidence to date supports the gravitational settling timescales of Koester & Wilken (2006) and episodic accretion from G29-38's debris disk. Furthermore, we have detected evidence for time-variable accretion with a timescale $\leq$ 24 hours, and typical variability of $\sim$4% during the 100 days of our autumn 2007 monitoring campaign.

1. Introduction

The presence of heavy elements in the photospheres of many hydrogen-dominated (DA) white dwarfs (WDs) runs counter to our expectation. The high surface gravities of WDs cause their atmospheres to become highly stratified, and for $T_{\rm eff} < 22{,}000$ K, heavy elements should settle out of the photosphere on timescales of 10^{-2} to 10^{6} years (Dupuis et al. 1992; Koester & Wilken 2006), depending on the mass and surface temperature of the white dwarf. Yet most WDs are old, and almost every known DAZ, as these stars are now called (Sion et al. 1983), has been cooling as a WD for at least 10^{8} years. DAZ stars must be currently accreting heavy elements. We have argued elsewhere (von Hippel et al. 2007) that most DAZs, and perhaps all, are accreting from circumstellar dust disks, rather than from the interstellar medium. We will not reconsider that argument here, as our focus is on G29-38, a DAZ with a known dust disk. (Where the dust disk is detected as an infrared excess, we abbreviate the system DAZd, DZd, DBZd, etc.) In this paper, we will focus on the accretion process from the disk to the WD's photosphere.

1.1. What is reasonably understood

We understand that the major constituent of WD disks is silicates (Reach et al. 2005; Farihi, et al. 2009), and that this material is accreted from disks onto WDs at a rate of order 10^{-17} $M_\odot$ yr^{-1} (Koester & Wilken 2006, with the rate corrected for circumstellar rather than ISM accretion). This material is refractory-rich, C-poor (Jura 2006; Zuckerman et al. 2007), and grossly underabundant in H and He (Reach et al. 2005; Gänsicke et al. 2006). We do not expect the luminosity, radius, or effective temperature of an accreting WD to be noticeably affected, since the accretion luminosity is only of order 10^{-10} $L_\odot$.

1.2. What is not understood

Theory substantially lags observation for DZd systems. At present, we do not know how these disk were formed, though we believe that the tidal destruction of asteroids or other minor planetary bodies played a key role (Jura 2003). We do not know how the complex set of dynamical interactions that take place after the mass-losing AGB phase could yield so many DZd systems, especially if this mechanism is responsible for all DZs, which account for $\sim$ 25% of all WDs (Zuckerman & Reid 1998). If we could understand the dynamical history of DZd systems, we would probably discover important new constraints on the prevalence and architecture of planetary systems. We may also learn more about AGB mass loss.

The WD disks themselves are poorly understood. While an optically thick, physically thin disk model approximately fits many of these systems (von Hippel et al. 2005; Jura, Farihi, & Zuckerman 2007), other systems do not fit this simple model (Jura et al. 2007). Nor do we know if the disk evolves through viscous dissipation, Poynting-Robertson drag, or some form of magnetic instability. Without this knowledge, we do not know if these disks start out as relatively massive entities that erode with time, or whether they have short lifetimes but are often replaced by new infalling material.

We also do not know the dominant accretion mechanism that moves material from the disks to the WD photospheres. None of the DAZd WDs are known magnetics, though magnetic fields could play a key role. At present, there are only two constraints on the accretion mechanism, besides the accretion rate itself. One is the preliminary result reported by Montgomery, Thompson, & von Hippel (2008) that G29-38 is accreting onto its equator, rather than uniformly, or onto its poles. The other is that G29-38 is accreting episodically (von Hippel & Thompson 2007). We follow up on the latter discovery in this conference paper.

2. The case for metal-line variability in G29-38

The case for metal-line variability in G29-38 was first made by von Hippel & Thompson (2007), then questioned by Debes & Lopez-Morales (2008). Here we present our further efforts to demonstrate that G29-38 is accreting in a time-variable manner, and characterize that variation.

In the spirit of a conference, our data and analysis represent a work in progress. Indeed, as we submit this article, we are still acquiring more data on G29-38's Ca and Mg abundances at both Gemini North and the Hobby-Eberly Telescope.

Figures 1 and 2 presents our preliminary Ca and Mg EWs from our autumn 2007 monitoring program. Each data point has two sets of error bars. The inner set of error bars represent the 1σ scatter in the equivalent width (EW) measurement due to Poisson noise in the spectrum and any sky subtraction issues. The outer set of error bars includes these effects, as well as a conservative 1σ estimate of surface temperature variations affecting the line strengths during the exposure. We have not yet taken continuum measurement errors into account, which we estimate could increase the total error bar by as much as 50%. Assuming our errors are approximately correct, the distribution of points is completely inconsistent with constant Ca and Mg line strengths, and therefore inconsistent with steady-state accretion from the disk.

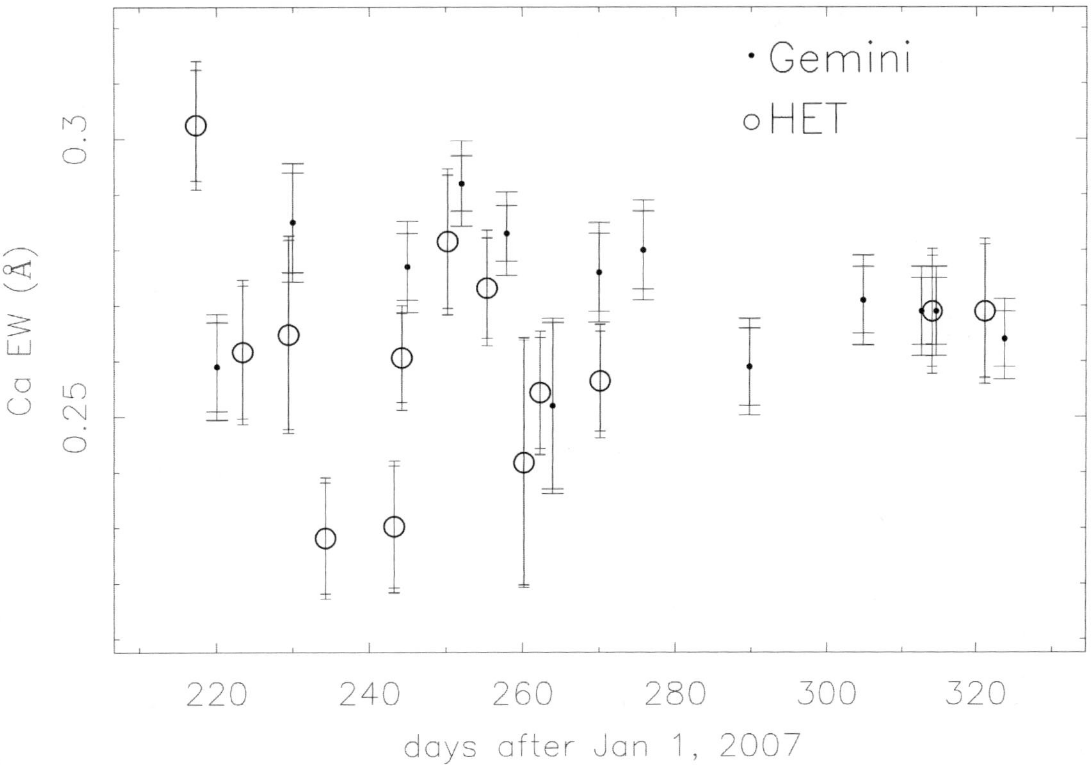

Figure 1. Ca EW as a function of time. The normalization has been adjusted upward by 4.7% for the HET data, which compensates for differences in how our technique measures EW between the moderate resolution Gemini and high resolution HET data. These were our best error estimates as of the conference. On-going analysis indicates that there may be an increase in these errors by as much as 50% due to noise in the continuum.

2.1. Could rotation with spots cause Ca/Mg EW variability?
G29-38 is rotating with a velocity, $v\sin(i)$ =11–28 km s^{-1} (Berger et al. 2005). This corresponds to a rotation period of 900–2400 s or 2000–5000 s, assuming a radius consistent with $\log(g) = 8.14$ (Bergeron et al. 1995) or 7.9 (Koester et al. 2005), respectively. Our VLT and Keck time-series observations cover 4.72 and 6.14 hours, respectively, so if the star had a strong spot, any metal-line modulation caused by the spot would have to have been meaningfully longer than 6 hours, which is inconsistent with the measured rotation rate. Additionally, the Ca and Mg variations should be modulated on the same timescale, for which we see no evidence.

3. The data at face value: implications
In the following, we assume that 1) the metals are accreting onto G29-38 uniformly over the surface, 2) gravitational settling theory is correct, and 3) that accretion never stops, but may decrease or increase. While item #1 may be incorrect (see Montgomery et al. 2008), our calculations are meant to serve only as a guideline. Further refinements, of the order of the area of the accretion surface divided by the total surface area of the WD, can be made once we learn more about the accretion geometry. Since there have been no observational tests of item #2 to date, with the exception of the general consistency of the observed variability (von Hippel & Thompson 2007) at an appropriate timescale, we assume the timescales of Koester & Wilken (2006). Item #3 just states that we do not assume that accretion ever shuts off, and therefore a

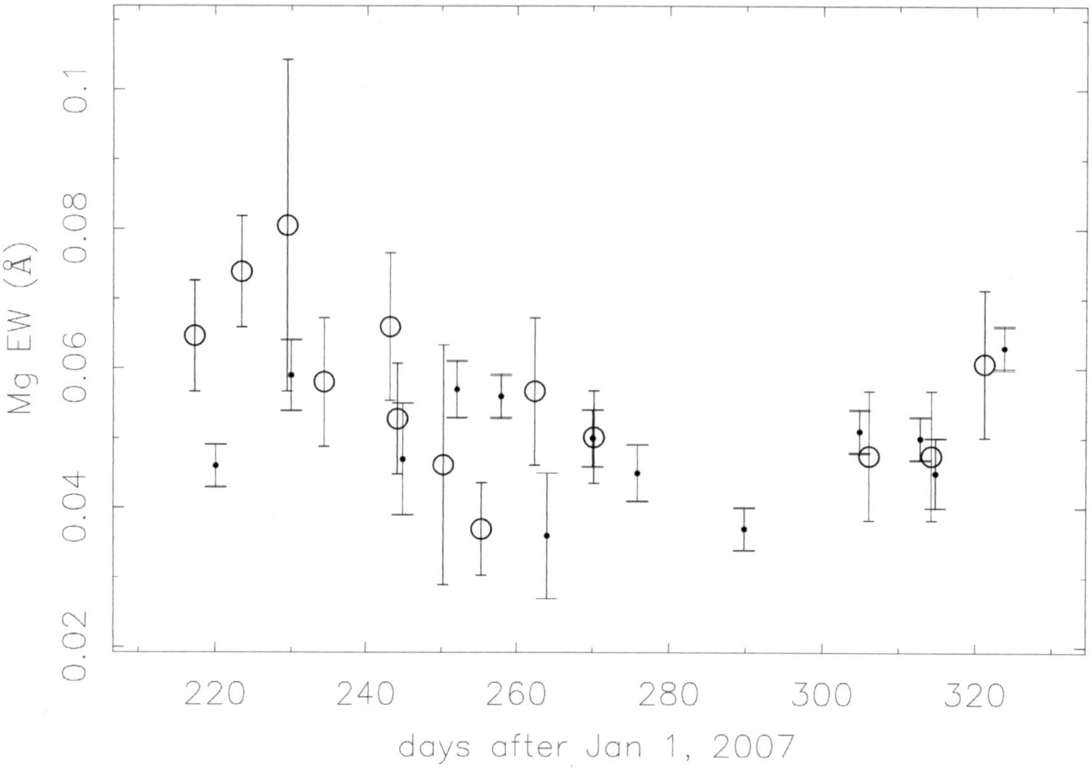

Figure 2. Mg EW as a function of time. The symbols are the same as in Figure 1. The normalization has been adjusted upward by 32% for the HET data.

decrease in photospheric abundances provides only an upper limit to the gravitational settling timescale.

3.1. Decreases in Ca & Mg line strengths

Since we have not yet fully inter-calibrated the HET and Gemini EWs, we restrict our comparisons of EW variations to be within either the HET or Gemini data sets. We find three instances where the Ca EW declines, two within the HET data and one within the Gemini data. In those three cases, we find that the timescale for Ca depletion is ≤ 32.4, ≤ 39.8, and ≤ 52.3 days. We find one such instance involving 3 epochs for the Mg EWs within the Gemini data where we find the timescale for Mg depletion is ≤ 66 days. The calculations of Koester & Wilken predict settling timescales (depending on which $\log(g)$ value is used) of ~13–23 days for Ca and ~20–36 days for Mg, respectively. Our data support these timescales.

3.2. Increases in Ca & Mg line strengths

The fastest rise in accretion occurred between epochs 5 & 6 in the HET data. This was an increase from 220 ± 11 to 249 ± 8 mÅ in 0.99 days, or 13% (though only a marginal 2.1σ result). If we compare this relative accretion rate change with the theoretical accretion rate (Koester & Wilken 2006, again reduced by 50 × assuming G29-38 is only accreting metals, not the Solar mixture), this corresponds to an increase of ~130 metric tons per second. Assuming a porous silicate structure with density ~2 g cm^{-3}, this corresponds to accreting one extra 2.5 m radius rock per second or one extra 112 m boulder in the unobserved 24 hour period. We do not know, however, if this accretion takes place quickly enough to happen in the solid phase, or whether this material is slowly brought in as gas.

3.3. How variable was accretion (while we were looking)?
The mean Ca EW difference between adjacent epochs (e.g., epoch1 − epoch2 or epoch4 − epoch5) was 13 mÅ for the Gemini data, whereas the average error bar was 7 mÅ. Subtracting these numbers in quadrature yields an average variation of 11 mÅ, or 4%, between epochs. (Increasing the error bars by 50% would mean an average variation of 8 mÅ, or 3%.) The HET data also yield an average variation of 11 mÅ. The maximum observed variation was 31 mÅ (11%) in the Gemini data and 39 mÅ (16%) in the HET data.

The mean difference between adjacent epochs for Mg from the Gemini data was 10 mÅ with an average error of 4 mÅ, or a result of 9 mÅ (18%). The largest Mg change was 20 mÅ (41%), though Mg is weaker and more difficult to measure, and therefore the results are not as firm..

If we just use the (more conservative) Ca numbers, then G29-38 had typical accretion variations of only 4% during this period. If we imagine the accretion process to be a series of discrete rocks or dust blobs all of the same mass hitting the surface, then we can estimate the number and mass of rocks/blobs hitting the surface with simple Poisson statistics. The 4% scatter we see would imply $\sim(4\%)^{-2} = 625$ discrete Ca sources per typical inter-epoch period of 8 days. More likely, a power-law source of rocks or dust blobs is responsible, rather than a Poisson distribution, and this number would be characteristic of the larger accreted rocks/blobs.

4. Conclusions

The convincing measurement of time-variable metal-line strengths in any DAZ would indicate both that the timescales for gravitational settling theory are broadly correct and that accretion from the dust disk is episodic. The former is important, since it is the primary motivator for the interpretation of on-going accretion and the mystery of the DZ phenomenon. The latter may offer some insights into the accretion physics, which at present are not even sketched out. We are compiling increasing evidence for time-variable metal-line strength variations in the best studied DAZ, G29-38. Our evidence to date broadly supports the gravitational settling timescales of Koester & Wilken (2006) and episodic accretion from G29-38's debris disk. Furthermore, we have detected evidence for time-variable accretion with a timescale $\leq$ 24 hours, and typical variability of $\sim$4% during the 100 days of our autumn 2007 monitoring campaign.

Acknowledgments

This material is based upon work supported by the National Aeronautics and Space Administration under Grant No. NAG5-13070 issued through the Office of Space Science. SET thanks the Crystal Trust for their financial support.

References

Berger L Koester D Napiwotzki R Reid I N & Zuckerman B 2005 *A&A* **444**, 565
Bergeron P Wesemael F Lamontagne R Fontaine G Saffer R A & Allard N F 1995 *ApJ* **449** 258
Debes J H & Lopez-Morales M 2008 *ApJ* **677** L43
Dupuis J Fontaine G Pelletier C & Wesemael F 1992 *ApJS* **82** 505
Farihi J Zuckerman B & Jura M 2009 *ApJ* submitted
Gänsicke B T Marsh T R Southworth J & Rebassa-Mansergas A 2006 *Sci* **314** 1908
Jura M 2003 *ApJ* **584** L91
Jura M 2006 *ApJ* **653** 613
Jura M Farihi J & Zuckerman B 2007 *ApJ* **663** 1285
Jura M Farihi J Zuckerman B & Becklin E E 2007 *AJ* **133** 1927
Koester D Rollenhagen K Napiwotzki R Voss B Christlieb N Homeier D & Reimers D 2005 *A&A* **432** 1025
Koester D & Wilken D 2006 *A&A* **453** 1051
Montgomery M H Thompson S E & von Hippel T 2008 *ApJL* in press
Reach W T Kuchner M J von Hippel T Burrows A Mullally F Kilic M & Winget D 2005 *ApJ* **635** L161

Sion E M Greenstein J L Landstreet J D Liebert J Shipman H L & Wegner G A 1983 *ApJ* **269** 253
von Hippel T Kuchner M J Kilic M Mullally F & Reach W T 2007 *ApJ* **662** 544
von Hippel T & Thompson S E 2007 *ApJ* **661** 477
Zuckerman B & Reid I N 1998 *ApJ* **505** L143
Zuckerman B Koester D Melis C Hansen Brad M & Jura M 2007 *ApJ* **671** 872

Journal of Physics: Conference Series **172** (2009) 012060 doi:10.1088/1742-6596/172/1/012060

Spectral synthesis of circumstellar disks - application to white dwarf debris disks

S D Hügelmeyer[1], S Dreizler[1], D Homeier[1] and P Hauschildt[2]

[1]Institut für Astrophysik, Georg-August-Universität Göttingen, Friedrich-Hund-Platz 1, 37077 Göttingen, Germany

[2]Hamburger Sternwarte, Gojenbergsweg 112, 21029 Hamburg, Germany

E-mail: shuegelm@astro.physik.uni-goettingen.de

Abstract. Gas and dust disks are common objects in the universe and can be found around various objects, e.g. young stars, cataclysmic variables, active galactic nuclei, or white dwarfs. The light that we receive from disks provides us with clues about their composition, temperature, and density. In order to better understand the physical and chemical dynamics of these disks, self-consistent radiative transfer simulations are inevitable. Therefore, we have developed a 1+1D radiative transfer code as an extension to the well-established model atmosphere code PHOENIX. We will show the potential of the application of our code to model the spectra of white dwarf debris disks.

1. Introduction

The detection of infrared excesses in association with the metal rich white dwarfs GD 362 and G29-38 has introduced the picture of debris disks around these stars. The high metal content observed in these evolved stars' atmospheres requires a steady or recent source of enriched matter, since the metals could only survive depletion by gravitational settling for a very short time. This source is believed to be provided by accretion of dusty material from a disrupted solid body, such as an asteroid. The IR excess is a strong indication for the formation of an accretion disk. To understand the origin of the dust, as well as to constrain the accretion rate and thus models for the gravitational diffusion of heavy elements in the atmosphere, comparison of observations with model spectra are necessary.

2. Standard accretion model

We adopt the standard accretion model for geometrically thin disks, i. e. height $H \ll$ radius R (Shakura & Syunyaev 1973). Matter is assumed to rotate with Kepler velocity and viscous shear decelerates inner and accelerates outer parts leading to accretion of matter and outward transportation of angular momentum. Turbulent cells smaller than the disk height H are the proposed origin of kinematic viscosity. The height averaged kinematic viscosity is usually described by

$$\nu = \alpha c_{\mathrm{s}} H \quad , \tag{1}$$

where $0 \leq \alpha \leq 1$ is the angular momentum transfer efficiency.

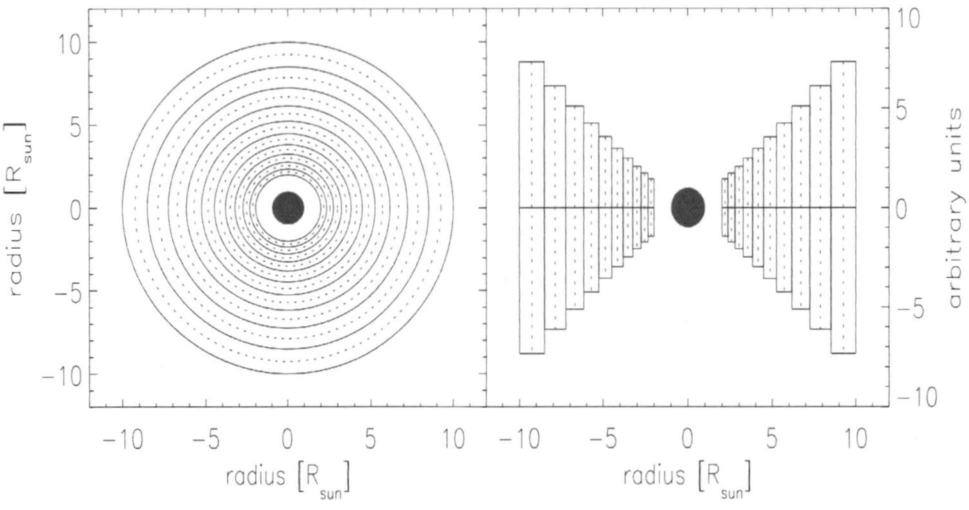

Figure 1. Disk structure as adopted for our numerical calculations. Radial disk ring extensions are chosen so that the structure varies only little over ring width. The disk is thin ($H \ll R$) so that vertical and radial structure can be separated.

3. Model calculations

We use a 1+1D disk ring structure approximation. The disk is divided into rings and each ring is assumed to be plane-parallel and to have constant physical properties from its inner to its outer radius. Radiative transfer and disk structure are calculated vertically from the disk's midplane to the top layer. Gas and dust are in chemical equilibrium and dust is assumed not to settle in the disk.

Basic program input parameters are the radius $R_\star$ and mass $M_\star$ of the central star, the distance R between star and disk ring, the mass accretion rate $\dot{M}$ in the disk, and the Reynolds number Re as a measure for the mean kinematic viscosity $\nu = \sqrt{GM_\star R}/Re$.

The iterative calculation of a disk ring model atmosphere starts by either constructing a grey start model for disks (after Hubeny 1990) or using a given PHOENIX model. We have further adopted the following conservation and constraint equations:

Hydrostatic equilibrium: This equation has to be used in a form to account for the varying surface gravity, i. e.

$$\frac{dP}{dm} = \frac{GM_\star}{R^3} z \quad . \tag{2}$$

Energy conservation: Unlike ordinary stellar atmospheres, mechanical energy is released in disk atmospheres in every layer. Therefore, the flux is not conserved and we have to consider energy equilibrium of the form

$$E_{\rm mech} = E_{\rm rad} + E_{\rm conv} \quad . \tag{3}$$

Irradiation: Also irradiation from the central star can be considered. As input source serves one or a combination of black body spectra of given effective temperatures. A PHOENIX spectrum can also be used as input. Discrete rays that receive radiation from the star are determined and assigned a fraction of the flux incident to the disk. The irradiation geometry is shown in Figure 2.

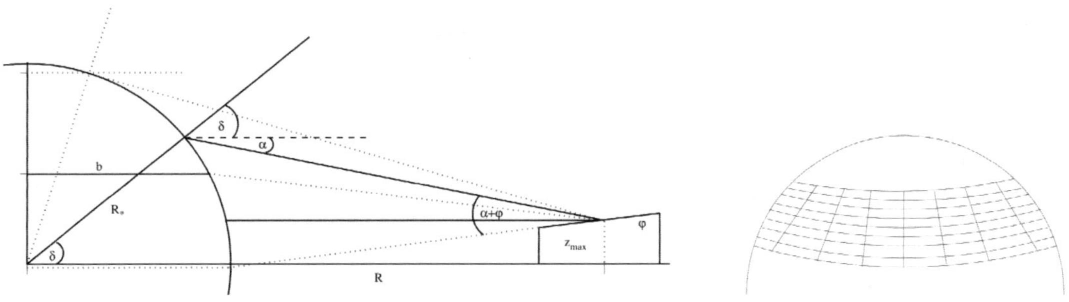

Figure 2. *Left*: Star – disk irradiation geometry. *Right*: The star's surface fraction is subdivided in sections and the irradiation flux is calculated for each ray considering limb darkening.

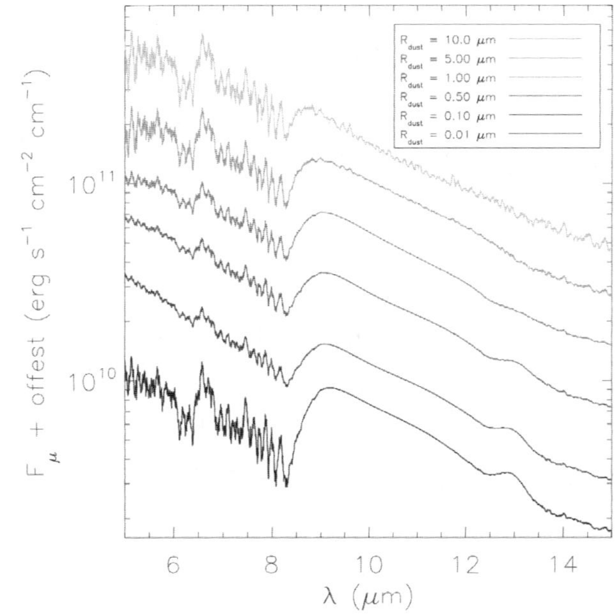

Figure 3. Spectra in the range of the characteristic 10 μm silicate feature for different mean dust grain sizes. For these calculations we have assumed $M_\star = 0.6\ M_\odot$, $R_\star = 0.01\ R_\odot$, $R = 100\ R_\star$, and $\dot{M} = 10^{-11}\ M_\odot\ \mathrm{yr}^{-1}$. For this setup, the smallest grain size produces the most prominent 13 μm feature which is also clearly visible in the *Spitzer* spectrum

4. Results

We have calculated a simple disk model consisting of three disk rings. The rings are in hydrostatic equilibrium and the mass accretion rate is set according to Becklin et al. (2005) to $\dot{M}_\star = 10^{11}\ \mathrm{g\ s}^{-1} \approx 2 \cdot 10^{-15}\ M_\odot\ \mathrm{yr}^{-1}$. This small accretion rate leads to small surface densities ($\doteq$ disk mass) and the disk rings are optically thin in the continuum. The vertical temperature distribution is almost constant and only "artificially" brought to a value reasonably in agreement with the blackbody temperature fit to the infrared excess by setting an unrealistically high viscosity value.

In order to estimate the mean dust grain size, we have performed various computations varying this parameter. This allows us to get an idea of the influence on the strength of the silicate feature at 10 μm and the 13 μm emission. We have varied the mean dust grain size between 0.01 μm and 10 μm. The resulting spectra of these simulations are shown in Figure 3. For our refined models of GD 362 we have chosen a mean dust grain size of 0.01 μm.

Since the disk slabs are optically thin, the continuum flux is in a reasonable approximation

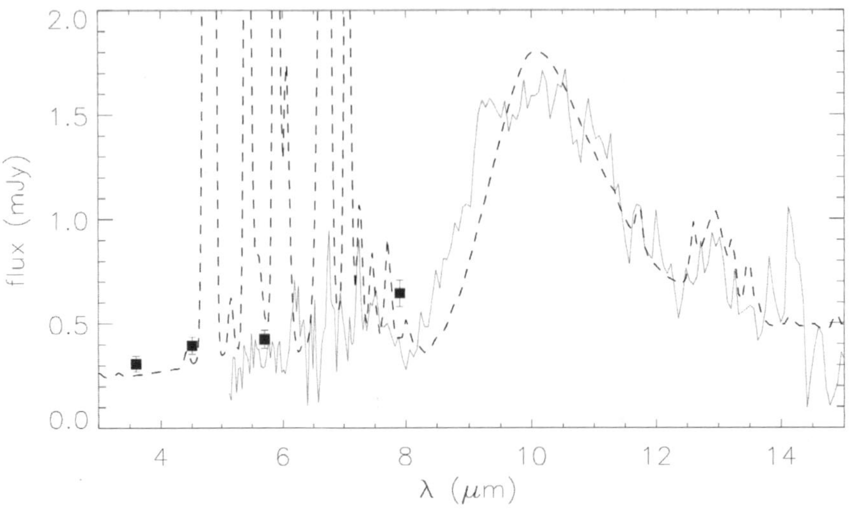

Figure 4. *Spitzer* IRS spectrum (grey line; Farihi, private communication) and photometry (squares with error bars; Jura et al. 2007) of GD 362. Our model fit is plotted as dashed line (disk+star) and reproduces the 10 μm silicate and the 13 μm feature fairly well.

linear with the optical depth. Therefore, the fitting of the models' continuum to the *Spitzer* spectrum can be adjusted by varying the mass accretion rate. We simply used a fudge factor to account for the missing flux.

In Figure 4 we show the co-added models folded to the resolution of the instrument. The innermost ring is directly irradiated and heated up by the white dwarf and is assumed to be isothermal (700 K), thus shielding the disk parts behind it. The inner ring extends from 9.3−10 $R_\star$. The following three rings go from 10−70 $R_\star$ and have almost isothermal temperature structures between 670 K and 500 K.

5. Discussion

We have shown that our disk model atmosphere code is in principal applicable to the modelling of white dwarf debris disks. Our models calculated for the parameters of GD 362 show a reasonable fit to IR observations of the objects. Masses and accretion rates of WD debris disks are however small and therefore make the model calculations challenging.

Acknowledgments

SDH is supported by a scholarship of the DFG Graduiertenkolleg 1351 "Extrasolar Planets and their Host Stars"

References

Becklin E E Farihi J Jura M et al. 2005 *ApJ* **632** L119
Hubeny I 1990 *ApJ* **351** 632
Jura M Farihi J and Zuckerman B 2007 *ApJ* **663** 1285
Shakura N I & Syunyaev R A 1973 *A&A* **24** 337

Preliminary XCOV26 Results for EC14012-1446

J L Provencal[1,2], S Thompson[1,2], M Montgomery[2,3], A Kanaan[4], H L Shipman[1,2], J Dalessio[1,2], D Childers[5], C Clemens[6], R Rosen[7], P Henrique[3], A Kim[2], W Strickland[8], D Chandler[8], B Walter[8], T K Watson[9], B Castanheira[10], M Wood[11], S Vennes[11], S O Kepler[12], M Reed[13], A Nitta[14], S J Kleinman[14], T Brown[15], S -L Kim[16], D Sullivan[17], Wen-Ping Chen[18], M Yang[18], Chia-You Shih[18], X J Jiang[19], A V Sergeev[20], A Maksim[20], R Janulis[21], H O Vats[22], K S Baliyan[22], S Zola[23,24], A Baran[23], M Winiarski[23,24], W Ogloza[23,24], M Paparo[25], Z Bognar[25], P Papics[25], D Kilkenny[27], R Sefako[26,27], D Buckley[26], N Loaring[26], A Kniazev[26], R Silvotti[28], S Galleti[28], G Handler[10], T Nagel[29], G Vauclair[30], N Dolez[31], J R Fremy[31], J Perez[32], J M Almenara[32], L Fraga[4]

[1]University of Delaware, Department of Physics and Astronomy Newark, DE 19716; jlp@udel.edu
[2]Delaware Asteroseismic Research Center, Mt. Cuba Observatory, Greenville, DE 19807
[3]Department of Astronomy, University of Texas, Austin, TX 78712; mikemon@rocky.as.utexas.edu
[4]Departamento de Física Universidade Federal de Santa Catarina, C.P. 476, 88040-900, Florianópolis, SC, Brazil; ankanaan@gmail.com
[5]Department of Math and Science, Delaware County Community College, 901 S. Media Rd, Media, PA 19063; dpc@udel.edu
[6]University of North Carolina 288 Phillips Hall, Chapel Hill, NC 27599; clemens@physics.unc.edu
[7]NRAO Green Bank, WV 24944; rachel.rosen@gmail.com
[8]Meyer Observatory and Central Texas Astronomical Society, 209 Paintbrush, Waco, TX 76705; chandler@vvm.com
[9]Southwestern University, Georgetown, TX; tkw@sousthwestern.edu
[10]Institut für Astronomie Universität Wien, Türkenschanzstrasse 17, 1180, Austria; handler@astro.univie.ac.at
[11]Florida Institute of Technology, Dept. of Physics & Space Sciences, 150 W Univ. Blvd, Melbourne, FL 3290; wood@fit.edu
[12]Instituto de Fíisica UFRGS, C.P. 10501, 91501-970 Porto Alegre,RS, Brazil; kepler@if.ufrgs.br
[13]Missouri State University and Baker Observatory, 901S. National, Springfield, MO 65897; MikeReed@missouristate.edu
[14]Gemini Observatory, Northern Operations Center, 670 North A'ohoku Place, Hilo, HI 96720; atsuko.nittakleinman@gmail.com
[15]Las Cumbres Observatory Global Telescope Network, Inc. 6740 Cortona Dr. Suite 102 Santa Barbara CA 93117; tbrown@lcogt.com
[16]Korea Astronomy and Space Science Institute, Daejeon 305-348, Korea; slkim@kasi.re.kr
[17]School of Chemical & Physical Sciences, Victoria University of Wellington, P.O. Box 600, Wellington, New Zealand;denis.sullivan@vuw.ac.nz
[18]Lulin Observatory, National Central University, Taiwan; wchen@astro.ncu.edu.tw
[19]National Astronomical Observatories, Academy of Sciences, Beijing 100012, PR China; xjjiang@bao.ac.cn

[20]Ukrainian National Academy of Sciences, Main Astronomical Observatory, Golosiiv, Kiev 022 252650; sergeev@terskol.com
[21]Institute of Theoretical Physics and Astronomy, Vilnius University, Vilnius, Lithuania; jr@itpa.lt
[22]Physical Research Laboratory, Ahmedabad 380009, India
[23]Mount Suhora Observatory, Cracow Pedagogical University, Ul. Podchorazych 2, 30-084 Krakow, Poland; zola@astro1.as.ap.krakow.pl
[24]Astronomical Observatory, Jagiellonian University, ul. Orla 171, 30-244 Cracow, Poland
[25]Konkoly Observatory, P.O. Box 67, H-1525 Budapest XII, Hungary; paparo@konkoly.hu
[26]South African Astronomical Observatory, PO Box 9, Observatory 7935, South Africa
[27]Department of Physics, University of the Western Cape, Private Bag X17, Belville 7535, South Africa
[28]INAF-Osservatorio Astronomico di Capodimonte, via Moiariello 16, 80131 Napoli, Italy
[29]Institut für Astronomie und Astrophysik, Universiät Tübingen, Sand 1, 72076 Tübingen, Germany; nagel@astro.uni-tuebingen.de
[30]Laboratoire d'Astrophysique de Toulouse-Tarbes, Université de Toulouse, CNRS, 14 avenue Edouard Belin, F314000 Toulouse, France; gerardv@srvdec.obs-mip.fr
[31]Observatoire de Paris, LESIA, 92195 Meudon, France
[32]Instituto de Astrofisica de Canarias, 38200 La Laguna, Tenerife, Spain

Abstract.
EC14012-1446 is a hydrogen atmosphere (DA) white dwarf pulsator. Its rich pulsation spectrum displays a range of excited modes with complex multiplet structure, in addition to numerous combination frequencies. In April 2008, EC14012-1446 was the primary target of XCOV26. We obtained over 300 hrs of nearly continuous high speed photometry with the goal of using the nonlinear pulse shapes to empirically determine the parameters of the convection zone. The Fourier transform (FT) of the light curve contains power between 1000 to 4000 μHz, with the dominant peak at 1234 μHz. We find 13 independent frequencies distributed in 8 modes, as well as a myriad of combination frequencies. In the following, we present preliminary results and lay the groundwork for future investigation leading to light curve fitting of EC14012-1446.

1. Introduction

Every star will meet one of three ends as it approaches the limits of its evolution. If the star is massive, the events triggered by the exhaustion of fuel in the stellar core will lead to a black hole. If the star is of intermediate mass, the product will be a neutron star. If the star is low mass, a definition that includes over 90% of all stars currently on the main sequence, the end product will be an electron-degenerate white dwarf.

White dwarfs play important roles in our understanding of the Galaxy and the universe. The structure and composition of white dwarfs contain records of stages of stellar evolution. The chemical evolution of the Galaxy is traced through the generations of stars formed from contaminated material ejected by pre-white dwarfs as they evolve through the planetary nebula phase. The white dwarf mass distribution and the luminosity function are used to examine the Galaxy's overall formation history. Type I supernovae, in which an accreting white dwarf undergoes a thermonuclear event, are used as distance indicators demonstrating the acceleration of the universe.

Asteroseismology allows us to peer beneath the photospheres of white dwarf stars and reveal the secrets of their interior structure and composition. Pulsating white dwarfs are only unique in their temperatures, as any white dwarf cools, it passes through temperature regimes where oscillations may be excited. The oscillations reveal information about basic physical parameters, such as mass, rotation rate, internal transition profiles, and compositional structure. This information (see for example: Nather et al. 1990; Winget et al. 1991; Winget et al. 1994; Kepler et al. 2000; Kanaan et al. 2005; Provencal et al. 2008) can be applied to the white dwarf

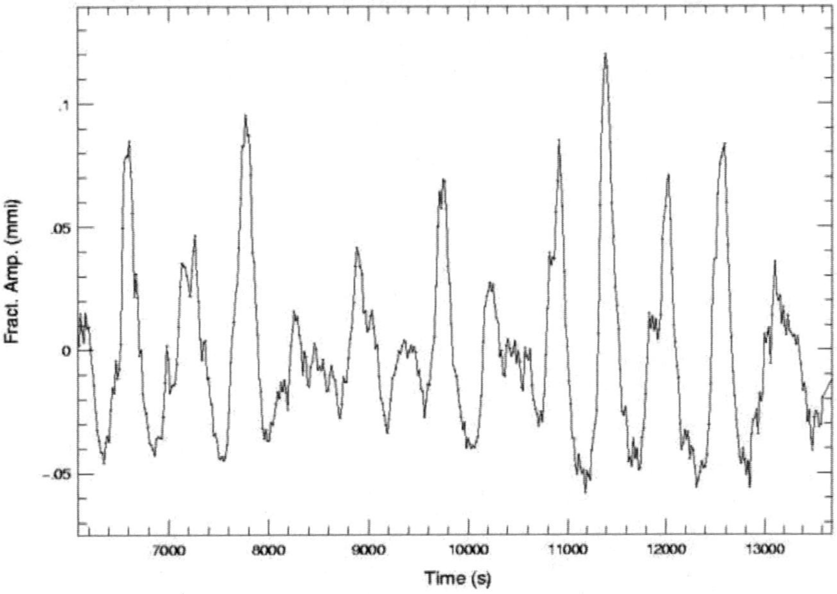

Figure 1. A portion of EC14012-1446-s light curve obtained with SOAR, demonstrating the light curve quality required for Montgomery's fitting technique. The complete light curve can be seen at www.physics.udel.edu/darc.

population as a whole, providing important constraints on a wide range of fields of astrophysical importance.

Asteroseismology is now expanding its focus to attack the fundamental physical process of convection. Convection is one of the largest sources of theoretical uncertainties in our understanding of stars and other natural phenomena. Montgomery (2005) shows how precise observations of variable star light curves can be used to observationally derive the depth of the pulsator's convection zone and its sensitivity to changes in temperature. The Whole Earth Telescope and the Delaware Asteroseismic Research Center are currently conducting a project to map convection across the white dwarf instability strips. The criteria for selecting candidate objects include nonlinear pulsations and a fairly bright target. The ZZ Ceti EC14012-1446 was chosen as a primary target for XCOV26 in April 2008 based on its suitability as a candidate for light curve fitting. In the following, we present the resulting data set, our preliminary analysis of the the light curves and Fourier transforms (FTs), and list the identified frequencies. Our discussion will lay the groundwork for the success of convective light curve fitting with EC14012-1446.

2. Observations

XCOV26 spanned March 25 - April 18 2008, with the densest coverage occurring April 1-14. Our goals were twofold for EC14012-1446: provide accurate frequency identifications, and obtain several nights of very high signal to noise light curves to be combined with the frequency information as required for light curve fitting. Over 20 telescopes participated in the complete run, including SOAR (4 m) and SALT (10 m). EC14012-1446 was the primary target. This DAV ($T_{eff} = 11,300$ K (Koester et al. 2001)) is a multiperiodic high amplitude pulsator discovered by

Journal of Physics: Conference Series **172** (2009) 012061 doi:10.1088/1742-6596/172/1/012061

Table 1. EC14012-1446 Journal of Site Observations

Telescope	Instrument	Length
Hawaii 2.2 m	CCD	6.2
SAAO 1.0 m	CCD	77
Tenerife 0.8 m	CCD	6
McDonald 2.1 m	CCD	32
Loia 1.52 m	CCD	1.1
CTIO 0.9 m	CCD	70
SOAR 4.1 m	CCD	30
SARA 1.0 m	CCD	8
SALT 10.0 m	CCD	2.4
Mt. John 1 m	CCD	6.6
BOAO 1.8 m	CCD	20
Lulin 1.0 m	CCD	10
KPNO 2.1 m	CCD	15
LAS Cumbres 2.0 m	CCD	20
Mt. Abu 1.2 m	CCD	4

Stobie et al. (1995) and observed extensively by Hander et al. (2008). Over 146,000 data points were obtained during XCOV26 by 24 telescopes distributed around the globe, achieving over 80% coverage during the central 5 days of the run. The Journal of Observations for EC14012-1446 is given in Table 1. In the interest of space, we present the total observing time for each site, rather than the individual run names. The individual runs can be found at www.physics.udel.edu/darc.

A goal of all Whole Earth Telescope (WET) runs is to obtain as uniform a data set as possible. XCOV26 was the first full WET run comprised solely of CCD photometers. We minimize bandpass issues by using CCDs with similar detectors and equipping each CCD with a BG40 or S8612 filter to normalize wavelength response and reduce extinction effects. Details of instruments and cycle times can be found at www.physics.udel.edu/darc. We followed the reduction procedures outlined in Provencal et al. (2008). Each reduced light curve consists of a list of times and fractional intensities. Finally, we combined the individual light curves to produce the complete data set. Figure 1 shows a light curve obtained by the SOAR telescope. We obtained 5 nights of SOAR observations and $\approx$2 hrs of SALT observations. These lightcurves easily achieve the signal to noise required by Montgomery's technique.

3. The Fourier Transform

The XCOV26 Fourier Transform (FT) is given in Figure 2. Following the guidelines and procedures outlined in Provencal et al. (2008), we identify a total of 70 frequencies, including 13 independent frequencies distributed in 10 modes and numerous combination frequencies (Table 2). The main frequencies are at 1633.905 and 1887.405 μHz. We find most of the modes to contain a single dominant frequency. Only the 1887.405 μHz mode exhibits triplet structure, with a splitting of 3.8 μHz. The remaining multiplet splittings do not conform to the predictions of frequency splitting in the limit of slow rotation. We do not find and so cannot confirm the multiplet splittings from Handler et al. (2008).

However, the period spacings provide some information. The two dominant modes, at 1633.905 and 1887.405 μHz, are separated in period by 82 seconds, indicating that these are not consecutive radial overtones. EC14012-1446's mode distribution is difficult to understand

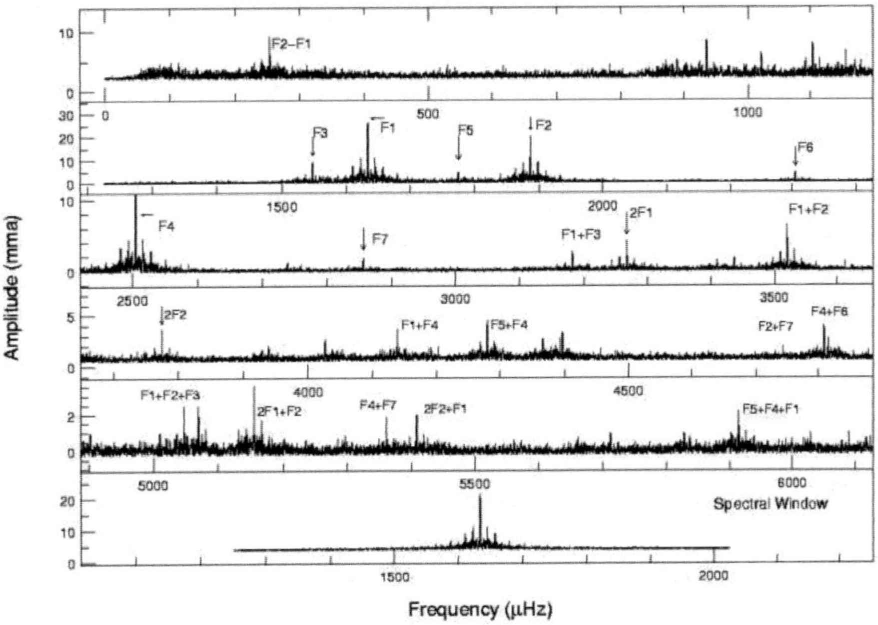

Figure 2. XCOV26 Fourier Transform of EC14012-1446. We find 13 independent frequencies distributed in 8 modes, as well as numerous combination frequencies. The spectral window is given in the last panel.

assuming consecutive overtones of a single l value. We do find some early indications of period spacings of ≈34 seconds, as well as ≈20 seconds. Based on these indications, and given the limitations on log g imposed by spectroscopic evidence, we believe that EC14012-1446's FT contains a mixture of l=1 and l=2 modes. Detailed identifications are still underway.

Handler et al. (2008) reports that EC14012-1445's pulsation modes undergo amplitude and frequency variation similar to that reported in the pulsating DBV GD358 (Provencal et al. 2008). Figure 3 shows EC14012-1446's FTs from 2004, 2007, and XCOV26 (2008). In both stars, the modes with higher radial overtones (higher frequency) show pronounced amplitude variability accompanied by frequency changes. Note in particular the slight frequency shift associated with the 1633 μHz mode, the complete disappearance of power at 1300 and 1400 μHz in the 2008 FT, and the overall lower amplitudes present in the 2008 FT.

Our goal of light curve fitting to determine convective parameters requires that we understand EC14012-1446's stability over the length of XCOV26. We divided the data set into two equal chunks and calculating the FT of each chunk (Figure 4). For the dominant peaks, the frequencies are consistent to within measurement error and the amplitudes remain stable to within 3σ. The differences between each FT are explained by variation in window structure and resolution from chunk to chunk. We are confident that EC14012-1446 was fairly stable over the length of XCOV26.

4. Conclusions

The 2008 XCOV26 observations were obtained with the goal of using EC14012-1446's nonlinear lightcurve to characterize its convection zone. XCOV26 reached both of its observational goals,

Table 2. Table of Independent Frequencies

ID	Frequency (μHz)	Amplitude (mma)
F3	1548.144 ± 0.001	7.9 ± 0.1
	1521.570 ± 0.002	2.2 ± 0.2
F1	1633.905 ± 0.001	25.6 ± 0.1
	1624.017 ± 0.003	3.1 ± 0.2
F5	1775.168 ± 0.002	4.5 ± 0.2
F2	1883.601 ± 0.003	1.6 ± 0.2
	1887.405 ± 0.001	20.6 ± 0.1
	1891.141 ± 0.002	3.7 ± 0.2
F6	2304.730 ± 0.001	4.0 ± 0.1
F4	2504.896 ± 0.001	12.7 ± 0.1
	2508.062 ± 0.002	2.1 ± 0.2
F8	2738.108 ± 0.002	1.2 ± 0.2
F7	2856.161 ± 0.002	2.0 ± 0.2

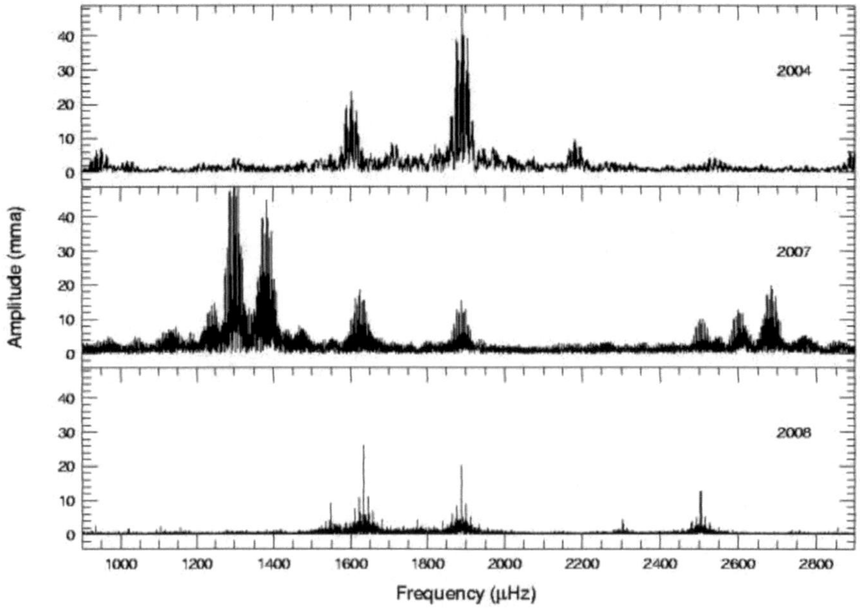

Figure 3. FTs of EC14012-1446 from 2004, 2007, and 2008. EC14012-1446 displays frequency and amplitude changes similar to GD358.

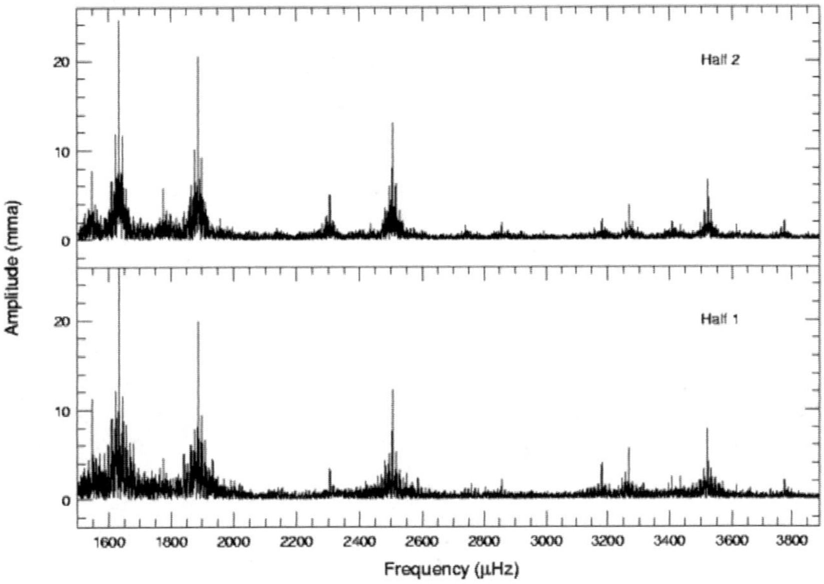

Figure 4. FTs of EC14012-1446 during the first half of XCOV26 (top panel) and second half (bottom panel). EC14012-1446 was fairly stable of the length of the run.

providing us with accurate frequencies for EC14012-1146 and obtaining several nights of high signal to noise light curves suitable for light curve fitting. Our asteroseismic analysis of this data set is still in its early stages, but promises to reveal a great deal of interesting information about EC14012-1446's pulsational behavior. We have identified over 70 frequencies present in the FT, with 13 independent frequencies distributed in 8 modes. While our identifications parallel the results of Handler et al. (2008), EC14012-1446 presented a somewhat different appearance during XCOV26. We find most modes to be dominated by a single frequency and lacking clear triplet structure. In addition, the period spacings seem to indicate that EC14012-1446 pulsates with both l=1 and l=2 modes. These results complicate the l and m identifications necessary for light curve fitting. We are proceeding along a number of fronts, including an examination of the combination frequencies and an analysis of period spacings for all of EC14012-1446's excited modes.

References

Handler G Romero-Colmenero E Provencal J L Sanchawala K et al. 2008 *M.N.R.A.S.* **388** 1444
Kanaan A Nitta A Winget D E et al. 2005 *A&A* **432** 219
Kepler S O Mukadam A Winget D E Nather R E T S et al. *ApJL* **534** L185
Koester D et al. 2001 *A&A* **378** 556
Montgomery M H 2005 *ApJ* **633** 1142
Nather R E Winget D E & Clemens J C 1990 *ApJ* **361** 390
Provencal J L et al. 2008 *ApJ* in press
Stobie R S O'Donoghue D Ashley R et al. 1995 *M.N.R.A.S.* **272** L21
Winget D E Nather R E Clemens J C Provencal J L et al. 1991 *ApJ* **378** 326
Winget D E Nather R E Clemens J C Provencal J L et al. 1994 *ApJ* **430** 839

Journal of Physics: Conference Series **172** (2009) 012062 doi:10.1088/1742-6596/172/1/012062

A Photometric Analysis of ZZ Ceti Stars: A Parameter-Free Temperature Indicator?

P Bergeron[1], S K Leggett[2] and H C Harris[3]

[1] Département de Physique, Université de Montréal, C.P. 6128, Succ. Centre-Ville, Montréal, Québec H3C 3J7, Canada

[2] Gemini Observatory, Northern Operations Center, 670 North A'ohoku Place, Hilo, Hawaii 96720, USA

[3] US Naval Observatory, Flagstaff Station, Flagstaff, Arizona 86001, USA

E-mail: bergeron@astro.umontreal.ca, sleggett@gemini.edu, hch@nofs.navy.mil

Abstract. We present a model atmosphere analysis of optical VRI and infrared JHK photometric data of about two dozen ZZ Ceti stars. We first show from a theoretical point of view that the resulting energy distributions are not particularly sensitive to surface gravity or to the assumed convective efficiency, a result which suggests a parameter-free effective temperature indicator for ZZ Ceti stars. We then fit the observed energy distributions with our grid of model atmospheres and compare the photometric effective temperatures with the spectroscopic values obtained from fits to the hydrogen line profiles. Our results are finally discussed in the context of the determination of the empirical boundaries of the ZZ Ceti instability strip.

1. Introduction

The atmospheres of hydrogen-line (DA) white dwarfs become convective below roughly $T_{\rm eff} \sim 17,000$ K, although it is only when they have cooled below $\sim 12,000$ K that a significant fraction of the total flux is transported by convection. At that point, the bottom of the convection zone sinks deeper into the star until it reaches a maximum depth of $\Delta M/M_\star \sim 10^{-6}$ near $T_{\rm eff} \sim 5000$ K. Below 9000 K or so, convection in hydrogen-atmosphere white dwarfs becomes adiabatic, and the thermodynamic stratification is then specified by the value of the adiabatic gradient and the hydrostatic equilibrium equation. As such, the convective flux becomes almost completely independent of the assumed convective efficiency. Above that temperature, however, the structure of the atmosphere and the deeper stellar envelope depend sensitively on the way convection is treated in the model calculations. For instance, Bergeron et al. (1992) have shown that in the range $T_{\rm eff} = 10,000 - 14,000$ K, the predicted energy distributions and line profiles are strongly affected by the parameterization of the mixing-length theory, the most commonly used formalism that can take into account convective energy transport in model atmosphere calculations.

This is unfortunately the precise range of effective temperature where the DA white dwarf pulsators, or ZZ Ceti stars, are found. According to the spectroscopic analysis of Gianninas et al. (2006, see also these proceedings), ZZ Ceti stars occupy a trapezoidal region in the $T_{\rm eff} - \log g$ plane, with an empirical blue edge around 12,400 K and a red edge near 11,000 K, although both edges show a strong dependence on surface gravity. It must be stressed, however, that the atmospheric parameter determinations of ZZ Ceti stars depend strongly on the assumed convective efficiency in the model atmosphere calculations. Moreover, this efficiency cannot be easily estimated from first principles. To overcome this situation, Bergeron et al. (1995) have attempted to calibrate the convective efficiency in the atmospheres of DA stars by comparing effective temperatures obtained from fits to Balmer line profiles with those derived from UV energy distributions. The authors convincingly demonstrate that the so-called ML2/α = 0.6 parametrization of the mixing-length theory provides the best internal consistency between optical and UV effective temperatures, as well as trigonometric parallaxes, V magnitudes, and gravitational redshifts. This has been the parametrization used in all model atmosphere calculations since then.

In this paper, we present an independent method for estimating the internal consistency of this parametrization by fitting optical VRI and infrared JHK photometric energy distributions of ZZ Ceti stars with detailed model atmosphere calculations. In particular, we show that this technique provides a temperature indicator for ZZ Ceti stars that is independent of the assumed convective efficiency.

2. A Parameter-Free Temperature Indicator

Our fitting technique relies on the approach originally developed by Bergeron et al. (1997) in their study of cool white dwarfs. Briefly, the atmospheric parameters for each star are obtained by converting the optical and infrared photometry into observed fluxes, and by comparing the resulting spectral energy distributions with those predicted from model atmosphere calculations. The first step is accomplished by transforming the magnitudes into average stellar fluxes f_λ^m received at Earth using the calibration of Holberg et al. (2006). The observed and model fluxes, which depend on $T_{\rm eff}$ and $\log g$, are related by the equation $f_\lambda^m = 4\pi\ (R/D)^2\ H_\lambda^m$ where R/D is the ratio of the radius of the star to its distance from Earth, and H_λ^m is the Eddington flux, properly averaged over the corresponding filter bandpass. We finally minimize the difference between observed and predicted fluxes at all bandpasses using our standard Levenberg-Marquardt minimization procedure. Only $T_{\rm eff}$ and the solid angle $\pi\ (R/D)^2$ are considered free parameters.

As a theoretical exploration of our method, we show in Figure 1 the predictions from our model atmospheres for various values of $T_{\rm eff}$, $\log g$, and convective efficiency. Monochromatic Eddington fluxes are shown together with the average over the optical $UBVRI$ and infrared JHK filter bandpasses. One can already notice that if the U filter, and to a lesser extent, the B filter are omitted from the fits, the predicted energy distributions (normalized at V) do not depend on the assumed value of $\log g$ (middle panel), and more interestingly, they do not depend on the assumed parametrization of the mixing-length theory either (top panel; the ML1, ML2, and ML3 prescriptions are described in detail in Bergeron et al. 1995). Fortunately, the energy distributions remain sensitive to variations of effective temperature (bottom panel). Hence, by restricting our analysis to optical VRI and infrared JHK photometry, we obtain an *independent temperature scale* for ZZ Ceti stars that does not depend on the assumed value of $\log g$ (or mass) or convective efficiency.

3. Observational data

Since 2001, we have been obtaining time-averaged optical and infrared photometric observations of ZZ Ceti stars. Our data for 24 ZZ Ceti white dwarfs are summarized in Table 1. The optical photometry were secured with a CCD detector attached to the 1.0 m R/C reflector at the U.S. Naval Observatory in Flagstaff. Data were calibrated by observing standards from Landolt (1992) throughout each night and determining nightly extinction and color terms. When possible, stars were observed on two to three different nights to average out their variability. We measured the V magnitude for all 24 objects in our sample, and R and I magnitudes for only 9 and 20 objects, respectively. The infrared JHK photometry was secured for 19 objects

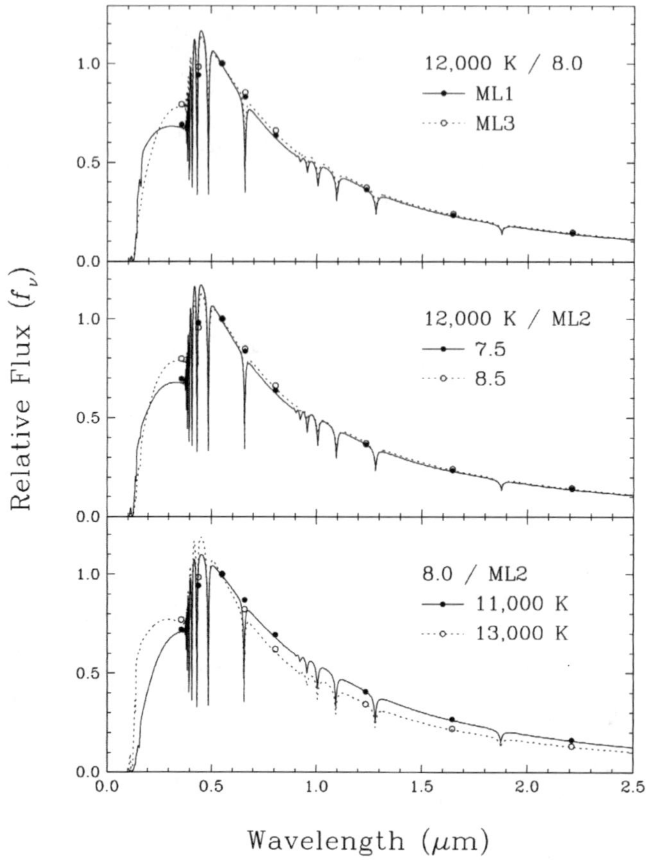

Figure 1. Energy distributions for white dwarfs in the ZZ Ceti temperature range for various values of $T_{\rm eff}$, $\log g$, and convective efficiency. Also shown (open and filled circles) are the integrated fluxes over the optical *UBVRI* and infrared *JHK* filter bandpasses. All distributions are normalized at *V*.

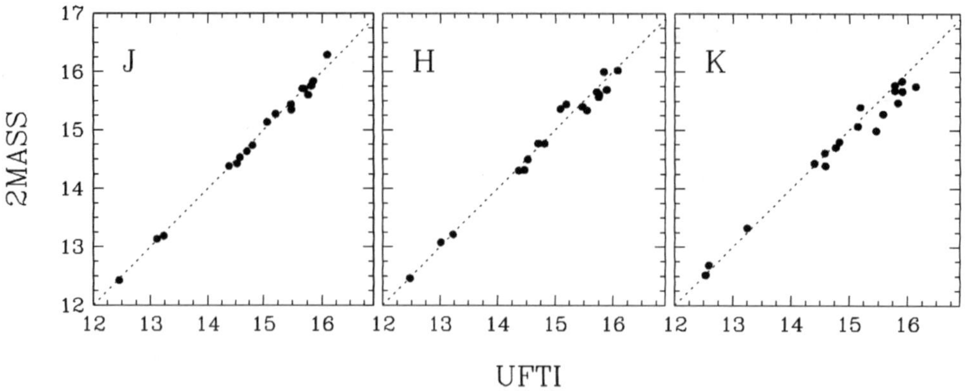

Figure 2. Comparison of infrared photometry for 17 ZZ Ceti stars in our sample that have both UFTI and 2MASS *JHK* measurements.

Table 1. Optical and infrared photometry of ZZ Ceti stars.

				UFTI			2MASS		
Name	V	R	I	J	H	K	J	H	K
Ross 548	14.16			14.37	14.36	14.40	14.38	14.30	14.43
KUV 02464+3239	16.03		16.03				15.97	15.92	15.44
HL 76	14.99		15.05	15.06	15.09	15.15	15.13	15.36	15.06
G38-29	15.59			15.70	15.75	15.78	15.70	15.56	15.77
HS 0507+0435B	15.33		15.39	15.47	15.55	15.58	15.43	15.33	15.27
GD 66	15.55		15.61	15.67	15.72	15.78	15.70	15.65	15.68
KUV 08368+4026	15.60	15.62	15.64	15.85	15.89	15.90	15.84	15.69	15.84
GD 99	14.51	14.53	14.52	14.70	14.70	14.77	14.63	14.77	14.70
G117-B15A	15.46	15.51	15.51	15.77	15.76	15.83	15.59	15.61	15.47
KUV 11370+4222	16.55	16.57	16.56	16.78	16.76	16.68			
G255-2	15.97	15.95	16.00				16.00	15.87	16.28
GD 154	15.26	15.24	15.29	15.48	15.47	15.46	15.34	15.39	14.98
G238-53	15.46	15.52	15.52				15.68	15.41	15.30
EC 14012-1446	15.66	15.66	15.69	15.82	15.84	15.91	15.75	16.00	15.67
GD 165	14.27	14.33	14.34	14.57	14.52	14.58	14.53	14.49	14.61
PG 1541+651	15.54		15.58				15.60	15.91	15.42
G226-29	12.24			12.45	12.47	12.53	12.42	12.46	12.52
G207-9	14.61		14.63	14.80	14.81	14.83	14.73	14.76	14.79
G185-32	12.98			13.23	13.23	13.25	13.18	13.21	13.32
GD 385	15.10		15.12	15.21	15.19	15.19	15.27	15.44	15.39
G232-38	16.83		16.79				16.34	15.97	16.01
PG 2303+243	15.26		15.34				15.49	15.41	15.70
G29-38	13.04		13.01	13.11	13.01	12.58	13.13	13.07	12.68
G30-20	16.07		16.04	16.09	16.08	16.14	16.29	16.02	15.75

using the U.K. Infrared Telescope (UKIRT) Fast-Track Imager (UFTI, Roche et al. 2003). The camera's filters are on the Mauna Kea Observatories photometric system (Tokunaga et al. 2002), and the data were calibrated using standards from Leggett et al. (2006). Since the beginning of our survey, 2MASS photometry has also become available for 23 objects in our sample, and this data is reported in Table 1 and included in our analysis below. The UFTI and 2MASS JHK photometric observations for the 17 objects in common are compared in Figure 2. Both data sets agree extremely well, with the exception perhaps of the K filter for faint stars where our UFTI photometry is arguably better.

4. Photometric analysis

As discussed above, the magnitudes from Table 1 are converted into average fluxes using the appropriate filter transmission functions for each photometric system. Similarly, the model monochromatic fluxes are also averaged over the same filter sets. The model atmospheres used in this analysis are slightly different from those used by Bergeron et al. (1995) and in all previous analyses of ZZ Ceti stars by the Montreal group (Gianninas et al. 2006 and references therein). We are now making use of the extensive Stark broadening profiles calculated by Lemke (1997), while our earlier calculations relied on similar calculations by Schöning & Butler (1991, private communication) but for temperatures and densities characteristic of hot subdwarfs stars, which

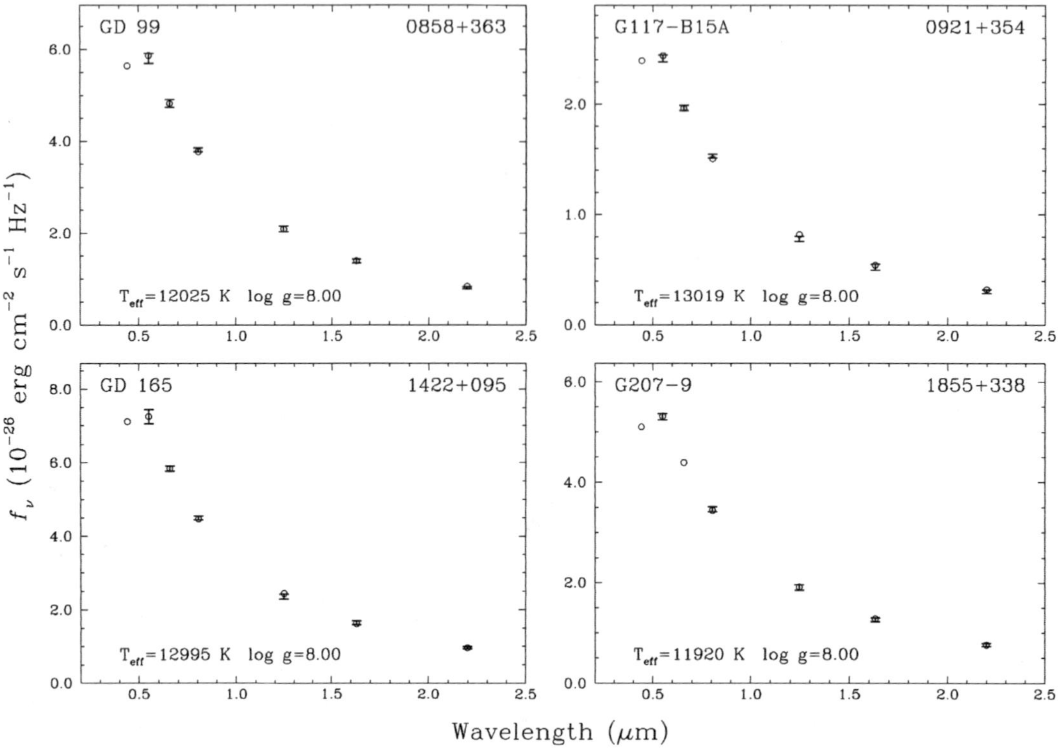

Figure 3. Sample fits to the energy distributions of ZZ Ceti stars. The VRI and JHK photometric observations are represented by error bars while the model fluxes are shown as open circles.

often required extrapolations in the higher density regime of white dwarf atmospheres. With these new models in hand, we redid the analysis of Bergeron et al. (1995) and found that a slightly more efficient version of the mixing-length theory, namely ML2 with $\alpha = 0.7$ (instead of 0.6), provides the best internal consistency between effective temperatures obtained from optical and UV spectra. Sample fits to the VRI and JHK energy distributions of four ZZ Ceti stars in our sample using these models are displayed in Figure 3. We assume in this figure a value of $\log g = 8.0$ for all objects.

The photometric effective temperatures for all ZZ Ceti stars in our sample are compared in Figure 4 with spectroscopic temperatures obtained from fits to the hydrogen Balmer lines using the so-called spectroscopic technique (see Bergeron et al. 1995 and references therein); the optical spectra are taken from the analysis of Gianninas et al. (2006). Both infrared data sets (UFTI and 2MASS) are considered independently in Figure 4. We also assume for our photometric fits a value of $\log g = 8.1$ for all objects, which corresponds to the average spectroscopic value of our sample ($\langle \log g \rangle = 8.143$, $\sigma = 0.166$). A variation of $\log g$ by ± 0.25 dex from the mean value changes the photometric temperatures by only 200 K, on average, as expected from the results displayed in the middle panel of Figure 1. Overall, the agreement observed in Figure 4 is satisfactory, although there are some objects with photometric temperatures significantly larger than the spectroscopic estimates. For these stars, however, we also notice that the photometric uncertainties appear larger. The object all the way to the left in the 2MASS panel is G29-

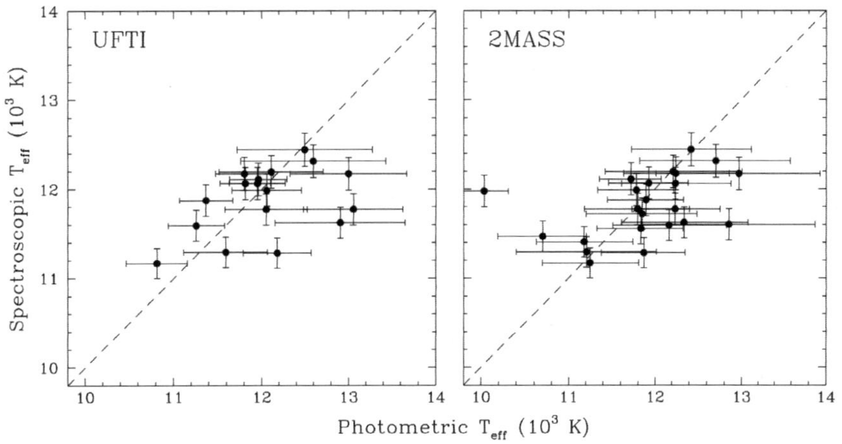

Figure 4. Comparison of photometric and spectroscopic effective temperatures for the two sets of infrared photometry used in our analysis. Model atmospheres calculated with the ML2/$\alpha = 0.7$ version of the mixing-length theory are used to obtain both estimates of $T_{\rm eff}$.

38, a ZZ Ceti star with an infrared excess due to the presence of a dust disk, which affects our photometric temperature determination (the corresponding measurement with the UFTI photometry is outside the temperature range displayed here).

The photometric temperatures in Figure 4 are spread over a somewhat wider range of values than the spectroscopic temperature determinations. This can be explained by the fact that the photometric method is less accurate than the spectroscopic method in this particular range of effective temperatures. For instance, a difference of $\Delta T_{\rm eff} = 2000$ K yields energy distributions (bottom of Fig. 1) that differ less than the predicted strengths of the Balmer line profiles for the same difference in $T_{\rm eff}$.

Of greater interest in the context of our study is the comparison of photometric and spectroscopic effective temperatures for different versions of the mixing-length theory. This is shown in Figure 5 for the ML1, ML2, and ML3 parameterizations (in what follows we consider the UFTI and 2MASS infrared data as independent measurements and simply merge the samples). Although the spectroscopic $T_{\rm eff}$ values vary over a range of nearly 4000 K going from ML1 (less efficient version) to ML3 (more efficient version), the photometric temperatures remain nearly unchanged, as anticipated from our results shown in the top panel of Figure 1. Also, for both temperature estimates to agree, one must rely on a parametrization which is less efficient than the ML2/$\alpha = 1.0$ of the mixing-length theory (used to produce Fig. 5), in agreement with the lower value of $\alpha = 0.7$ adopted in the comparison shown in Figure 4, which is reproduced in Figure 6 by combining both UFTI and 2MASS data sets. These results add additional confidence to our conclusion that the atmospheric convective efficiency in DA white dwarfs is fairly well constrained and understood.

Despite these encouraging results, there are still some stars in Figure 6 for which the differences in photometric and spectroscopic temperatures are uncomfortably large. One example labelled A in the figure is G117-B15A. The spectroscopic solution yields $T_{\rm eff} = 11,770$ K and $\log g = 8.04$ while the photometric solution with the UFTI infrared data is about 1250 K hotter. On the other hand, the photometric solution obtained with the 2MASS data is only

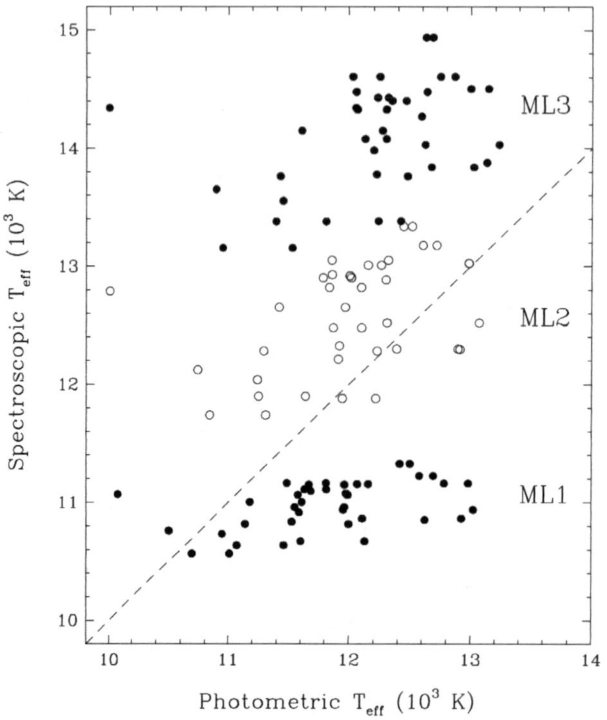

Figure 5. Comparison of photometric and spectroscopic effective temperatures for different versions of the mixing-length theory.

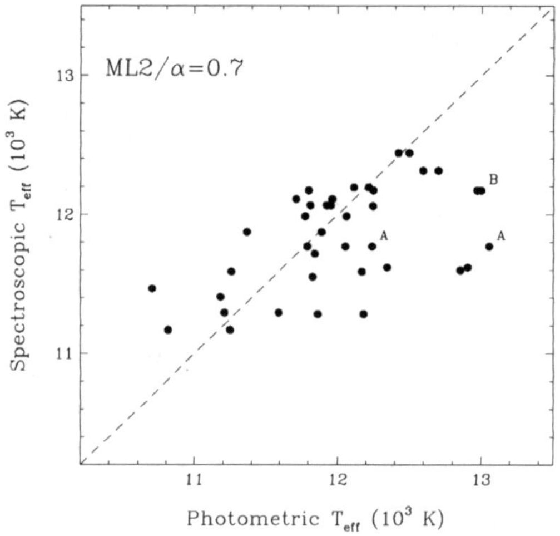

Figure 6. Comparison of photometric and spectroscopic effective temperatures obtained with model atmospheres calculated with ML2/α = 0.7. A value of $\log g$ = 8.1 was used for the photometric determinations. The objects labeled in the figure and discussed in the text correspond to (A) G117-B15A and (B) GD 165.

400 K hotter. We believe this reflects the difficulty with obtaining precise $T_{\rm eff}$ measurements for individual objects from photometry in this particular range of effective temperature due mostly to the sensitivity of the method, but also to problems with the calibration of the synthetic photometry. An alternative, perhaps more exotic explanation is provided by the second object, GD 165, labelled B in Figure 6. For this star, the spectroscopic solution yields $T_{\rm eff} = 12,170$ K and $\log g = 8.13$. In this case, however, both solutions with the UFTI and 2MASS photometry agree within 90 K, with an average value of $T_{\rm eff} = 12,950$ K, which is nearly 800 K hotter than the spectroscopic solution. Interestingly enough, had we used the ML2/$\alpha = 1.0$ version of the mixing-length theory, both spectroscopic and photometric temperatures would have been in perfect agreement (the spectroscopic temperature increases to a value of $T_{\rm eff} = 13,030$ K in this case). Since GD 165 is one of the hottest ZZ Ceti star, perhaps our results is suggesting that convective energy transport is more efficient near the blue edge of the ZZ Ceti instability strip.

5. Conclusion

We have successfully demonstrated that energy distributions of ZZ Ceti stars built from optical VRI and infrared JHK photometric data provide a parameter-free and independent temperature scale for these stars that does not depend on the assumed value of $\log g$ or convective efficiency. Although we believe that the effective temperatures measured from photometry are less accurate for *individual* objects than temperatures obtained from the standard spectroscopic technique, our results nevertheless show that a parameterization of the mixing-length theory with ML2/$\alpha = 0.7$ provides, on average, an excellent internal consistency between photometric and spectroscopic temperatures. In particular, both methods yield consistent values for the boundaries of the ZZ Ceti instability strip between roughly $T_{\rm eff} = 11,000$ K and 12,500 K.

Acknowledgments

This work was supported in part by the NSERC Canada and by the Fund FQRNT (Québec). P. Bergeron is a Cottrell Scholar of Research Corporation for Science Advancement. The United Kingdom Infrared Telescope is operated by the Joint Astronomy Centre on behalf of the Science and Technology Facilities Council of the U.K.

References

Bergeron P Ruiz M T and Leggett S K 1997 *ApJ Suppl* **108** 339
Bergeron P Wesemael F and Fontaine G 1992 *ApJ* **387** 288
Bergeron P Wesemael F Lamontagne R Fontaine G Saffer R A and Allard N F 1995 *AJ* **449** 258
Gianninas A Bergeron P and Fontaine G 2006 *AJ* **132** 831
Holberg J B and Bergeron P 2006 *AJ* **132** 1221
Landolt A U 1992 *AJ* **104** 340
Leggett S K et al. 2006 *MNRAS* **373** 781
Lemke M 1997 *A&A Suppl* **122** 285
Roche P F et al. 2003 *SPIE* **4841** 901
Tokunaga A T Simons D A and Vacca W D 2002 *PASP* **114** 180

Journal of Physics: Conference Series **172** (2009) 012063 doi:10.1088/1742-6596/172/1/012063

Frequency variations in the PG 1159 pulsator PG 0122+200

Gerard Vauclair[1] **and Jian-Ning Fu**[2]

[1]Laboratoire d'Astrophysique de Toulouse-Tarbes, Université de Toulouse, CNRS, 14 avenue E. Belin, 31400 Toulouse, France

[2]Department of Astronomy, Beijing Normal University, 19 road Xinjiekouwai, 100875 Beijing, China

E-mail: gerardv@ast.obs-mip.fr

Abstract. PG 0122+200 is the coolest PG 1159 pulsator. On theoretical basis, its cooling time scale was predicted to be dominated by neutrino emission. The measurement of the frequency variation (or $\dot{P}$) for a number of its pulsation modes would then provide an empirical estimate of the yet uncertain neutrino emission rate. To test this prediction, we reanalysed the time-series obtained during the seven observing campaigns spanning a time interval of 19 years. We determine the time variations for the seven largest amplitude modes. This includes two complete triplets and one component of a third triplet. We find that PG 0122+200 exhibits much larger frequency variations and on considerably shorter time scales than the ones predicted by cooling theory. We suggest that these variations are not dominated by the star cooling but instead by nonlinear coupling induced by rotation.

1. PG 0122+200, the coolest PG 1159 pulsator

The PG 1159 pulsator (or GW Vir variable star) PG 0122+200 is presently the coolest member of its class with a spectroscopic determination of $T_{\rm eff}$= 80 kK. It defines consequently the red edge of the GW Vir stars instability strip. That was a first justification for obtaining a better knowledge of its global parameters and of its internal structure. In addition, it was shown that its evolution time scale could be dominated by neutrino cooling (O'Brien et al. 1998). The prospect to get a direct empirical determination of the neutrino emission rate through a measurement of the rate of change of the pulsation periods, $\dot{P}$, was a second motivation to follow up the star. For these reasons, a number of observing campaigns were organized, starting in 1986 with a monosite campaign, followed by six multisite campaigns in 1990, 1996, 1999, 2001, 2002 and 2005.

As most of the other pre-white dwarf and white dwarf pulsators, PG 0122+200 exhibits time variations of its pulsation modes amplitudes. Only a few pulsation modes show up at a given season. For instance, all the components of the multiplets generated by rotation (rotational splitting) are not generally seen during the same observing season as one or two of them fall below the detection limit of this particular run. The reconstruction of the whole multiplets is required in order to get a safe identification of the degree of the modes (ℓ=1 for triplets or ℓ=2 for quintuplets etc). It took a number of observing campaigns to finally get enough modes for a proper identification and to constrain the model of PG 0122+200.

The star shows 23 pulsation modes with periods between 336 s and 612 s, distributed in seven triplets and two single frequencies. All the pulsation modes are ℓ=1 modes (Fu et al. 2007). The identification and frequency of the modes were used to constrain the model of the star with high accuracy (Corsico et al. 2007). The global parameters of PG 0122+200 derived from this asteroseismological study are: $M/M_{\odot} = 0.556^{+0.009}_{-0.014}$, $T_{\rm eff} = 81.54\ kK^{+0.8}_{-1.4}$, $\log\ g = 7.65^{+0.02}_{-0.07}$, $\log(L/L_{\odot}) = 1.14^{+0.02}_{-0.04}$, $M_{\rm env}/M_{\odot} = 0.019 \pm 0.006$. The deduced distance of PG 0122+200 is 614^{+58}_{-32} pc and the mean rotation period derived from the rotational splitting measured in the seven triplets is 1.55 days. The accuracy on the global parameters of PG 0122+200 deduced from asteroseismology is significantly improved compared to previous spectroscopic determinations. Since the rate of neutrino production is sensitive to the total mass, effective temperature and luminosity of the star, determining those parameters with enough accuracy was a pre-requirement for estimating the importance of the neutrino emission on the cooling of the star.

2. What does the cooling theory predict?

According to the theoretical predictions of O'Brien & Kawaler (2000), the evolution of a pre-white dwarf corresponding to PG 0122+200, with a typical mass of 0.6 $M_{\odot}$, should be dominated by cooling due to neutrino emission. At a $T_{\rm eff}$= 80 kK, the ratio of the neutrino luminosity, L_{ν}, on the photon luminosity, L_{γ}, is $\approx$ 1.6 for a "normal" neutrino emission rate, as estimated from Itoh et al. (1989, 1992). However, since the neutrino emission rate is rather uncertain, O'Brien & Kawaler (2000) estimated the influence of a factor 3 uncertainty of this rate on the ratio L_{ν}/L_{γ}. They find that this ratio may be in the range 1.2–2.0. This uncertainty translates into a 45% uncertainty on the age determination of the model. Similarly, this uncertainty on the neutrino emission rate results in a factor of $\approx$4 difference in the predicted $\dot{P}$ in the range of periods observed in PG 0122+200 (O'Brien & Kawaler 2000). Inversely, if one could measure $\dot{P}$ for a number of pulsation modes, assuming that they are really measuring the cooling rate of the star, one could constrain empirically the value of the neutrino emission rate.

The $\dot{P}$ values for the modes observed in PG 0122+200 have been obtained by Corsico et al. (2007) for the best fit model of the star, which is computed with the "normal" neutrino emission rate. They correspond to cooling timescales in the range 6–12 Myr. The period variations expected on the $\approx$15 y covered by the multisite campaigns (19 yr if one includes the 1986 monosite campaign) would not exceed 6 $\times 10^{-4}$ s to 1.5 $\times 10^{-3}$ s, depending on the mode. In frequency, this corresponds to variations of only a few nHz. With the frequency determination accuracy presently achieved by the best quality multisite campaigns, we estimate that it would require $\approx$ 50 yr to obtain a frequency variation detection at the 3σ level.

3. What the star does?

We have reanalysed the time-series obtained during all the campaigns in an homogeneous way, using the Period04 software (Lenz & Breger 2005). The frequencies, amplitudes and phases were obtained by nonlinear least square fits. Realistic uncertainties on the frequencies were estimated by Monte-Carlo simulations. We were able to determine the frequency variations for the seven largest amplitude modes; this includes the two complete triplets at 2221-2224-2228 μHz and at 2490-2493-2497 μHz and the prograde component of a third triplet at 2973 μHz. All these modes are not always seen in the data except for the dominant mode at 2497 μHz (400 s), which is the prograde component within its triplet. Fig. 1 shows the result for this mode. The unpublished result of the additional 2005 multisite campaign is shown for completness so that the frequency variation is shown on a total time interval of 19 years. It shows clearly that the frequency variation for this mode is much larger than predicted from cooling theory (a few tenths of μHz instead of a few nHz). The time scale of the variation is also much shorter than the expected

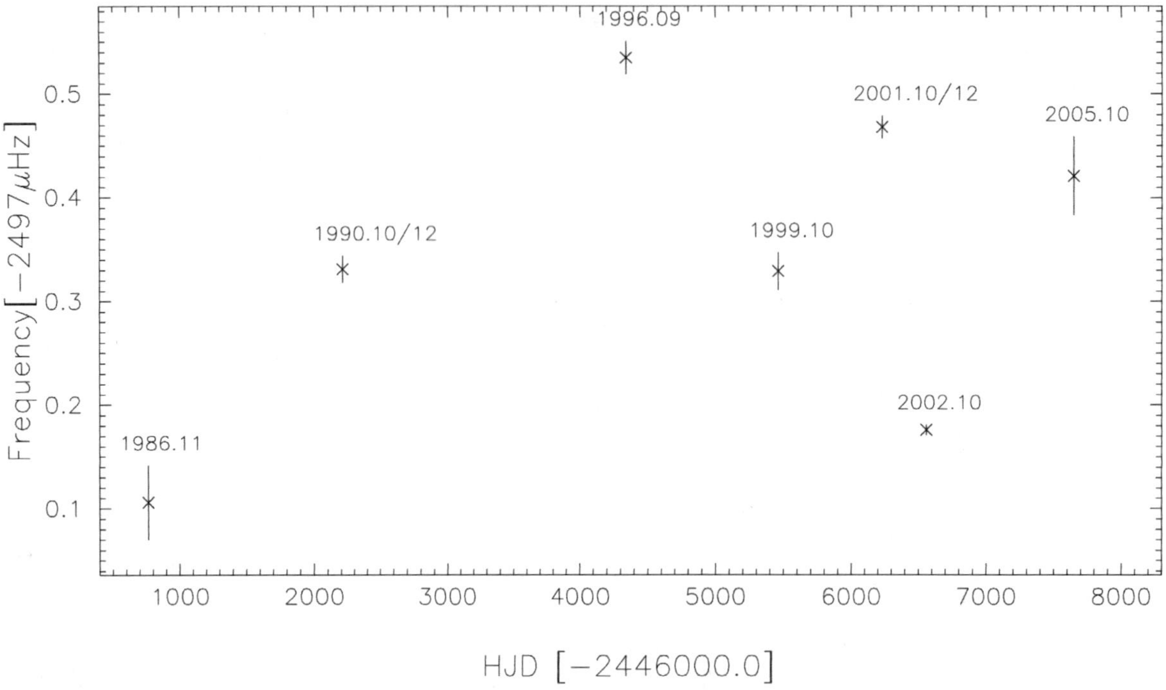

Figure 1. Frequency variation of the largest amplitude mode in PG 0122+200: the frequency of the 2497 μHz mode, after subtraction of 2497 μHz, is plotted as a function of time. Each data point is marked with the date of the seasonal observing campaign as (year.month). All observing campaigns were multisite campaigns except the first one in 1986. Error bars are estimated from Monte-Carlo simulations.

cooling time scale, of the order of years instead of a few Myr. Each of the other six modes, not shown here, exhibits frequency variations of its own.

4. Conclusions and interpretation

The seven largest amplitude modes of pulsation in PG 0122+200 show much larger frequency variations and on a considerably shorter timescale than expected would they result from the cooling of the star, whatever would be the impact of the neutrino emission efficiency on the cooling rate. The disagreement is by orders of magnitude and the frequency variations are not a monotonic function of time. This cannot be accounted for by uncertainty on the neutrino emission rate. We conclude that the frequency time variations observed in PG 0122+200 is not dominated by the star cooling. One can exclude the influence of a companion orbiting the star (brown dwarf or planet) since the frequency variations observed for different modes are not periodical (as far as one can say from at most seven data points), are not in phase and have not the same amplitude.

The analysis of the period changes in PG 1159-035, the well studied prototype of the PG 1159 pulsators, shows quite similar results (Costa & Kepler 2008). In the case of PG 1159−035, the period variations of a larger number of pulsation modes were obtained, with different behaviour from mode to mode. Interestingly, the analysis of the both stars PG 1159−035 and PG 0122+200 span a time interval of 19 years and in both cases, the dominant mode is the prograde component of a ℓ=1 triplet. However, the dominant mode in PG 1159−035, at 516 s, shows a quite linear increase with time, while the dominant mode in PG 0122+200, at 400 s, exhibits a more complex time variation.

We interpret the observed frequency variations as the signatures of nonlinear mode coupling due to rotation, as suggested by Goupil et al. (1998). In this case, the second-order effect of rotation produces a frequency mismatch within the triplets induced by rotation for the ℓ=1 modes. The sum of the frequencies of the m= + 1 and m= - 1 components differs from twice the frequency of the m= 0 component within a triplet. The coupling between the components of the triplets is then governed by the ratio of the frequency mismatch by the growth rate of the mode. Three different regimes may appear according to the value of this ratio. When it is small, the triplet is in a frequency lock regime. When the ratio is large, the triplet follows the linear behaviour and the frequencies are the ones predicted by linear theory. For intermediate value, both the frequencies and the amplitudes of the components are time modulated. In the case of PG 0122+200, our results suggest that all the modes are in the intermediate regime. In the case of PG 1159−035, different modes are in different regimes and the dominant mode at 516 s seems to be unaffected by nonlinear coupling. A better knowledge of the regime followed by different modes and a determination of the typical modulation time scales for those modes which are in the intermediate regime would give important insight on their growth rates.

Acknowledgments

The authors acknowledge the support from the National Natural Science Foundation of China (NSFC), through the Grants 10673001, 10778601 and 10878007.

References

Corsico A Miller-Bertolami M Althaus L et al. 2007 *A&A* **475** 619
Costa J E S & Kepler S O 2008 *A&A* **489** 1225
Fu J-N Vauclair G Solheim J-E et al. 2007 *A&A* **467** 237
Goupil M-J Dziembowski W Fontaine G 1998 *Balt.Astr.* **7** 21
Itoh N Adachi T Nakagawa M et al. 1989 *ApJ* **339** 354
Itoh N Mutoh H Hikita A et al. 1992 *ApJ* **395** 622
Lenz P & Breger M 2005 *Con. Ast.* **146** 53
O'Brien M S & Kawaler S D 2000 *ApJ* **539** 372
O'Brien M S Vauclair G Kawaler S D et al. 1998 *ApJ* **495** 458

Journal of Physics: Conference Series **172** (2009) 012064 doi:10.1088/1742-6596/172/1/012064

New Pulsating ZZ Ceti Stars

S O Kepler[1] and B G Castanheira[1,2]

[1]Instituto de Física, Universidade Federal do Rio Grande do Sul, Porto Alegre, RS, Brazil
[2]Institut für Astronomie, Türkenschanzstr. 17, A-1180 Wien, Austria

E-mail: kepler@if.ufrgs.br

Abstract. Over the last four years the number of known pulsating DA white dwarf stars increased from 32 to a total of 147; 87 are white dwarf stars identified by the Sloan Digital Sky Survey and selected from fits of synthetic spectra from model atmospheres to the S/N<30 optical spectra. With time series photometric observations at the SOAR 4.1m telescope, we found 42 new pulsators, extending the instability strip in terms of effective temperature and surface gravity, but mainly in terms of very low amplitude pulsators. The discovery of 1.5 mma pulsators raises the necessity to re-observe all stars previously classified as non-pulsators close to or inside the instability strip, as their detection limits, in general, were above 3 mma. The real extent of the instability strip and its possible contamination with non-pulsators, specially at the borders, are still open questions.

1. Introduction

White dwarfs with hydrogen atmospheres comprise at least 83% of all known spectroscopically identified white dwarf stars. Observationally, they start to pulsate when they cool down to around $T_{\rm eff} \simeq 12\,300$ K, defining the ZZ Ceti instability strip. The position of this strip has a small dependency on the stellar mass, such that more massive stars start to pulsate at an effective temperature slightly higher than less massive stars. According to the current best models, pulsations start when the hydrogen partial ionization zone at the bottom of a very shallow convection zone reaches a thermal timescale around 100 s. The partial ionization zone has higher opacity than the hotter ionized state, hindering the energy transport and the star starts to pulsate in non-radial g-modes. At the blue edge of the instability strip, we observe low amplitude pulsating stars with periods close to the 100 s thermal timescale and showing fractional amplitudes of the order of a few mili-modulation-amplitude (mma) (Winget and Kepler 2008). Theory predicts that, as the convection zones deepens with the subsequent cooling, the thermal timescale increases and the pulsation periods and average pulsation amplitude also increases. This is in agreement with the observations of cooler stars, that pulsate with higher amplitudes and longer periods. After cooling by $\simeq 1500$ K, the models show that the surface convection layer becomes so large that pulsations subside. Mukadam et al. (2004b, 2006) and Castanheira et al. (2007) have detected stars with small amplitudes close to the red edge of the instability strip, sampling the stage when stars are stopping to pulsate.

2. SDSS Variables

In 2004 Kleinman (2004) started to identify a few thousand new white dwarf stars with the Sloan Digital Sky Survey (SDSS) and to measure their effective temperature and surface gravity by

Journal of Physics: Conference Series **172** (2009) 012064 doi:10.1088/1742-6596/172/1/012064

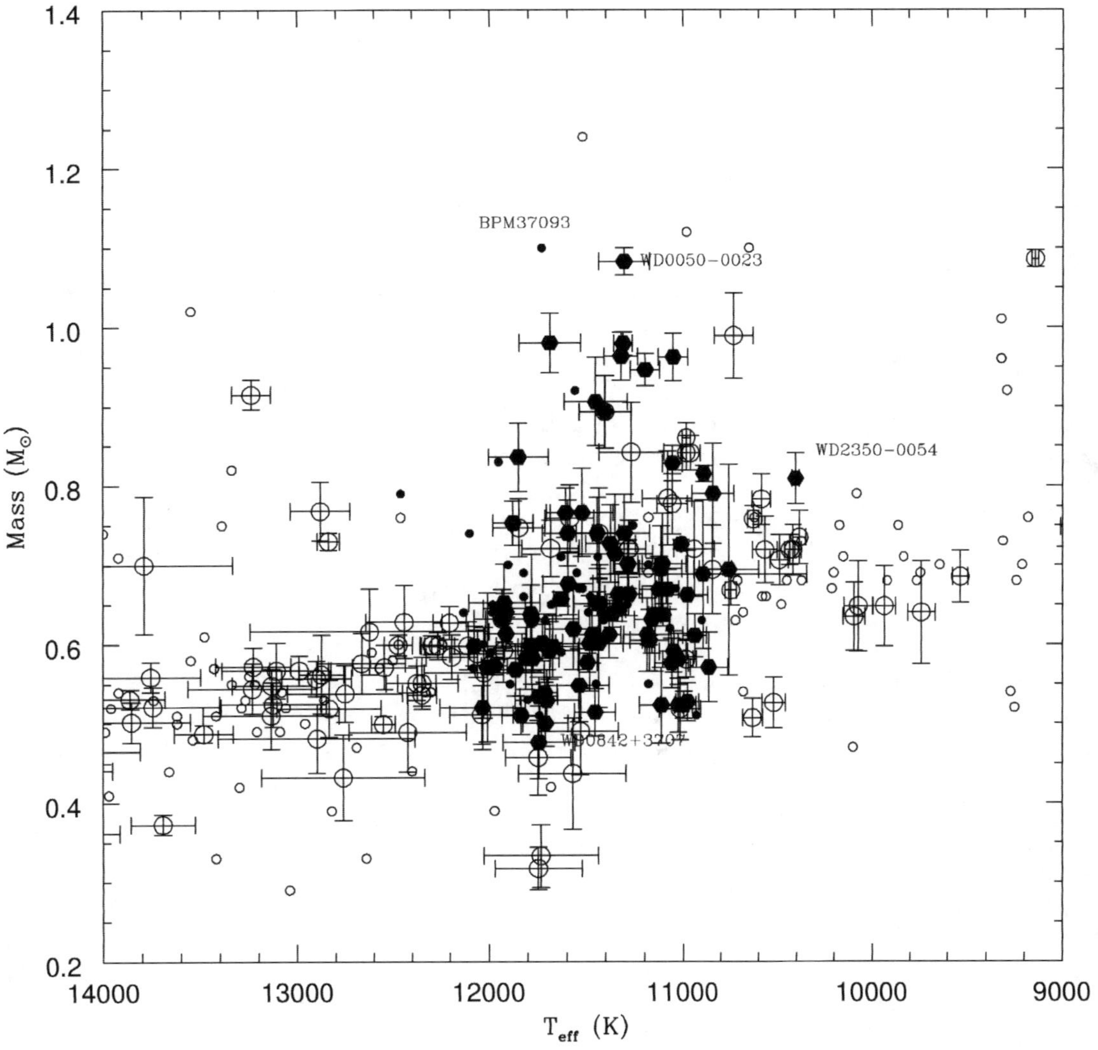

Figure 1. The ZZ Ceti instability strip showing variables as filled circles and non-variables as open circles. The cooler variable SDSS J235040.74-005430.8, with $T_{\rm eff}$ = 10 370 ± 20 K from two independent spectra, remains a challenge to explain. The smaller points are from the bright sample of Gianninas et al (2006). Most of the stars for which previous studies did not detect pulsations had limits around 5 mma, but our discovery of pulsators with amplitudes as low as 1.5 mma requires they all be re-observed to lower the detection limits.

fitting the signal-to-noise rate ≤ 30 optical spectra to synthetic spectra calculated from model atmospheres by Koester (2008). Mukadam (2004a), Mullally (2005), Kepler(2005), Castanheira et al (2006), (2007) selected the stars with effective temperature close to the ZZ Ceti instability strip from these SDSS candidates and obtained time series photometry using CCD cameras on the 2.1 m telescope at McDonald Observatory, the 1.6 m telescope at Observatório Pico dos Dias, and the 4.1 m telescope SOAR, to search for variability. They found that that most stars with temperatures in the range $12\,300\text{ K} \leq T_{\text{eff}} \leq 10800$ K do pulsate. Some of the new pulsators show pulsations with fractional amplitudes of only 1.5 mma.

3. Results and Discussion

We have continued this search with the 4.1 m telescope SOAR using white dwarf candidates selected by Scot Kleinman and Atsuko Nitta from the subsequent SDSS data releases (Eisenstein et al 2006, Adelman-McCarthy 2007, 2008) and their fits to new and improved synthetic spectra by Detlev Koester. We discovered low amplitude pulsations in stars previously classified as Not-Observed-to-Vary by Mukadam (2004a) and Mullally et al. (2005), because with a larger telescope we could reach smaller amplitude limits and find previously undetected pulsations, as in Castanheira et al (2007).

Kepler at al. (2006) obtained S/N$\geq$70 optical spectra of a few of the white dwarfs discovered by the SDSS to compare with the low S/N SDSS spectra and demonstrated the internal uncertainties from SDSS spectra are underestimated, but only by $\Delta T_{\text{eff}} \simeq 320$ K, and $\Delta \log g \simeq 0.24$ dex. We have continued to observe other stars with Gemini over the last two years and obtain similar results.

Our main goal has been to use the observed pulsation modes to measure the structure of the white dwarfs through seismology (Kim et al 2008, Castanheira et al. 2008). We also aim to determine if pulsation is a normal stage that all white dwarf stars pass as they cool down (Gianninas et al. 2005, 2006, 2007, Castanheira et al 2007). As the detection limit for pulsations in previous studies were larger than the now known lowest amplitude pulsator, we do not know at present if there are non-pulsators inside the instability strip or even the real extent of the instability strip.

The discovery of a substantial number of pulsators is necessary to map the instability strip, not only with T_{eff}, but also in terms of its mass dependency. Our discoveries of massive pulsators, shown in Figure 1, are steps towards such coverage. ZZ Ceti stars are useful tools in the study of high density physics, specially considering that the massive pulsators crystallize inside the DAV instability strip (Kanaan et al 2005).

References

Adelman-McCarthy J K et al. 2007 *ApJS* **172** 634
Adelman-McCarthy J K et al. 2008 *ApJS* **175** 297
Bischoff-Kim A, Montgomery M H and Winget D E 2008 *ApJ* **675** 1505
Castanheira B G and Kepler S O 2008 *MNRAS* **385** 430
Castanheira B G et al. 2007 *A&A* **462** 989
Castanheira B G et al. 2006 *A&A* **450** 227
Eisenstein D J et al. 2006 *ApJS* **167** 40
Gianninas A Bergeron P and Fontaine G 2005 *ApJ* **631** 1100
Gianninas A Bergeron P and Fontaine G 2006 *AJ* **132** 831
Gianninas A Bergeron P and Fontaine G 2007 *ASPC* **372** 577
Kanaan A et al. 2005 *A&A* **432** 219
Kepler S O Kleinman S J Nitta A Koester D Castanheira B G Giovannini O Costa A F M and Althaus L 2007 *MNRAS* **375** 1315
Kepler S O Castanheira B G Costa A F M and Koester D 2006 *MNRAS* **372** 1799
Kepler S O Castanheira B G Saraiva M F O Nitta A Kleinman S J Mullally F Winget D E and Eisenstein D J 2005 *A&A* **442** 629

Kleinman S J et al. 2004 *ApJ* **607** 426
Koester D 2008 *arXiv:0812.0482*
Mullally F Thompson S E Castanheira B G Winget D E Kepler S O Eisenstein D J Kleinman S J and Nitta A 2005 *ApJ* **625** 966
Mukadam A S Montgomery M H Winget D E Kepler S O and Clemens J C 2006 *ApJ* **640** 956
Mukadam A S Winget D E von Hippel T Montgomery M H Kepler S O and Costa A F M 2004 *ApJ* **612** 1052
Mukadam A S et al. 2004 *ApJ* **607** 982
Winget D E and Kepler S O 2008 *ARA&A* **46** 157

Journal of Physics: Conference Series **172** (2009) 012065 doi:10.1088/1742-6596/172/1/012065

Dynamic masses for the close PG1159 binary SDSS J212531.92−010745.9

S Schuh[1], **B Beeck**[1] **and T Nagel**[2]

[1]Institute for Astrophysics, Georg-August-University of Göttingen, Friedrich-Hund-Platz 1, 37077 Göttingen, Germany

[2]Institute for Astronomy and Astrophysics, Kepler Center for Astro and Particle Physics, Eberhard-Karls-University of Tübingen, Sand 1, 72076 Tübingen, Germany

E-mail: `schuh@astro.physik.uni-goettingen.de`

Abstract. The evolutionary scenarios which are commonly accepted for PG 1159 stars are mainly based on numerical simulations. These stellar evolution scenarios have to be tested and calibrated with real objects with known stellar parameters. One of the most crucial but also quite uncertain parameters is the stellar mass. PG 1159 stars have masses between 0.5 and 0.8 solar masses, as derived from asteroseismic and spectroscopic determinations. Such mass determinations are, however, themselves model-dependent. Moreover, asteroseismologically and spectroscopically determined masses deviate systematically for PG 1159 stars by up to 10 %. SDSS J212531.92-010745.9 is the first known PG 1159 star in a close binary with a late main sequence companion allowing a dynamical mass determination. The system shows flux variations with a peak-to-peak amplitude of about 0.7 mag and a period of about 6.96 h. In August 2007, 13 spectra of SDSS J212531.92-010745.9 covering the full orbital phase range were taken at the TWIN 3.5 m telescope at the Calar Alto Observatory (Almería, Spain). These confirm the typical PG 1159 features seen in the SDSS discovery spectrum, together with the Balmer series of hydrogen in emission (plus other emission lines), interpreted as signature of the companion's irradiated side. A radial velocity curve was obtained for both components. Using co-added radial-velocity-corrected spectra, the spectral analysis of the PG 1159 star is being refined. The system's lightcurve, obtained during three seasons of photometry with the Göttingen 50 cm and Tübingen 80 cm telescopes, was fitted with both the `NIGHTFALL` and `PHOEBE` binary simulation programs. An accurate mass determination of the PG 1159 component from the radial velocity measurements requires to first derive the inclination, which requires light curve modelling and yields further constraints on radii, effective temperature and separation of the system's components. From the analysis of all data available so far, we present the possible mass range for the PG 1159 component of SDSS J212531.92-010745.9.

1. The SDSS J212531.92−010745.9 system: a close PG 1159 binary

1.1. Evolutionary scenarios and masses for PG 1159 stars

PG 1159 stars are hot hydrogen-deficient (pre-)white dwarfs with effective temperatures between 75 000 and 200 000 K, and $\log(g/\mathrm{cm\,s^{-2}}) = 5.5$–$8.0$ (Werner 2001). They are in the transition between the asymptotic giant branch (AGB) and cooling white dwarfs. Spectra of PG 1159 stars are dominated by absorption lines of He II, C IV and O VI. Roughly half of them have a planetary nebula. Current theory suggests (e. g. Werner 2001) that they are the outcome of a late or very late helium-shell flash, a phenomenon that drives the currently observed fast evolutionary rates of three well-known objects (FG Sge, Sakurai's object, V605 Aql). Flash-induced envelope

Journal of Physics: Conference Series **172** (2009) 012065 doi:10.1088/1742-6596/172/1/012065

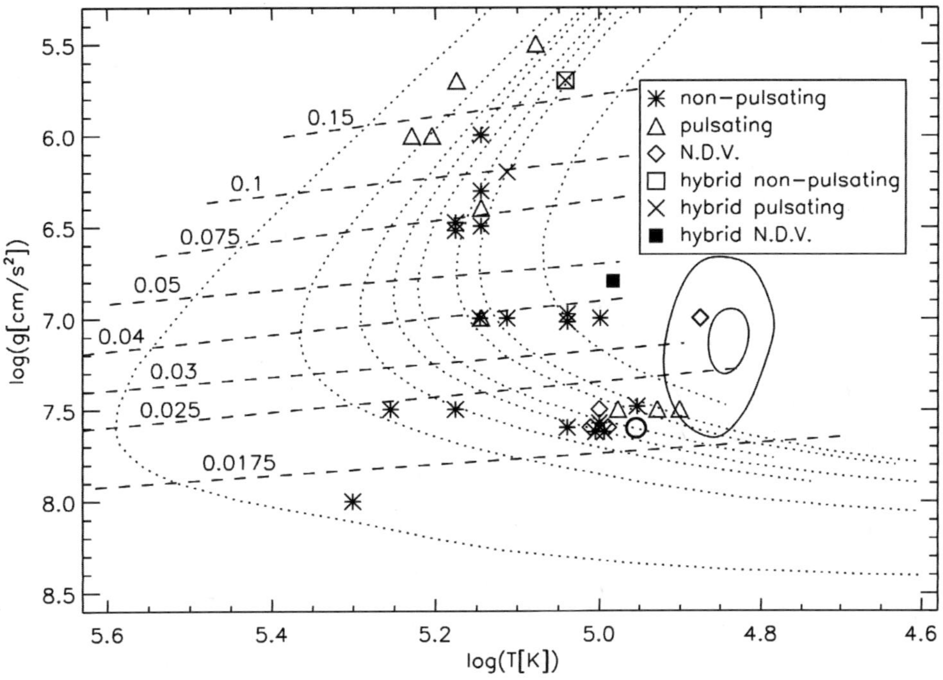

Figure 1. Positions of PG 1159 stars with known $T_{\rm eff}$ and $\log g$ in the $\log T_{\rm eff}$-$\log g$-diagram. The current state-of-the-art post-AGB evolutionary tracks, first presented by Miller Bertolami et al. (2006a), as well as the positions of known PG1159 stars, are adopted from Figure 3 in Miller Bertolami & Althaus (2006b, dotted lines). The masses correspond to evolutionary sequences for 0.87, 0.664, 0.609, 0.584, 0.564, 0.542, 0.53, and 0.512 $M_\odot$ from left to right. We also give lines of constant radii for the theoretical tracks (dashed lines with annotations in $R_\odot$). The position of SDSS J212531.92−010745.9 found by Nagel et al. (2006) is shown as a circle; the position we find from our new spectral analysis (see section 2.2) is indicated by the error ellipse.

mixing produces a H-deficient stellar surface. The photospheric composition then essentially reflects that of the region between the H- and He-burning shells in the precursor AGB star. The He-shell flash forces the star back onto the AGB. The subsequent, second post-AGB evolution explains the existence of Wolf-Rayet central stars of planetary nebulae and their successors, the PG 1159 stars. These in turn are the progenitors of the helium rich WD sequence. Currently, 40 PG 1159 stars are known, Figure 1 shows their positions in a $\log T_{\rm eff}$-$\log g$-diagram.

Recent evolutionary models for PG 1159 stars (Miller Bertolami & Althaus 2006b, Werner & Herwig 2006) lead to spectroscopic masses which still suffer from serious uncertainties concerning the efficiency of the overshooting process in the He-flash driven convection zone during the thermal pulse phase on the AGB. This efficiency is represented in the models with an a priori unknown free parameter. The other ingredient for the determination of spectroscopic masses, the stellar atmosphere models, have to deal with uncertainties in the line broadening theory, so lacking reliable physical input there makes the determination of $\log g$ particularly difficult.

The overall resulting uncertainties in model masses manifest themselves as discrepancies between current asteroseismological masses and spectroscopic masses, as is evident from Table 3 in Werner & Herwig (2006; an updated version has been adopted as Table 1 in Schuh & Nagel 2007). The determination of a dynamical mass as an alternative method and test case for the models is currently restricted to only one suitable object: SDSS J212531.92−010745.9.

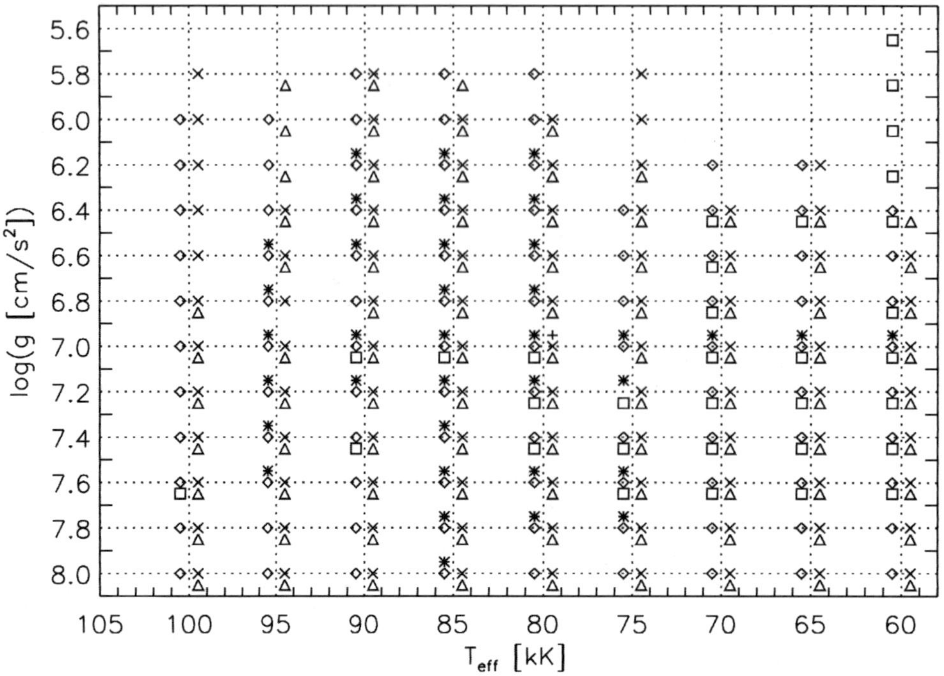

Figure 2. The model atmospheres were calculated with `NGRT` (Werner et al. 2003). Shown are the models calculated by end of October 2008 in the $\log g - T_{\rm eff}$-plane for six different C-abundances represented by different plot symbols (squares: C/HE=0.01, triangles: C/HE=0.03, diamonds: C/HE=0.05, crosses: C/HE=0.07, asterisks: C/HE=0.10, plus signs: C/HE=0.13). Other elemental abundances (all by number) are fixed at N/HE=0.01 and O/HE=0.01. The grid is still being completed: The whole range from $T_{\rm eff} = 60\ldots100$ kK in steps of 5 kK and from $\log(g/\mathrm{cm\,s^{-2}}) = 5.6\ldots8.0$ in steps of 0.2 dex will be filled with models in all six C-abundances by December 2008. Model spectra were calculated from the atmosphere models using `profile` and were convolved with a gauss profile to match the spectral resolution of 1.2 Å of the Calar Alto TWIN Spectograph in the configuration used.

1.2. Discovery of one PG 1159 close binary among many SDSS white dwarf binaries

The important role of binaries containing white dwarfs, for dynamical mass and other parameter determinations, is evident from the number of publications devoted to the topic in these proceedings. The contributions by Heller et al. (2009), Rebassa-Mansergas et al. (2009), Schreiber et al. (2009), and further contributions on specific individual systems, reflect the remarkable increase in number of known white dwarfs with companions that the Sloan Digital Sky Survey (SDSS) has brought about. Among the PG 1159 stars, however, SDSS J212531.92−010745.9 is the only known close binary system (Nagel et al. 2006). Eisenstein et al. (2006) also list it as a spectroscopically confirmed white dwarf from the SDSS.

The spectrum of SDSS J212531.92−010745.9 shows significant features that are typical for PG 1159 stars, for example the strong C IV absorption lines at 4650 – 4700Å and He II at 4686Å. Furthermore, the spectrum shows the Balmer series of hydrogen in emission. We have shown from our extensive follow-up photometry and modelling that this is due to a close cool companion which is heated up by irradiation from the hydrogen-deficient PG 1159 star. Thus, the radial velocities for both stars can be measured in this double-lined system.

Nagel et al. (2006) first reported a photometric period of 6.95573(5) h with an amplitude

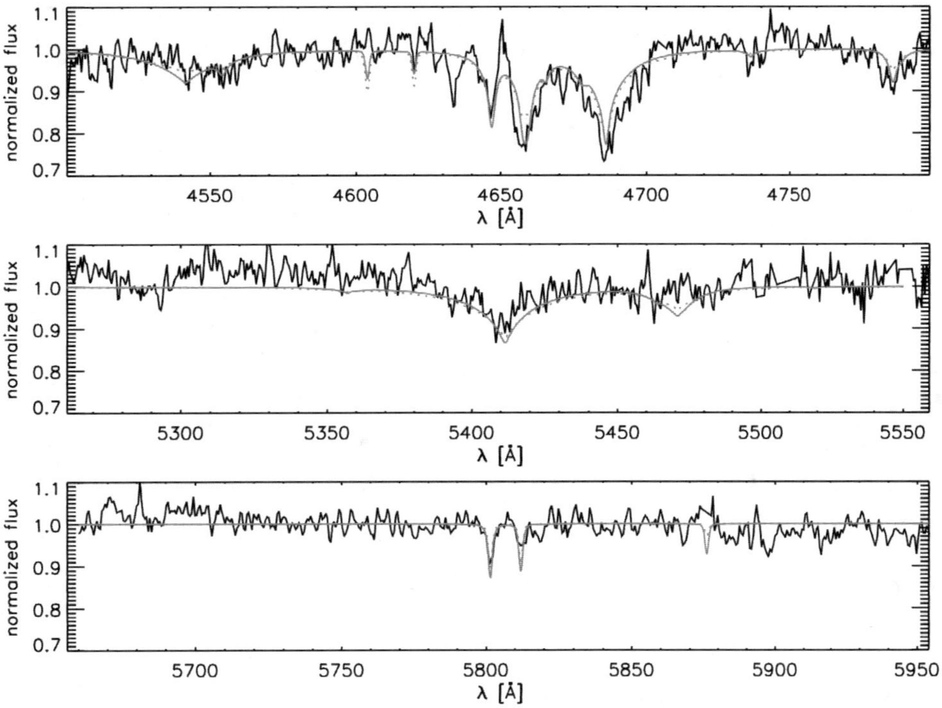

Figure 3. Cross-correlating specific spectral features of the Calar Alto spectra radial velocities were deduced. The three panels show three distinctive PG1159 features in the median spectrum (solid black line) along with the two model spectra (grey lines), one of which (dotted line) belongs to the model atmosphere with $\log (g/\mathrm{cm\,s^{-2}}) = 7.60$ and $T_{\mathrm{eff}} = 90000\,\mathrm{K}$ and C/He = 0.05 and was found as first preliminary result by Nagel et al. (2006). The solid grey line is the model spectrum with the highest cross-correlation value found by the quantitative spectral analysis with the extended model grid ($\log (g/\mathrm{cm\,s^{-2}}) = 7.1$, $T_{\mathrm{eff}} = 72500\,\mathrm{K}$, C/He = 0.07). The lowermost panel shows the narrow C IV lines used for the radial velocity determination of the PG1159 star.

of 0.354(3) mag which represents the orbital period of the binary system. From an initial comparison of the SDSS archive spectrum (Data Release 4) with NLTE model spectra they derived, as preliminary results, an effective temperature of 90 000 K, $\log (g/\mathrm{cm\,s^{-2}}) \sim 7.60$ and the abundance ratio C/He $\sim$ 0.05 for the PG 1159 component. They simulated the light curve of the binary system using `NIGHTFALL`, showing that in in addition to the constant contribution from the PG 1159, the irradiated side of the companion leads to a periodic brightening due to the reflection effect. For the inclination they obtained $(70 \pm 5)\,^{\circ}$. The system is not eclipsing and has not been found to be pulsating. While their solution for a mean radius of $(0.4 \pm 0.1)\,R_{\odot}$, a mass of $(0.4 \pm 0.1)\,M_{\odot}$ and a temperature of the irradiated surface of about 8 200 K for the companion was in good agreement with the observed light curve, a manifold of solutions remains equally plausible as long as we cannot reliably and indepedently fix the PG 1159 mass and radius.

We note that in the SDSS Data Release 6 catalogue Adelman-McCarthy et al. (2008) quote a proper motion of $+5(3)\,\mathrm{mas\,yr^{-1}}$ in RA and $+4(3)\,\mathrm{mas\,yr^{-1}}$ in DEC, a value for the redshift corresponding to $138(60)\,\mathrm{km\,s^{-1}}$, and Sloan colours of u'=17.144(0.009) mag, g'=17.538(0.005) mag, r'=17.757(0.006) mag, i'=17.794(0.008) mag, z'=17.828(0.020) mag.

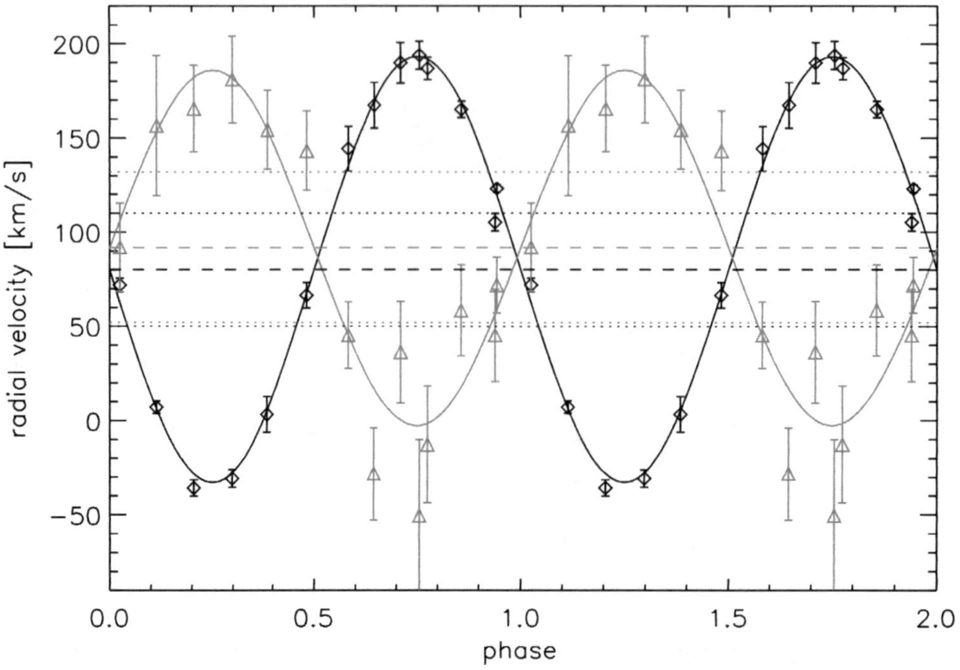

Figure 4. Radial velocity curves for the PG 1159 star (triangles, light gray) and the companion (diamonds, black) determined from spectra obtained with TWIN at Calar Alto. The best fit sine curves yield semi-amplitudes of $k_1 = 94.3 \pm 15.0\,\mathrm{km\,s^{-1}}$ for the PG 1159 (light gray sine curve) and $k_2 = 113.0 \pm 3.0\,\mathrm{km\,s^{-1}}$ for the the companion (black sine curve). The zero points of the two velocity curves are shown as dashed lines, the dotted lines indicate their uncertainties. The shift of the zero points relative to each other by about $11\,\mathrm{km\,s^{-1}}$ could be explained by gravitational redshift of the PG 1159 star but is not significant.

1.3. Orbit ephemeris from photometric observations

From time-series photometry of SDSS J212531.92−010745.9 performed in 2005, 2006 and 2007 with the Tübingen 80 cm and the Göttingen 50 cm telescopes, Schuh et al. (2008b) find the ephemeris of the predicted *maxima* times to be $\mathrm{HJD} = 2454055\overset{\mathrm{d}}{.}2134(4) + 0\overset{\mathrm{d}}{.}289822(2) \cdot E$. Further multi-colour observations are available, which show that the photometric amplitude in the optical is roughly 35% smaller in the blue than in the red.

2. Analysis of the Calar Alto 3.5 m TWIN spectroscopic data

2.1. Observational data and stellar atmosphere model grid

In 2007 we obtained time-resolved spectra with the TWIN spectrograph at the 3.5 m telescope at Calar Alto. The observations have been described in Schuh et al. (2008a): in the double-lined spectrum both sets of lines vary as expected with the orbital phase. The series of phase-resolved spectra can be used to derive radial velocity curves on the one hand, and to refine the spectral analysis (originally done on one SDSS spectrum) on the other hand. In order to do a spectral analysis of the PG 1159 component, a suitable grid of stellar model atmospheres is being calculated using `NGRT` (see Figure 2 for details).

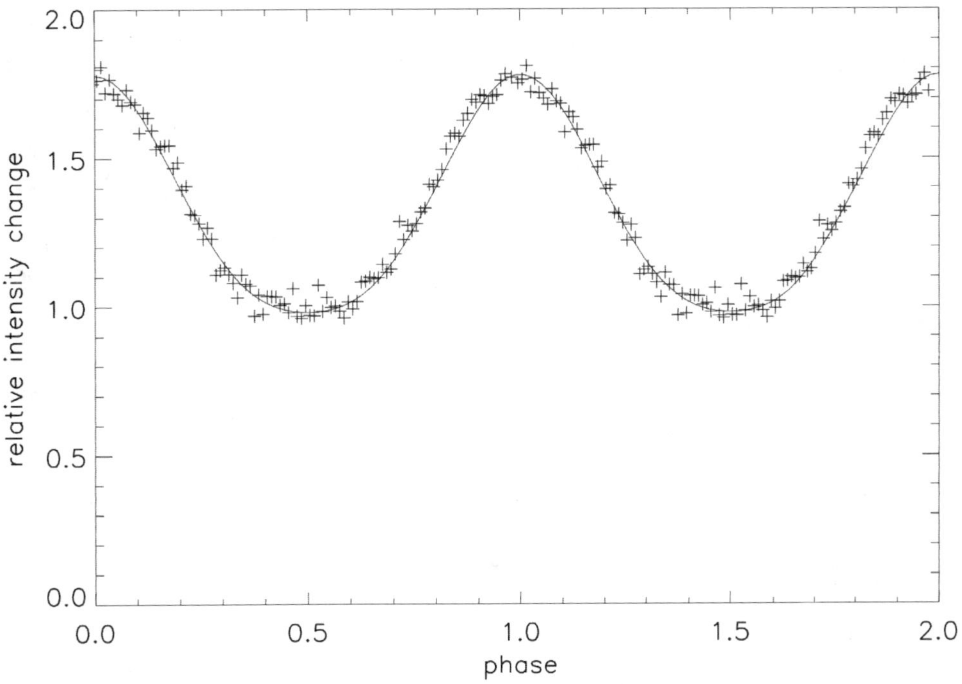

Figure 5. The observed light curve of all nights from 2005 to 2007 folded onto the orbital period and the simulated light curve of a binary system, consisting of a PG 1159 star and a heated M dwarf, calculated with PHOEBE for $m_1 = 0.55\,M_\odot$, $\log(g/\mathrm{cm\,s^{-2}})=7.1$, $R_1 = 0.035\,R_\odot$(black line).

2.2. Spectral analysis of the PG 1159 component

The spectra were wavelength shifted according to the radial velocity solution shown in Figure 4 and co-added, however without correcting for the phase-variable continuum contribution by the irradiated side of the companion yet, for a spectral analysis with our model grid. The median spectrum of the PG 1159 component was normalised to the continuum and cross-correlated with each normalised spectrum of the model grid. This cross-correlation was restricted to wavelength ranges where at least one of the model spectra shows significant spectral features (minimum line depth is a free parameter of this restriction; usually 1% was chosen, 3%, however, yields similar results) and where the mean companion spectrum shows no significant spectral features (the maximum line height of emission lines is another free parameter for the restriction — possible red shift differences between the two components are taken into account). Different choices of the free parameters show a stable maximum in the cross-correlation values for the median spectrum at an effective temperature of about (72500±5000) K, $\log(g/\mathrm{cm\,s^{-2}})$ of about 7.1±0.5 and C/He of 0.07 ± 0.03. The error bars quoted are very preliminary by-eye values. Two spectra from the model grid are shown in comparison to three spectral regions containing lines of the PG 1159 component of SDSS J212531.92−010745.9 in Figure 3. Combing the quantitative with the by-eye spectral analysis, we currently constrain the stellar parameters of the PG 1159 component to the following cuboid in parameter space: $T_{\mathrm{eff}} = 65000\,\mathrm{K} \ldots 90000\,\mathrm{K}$, $\log(g/\mathrm{cm\,s^{-2}}) = 6.50 \ldots 8.00$, C/He = 0.04 ... 0.10. These preliminary values were obtained as part of the diploma thesis work by Beeck (2009) and will be finalised when the model grid has been completed and the quantitative spectral analysis has been optimised.

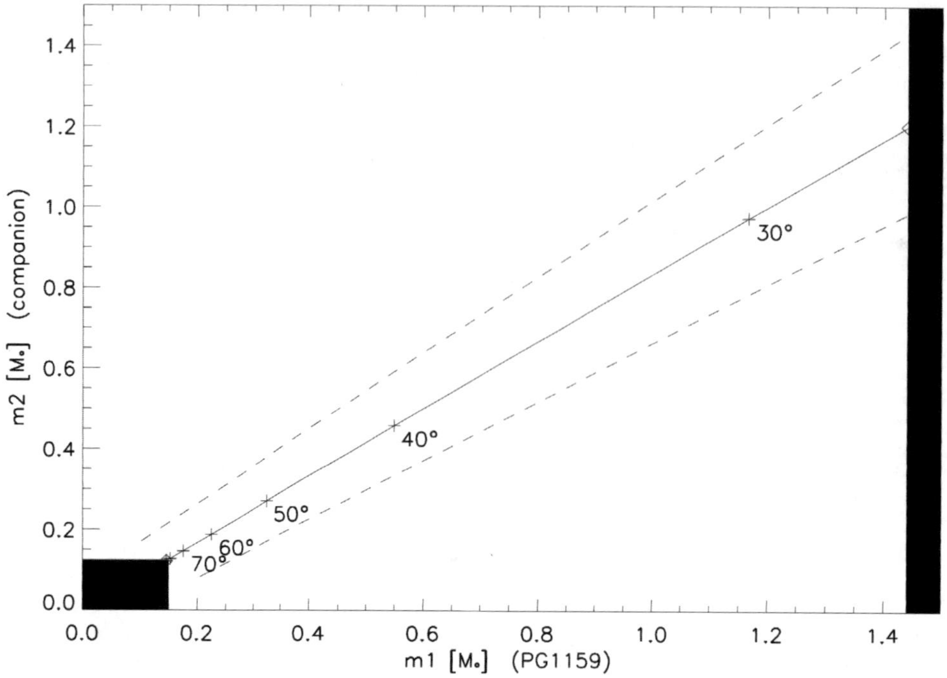

Figure 6. Using the photometric period and the mass ratio from the radial velocity amplitudes, the masses can be obtained as a function of the inclination. For the radial velocity curve solution in Figure 4, the (m_1,m_2) pairs lie on the line (error range: dashed) shown above, with $m_{1,\mathrm{min}} = 0.15\,M_\odot$ and $m_{2,\mathrm{min}} = 0.12\,M_\odot$. The dark areas mark forbidden ranges of the inclination: $m_1 < 1.44\,M_\odot$ requires $i > 20°$, and the fact that the system is not eclipsing sets a limit of $i < 80°$, implying $m_1 > 0.15\,M_\odot$; more stringent constraints are clearly required. A comparison with the evolutionary tracks in Figure 1 (although these may not be appropriate for a presumed PCEB system) favours a higher inclination (lower masses) than assumed in Figure 5.

2.3. Radial velocity curves

The radial velocity curves derived by Beeck (2009) are shown in Figure 4. Due to the individual exposure times of 30 min the spectral lines are somewhat smeared by the orbital motion, adding to the uncertainties in the determination of the radial velocities. The ratio of projected semi-amplitudes yields $k_1/k_2 \approx m_1/m_2 \approx 1.2$, allowing a range of mass ratios $\approx 1.0 \ldots 1.5$. Although we find a shift of the zero point of the PG 1159's velocity curve with regard to the companion's velocity curve zero point of the order of $11\,\mathrm{km\,s^{-1}}$ (Figure 4) which could be caused by the gravitational redshift of the PG 1159 star, this result is not formally significant. The expected gravitational redshift for a star with a mass of 0.5 - 0.8 $M_\odot$ and a radius of 0.02 - 0.15 $R_\odot$ would be of the order of 4 to 25 $\mathrm{km\,s^{-1}}$. Better data would be required to deduce reliable constraints for the combination of mass and radius of the PG 1159 star from a gravitational redshift measurement.

3. Phase resolved light curve and spectral modelling

3.1. Light curve modelling with NIGHTFALL and PHOEBE

With the know orbital period P and projected radial velocities k_1, k_2 measured from spectroscopy, in principle the only value the light curve modelling still needs to provide is the inclination i to obtain the mass:

$$m_1 \sin^3 i \ \ (k_1, k_2, P) = \frac{P}{2\pi G} \frac{(k_1 + k_2)^3}{1 + \frac{k_1}{k_2}}$$

(similarly for m_2). At an assumed value for the inclination of 40°, the photometric variation can be fit using PHOEBE as shown in Figure 5; further independent considerations to constrain the inclination are not overwhelmingly helpful (Figure 6). The best value will be determined from a comprehensive parameter study, including a comparison of results from NIGHTFALL and PHOEBE (Beeck 2009).

3.2. Challenges in the modelling approach and additional constraints

All the light curve solutions are obtained under the simplifying assumptions of bound rotation and a circular orbit. Further complications arise from the choice of the albedo value in particular for the companion which is unknown and possibly larger than 1 in the relevant wavelength range due to redistribution, large uncertainties in the physical modelling of reflection effect, and a formal limitation inherent to PHOEBE where the maximum $T_{\rm eff}$ currently available is 50 000 K.

Constraints for the many parameters governing a light curve simulation will come from the spectral analysis not only of the PG 1159, but also from irradiated spectra calulated using PHOENIX to model the hot side of the companion. A particularly strong additional constraint would come from a good determination of the gravitational redshift for the PG 1159 star.

Acknowledgments

We would like to thank B. Gänsicke and M. Schreiber for pointing out to us the interesting SDSS spectrum of this object. We also thank S. Dreizler and S. Hügelmeyer for their help in setting up the stellar atmosphere model calculations, and K. Werner and M. Miller-Bertolami for inspiring discussions. Thanks to I. Traulsen and many other observers for supporting the photometric observations in Göttingen and Tübingen. The spectroscopy is based on observations collected in service mode by Javier Aceituno and Ulrich Thiele for our program H07-3.5-038 at the Centro Astronómico Hispano Alemán (CAHA) at Calar Alto, operated jointly by the Max-Planck Institut für Astronomie and the Instituto de Astrofísica de Andalucía (CSIC). Travel to the 16th White dwarf workshop where this contribution was presented has been subsidised by the Deutsche Forschungsgemeinschaft (DFG) under grant number KON 856/2008 SCHU 2249/4-1.

References

Adelman-McCarthy J K et al. 2008 *APJS* **175** 297
Beeck B 2009 University of Göttingen *Diploma Thesis* in prep.
Eisenstein D J et al. 2006 *APJS* **167** 40
Heller R et al. 2009 *these proceedings*
Miller Bertolami M M Althaus L G Serenelli A M & Panei J A 2006a *A&A* **449** 313
Miller Bertolami M M & Althaus L G 2006b *A&A* **454** 845
Nagel T Schuh S Kusterer D-J Stahn T Hügelmeyer S D Dreizler S Gänsicke B T & Schreiber M R 2006 *A&A* **448** L25
Rebassa-Mansergas A et al. 2009, *these proceedings*
Schreiber M R et al. 2009 *these proceedings*
Schuh S & Nagel T 2007 in "Proc. of 15th European Workshop on White Dwarfs" *ASPC* **372** 491
Schuh S Nagel T Traulsen I & Beeck B 2008a in "Hydrogen-Deficient Stars" *ASPC* **391** 133
Schuh S Traulsen I Nagel T Reiff E Homeier D Schwager H Kusterer D-J Lutz R & Schreiber M R 2008b *Astronomische Nachrichten* **329** 376
Werner K 2001 *Ap&SS* **275** 27
Werner K Deetjen J L Dreizler S Rauch T Barstow M A & Kruk J W 2003 in "White Dwarfs" *NATO ASIB Proc.* **105** 117
Werner K & Herwig F 2006 *PASP* **118** 183

Journal of Physics: Conference Series **172** (2009) 012066 doi:10.1088/1742-6596/172/1/012066

Pulsations in carbon-atmosphere white dwarfs: A new chapter in white dwarf asteroseismology

G Fontaine[1], P Brassard[1], P Dufour[2], E M Green[2] and J Liebert[2]

[1] Département de Physique, Université de Montréal, C.P. 6128, Succ. Centre-Ville, Montréal, Québec, Canada H3C 3J7

[2] Department of Astronomy and Steward Observatory, University of Arizona, 933 North Cherry Avenue, Tucson, AZ 85721, USA

E-mail: fontaine@astro.umontreal.ca; brassard@astro.umontreal.ca; dufourpa@as.arizona.edu; bgreen@as.arizona.edu; liebert@as.arizona.edu

Abstract. We present some of the results of a survey aimed at exploring the asteroseismological potential of the newly-discovered carbon-atmosphere white dwarfs. We show that, in certains regions of parameter space, carbon-atmosphere white dwarfs may drive low-order gravity modes. We demonstrate that our theoretical results are consistent with the recent exciting discovery of luminosity variations in SDSS J1426+5752 and some null results obtained by a team of scientists at McDonald Observatory. We also present follow-up photometric observations carried out by ourselves at the Mount Bigelow 1.6-m telescope using the new Mont4K camera. The results of follow-up spectroscopic observations at the MMT are also briefly reported, including the surprising discovery that SDSS J1426+5752 is not only a pulsating star but that it is also a magnetic white dwarf with a surface field near 1.2 MG. The discovery of g-mode pulsations in SDSS J1426+5752 is quite significant in itself as it opens a fourth asteroseismological "window", after the GW Vir, V777 Her, and ZZ Ceti families, through which one may study white dwarfs.

1. Introduction

Dufour et al. (2007) reported recently on the unexpected discovery of a new type of white dwarfs, those with carbon-dominated atmospheres, also known as Hot DQ stars. These are very rare, and Dufour et al. (2007) wrote on the discovery of nine of those, out of a total of several thousands white dwarfs already identified spectroscopically. The Hot DQ stars were, in fact, uncovered within the framewok of the SDSS project, which has led to many other surprises. Among other characteristics, the C-atmosphere white dwarfs bunch together in a narrow range of effective temperature, between about 18,000 K and 23,000 K (Dufour et al. 2008a).

It has been known for quite some time (see, e.g., Fontaine & Van Horn 1976) that models of C-atmosphere white dwarfs in this temperature range possess superficial convection zones that bear strong similarities with those found in H- and He-atmosphere stars. Given that the newly-found C-atmosphere white dwarfs are sandwiched between the V777 Her and ZZ Ceti instability strips, and given that convection plays a key role in the excitation of pulsation modes in these pulsators, it follows that the prospects of finding unstable models of Hot DQ stars looked good at the outset.

2. Stability survey of models of Hot DQ white dwarfs

In an initial effort, three of us carried out an exploratory stability survey of models of carbon-atmosphere white dwarfs using a full nonadiabatic approach, and this was reported in Fontaine, Brassard & Dufour (2008). Our theoretical survey has revealed that g-modes can indeed be driven in models of Hot DQ stars. However, we found that only those stars with a sufficiently large amount of He ($X(\mathrm{He}) > 0.25$) in their C-rich envelope mixture could pulsate *in the range of effective temperature where the real C-atmosphere white dwarfs are found.*

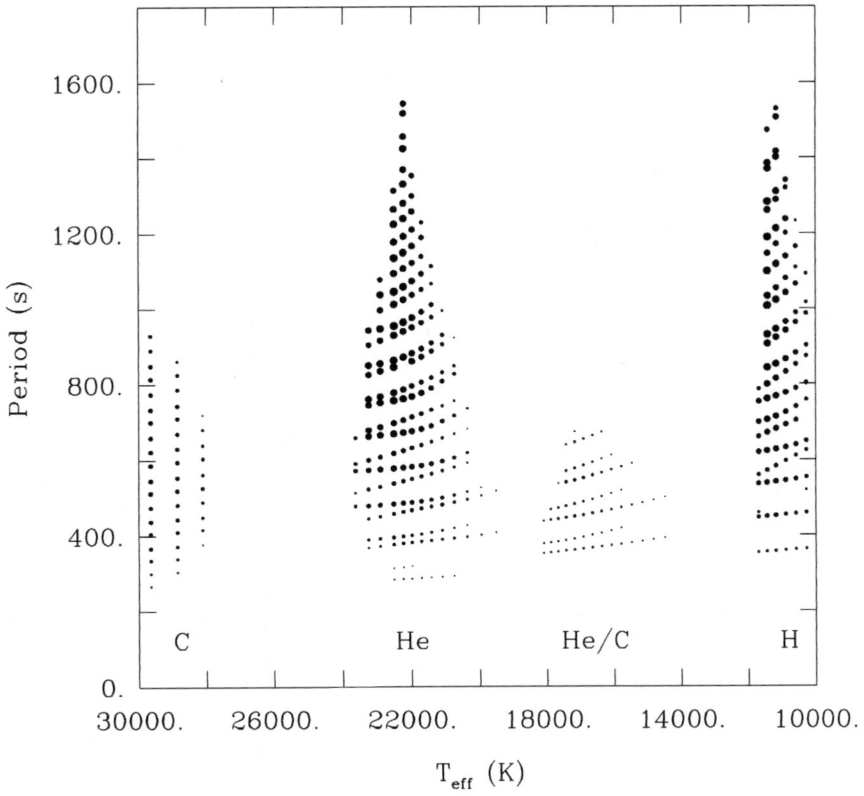

Figure 1. Predicted spectra of excited dipole g-modes computed from four distinct evolutionary sequences, each characterized by a total mass of 0.6 $M_\odot$, but with a different envelope composition: pure C, pure He, $X(\mathrm{He}) = X(\mathrm{C}) = 0.5$, and pure H, from left to right. The so-called ML2/α=0.6 version of the mixing-length theory was used in these calculations. Each dot gives the period of an unstable mode, and its size represents a logarithmic measure of the modulus of the imaginary part σ_I of the complex eigenfrequency. The bigger the dot, the more unstable the mode.

In this connection, Figure 1 displays some revealing results. It depicts the locations of theoretical instability strips for evolving 0.6 $M_\odot$ white dwarf models with different envelope compositions. Along with the usual V777 Her (pure He) and ZZ Ceti (pure H) instability strips, one can recognize the red edge of the pulsating pure C envelope white dwarf models. In fact, the pure C instability strip extends all the way up to the GW Vir regime as described at length in

Quirion et al. (2007). Given that the true red edge is hotter than the ~28,000 K value found in our survey (and see Fontaine & Brassard 2008), it follows that pure C-atmosphere white dwarfs cannot pulsate in the range of effective temperature where the Hot DQ stars congregate. This is contrary to the arguments put forward by Montgomery et al. (2008).

On the other hand, models with a mixed He and C envelope composition can also pulsate, but in different temperature intervals. For instance, Figure 1 illustrates a new instability strip between the V777 Her and the ZZ Ceti domains associated with white dwarf models with a mixed envelope composition specified by $X(\mathrm{He}) = X(\mathrm{C}) = 0.5$. Naively, one could have expected to find such a strip in between the pure C and pure He strips, but structural differences in the mixed envelope composition models explain why this is not so (see Fontaine et al. 2008 for more detailed explanations on this). In brief, the survey of Fontaine et al. (2008) revealed that some Hot DQ white dwarfs can indeed undergo low-order, low-degree g-mode pulsational instabilities, provided that the surface gravity is larger than average and a substantial amount of He is present in the C-rich envelope mixture.

Figure 2 illustrates the details of the driving/damping region in a Hot DQ model defined by $\log g = 8.0$, $T_{\rm eff} = 17{,}000$ K, and with an envelope composition specified by $X(\mathrm{He}) = X(\mathrm{C}) = 0.5$. While this is cooler than the coolest C-atmosphere white dwarf known, there is a compensation effect related with the gravity such that a similar behavior is observed in a hotter but higher-gravity model as described in Fontaine et al. (2008). Out of the many g-modes found excited in the retained model, we have singled out a representative one with indices $k = 8$ and $\ell = 1$. It has a period of 489.1 s. The figure illustrates a driving/damping situation comparable to the cases of the V777 Her and ZZ Ceti pulsators, but is more complicated because of the presence of two maxima in the opacity distribution instead of a single peak in these other pulsators. In the present case, both opacity maxima (caused by two distinct partial ionization zones in the envelope mixture) are "active" in the sense that they both contribute to the driving/damping process. What we can observe from the plot is that the regions on the descending side (going in from the surface) of an opacity bump contribute locally to driving, while the deeper adjacent zones, where the opacity plummets to relatively low values, contribute instead to damping. In the model, the two opacity bumps are relatively close to each other and are part of a single convection zone. The damping region between the two bumps is then relatively narrow and the overall work integral comes out positive, meaning that the mode is globally excited.

In parallel with our theoretical investigations, but totally independently, Montgomery et al. (2008) carried out an observational search for luminosity variations in six Hot DQ stars accessible to observations in the winter of 2008. They were able to report the very exciting discovery that SDSS J1426+5752, one of the Hot DQ's found by Dufour et al. (2007), pulsates in, at least, one mode with a period of 417 s, thus establishing the existence of a fourth type of pulsating white dwarf. Full credit should be given to Montgomery et al. (2008) for this important breakthrough. In addition, they reported that no luminosity variations were found, to the limit of detection, in the five other stars in their sample.

Concerning their work, we have to point out, however, that the theoretical arguments put forward by Montgomery et al. (2008) to "predict" pulsational instabilities in Hot DQ stars are fallacious. In particular, the thermal timescale argument that they used is a necessary but *not* a sufficient condition for instability. Only full nonadiabatic calculations such as those carried out by Fontaine et al. (2008), for example, can lead to the final verdict as whether or not a pulsation mode is unstable in a given stellar model. Montgomery et al. (2008) did not carry out nonadiabatic calculations and could not, therefore, conclude about the stability of Hot DQ models. Fortunately, this did not prevent them from going to the telescope and discovering the first pulsating C-atmosphere white dwarf. In the end, this is really what matters.

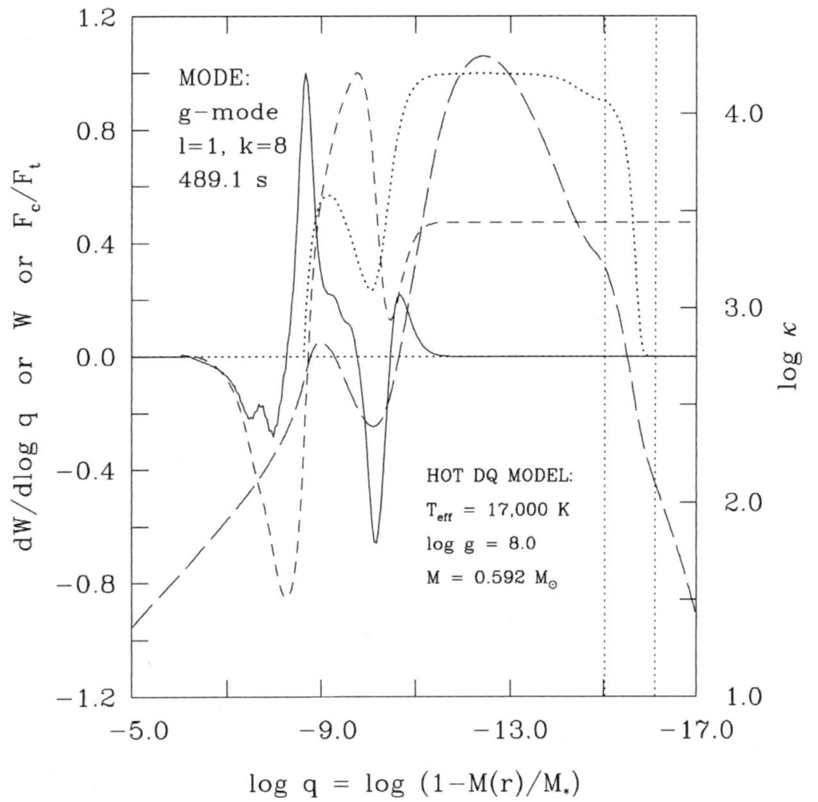

Figure 2. Details of the driving/damping process for a typical g-mode excited in a 17,000 K, ML2 model of a Hot DQ white dwarf with a envelope composition $X(\mathrm{C}) = X(\mathrm{He}) = 0.5$ and a gravity $\log g = 8.0$. The solid curve shows the integrand of the work integral of the mode as a function of fractional mass depth. The dashed curve shows the running work integral, from left to right, toward the surface of the model. The dotted curve shows the ratio of the convective to total flux. The long-dashed curve gives the run of the Rosseland opacity, to be read on the RHS ordinate axis. The maximum in the opacity profile, located at $\log q \simeq -12.41$ and corresponding to a temperature $T \simeq 1.045 \times 10^5$ K, is caused by the partial ionization of He II, C III, and C IV in the envelope mixture. The secondary maximum, located at $\log q \simeq -8.98$ and corresponding to a temperature $T \simeq 1.177 \times 10^6$ K, is caused by the partial ionization of C V, and C VI. The vertical dotted line on the left (right) gives the location of the base of the atmosphere at optical depth $\tau_R = 100$ (of the phoptosphere at $\tau_R = 2/3$).

3. Follow-up photometric observations of SDSS J1426+5752

Following this discovery, Green, Dufour & Fontaine (2008, in preparation) undertook follow-up wide band photometric observations of SDSS J1426+5752 at the Steward Observatory 1.6-m telescope on Mount Bigelow with the help of the new Montréal 4K×4K CCD camera (Mont4K), a joint venture between the University of Arizona and the Université de Montréal. Some 106 h of observations were obtained on this rather faint star (g = 19.2). Figure 3 illustrates one of the nightly light curves obtained by Green et al., leaving no doubt as to the variability of

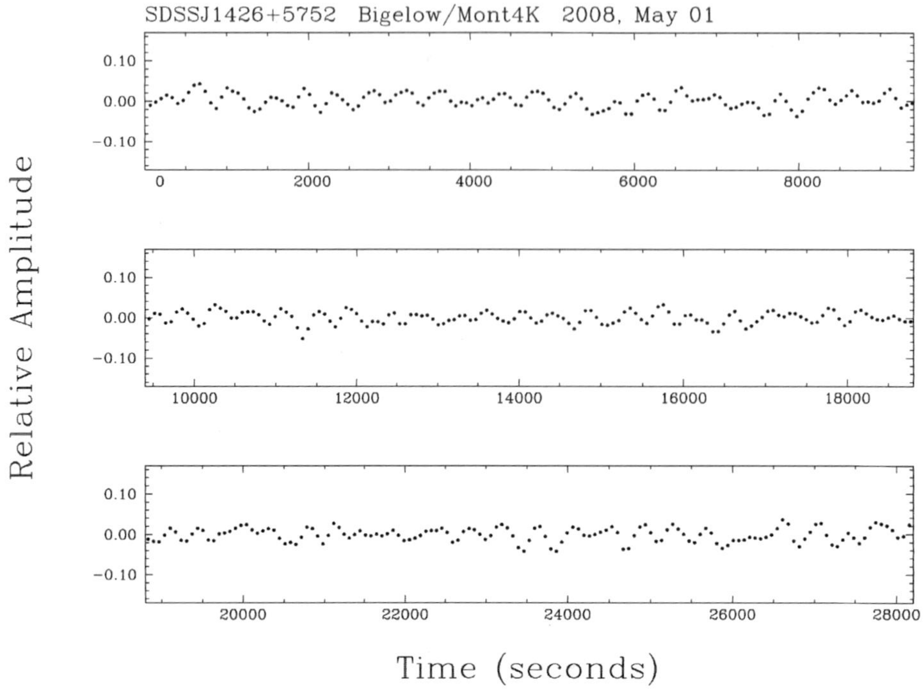

Figure 3. Sample broadband light curve of SDSS J1426+5752 obtained with the Mont4K CCD camera mounted on the Steward Observatory 1.6-m Kuiper telescope through a Schott 8612 filter. Each plotted point corresponds to an effective sampling time of 67 s.

SDSS J1426+5752. The Fourier amplitude spectrum for the full data set confirms the presence of a dominant pulsation with a period of 417 s and of its first harmonic as first reported by Montgomery et al. (2008). In addition, it also reveals the likely presence of an additional pulsation with a period of 319 s at the 4.9 sigma level. Hence, with at least two independent periodicities uncovered, SDSS J1426+5752 can be considered as a multiperiodic pulsator like the other types of pulsating white dwarfs.

4. Follow-up spectroscopic observations of SDSS J1426+5752

In view of the rather poor SDSS spectrum available for SDSS J1426+5752, follow-up spectroscopic observations were pursued by Dufour et al. (2008b) using both the MMT and one of the Keck telescopes. The objective was, firstly, to obtain a sufficiently good spectrum for detailed atmospheric modeling and, secondly, to search for the presence of He required to account for the observed pulsational instabilities according to the nonadiabatic calculations of Fontaine et al. (2008). The spectral analysis of the improved spectra readily revealed the presence of a substantial amount of helium in the atmosphere of SDSS J1426+5752, an abundance comparable to that of carbon by mass fraction. This is in line with the expectations of nonadiabatic pulsation theory which require an important He "pollution" in the atmosphere/envelope of SDSS J1426+5752 for it to pulsate at its current effective temperature. In this context, Dufour et al. (2008a) also showed that the five other objects found not to vary by Montgomery et al. (2008) are, contrary to SDSS J1426+5752, not expected to vary.

To add to this small success, an unexpected surprise came out of the follow-up spectroscopic

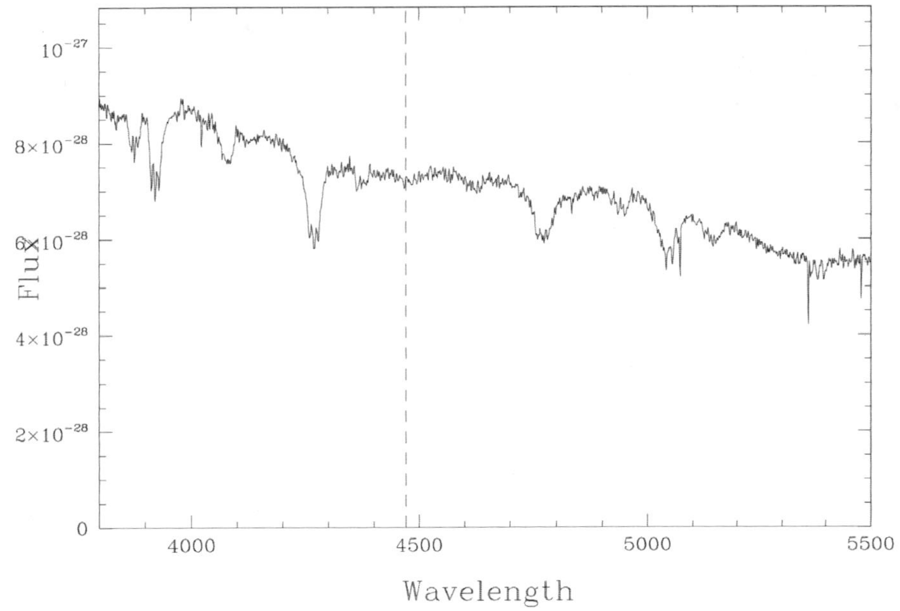

Figure 4. Optical spectrum of SDSS J1426+5752 obtained at the Multiple Mirror Telescope. The strong carbon lines are obvioulsy affected by Zeeman splitting due to the presence of a strong magnetic field. The same is true of the faint HeI 4471 line. Although relatively weak, that line implies an atmospheric helium abundance comparable to that of carbon in that star.

observations of Dufour et al. (2008b). It was found that the strong carbon lines seen in the spectrum of SDSS J1426+5752 feature Zeeman splitting, a structure that could not be seen in the original noisy SDSS spectrum. Figure 4 illustrates our MMT spectrum. The observed splitting between the π and σ components implies a large scale magnetic field of about 1.2 MG. Hence, SDSS J1426+5752 is both a pulsating and a magnetic white dwarf. As there is no sign of binarity in either the light curve or the optical spectrum (although the phase coverage has been quite limited), SDSS J1426+5752 is most likely the first example of an isolated pulsating white dwarf with a large detectable magnetic field. As such, it is the white dwarf equivalent of an roAp star. Interestingly, SDSS J1426+5752 could have been itself an roAp star in its distant past.

5. Conclusion

It remains to be seen if other stars similar to SDSS J1426+5752 will be found or if it will remain an isolated "freak". If we adopt a conservative point of view, it takes at least *two* members to define a "class", so we should perhaps refrain from referring to it as the prototype of a class for the time being. Note, however, that at the time of writing, rumors have it that two more Hot DQ stars may also be pulsating objects. Be it as it may, SDSS J1426+5752 is certainly different from the other kinds of pulsating white dwarfs that we know of (GW Vir, V777 Her, and ZZ Ceti). It is hoped that further data releases from the SDSS project might reveal siblings of this fascinating star. A fourth asteroseismological window has thus been opened through which one can further study the properties of white dwarf stars.

Journal of Physics: Conference Series **172** (2009) 012066 doi:10.1088/1742-6596/172/1/012066

References

Dufour P Fontaine G Liebert J Schmidt G D and Behara N 2008a *ApJ* **683** 978
Dufour P Fontaine G Liebert J Williams K A and Lai D K 2008b *ApJ* **683** L167
Dufour P Liebert J Fontaine G and Behara N 2007 *Nature* **450** 522
Fontaine G and Brassard P 2008 *PASP* **120** 1043
Fontaine G Brassard P and Dufour P 2008*ApJ* **483** L1
Fontaine G and Van Horn H M 1976 *ApJS* **31** 467
Montgomery M H Williams K A Winget D E Dufour P DeGennaro S and Liebert J 2008 *ApJ* **678** L51
Quirion P-O Fontaine G and Brassard P 2007 *ApJS* **171** 219

Journal of Physics: Conference Series **172** (2009) 012067 doi:10.1088/1742-6596/172/1/012067

Whole Earth Telescope Observations of the DAVs R808 and G38-29

Susan E Thompson[1,2], J L Provençal[1,2], A Kanaan[3], M H Montgomery[1,4], A Bishoff-Kim[5], H Shipman[1,2] and the WET team

[1]Delaware Asteroseismic Research Center, Mt. Cuba Observatory, Greenville, DE 19807
[2]Department of Physics and Astronomy, University of Delaware, Newark, DE, 19716
[3]Universidade Federal de Santa Catarina, C.P. 476, 88040-900, Florianópolis, SC, Brazil
[4]Department of Astronomy, University of Texas, Austin 78712
[5]Georgia College and State University, Milledgeville, GA 31061

E-mail: sthomp@physics.udel.edu

Abstract. The Whole Earth Telescope, under operation by the Delaware Asteroseismic Research Center, have obtained multi-site observations of two pulsating white dwarf stars. The DAVs R808 and G29-38 both show an abundance of excited modes, mostly clustered around 1050μHz. We present the Fourier analysis of both multi-site campaigns, present the measured periods, discuss their combination modes, and show that amplitude modulation is present in these stars.

1. Introduction

The Delaware Asteroseismic Research Center (DARC) sponsored two multi-site campaigns of DAVs prior to the White Dwarf Workshop in Barcelona; a small campaign in November 2007 of G 38-29 and a full Whole Earth Telescope run in April 2008 of EC14012-1446, PG1159-035, and R808. The main purpose of these observations were to 1) accurately measure the light curve in order to map convective timescale across the DAV instability strip (Montgomery, 2005) 2) measure the frequencies for the purposes of asteroseismology and 3) for PG1159-0035 to measure the $\dot{P}$.

The convection zone distorts the pulsations, creating pulse shapes with sharp peaks and shallow troughs. By measuring the shapes of these pulses we can ultimately determine the depth of the convection zone and how it changes with temperature. This technique is most effective on stars with large amplitudes where the nonlinear effects introduced by the convection zone are largest. Though the technique is most easily applied to pulsators with only a couple modes, analysis of GD 358 (Montgomery, 2008; Provençal, 2008), with its large, complicated set of frequencies, has shown that the convective fitting technique may be applied to a much larger sample of stars.

The first step for both convective fitting and asteroseismological analysis is to measure the pulsation modes on the star. Here we present the preliminary analysis of R808 and G38-29. We present the abundance of modes found for each, discuss how combination modes may help determine ℓ for some modes, and demonstrate that amplitude modulation plays an important role in the light curves of these stars. The large set of frequencies on each star may be ideal for asteroseismological studies and may complicate the measuring of their convection zones.

2. Light curves of R808 and G38-29

We analyzed time series photometry on R808 between April 4 and April 17, 2008 collected by 13 different telescopes. Similarly time series photometry of G38-29 was collected by 9 telescopes between November 5 and November 16, 2007 (Table 1). For both runs the reductions of the CCD images include, bias/dark subtraction and flat-fielding. We used IRAF routines to perform the aperture photometry, `Wqed` (Thompson & Mullally, 2008) to clean the light curve and apply the barycentric correction and Period04 (Lenz & Breger, 2005) for Fourier analysis.

Both were chosen for their large, nonlinear pulse shapes, in the hopes they may be ideal targets for measuring convection via light curve fitting. Figure 1 shows an example of both light curves. Each shows the nonlinear pulse shapes and beating between the various modes.

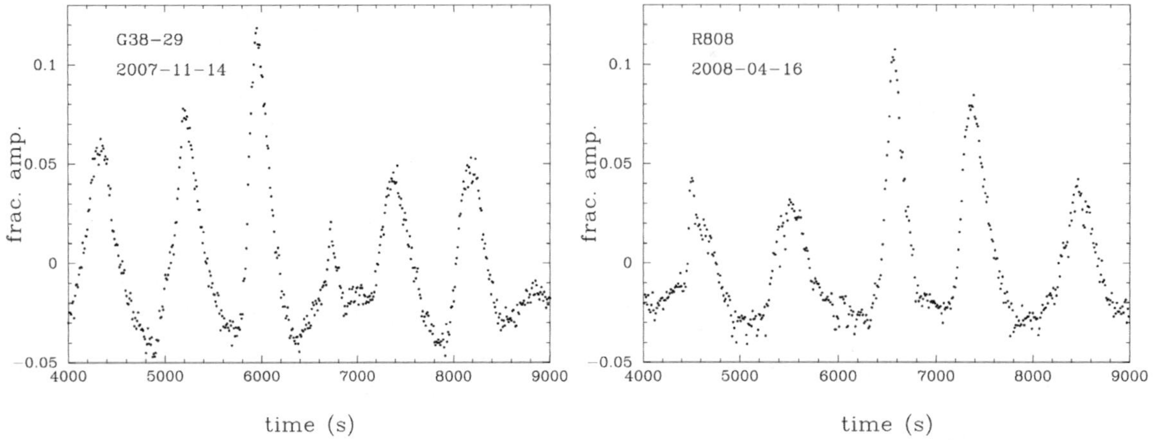

Figure 1. A portion of the light curves of R808 (taken by Kitt Peak's 2.1 m) and G38-29 (taken by McDonald's 2.1 m). Each shows evidence of the nonlinear pulse shapes, with peaky tops and shallow troughs typical of large amplitude DAV light curves.

3. FTs and Periods

The Fourier transform for G38-29 and R808 are presented in Figure 2. By sequentially selecting the largest peak in the Fourier transform, we do a nonlinear fit at the selected frequencies and remove the fit from the light curve. With this process we found 24 frequencies in G38-29 and 29 in R808. Each mode is 4 times the average power of the surrounding residual peaks. However, be cautious when considering the low amplitude modes near 1000μHz in both stars. The one day alias peak is 57% and 38% of the amplitude of the mode for G38-29 and R808 respectively. Thus, with closely spaced, low amplitude modes (as around 1000μHz), the highest peak is not always the real signal. When presented with several peaks of similar amplitude we selected those modes that, once removed, left the residual FT with the least signal. Modes identified in each star are shown in Table 2 and 3.

3.1. Combination Modes

Combination modes and harmonics are common to DAVs as a result of the nonlinear pulse shape. These are not physical standing waves of the white dwarf that could help reveal the internal structure of the star. The convection zone changes depth and it strongly attenuates the modes as the temperature changes on the star. As a result, it causes the pronounced nonlinearities and combination modes. Both G38-29 and R808 show peak-to-peak variations of 10%, however only G38-29 shows obvious combination modes. R808 instead shows an abundance of low amplitude

Journal of Physics: Conference Series **172** (2009) 012067 doi:10.1088/1742-6596/172/1/012067

pulsations that constructively interfere to create large amplitude pulsations. The combination modes on R808 are presumably present, but simply with too low amplitude to be detected when compared to the 1 mma noise level in the FT.

The ratio of the amplitudes between the combination mode and its parents, R_c (van Kerkwijk et al., 2000; Wu, 2001) has been used as a diagnostic of ℓ in the low amplitude DAVs (Yeates et al., 2005). This method is complicated by the fact that R_c also depends on the inclination and the azimuthal order (m) of the mode. We do not have space here to do a full combination mode analysis. However we note that for G38-29, the harmonic of F15 is much lower than the harmonic of F07, as one would expect if F15 is an ℓ=1, m=0 and F07 is an ℓ=2, m=0. Notice also that F10+F15 and F12+F15 have R_c values similar to 2F15, as one would expect for the combination of two ℓ=1 modes. The R_c values for R808 are very large, implying large ℓ, or more likely, mis-identification as combination modes.

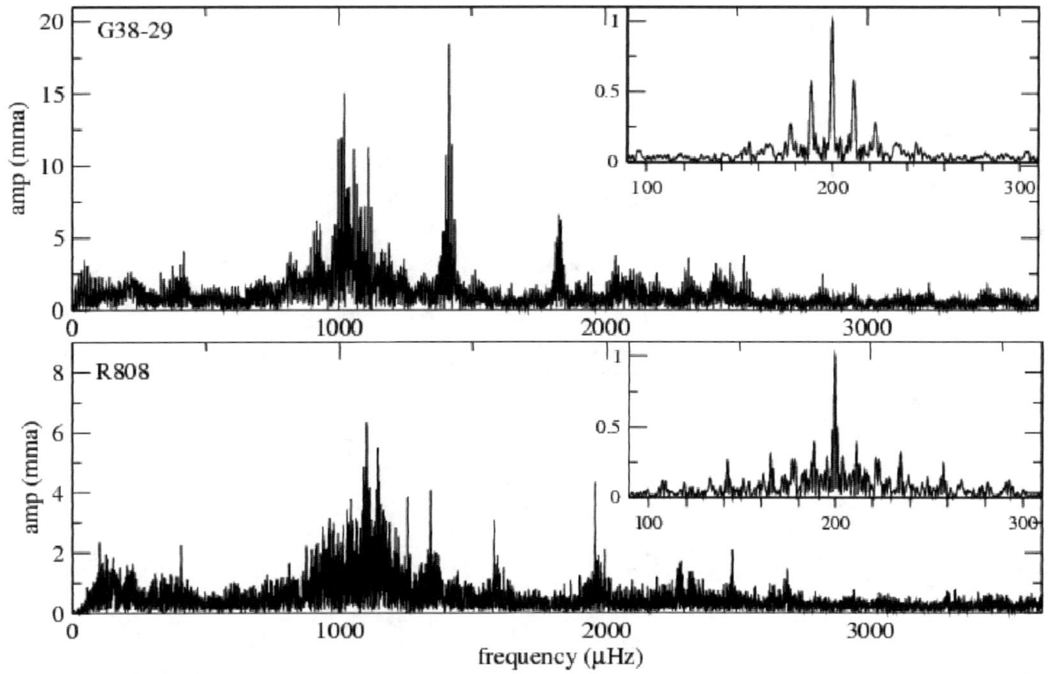

Figure 2. The Fourier transform of G38-29 and R808 and their windows (upper right plots). Both show an abundance of power near 1000 μHz. Many of the shorter period modes in G38-29 are combination modes, while few of the significant peaks in R808 are combinations. See Tables 2 and 3 for the measured periods.

4. Amplitude Modulation

Both R808 and G38-29 show significant variations in the amplitude of the modes during the run. See Figures 3 and 4 for plots of amplitude versus time for a sampling of modes. In each

Journal of Physics: Conference Series **172** (2009) 012067 doi:10.1088/1742-6596/172/1/012067

case the amplitudes were measured by fitting the modes listed in this paper to 5.5 day long segments of each light curve, long enough to resolve modes as close as 2.1μHz. For R808, the larger amplitude variations occur for the longer period modes where an abundance of modes have been excited. For G38-29 there is no basic trend with period; one of the shortest periods, 546 s, shows a rise in amplitude from 4-10 mma over the course of 10 days.

Either Figures 3 and 4 represent true variations in the amplitudes on time scales of days, or nearby modes exist that are not resolved in the 2-3 week time span of each data set. If these amplitude modulations were due to beating, we would expect sinusoidal variations in amplitude. As that is not what we observe, these are likely intrinsic amplitude modulations. However, amplitude modulation exemplifies the limitations of Fourier analysis; prewhitening by a constant amplitude when the mode's amplitude is actually varying will leave significant signal that can easily be interpreted as a true stellar pulsation.

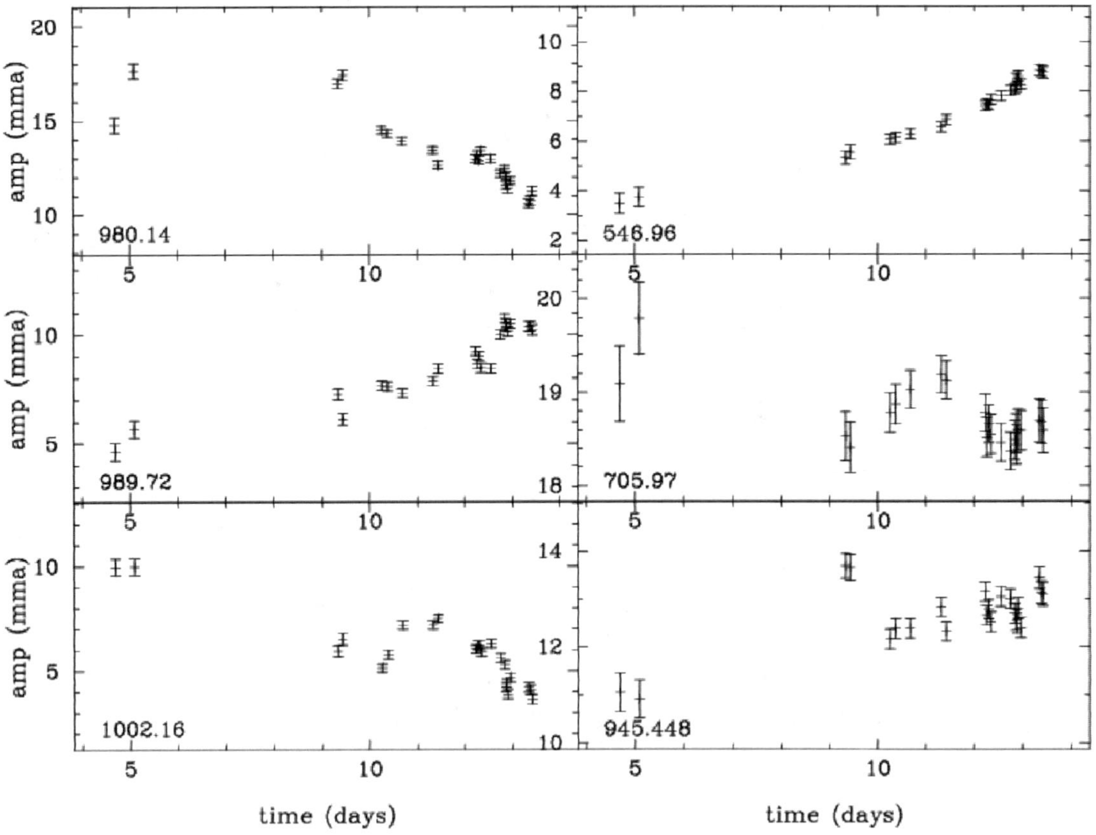

Figure 3. Mode amplitude vs. time for select modes of G38-29. Some modes show significant variations (980.14 s and 546.96 s) in amplitude while others show almost none (705.97 s). The first two points represent data taken prior to the November campaign, not included in the Fourier analysis of this paper.

5. Conclusions

We have observed two DAVs with large pulses and a plethora of modes. As each mode probes different layers of the star, these are good candidates for asteroseismology if we can be confident

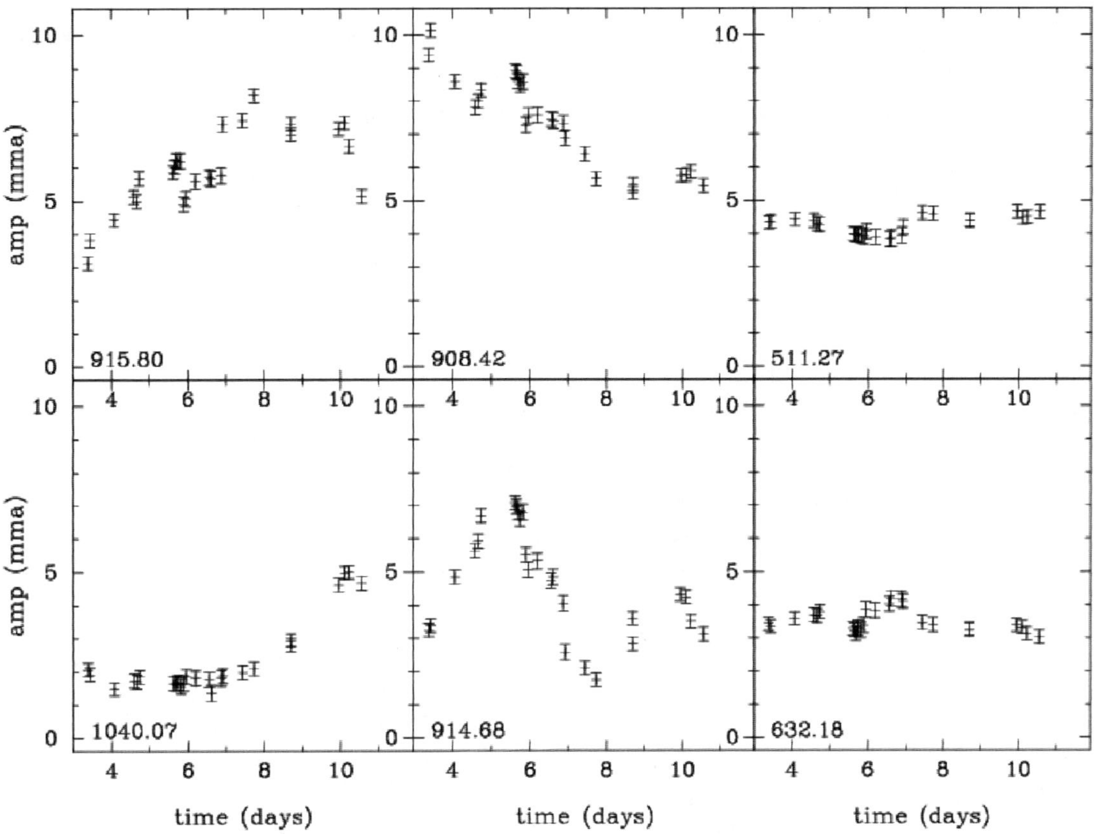

Figure 4. Mode amplitude vs. time for select modes of R808. Shorter period modes clearly show less modulation than longer periods.

that the modes we identify are due to the star's inherent structure, and are not due to changes in long term variations in mode amplitudes, or nonlinear effects on the star (combination modes). These data emphasize the need to explore different observing and/or analysis techniques to positively identify the physical pulsations from the nonlinear effects. We plan to apply the convection zone fitting technique on both of these stars if asteroseismological models can provide sufficient constraints on the ℓ and m values of the observed modes present in G38-29 and R808.

References

Lenz P & Breger M 2005 *Communications in Asteroseismology* **146** 53

Montgomery M H 2005 *ApJ* **633**, 1142

Montgomery M H 2008 *Communications in Asteroseismology*, **154**, 38

Thompson S E & Mullally F 2008 *White Dwarf Workshop*, Barcelona http://physics.udel.edu/darc/wqed

Provençal J L 2008, *ApJ* accepted

van Kerkwijk M H Clemens J C & Wu Y 2000 *Mon. Not. Royal Ast. Soc.* **314** 209

Wu Y 2001 *Mon. Not. Royal Ast. Soc.* **323** 248

Yeates C M Clemens J C Thompson S E & Mullally F 2005 *ApJ* **635** 1239

Journal of Physics: Conference Series **172** (2009) 012067 doi:10.1088/1742-6596/172/1/012067

Table 1. Telescopes that observed R808, April 2008 and G38-29, November 2007.

Observatory	Longitude (° East)	Length (Hours)
R808		
OHP	5.7	1.7
Loiano	11.3	6.48
Vienna	16.3	13.72
Konkoly	19.9	11.59
Suhora	20.1	32.90
Moletai	25.5	1.25
Terskol	42.5	4.00
BAO	116.3	5.09
BOAO	128.9	9.09
Kitt Peak	248.4	9.32
SARA	248.4	20.39
PJ Meyer	262.3	49.25
Mount Cuba	284.4	5.50
G38-29		
Konkoly	19.9	11.50
Suhora	20.1	22.82
MIRO	72.4	27.2
BOAO	128.9	34.54
Faulkes North	203.7	55.70
U. Hawaii	204.5	68.05
Apache Point	247.5	11.15
McDonald	257.0	42.71
Mount Cuba	284.4	6.67

Journal of Physics: Conference Series **172** (2009) 012067 doi:10.1088/1742-6596/172/1/012067

Table 2. Properties of the pulsation modes of G38-29 as measured with the function A $\sin(2\pi f(t + t_0))$. Error in frequency is $\sim$0.02 μHz, error in amplitude is $\sim$0.2 mma. The initial barycentric Julian ephemeris day for this data set is 2454410.47025381.

ID	f	Period	A	t_0	R_c
	(μHz)	(sec)	(mma)	(sec)	
F01	917.948	1089.39	5.20	126.95	–
F02	920.169	1086.76	3.92	129.55	–
F03	924.369	1081.82	5.04	886.08	–
F04	984.107	1016.15	5.79	643.03	–
F05	997.841	1002.16	7.14	562.24	–
F06	1010.39	989.719	10.04	649.00	–
F07	1020.26	980.141	11.43	173.88	–
F08	1037.78	963.593	4.58	800.31	–
F09	1039.42	962.077	8.09	56.091	–
F10	1057.70	945.448	12.34	788.88	–
F11	1083.83	922.657	5.94	863.09	–
F12	1111.15	899.971	10.59	438.43	–
F13	1189.92	840.390	5.19	50.605	–
F14	1409.98	709.232	6.03	558.67	–
F15	1416.49	705.970	18.44	343.07	–
F16	1821.69	548.939	4.83	425.65	–
F17	1828.29	546.960	6.97	208.18	–
F18	1835.10	544.928	6.12	331.45	–
F19	2312.92	432.354	3.57	235.77	–
F20	2419.51	413.307	3.07	297.05	–
F18-F15	418.185	2391.28	4.11	1131.97	18.2
F17-F14	"	"	"	"	48.9
F17-F06	818.380	1221.93	3.70	937.64	26.4
2F07	2040.54	490.067	3.72	19.88	28.5
F07+F15	2436.80	410.375	2.80	203.40	6.6
F10+F15	2474.26	404.162	3.46	1.457	7.6
F12+F15	2527.72	395.614	3.63	266.67	9.3
2F15	2832.61	353.031	2.55	352.51	7.5
F15+F16	3238.14	308.819	2.07	22.86	11.6
F14+F17	"	"	"	"	24.6

Journal of Physics: Conference Series **172** (2009) 012067 doi:10.1088/1742-6596/172/1/012067

Table 3. Properties of the pulsation modes of R808 as measured with the function A $\sin(2\pi f(t+t_0))$. Error in frequency is ~0.02 μHz, error in amplitude is ~0.2 mma. The initial barycentric Julian ephemeris day for this data set is 2454555.53472367.

ID	f (μHz)	Period (sec)	A (mma)	t_0 (sec)	R_c
F01	406.652	2459.1	2.22	670.43	–
F02	874.16	1143.96	2.50	502.95	–
F03	916.516	1091.09	2.36	403.53	–
F04	937.444	1066.73	2.21	604.71	–
F05	959.649	1042.05	2.89	908.71	–
F06	961.476	1040.07	3.34	899.31	–
F07	988.736	1011.39	2.54	623.36	–
F08	1041.1	960.527	3.68	704.48	–
F09	1049.98	952.398	3.36	658.35	–
F10	1084.01	922.504	3.40	727.11	–
F11	1091.94	915.803	5.54	742.81	–
F12	1093.27	914.683	3.94	847.06	–
F13	1097.05	911.534	3.16	702.63	–
F14	1100.81	908.422	7.59	819.87	–
F15	1112.71	898.707	3.57	785.50	–
F16	1138.33	878.479	3.62	352.26	–
F17	1142.67	875.146	3.73	620.38	–
F18	1162.48	860.227	3.48	664.51	–
F19	1186.65	842.707	2.81	409.19	–
F20	1255.88	796.253	3.97	369.707	–
F21	1342.07	745.12	3.97	69.106	–
F22	1581.83	632.179	3.41	473.715	–
F23	1589.25	629.228	1.88	8.246	–
F24	1955.93	511.266	4.49	35.256	–
F25	2472.45	404.457	1.99	225.66	–
F17-F08	101.555	9846.84	2.38	3881.2	86.7
2F16	2276	439.368	1.71	133.53	130.7
F14+F22	2682.77	372.749	1.5	131.561	29.0

Seismological studies of ZZ Ceti stars

B G Castanheira[1,2] and S O Kepler[2]

[1]Institut für Astronomie, Türkenschanzstr. 17, A-1180 Wien, Austria
[2]Departamento de Astronomia Universidade Federal do Rio Grande do Sul, Av. Bento Gonçalves 9500, Porto Alegre 91501-970, RS, Brazil

E-mail: castanheira@astro.univie.ac.at

Abstract. We calculate an extensive adiabatic model grid for pulsating white dwarfs with H dominated atmospheres, the ZZ Ceti stars. We developed a new approach for asteroseismology, using the relative observed amplitudes as weights, and compared the computed modes with the observed ones for the class of ZZ Ceti stars. We measure the H layer mass for 83 stars and found an average of $M_H/M_* = 10^{-6.3}$, which is thinner than the predicted value of $M_H/M_* = 10^{-4}$. Our results indicate that the stars lose more mass during their evolution than previously expected.

1. Introduction

White dwarfs are the final evolutionary stage of almost all stars (95–98%, e.g., Fontaine et al. 2001). Their evolution is basically dominated by cooling. As they cool down, because of the κ-γ mechanism and/or convection driving, white dwarfs pulsate in three different instability strips, depending on the chemical element that drives pulsation. The 143 currently known DAVs (or ZZ Ceti stars) have hydrogen atmospheres and are confined in an observational instability strip between 10 800 and 12 300 K. Recent observations (Castanheira et al. 2007 and Kepler et al. 2006) indicate that pulsation is a phase every white dwarf goes through, i.e., the properties we measure for pulsators can be applied to all white dwarfs. The other two classes are the DBVs, with helium-rich atmospheres, and the DOVs, the hotter pulsators, with roughly a dozen of known pulsators in each class. Their temperatures range from 20 000 to 30 000 K and 75 000 to 170 000 K, respectively.

Asteroseismology is the most powerful tool to study stellar interiors and evolution. Pulsations probe the internal structure of the stars, as every mode is an independent measurement. Even from just a few pulsation modes we can estimate fundamental stellar properties, such as stellar mass (Castanheira & Kepler 2008). Moreover, when a large number of modes is observed, the structure of the star can be determined in detail (e.g., Metcalfe 2005). Pulsating white dwarfs constitute the largest population of variable stars, even though we can only detect the nearest ones because of their intrinsic faintness. Despite the relatively small number of known pulsators, their studies have extremely important astrophysical implications.

The ZZ Ceti stars can be separated in three groups: hot, intermediate and cool. The hot ZZ Ceti stars have only a few modes excited to visible amplitudes, all of them with short periods (smaller than 300 s), small amplitudes (from 1.5 to 20 mma), and the power spectrum does not change substantially from season to season. These stars define the blue edge of the ZZ Ceti instability strip. The standard picture is that pulsations start when the partial ionization zone of hydrogen deepens into the envelope. As the stars cool down, the depth of the convective zone increases, as well as its size. Around 11 500 K there is a sudden deepening of the convection zone in the models; the observed pulsations change in character, with more modes with long and higher amplitudes excited. At temperatures slightly lower than 11 000 K, the stars stop to pulsate, defining the red edge of the ZZ Ceti instability strip. The high amplitudes observed on the cool part of the strip decreases, consistent with an increase of the depth of the convective zone (Brickhill 1991). The cool ZZ Ceti stars show a very rich pulsational spectra of high amplitude modes (up to 30%) and longer periods (up to 1500 s). The modes interfere with each other, causing dramatic changes in the pulsational spectra. There are also a few red edge stars with long periods of low amplitude, representing the stage when the star is ceasing to pulsate. The intermediate ZZ Ceti stars have characteristics between the hot and the cool groups.

2. The seismological models

2.1. Calculating the models

As the internal structure of white dwarfs with $T_{\rm eff} \sim 12\,000$ K has been sufficiently modified due to cooling, stellar contraction, and chemical diffusion, a hot polytrope static model is equivalent to self-consistent models (e.g. Wood 1990).

We used the White Dwarf Evolutionary Code (Lamb & van Horn 1975, Wood 1990) to evolve the starting model until the desired temperature. The equation of state for the core of our models is from Lamb (1974) and for the envelopes from Fontaine, Graboske & van Horn (1977). We used the updated opacity OPAL tables (Iglesias & Rogers 1996), neutrino rates of Itoh et al. (1996), and ML2/α=0.6 MLT (Bohm & Cassinelli 1971). The core evolution calculations are self-consistent, but the envelope was treated separately. The core and the envelope are stitched together, fulfilling the boundary conditions in the interface. The transitions between the layers are consistent with time diffusion, following Althaus et al. (2003).

2.2. Grid dimensions

We calculated an extensive adiabatic model grid for the pulsation modes, varying four quantities: $T_{\rm eff}$, M, $M_{\rm H}$, and $M_{\rm He}$. The whole grid was generated from a starting polytrope of $M = 0.6\,M_\odot$, using homology transformations to obtain models with masses from 0.5 to 1.0 $M_\odot$, in steps of 0.005 $M_\odot$, initially. The effective temperature ($T_{\rm eff}$) varies from 10 600 to 12 600 K, in steps of 50 K. The upper H and He layer mass values are $10^{-4}\,M_*$ and $10^{-2}\,M_*$, respectively; above these limits, nuclear reactions would excite g-mode pulsations in the PNNVs by the ϵ mechanism, which are not observed (Hine 1988). The lower H and He layer mass values are $10^{-9.5}\,M_*$ (the minimum H amount to be detected in a DA white dwarf spectra) and $10^{-3.5}\,M_*$, respectively.

2.3. Core composition: C/O

Varying the core composition introduces three parameters in the fit: the abundance itself from the uncertain C(α,γ)O reaction rate (Metcalfe 2005) and the two extreme points of the function that better describes the transition zone. We chose a homogeneous core of 50:50 C/O, for simplicity. We calculated a representative model of a star in the middle of the instability strip, $T_{\rm eff}$=11 600 K, with mass close to the DA white dwarfs mean distribution, $M = 0.6M_\odot$ (e.g., Kepler et al. 2007), and canonic values for H and He layer masses, $M_{\rm H} = 10^{-4}M_*$ and $M_{\rm He} = 10^{-2}M_*$. The change for the three first $\ell = 1$ overtones is only a few seconds when the O core abundance (C is the missing quantity) varies from 0% to 90%!

We also investigated the shape of the transition zones when a Salaris-like profile is used (Salaris et al. 1997). The transition zones between the core and the He outer envelope are in different positions; therefore, the trapped modes in these cavities are different. Because the ZZ Cetis pulsate with only a few modes, we decided to use a fixed homogeneous C/O 50:50 core to decrease the number of free parameters in the fit, but still be consistent with the reaction

Journal of Physics: Conference Series **172** (2009) 012068 doi:10.1088/1742-6596/172/1/012068

rate uncertainty. The price payed for this choice is that the He layer mass determinations are uncertain. The differences in the shape of the transition zone can be compensated by changes in the thickness of the He layer. The Salaris profile introduces a more complicated than a linear decrease of the O profile in the outer layers (see Fig. 2. of Metcalfe 2005). This effect potentially changes the calculated trapped modes in the models.

3. Seismology of individual stars: testing for G117-B15A

Since the discovery of its variability (McGraw & Robinson 1976), G117-B15A was considered the most regular ZZ Ceti. Due to its mode stability, Kepler et al. (2005) measured the rate of change of the main mode at 215 s, with a precision better than 10^{-15} s/s. G117-B15A is the most precise optical clock known.

G117-B15A pulsates with only three independent low amplitude modes, showing also the harmonic of the main mode and linear combinations between the modes (Kepler et al. 1982). In table 1 we list the average values for all periodicities detected in observations with the 2.1 m telescope at McDonald Observatory.

Table 1. All detected periodicities for G117-B15A, with mode labeled as well as the linear combinations and the harmonic of the main mode, which are not fundamental modes of the star.

Period (s)	Amplitude (mma)	identification
215.20	17.36	f1
270.46	6.14	f2
304.05	7.48	f3
107.7	1.65	2×f1
126.2	1.40	f1+f3
119.8	1.30	f1+f2

We used G117-B15A as a test of our seismological analysis for two main reasons. First, this star shows a very simple pulsation spectrum, but more than one mode is excited. Second, there are published seismological studies, which can be used to test the validity of our approach.

Our seismological study starts with a literature search for all available information about the star. Besides more than 30 years of time series photometry (Kepler et al. 2005), there are independent measurements of the atmospheric properties ($T_{\rm eff}$ and $\log g$) of G117-B15A, from different techniques (Bergeron et al. 2004, Koester & Allard 2000, and Wegner & Reid 1991). The differences among the published values can be explained by the intrinsic degeneracy between $T_{\rm eff}$ and $\log g$, which allows more than one combination of solutions.

As our models do not account for the excitation of linear combinations nor harmonics, it is very important to be sure that only normal modes are being used in the fit. For G117-B15A, the modes are identified in table 1 as f_1, f_2, and f_3.

To do seismology, we compared the observed modes ($P_{\rm obs}$) with the computed ones ($P_{\rm model}$) by adding the square of the differences, similar to a χ^2 fit. We choose to weight the periods by pulsation energy, which is proportional to the observed amplitude squared ($w_P \propto A^2$), minimizing the quantity S:

$$S = \sum_{i=1}^{n} \sqrt{\frac{[P_{\rm obs}(i) - P_{\rm model}]^2 \times w_P(i)}{\sum_{i=1}^{n} w_P(i)}} \qquad (1)$$

where n is the number of observed modes and w_P is the weight given to each mode.

Even though our model grid uses the adiabatic approximation and does not provide theoretical amplitudes, our choice to weight our fits is to guarantee that the fit will be always dominated by high amplitude modes. For G117-B15A, the periods and their respective normalized weights are 215.20 s, 270.46 s, and 304.05 s, and 0.13, 1.00, and 0.67.

At first, we have tried $\ell = 1$ for all modes. This choice is supported by previous observations of Robinson et al. (1995) and Kotak, van Kerkwijk, & Clemens (2004). We would have tried higher ℓ values if no solution was found, despite the fact that all chromatic amplitude changes to date indicate the modes are more likely $\ell = 1$ than higher ℓ for all ZZ Ceti stars (e.g., Kepler et al. 2000).

By comparing the observations and the models, we found several families of solution below the cut $S < 1.8$ s (the quadratic sum of the relative uncertainties). Figure 1 shows the possible combinations of $T_{\rm eff}$, M, $M_{\rm H}$, and $M_{\rm He}$ below the S cut off, and the minimum of each family is listed in table 2.

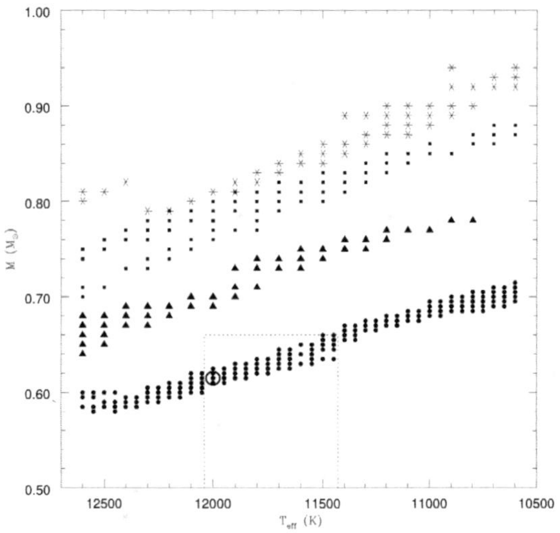

Figure 1. Results from comparison between the pulsation modes of the star G117-B15A and the models. The circles are the solutions for $M_{\rm He} = 10^{-2}M_*$, the triangles for $M_{\rm He} = 10^{-2.5}M_*$, the squares for $M_{\rm He} = 10^{-3}M_*$, and the asterisks for $M_{\rm He} = 10^{-3.5}M_*$. The dotted line box limits the region of the independent temperature and mass determinations ($\pm 1\sigma$) listed in table 2 and the open circle shows the location of the minimum for this family of solutions.

Table 2. Seismological solutions for G117-B15A: absolute minima for each possible family of solution in figure 1.

Symbol in fig. 1	$T_{\rm eff}$ (K)	M ($M_\odot$)	$-\log M_{\rm H}$	$-\log M_{\rm He}$	S (s)	Modes (ℓ, k)
Circles	**12 000**	**0.615**	**7**	**2**	0.67	215.3(1,1), 275.0(1,2) 302.1(1,3)
Triangles	11 500	0.75	5	2.5	0.97	215.1(1,2), 265.1(1,3) 308.0(1,4)
Squares	12 600	0.71	7.5	3	0.73	215.4(1,1), 266.0(1,2) 301.6(1,3)
Asterisks	11 500	0.85	8.5	3.5	0.19	215.2(1,1), 271.5(1,2) 303.3(1,4)

Zhang, Robinson, & Nather (1986) derived the following expression to calculate the uncertainties in the fit parameters: $\sigma^2 = \frac{d^2}{S - S_0}$, where d is the imposed step, S_0 is the absolute minimum, and S is the local minimum in a d difference between the quantities.

Journal of Physics: Conference Series **172** (2009) 012068 doi:10.1088/1742-6596/172/1/012068

The mean uncertainties for the quantities we fit are $\sigma_{T_{\rm eff}} \sim 50\,{\rm K}$, $\sigma_M \sim 0.005\,M_\odot$, $\sigma_{M_{\rm H}} \sim 10^{-0.5}\,M_*$, and $\sigma_{M_{\rm He}} \sim 10^{-0.5}\,M_*$, of the order of the spacings in the model grid. These values are the typical uncertainties in our seismological studies.

The final step is to identify which of the possible solutions correlates with the atmospheric determinations. From external determinations, mass should be between $0.46\,M_\odot$ to $0.66\,M_\odot$, which excludes all seismological high mass solutions; only the first solution in table 2 is consistent with the previous atmospheric determinations. In the same way, the spectroscopic temperature is within the range 11 430 to 12 040 K, which also agrees with this solution. Our seismological solution indicates that $T_{\rm eff}$ is hotter than determined by Bergeron et al. (2004), but consistent with the value from the UV spectra (Koester & Allard 2000).

Alternatively to our procedure, Bradley's (1998) first step is to select seismological models consistent with spectroscopic mass and temperature. He uses a standard model with $M_{\rm He} = 10^{-2}\,M_*$ and C/O nominal profile. Afterwards, he adjusts $M_{\rm H}$ to obtain a mode with $\ell = 1$, $k = 1$ or 2 close to 215 s. Then, he refines $T_{\rm eff}$ and $M_{\rm H}$ to bring the 270 s mode into agreement. The next step is to use the 304 s mode to adjust C/O profile. Finally, he refines the adjustment even more and/or applies small changes to $M_{\rm H}$, $M_{\rm He}$, and in the core structure. He found that the mass of the H layer is either $M_{\rm H} = 10^{-4}\,M_*$ or $10^{-7}\,M_*$, depending if the 215 s $\ell = 1$ mode is $k = 2$ or 1, respectively.

The code he used to calculate his models is basically the same as ours. However, his version used the trace element approximation, while our updated version uses a parameterization that mimics the results of time-dependent diffusion calculations (Althaus et al. 2003) to describe the transition zones. The biggest advantage of our seismological recipe is that we can explore all possibilities, avoiding local minima and error propagation if the spectroscopic determinations are uncertain.

Benvenuto et al. (2002) used realistic He, C, and O profiles, calculated from previous stellar evolutionary phases, obtaining $T_{\rm eff} = 11\,800\,{\rm K}$, $M = 0.525\,M_\odot$, and $M_{\rm H} = 10^{-3.83}\,M_*$, for G117-B15A. Their $T_{\rm eff}$ and mass are consistent with Koester & Allard (2000) determinations by fitting the UV spectrum.

Our seismological study of G117-B15A is in agreement with other approaches, encouraging us to apply the same technique to other ZZ Cetis.

4. Seismology of ZZ Ceti stars by groups

The motivation of grouping the stars is to use the similarities to lower the uncertainties. For stars with similar temperature, mass, and excited modes, it is reasonable to expect that their internal structure would be similar as well. Our approach is akin to having several independent measurements for one particular star.

Our first task was the identification of all periodicities known for all ZZ Ceti stars. We re-analyzed the light curves of the stars observed at the 2.1 m telescope at McDonald Observatory for other projects, like the search for new ZZ Ceti stars conducted by Mukadam et al. (2004), Mullally et al. (2005), and Castanheira et al. (2006), and the search for extra-solar planets orbiting ZZ Ceti stars with stable modes (Mullally et al. 2008), and also for the new ZZ Ceti stars discovered with the 4.1 m telescope SOAR and the 1.6 m telescope at Observatório Pico dos Dias (Kepler et al. 2005b, Castanheira et al. 2006, 2007). The other ZZ Ceti periods were obtained from the literature.

We separate the blue edge ZZ Ceti stars according to the excited mode with highest amplitude. We compared the observed modes with our model grid in the same way as for G117-B15A. We searched for common properties to characterize a particular group, as each group is a specific evolutionary stage in the white dwarf cooling.

83 ZZ Ceti stars have been selected from the SDSS sample. SDSS spectra for $g \leq 18$ stars have S/N~ 30, and therefore, temperature and mass are reasonably well determined (Kepler et

al. 2006). However, this is not the case for stars fainter than $g = 18$. Bergeron et al. (1995, 2004) followed by Gianninas, Bergeron, & Fontaine (2005) discussed that only with spectra S/N$\geq$70 one can obtain precise temperature ($\Delta T_{\rm eff} \simeq 300\,$K) and gravity determinations ($\Delta \log g \simeq 0.1$). Their uncertainties are the external estimates, by fitting the H line profiles of duplicate spectra. The published uncertainties from SDSS spectra are the internal values, obtained by fitting the whole spectrum, combined with colors (see Eisenstein et al. 2006 for detailed explanation). The external uncertainties by comparing duplicated SDSS spectra are of order of 300 K (Kleinman et al. 2004, Eisenstein et al. 2006, Kepler et al. 2007). Kepler et al. (2006), comparing SDSS and Gemini (S/N$\geq$ 70) spectra, found $\Delta T_{\rm eff} \simeq 320\,$K, systematically lower in SDSS and $\Delta \log g \simeq 0.24$ dex, systematically larger in SDSS.

Koester & Allard (2000) used UV spectra instead of optical, which results in a higher mass uncertainty, but much better temperature determination. An additional external uncertainty comes from the differences in the model grids from Kiel (e.g. Koester et al. 2001) and Montreal (e.g. Bergeron et al. 2004), which is around 200 K, using a similar line profile fitting technique (Bergeron et al. 2004).

We used the external temperature and mass determinations as a guide for the search of the best among all the possible families of seismological solutions.

5. Results and discussions

We have built an extensive model grid, which calculates all possible modes that can be excited for a given internal structure at certain temperatures. We have also developed an independent technique of model fitting to compare the observed to the calculated periods. In our approach, we used the external temperature and mass determinations to guide the seismological solutions, but we never limited the search to the uncertainty range of the spectroscopic determinations. G117-B15A was the best star to test our seismological approach.

In our seismological study of a few ZZ Ceti stars, seismology was proven to be really a powerful tool in the study of stellar evolution. Even for stars with few excited modes, it is possible to determine some characteristic of their interior. For G226-29, with only one detected mode, it was possible to restrict the mass to be above $0.7\,M_{\odot}$. Combined with reliable values for temperature and mass, three internal parameters could be determined, because the modes do not show an asymptotic behavior, i.e., they are more sensitive to the structure of the star.

The study of G185-32 was the motivation to include the relative amplitudes to give weights to the observed periods in the fits. The idea is that the high amplitude modes should be present in the best models. Our conclusion is that the observed amplitudes should be taken into account, even when one calculates the modes from an adiabatic model.

In the red edge, we studied HL Tau76 and G29-38. For the first, the period spacing was a very useful tool, especially because of the asymptotic behavior of the modes when they are larger. This analysis was consistent with the relative amplitudes between the $\ell = 1$ and 2 modes. In the case of G29-38, we suspected that the spectroscopic solutions were wrong, but we have obtained the same temperature and mass from our seismological study. It was a nice surprise that spectroscopy was right, agreeing with seismology, and that our prejudice that G29-38 should be in the red edge, wrong.

We also studied BPM37093, a high mass white dwarf with $1.1\pm0.05\,M_{\odot}$ (e.g., Bergeron et al. 2004), that has a significant crystallized portion. Even though it was not possible to determine the He layer mass, we could constrain the total stellar masses as high ($M > 1\,M_{\odot}$), temperatures between 11 600 and 11 800 K (only 200 K of external uncertainty), and H layer mass between $10^{-6}\,M_*$ and $10^{-7}\,M_*$, consistent with seismological determinations from Althaus et al. (2003). Our seismological determinations for the high mass red edge $T_{\rm eff}$ is also in theoretical and observational agreement with this to be hotter than the low mass red edge.

We found that there are many stars in the red edge that need to be re-observed with time-resolved photometry to detect more of the excited modes. This will allow us to study their interiors. We have demonstrated that it is possible to do seismology, even when a few modes are detected and/or reliable temperature and mass is available, with a minimum of four parameters. More precise spectroscopy will allow better determinations of the atmospheric parameters of the faint stars, providing two extra measurements. For many years, the red edge stars were relegated to oblivion, because we did not know how to study them.

Another important conclusion is that the H layer mass is not dependent on temperature (see Figure 2), according to Kolmogorov–Smirnov and correlation of coefficients tests. Therefore, there is no evidence for accretion or loss of the external layers, as it happens for Miras, for example, as the H layer mass does not vary univocally with temperature (or age). The mass loss of the external layer could come from the lack of reflectivity of the wave in the external layers, as calculated by Hansen, Winget & Kawaler (1985).

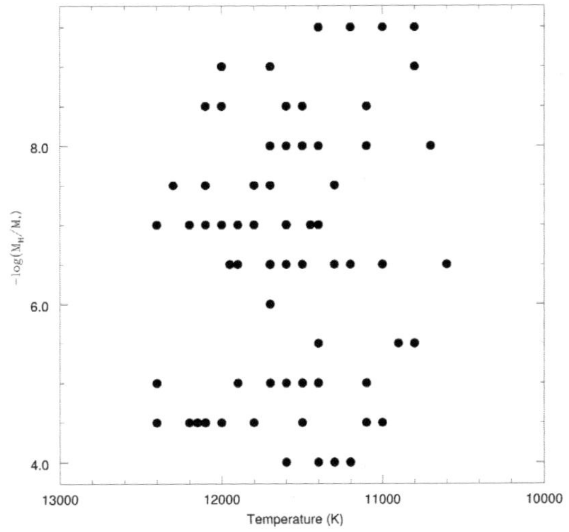

Figure 2. H mass vs. $T_{\rm eff}$, showing that there is no evidence of accretion nor loss of the external layers, as the ZZ Ceti evolves in the instability strip.

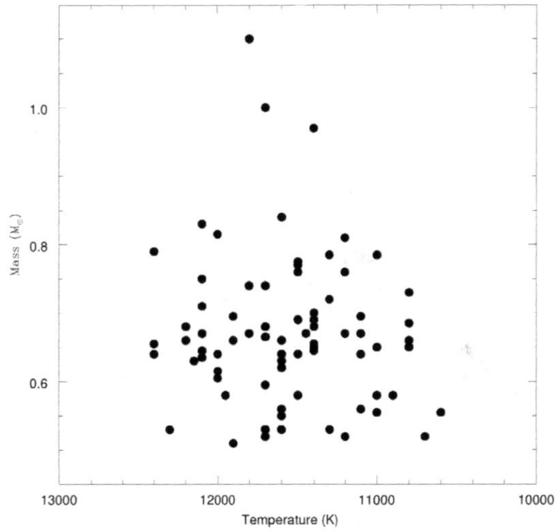

Figure 3. ZZ Ceti instability strip derived from seismology of these stars. We have included just the best solution for each studied star.

The mean value for the H layer mass is $10^{-6.3\pm1.6}\,M_*$, which is different than the canonical value of $10^{-4}\,M_*$, from evolutionary calculations. This result indicates that some white dwarfs, even if their masses are near the most probable value, might have formed with H mass several orders of magnitude smaller than the value predicted by theory, i.e., it is probable that the mass loss during their evolution was, in fact, more efficient than assumed by the models.

Using the seismological results, we derived an auto–consistent instability strip (see Figure 3), which includes the stars from the bright sample and the SDSS stars. This is the first time that the ZZ Ceti stars are studied as a group, by seismology.

We have done the first large seismological analysis of the ZZ Ceti stars as a group, studying 83 stars; before, only 12 ZZ Ceti stars had been studied seismologically. Even though we used the spectroscopic determinations as a guide, we only restricted the seismological solution to the range of spectroscopic parameters if there were not enough modes detected, avoiding local minima to be mistaken as global. After 40 years since the discovery of the first ZZ Ceti star (Landolt 1968), we finally extracted information about this class as a whole.

Acknowledgments
We acknowledge support from the CNPq-Brazil and B. G. C. acknowledges the support from Österreichische Forschungsgemeinschaft. We also acknowledge Travis S. Metcalfe for making available his scripts to calculate the model grid used in this work and Agnes Bischoff-Kim for sharing her version of the model with the Salaris profile. This work has been done with observations from the SOuthern Astrophysical Research telescope, a collaboration between CNPq-Brazil, NOAO, UNC, and MSU.

References

Althaus L G Serenelli A M Córsico A H & Montgomery M H 2003 *A&A* **404** 593
Benvenuto O G Córsico A H Althaus L G & Serenelli A M 2002 *MNRAS* **332** 399
Bergeron P Wesemael F Lamontagne R et al. 1995 *ApJ* **449** 258
Bergeron P Fontaine G Billères et al. 2004 *ApJ* **600** 404
Bohm K H & Cassinelli J 1971 *A&A* **12** 21
Bradley P A 1998 *ApJS* **116** 307
Brickhill A J 1991 *MNRAS* **252** 334
Castanheira B G Kepler S O Mullally F et al. 2006 *A&A* **450** 227
Castanheira B G & Kepler S O 2008 *MNRAS* **385** 430
Castanheira B G 2007 *Ph.D. Thesis* UFRGS-Brazil
Eisenstein D J Liebert J Harris H C et al. *ApJS* **167** 40
Gianninas A Bergeron P & Fontaine G 2005 *ApJ* **631** 1100
Hansen C J Winget D E & Kawaler S D 1985 *ApJ* **297** 544
Hine B P A 1988 *Ph. D. Thesis* UT-Austin
Iglesias C A & Rogers F J 1996 *ApJ* **464** 943
Itoh N Hayashi H Nishikawa A & Kohyama Y 1996 *ApJS* **102** 411
Kepler S O Nather R E McGraw J T & Robinson, E L 1982 *ApJ* **254** 676
Kepler S O Robinson E L Koester D et al. 2000 *ApJ* **539** 379
Kepler S O Costa J E S Castanheira B G et al. 2005a *ApJ* **634** 1311
Kepler S O Castanheira B G Saraiva M F O et al. 2005b *A&A* **442** 629
Kepler S O Castanheira B G Costa A F M & Koester D 2006 *MNRAS* **372** 1799
Kepler S O Kleinman S J Nitta A et al. 2007 *MNRAS* **375** 1315
Kleinman S J Harris H C Eisenstein D J et al. 2004 *ApJ* **607** 426
Koester D & Allard N F 2000 *Baltic Astronomy* **9** 119
Koester D Napiwotzki R Christlieb N et al. 2001 *A&A* **378** 556
Kotak R van Kerkwijk M H & Clemens J C 2004 *A&A* **413** 301
Lamb D Q 1974 *ApJL* **192** L129
Lamb D Q & van Horn H M 1975 *ApJ* **200** 306
Landolt A U 1968 *ApJ* **153** 151
McGraw J T & Robinson E L 1976 *ApJL* **205** L155
Metcalfe T S 2005 *MNRAS* **363** L86
Mukadam A S Mullally F Nather R E et al. 2004 *ApJ* **607** 982
Mullally F Thompson S E Castanheira B G et al. 2005 *ApJ* **625** 966
Mullally F Winget D E Degennaro S et al. 2008 *ApJ* **676** 573
Robinson E L Mailoux T M Zhang E et al. 1995 *ApJ* **438** 908
Salaris M Dominguez I Garcia-Berro E et al. 1997 *ApJ* **486** 413
Wegner G & Reid I N 1991 *ApJ* **375** 674
Wood M A 1990 *Ph.D. Thesis* UT-Austin
Zhang E H Robinson E L & Nather R E 1986 *ApJ* **305** 740

Singing and Dancing White Dwarfs

Anjum S Mukadam[1], Paula Szkody[1], Boris T Gänsicke[2] and Atsuko Nitta[3]

[1] Department of Astronomy, University of Washington, Seattle, WA 98195, USA
[2] Department of Physics, University of Warwick, Coventry, CV4 7AL, UK
[3] Gemini Observatory, Hilo, HI 96720, USA

E-mail: anjum@astro.washington.edu

Abstract. Accreting white dwarfs have recently been shown to exhibit non-radial pulsations similar to their non-interacting counterparts. This allows us to probe the interior of the accreting white dwarf using seismology, and may be the only way to determine masses for non-eclipsing cataclysmic variables. Improving our understanding of accreting white dwarfs will have implications for models of supernovae Type Ia. Pulsating white dwarfs in cataclysmic variables are also useful in establishing the effects of accretion on pulsations.

A search for nonradial pulsations among suitable candidates has led to the discovery of twelve such systems known to date. With the goal of establishing an instability strip (or strips) for these pulsating accretors, we acquired HST ultra-violet time-series spectroscopy of six pulsating white dwarfs in cataclysmic variables in 2007 and 2008. This approach enables us to measure the effective temperature of the white dwarf using the co-added spectrum, and to simultaneously characterize the pulsations. We also intended to constrain the pulsation mode identification by comparing the ultra-violet amplitudes to those from near-simultaneous ground-based photometry. Our preliminary results indicate a broad instability strip in the temperature range of 10500–15400 K.

1. Introduction to accreting white dwarf pulsators

Cataclysmic variables are close binary systems in which a late-type star (secondary) fills its Roche lobe and transfers mass through the inner Lagrangian point to a white dwarf (primary). GW Librae was the first cataclysmic variable discovered in 1998 with an accreting white dwarf showing photometric variations consistent with nonradial g-mode pulsations (van Zyl et al. 2000, 2004), typically observed in non-interacting white dwarf stars. This discovery has opened a new venue of opportunity to learn about the stellar parameters of accreting variable white dwarfs using asteroseismic techniques (Townsley, Arras, & Bildsten 2004). A unique model fit to the observed periods of the variable white dwarf can reveal information about the stellar mass, core composition, age, rotation rate, magnetic field strength, and distance (see the review paper Winget 1998). There are now twelve accreting pulsating white dwarfs known to date.

The optical spectrum of an accreting pulsator includes prominent broad absorption lines from the white dwarf as well as the central narrow emission features from the accretion disk. When the orbital period of a cataclysmic variable is ~80-90 min, it is near the evolutionary orbital period minimum, where the rate of mass transfer is very low $\sim 10^{-13}\,M_{\odot}/yr$ (e.g. Gänsicke et al. 2006). The low rates of mass transfer allow a larger surface area of the white dwarf to be visible; 90% of the optical light is dominated by the white dwarf in these systems.

Accreting pulsators have probably undergone a few billion years of accretion and thousands of thermonuclear runaways. Studying these systems will allow us to address the following questions: to what extent does accretion affect the white dwarf mass, temperature, and composition and how efficiently is angular momentum transferred into the core of the white dwarf. White dwarf rotation rates constrain models of Type Ia supernovae (e.g. Piersanti et al. 2003), and also play a role in the formation of millisecond pulsars and gamma-ray bursts from collapsing white dwarfs (Yoon & Langer 2005). Accreting white dwarfs with sufficient angular momentum may be detectable sources of gravitational waves with frequencies of 0.1-1.0 Hz by the Laser Interferometer Space Antenna (Yoon & Langer 2004). Accreting pulsators in non-eclipsing cataclysmic variables may prove to be our only opportunity to obtain meaningful mass constraints for the primary white dwarfs. Constraining the population, mass distribution, and evolution of accreting white dwarfs is also important for studying supernovae Type Ia systematics.

2. Theoretical background: expected instability strip(s) for accreting pulsators

Arras et al. (2006) investigate the temperature range in which models of accreting white dwarfs with a wide range of masses and Helium enrichment from the donor star would be pulsationally unstable. They find a H/HeI instability strip for accreting model white dwarfs with a blue edge near $\leq$12000 K for a 0.6 $M_\odot$ star. The blue edge shifts to hotter (cooler) temperatures by about 2000 K for a factor of 10 increase (decrease) in gravity; we can expect the blue edge at 14000 K corresponding to log $g = 9$ and at 10000 K for log $g = 7$. This theoretical instability strip is similar to the ZZ Ceti instability strip. Non-interacting hydrogen atmosphere (DA) white dwarfs are observed to pulsate in a narrow instability strip located within the temperature range 10800–12300 K for log $g \approx 8$ (e.g. Bergeron et al. 1995, 2004; Koester & Holberg 2001; Gianninas et al. 2005), and are also known as the ZZ Ceti stars.

For accreting model white dwarfs with a high He abundance (> 0.38), Arras et al. (2006) find an additional hotter instability strip at $\approx$15000 K due to HeII ionization. The boundaries of this intermediate instability strip depend on the Helium abundance and the mass of the accreting white dwarf. For a He abundance higher than 0.48, models of the H/HeI and HeII instability strips essentially merge. Arras et al. (2006) expect that there are thus two distinct instability strips for accreting model white dwarfs with a He abundance between about 0.38 and 0.48.

3. Theoretical background: nature of white dwarf pulsations

Each nonradial pulsation mode in a white dwarf can be described with a unique set of indices, similar to the quantum numbers that describe the state of a bound electron in the spherical electrostatic potential of a nucleus. Mode identification is essential in asteroseismology as we cannot determine the inner structure of variable stars without mode-indices. The nonradial luminosity variations observed in white dwarfs are almost entirely due to temperature fluctuations (Robinson, Kepler & Nather 1982). The stellar surface is divided into zones of higher and lower effective temperature, depending on the degree of spherical harmonic ℓ. Since we cannot resolve the stellar disk, geometric cancellation results in smaller observable amplitudes for these modes. At ultra-violet (UV) wavelengths, the increased limb darkening decreases the contribution of zones near the limb. Hence modes with $\ell = 3$ are canceled less effectively in the UV compared to the low $\ell \leq 2$ modes. However ℓ=4 modes do not show a significant change in amplitude as a function of wavelength. Robinson et al. (1995) suggested a new mode identification technique based on this effect. We obtained optical photometry on the accreting pulsators, nearly simultaneous with the HST UV data, with the intention of constraining the mode identification of the observed pulsation periods.

Journal of Physics: Conference Series **172** (2009) 012069 doi:10.1088/1742-6596/172/1/012069

4. Data acquisition & reduction

We acquired five orbits of HST UV time-series spectroscopy for each accreting pulsator using the Solar Blind Channel (SBC) on the Advanced Camera for Surveys (ACS). With exposure times of approximately 61 s and a dead time of 40 s, yields a time resolution of 101 s. Our wavelength coverage is approximately 1200 to 1960 Å. The dispersion varies from about 1.5 Å/pixel at the far blue end to about 25 Å/pixel at the red end. The first orbit of each target has typically 26 integrations (due to the initial overhead of setting up) and the remaining four orbits contain 29 integrations; each light curve has $\sim$142 points.

We analyzed the time-series spectra using the software package aXe1.4. Summing the individual spectra yields a high S/N total spectrum, while summing over wavelength yields light curves that reveal the pulsation characteristics of the targeted accreting white dwarf. We used a fixed extraction width of $\pm$17 pixels, corresponding to $\pm$0.5 arcsecond from the dispersion axis, to obtain the co-added spectrum for each accreting pulsator. This gives us a uniform basis for flux calibration. To extract the light curves (differential photometry), we determined the optimal extraction width and wavelength range that gave us the least noise. We converted the mid-exposure times of the individual spectra to Barycentric Coordinated Time (TCB; Standish 1998). We computed the Discrete Fourier Transforms (DFTs) of the extracted light curves and conducted a least squares analysis of all those that showed signs of variability. The detected orbital and pulsation-related periods are shown in Table 1 for the observed systems with the corresponding wavelength range we used to determine the light curves.

Table 1. Observed periodicities of the accreting pulsators from HST time-series spectroscopy

Accreting Pulsator	UT Date of Observation	Prism	Wavelength Range (Å)	Observed Periods
PQ And	13 September 2007	PR110L	1200–1820	
SDSS J0745+4538	1 November 2007	PR110L	1240–1910	
SDSS J0919+0857	14–15 November 2007	PR130L	1226–1955	40.75 min
REJ 1255+266	14 May 2008	PR110L	1340–1885	
SDSS J1339+4847	25 January 2008	PR130L	1245–1955	210.3 s, 229.6 s
SDSS J1514+4549	8 May 2008	PR130L	1324–1935	

5. Modeling the HST spectra

We model the HST UV spectra of these cataclysmic variables as a combination of the underlying white dwarf, the accretion disk, and a source of emission lines (e.g. Gänsicke et al. 2005). We use Hubeny white dwarf LTE models TLUSTY195 and SYNSPEC46 (Hubeny & Lanz 1995; Lanz & Hubeny 1995), consisting of a grid of temperatures, gravities, and abundances. Fixing the mass of the white dwarf to about 0.6 $M_\odot$ corresponding to log g=8, and the solar abundance at 0.01, we require that the best-fit white dwarf model must match the UV continuum shape, especially the broad Lyα line, and its optical flux must fall below the observed optical fluxes. In the absence of reliable models of a non-steady state disk at low mass transfer rates, we model the disk as a simple black body, adding in emission lines as Gaussian functions. Our temperature determinations for the observed accreting white dwarfs are shown in Table 2; we also indicate whether they were pulsating at the time of acquisition of the spectrum or not. We also include the best-fit temperatures of GW Librae, V455 And, SDSS J0131-0901, SDSS J1610-0102, and SDSS J2205+1155 for completeness, calculated using a similar approach (Araujo-Betancor et al. 2005; Szkody et al. 2007). Defining the instability strip for accretors requires a statistically significant sample, but our preliminary results suggest a broad strip in the range of 10500–15400 K.

Table 2. Temperatures of the accreting white dwarfs from HST Spectroscopy

Accreting White Dwarf	T_{eff} (K)	Pulsating During Spectroscopy
V455 And	10500 ± 1000	Yes
PQ And	12000 ± 1000	Ambiguous
REJ 1255+266	12000 ± 1000	Ambiguous
SDSS J1339+4847	12500 ± 1000	Yes
SDSS J0919+0857	13500 ± 1000	No
SDSS J1610-0102	14500 ± 1000	Yes
SDSS J0131-0901	14500 ± 1000	Yes
SDSS J2205+1155	15000 ± 1000	Yes
GW Librae	15400 ± 1000	Yes
SDSS J0745+4538	16500 ± 1000	No

6. Data analysis for observed accreting pulsators

6.1. PQ And

The HST UV light curves we extracted for PQ And do not reveal any sign of pulsations (see Figure 1). However simultaneous optical data acquired on September 13 reveal two periods at 1285±10 s and 2337±30 s with amplitudes close to 20 mma. The 2337 s period is beyond the typical range of ZZ Ceti pulsation periods of 70–1400 s, and may well be flickering noise as it does not show up in optical observations from September 15 and 19. The optical data from September 15 show a period at 1309±13 s, while that from September 19 show a period at 1301±13 s. Given the large uncertainties of these poor time resolution data (exposure time ~180 s) acquired using 1 m class telescopes, the 1300 s period appears to be persistent for 6 days and is unlikely to be caused by flickering. We expect that for ℓ=1 or ℓ=2 modes, typically observed in pulsating white dwarfs, the UV amplitudes should be significantly higher than the optical amplitudes. The absence of the 1300 s from the HST UV data suggests that it is either a high ℓ mode or not related to nonradial pulsation.

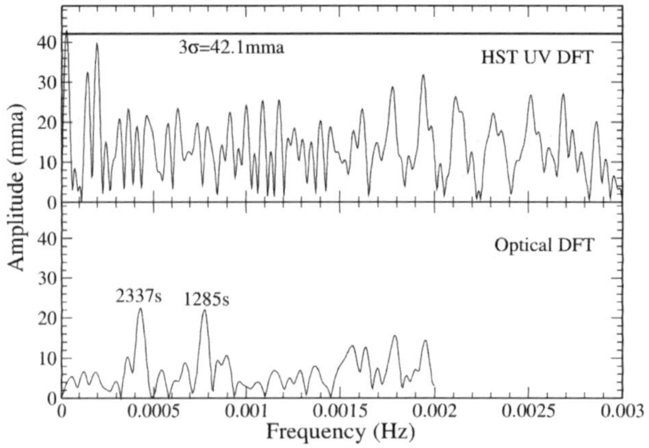

Figure 1. Comparing the HST UV DFT to simultaneous optical data on PQ And.

6.2. SDSS J0745+4538

The HST UV data acquired on 1 November 2007 and the optical data acquired on four consecutive nights from 30th October to 2nd November 2007 reveal no sign of the large amplitude pulsations (see Figure 2 below), previously reported by Mukadam et al. (2007). The optical light curves prominently display the orbital period of about 85.8 min with harmonics. We found that this star had undergone an outburst around October 2006 when its brightness increased by five orders of magnitude; this event was recorded by the Catalina Sky Survey (Figure 3). The white dwarf is heated to high temperatures well beyond the instability strip during such an outburst, and we expect the pulsations to cease. Godon et al. (2006) obtained a far UV spectrum of the dwarf nova WZ Sge three years after its superoutburst of 2001, and found the white dwarf was still ~1500 K above its quiescent temperature. Light curves of SDSS J0745+4538 may reveal pulsations again by 2010.

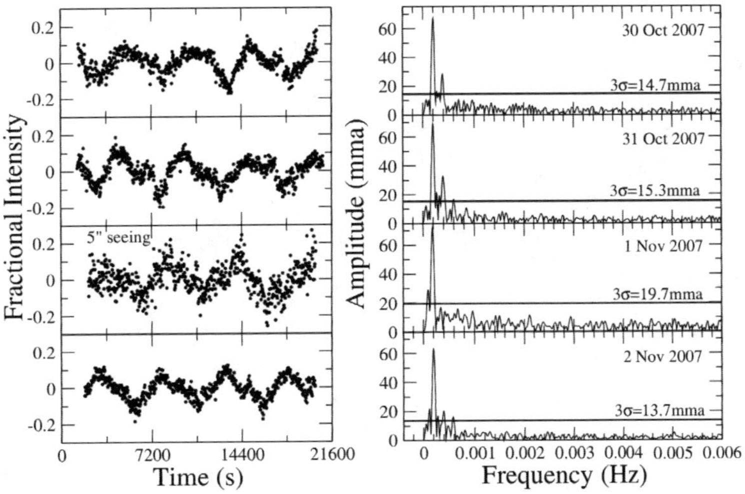

Figure 2. Optical data obtained on SDSS J0745+4538 reveal the cessation of previously observed pulsations in the accreting white dwarf.

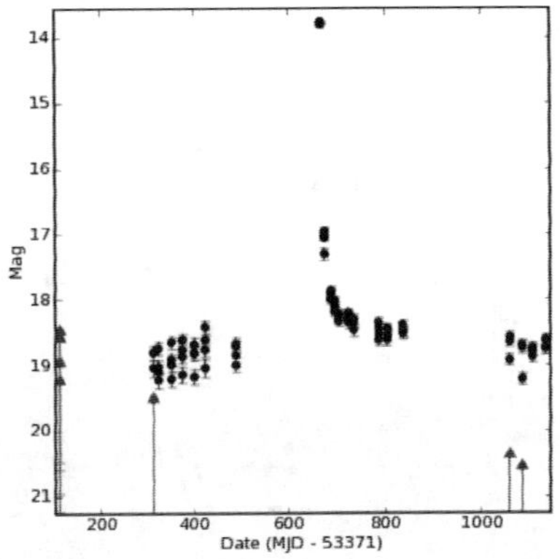

Figure 3. The Catalina Sky Survey captured the Fall 2006 outburst of SDSS J0745+4538.

Journal of Physics: Conference Series **172** (2009) 012069 doi:10.1088/1742-6596/172/1/012069

6.3. SDSS J0919+0857

The HST UV observations acquired on SDSS J0919+0857 during 14–15 November 2007 reveal only a harmonic of the orbital period at 40.75 min (Figure 4). The DFT shows no signs of the previously observed nonradial pulsation period near 260 s (Mukadam et al. 2007). The optical data we acquired on 12 November 2007 also do not reveal any periods in the vicinity of 260 s above 4 mma. We are at a loss to comprehend why pulsations may have disappeared from this star. It is possible that due to accretion heating, the temperature of the white dwarf changed slightly, causing the white dwarf to move just beyond the blue edge of the instability strip.

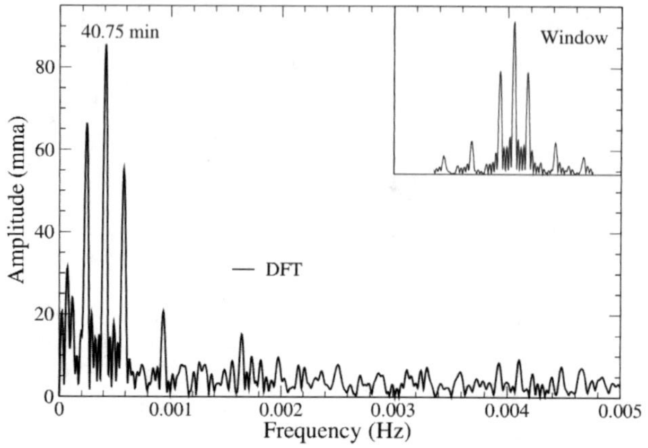

Figure 4. The HST UV SDSS J0919+0857 data reveal the orbital modulation at 40.75 min. The window function shown is the DFT of a single noiseless sinusoid with the same time sampling as the data.

6.4. REJ 1255+266

Our HST UV data on REJ 1255+266 reveal only a long period at 1.9 hr, while the optical data reveal two shorter periods at 582.1 s and 654.5 s, both with similar amplitudes of 12–13 mma (Figure 5). This is the second case where short periods detected in the optical are not present in the UV data, suggesting that either we are dealing with high ℓ modes or not dealing with nonradial pulsations at all.

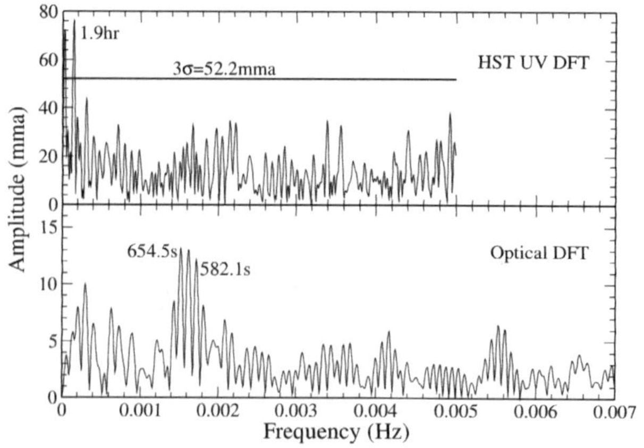

Figure 5. The HST UV data on REJ 1255+266 reveal no periods, although the optical data reveal two periods at 582.1 s and 654.5 s.

6.5. SDSS J1339+4847

The HST UV observations of SDSS J1339+4847 from 25 January 2008 reveal both high and low frequency variations (Figure 6). We observe a long period of 7.4 hr along with short periods at 229.6 s and 210.3 s, both close to the Nyquist frequency. We acquired optical data on SDSS J1339+4847 on 23 January 2008 for 1.5 hr. These data reveal the orbital period at 83.2 min, besides another long period at 1539 s above the 3σ limit. We suspect that the 1539 s period may have been caused by flickering as it lies beyond the typical ZZ Ceti pulsation period range of 70–1400 s. It is possible that the amplitudes of the 229.6 s and 210.3 s observed in the UV are below the detection limit in the optical. We expect typical pulsations modes with $\ell \leq 2$ to have significantly lower optical amplitudes compared to those in the UV. Both these modes cannot be the same ℓ if they are both real.

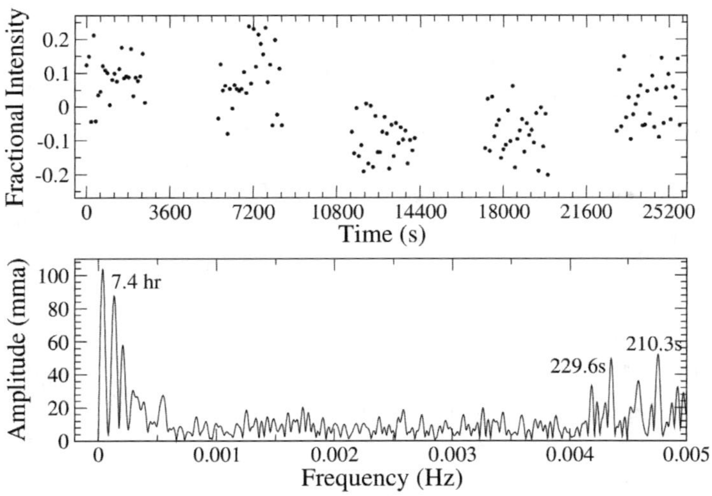

Figure 6. We show the HST UV light curve (top panel) and DFT (bottom panel) of SDSS J1339+4847 that reveal a long period at 7.4 hr besides two short periods close to the Nyquist frequency at 229.6 s and 210.3 s.

6.6. SDSS J1514+4549

We acquired HST observations of SDSS J1514+4549 on 8 May 2008 and optical observations on 7 May 2008; neither the UV nor the optical data reveal any periodicities. Since there is only a scarce amount of data published on SDSS J1514+4549(see Nilsson et al. 2006), we strongly suggest that long-term observations of this object are necessary to test whether it is a genuine pulsator.

7. Discussion of results

Szkody et al. (2007) used HST UV observations to determine the temperatures of three accreting pulsators, all of which showed distinct pulsation periods. Contrary to those well behaved systems, our preliminary results for six accreting pulsators observed similarly show peculiar results. We can dismiss SDSS J1514+4549 from further analysis as it may not actually be a pulsator. We expect that after its outburst in October 2006, SDSS J0745+4538 is still heated and will probably not reveal any pulsations until 2010. Two of the accreting pulsators PQ And and REJ 1255+266 revealed periods within the ZZ Ceti range in the optical, but none in the HST UV data. It is possible that these pulsators have a selection effect biased towards high ℓ

modes, or we are mistaking other variability in the system for nonradial pulsations. The pulsator SDSS J1339+4847 reveals two short periods in the UV data, but these are not visible above the noise in the optical DFT. We cannot easily explain the non-variability of SDSS J0919+0857, and expect that perhaps accretion heating may have driven the star just outside the instability strip. Note that the complete disappearance of pulsations has not been observed for any well-established ZZ Ceti star. The cessation of pulsations must not be confused with the phenomenon of amplitude modulation that causes the pulsation spectrum to change from time to time.

Our conclusion is that we need to monitor each accreting white dwarf for several years to improve our understanding of the variability of these systems, and to truly establish the subset that are accreting pulsators from amongst those that have been suggested to be nonradial pulsators.

Acknowledgments

We thank the National Science Foundation for the grant AST-0607840 and the Space Telescope Science Institute for the grants HST GO-10233.01-A and HST GO-11163.01-A, that financially supported this project. Based on observations made with the NASA/ESA Hubble Space Telescope, obtained at the Space Science Institute, which is operated by the Association of Universities for Research in Astronomy, Inc., under NASA contract NAS5-26555 and on observations from the Apache Point Observatory 3.5 m telescope, which is owned and operated by the Astrophysical Research Consortium.

References

Araujo-Betancor S et al. 2005 *A&A* **430** 629
Arras P Townsley D M & Bildsten L 2006 *ApJ* **643** L119
Bergeron P Fontaine G Billères M Boudreault S & Green E M 2004 *ApJ* **600** 404
Gänsicke B T Szkody P Howell S B & Sion E M 2005 *ApJ* **629** 451
Gänsicke B T et al. 2006 *MNRAS* **365** 969
Gianninas A Bergeron P & Fontaine G 2005 *ApJ* **631** 1100
Godon P Sion E M Cheng F Long K S Gänsicke B T & Szkody P 2006 *ApJ* **642** 1018
Hubeny I & Lanz T 1995 *ApJ* **439** 875
Koester D & Holberg J B 2001 ASP Conf Ser 226: *12th European Workshop on White Dwarfs* **226** ed J L Provencal H L Shipman et al. (San Francisco: Astronomical Society of the Pacific) p299
Lanz T & Hubeny I 1995 *ApJ* **439** 905
Mukadam A S Gänsicke B T Szkody P Aungwerojwit A Howell S B Fraser O J & Silvestri N M 2007 *ApJ* **667** 433
Nilsson R Uthas H Ytre-Eide M Solheim J -E & Warner B 2006 *MNRAS* **370** L56
Piersanti L Gagliardi S Iben I J & Tornambé A 2003 *ApJ* **583** 885
Robinson E L Kepler S O & Nather R E 1982 *ApJ* **259** 219
Robinson E L et al. 1995 *ApJ* **438** 908
Standish E M 1998 *A&A* **336** 381
Szkody P et al. 2007 *ApJ* **658** 1188
Townsley D M Arras P & Bildsten L 2004 *ApJ* **608** L105
van Zyl L Warner B O'Donoghue D Sullivan D Pritchard J & Kemp J 2000 *Baltic Astronomy* **9** 231
van Zyl L et al. 2004 *MNRAS* **350** 307
Winget D E 1998 *Journal of Physics: Condensed Matter* **10** 11247
Yoon S -C & Langer N 2004 *A&A* **419** 623
Yoon S -C & Langer N 2005 *A&A* **435** 967

Journal of Physics: Conference Series **172** (2009) 012070 doi:10.1088/1742-6596/172/1/012070

The long term period stability of the hot DBV white dwarf EC 20058−5234

Denis J Sullivan

School of Chemical & Physical Sciences, Victoria University of Wellington, P Box 600, Wellington 6012, New Zealand

E-mail: `denis.sullivan@vuw.ac.nz`

Abstract. Since its discovery over a decade ago, the helium atmosphere pulsator EC 20058−5234 (QU Tel) has been studied both spectroscopically, and also extensively using the techniques of time-series photometry. Model atmosphere fits to quality spectra obtained with one of the Magellan telescopes have confirmed its status as the hottest know DBV, so it currently defines the blue edge of the DBV instability strip. Extensive time-series photometry (primarily from Mt John in NZ but also including a Whole Earth Telescope run) clearly demonstrates that this white dwarf is a very stable low amplitude pulsator. This is consistent with its position at or near the blue edge of the DBV instability strip. However, of perhaps greater significance is the possibility of employing this period stability to look for a period change that can be sourced to the predicted neutrino-dominated cooling of the hot white dwarfs. This paper provides an update on this work.

1. Introduction

The white dwarf EC 20058−5234 (QU Tel, henceforth simply EC 20058) was discovered to be a variable by Koen et al. (1995) in 1994. Their work enabled its classification as a helium atmosphere (DBV) pulsator and identified 8 pulsation frequencies. A Whole Earth Telescope (WET) campaign in 1997, combined with a week of single-site photometry in 2004, resulted in nearly 180 hours of quality time-series photometry, which revealed a total of 18 pulsation frequencies for EC 20058 (Sullivan et al. 2008). Comparison with suitable models led to the identification of 11 individual pulsation modes.

The EC 20058 light curve is dominated by the presence of two dominant pulsation modes that produce approximately equal modulation amplitudes a little less than 1% ($\sim$ 8 mma); they have periods of 257 s and 281 s respectively. The other frequencies in the light curve have smaller amplitudes. The WET run, combined with subsequent single-site observing at Mt John Observatory in NZ (eg Sullivan et al. 2007) indicated that EC 20058 was a stable low-amplitude pulsator (see Figure 1), and thus was a suitable object for endeavouring to measure an evolutionary period change using photometry obtained over a number of seasons. We report briefly here on the motivation for this work and the current state of the play.

2. Period stabilities of the pulsating white dwarfs

Two classes of pulsating white dwarfs (DAV and DBV) and one class of pulsating pre-white white dwarf (DOV) have been identified. For brevity we will simply refer to all three classes as pulsating

Journal of Physics: Conference Series 172 (2009) 012070 doi:10.1088/1742-6596/172/1/012070

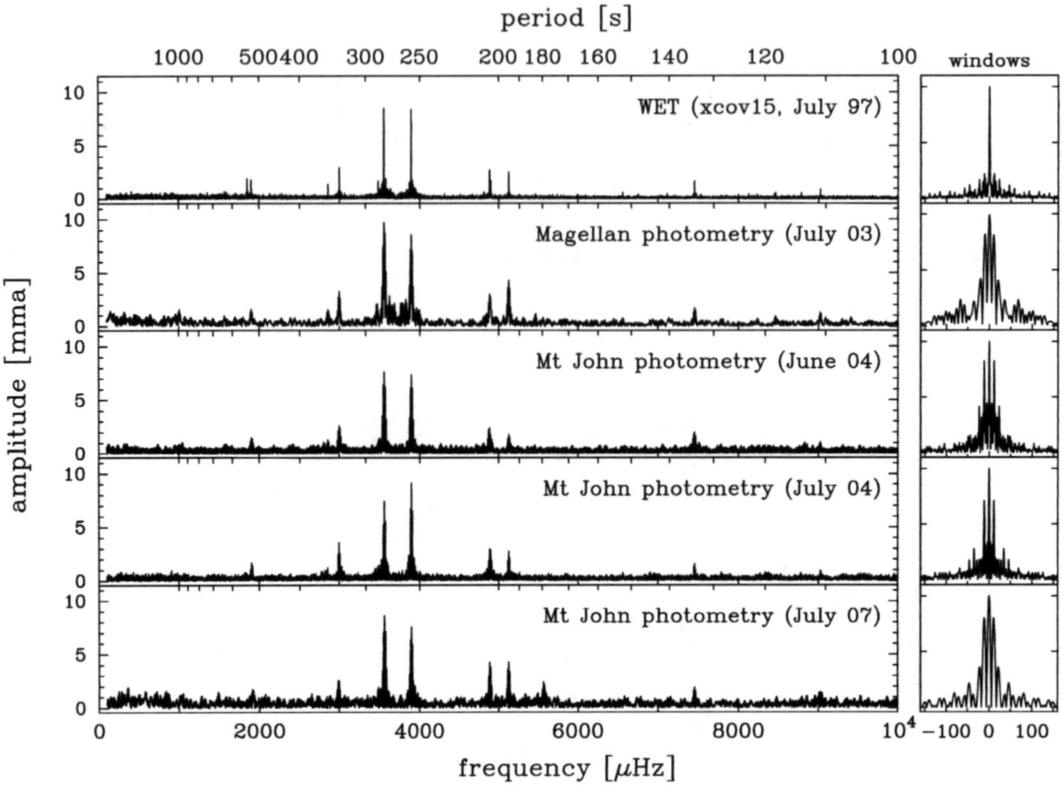

Figure 1. A selection of discrete Fourier transforms and the DFT windows for some of the EC 20058 time-series photometry, covering the period from the 1997 Whole Earth Telescope run to July 2007. The vertical axes use units of millimodulation amplitude (mma) whereby 10 mma corresponds to a 1% modulation and the horizontal axes employ μHz. The period structure for this pulsator has proven to be quite stable, although there is some variation in the actual pulsation amplitudes from season to season. Although the existence of some combination frequencies suggests some nonlinearity in the generation of the light curve modulations it is obviously quite small; this is consistent with the folded light curves in Figure 3

.

white dwarfs (WDs). The very hot object PG 1159−035 ($T_{eff} \sim 140$ kK) is the prototype for the DOV class, and it has been extensively studied (eg Costa et al. 2008) since its discovery. The prototypes for the other two types are GD 358 ($T_{eff} \sim 25$ kK) for the DBV class and ZZ Ceti ($T_{eff} \sim 12$ kK) for the DAV class. For all three classes, the onset of pulsation is believed to arise as a result of evolutionary cooling of the respective stellar structures through temperature-dependent cooling strips which leads to partial ionization zones for the most abundant surface elements (Winget & Kepler 2008).

For the DAVs, partial ionization in the H surface layer drives the pulsation, while it is partial ionisation of He ions in the surface layers that is responsible in the DBVs. For the pre-white dwarf H-deficient DOVs, the situation is less certain and more complicated, but it is appears clear that the main drivers of pulsation are the C and O partial ionization zones in the surface layers (Winget & Kepler 2008). This explains the significant difference in temperature between the respective instability strips.

Due to the compact mechanical structure of white dwarfs, the pulsation modes are expected to be very stable in period, and this is particularly true for the largely degenerate lower temperature DBV and DAV objects. The available observational material supports this prediction. However, extremely small period change rates ($\dot{P} = dP/dt$) are expected from cooling of the stellar material (a period increase or positive $\dot{P}$) and from contraction of the stellar structure (a period decrease of negative $\dot{P}$). The results from model calculations can be represented in the form (Winget, Hansen & Van Horn 1983)

$$\frac{\dot{P}}{P} = -a\frac{\dot{T}}{T} + b\frac{\dot{R}}{R}$$

where the a and b are positive constants of the order of unity. Order of magnitude estimates for $\dot{P}$ for the three classes of pulsating white dwarfs are given in Table 1.

Table 1. Theoretical $\dot{P}$ values for the three different classes of pulsating white dwarf stars

Type	$\dot{P}$	Comment
DOV	$\sim +10^{-11}\,\mathrm{s\,s^{-1}}$	If $\dot{T}$ dominates, but $\dot{R}$ important
DBV	$\sim +10^{-13}\,\mathrm{s\,s^{-1}}$	$\dot{R}$ unimportant
DAV	$\sim +10^{-15}\,\mathrm{s\,s^{-1}}$	$\dot{R}$ unimportant

Note that there are two orders of magnitude difference between the predicted $\dot{P}$s for the three classes of pulsator, and the very hot DOVs are expected to exhibit the largest change. Also note that for these latter objects both cooling and contraction should make significant contributions, but for both the DBVs and DAVs the near complete degeneracy of the core electrons providing mechanical support means that any radius change for the white dwarf is negligible for our purposes.

3. Observed white dwarf period stabilities

Table 2 summarises the relevant data for measured $\dot{P}$ values. As one might expect, the rapidly cooling DOV object, PG 1159, provided the first measurement. By analysing two-site time-series photometry obtained on PG 1159 in the early 1980s, Winget et al. (1985) actually 'detected' a period *decrease* for the 516 s dominant pulsation mode of the order of $10^{-11}\,\mathrm{s\,s^{-1}}$, which conflicted with the predicted period increase of the same order. Furthermore, using additional photometry over an extended time base including Whole Earth Telescope (WET) multi-site data, Winget et al. (1991) 'confirmed' this decrease and provided a more precise estimate of $-2.49 \pm 0.06 \times 10^{-11}\,\mathrm{s\,s^{-1}}$ for the $\dot{P}$ of the 516 s mode.

Although atmospheric contraction is important for the DOVs, the models predict that the effects of cooling (largely due to energy loss from core neutrino emission) should have a larger impact on the pulsation periods; hence this should lead to an observable period increase. Attempts to explain the surprising measured period decrease centred around the idea of mode trapping (Kawaler & Bradley 1994), in which the properties of particular pulsation modes are affected by resonances in the assumed stratified chemical layers of the star. The idea is that for some modes this allows the relatively less degenerate envelope, which is undergoing some contraction, to dictate the period change and hence the negative impact of a shrinking radius on the $\dot{P}$ can e xceed the positive impact of a temperature reduction.

However, subsequently, armed with more data in the form of several multi-site WET data sets, Costa et al. (1999) convincingly demonstrated that the 516 s mode in PG 1159 was actually

Table 2. A summary of relevant measured $\dot{P}$ values for pulsating white dwarf stars. The methods used are discussed in the text and are the phase cycle count procedure (O−C), a direct comparison of seasonal period measurements and a combination of both these methods. The blank row in the PG 1159 category indicates there a number of other periods that have measured $\dot{P}$ values for this star.

Object	Type	Mode	$\dot{P}$ ($\mathrm{s\,s^{-1}}$)	Method	Reference
PG 1159	DOV	516 s	$-1.21 \pm 0.06 \times 10^{-11}$	O−C	Winget et al. (1985)
		516 s	$-2.49 \pm 0.06 \times 10^{-11}$	O−C	Winget et al. (1991)
		516 s	$+13.0 \pm 2.6 \times 10^{-11}$	direct	Costa et al. (1999)
		516 s	$+13.07 \pm 0.03 \times 10^{-11}$	O−C	Costa et al. (1999)
		539 s	$-0.82 \pm 0.04 \times 10^{-11}$	O−C	Costa et al. (1995)
		516 s	$+13.146 \pm 0.003 \times 10^{-11}$	combn	Costa & Kepler (2008)
		539 s	$-0.34 \pm 0.02 \times 10^{-11}$	combn	Costa & Kepler (2008)
		...	...	...	Costa & Kepler (2008)
GD 358	DBV	...	(unstable modes)	...	Kepler et al. (2003)
EC 20058	DBV	257 s	(promising)	O−C	Sullivan et al. (2008)
		281 s	(promising)	O−C	Sullivan et al. (2008)
G 117−B15A	DAV	215 s	$+3.6 \pm 0.8 \times 10^{-15}$	O−C	Kepler et al. (2005)
ZZ Ceti	DAV	213 s	$\leq 5.5 \pm 1.9 \times 10^{-15}$	O−C	Mukadam et al. (2003)
L 19−2	DAV	192 s	$\leq 3.0 \times 10^{-14}$	O−C	O'Donoghue ... (1987)

undergoing a period increase in agreement with model predictions. The superior quality of the WET data sets and the relatively large period change rate in PG 1159 allowed them to directly detect a $\dot{P}$ of $+13 \pm 3 \times 10^{-11}\ \mathrm{s\,s^{-1}}$ for the 516 s mode by calculating the best periods for four seasonal data sets over an interval of 10 years; they then improved the precision for this correct value to yield $\dot{P} = +13.07 \pm 0.03 \times 10^{-11}$ by employing the O−C method. But there still remained some mystery as this new measured value was an order of magnitude larger than the values inferred from realistic models.

It is something of a cautionary tale to understand the key factors in this change of sign for the measured $\dot{P}$ value of the 519 s mode in PG 1159. Unquestionably, these are difficult measurements. The earlier measurements of Winget et al. employed the sensitive O−C method, in which the phases of the selected frequency in each photometric data set are established by least squares fitting and then an accurate period is used to construct an 'observed-minus-calculated' time of maxima for each data set as a function of the number of elapsed periods or cycles. If this plot is linear then a correction to the starting period can be obtained, but if there is some curvature then this indicates a non zero $\dot{P}$ value. The problem with this method is that any cycle-count errors due to missing data between the photometry sets (and a possible large number of elapsed cycles) leads to erroneous conclusions.

Even though PG 1159 is a relatively stable pulsator, it is also a rich multi-periodic pulsator with many different frequencies present in the light curve – nearly 200 at the last count (Costa et al. 2008). Underestimating the uncertainties in establishing the times of maxima (ie the phases) for the different data sets appears to underly the identification of an erroneous $\dot{P}$ in the earlier work. The most recent work on this pulsator (Costa & Kepler 2008) has yielded $\dot{P}$ values for no less than 27 of the periods and with 11 of them having period decreases. It is clear that this object is rather complex, and mode trapping is definitely required to understand the variation in the sign of the $\dot{P}$ values among the modes. The order of magnitude discrepancies between the measured $\dot{P}$s and predicted ones also need to be explained.

At face value, one would expect the DBVs to be the next class of pulsating white dwarf to yield a $\dot{P}$ measurement. The class prototype, GD 358, is the brightest one of these objects and has been studied extensively, including via several WET runs. However it is a large amplitude and relatively unstable pulsator (Kepler et al. 2003), which is consistent with it being near the red edge of the DBV instability strip: it has proven to be an unsuitable candidate for evolutionary $\dot{P}$ measurements. On the other hand, EC 20058 is a small amplitude stable pulsator (and it currently defines the blue edge of the instability strip); it looks to be a much better prospect for an evolutionary measurement, as discussed in the last section.

The DAV objects have to date provided one $\dot{P}$ measurement that is in agreement with model predictions and one upper limit that is close to these values (see Table 2). Quality time-series photometry over a 30-year baseline for G 117−B15A has enabled Kepler et al. (2005) to determine a value of $\dot{P} = +3.6 \pm 0.8 \times 10^{-15}$ for this object, which is a very fine achievement. A less comprehensive data set has been used by Mukadam et al. (2003) to establish the upper limit for the object ZZ Ceti given in Table 2. A less impressive upper limit for the DAV L 19−2 has been determined by O'Donoghue and Warner (1987), but note that this author has archival data that should significantly improve this limit.

For both the DBV and DAV classes the expected $\dot{P}$ values for the periods are too small to usefully employ the direct measurement of seasonal periods to detect a change in their values or even provide a check on other methods. One must rely on the O−C method and be very vigilant for possible cycle count errors. In view of the PG 1159 saga described above, extreme caution is warranted. In some ways, establishing an upper limit that is consistent with model predictions is an easier task, and this has been the case for the DAVs, until the recent positive result for G 117−B15A.

4. Neutrino cooling of white dwarfs

Neutrino emission from white dwarfs cores has been included in relevant models since the 1960s following earlier developments in weak interaction physics. Its effect largely drives the thermal behaviour of the hot stages of the white dwarf cooling sequence, including effective temperatures down to the DBV instability strip for white dwarfs with typical masses (Winget et al. 2004).

Although there are a number of possible mechanisms, such as plasmon neutrinos, bremsstrahlung neutrinos and photoneutrinos, the underlying processes have a common feature: they all involve the transformation of a virtual electron-positron pair into a neutrino-antineutrino pair, rather than recombination back into a photon. Since this transformation is mediated by the weak interaction, it is unlikely in the extreme: in round figures, one expects to see one neutrino-antineutrino pair in every $\sim 10^{19}$ events. However, in the regime of interest in the long-lived gravitationally bound hot dense plasmas at the core of a white dwarf, this process is nevertheless predicted to be significant and lead to a one-way leakage of energy from the white dwarf that adds to the emission of photons from the atmosphere.

The key steps are as follows. Electron-positron pairs are produced from the thermally generated photons. But these must be virtual particles as there is insufficient energy in the $\sim$ keV photons (even in the high-energy thermal tail) to provide the 1.02 MeV rest mass energy of the pair. Overwhelmingly, these pairs annihilate back into a photon within the constraints of the energy-time uncertainty principle; but *'every now and then'* either an exchange of a charged $W^{\pm}$ vector boson or the creation of a neutral Z^0 leads to a transformation into a *real* neutrino-antineutrino pair (eg Misiaszek et al. 2006). Feynman diagrams depicting the relevant processes are provided in Figure 2.

Since the rest mass-energy of the final state neutrinos is significantly less than the typical photon energy (not long ago it was thought to be zero), satisfying energy conservation is not a

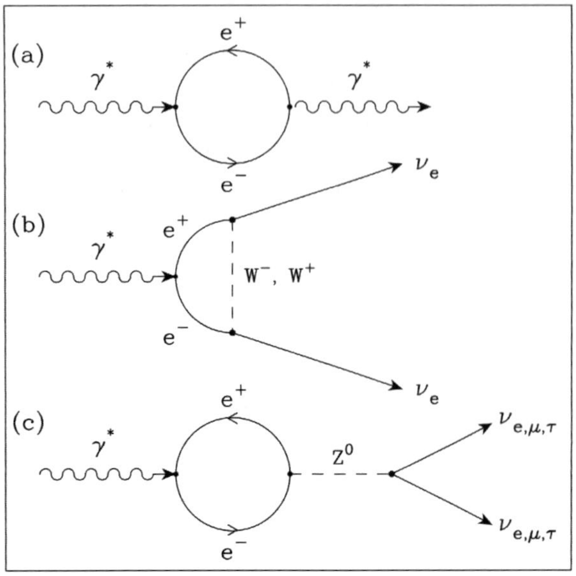

Figure 2. Feynman diagrams representing the three relevant plasmon transformation processes in the hot plasmas in the cores of white dwarfs. Schematic (a) represents the creation and subsequent annihilation of virtual e^+e^- pairs from plasmons with insufficient energy to create real particles. Schematics (b) and (c) represent the possible processes involving the weak force that lead to the creation of real $\nu\bar{\nu}$ pairs from these virtual e^+e^- pairs. Process (b) involves an exchange of a charged W^+ or W^- boson, while (c) features the neutral Z^0 boson. Either way, these latter transformations result in a one-way leakage of energetic neutrinos from the core.

problem. However, overall momentum conservation must also be satisfied and interaction with another external body is required: in the plasmon process, coupling to the collective motions of the plasma performs this task. And it is the plasmon process that dominates neutrino production in white dwarf cores (Kantor & Gusakov 2007, Itoh et al. 1996).

5. The potential for EC 20058

As mentioned previously, EC 20058 has proven to be a very stable low amplitude pulsator. The two dominant modes (257 s and 281 s) look to be the best candidates for measuring $\dot{P}$ values, and their frequencies are resolved in several hours of quality time-series photometry. However, the 281 s periodicity has two adjacent satellite frequencies separated from the main peak by 70 μHz and 81 μHz respectively (Sullivan et al. 2008), so longer runs are required to properly isolate this frequency. Hence the 257 s mode offers the best possibilities.

In addition to single-site data obtained from Mt John in NZ over a number of seasons, the Clay 6.5 m Magellan telescope has been used (Sullivan et al. 2007) to obtain both time-series photometry (July 2003) and time-series spectroscopy (July 2004). A DFT for the Magellan photometry is given in Figure 1.

Figure 3 demonstrates in an alternative way to the DFT plots that EC 20058 exhibits little sign of nonlinearities in the light curve variations. Both the 2003 Magellan photometry and July 2004 Mt John photometry have been folded at the two periods corresponding to the dominant modes in order to obtain pulse shapes for these frequencies. A sine function has then been least squares fitted to each data set. It is clear that there is no evidence of a significant departure from a sinusoidal pulse shape in all four cases, in agreement with the absence (except for some small peaks) of harmonic content in the WET data set (Sullivan et al. 2008).

Model fitting of the 2004 integrated Magellan spectra by Detlev Koester yielded an effective surface temperature for EC 20058 nearly as high as 31 kK with no assumed H component in the atmosphere, and a value of $\sim$ 29 kK if the atmosphere contains a small (undetectable) H component (Sullivan et al. 2007). Hence, this object is quite a hot DBV.

We also have a firm idea of the mass of this star. Both the $\log g$ values obtained from the spectral fitting and the asteroseismic analysis presented in Sullivan et al. (2008) are consistent with a mass for EC 20058 of about 0.55 $M_\odot$. So, the combination of a high effective temperature and a typical white dwarf mass means that the evolutionary period change in this pulsator should

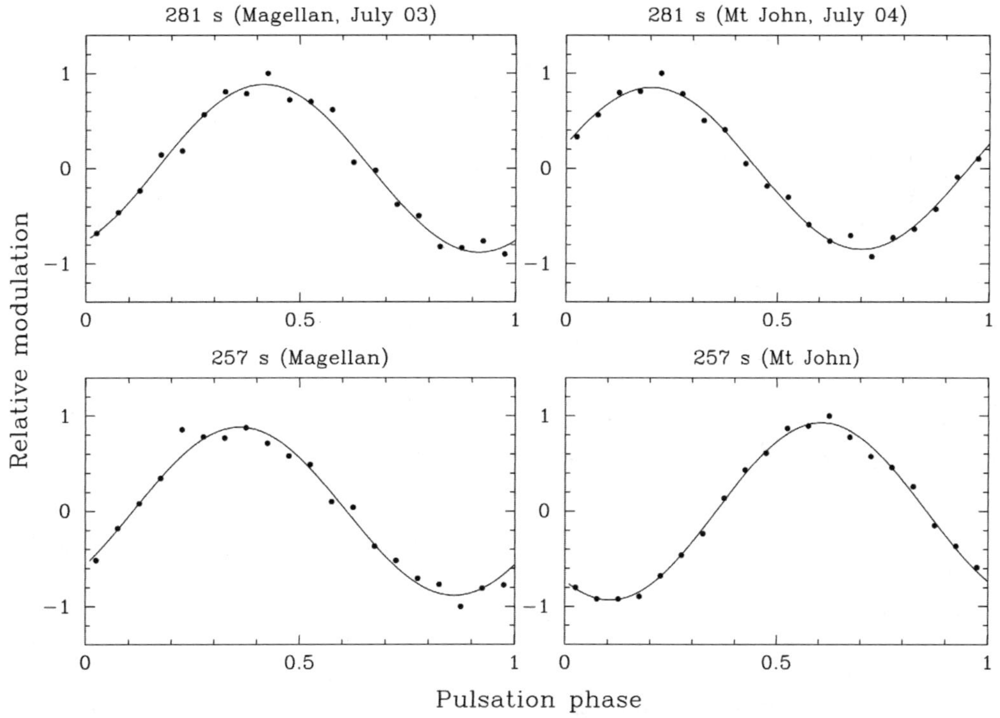

Figure 3. Folded light curves and sinusoidal fits for the two dominant modes in EC 20058 derived from two consecutive nights of Magellan photometry in July 2003 and a good quality run of 6 nights at Mt John Observatory in July 2004. All four folded light curves are consistent with sinusoidal pulse shapes.

be driven primarily by the core neutrino flux, as this energy loss mechanism is expected to exceed that due to the surface photon flux by a factor of two or so. Given the lack of complications arising from model-dependent contraction rates with this object, a measurement of a $\dot{P}$ can be used to not only directly check the predicted neutrino emission rates, but also verify the actual existence of these weak interaction processes in the untested low-energy regime (Winget et al. 2004).

It is true that a positive $\dot{P}$ was eventually detected for the dominant 516 s mode in the DOV PG 1159, and that neutrino cooling undoubtedly plays a significant role here. However, modelling uncertainties, including the effects of contraction, the measurement of both period increases *and decreases* for different modes and $\dot{P}$ values an order of magnitude more than 'best' model values, make this theory-observation confrontation somewhat complicated and uncertain.

Figure 4 schematically illustrates the extended photometric data sets available for the EC 20058 $\dot{P}$ measurements. Most of the photometry has been obtained from Mt John in NZ, but we also have seasonal data obtained from Chile over the last few years (Provencal 2008). We aim to soon converge on a $\dot{P}$ conclusion, with particular emphasis on the epoch between 2004 and 2008 underwriting the measurement.

It is important to note that due to the expected $\dot{P}$ being $\sim 10^{-13}\,\mathrm{ss}^{-1}$ the direct method of searching for a period change that has been so important in the case of PG 1159 is not useful here, as it is not sensitive enough. For the DOV objects the expected period change of $\sim 10^{-11}\,\mathrm{ss}^{-1}$ translates into a $|\Delta P| \approx 3 \times 10^{-3}$ s or 3 msec in 10 years. Costa et al. (1999), using quality data sets, were actually able to measure for the interval 1983 to 1993, an unambiguous change of $\sim$

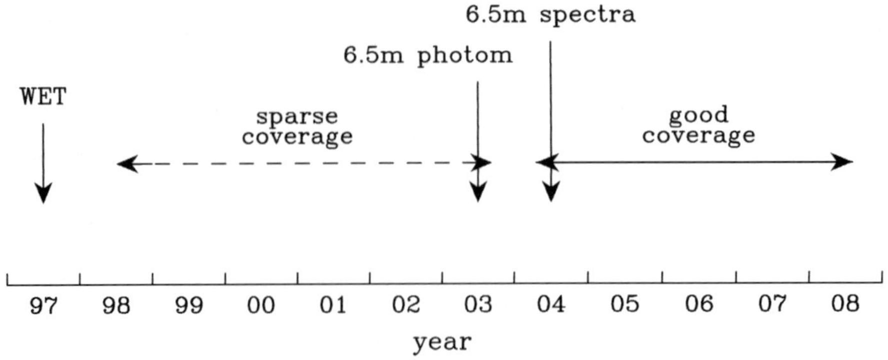

Figure 4. A schematic representation showing the time span for the currently available EC 20058 photometry data sets that are suitable for estimating $\dot{\mathrm{P}}$ values.

40 msec in the period of the 516 s mode of PG 1159. The theoretical equivalent figure for the DBV EC 20058 over a 10-year interval is $|\Delta P| \sim 3 \times 10^{-5}$ s. Given the quality of the available data sets and the real uncertainties in the seasonal period determinations, there is no hope of measuring such a change directly.

We will need to identify a correct value for $\dot{P}$ on the basis of the O−C method alone. An important part of this process is establishing realistic uncertainties for the phase fitting. We are employing procedures similar to the Monte Carlo methods described by Costa et al. (1999) and Costa and Kepler (200) to achieve this.

References

Costa J E S and Kepler S O (2000) *Baltic Astron.* **9** 415
Costa J E S and Kepler S O 2008 *A&A* **489** 1225
Costa J E S Kepler S O and Winget D E 1995 *Baltic Astron.* **4** 334
Costa J E S Kepler S O and Winget D E 1999 *ApJ* **522** 973
Costa J E S et al. 2008 *A&A* **477** 627
Itoh N Hayashi H Nishikawa A Kohyama Y 1996 *ApJS* **102** 411
Kantor E M and Gusakov M E 2007 *MNRAS* 381 1702
Kawaler S D and Bradley P A 1994 *ApJ* **427** 415
Kepler S O et al. 2003 *A&A* **401** 639
Koen C O'Donoghue D Stobie R S Kilkenny D and Ashley R 1995 *MNRAS* **277** 913
Mukadam A S et al. 2003 *ApJ* **594** 961
O'Donoghue D and Warner B 1987 *MNRAS* **228** 949
Provencal J H (private communication)
Sullivan D J 2005 *ASP Conf. Ser.* **334** 495
Sullivan D J et al. 2007 *ASP Conf. Ser.* **372** 629
Sullivan D J et al. 2008 *MNRAS* **556**
Winget D E et al. 1991 *ApJ* **378** 326
Winget D E Hansen and Van Horn 1983 *Nature* **303** 781
Winget D E and Kepler S O 2008 ARA&A **46** 157
Winget D E Kepler S O Robinson E L Nather R E and O'Donoghue D 1985 *ApJ* **292** 606
Winget D E Sullivan D J Metcalfe T S Kawaler S D and Montgomery M H 2004 *ApJ* **602** L109

Journal of Physics: Conference Series **172** (2009) 012071 doi:10.1088/1742-6596/172/1/012071

The white dwarf in dwarf nova SDSS J080434.20+510349.2: Entering the instability strip?

E Pavlenko[1,2]

[1]Crimean astrophysical observatory, Crimea 98409, Ukraine
[2]Tavrida National Vernadsky University, Crimea 95007, Ukraine

E-mail: pavlenko@crao.crimea.ua

Abstract. SDSS J080434.20+510349.2 is a WZ Sge type binary that displayed a rare outburst in 2006 (Pavlenko et al. 2007). During the long-lasting tail of the late stage of the outburst, the binary showed a two-humped or four-humped profile of the orbital light modulation. The amplitude of the orbital light curve decreased while the mean brightness decreased; moreover, that occurred $\sim$ 10 times faster during the fast outburst decline with respect to the late quiet state of slow outburst fading. There were no white dwarf pulsations detected in this system, neither 1 - 1.5 months prior to the outburst, nor in 1.5 - 2 months after the 2006 outburst. However, strong non-radial pulsations with period 12.6 minutes and a mean amplitude of 0.05^m were first detected in the V band with the 2.6-m Shajn mirror telescope of the Crimean astrophysical observatory, $\sim$ 8 months after the outburst. The evolution of pulsations over two years, in 2006 - 2008, is considered. It is supposed that pulsations first appeared when the cooling white dwarf (after the outburst) entered the instability strip, although the possibility of temporary lack of pulsations at some occasions could not be excluded.

1. Introduction

Twelve cataclysmic variables (CVs) with orbital period close to the evolutional orbital period minimum, containing a pulsating white dwarf (WD), were known up to now (Mukadam et al. this volume). SDSS J080434.20+510349.2 (hereafter SDSS J0804) is the 13th system showing such pulsations (Pavlenko and Malanushenko 2009) which are generally believed to be due to non radial g-mode pulsations of the white dwarf (Warner and Robinson 1972). P. Szkody et al. reported that there are 4 dwarf novae (PQ And, GW Lib, V455 And, REJ 1255+26) containing an accreting WD pulsator that have undergone outbursts (paper available at www.narit.or.th/conference-prcsa2008/). SDSS J0804 (Pavlenko et al. 2007) is the 5th binary of such type. SDSS J0804 is a WZ Sge type star that displayed an outburst in 2006 with 11 rebrightenings. Superhumps, that have been observed both during the main outburst and the rebrightenings, were replaced later on by the orbital two-humped light variations (Zharikov et al. 2008; Pavlenko 2007) typical of the WZ Sge stars in quiescence. The phenomenon of a second series of rebrightenings or "mini-outbursts", with longer recurrence time and less amplitude, was found by Zharikov et al. (2008). In quiescence before outburst, the optical spectrum of SDSS J0804 displayed a blue continuum with broad absorption lines from a white dwarf, surrounding the double-peaked Balmer emission lines formed in the accretion disk (Szkody et al. 2006). These authors also noted the particular behavior of the binary during one night in 2005: quiet state - sudden increase of brightness (0.5^m in a few minutes) accompanied by increase of amplitude

Journal of Physics: Conference Series **172** (2009) 012071 doi:10.1088/1742-6596/172/1/012071

from 0.05^m to 0.2^m, that was never detected after the outburst (Zharikov et al. 2008). Despite J0804 was a promising CV to search for the WD non-radial g-mode pulsations, it did not show them before the 2006 outburst (Szkody et al. 2006). We had pointed attention cto the short-term light variations, with the most significant 12.6 min. oscillation in 2008, January 1, when its amplitude at some pulses was commensurable with orbital light modulation $\sim 0.05^m$ (Pavlenko and Malanushenko 2009). The results of searching for possible pulsations in the available data base of our observations, both before and after the 2006 outburst, in order to detect their first appearance, is presented here.

2. Observations

Taking into account the small amplitude of short-term variations of SDSS J0804, the analysis was restricted by using the most precise CCD data collected from 14 nights in 2006 - 2008, ensuring the accuracy of a single brightness measure to better than $0.01^m - 0.015^m$, time resolution of a few minutes for the multicolor observations and 20 - 30 seconds for observations in white light. They were obtained with the 2.6-m Shajn telescope of the Crimean astrophysical observatory (CrAO) using the CCD FLI 1001E in Johnson B, V, R bands or in white light reduced to R. All measurements were made with respect to the comparison stars pointed by Zharikov et al. (2008). The photometric standards in cluster M67 (Mendoza 1967) were observed as well and used to measure the V and R values of comparison stars. Other details of observations are given in Pavlenko and Malanushenko (2009). The time resolution of observations was different and varied from a few minutes for multicolor observations to 15 - 20 seconds for observations in white light.

3. Outburst and orbital light curves

The light curve of SDSS J0804 for 2006 - 2008 is given in Fig.1, including one night prior to the outburst. The dense part of points corresponds to rebrightenings. After the end of rebrightenings, the rapid brightness decline continued and has been observed during $\sim$ 3 - 4 weeks. Then the star began to fade very slowly. We first found SDSS J0804 in the 2006 outburst (Pavlenko et al. 2007) at the end of the outburst plateau (two days before the plateau end). Assuming the possible plateau duration to be 2 - 3 weeks one could expect the start of the plateau outburst at JD $\sim$ 2453780 - 2453787. c

The color-index V-R was estimated for several dates during the different stages of the outburst. For JD 2453856-2453858 (fast brightness fading after rebrightenings), mean V = 16.4^m, V-R= 0.17^m; for JD 2454479 (late fading stage), mean V = 17.26^m and V-R = 0.11^m.

The dramatic change of the nightly light curve profile has been detected during the fast stage of the outburst fading in the three weeks after rebrightenings were finished (see Fig. 2). To make sure that this modulation is really the orbital one, the data were folded on the orbital period using the ephemeris MinI = HJD 2453744.37 + 0.0590048(3)E (Pavlenko and Malanushenko 2009). One could see that the light curve is already a two-humped one, but the heights of the neighbor humps are 0.3^m and 0.05^m for (a), 0.3^m and 0.1^m for (b) and 0.2^m and 0.2^m for (c). Both minima are observed close to phases 0.0 and 0.5. While the slightly deeper minimum in (a) and (b) cases falls into phase 0.0, the deepest minimum in case (c) falls into phase 0.5. Note the sharp eclipse-like minimum at phase near 0.0 for (a) and (b) cases. The two-humped light curve and the perfect fitting to the ephemeris are convincing arguments for considering this modulation as the orbital one.

The orbital light curves for later stages of the outburst display small amplitude and show diversity of their two-humped profiles. In Fig. 3 typical examples of the orbital light curves obtained 195 - 775 days after the end of rebrightenings are shown. Some of them (a) display two near-equal sine-like waves per period, another - W UMa-like - nceighbor humps with round

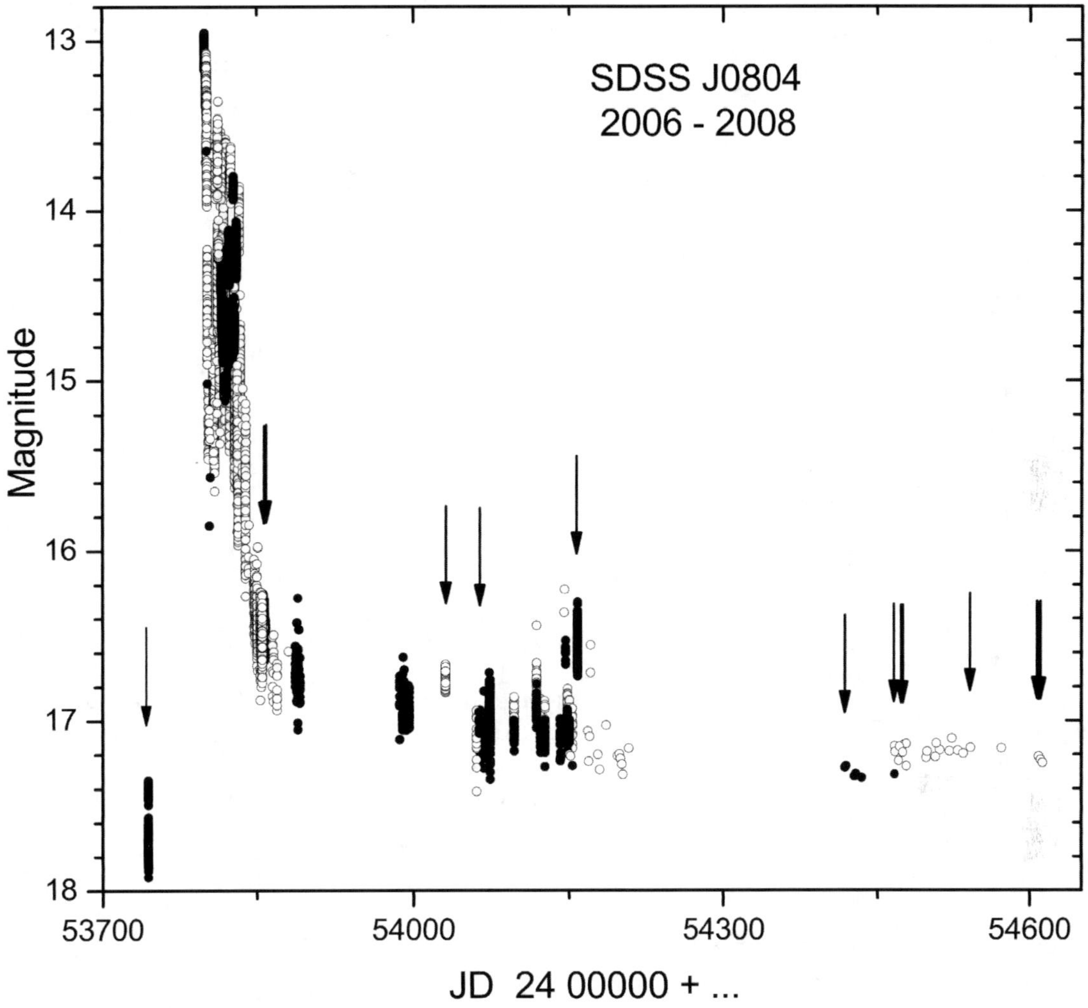

Figure 1. Light curve of the outburst of SDSS J0804. The V and R data are marked by filled and open circles respectively. Part of the data are averaged an expressed as one or two points per night. Arrows point to the data used for the subsequent periodogram analysis.

maxima and sharp minima (c). The curves (b) and (d) demonstrate rather four-humped than two-humped light curves (or splitting of the humps with unequal heights).

The dependence of the amplitude of the orbital light curves on the mean brightness is shown in Fig. 4. The amplitudes decrease when brightness decreases, but with a different rate. It is possible to pick out three different dependencies: the first one corresponds to the data shown in Fig. 2 (they are shown by the dashed line) - at that time the dependence is the sharpest; the second one refers to the data of the quiet long-lasting fading (dash-dotted line). Note that the amplitude decrease (expressed in intensities per unit of intensity) was $\sim$ 10 times faster during the fast outburst decline than during the slow fading. The third dependence corresponds to the data falling into the vicinity of the maxima of the two mini-outbursts (dotted line). These amplitudes themselves are slightly bigger than the amplitude apart the mini-outbursts.

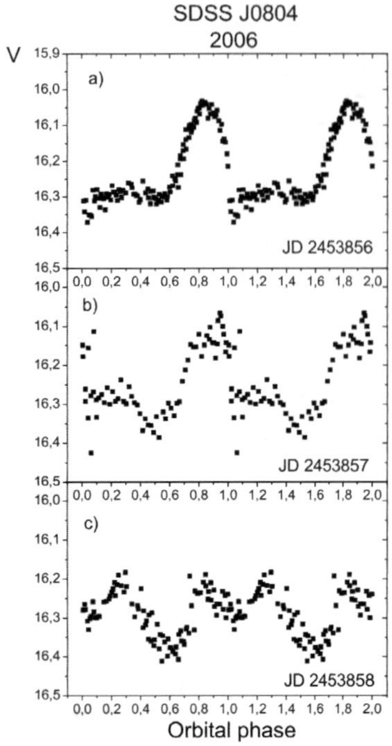

Figure 2. Example of data for April 30, May 01 and May 02, 2006 folded on the orbital period. For clarity data are plotted twice.

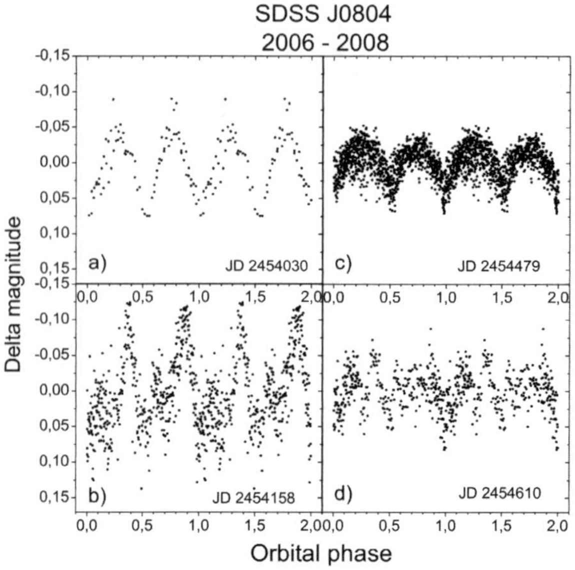

Figure 3. Example of data for October 21, 2006; February 26, 2007; January 13, 2008 and May 23, 2008 folded on the orbital period. For clarity data are plotted twice.

4. Analysis of the short-term variations

The white dwarf pulsations in CVs are difficult to search for, because these variations could be contaminated by other multiperiodical and quasi-periodical signals connected with orbital and rotational motion, instability in the accretion stream and disk. The orbital light modulation profile itself also could vary from night to night as it was described above. The investigation of the short-term light variations is performed for three data sets: 1) before outburst, 2) during the fast outburst decline $\sim$ 3 weeks after the end of rebrightenings and 3) during the long tail of the outburst.

The light curve for 2006, January 8 (1 - 1.5 monthsc before outburst) is shown in Fig. 5 (a). One could see the jump-like brightness decrease from $V = 17.4^m$ to $V = 17.7^m$, which occured during $\sim$ 9 minutes. A similar behavior had been observed by Szkody et al. (2006) also before outburst at JD 2453380, but in the "opposite direction": they detected a sharp brightness increase of 0.5^m. The Fourier transform (FT) for the initial data, corresponding to the lower linear part of the light curve, has been calculated (see Fig.5 (b)). The most significant peaks point to one quarter and one half of the orbital period. After the orbital wave subtraction, the FT was calculated again for residuals. The ISDA package was used for calculations (Pelt 1992). The amplitude periodogram is shown in Fig. 5 (c), where the amplitudes are normalized according to Pelt (1992). No significant peaks were seen on that periodogram (Fig. 5 (c)).

The most dense data from the second data set (JD 2453856) were selected for the analysis. The FT for original data is presented in Fig. 6 (a), where the significant peaks point to the orbital period and its twice value. FT for the data after orbital period is removed is shown in

Journal of Physics: Conference Series **172** (2009) 012071 doi:10.1088/1742-6596/172/1/012071

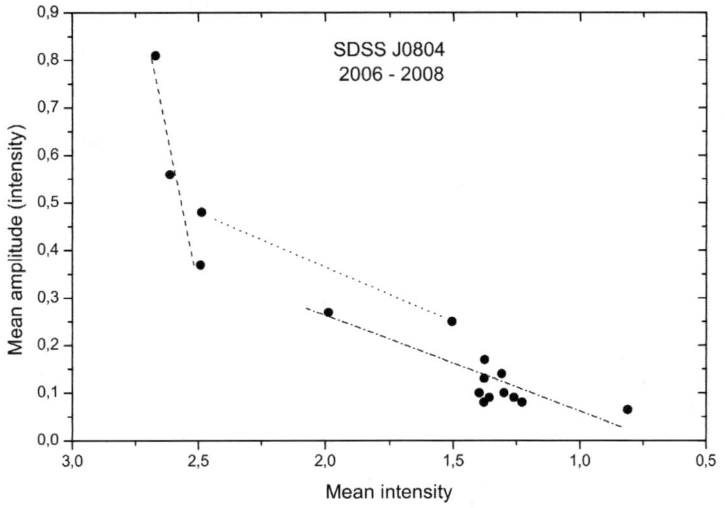

Figure 4. Dependence of the orbital amplitude on mean brightness expressed as intensity (arbitrary values). Dashed line is drawn through the data of April 30, May 01 and May 02, corresponding to the rapid outburst decline after the end of the rebrightenings. Dash-dotted line is drawn through the data belonging to the "long tail" of the outburst and before one (last point), and dotted line is drawn through the data from two mini outbursts.

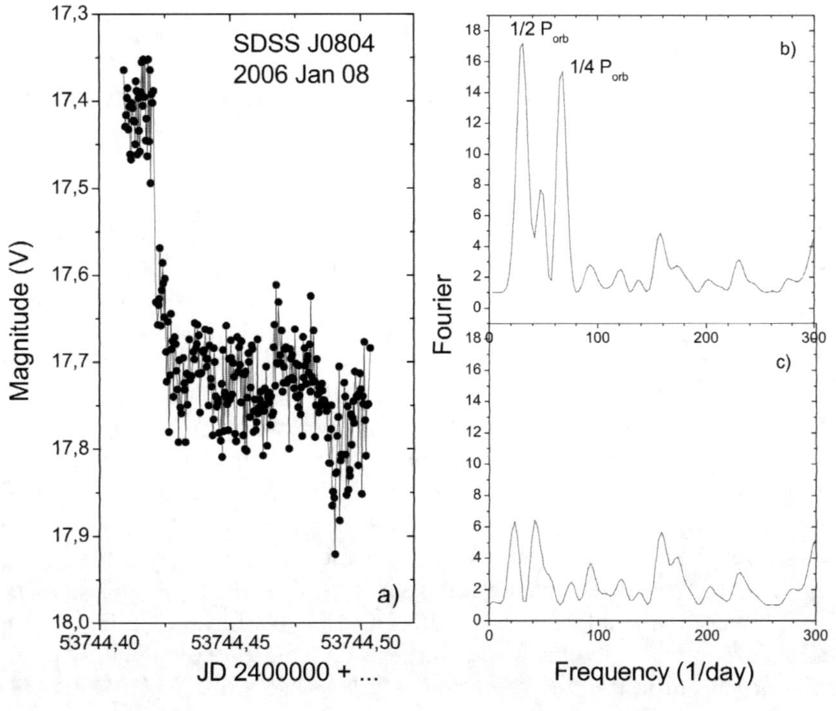

Figure 5. a) The light curve for 2006, Jan 08; b) Fourier transform for the initial data; c) Same as b) but with orbital period removed.

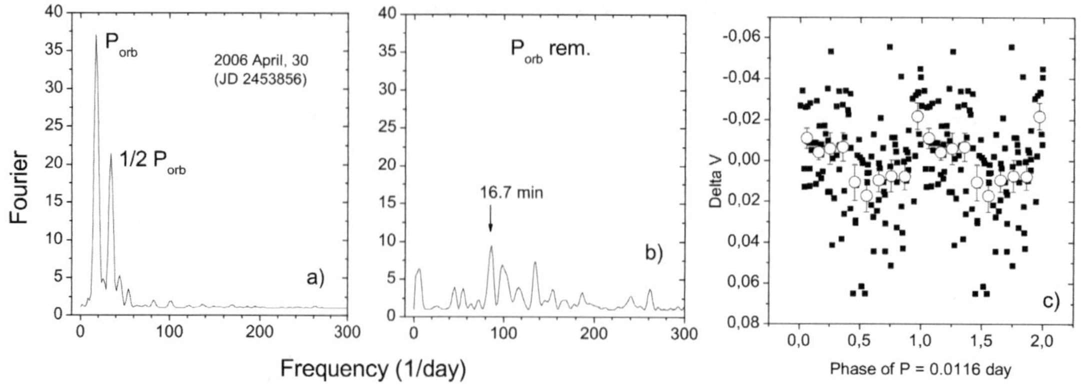

Figure 6. The Fourier transform for the initial data of 2006, April 30 (a), those for the data after orbital period is removed (b) and data folded on the best period for residual spectrum 0.01156 day (16.7 min). For clarity data are plotted twice.

Fig. 6 (b). There are a few peaks of low significance, the most prominent one pointing to the frequency 86 day^{-1} (16.7 min.). This corresponds to the light variations with a mean amplitude of 0.025^m (see Fig. 6 (c)). In Fig. 7 (a - k) the original light curves for the data from the third set of observations are shown. For each data the FT after orbital period subtraction was calculated, and the corresponding periodograms are shown in Fig. 7 (l - u). One could see that every periodogram (with exception of (m)) shows a sequence of significant peaks. These series concentrate within 40 - 150 day^{-1}. The most stable pulsation corresponds to 114 day^{-1} (12.6 min.) and its twice value 57 day^{-1}. The peak at 68 day^{-1} could be caused by the four-humped structure of the orbital light curve profile, because it coincides just with 1/4 P_{orb}. The less stable peak could be seen at frequency near 74 day^{-1} and its twice value 148 day^{-1}. Many of peaks are probably not connected with WD pulsations.

5. Discussion

The orbital light curves of SDSS J0804 had typically a two-humped profile, often displaying the splitting of one or both humps. The color-index V-R continued to decrease with time during the long brightness decay, indicating the decrease of the accretion disk contribution to the total light and, so, the increase of the white dwarf contribution. The source J0804 is localized in the Galaxy in a region of low interstellar extinction; its E(V-R) could be calculated following Schlegel et al. (1998), and it is roughly E(V-R) = $\sim 0.037^m$, so the corrected color index is $(V - R)_0 = 0.073^m$. That corresponds to a black-body temperature $\sim$ 15000° K with accuracy of a few of thousands K. Taking into account that $\sim$ 2 years after the outburst the disk could also still contribute to the total light, one could expect that the temperature of the white dwarf was hotter than 15000° K.

The dependence of the orbital amplitude on the mean brightness could be caused by the decreasing brightness of the bright tail of the accretion disk, due to the decrease of the mass transfer rate over the source, that occured $\sim$ 10 times faster during the fast outburst decline, with respect to the late quiet state of slow outburst fading. Such dependence is obviously caused by the decreasing of the initially more enhanced mass transfer rate; decreasing itself was $\sim$ 10 times faster during the fast outburst decline (after the end of rebrightenings) than during the late stage of the outburst.

The variable and sometimes four-humped orbital modulation points to a complex and variable

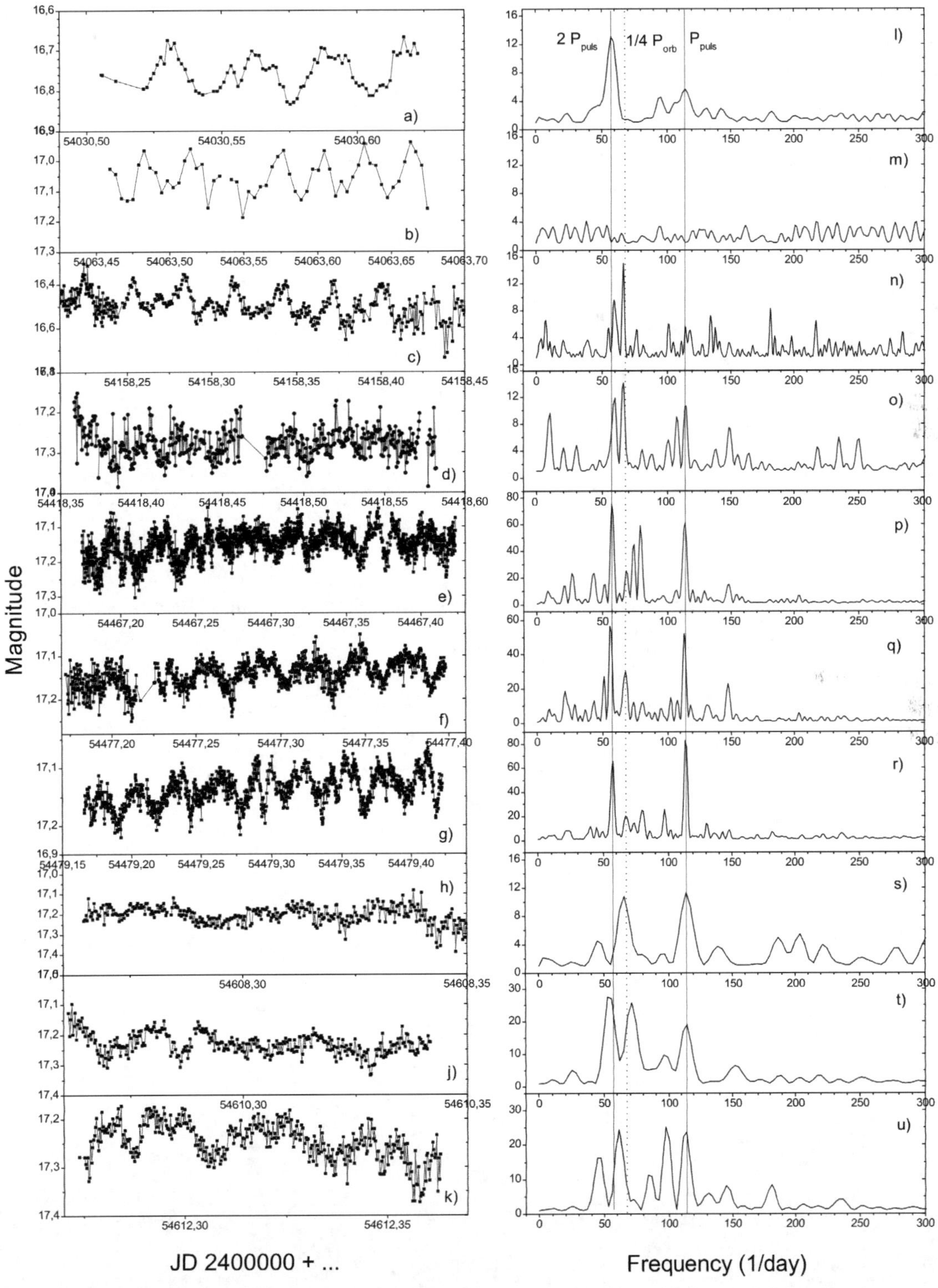

Figure 7. The light curves (a - k) and corresponding Fourier transforms (l - u) for the observations of SDSS J0804 in 2006 - 2008. The solid lines are drawn at the frequencies 57 and 114 day^{-1}, while the dotted one at 68 day^{-1}.

morphology of the accretion disk.

We did not find any significant pulsation during the period 1 - 1.5 months before the outburst. The first detection of the most stable 12.6-min. pulsations was $\sim$ 8 months after the expected start of the outburst. It is not clear whether the lack of pulsations of JD 2454063 was due to their temporary disappearance, or was caused by insufficient data statistics. It is already known that the pulsators in dwarf novae could stop their pulsations by some reason, contrary to what occurs in the ZZ Ceti stars (Southworth et al. 2008). So, the lack of pulsations in SDSS J0804 in two occasions before outburst at JD 2453384 (Szkody et al. 2006) and at JD 2453856 could not be used to claim that before the outburst the white dwarf never pulsated. However it is possible to suggest that compressional heating during the outburst and further fast cooling entered the white dwarf of SDSS J0804 into the instability strip.

It is also impossible to be confident on the 16.7 min. periodicity of JD 2453856 as that of the white dwarf pulsations, because the periodogram is different from those shown in Fig. 7.

Acknowledgments

This work was partially supported by the grant F 25.2/139 of FRSF and CosmoMicrophysics program 5-20 of the Ukrainian National academy of science. I am also grateful to the SOC and LOC for the hospitality during the Conference.

References

Mendoza E E 1967 *BOOT* **4** 149

Pavlenko E P et al. 2007 *15th European Workshop On White Dwarfs* **372** Eds Ralf Napiwotzki and Matthew R Burleigh. (San Francisko: ASP Conference Series) 511

Pavlenko E P 2007 *Odessa Astronomical Publications* **20** 168

Pavlenko E P and Malanushenko V P 2009 *Kinematics and Physics of Celestial Bodies* **25** 72

Pelt Ja. 1992 *Irregularly Spaced Data Analysis, User Manual and Program Package* (Helsinki: Helsinki University) 267

Schlegel D I Finkbeiner D P and Davis M 1998 *ApJ* **500** 525

Southworth J Townsley D M and Gänsicke B T 2008 *MNRAS* **388** 709

Szkody P Henden A Agueros M et al. 2006 *AJ* **131** 973

Warner B and Robinson E L 1972 *Nature* **239** 2

Zharikov S V et al. 2008 *A&A* **486** 505

Journal of Physics: Conference Series **172** (2009) 012072 doi:10.1088/1742-6596/172/1/012072

Asteroseismic probing of internal rotation in hot B subdwarf stars: Testing spin-orbit synchronism in two close binary systems

V Van Grootel[1,2], S Charpinet[1], G Fontaine[2], P Brassard[2], and D Reese[3]

[1]Laboratoire d'Astrophysique de Toulouse-Tarbes, Université de Toulouse, CNRS, 14 Av. E. Belin, F-31400 Toulouse, France

[2]Département de Physique, Université de Montréal, C.P. 6128, Succ. Centre-Ville, Montréal, QC H3C 3J7, Canada

[3]Department of Applied Mathematics, University of Sheffield, Hounsfield Road, Sheffield, S3 7RH, UK

E-mail: valerie.vangrootel@ast.obs-mip.fr; scharpin@ast.obs-mip.fr; fontaine@astro.umontreal.ca; brassard@astro.umontreal.ca; D.Reese@sheffield.ac.uk

Abstract. We present internal rotation profiles derived from asteroseismology for the hot pulsating B subdwarf stars PG 1336-018 and Feige 48. These two pulsators are primaries of close binary systems of known orbital period and, therefore, provide laboratories to test, for the first time, spin-orbit synchronization as a function of depth. We show that PG 1336-018 and Feige 48 clearly rotate as solid bodies with periods equal to their orbital periods from the surface down to at least $\sim$ 0.5 and $\sim$ 0.3 their radius, respectively. Deep tidal locking has therefore developed within the relatively short lifetime of these stars ($\sim 10^8$ yr).

1. Introduction

Subdwarf B (sdB) stars are hot ($T_{\rm eff}$ = 20,000−40,000 K) and compact (log g = 5.2 − 6.2) evolved Extreme Horizontal Branch stars (EHB). They are composed of convective helium-burning cores, surrounded by radiative helium mantles and very thin radiative hydrogen-rich envelopes ($M_{\rm env} < 0.02\ M_\odot$, while the total mass is around 0.5 $M_\odot$). They live about 1.5×10^8 yr on the EHB and then evolve, after core-helium exhaustion, directly toward the white dwarf cooling sequence (Dorman et al. 1993). Subdwarf B stars host two groups of non-radial, multiperiodic pulsators. Rapid oscillations (80 − 600 s) in the so-called EC 14026 stars are usually identified to low-degree, low-order p-modes. The longer periods (0.75 − 2 h) in the PG 1716+426 stars are due to low-degree, mid-order g-modes. The presence of excited pulsation modes in both type of pulsators is caused by a classic κ-effect associated with an opacity bump due to partial ionization of heavy metals, especially iron, locally enhanced by radiative levitation at work in the envelope of these stars (Charpinet et al. 2001).

A significant fraction of sdB stars reside in close binaries, with orbital periods from hours to days and white dwarfs or M dwarf companions (e.g. Maxted et al. 2001, Green et al. 2001). The Feige 48 system is made of a short-period pulsating sdB star and an unseen companion (most likely a white dwarf), with an orbital period of 9.024 ± 0.072 h (O'Toole et al. 2004).

PG 1336−018 ($P_{\rm orb}$ = 2.42438 h; Kilkenny et al. 2000) is one of the very few *HW Vir*-type sdB + dwarf M close eclipsing binaries, and the only known – to date – that exhibits short-period pulsations for its sdB component.

Tidal forces in binary systems, among other long-term effects such as circularization and alignment of the rotation axis to the normal of the orbit, tend to synchronize the rotation of the two stars with the orbital motion. Theoretical frameworks on tidal interaction have been developed in the last decades essentially by Zahn (1975, 1977, and references therein), where turbulent dissipative processes and radiative damping are invoked. Another model based on large-scale hydrodynamical currents was proposed by Tassoul & Tassoul (1992, and references therein), but the question of its validity is still under debate given its free parameter dependence. The theoretical synchronization times can differ by orders of magnitude depending on the physical mechanism invoked, especially in the case of hot stars with radiative envelopes (such as sdB stars), where tidal forces are less efficient for synchronization. Anyway, the confrontation of theory with observations, through the traditional photometric or spectroscopic techniques, only deals with the surface layers. The synchronization level reached in the inner parts, by transport of the angular momentum from the surface (Goldreich & Nicholson 1989), is only accessible through asteroseismology. It therefore constitutes a unique opportunity to test the theory of stellar synchronization and angular momentum transport as a function of depth.

2. Inclusion of rotation in the forward modeling approach

The group of rapidly pulsating sdB stars has proved its potential for performing objective asteroseismic analyses (see Fontaine et al. 2008 for a recent review). The method implements to so-called forward modeling approach, built on the requirement of global optimization: theoretical pulsation spectra computed from sdB models must match *all* the observed periods simultaneaously. The goodness of the fit is evaluated through a merit function defined as

$$S^2 = \sum_{i=1}^{N_{\rm obs}} \Big(\frac{P^i_{\rm obs} - P^i_{\rm th}}{\sigma_i}\Big)^2 \tag{1}$$

where $N_{\rm obs}$ is the number of observed periodicities. The method performs a double-optimization procedure in order to find the minima of the merit function, which constitutes the potential asteroseismic solutions (see details in Charpinet et al. 2005b). The codes have been recently improved to incorporate the effect of the rotation of the star, which lifts the $(2l + 1)$- fold degeneracy of eigenfrequencies of a perfectly spherically symetric star (Van Grootel al. 2008). Assuming an internal rotation law $\Omega(r)$, the rotational multiplets are calculated, with the perturbative method to first order, by

$$\sigma_{klm} = \sigma_{kl} - m \int_0^R \Omega(r) K_{kl}(r) dr \quad ; \quad K_{kl} = \frac{\xi_r^2 - [l(l+1) - 1]\xi_h^2 - 2\xi_r\xi_h}{\int_0^R [\xi_r^2 + l(l+1)\xi_h^2]\rho r^2 dr} \rho r^2 \tag{2}$$

where K_{kl} is the rotational kernel. The optimal solution gives the structural parameters of the star ($T_{\rm eff}$, $\log g$, M_*, etc.) and, of utmost interest here, the internal dynamics $\Omega(r)$ as a function of the radius of the star.

The effects of higher orders on rotation have recently been evaluated for polytropic ($N = 3$) models of sdB stars, with a full (nonperturbative) treatment of stellar rotation developed by one of us (Reese et al. 2006). The main results show that these higher-order perturbation effects due to rotation and tidal deformation of the star cannot affect in a significant way the proposed asteroseismic solution, for Feige 48 as well as for PG 1336−018, at the present level of accuracy (further details are given in Charpinet et al. 2008).

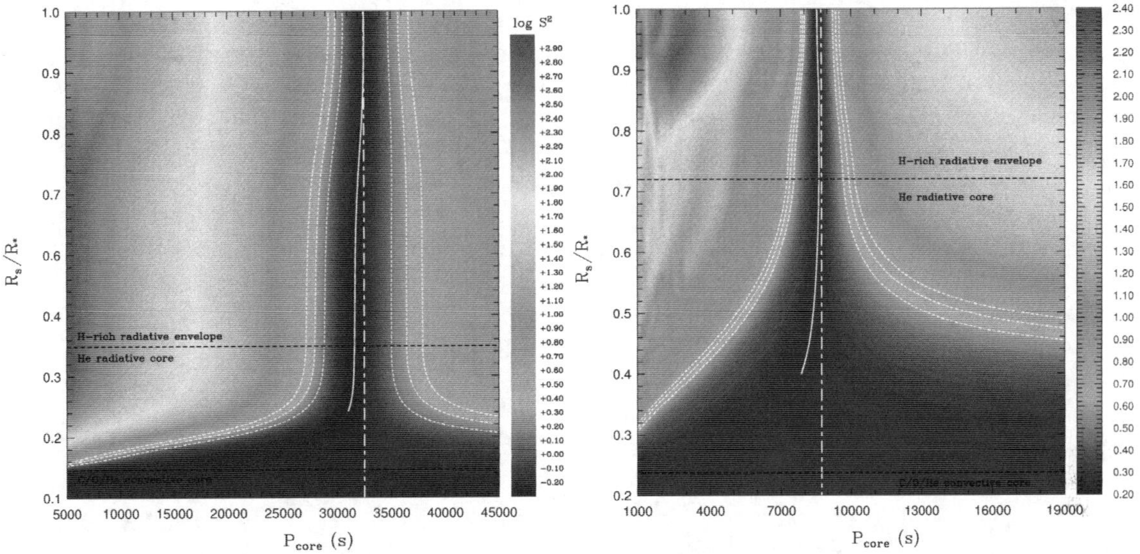

Figure 1. Seismic internal rotation profile of the sdB star in Feige 48 (left; $P_{\rm orb} = 9.024$ h) and PG 1336−018 (right; $P_{\rm orb} = 2.424$ h). This S^2 map (on a logarithmic scale) shows the quality of fit to the observed pulsation periods as a function of the parameters $P_{\rm core}$ and R_S/R_*. The continuous white line indicate the minima of the merit function. The white dot-dashed vertical lines indicates the orbital period. White dotted-line contours indicates the 1-σ, 2-σ et 3-σ confidence level relative to the best-fit solution. The transitions between the H-rich envelope and the radiative He core; and between the radiative He core and the convective C-O-He core are also indicated.

3. Asteroseismic tests of spin-orbit synchronism as a function of depth

The pulsating sdB star in the Feige 48 system exhibits nine pulsation periods in the range 343 – 383 s, as observed in white light photometry at the 3.6-m CFHT during six nights in June 1998 (Charpinet et al. 2005a). Three groups of modes can naturally be constructed from this pulsation spectrum, as components of rotational multiplets approximately evenly distributed in frequency with a mean spacing of about ~ 28 μHz. These nine pulsation periods are used in the optimization procedure in order to find the minima of the merit function, assuming several internal rotation laws. First, the hypothesis of a solid-body rotation ($\Omega(r) = \Omega =$ constant) is tested, and the star rotation period $P_{\rm rot} = 2\pi/\Omega$ therefore constitutes a free parameter in the optimization procedure. This leads to the determination of the structural parameters of the star and to a rotation period $P_{\rm rot} = 9.028 \pm 0.48$ h (details can be found in Van Grootel al. 2008), in excellent agreement with the orbital period of the system $P_{\rm orb} = 9.024 \pm 0.072$ h measured from RV variations. This result strongly suggests that the sdB star is tidally locked in the Feige 48 system.

To investigate further this question, the hypothesis of differential rotation is tested by dividing the star in two regions that rotate each independently as solid structures. The transition between the two layers is allowed to vary from 0.1 to 1.0 R_*, while the structural parameters are fixed to their optimal values. The surface rotation is fixed to the value of 32,500 s, and the core rotation period $P_{\rm core}$ can vary, as shown in Fig. 1 (left panel), from 5,000 to 45,000 s. All the merit functions are shown in Fig. 1, with a gray scale in logarithmic units. The optimal core rotation period $P_{\rm core}$ (continuous white line) is not significantly different from the surface rotation period (equal to the orbital period, vertical dot-dashed line) in most of the star: the sdB component in the Feige 48 system is tidally locked and rotates as a solid-body from the surface down to ~ 0.22

R_* at least. The dark-gray valley enlarges significantly under this limit, which translates the insensitivity of the pulsation modes to these deep regions. Nonetheless, a very fast core rotation can be rejected from Fig. 1, as it would lead to much poorer merit functions.

The same exercise is carried out with the 25 pulsation periods in the frequency spectrum of the sdB star in the PG 1336−018 system (Kilkenny et al. 2003). Details on the asteroseismic analysis is reported in Charpinet et al. (2008), and the internal rotation profile is shown in Fig. 1 (right panel). The sdB star is also tidally locked, down to $\sim 0.55\ R_*$ at least. Again, the dynamics of deeper regions (under the H-rich envelope much thinner in this case compared to the envelope of Feige 48), cannot be probed by the p-modes in action here.

4. Conclusion

We have demonstrated, for the first time by asteroseismic means, that spin-orbit synchronism is reached in most of two short-period pulsating sdB stars residing in close binary systems, namely Feige 48 and PG 1336−018. Both rotate as solid bodies with periods equal to their orbital periods from the surface down to at least $\sim$ 0.22 and 0.55 of their radius, respectively. The rotation of deeper layers cannot be inferred with the type of modes observed in short-period pulsating sdB stars. This observed synchronization as a function of depth, achieved within $\sim 1.5\times10^8$ yr (the lifetime of a sdB star on the EHB), could provide constraints to tidal evolution theories, concerning particularly the angular momentum transport from the surface to the center. In a next step, the more internal dynamics of sdB stars residing in binaries could be probed by the deep g-modes of long-period pulsating sdB stars.

Acknowledgments

This work made extensive use of the computing facilities offered by the Calcul en Midi-Pyrénées (CALMIP) project and by the Centre Informatique National de l'Enseignement Supérieur (CINES), France. Numerical experiments presented in this paper were also carried out using the Grid'5000 experimental testbed, an initiative from the French Ministry of Research through the ACI GRID incentive action, INRIA, CNRS and RENATER, and other contributing partners (see https://www.grid5000.fr). This work was also supported in part by the NSERC of Canada. G.F. gratefully acknowledges the contribution of the Canada Research Chair Program.

References

Charpinet S Fontaine G Brassard P 2001 *PASP* **113** 775
Charpinet S Fontaine G Brassard P et al. 2005a *A&A* **443** 251
Charpinet S Fontaine G Brassard P et al. 2005b *A&A* **437** 575
Charpinet S Van Grootel V Reese D et al. 2008 *A&A* **489** 377
Dorman B Rood R T & O'Connell R W 1993, *ApJ* **419** 596
Fontaine G Brassard P Charpinet S et al. 2008 *ASPC* **392** 231
Goldreich P & Nicholson P D 1989, *ApJ* **342** 1079
Green E M Liebert J & Saffer R A 2001 *ASPC* **226** 192
Kilkenny D Keuris S Marang F et al. 2000 *The Observatory* **120** 48
Kilkenny D Reed M D O'Donoghue D et al. 2003 *MNRAS* **345** 834
Maxted P Heber U Marsh T R & North R C 2001 *MNRAS* **326** 1391
O'Toole, S. J., Heber, U. & Benjamin, R. A. 2004, A&A, **422**, 1053
Reese D Lignières F & Rieutord M 2006 *A&A* **455** 621
Tassoul J-L & Tassoul M 1992 *ApJ* **395** 259
Van Grootel V Charpinet S Fontaine G & Brassard P 2008 *A&A* **483** 875
Zahn J-P 1975 *A&A* **41** 329
Zahn J-P 1977 *A&A* **57** 383

Doubling the Number of Pulsating DB White Dwarfs

Atsuko Nitta[1,2], S J Kleinman[1], J Krzenski[3], S O Kepler[4], T S Metcalfe[5], Anjum S Mukadam[6], F Mullally[7], R E Nather[8], D Sullivan[9], Susan E Thompson[10,11] and D E Winget[8]

[1]Gemini Observatory, 670 N A'ohoku Pl., Hilo, HI 96720
[2] Subaru Telescope, 650 N A'ohoku Pl., Hilo, HI 96720
[3] Mount Suhora Observatory, Cracow Pedagogical University, Cracow, Poland
[4] Inst de Fsica UFRGS Campus do Vale CP 15051 BR Porto Alegre, RS 91501-970 Brazil
[5] High Altitude Observatory, National Center for, Atmospheric Research, P.O. Box 3000, Boulder CO 80307 U.S.A.
[6] Dept. of Astronomy, University of Washington, 3910 15th Ave NE, Seattle WA 98195 U.S.A.
[7] Deptartment of Astrophysical Sciences, Princeton University, Princeton, NJ, 08544 U.S.A.
[8] Department of Astronomy, University of Texas at Austin, Austin, TX, U.S.A.
[9] School of Chemical & Physical Sciences, Victoria University of Wellington, Wellington, NZ
[1]0 Department of Physics and Astronomy, University of Delaware, Newark, DE
[1]1Delaware Asteroseismic Research Center, Mt. Cuba Observatory, Greenville, DE

E-mail: anitta@gemini.edu

Abstract. We are searching for new pulsating DB white dwarf stars (DBVs) based on the newly found white dwarf stars from the spectra obtained by the Sloan Digital Sky Survey. DBVs pulsate at hotter temperature ranges than their better known cousins, DAVs or ZZ Ceti stars. Since the evolution of white dwarf stars is characterized by cooling, asteroseismological studies of DBVs give us opportunities to study white dwarf structure at a different evolutionary stage than the DAVs. The hottest DBVs are thought to have neutrino luminosities exceeding their photon luminosities (Winget et al. 2004), a quantity measurable through asteroseismology. Therefore, they can also be used to study neutrino physics in the stellar interior. At the time of the meeting, we reported on the nine new DBVs, doubling the number of previously known DBVs. Here we report the new nine pulsators' lightcurves and power spectra.

1. Fourier Transform and the DBV instability stirp

The more detail description of this work have been recently published (Nitta et al. 2009). The table shows the list of objects we have observed at the time of the meeting. In the status section, we note which ones are new DBVs and which ones are observed not to vary with their amplitude limits. (1) means we only have one observation of the object. Beating of multiple modes and amplitude modulation can make a pulsator appear as a non-pulsator. Therefore, we try to observe each object at least two separate times. We also observe a new pulsator two different times to ensure we have found a real pulsator

This work is on-going and we are happy to report that since the time of the meeting, we have discovered three new DBVs, although two of them does not have confirmation observations yet. We also were granted observing time on Gemini telescopes to get high signal to noise spectra of the nine new DBVs reported here. So we hope to have better physical parameters in the near

Journal of Physics: Conference Series **172** (2009) 012073 doi:10.1088/1742-6596/172/1/012073

Table 1. Here is the list of DBs we observed by the time of the meeting. The top section of the table details the objects that showed variability during at least one observation. Separated by a double horizontal line, the second half of the table lists the objects for which we have not (yet?) seen significant variability. In the status section, we note new variable objects by "DBV". For objects in which we have not detected variability, we give the amplitude limit in mma in the status section. If we have only observed an object once, then we add a "(1)". Due to lack of observing time and a large number of candidates, we have yet been able to observe all DBV candidate objects, nor all these a second time. The physical parameters here come from fitting SDSS DR6 spectral data with a denser, but otherwise consistent, model grid than used in the DR4 WD catalog.

Object SDSS J	Plate	Fiber	MJD	g [mag]	Teff [K]	σ_{Teff} [K]	logg	σ_{logg}	Status
034153.03 − 054905.8	462	506	51909	18.25	25087	524	8.02	0.062	DBV
085202.44 + 213036.5	2280	604	53680	18.50	25846	6361	8.02	0.056	DBV
094749.40 + 015501.8	480	520	51989	19.95	23453	1659	8.13	0.192	DBV
104318.45 + 415412.5	1361	155	53047	18.95	26291	919	7.77	0.138	DBV(1)
122314.25 + 435009.1	1371	205	52821	18.98	23442	1069	7.84	0.127	DBV
125759.03 − 021313.3	338	436	51694	19.16	25820	1296	7.57	0.151	DBV
130516.51 + 405640.8	1458	21	53119	17.46	24080	414	8.14	0.056	DBV(1)
130742.43 + 622956.8	783	513	52325	18.83	23841	913	8.14	0.097	DBV(1)
140814.63 + 003838.9	302	490	51688	19.19	26073	1227	7.98	0.117	DBV
001529.74 + 010521.3	389	530	51795	18.94	34379	1079	7.96	0.163	8.20(1)
031609.12 − 062556.8	459	605	51924	19.97	24478	2520	7.96	0.222	17.0(1)
081904.19 + 354255.8	826	422	52295	18.22	22540	867	8.18	0.079	4.80(1)
085950.29 − 000339.6	469	49	51913	20.19	23729	2391	8.12	0.291	13.3(1)
090409.03 + 012740.9	470	442	51929	17.96	23183	533	7.95	0.062	4.28
090456.11 + 525029.8	552	547	51992	18.95	37584	953	7.99	0.091	10.1(1)
092200.97 + 000834.3	474	388	52000	18.56	22581	769	8.10	0.074	7.56(1)
095256.68 + 015407.6	481	513	51908	17.50	32920	323	8.16	0.041	4.84(1)
095455.11 + 440330.3	942	275	52703	18.18	20072	368	8.29	0.064	5.85(1)
095649.55 + 010812.4	481	20	51908	20.48	17125	1257	7.37	0.261	13.0(1)
101131.88 + 050729.3	574	331	52355	18.97	24301	984	7.71	0.115	8.98(1)
101502.95 + 464835.3	944	328	52614	18.61	23312	830	8.01	0.076	7.24(1)
105929.60 + 554039.2	908	317	52373	18.47	24742	571	8.17	0.101	8.46(1)
122241.27 − 003614.4	288	63	52000	18.10	24023	676	8.21	0.073	4.66(1)
131148.49 + 053847.6	850	522	52338	17.65	20249	268	8.30	0.041	11.7(1)
133215.93 + 640656.2	603	118	52056	18.41	21365	1694	7.99	0.097	9.73(1)
135610.32 − 002230.6	301	232	51641	19.38	18584	397	8.20	0.149	13.1(1)
141258.17 + 045602.2	583	432	52055	17.35	30343	329	7.97	0.038	2.88
231324.24 − 001636.9	381	72	51811	19.83	19588	1987	7.93	0.298	3.19
235322.16 + 002653.8	386	549	51788	19.71	25012	1800	8.15	0.203	13.0(1)

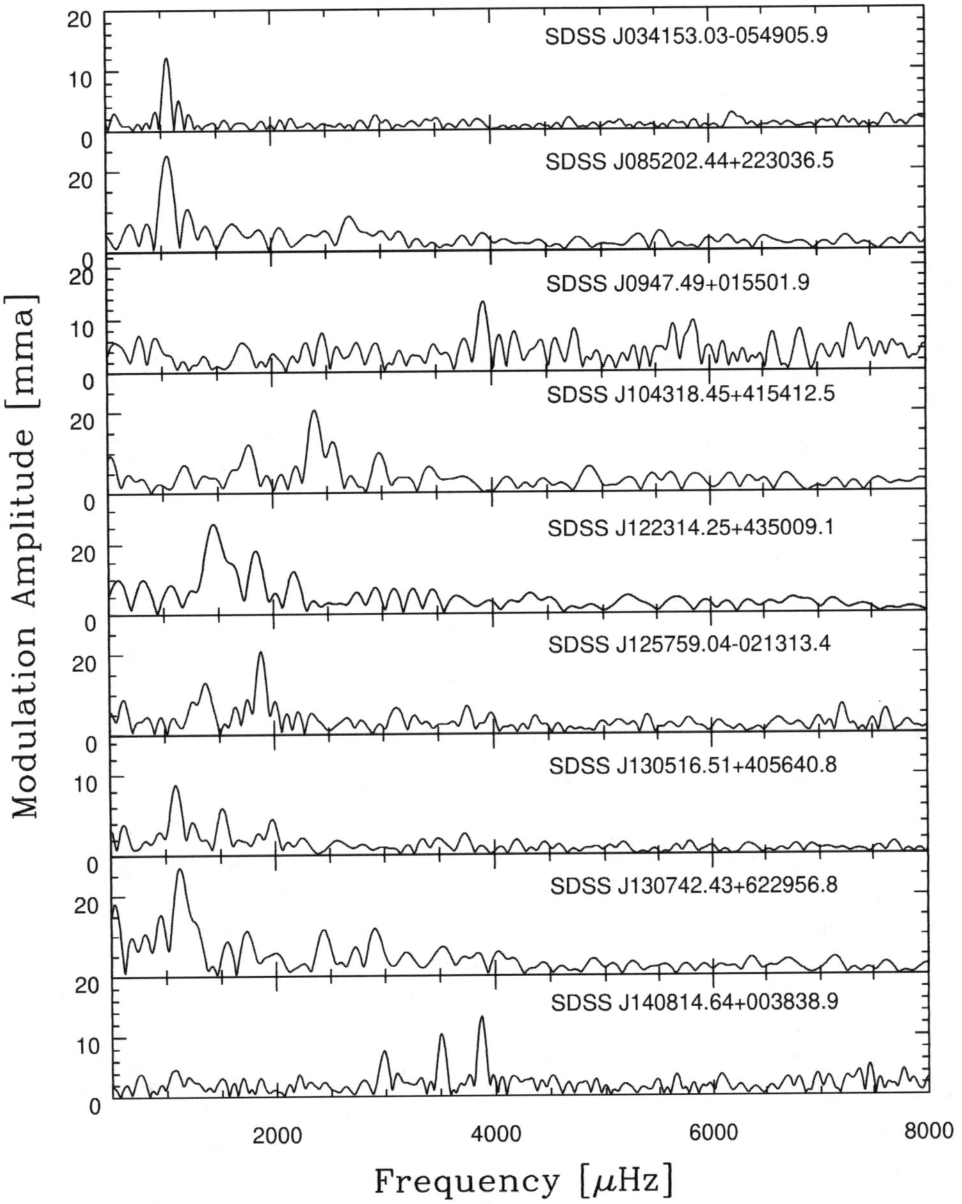

Figure 1. Fourier transform of the 9 new DBVs. SDSS J0947.49+015501.9's FT, perhaps, is not as visually convincing as other new DBVs shown here. The FT of the second observation of this target also shows the largest peak at a consistent frequency as the data shown here with similar significance.

feature. That will help the error bars in Figure 2 to reduce so we have better picture of the DBV instability strip.

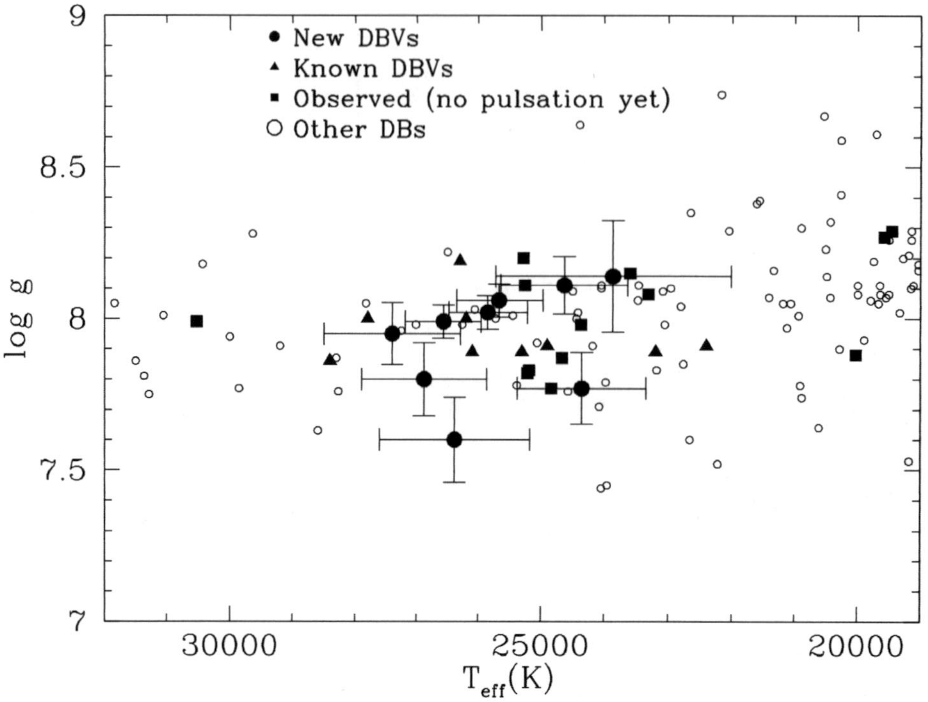

Figure 2. The effective temperatures and surface gravities of the SDSS DBs (our latest fits to the DR6 SDSS spectra) around the DBV instability strip, along with the previously known DBVs (with their physical parameters taken from Beauchamp et al. 1999). Circle dots show the new DBVs and triangles, the previously-known DBVs. The square show the non-pulsating DBVs we have observed and the hollow dots show the rest of the DBs in the WD catalog (Eisenstein, et al. 2004). Most of the DBs observed not to pulsate so far have such high amplitude limits that we cannot tell if they are truly non-pulsators or not. In order to better characterize the instability strip, we need better determinations of the physical parameters for some of the DBs (which requires better signal-to-noise spectra) and better variability amplitude limits (1mma or better). We have been granted observing time on Gemini to observe the nine new DBVs so we should have better estimates of their parameters in the near future. So far we have not found any pulsator hotter than EC20058 and hence the best chance of determining the neutrino production rates still lies with this star.

References

Beauchamp A Wesemael F Bergeron P Fontaine G Saffer R A Liebert J Brassard P 1999 *ApJ* **516** 887
Eisenstein Daniel J et al. 2006, *AJ* **132** 676
Nitta A et al. 2009 *ApJ* **690** 560
Winget D E Sullivan D J Metcalfe T S Kawaler S D Montgomery M H 2004 *ApJ* **602** 109

Journal of Physics: Conference Series **172** (2009) 012074 doi:10.1088/1742-6596/172/1/012074

Watching ZZ Ceti evolve

Anjum S Mukadam[1], Agnes Kim[2], Oliver Fraser[1], D E Winget[3], S O Kepler[4], D J Sullivan[5], D Reaves[3], E L Robinson[3], T von Hippel[3], F Mullally[6], H Shipman[7], S E Thompson[7], N M Silvestri[1] and R I Hynes[8]

[1] Department of Astronomy, University of Washington, Seattle, WA 98195, USA
[2] Georgia College & State University, Milledgeville, GA 31061, USA
[3] Department of Astronomy, University of Texas at Austin, Austin, TX 78759, USA
[4] Universidade Federal do Rio Grande do Sul, Porto Alegre 91501-970, RS, Brazil
[5] Victoria University of Wellington, P. O. Box 600, Wellington, New Zealand
[6] Department of Astrophysical Sciences, Princeton University, Princeton, NJ 08544, USA
[7] Department of Physics & Astronomy, University of Delaware, Newark, DE 19716, USA
[8] Louisiana State University, Baton Rouge, LA 70803, USA

E-mail: anjum@astro.washington.edu

Abstract. We report preliminary results from our analysis of the stability of periods observed in the pulsating hydrogen atmosphere white dwarf ZZ Ceti (R548) based on observations that span 37 years from 1970 to 2007. We determine the rate of change of period with time to be $dP/dt = (0.8 \pm 1.9) \times 10^{-15}$ s/s using the O-C method and $dP/dt = (4.3 \pm 1.2) \times 10^{-15}$ s/s using the direct non-linear least squares fit NLSPDOT for the dominant period 213.13260643 s after correcting for proper motion. We do not claim either of these values as a measurement at this time, but hope to arrive at a conclusive result in the near future with more observations. The reduced uncertainty for both methods shows the improvement we obtained over the previous evolutionary constraint on ZZ Ceti (Mukadam et al. 2003). These dP/dt values are consistent within uncertainties with the measurement of $dP/dt = (3.57 \pm 0.82) \times 10^{-15}$ s/s for the period 215.2 s observed in another pulsating white dwarf G117-B15A (Kepler et al. 2005). Using the 213 s triplet spacing of 4 μHz, we compute the rotation period of ZZ Ceti to be 1.5 days.

1. Introduction

Non-interacting or single hydrogen atmosphere (DA) white dwarfs pulsate in a narrow instability strip located within the temperature range 10800–12300 K for log $g \approx 8$. These DA variables (DAVs) are also called the ZZ Ceti stars, with pulsation periods in the range 70–1400 s, consistent with nonradial g-mode pulsations. Even though HL Tau 76 was the first pulsating white dwarf to be discovered, ZZ Ceti (R 548) is the prototype of the class because it was the first pulsator to be resolved. We show in subsequent sections that we have truly been able to resolve the pulsation spectrum of ZZ Ceti only now.

2. Why measure the evolutionary drift rates of the periods observed in ZZ Ceti?

Evolutionary cooling of the ZZ Ceti stars dictates the rise in their periods as a result of the increasing degeneracy (Winget, Hansen & van Horn 1983; Kepler et al. 2000). Our motivation in measuring the rate of change of period with time for ZZ Ceti (R548) is to ultimately measure the cooling rate of the star. Measuring the cooling rates of white dwarfs proves helpful in calibrating the white dwarf cooling curve, reducing some of the theoretical uncertainty in white dwarf cosmochronometry (e.g. Winget et al. 1987; Hansen et al. 2002). We can also use the cooling rates of these stars to study exotic particles such as axions (Isern, Hernanz, & Garcia-Berro 1992; Bischoff-Kim et al. 2008).

3. Constructing an O-C diagram of ZZ Ceti with a doublet fit to the 213 s mode

The O-C technique can be used to improve the period estimates for any periodic phenomenon. The O stands for the observed value of the time of maximum (or time of zero) for a cycle or an epoch E that occurs in a data set. The C stands for its calculated value or ephemeris assuming an ideal stable clock. A non-linear trend in the O-C diagram shows that our measure of the period is changing. We set out to measure the rate of change of period with time (dP/dt) using the O-C technique for the dominant mode at 213 s, which we had so far believed to be a doublet. We found $dP/dt = (3.1 \pm 2.5) \times 10^{-15}$ s/s for the 213.13260643 s period based on 37 years of data from 1970 to 2007. We also determined $dP/dt = (2.6 \pm 2.8) \times 10^{-15}$ s/s for the 212.76842927 s period. We show the resultant O-C diagrams in Figure 1.

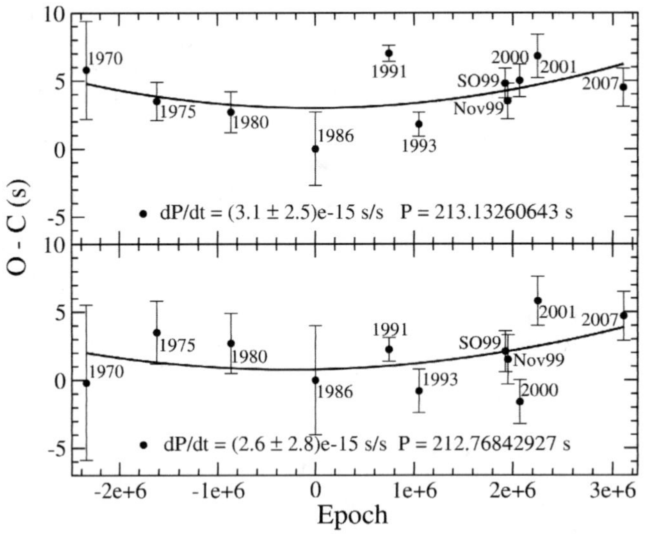

Figure 1. We show the O-C diagram for ZZ Ceti fitting a doublet to the 213 s mode

4. New observations with the highest signal-to-noise ratio for ZZ Ceti data

We acquired time-series photometry on ZZ Ceti from the 26th of August until the 7th of September 2007 using the prime focus CCD photometer Argos (Nather & Mukadam 2004) at the 2.1 m telescope at McDonald Observatory. These data constitute the highest signal-to-noise season of observations on ZZ Ceti which has mostly been observed using 1 m class telescopes and instruments based on photo-multiplier tubes with 35% intrinsic quantum efficiency. These high signal-to-noise data have allowed us to discover without ambiguity that the 213 s and 274 s modes are both triplets. Although we had checked for the existence of triplets previously, we could not conclusively prove their existence until now. We expect that both the 213 s and 274 s triplets are rotationally split ℓ=1 modes. We measure the 213 s triplet spacing to be about 4 μHz (see Figure 2), and compute the rotation period of ZZ Ceti to be 1.5 days.

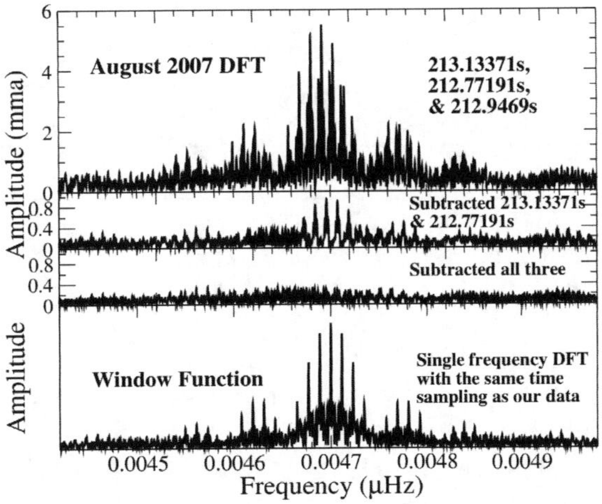

Figure 2. Our highest signal-to-noise data on ZZ Ceti from 2007 demonstrates that the 213 s mode is a triplet.

We revised the O-C diagram with new phases obtained by simultaneously fitting all frequencies of the 213 s triplet (see Table 2). The reduction in uncertainties obtained for the new dP/dt values shows the improvement obtained with a triplet fit (see Figure 3).

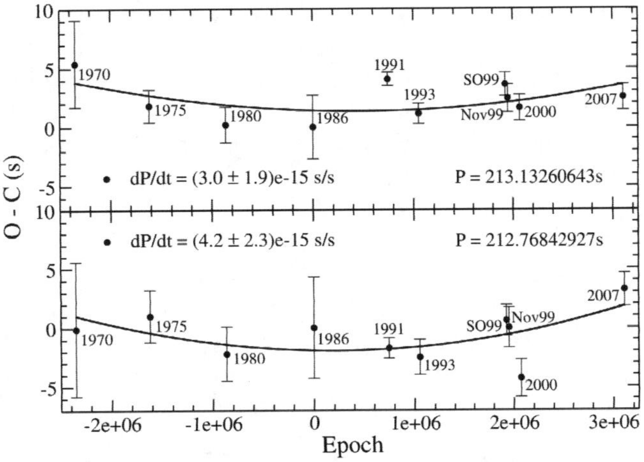

Figure 3. We show the O-C diagram for ZZ Ceti fitting a triplet to the 213 s mode

Table 1. Best Fit dP/dt values obtained

Period (s)	dP/dt from O-C Method (s/s)	dP/dt from NLSPDOT (s/s)
$213.13260643 \pm 0.00000034$	$(0.8 \pm 1.9) \times 10^{-15}$	$(4.3 \pm 1.2) \times 10^{-15}$
$212.76842927 \pm 0.00000040$	$(2.0 \pm 2.3) \times 10^{-15}$	$(3.3 \pm 1.9) \times 10^{-15}$

Table 2. O-C Tables for P = 213.13260643 s (left) and P = 212.76842927 s (right)

Epoch	O-C (s)	Epoch	O-C (s)
−2346428	5.4 ± 3.7	-2350444	−0.1 ± 5.7
−1617531	1.8 ± 1.4	-1620300	1.0 ± 2.2
−862740	0.2 ± 1.5	-864217	−2.2 ± 2.3
0	0 ± 2.7	0	0 ± 4.3
743875	4.06 ± .57	745148	−1.73 ± 0.85
1049404	1.12 ± .85	1051200	−2.5 ± 1.5
1924342	3.6 ± 1.0	1927636	0.6 ± 1.3
1949381	2.4 ± 1.2	1952718	0 ± 1.7
2067847	1.6 ± 1.1	2071386	−4.3 ± 1.6
3104809	2.5 ± 1.1	3110124	3.2 ± 1.4

5. Theoretical prediction of dP/dt based on the best-fit seismological model

We performed asteroseismological fits using the five periods observed in ZZ Ceti at 213 s, 274 s, 333 s, 318 s, and 187 s. We allowed the effective temperature to be within the range of 11700–12200 K, and fixed the helium layer mass to be Log M_{He}=-2.0. We constrained the stellar mass to be within the range 0.55–0.58 $M_{\odot}$, and the hydrogen layer mass to be Log M_H=-5.0.

We used models with ramp core profiles (see Bischoff-Kim et al. 2008), varying two core parameters: the edge of the homogeneous core where the oxygen abundance begins to drop off and secondly the central oxygen abundance. The latter is strongly constrained to 60-80% in the best fit model, and correlated with the location of the edge of the homogeneous core. Our best core parameter fits with uncertainties less than 1 s consist of a central oxygen abundance of 60–80%. Our best prediction of the rate of period change with time is dP/dt = $(8.3 \pm 0.046) \times 10^{-15}$ s/s assuming the 213 s mode is ℓ=1.

The observational results from both the O-C method and the nonlinear least squares fit NLSPDOT are not consistent with the model fit dP/dt value. Invoking weakly interacting particles like axions will only increase the discrepancy between model and observations. While our models are presently unable to reproduce the observed dP/dt value for ZZ Ceti, the empirical value should serve to constrain model white dwarfs in the end.

Acknowledgments
We thank NSF for the grant AST-0607840 for funding the new observations we report here.

References

Bischoff-Kim A Montgomery M H & Winget D E 2008 *ApJ* **675** 1505
Bischoff-Kim A et al. 2008 *ApJ* **675** 1512
Hansen, B M S et al., 2002 *ApJ* **574** L155
Isern J Hernanz M & Garcia-Berro E 1992 *ApJ* **392** 23
Kepler S O et al. 2000 *ApJ* **534** L185
Kepler S O et al. 2005 *ApJ* **634** 1311
Mukadam A S et al. 2003 *ApJ* **594** 961
Nather R E & Mukadam A S 2004 *ApJ* **605** 846
Winget D E et al. 1983 *Nature* **303** 781
Winget D E et al. 1987 *ApJ* **315** L77

Revisiting the theoretical DBV (V777 Her) instability strip: the MLT theory of convection

A H Córsico[1], L G Althaus[1], M M Miller Bertolami[1] and E García–Berro[2,3]

[1]Facultad de Ciencias Astronómicas y Geofísicas, Universidad Nacional de La Plata, Paseo del Bosque S/N, (1900) La Plata, Argentina

[2]Departament de Física Aplicada, Escola Politècnica Superior de Castelldefels, Universitat Politècnica de Catalunya, Av. del Canal Olímpic, s/n, 08860 Castelldefels, Spain

[3]Institute for Space Studies of Catalonia, c/Gran Capità 2–4, Edif. Nexus 104, 08034 Barcelona, Spain

E-mail: acorsico@fcaglp.unlp.edu.ar

Abstract. We reexamine the theoretical instability domain of pulsating DB white dwarfs (DBV or V777 Her variables). We performed an extensive g-mode nonadiabatic pulsation analysis of DB evolutionary models considering a wide range of stellar masses, for which the complete evolutionary stages of their progenitors from the ZAMS, through the thermally pulsing AGB and born-again phases, the domain of the PG1159 stars, the hot phase of DO white dwarfs, and then the DB white dwarf stage have been considered. We explicitly account for the evolution of the chemical abundance distribution due to time-dependent chemical diffusion processes. We examine the impact of the different prescriptions of the MLT theory of convection and the effects of small amounts of H in the almost He-pure atmospheres of DB stars on the precise location of the theoretical blue edge of the DBV instability strip.

1. Introduction

Variable DB white dwarfs (also called DBV or V777 Her stars) are g-mode nonradial pulsators with periods ranging from 200 to 1000 s. They are characterized by He-rich atmospheres, possibly contaminated with small impurities of H. Their pulsations are thought to be triggered by the $\kappa - \gamma$ mechanism acting on the partial ionization of He at the base of the outer convection zone. The observed instability strip of the DBV stars is located between $T_{\rm eff} \approx 28\,400$ K and $T_{\rm eff} \approx 22\,500$ K (Winget & Kepler 2008). All of the published stability analysis of DB models — see, for instance, Bradley & Winget (1994) and Beauchamp et al. (1999) — clearly indicate a strong dependence of location of the *theoretical* blue (hot) edge of the DBV instability strip with the convective efficiency adopted in the envelope of the stellar models. At present, there is no general consensus between different authors about what is the right convective efficiency of DB models in order to fit the observed blue edge. On the other hand, there exists an additional uncertainty in the definition of the temperature scale of the DB stars, and consequently in the location of the *observed* blue edge, due to the strong dependence with the convective efficiency adopted in the model atmosphere fits. Finally, another difficulty is the uncertainty in the surface parameters that results from the presence of undetectable amounts of H in the atmospheres of DB stars (Beauchamp et al. 1999). In this work, we explore the impact of different convective

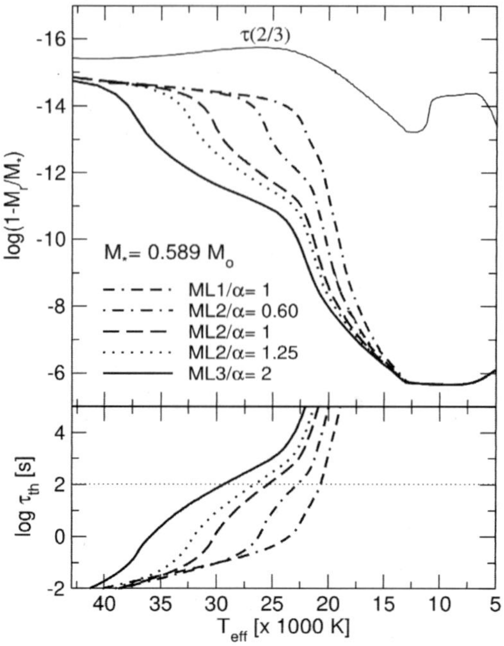

Figure 1. The extent in mass of the outer convection zone in terms of $T_{\rm eff}$ for different prescriptions of the MLT theory of convection (upper panel), and the run of the thermal timescale at the basis of the outer convective zone (lower panel).

efficiencies and of small amounts of H in the atmospheres of DB stars on the precise location of the theoretical blue edge of the DBV instability strip. We employ the Mixing Length Theory (MLT) of convection to model the outer convection zone of our models. We also examine the dependence of the blue edge with the stellar mass. To this end, we perform fully nonadiabatic pulsation calculations of up-to-date DB white dwarf evolutionary models with a wide range of stellar masses.

2. Stellar structure, evolution and pulsation modeling

We employ a new set of fully evolutionary DB white dwarf models that descend from the post-born again PG1159 models (Miller Bertolami & Althaus 2006). Specifically, we employ the `LPCODE` evolutionary code (Althaus et al. 2005) to compute the evolution of PG1159 models (H-deficient, He-, C-, and O-dominated atmospheres) towards the DB regime (almost He-pure atmospheres) taking into account the time-dependent microscopic diffusion of ^{4}He, ^{12}C, ^{13}C, ^{14}N, ^{16}O and ^{22}Ne. The metallicity is assumed to be $Z = 0$ in the metal-free He-rich envelopes and $Z \leq 0.02$ otherwise. The He-rich envelopes of our models are characterized by a double-layered chemical structure built up by chemical diffusion. We perform nonadiabatic pulsation computations of nonradial g-modes for $\ell = 1, 2$ with the code described in Córsico et al. (2006). We analyze the pulsation stability of DB white dwarf models with effective temperatures between 35 000 K and 17 000 K and stellar masses of $0.530, 0.542, 0.556, 0.565, 0.589, 0.609, 0.664, 0.741$ and $0.870 M_\odot$. We model convection in the envelope of our models in the framework of the MLT theory. We employ the following prescriptions of the MLT, in order of increasing efficiency: ML1/α = 1, ML2/α = 0.60, ML2/α = 1, ML2/α = 1.25, and ML3/α = 2, α being the characteristic mixing length (Tassoul et al. 1990).

3. Results

The evolution and extent of the outer convective zone of our DB white dwarf sequences during the DBV instability strip is shown in the upper panel of Fig. 1 for the different treatments of

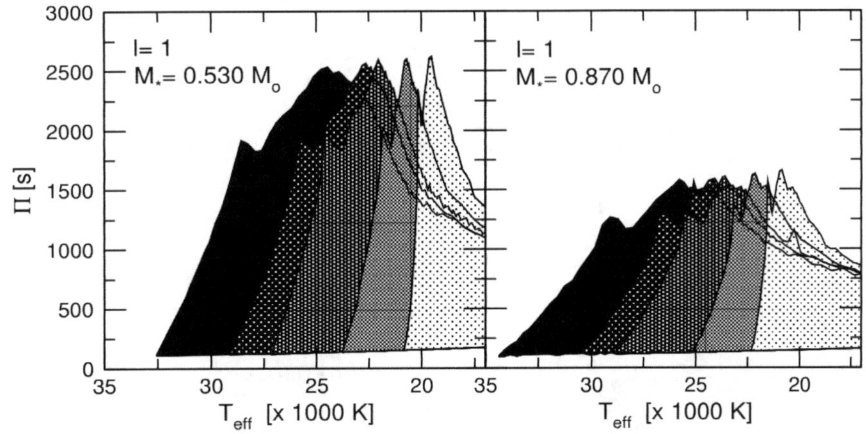

Figure 2. The instability domains of dipole modes for different MLT prescriptions (from left to right: ML3/$\alpha = 2$, ML2/$\alpha = 1.25$, ML2/$\alpha = 1$, ML2/$\alpha = 0.60$, and ML1/$\alpha = 1$).

the MLT. Note that the base of the outer convective zone notably deepens with cooling, being this effect more pronounced in the case of more efficient MLT prescriptions in the proximity of the blue edge ($T_{\rm eff} \approx 29\,000$ K). In the lower panel we display the run of the thermal timescale at the basis of the outer convective zone. DB models are expected to became g-mode pulsationally unstable when $\tau_{\rm th} \approx 100$ s. This would happen first for the highest convective efficiency (ML3/$\alpha = 2$) models. This is confirmed through full nonadiabatic computations.

Fig. 2 displays the instability domains in the $T_{\rm eff} - \Pi$ diagram, corresponding to the sequence of $0.530 M_{\odot}$ (left) and the sequence of $0.870 M_{\odot}$ (right), for the different MLT prescriptions. On the basis of this figure, we can outline several trends: (1) there is a strong dependence of the longest excited periods with the stellar mass, being larger for less massive models; (2) the shorter excited periods, on the contrary, do not exhibit any dependence with M_*; (3) regarding convection, there is no dependence of the longest (nor the shorter) excited periods with the particular MLT prescription adopted; (4) the blue edge is strongly sensible to the convective efficiency, being much hotter for efficient convection (that is, for ML3 and ML2/$\alpha = 1.25$); and (5) the blue edge is hotter for more massive DB models.

Now, we examine how our theoretical blue edges of the instability strip compare with the location of the observed pulsating DB white dwarfs. We consider the eight DBV stars considered in Beauchamp et al. (1999) , plus the nine DBVs recently discovered in the SDSS (Nitta et al. 2009). We consider only the case of atmospheres devoided of H. Fig. 3 shows our complete set of DB evolutionary sequences (dashed lines) on the $T_{\rm eff} - \log g$ diagram, along with the blue edge for the different MLT prescriptions. Note that there is a clear dependence of the blue edge with the stellar mass. It is apparent that only the ML2/$\alpha = 1.25$ prescription is able to account for the location of all known DBVs.

4. The effects of H on the location of the blue edge

At present, it is a well known observational fact that the atmospheres of DB white dwarf stars exhibit small traces of H, attributed mainly to accretion from the ISM. A modest accretion rate of about $10^{-19} - 10^{-21} M_{\odot}$/yr would be enough to explain the presence of H in the DBs. The effective temperature of DB stars as derived from model atmospheres that contain H are considerably lower than in the cases in which H is neglected (Beauchamp et al. 1999; Castanheira et al. 2006; Voss et al. 2007). It is expected also that a non-negligible shift of the theoretical blue edge should result when the He-rich envelopes of the equilibrium models are contaminated with small impurities of H. To test this possibility, we have restricted ourselves to the sequences of $M_* = 0.530 M_{\odot}$ and $M_* = 0.741 M_{\odot}$. For these two values of the stellar mass, we have explored the cases in which $X_{\rm H} = 0.0001, 0.001$ and 0.01, corresponding

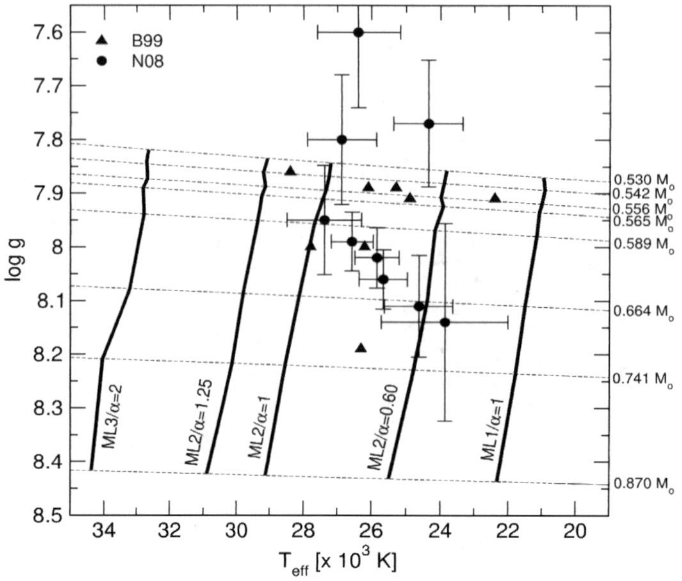

Figure 3. The blue edge of the DBV instability strip for different recipes of the MLT.

to $\log(n_{\rm H}/n_{\rm He}) = \log(4X_{\rm H}/X_{\rm He}) = -3.4, -2.4$ and -1.44, respectively. We found that if $X_{\rm H} = 0.0001$, there is no appreciable effects on the location of the blue edge. However, if $X_{\rm H} = 0.001$ the blue edge is shifted to lower effective temperatures by ~ 800 K (for ML3/$\alpha = 2$) and ~ 300 K (for ML1/$\alpha = 1$). Finally, if $X_{\rm H} = 0.01$ we obtain a shift of ~ 3000 K (for ML3/$\alpha = 2$) and ~ 1200 K (for ML1/$\alpha = 1$).

5. Conclusions

The results of the present work indicate that the best agreement between the location of the DBV stars previously studied in Beauchamp et al. (1999) with surface parameters from H-free model atmospheres, and the predictions of the nonadiabatic pulsation calculations is provided by the case of ML2/$\alpha = 1.25$. Notably, the $T_{\rm eff}$ and $\log g$ previously derived in Beauchamp et al. (1999) are obtained with model atmospheres that assume ML2/$\alpha = 1.25$. Our ML2/$\alpha = 1.25$ blue edge is consistent also with the SDSS DBV stars reported in Nitta et al. (2009), although their surface parameters are derived from model atmospheres with ML2/$\alpha = 0.60$. The present work constitutes the first phase of a thorough pulsation study (adiabatic and nonadiabatic) of DBV stars on the basis of a new generation of fully evolutionary DB models.

Acknowledgments

We thank S. O. Kepler for providing us the $T_{\rm eff}$ and $\log g$ values of the recently discovered SDSS DBV stars. Part of this work was supported by PIP 6521 grant from CONICET, by MEC grant AYA05–08013–C03–01, by the European Union FEDER funds and by the AGAUR.

References

Althaus L G Serenelli A M Panei et al. 2005 *A&A* **435** 631
Beauchamp A Wesemael F Bergeron P et al. 1999 *ApJ* **516** 887
Bradley P A & Winget D E 1994 *ApJ* **421** 236
Castanheira B G Kepler S O Handler G & Koester D 2006 *A&A* **450** 331
Córsico A H Althaus L G & Miller Bertolami M M 2006 *A&A* **458** 259
Miller Bertolami M M & Althaus L G 2006 *A&A* **454** 845
Nitta A Kleinman S J Krzesinski J et al. 2009 *J. of Phys.: Conf. Ser.*, in press
Tassoul M Fontaine G & Winget D E 1990 *ApJS* **72** 335
Voss B Koester D Napiwotzki R Christlieb N & Reimers D 2007 *A&A* **470** 1079
Winget D E & Kepler S O 2008 *ARA&A* **46** 157

16th European White Dwarfs Workshop IOP Publishing
Journal of Physics: Conference Series **172** (2009) 012076 doi:10.1088/1742-6596/172/1/012076

SPA, "The Stellar Photometry Assistant"

J Dalessio, J L Provencal and H L Shipman

The University of Delaware and Mt. Cuba Astronomical Observatory (DARC), Department of Physics and Astronomy, 104 The Green, 217 Sharp Laboratory, Newark, DE 19716, USA

E-mail: `dalessio@udel.edu`

Abstract. SPA is a stand alone software package for high speed photometry reduction and analysis. The goal of SPA is to be simple, powerful and intuitive. SPA was born out of complications studying the pulsating DB white dwarf EC20058-5234 (QuTel) due to the proximity of its companions. SPA addresses the Whole Earth Telescope's (Nather et al. 1990) demand for large scale rapid data reduction from multiple sites. SPA is being developed in MATLAB by the Delaware Asteroseismic Research Center (DARC) in collaboration with the University of Delaware and the Mount Cuba Astronomical Observatory.

1. Program Features

1.1. Simplifying the WET Pipeline

During a Whole Earth Telescope (WET) observation, incoming data is processed as quickly as possible. Rapid reduction and analysis allows the control center to efficiently allocate observing time to WET targets, and hence maximize the quality and quantity of the science. When working with images from many different acquisition systems and CCDs, problems are encountered when extracting the necessary quantities from the image sets, i.e. the Julian Day of each image. These problems are exacerbated by site specific issues, such as systematic timing errors and non-standard FITS keywords. Though the WET control center is typically manned continuously, detailed knowledge of IRAF and the reduction scripts (Kanaan et al. 2002) is often required to process problematic incoming image sets. This means that some data is not processed immediately, or excessive man-hours are required to reduce certain data sets. This hampers the ability for WET headquarters to make critical real time decisions. A goal of SPA is to alleviate these problems. SPA includes site specific scripts which, when called, pre-process the images, correct timing errors and translate non-standard FITS keywords.

1.2. Crowded Field Photometry

QuTel has been a target of seasonal observations since its discovery in the mid-1990s (Koen et al. 1995), including a WET observation (Sullivan et al. 2008). The pulsation modes of QuTel appear stable, which could allow a measurement of the rate of period change ($\dot{P}$) for each mode. A measurement of $\dot{P}$ could reveal information about neutrino/axion emission (Bischoff-Kim, 2008) . However, QuTel has two nearby companions, one of which lies well within efficient aperture sizes. In order to obtain precise measurements of each mode's phase, and hence $\dot{P}$, along with the general goal of reducing noise, advanced data reduction techniques such as growth curve photometry (Stetson 1987) may be more suitable than synthetic aperture photometry. The modular nature of SPA allows seem-less integration of reduction techniques more powerful than synthetic aperture photometry. Work is currently underway to incorporate growth curve photometry and various other methods involving calculation of the point spread function (PSF).

1.3. Rapid Data Reduction
Due to the time critical nature of WET reductions, and to reduce the manpower required at headquarters, the reduction algorithms of SPA are designed with speed in mind. For example, instead of reading an entire FITS frame, SPA will only read the needed sections of each image. This has decreased the time it takes to reduce an image by over an order of magnitude (for a 1024×1024 image).

1.4. Multiple Interfaces
The core of SPA lies in object oriented MATLAB classes. This encourages a modular structure, an important feature when data formatting and reduction techniques are subject to change. This also separates the user interface from the core components of the program, allowing the development of graphical user interfaces, or console interfaces without any modifications to the core code. Graphical interfaces can provide simple tools for beginner users, while the console allows complete control for more advanced users.

1.5. Open Source Development
SPA is being developed under the Academic Free License, i.e. the source code will be available. The authors, however, will discourage repackaging of the program. Instead it is suggested that any additional modules be implemented (after rigorous testing) into main branch of development. Those interested in developing SPA are encouraged to contact the authors.

1.6. Portability
SPA is compatible with R2007a and higher MATLAB releases. This means any computer, independent of operating system, can run SPA so long as MATLAB or the MATLAB component runtime can be installed. Resting on the shoulders of MATLAB is an advantage over relying on open-source software which may not be well maintained in the future. This ensures longevity, and defers system compatibility issues to the developers of MATLAB. Effort may also be taken to increase SPA's compatibility with open source MATLAB alternatives such as Octave.

2. Testing the Effectiveness of SPA
IRAF scripts developed by A. Kanaan (Kanaan et al. 2002) are being used to test the reliability and effectiveness of SPA. Synthetic aperture photometry produces similar light curves. Differences in the light curves are marginal and are likely due to variations in the methods chosen to round pixels or center an object. These tests are preliminary and a more rigorous comparison will be underway shortly.

Acknowledgments
The authors would like to thank the following entities: The Delaware Asteroseismic Research Center and Mount Cuba Astronomical Observatory. The authors would also like to thank the following individuals: Antonio Kanaan for sharing his knowledge of aperture photometry, and Fergal Mullally for encouraging felicitous coding.

References
Bischoff-Kim A 2008 *CoAst* **154** 16
Kanaan A Kepler S O & Winget D E 2002 *A&A* **389** 896
Koen C O'Donoghue D Stobie R S Kilkenny D & Ashley R 1995 *MNRAS* **277** 913
Nather R E Winget D E Clemens J C Hansen C J & Hine B P 1990 *ApJ* **361** 309

Journal of Physics: Conference Series **172** (2009) 012076 doi:10.1088/1742-6596/172/1/012076

Stetson P B 1987 *PASP* **99** 191
Sullivan D J et. al. 2008 *MNRAS* **387** 137

Journal of Physics: Conference Series **172** (2009) 012077 doi:10.1088/1742-6596/172/1/012077

Nonadiabatic Asteroseismology of GW Vir Stars

Pierre-Olivier Quirion[1], **Gilles Fontaine**[2] **and Pierre Brassard**[2]

[1]Institut for Fysik og Astronomi, Aarhus Universitet Ny Munkegade, Bygn. 1520, 8000 Århus-C

[2]Département de physique Université de Montréal, C.P. 6128, Succ. centre-ville, Montréal (Québec) H3C 3J7

E-mail: `olivier@phys.au.dk`

Abstract. We present results from nonadiabatic stability calculation to determine the position of the GW Vir band in the $\log g - T_{\rm eff}$ diagram. With this method, we calculate the position of seven GW Vir stars. The results are consistent with and complementary to the spectroscopic determination of atmospheric parameters and the chemical compositions of these stars except for the newly discover GW Vir star VV 47. This discrepancy is discussed.

1. Astrophysical Context

The GW Vir stars are extremely hot, compact stars that pulsate in gravity modes. They include at least two different spectral types: the hot H-deficient Wolf-Rayet central stars of planetary nebulae or [WCE] stars, and their descendants, the PG 1159 stars. The multiperiodic luminosity variations observed in GW Vir stars are due to a classic κ-mechanism associated with the ionization of the K-shell electrons of carbon and oxygen, two of the main atmospheric/envelope constituents in these objects.

Quantitative spectroscopy, based on detailed Non-LTE model atmospheres and synthetic spectra, has revealed that GW Vir stars occupy a wide domain in the surface gravity−effective temperature plane, with objects in the ranges $5.5 \leq \log g \leq 7.5$ and $80,000$ K $\leq T_{\rm eff} \leq 170,000$ K. In addition, the inferred atmospheric abundances vary largely from one star to another, and the main atmospheric constituents are found in the intervals, $0.33 \leq$ X(He) ≤ 0.76, $0.16 \leq$ X(C) ≤ 0.55, and $0.00 \leq$ X(O) ≤ 0.17. We refer the reader to the excellent review by Werner & Herwig (2006) for more details on the chemical compositions of these stars. Unfortunately, at the very high $T_{\rm eff}$ of the GW Vir regime, the uncertainties on the derived values of $T_{\rm eff}$ and $\log g$ remain relatively large with typical values of 5 to 10% for $T_{\rm eff}$, and 0.5 dex for $\log g$. We argue here that it is possible to infer the atmospheric parameters of GW Vir pulsators with a higher accuracy by exploiting nonadiabatic asteroseismology. This is based on a comparison of the observed ranges of excited periods with ranges predicted from nonadiabatic models.

2. Models and Results

We know that the instability band of GW Vir stars is strongly correlated with its location in the $\log g - T_{\rm eff}$ diagram (Quirion et al. 2007). The range of the instability bands depend mainly on surface gravity for the "fluffy" stars of lower gravities and is a function of temperature and gravity for the more compact/evolved objects of higher gravities. We thus expect an especially

Journal of Physics: Conference Series **172** (2009) 012077 doi:10.1088/1742-6596/172/1/012077

Table 1. Comparison between spectroscopic results, our nonadiabatic asteroseismology calculation and the observations.

Star	Spectroscopic Fit			Nonadiabatic Asteroseismology			Observable
	$\log g$ $\pm.5$ dex	$T_{\rm eff}$ [kK] $\pm<10\%$	Periods [s]	$\log g$	$T_{\rm eff}$ [kK]	Periods [s]	Periods [s]
NGC 246	5.70	150	514−11453	~ 5.75	−	604−4477	1464−1842
RX J2117.1+3412	6.00	170	635−2202	~ 6.10	180 ± 10	178−779	694−1530
PG 1159−035	7.00	140	247−623	$6.80 \pm .05$	142 ± 3	336−987	339−982
VV 47	7.00	130	216−891	$6.10 \pm .10$	130 ± 20	235-3531	261−4310
PG 2131+066	7.50	95	197−650	$7.25 \pm .25$	−	118−508	339−508
PG 1707+427	7.50	85	188−765	$7.35 \pm .05$	81 ± 5	224−960	336−942
PG 0122+200	7.50	80	198−847	$7.50 \pm .30$	−	330−602	336−612

good constraint on surface gravity for GW Vir stars of low gravity, and both good fit on gravity and temperature for the more compact object.

We use the test function S, which depends on P_{min} and P_{max}, the shortest and longest excited periods, to measure the goodness of our fit and to find the best matching model for the observed GW Vir stars.

$$S^2 = \left(\frac{P_{\rm max}^{\rm obs} - P_{\rm max}^{\rm mod}}{\sigma_{\rm max}}\right)^2 + \left(\frac{P_{\rm min}^{\rm obs} - P_{\rm min}^{\rm mod}}{\sigma_{\rm min}}\right)^2 \qquad (1)$$

where the weight $\sigma_i = 1\%\ P_i^{\rm obs}$ is chosen to be proportional to the observed period boundaries.

An other important parameter fixing the instability band is the chemical composition in the envelope of the star. The composition used for our nonadiabatic calculation are taken from the spectroscopic measurements. In that sense the results shown here are a consistency check between spectroscopic and photometric measurements. A correct interpretation our calculations should help us understanding if photometric or spectroscopic results need to be revisited.

We explicitly show our results for two stars. One which gives perfect agreement, PG 1707+427, and one where there is no consistency between the spectroscopy and the photometry, VV 47. We also present in Table. 1 results for other stars.

The consistency check for PG 1707+427 (Figure 1) is perfect. In addition, asteroseismology provides a better constraint on both the gravity and the effective temperature (see Table 1). Nonadiabatic calculation finds the best fit at $\log g = 7.4$ and $T_{\rm eff} = 79,000$ K. The test function S reaches a minimum value of 0.3. At that point, the instability band of the model extends from 224 s to 960 s. This GW Vir is compact enough to let the effective temperature have a significant effect on the instability band. In this high gravity regime, we can also improve the determination of $\log g$. The analysis of that star is a spectroscopic and photometric success!

For VV 47, if we calculate the instability range at the spectroscopic value ($\log g = 7.00$ and $T_{\rm eff} = 130,000$ K) we find a band of 216 s to 891 s. However, find a better fit outside the estimated spectroscopy error, at $\log g = 6.3$ and $T_{\rm eff} = 103,000$ K (Figure 2). At this point, the instability band of the model extends from 235 s to 3531 s. The "best fit valley" of the S function around $\log g = 6.0$ is quite far from the spectroscopic value. This inconsistency is a hint that the composition ratio and/or the temperature of VV 47 is to be revised. This poor correspondence between spectroscopy results and nonadiabatic asteroseismology is not really surprising when one consider that in their spectroscopic analysis, Rauch & Werner (1995) stress that their analysis of VV 47 was uncertain. We hope that the recent discovery of pulsation in

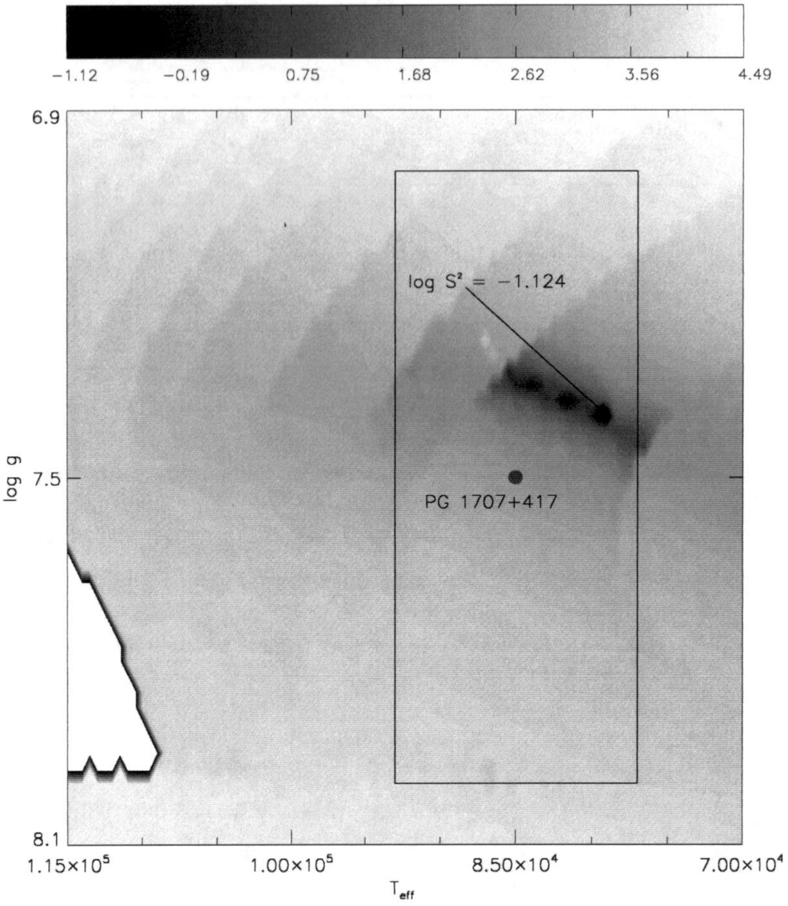

Figure 1. Position of PG 1707+427 in the $\log g - T_{\rm eff}$ diagram as fixed by spectroscopy is shown by the black dot. The error range is depicted by the black rectangle.

VV 47 and the inconsistency shown in Figure 2 will revive the interest in what looked, in 1995, to be one of the most compact PG 1159 star with a planetary. We notice that the value of $\log g$ obtained with nonadiabatic calculations would replace VV 47 along with the other PG 1159 and [WCE] stars which display a planetary nebulae.

If we examine results of Table 1 we see that nonadiabatic asteroseismology calculation do not always constrain temperature, this can be caused by the model intrinsic low dependency on temperature or by unresolved photometric features. For all star, nonadiabatic asteroseismology gives values of the surface gravity that is equivalent or more precise than what is achieved by spectroscopy alone.

Journal of Physics: Conference Series **172** (2009) 012077 doi:10.1088/1742-6596/172/1/012077

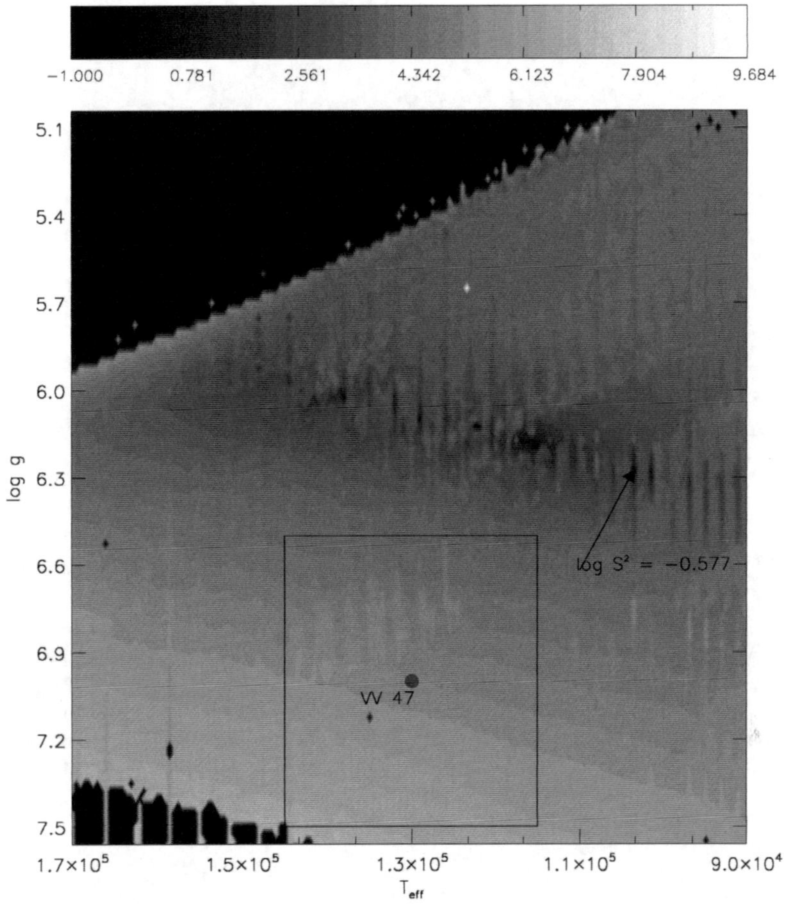

Figure 2. Position of VV 47 in the $\log g - T_{\rm eff}$ diagram (black dot) as fixed by spectroscopy with canonical GW Vir error bars (black rectangle).

3. Conclusion

Nonadiabatic asteroseismology is more than a consistency check over different GW Vir analysis techniques. It give a significant improvement to GW Vir stars surface gravity, and in some case the their effective temperature. This helps us establish these stars absolute luminosity and boost the understanding of physical processes leading from the post-AGB to the white dwarf track.

References

Quirion P-O Fontaine G & Brassard, P 2007 *ApJS* **171** 219
Rauch T & Werner K 1995 *LNP* **443** 186
Werner K & Herwig F 2006 *PASP* **118** 183

Journal of Physics: Conference Series **172** (2009) 012078 doi:10.1088/1742-6596/172/1/012078

Seismological constraints on the high-gravity DOV stars PG2131+066 and PG 1707+427

A H Córsico[1], L G Althaus[1], M M Miller Bertolami[1] and E García–Berro[2,3]

[1]Facultad de Ciencias Astronómicas y Geofísicas, Universidad Nacional de La Plata, Paseo del Bosque S/N, (1900) La Plata, Argentina
[2]Departament de Física Aplicada, Escola Politècnica Superior de Castelldefels, Universitat Politècnica de Catalunya, Av. del Canal Olímpic, s/n, 08860 Castelldefels, Spain
[3]Institute for Space Studies of Catalonia, c/Gran Capità 2–4, Edif. Nexus 104, 08034 Barcelona, Spain

E-mail: acorsico@fcaglp.unlp.edu.ar

Abstract. A seismological study of the pulsating PG1159 stars PG2131+066 and PG 1707+427 is presented. We perform extensive adiabatic computations of g-mode pulsation periods of PG1159 evolutionary models with stellar masses ranging from 0.530 to 0.741 $M_{\odot}$. We constrain the stellar mass of PG2131+066 and PG 1707+427 by comparing the observed period spacing of each star with the theoretical asymptotic period spacings and with the average of the computed period spacings. We also employ the individual observed periods to find representative seismological models for both stars.

1. Introduction

Pulsating PG1159 stars (also called GW Vir or DOV stars) are very hot hydrogen-deficient post-Asymptotic Giant Branch stars with surface layers rich in He, C and O (Werner & Herwig 2006) exhibiting multimode pulsations (non-radial g-modes) with periods ranging from 5 to 50 minutes (Winget & Kepler 2008). It is generally accepted that these stars have their origin in a born-again episode induced by a post-AGB He thermal pulse (Althaus et al. 2005). Recently, considerable observational effort has been invested to study pulsating PG1159 stars. Particularly noteworthy are the observational efforts devoted to RXJ 2117.1+3412 (Vauclair et al. 2002), PG 0122+200 (Fu et al. 2007) and PG 1159−035 (Costa et al. 2008). On the theoretical front, recent important progress in the numerical modeling of PG1159 stars (Althaus et al. 2005; Miller Bertolami & Althaus 2006) has paved the way for unprecedented seismological inferences for the mentioned stars (Córsico et al. 2007a; Córsico et al. 2007b; Córsico et al. 2008). Here, we present a seismological study of the high-gravity pulsating PG1159 stars PG 2131+066 and PG 1707+427, two stars that have been intensively scrutinized with the Whole Earth Telescope.

Our pulsation analysis is based on a new generation of stellar models that take into account the complete evolution of PG1159 progenitor stars with initial masses on the ZAMS ranging from 1 to 3.75 $M_{\odot}$ (Althaus et al. 2005; Miller Bertolami & Althaus 2006). All the post-AGB evolutionary sequences were computed using the LPCODE evolutionary code and were followed through the very late thermal pulse and the resulting born-again episode that gives rise to the H-deficient, He-, C- and O-rich composition characteristic of PG1159 stars. The masses of

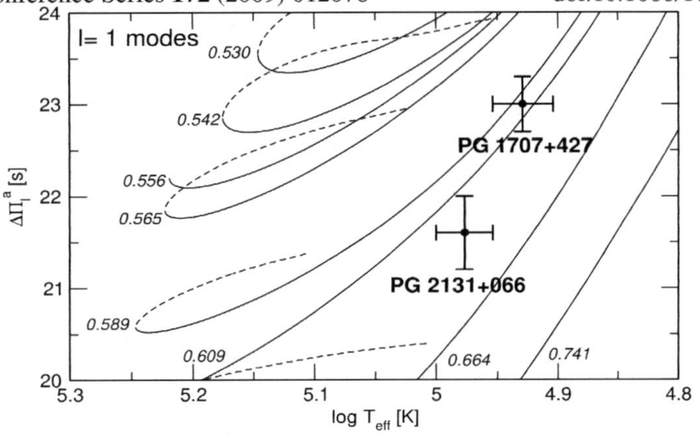

Figure 1. The dipole ($\ell = 1$) asymptotic period spacing in terms of the effective temperature for different stellar masses (in solar units). Also shown is the location of PG 2131+066 and PG 1707+427.

the resulting remnants are 0.530, 0.542, 0.556, 0.565, 0.589, 0.609, 0.664, and 0.741 $M_\odot$. We computed $\ell = 1$ g-mode adiabatic pulsation periods for these equilibrium models with the same numerical code and methods we employed in our previous works (Córsico & Althaus 2006).

2. Mass determination from the observed period spacing

Here, we constrain the stellar mass of PG 2131+066 and PG 1707+427 by comparing the asymptotic period spacing and the average of the computed period spacings with the *observed* period spacing. These approaches are based on the fact that the period spacing of PG1159 pulsators depends primarily on the stellar mass, and the dependence on the luminosity and the He-rich envelope mass fraction is negligible (Kawaler & Bradley 1994; Córsico & Althaus 2006). Fig. 1 displays the asymptotic period spacing for $\ell = 1$ modes as a function of the effective temperature for different stellar masses. Also shown in this diagram is the location of PG 2131+066, with $T_{\rm eff} = 95 \pm 5$ kK (Dreizler & Henber 1998) and $\Delta\Pi^{\rm O}_{\ell=1} = 21.6 \pm 0.4$ s (Reed et al. 2000), and PG 1707+427, with $T_{\rm eff} = 85 \pm 5$ kK (Dreizler & Heber 1998) and $\Delta\Pi^{\rm O}_{\ell=1} = 23.0 \pm 0.3$ s (Kawaler et al. 2004). The asymptotic period spacing is computed as in Tassoul et al. (1990). From the comparison between the observed $\Delta\Pi^{\rm O}_{\ell=1}$ and $\Delta\Pi^{\rm a}_{\ell=1}$ we found a stellar mass of 0.627 $M_\odot$ for PG 2131+066 and 0.597 $M_\odot$ for PG 1707+427. In Fig. 2 we show the run of average of the computed ($\ell = 2$) period spacings ($\ell = 1$) for PG 2131+066 (left panel), and PG 1707+427 (right panel) in terms of the effective temperature for all of our PG1159 evolutionary sequences. From the comparison between the observed $\Delta\Pi^{\rm O}_{\ell=1}$ and $\overline{\Delta\Pi_\ell}$, we found a stellar mass of 0.578 $M_\odot$ and 0.566 $M_\odot$ for PG 2131+066 and PG 1707+427, respectively. These values are 8.5% (for PG 2131+066) and 5.5% (for PG 1707+427) smaller than those derived through the asymptotic period spacing, showing once again that the asymptotic approach overestimates the stellar mass of PG1159 stars that, like PG 2131+066 and PG 1707+427, exhibit short and intermediate pulsation periods.

3. Constraints from the individual observed periods

In this approach we seek pulsation models that best matches the *individual* pulsation periods of PG 2131+066 and PG 1707+427. For both stars, we assume that all of the observed periods correspond to $\ell = 1$ modes (Kawaler et al. 1995; Kawaler et al. 2004). The goodness of the match between the theoretical pulsation periods ($\Pi^{\rm T}_k$) and the observed individual periods ($\Pi^{\rm O}_i$) is measured by means of a merit function, $\chi^2(M_*, T_{\rm eff})$, which takes into account the differences between the observed and theoretical periods squared. For each star, the PG 1159 model that shows the lowest value of χ^2 will be adopted as the "best-fit model". For PG 2131+066, we find a clear seismological solution corresponding to a model with $M_* = 0.589\, M_\odot$ and $T_{\rm eff} \approx 102$ kK. This model provides an excellent agreement between the theoretical and observed periods.

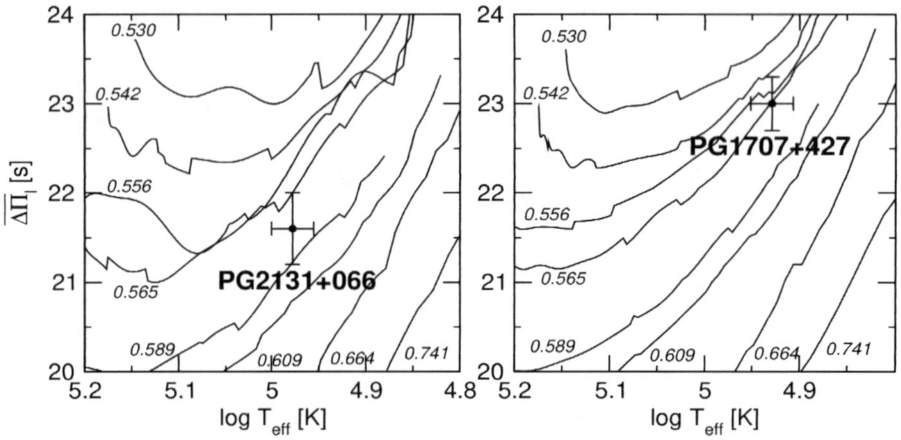

Figure 2. Same as Fig. 2, but for the average of the computed period spacings.

Table 1. The main characteristics of PG 2131+066. Column (1) corresponds to spectroscopic results from Dreizler & Heber (1998), column (2) presents results from previous pulsational studies (Kawaler et al. 1995; Reed et al. 2000) and column (3) provides the results from the seismological model of this work, respectively.

	(1)	(2)	(3)
$T_{\rm eff}$ [kK]	95 ± 5	—	$102.18^{+3.0}_{-2.8}$
M_* [$M_\odot$]	0.55 ± 0.1	0.608 ± 0.01	$0.589^{+0.020}_{-0.024}$
$\log g$ [cm/s^2]	7.5 ± 0.5	—	$7.63^{+0.12}_{-0.14}$
$\log(L_*/L_\odot)$	1.6 ± 0.3	1.41 ± 0.5	$1.57^{+0.07}_{-0.06}$
$\log(R_*/R_\odot)$	—	-1.73	$-1.71^{+0.06}_{-0.05}$
d [pc]	681^{+170}_{-137}	668^{+78}_{-83}	830^{+300}_{-224}
π [mas]	$1.47^{+0.4}_{-0.3}$	1.50 ± 0.2	$1.2^{+0.4}_{-0.3}$

This is quantitatively reflected by the average of the absolute period differences, $\overline{\delta\Pi_i} = 1.57$ and the root-mean-square residual $\sigma_{\delta\Pi_i} = 2.32$ s. For PG 1707+427, we find a unambiguous seismological solution for a model with $M_* = 0.542\,M_\odot$ and $T_{\rm eff} \approx 89.5$ kK, an effective temperature compatible with the spectroscopic determination. We found $\overline{\delta\Pi_i} = 1.75$ s and $\sigma_{\delta\Pi_i} = 1.99$ s.

The main features of our best-fit model for PG 2131+066 are summarized in Table 1. The total mass of the best-fit model ($M_* = 0.589\,M_\odot$) is in agreement with the value derived from the average of the computed period spacings ($M_* \sim 0.578\,M_\odot$), but at odds ($\sim 6\%$ smaller) with that inferred from the asymptotic period spacing ($M_* = 0.627\,M_\odot$). Also, the M_* value of our best-fit model is substantially larger than the spectroscopic mass of $0.53 \pm 0.1\,M_\odot$ (Dreizler & Heber 1998; Miller Bertolami & Althaus 2006) for PG 2131+066. A discrepancy between the seismological and the spectroscopic values of M_* is generally encountered among PG1159 pulsators (Córsico et al. 2007a; Córsico et al. 2007b). Until now, the seismological mass of PG 2131+066 has been about 11% larger ($\Delta M_* \approx 0.06\,M_\odot$) than the spectroscopic mass if we consider the early estimation for the seismological mass quoted in Reed et al. (2000). In light of the best-fit model derived in this paper, this discrepancy is slightly reduced to about 7% ($\Delta M_* \approx 0.04\,M_\odot$).

The main properties of our best-fit model for PG 1707+427 are shown in Table 2. The stellar mass of our best-fit model is $M_* = 0.542\,M_\odot$, in agreement with the value derived from

Table 2. Same as Table 1, but for PG 1707+427. Column (2) presents results from the pulsational study of Kawaler et al. (2004).

	(1)	(2)	(3)
$T_{\rm eff}$ [kK]	85 ± 4.5	—	$89.5^{+1.7}_{-1.8}$
M_* [$M_\odot$]	0.53 ± 0.1	0.57 ± 0.02	$0.542^{+0.014}_{-0.012}$
$\log g$ [cm/s^2]	7.5 ± 0.3	—	$7.53^{+0.09}_{-0.08}$
$\log(L_*/L_\odot)$	1.4 ± 0.3	1.36	1.40 ± 0.04
$\log(R_*/R_\odot)$	—	—	-1.68 ± 0.04
d [pc]	1300^{+1000}_{-300}	—	730^{+230}_{-175}
π [mas]	$0.77^{+0.2}_{-0.3}$	—	1.4 ± 0.4

the average of the computed period spacings ($M_* \sim 0.566\,M_\odot$), but at odds ($\sim 9\%$ lower) with that inferred from the asymptotic period spacing ($M_* = 0.597\,M_\odot$). On the other hand, we note that M_* for the best-fit model is in excellent agreement with the spectroscopic derivation ($0.542\,M_\odot$ versus $0.53\,M_\odot$). Until now, the seismological mass of PG 1707+427 has been more than 7% larger ($\Delta M_* \approx 0.04\,M_\odot$) than the spectroscopic mass if we adopt for the seismological mass the value found previously (Kawaler et al. 2004). In light of our best-fit model, this discrepancy is strongly reduced to about 2% ($\Delta M_* \approx 0.012\,M_\odot$).

4. Summary and conclusions

In this work we have estimated the stellar mass of PG 2131+066 and PG 1707+427 on the basis of the period-spacing information alone, and we have also been successful in finding seismological models for both stars from period-to-period fits. The present study closes our short series of seismological studies of PG1159 stars based on a grid of full evolutionary models characterized by thick He-rich envelopes, as dictated by canonical stellar evolution.

Acknowledgments

Part of this work was supported by PIP 6521 grant from CONICET, by MEC grant AYA05–08013–C03–01, by the European Union FEDER funds and by the AGAUR.

References

Althaus L G Serenelli A M Panei et al. 2005 *A&A* **435** 631
Córsico A H & Althaus L G 2006 *A&A* **454** 863
Córsico A H Althaus L G Miller Bertolami M M & Werner K 2007a *A&A* **461** 1095
Córsico A H Miller Bertolami M M Althaus L G Vauclair G & Werner K 2007b *A&A* **461** 1095
Córsico A H Althaus L G Kepler S O Costa J E S & Miller Bertolami M M 2008 *A&A* **478** 869
Costa J E S Kepler S O Winget D E et al. 2008 *A&A* **477** 627
Dreizler S & Heber U 1998 *A&A* **334** 618
Fu J N Vauclair G Solheim, J E et al. 2007 *A&A* **467** 237
Kawaler S D & Bradley P A 1994 *ApJ* **427** 415
Kawaler S D O'Brien M S Clemens J C et al. 1995 *ApJ* **450** 350
Kawaler S D Potter E M Vucković M et al. 2004 *A&A* **428** 969
Miller Bertolami M M & Althaus L G 2006 *A&A* **454** 845
Reed M D Kawaler S D & O'Brien M S 2000 *ApJ* **545** 429
Tassoul M Fontaine G & Winget D E 1990 *ApJS* **72** 335
Vauclair G Moskalik P Pfeiffer B et al. 2002 *A&A* **381** 122
Werner K & Herwig F 2006 *PASP* **118** 183
Winget D E & Kepler S O 2008, *ARA&A* **46** 157

16th European White Dwarfs Workshop IOP Publishing
Journal of Physics: Conference Series **172** (2009) 012079 doi:10.1088/1742-6596/172/1/012079

Pulsational instabilities in the hot sdO star SDSS J1600+0748: The key role of radiative levitation

G Fontaine[1], P Brassard[1], E M Green[2], P Chayer[3], S Charpinet[4], M Andersen[5] and J Portouw[2]

[1] Département de Physique, Université de Montréal, C.P. 6128, Succ. Centre-Ville, Montréal, Québec, Canada H3C 3J7

[2] Department of Astronomy and Steward Observatory, University of Arizona, 933 North Cherry Avenue, Tucson, AZ 85721, USA

[3] Space Telescope Science Institute, 3700 San Martin Drive, Baltimore, MD 21218, USA

[4] Laboratoire d'Astrophysique de Toulouse-Tarbes, Université de Toulouse, CNRS, 14 av. E. Belin, F-31400 Toulouse, France

[5] Spitzer Science Center, California Institute of Technology, Pasadena, CA 91125, USA

E-mail: fontaine@astro.umontreal.ca; brassard@astro.umontreal.ca; bgreen@as.arizona.edu; chayer@stsci.edu; scharpin@ast.obs-mip.fr; mortena@ipac.caltech.edu

Abstract. SDSS J1600+0748 is the only hot sdO star for which unambiguous multiperiodic luminosity variations have been reported so far. These rapid variations, with periods in the range from 60 s to 120 s, are best qualitatively explained in terms of pulsational instabilities, but the exact nature of the driving mechanism has remained a puzzle. Models of sdO stars with uniform metallicity, even those with artificially-boosted values of the metallicity Z, are unable to excite pulsation modes in the range of interest as demonstrated most eloquently by Cristina Rodríguez-López in her Ph.D. thesis at the Universidad de Vigo in 2007. We confirm her results here, but also show that the inclusion of radiative levitation in the equilibrium models changes the picture dramatically. We find indeed that p-mode pulsations with periods overlapping with the observed ones in SDSS J1600+0748 can be excited in models in which radiative levitation of iron is taken into account. This process provides the needed boost to the opacity driving mechanism. We conclude that radiative levitation is a likely essential ingredient in the excitation physics at work in this unique star.

1. Introduction

Woudt et al. (2006) reported the discovery of short-period (from ~60 s to ~120 s) multiperiodic luminosity variations in the relatively faint (g = 17.41) SDSS object J1600+0748. They also provided a SALT spectrum which strongly suggested that SDSS J1600+0748 is a spectroscopic binary made of a very hot sdO star and, most likely, a late-type main sequence companion. This came as a bit of a surprise as no driving mechanism able to excite short-period p-modes had been found in models of sdO stars as reported by Rodríguez-López in her Ph.D. work (and see Rodríguez-López 2006, 2007 for more details).

2. The nature of SDSS J1600+0748

In order to carry out an independent stability analysis of models of SDSS J1600+0748, we found it essential first to determine as accurately as possible its location in the log g – $T_{\rm eff}$ plane. This is because sdO stars are found in a very wide domain of the log g – $T_{\rm eff}$ plane, and because Woudt et al. (2006) have not modeled their SALT spectrum. Actually, this task is complicated by the fact that the optical spectrum of the sdO component is contaminated by the light of its companion in this binary system.

We thus gathered a very high-sensitivity (S/N $\sim$ 300) optical spectrum of SDSS J1600+0748 at the Steward Observatory 2.3-m telescope, the average of some 12.2 h of observations gathered over four nights in the spring of 2007. We next worked quite hard to deconvolve the polluted spectrum of the system, finding out in the process that the cool star companion is very likely a G0V star. We finally χ^2-fitted the cleansed spectrum of the sdO component using a grid of synthetic spectra calculated from NLTE H/He model atmospheres. This grid of models/spectra was briefly described in Charpinet et al. (2005), but was recently extended to effective temperatures reaching 100,000 K, thus fully covering the sdO, as well as the sdB, range of interest. The best model fit that we obtained gives log g = 5.93 $\pm$ 0.11, $T_{\rm eff}$ = 71,070 $\pm$ 2725 K, and log N(He)/N(H) = -0.85 ± 0.08 (internal errors of the fit only). Note that the largest source of uncertainties on these derived atmospheric parameters are systematic effects associated with our neglect of metals in the model atmospheres and synthetic spectra. Currently, those are impossible to quantify as our (low-resolution) spectrum reveals no metal lines in the sdO spectrum.

It is worthwhile to point out that, with an estimated surface gravity of log $g \simeq 5.93$ for the pulsator in SDSS J1600+0748, the possibility that the observed oscillations could be due to unstable g-modes as found in some models of low-mass DAO white dwarfs (Charpinet et al. 1997a) is completely ruled out. Indeed, Charpinet et al. (1997a) have predicted the existence of another type of pulsator associated with the ϵ-mechanism that becomes of importance in the late post-EHB phase of the evolution when a star "turns the bend" in the HR diagram. Such a model is then identified with a low-mass DAO white dwarf on its way to final cooling. When this happens, H-burning at the base of the H-rich envelope turns on again and is able to drive short-period g-modes. To excite pulsation modes through the ϵ-mechanism at the base of the H-rich envelope requires that the modes have large enough amplitudes there, which automatically exclude p-modes. To excite g-modes with periods as short as those observed in SDSS J1600+0748 through the same mechanism requires that the surface gravity be larger than log $g \sim 7$. Our spectral analysis rules out this value for SDSS J1600+0748. We note nevertheless that the jury is still out on the possibility that real pulsating DAO white dwarfs may exist.

3. Improved seismic models

One piece of physics that can potentially be very important, but was not included in the sdO models built by Rodríguez-López and collaborators, is radiative levitation. In order to investigate that possibility, we followed the same approach as that proposed by Charpinet et al. (1997b) in the construction of their so-called second-generation sdB models. In brief, a state of diffusive equilibrium is assumed between gravitational settling and radiative levitation of iron in the envelope of sdO star models. Several second-generation models consistent with the spectroscopic values of the surface gravity and effective temperature were then built and analyzed for pulsational stability. With the inclusion of radiative levitation, we find instabilities in models of J1600+0748 that match remarkably well those found in the real star. In contrast, models with uniform metallicity fail to produce unstable modes (see Figure 1). Details on our results can be found in Fontaine et al. (2008).

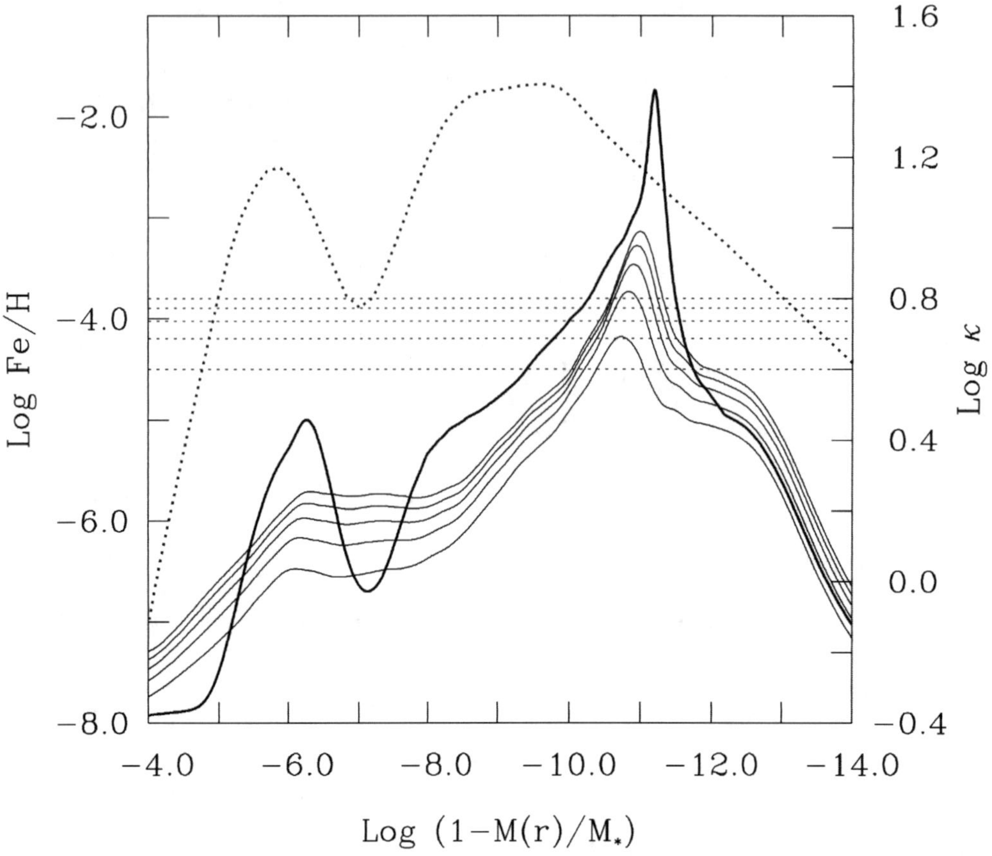

Figure 1. Iron-to-hydrogen number ratio (dotted curve) and Rosseland opacity (solid curve) profile in the envelope of representative stellar models. The thin curves refer to models with uniform metallicity specified by $Z = 0.02$, 0.04, 0.06, 0.08, and 0.10, from bottom to top. The thick curves refer to our reference model that takes radiative levitation into account. The largest bump in the opacity profile for each model is clearly larger and sharper in the latter case and produces pulsational instabilities.

References

Charpinet S Fontaine G Brassard P and Dorman B 1997a *ApJ* **489** L149
Charpinet S Fontaine G Brassard P et al. 1997b *ApJ* **483** L123
Charpinet S Fontaine G Brassard P et al. 2005 *A&A* **437** 575
Fontaine G Brassard P Green E M et al. 2008 *A&A* **486** L39
Rodríguez-López C Garrido R Moya A et al. 2006 *LNEA* **2** 167
Rodríguez-López C Ulla A and Garrido R 2007 *MNRAS* **379** 1123
Woudt P A Kilkenny D Ziestman E et al. 2006 *MNRAS* **371** 1497

Journal of Physics: Conference Series **172** (2009) 012080 doi:10.1088/1742-6596/172/1/012080

Testing the stationarity of white dwarf light-curves

L Molnár, Z Kolláth, E Plachy and M Paparó

Konkoly Observatory, P.O. Box 67. H-1525 Budapest, Hungary

E-mail: `lmolnar@konkoly.hu`

Abstract. Long period white dwarfs show changes in their frequency spectra from one observing season to another, i.e. their light-curves cannot be considered as stationary multiperiodic variations on long timescales. However, due to the complex frequency spectra of these stars and the narrow frequency spacing, it is still unknown, what the shortest time scale is, where real physical modulation exists. We present tests on artificial data, resembling the observations, using time-frequency distributions (TFDs), Fourier-analysis and the analytical signal method.

1. Introduction

Temporal modulations of light-curves provide a great deal of information on the nature of dynamics behind stellar pulsation. However, from a finite dataset it is usually very hard to conclude whether the signal is multiperiodic or a time varying one, with real amplitude and/or phase (frequency) variations. Furthermore, the decision whether a non-stationary signal is modulated due to nonlinear dynamics or some stochastic process, is even more challenging. Time-frequency representation (Cohen 1995) of a non-stationary signal yields information about characteristics of the dataset in the time-frequency plane. Then transient events or frequency/amplitude changes can be monitored easily with these tools. Time-frequency distributions are proved to be useful for analyzing non-stationary light-curves, e.g. semiregular variables (Buchler et al. 2004) or stellar activity (Kolláth and Oláh 2008). However, these methods can lead to false results when applied to multiperiodic variations with narrow frequency spacing, as the beating produces structures very similar to the modulationary one on the time-frequency plots.

Long term multisite observations provide very good frequency resolution, resulting in a good overall fit of the observations. But in some cases, single night observations deviate significantly from the multiperiodic curve fitted to the whole data. Whether it is a sign of ill frequency resolution or nonstationarity of the process, is another question hard to answer. The compatibility of single night data to a general model cannot be tested by Fourier transformation itself, if the length of the night is short compared to the timescale related to the frequency spacing. But it could be useful to investigate other methods, whether any fingerprints of nonstationarity can be obtained by them.

We have performed our tests based on published frequencies of pulsating white dwarf stars. We generated artificial light-curves using fixed frequencies and amplitudes. The frequencies have been derived from the WET (Whole Earth Telescope) observations of GD154 (Pfeiffer et al. 1996). For the sake of simplicity we concentrated on the frequency range 60-100 c/d, mainly on the structures around 73 c/d. We have to note that the narrowly spaced frequencies (e.g.

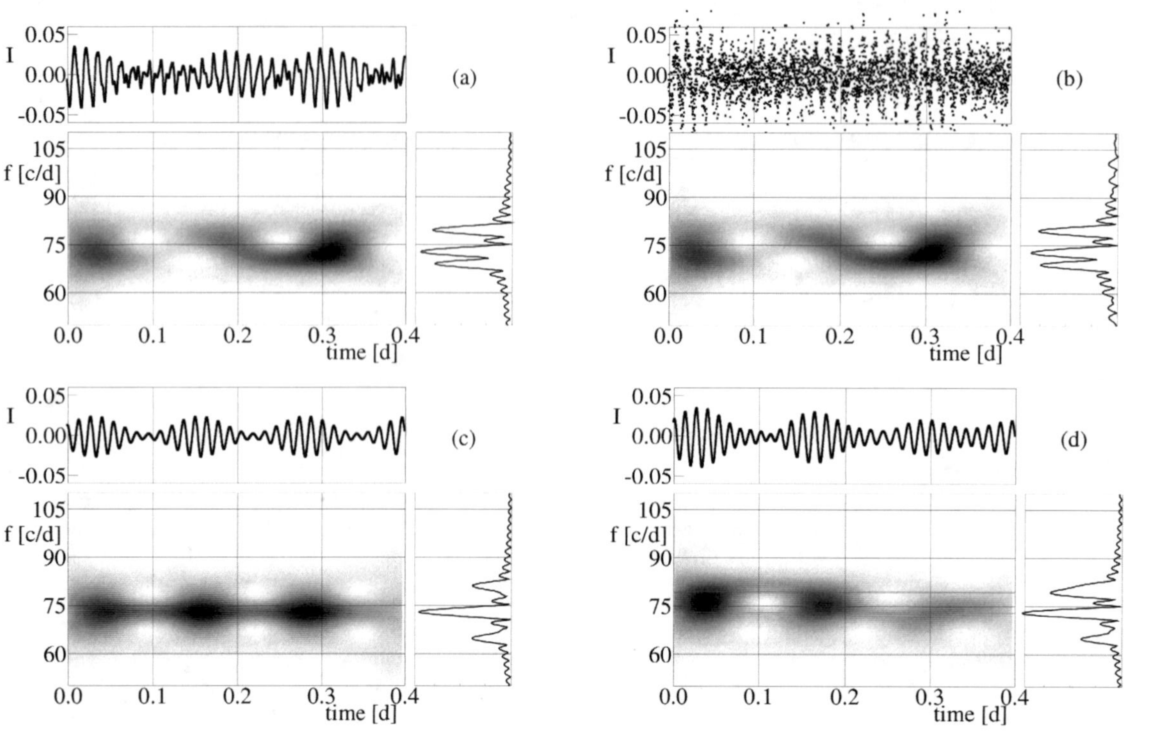

Figure 1. STFT distributions of different artificial data. Synthetic data with the published frequencies without noise *(a)* and with added noise *(b)*; Amplitude modulated (AM) signal *(c)*; and AM signal with a triplet around $\sim$ 79 c/d *(d)*.

a triplet at 73 c/d) practically act as a single frequency, as the time base of our tests are very small compared the characteristic time of the beating.

2. Time-Frequency Distributions

We calculated time-frequency distributions (TFDs) of different synthetic signals to check the characteristics of modulation and beats on single night observations. To construct TFDs, we used the TiFrAn (Time Frequency Analyzer) package, which has been developed at the Konkoly Observatory (see http://konkoly.hu/tifran). Different types of distributions have been integrated to this tool, for example Short-Term Fourier Transform (STFT), Choi-Williams- and Zhao-Atlas-Marks-Distributions (Cohen 1995).

A typical sample of synthetic variation calculated from the frequencies of GD154 is exhibited on panel *(a)* of Figure 1. With different randomized values of the phases the signal changes greatly, as light-curves do on different nights of observations. The Fourier-spectrum, right to the time-frequency distribution, clearly shows the three (groups of) frequencies. The TFD plot displays the traces of these peaks, but they are rather distorted by the modulation due to the close frequency spacing. TFDs have a very good noise rejection ratio, it is demonstrated on panel *(b)* of Fig. 1. Here we added a large Gaussian noise with a standard deviation of 0.02 to the synthetic data. Even with this amount of noise, the structure on the time-frequency plot remains unaltered, indicating that TFDs provide a good tool to compare noisy data with synthetic signals.

For comparison in panel *(c)* the same analysis of a single amplitude modulated (AM) signal is exhibited. For this tests we replaced the main triplet with an AM signal at 73 c/d. This simple test signal can be easily distinguished from the multiperiodic variation shown on panel

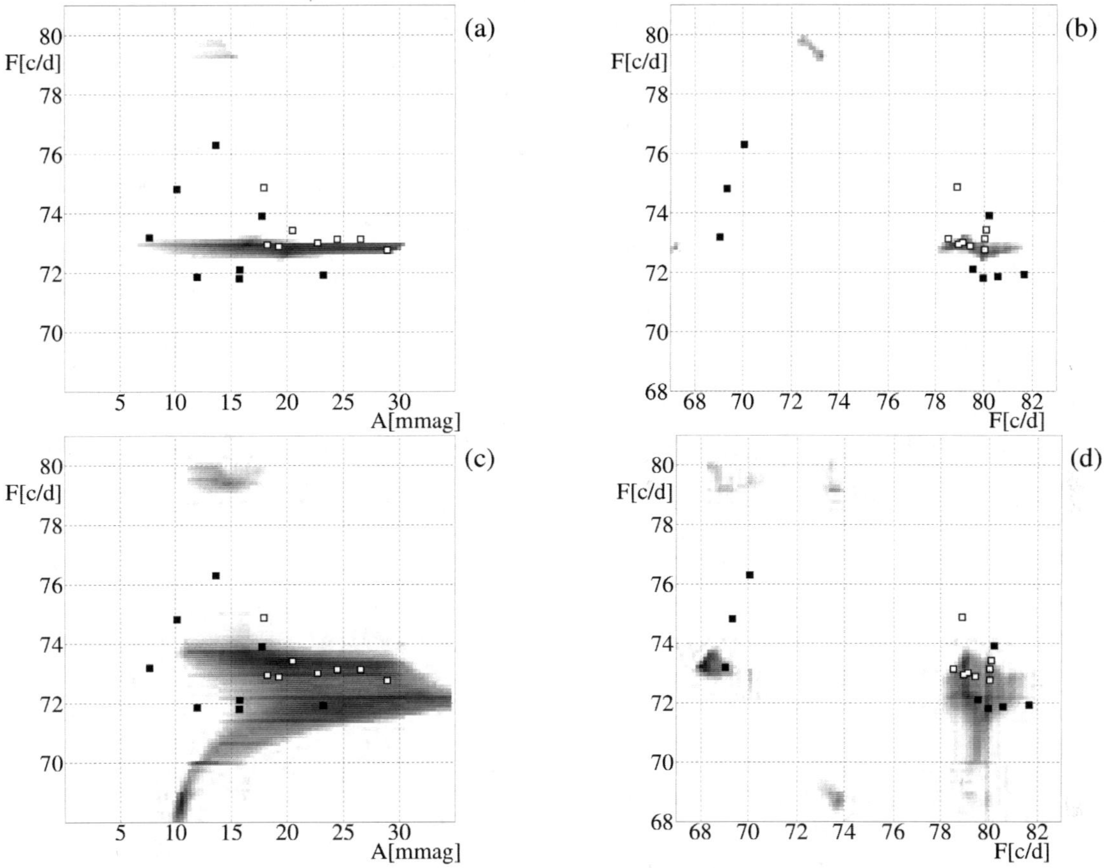

Figure 2. Probability distributions: main peak amplitudes *(a)*, frequencies of the two largest peaks *(b)*. Panel *(c)* and *(d)* are similar, with the 68 c/d frequency added. Open squares indicate the WET observation values (1991, partial), while solid squares show our observations (2006).

(a). However, if any component is included in the test signal in addition to the AM part, the resulting TFD looses the significant characteristic shown in panel *(c)*. This effect is clearly demonstrated on panel *(d)* of Fig. 1, where a triplet at 79 c/d is included in the synthetic variation. The datasets are 0.4 day long to be comparable to a single-night observation which is far too short to resolve the modulation of the triplet itself.

Another useful tool is to calculate the instantaneous frequencies and amplitudes by band filtered analytical signal (Kolláth and Buchler 2001). This method provides an efficient way to reconstruct a signal by a single $A(t)\exp(i\varphi(t))$ term, where $A(t)$ and $\varphi(t)$ are the time dependent instantaneous amplitude and phase. It has similar properties (e.g. in noise rejection) to time-frequency distributions, and the band filtered analytical signal method can be used together with TFDs to get quantitative information on modulations.

3. Monte-Carlo Simulations

Based on the hypothesis of stationary multiperiodic data, probability distribution functions of the parameters of single night observations can be constructed by Monte-Carlo simulation. We fixed the periods and amplitudes while phases were randomized. The simulation includes more than 100,000 artificial light-curves. This method allows us to define the probable parameter range of single-night observations, like the central frequency and amplitude of the largest Fourier-

Journal of Physics: Conference Series **172** (2009) 012080 doi:10.1088/1742-6596/172/1/012080

peak. Then the parameters of real observations can be compared to the original hypothesis. If the observed frequency and amplitude values fall outside the possible range, then that night is not compatible with the original assumption. Similar test can be performed using only the frequencies of the two largest peaks. In that case the results are independent from the amplitude scale. To minimize the effect of the side-lobes of the peaks in the Fourier-transform, we used a Gaussian windowing of the data. This preprocessing results in a better defined probability distribution of the Fourier peak parameters.

The published frequencies and amplitudes from WET observations of GD154 (Pfeiffer et al. 1996) have been used to construct the upper panels of Fig. 2, while the lower panels include an additional frequency at 68 c/d based on recent observations (Hürkal et al. 2005). The probability distribution functions of the largest Fourier-peak are shown in the left panels. The right panels display the possible ranges of the frequencies of the two largest peaks.

The WET observations are displayed with open squares on the figures. While the majority of the nights coincides with the most probable ranges, some observations are clearly show deviations from the assumed values. The additional frequency at 68 c/d helps in resolving the discrepancies. Preliminary results of recent observations of GD154 are also plotted by solid squares – these data suggest further departure from the original hypothesis.

4. Conclusion

It is an important question, whether white-dwarf light-curves are considered to be stable (stationary) on short timescales. We investigated several nonstandard methods to test if they are capable on tracing deviations from a multiperiodic variation in single night observations.

(i) Time-frequency distributions itself cannot show differences between multiperiodic and non-stationary time varying signals, if the timebase of the data is comparable to the modulation periods due to beats. However, due to the good signal-to-noise ratio in the TFD plots, it can be used to compare single night observations to the best fit of the whole dataset.

(ii) For continuous long timebase observations expected in the future from space-born observatories, these methods will be substantial tools to test the nonstationarity of the light variations.

(iii) With Monte-Carlo simulations based on the mathematical model of multisite observations, the probability distribution of the largest amplitudes and the corresponding frequencies in a single night can be calculated. Then the observed quantities, as compared to the probability distribution, provides information on the quality of the fit or on the validity of the stationarity of the data.

References

Buchler J R Kolláth Z and Cadmus R R 2004 *ApJ* **613** 532

Cohen L 1995 *Time-Frequency Analysis* (Englewood Cliffs, NJ: Prentice-Hall PTR)

Hürkal D Ö, Handler G, Steininger B A and Reed M D 2005 *ASPC* **334** 577

Kolláth Z and Oláh K 2008 *A&A* submitted

Kolláth Z and Buchler J R 2001 *Stellar Pulsation – Nonlinear Studies* (Astrophysics and Space Science Library Series vol 257) ed M Takeuti and D D Sasselov (Dordrecht: Kluwer Academic Publishers) pp 29-60

Pfeiffer B et al. 1996 *A&A* **314** 182-90

Wqed: A Lightcurve Analysis Suite

S E Thompson[1,2] and F Mullally[3]

[1]Delaware Asteroseismic Research Center, Mt. Cuba Observatory, Greenville, DE 19807, USA
[2]Department of Physics and Astronomy, University of Delaware, Newark, DE, 19716, USA
[3]Department of Astrophysical Sciences, Princeton University, Princeton, NJ 08544-1001 USA

E-mail: `sthomp@physics.udel.edu` & `fergal@astro.princeton.edu`

Abstract. `Wqed` is a suite of programs designed to clean lightcurves created by aperture photometry to allow more sophisticated analysis, like Fourier transforms, to be preformed. The data format for `Wqed` includes a header that allows users to keep all pertinent information about their data attached to the final light curve. `Wqed` was tested during the Whole Earth Telescope run on a DAV in 2008 and improved the ease and speed of data reduction. Visit http://physics.udel.edu/darc/wqed to download the current version of `Wqed` along with a full set of instructions on how to install and use the program.

1. Introduction

Analysis of astronomical time-series data is often complicated by atmospheric effects on the lightcurve. Clouds change the brightness of these stars, while extinction adds long-term trends. Removal of these effects is best done interactively. The Whole Earth Telescope (WET) conducts coordinated time-series observations of multiple targets using telescopes distributed across the globe, and requires a fast, light-weight, and easy to use software pipeline to reduce data as fast as it is obtained. In the early days, the WET used a DOS based program known as QED to reduce phototube data, but the increased popularity of CCD rendered this program obsolete.

Currently, the WET uses the IRAF based ccd_hsp package to extract raw lightcurves from CCD frames, and for the most recent WET run (XCov26, Thompson et al., this volume) we developed and successfully used a package for interactively removing atmospheric effects. This package, called `Wqed` (pronounced "Wicked"), allows the user to remove portions of the lightcurve, divide lightcurves, fit low order polynomials, and other useful tasks. While designed with multi-site data in mind, `Wqed` also conveniently serves the need of the single-site astronomer.

2. Usage

`Wqed` is designed to be quick to learn and easy to use. An intuitive interface based on the pgplot libraries allows the user to use the mouse to select and zoom in on regions of the lightcurve. A similar combination of mouse and keystrokes can remove irredeemably bad data from a lightcurve, divide the lightcurve of a target star by that of a reference to remove the effects of small cloud, fit and remove a polynomial to a lightcurve to remove long-term trends from the data, etc. Every action can be reversed using the undo function. Users of the old QED program will find the interface and design philosophy familiar.

Figure 1 shows a screenshot from a typical reduction. The screen is divided into three parts, information about the run extracted from the header (see § 3) is displayed at the top, the lightcurves are displayed in the middle, and at bottom, a message box provides feedback to the user. Each light curve is referenced by a channel number, listed on the left. The average counts of each channel are listed on the right. A list of available commands can be found by pressing the **?** key.

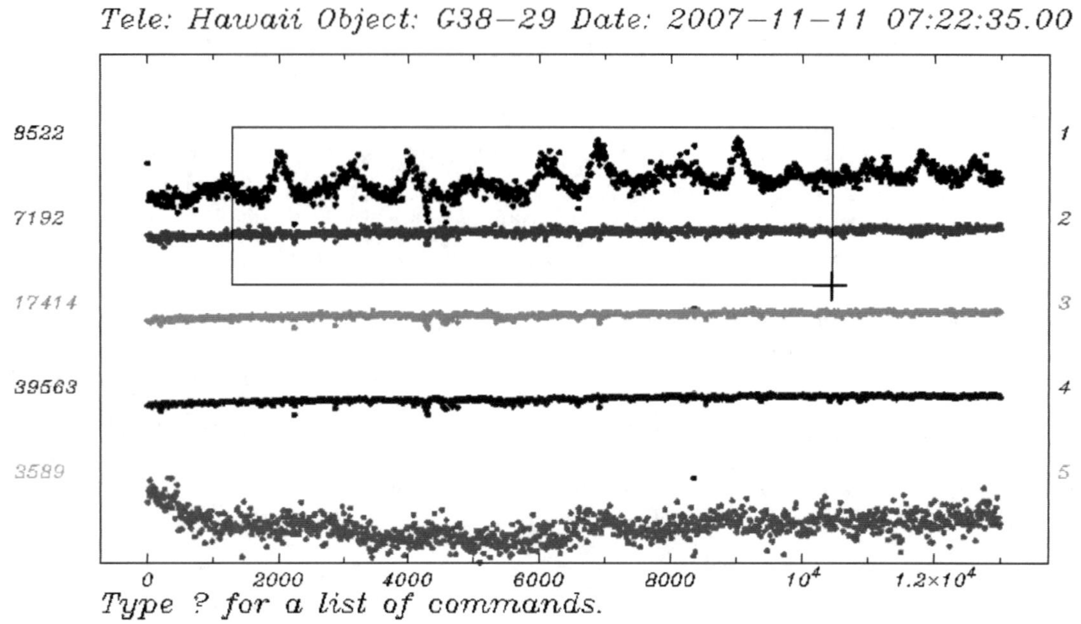

Figure 1. Screenshot of `Wqed` and a box for zooming. Five different light curves are shown and their channel numbers are listed on the right. The average counts in each light curve are listed on the left. When the user decides to zoom in, the key **z** is typed to form a box around the desired region.

Figure 1 shows the lightcurve of a variable white dwarf star in black, three reference stars, the bottom lightcurve shows the sky brightness during the observations. The following is a brief list of commands frequently used in `Wqed`.

2.1. Zooming

To examine a region of the lightcurve in greater detail, press the **z** key and select the region using the mouse.

2.2. Division

To divide the target star (in channel 1) by a bright reference star (in channel 2), one simply must type four key strokes: **1 / 2 <CR>**. To divide by the sum of two light curves, type **1 / 2 3 <CR>**.

2.3. Garbage

Because of clouds and cosmic rays, bad data points remain in the divided light curve. To garbage these points the user can create a box around these points by typing **g**. Similarly, to remove all points outside of the box, use the command **ctrl-g**.

2.4. Fit Polynomial
In differential astronomy of white dwarf stars, the reference star is usually redder than the target star. The result is some false curvature remaining in the divided lightcurve. To remove a low order polynomial from a lightcurve, the user must select the channel, type **f** and then the order of the polynomial to remove from the lightcurve, i.e type **1 f 2** to remove a 2nd order polynomial from channel 1.

2.5. Write Data
When ready to save the processed light curve, the user types **1 w e**; the entire first channel is written to a file with the extension **.lc1**. This file will have a header containing all the original header information, plus information about the `Wqed` reduction and the Bjed correction. The data will be in two columns, the first is time in seconds and the second is fractional amplitudes with a mean of zero.

3. Headers and Bjed Calculation
The `Wqed` package encourages the use of headers on all light curve files. These headers are simply ASCII text placed at the beginning of the file. Each line contains a key, value and a brief comment. A detailed header is invaluable for archival data, especially when working with lightcurves taken several years ago. The headers are flexible so the user may add any information necessary to describe the observations. Some values are necessary for `Wqed` to fully function, others are important for future astronomers to understand the analysis of these data. A program called `Whiff`, supplied with the `Wqed` package creates a suitable header by extracting relevant information from the first raw fits image of the series. Refer to Figure 2 for an example of a `Wqed` header.

With the inclusion of the object name, date and time of observation in the header, `Wqed` will automatically calculate the barycentric correction based on the method of Stumpff (1980). This implementation has been tested against the JPL DE405 ephemeris and agrees to better than 0.05 seconds. `Wqed` expects the date and initial UTC of your data to have the form **yyyy-mm-dd** and **hh:mm:ss.ss** respectively. The right ascension and declination of the object are determined from the object name. `Wqed` recognizes most pulsating white dwarf stars, and additional stars can easily be added by the user.

4. Conclusion
The `Wqed` package is an easy to use lightcurve manipulator that encourages the use of headers to keep track of the time, place and object of each observation. Those doing the data analysis for XCOV26, the 2008 WET run, (see Thompson et al. this proceedings) found the `Wqed` package very easy to learn, and helpful for data organization. `Wqed` was also recently used for the single-site observations that discovered two new DQVs (Barlow et al., 2008). `Wqed` comes with an extensive set of instructions on how to use and install the `Wqed` package. Several programs not described here are also provided with the package to plot and analyze variable star data.

We encourage anyone analyzing rapidly pulsating stars, such as white dwarf and subdwarf pulsators, to use `Wqed`. The program may be downloaded from the DARC web site at `http://physics.udel.edu/darc/wqed`.

Acknowledgments
We thank the WET for helping to test and develop this software. SET thanks the Crystal Trust for their financial support.

```
#Telescope      =Hawaii                          #Telescope Name
#Object         =G38-29                          #Object Name
#Date           =2007-11-11                      #Mid Exp. yyyy-mm-dd
#UTC            =07:22:35.00                     #Mid Exposure UTC
#Filter         =BG40                            #Filter Name
#Instrument     =Apogee Alta                     #Instrument used
#Exptime        =10.000000000000000              #Exposure Time (s)
#MJD            =54415.307349537                 #Mid Exposure MJD
#Observer       =set meo kt                      #Observer
#TelCode        =hawa                            #Short Tel. Code
#Runname        =hawa071111-07                   #Wet Run Name
#IrafMJD        =54415.3073494998                #ccd_hsp zero MJD
#OrigFile       =g3928_hawaii_20071110_7..star-sky #Contains Aperture Size
#WQED           =wqed v1.0.1 2008-05-30          #Version of Wqed
#Channel        =1                               #Outputted Channel
#WqedUser       =sthomp                          #User who ran wqed.
#RA             =04 20 18.00                     #From stars.dat (J2000)
#Dec            =+36 16 16.0                     #From stars.dat (J2000)
#Bjed           =2454415.81329600                #with leapsecs(Stumpff80)
#BjedCorr       =0.00003323                      #linear Bjed correction
#Leapsecs       =65.1840                         #Total number accounted
#TimeUnits      =secs                            #Units Time column
#Npoints        =1160                            #Number pts in data
```

Figure 2. An example header of a `Wqed` lightcurve. Headers are denoted by a # followed by a key. The value of the key comes after an equal sign. A comment to describe the key or value comes after a second #. The last ten lines of the header were added after running `Wqed`, including the Bjed correction. A program called `Whiff` is provided to create these headers from the first fits file of the time series.

References

Barlow B N, Dunlap B H, Rosen R & Clemens J C 2008, ArXiv:0810.2140

Stumpff P 1980, *A&A Suppl.*, **41**, 1

AUTHOR INDEX

AUTHOR INDEX

AUTHOR INDEX